The Wide World of Adhesion

Scrivener Publishing
100 Cummings Center, Suite 541J
Beverly, MA 01915-6106

Publishers at Scrivener
Martin Scrivener (martin@scrivenerpublishing.com)
Phillip Carmical (pcarmical@scrivenerpublishing.com)

The Wide World of Adhesion

A Festschrift in Honor of Dr. Kash Mittal on his 80th Birthday

Edited by

Anil N. Netravali

and

Frank M. Etzler

Scrivener Publishing

WILEY

This edition first published 2026 by John Wiley & Sons, Inc., 111 River Street, Hoboken, NJ 07030, USA and Scrivener Publishing LLC, 100 Cummings Center, Suite 541J, Beverly, MA 01915, USA

For more information about Scrivener publications please visit www.scrivenerpublishing.com.

Wiley Global Headquarters
111 River Street, Hoboken, NJ 07030, USA

For details of our global editorial offices, customer services, and more information about Wiley products visit us at www.wiley.com.

The manufacturer's authorized representative according to the EU General Product Safety Regulation is Wiley-VCH GmbH, Boschstr. 12, 69469 Weinheim, Germany, e-mail: Product_Safety@wiley.com.

Library of Congress Cataloging-in-Publication Data

ISBN 978-1-394-44950-7

Cover image: Courtesy of the Wikimedia Commons
Cover photo: Courtesy of the editors
Cover design by Russell Richardson

Set in size of 11pt and Minion Pro by Manila Typesetting Company, Makati, Philippines

Printed in the USA

10 9 8 7 6 5 4 3 2 1

Contents

Preface

In the broad field of materials there are many cases where two or more materials are bonded together to achieve properties that cannot be obtained from any single material. While in most cases these materials need to be bonded to each other permanently, in a few cases they need to be bonded temporarily so that they can be separated easily when needed. A good case where two materials need to be bonded permanently is fiber reinforced polymer composites. When the fiber/resin bonding is good, composites achieve high mechanical properties such as strength and stiffness. However, when fiber/resin bonding is weak for the same fiber/resin combination, both strength and stiffness of the composites are significantly reduced while the toughness is increased. This indicates the critical role fiber/resin bonding, i.e., adhesion plays in controlling the composite properties. On the other hand, a Post-it® note sticks to multiple surfaces but can be pulled out with ease as and when needed. While these are just two examples, adhesion is ubiquitous in the industrial world. It can be seen in fiber/resin in fiber reinforced composites, coatings/paints/varnishes in a variety of applications (on polymers, metals, wood, ceramics, and metallized plastics), lipstick and other cosmetics/skin patches, tissue joining after surgery, stickers, pressure-sensitive adhesives, etc. The list can go on and on.

Many questions arise from the two distinct cases mentioned above as well as other applications. How do you fine-tune adhesion between two disparate materials? What factors (chemical, mechanical, or others) control adhesion between the two materials? How does one measure adhesion or bond strength between the two materials? What is the effect of high temperature or other environmental conditions on adhesion? With better understanding of the interfacial characteristics and the factors that control adhesion, it should be possible to design the two materials that will result in required level of adhesion between them. This book is dedicated to Dr. Kash Mittal, a researcher devoted to 'adhesion science', an articulate teacher, and an outstanding editor committed to advancing the field of adhesion, on his 80th birthday. The book is primarily based on the Symposium, aptly named, "The Wide World of Adhesion" held at the American Chemical Society meeting in Chicago in August of 2022 in honor of Dr. Mittal's 75th birthday and brings together many experts and leading researchers from around the world who have made significant contributions to different aspects of the field of adhesion.

The book is divided into three parts after the introductory part. The introductory part consists of three write-ups: Dedication by editors Netravali and Etzler, A personal tribute to Dr. Mittal by Chibowski, and Half-a-century romance with adhesion science by Dr. Mittal himself. Parts 1, 2, and 3 consist of 23 technical papers (chapters) that not only provide in-depth discussions on various critical aspects of adhesion science and technology but clearly indicate the breadth of its use in diverse and critical applications. Part 1: Contact Angle, Wettability and Adhesion consists of chapters 1-4 by Perez-Huertas and Terpilowski, Rao and Saikia, Rossi *et al.*, and Tyowua *et al.*, respectively, which discuss various aspects of wettability and how the theory can be applied in disparate applications such as flow through porous media, release of active substances from ointments, and evaporation of wetting and non-wetting liquid drops. Part 2: Adhesion Aspects consists of chapters 5 to 14. In chapter 5 Luengo discusses controlling adhesion in dermatological applications. Chapter 6 by Jaiswal and Jaiswal discusses the role of surface topography in engineering interfacial adhesion. In chapter 7 Bui discusses critical adhesion aspects in color cosmetics. Chapters 8 and 9 by Daneshzand *et al.* and Chen, respectively, discuss adhesion of nanofiber mats and microplastic adhesion. Chapters 10 and 11 by Friedrich show how plasma technology can be used to improve polymer adhesion, whereas Chapter 12 by Rawal *et al.* shows how substrate surfaces can be engineered to promote adhesion. Thomas *et al.* in Chapter 13 discuss the results of interfacial studies in biocomposites and Chapter 14 by Netravali presents a brief review of micromechanical methods of interfacial shear strength measurement. Finaly Part 3: Adhesives and Adhesive Joints consists of chapters 15 to 23. In chapters 15 by Jarach *et al.* and 16 by Liang and Li, authors discuss the results of their research on switchable thermosets and pressure-sensitive adhesives, respectively. Chapters 17 by Dunky and 18 by Frihart present wood adhesives and soy-based adhesives prepared for wood bonding, respectively. Chapters 19 by Garcia *et al.* and 20 by Goyal and Bhatnagar present reviews of advances made in the fields of dental adhesives and bioadhesives, respectively. In Chapter 21 Tiwari *et al.* discuss surface treatments for multi-material adhesive bonding and in chapter 22 Ahmed and Panigrahi present a review of adhesives that are used specifically in high-temperature applications. Finally, chapter 23 by de Queiroz and Banea present various methods used to enhance mechanical performance of adhesive joints. All chapters are profusely illustrated and referenced.

The book should be of great interest to adhesionists or "stickologists" that work in academic, industry and government laboratories. as well as those researchers interested in bonding, packaging, printing, composite materials, construction and many other fields. The current overview presented in this book should provide R&D personnel in a broad set of industries such as automotive, aerospace, polymer composites, cosmetics, biomedical materials, dentistry, fiber manufacturing, textiles, medical, microelectronics, packaging, paints and varnishes, plastics, printing, and many others, ample information in designing, manufacturing and testing of products that involve bonding of two or more materials.

It is our great pleasure to thank those who made this book possible. First and foremost, we are very thankful to all contributing authors for their enthusiasm, sustained interest and cooperation in bringing this book to fruition. We are grateful for their willingness to share their cutting-edge research work and knowledge of the field of adhesion with the interested community. This book would not have been possible without their hard work. Many thanks to the unwavering support and keen interest from Martin Scrivener (Scrivener Publishing) in this book project. His efforts for giving this book a body form are much appreciated.

Anil N. Netravali
Jean and Douglas McLean Professor of Fiber Science (Emeritus), Cornell University
Frank M. Etzler
Professor of Pharmaceutical Science (Emeritus), Lake Erie College of Osteopathic Medicine

Dedication

This Festschrift is dedicated to Dr. Kashmiri Lal Mittal, affectionately known as Kash to all his friends and professional colleagues, on the occasion of his 80th birthday. Here is the truly inspiring story of this true gentleman, great teacher, scientist, and editor par excellence. Dr. Mittal was born into a middle-class business family in the small hamlet of Kilrodh, in the state of Haryana, India, on October 15, 1945. Kilrodh did not have electricity in those days, and the elementary school Dr. Mittal attended was far away. With no public or private transportation available, he had to walk that distance six days a week. By the time he could get home from school, it would already begin to get dark. With no electricity, he had to study during what little daylight was left or sit near an oil-lit lamp or a candle to read or do his homework. Winters were brutally cold, and summers were burning hot. However, it did not matter to Dr. Mittal how cold the winter days were and how hot the summer days felt; he simply concentrated on his studies. He was disciplined, inquisitive, and above all, highly motivated to study. He was ready to endure anything and everything. However, the worst thing that can happen to a person who loves reading happened to him. He was unfortunate to lose sight in his right eye at the young age of 13. But he was not the person (or the child) who would give up. He chose to work even harder to prove himself. He often quips: if I had both eyes, I could read double and produce double. As he graduated from high school, he received the highest accolades from his school.

After completing high school with high honors, Dr. Mittal joined the Nehru College, Jhajjar, of Punjab University. He earned his BSc degree in 1964 with flying colors. He stood 4th in a class of about 40,000 students who took that exam! That was not enough for Dr. Mittal. While his siblings got into business, he was determined to continue his education, learn more, and achieve more! He joined the Indian Institute of Technology Delhi (IIT-Delhi), Indian Ivy League equivalent, for his graduate studies in chemistry. Two years later, in 1966, he graduated at the top of his class, earning an MSc degree in Chemistry. The same year, he came to the US to carry out his Ph.D. research at the University of Southern California (USC), Los Angeles. There, he completed his Ph.D. in Colloid Chemistry in 1970 and began his long journey in the field of colloid and surfactant

Anil N. Netravali and Frank M. Etzler (eds.) The Wide World of Adhesion: A Festschrift in Honor of Dr. Kash Mittal on his 80th Birthday, (1–20) © 2026 Scrivener Publishing LLC

science, to which he has made significant contributions. His Ph.D. thesis was titled "Factors Influencing the Ultracentrifugal Stability of Oil-in-Water Emulsions". Within six years of receiving his Ph.D. from USC, he founded the international biennial symposium series "Surfactants in Solution" that has run uninterruptedly and smoothly, this year's meeting in Madrid marks the 50th anniversary.

After receiving his Ph.D., Dr. Mittal had a brief stint at the Pennsylvania State University as a postdoctoral associate before moving to the University of Pennsylvania (U Penn), Electrochemistry Laboratory of Prof. John O' M. Bockris as a postdoctoral associate. His co-authored paper at U Penn, "Mechanism of Activation in Electron Transfer in Solution," was published in Nature - Physical Sciences in 1971.

In 1972, Dr. Mittal joined IBM (San Jose, California) as a postdoctoral scientist, where he first began working in adhesion science and technology, the field that became his 'First Love'. Most of his friends recall him mentioning affectionately, 'he got stuck' to this field. In fact, since that time, he has literally stuck to this 'wonderful field of adhesion'. Perhaps it is the adhesion science that has stuck to him! Either way, he has excelled in the research and technology of adhesion for more than 5 decades. In that period, he edited and/or co-edited more than 75 books in this field alone! A list of his 154 edited books is presented as Appendix 1 in this dedication for the benefit of the readers. It should be mentioned here that many more books are at different stages of publication!

In 1974, Dr. Mittal moved to IBM in Poughkeepsie, New York, as a full-time IBM employee and worked in both IBM facilities in East Fishkill and Thornwood, New York, until 1994. At IBM, he was assigned to work on adhesion aspects as well as surface contamination and cleaning in microchip processing. He took the challenges wholeheartedly. He worked on many critical aspects of microchip manufacturing with respect to adhesion between various components, where obtaining clean surfaces was *sine qua non*, and was successful in improving the process efficiency. At IBM, he also got the opportunity to master various techniques and concepts, such as acid-base interactions, silanes, and plasma treatments, etc., and some new techniques, such as inelastic electron tunneling spectroscopy (IETS), XPS, STM, and AFM, that were used to modify and characterize surfaces and interfaces.

As he worked at IBM, not only did his knowledge of the field deepen, but he also developed an avid passion for the field of adhesion science. As he searched the literature, he quickly noticed that there was only one journal that was fully dedicated to adhesion science. He took this opportunity and soon founded the *Journal of Adhesion Science and Technology* (JAST).

Dr. Mittal remained the editor-in-chief of JAST until 2012. While at the helm of JAST, he was in touch with many top researchers in the field. He also had a knack for connecting with younger researchers and nurturing/encouraging them to do better research, and giving them an opportunity to publish in JAST. This is, in fact, how one of us (ANN) met Dr. Mittal. In June of 1987, the Greater New York Chapter of the American Vacuum Society organized '*Symposium on Polymer Interfaces and Adhesion*" at IBM Yorktown Heights. At the time, I was a postdoc at Cornell University and was asked by Prof. Sachse, with whom I worked, to present our research on fiber/resin interfaces. After my presentation, during the break, Dr. Mittal came to me and told me about JAST, which he had just founded, and asked me to consider submitting to it. I was thrilled since JAST was a perfect match for the research I was doing. As I started publishing papers in JAST, I realized some important characteristics of Dr. Mittal. He was never demeaning to the authors, no matter how bad the manuscript was. He kept asking for changes, additions, deletions, modifications, grammar corrections, and sometimes it could be a long list! However, all of that came with kind-heartedness and encouraging words! He could patiently go through revision after revision, but he never gave up on anyone. His perseverance and determination to get things right, both scientifically and linguistically, are simply unbelievable and unmatched! No wonder JAST quickly became the top journal in the field. It has been about 38 years since I met him, but since that time, we have truly 'bonded' as friends. In fact, we are so close that I consider him my elder brother!

While at IBM, two other things happened to Dr. Mittal. First, his job, along with working on adhesion issues, was to find ways to remove surface contaminants and to characterize clean surfaces, critical requirements in microchip processing. Second, he developed and taught courses for the IBM employees in Adhesion Science and Surface Chemistry. He trained many IBM employees through these courses. Along those lines, he also developed a full-fledged course on Adhesion Science and Technology. In the past 3-4 decades or so, this course was widely offered across the globe, and more than 5,000 individuals have taken it and have benefited from it. The course allowed him not just to satisfy his true desire to spread the 'Gospel' of 'Adhesion Science and Technology' around the globe but also to satisfy his passion for teaching. Later, Dr. Mittal also co-founded *MST Conferences, via* which he organized many conferences in the US and abroad.

Dr. Mittal remained Editor-in-Chief of JAST until 2012. In 2013, he initiated *Reviews of Adhesion and Adhesives* (RAA), another successful journal, and was its Editor-in-Chief until July 2022. Along with the journals,

Dr. Mittal also focused on editing books in Adhesion Science, Colloids and Surfactants, and Surface Cleaning, subjects where he is well known to the global research community. Besides these topics, Dr. Mittal has also worked on "polyimides", a high-temperature polymer that has been used in insulation for wiring and electronics in automotive and aerospace industries. It is also used as a material of choice in flexible circuits in electronics, as well as in medical devices such as implants, catheters, etc. Dr. Mittal has edited several books on polyimides as well.

Dr. Mittal's significant contributions have not gone unnoticed. He is listed in several biographical references: American Men and Women of Science, Who's Who in the East (USA), Men of Achievement, International Who's Who of Contemporary Achievement, Who's Who in Frontier Science and Technology, Who's Who in Technology Today, and Who's Who of Intellectuals.

Dr. Mittal has been a member of the editorial boards of numerous international journals. He has also published 80 research papers in refereed journals and chapters in books. A list of his papers is given in Appendix 2 in this dedication. He is the 1990 recipient of the Charles Dudley Award of the ASTM International and is a Robert L. Patrick Fellow of the Adhesion Society (USA). He is also the recipient of the 1995 Thomas D. Callinan Award of the Dielectric Science and Technology Division of the Electrochemical Society. The crowning glory of his career was when he was recognized for his contributions by the worldwide adhesion community by organizing the 1st International Congress on Adhesion Science & Technology in his honor on his 50th birthday in Amsterdam, The Netherlands, in 1995, where 235 papers were presented by researchers from 38 countries. Also, on this occasion, Prof. P. G. de Gennes (Nobel laureate in Physics, 1991) presented him with a plaque with the inscription "*I, on behalf of the global adhesion scientific community, would like to recognize him for his commitment and dedication to the advancement of adhesion science and technology and his exceptional achievements. All in the adhesion world would like to thank Dr. Mittal for his outstanding leadership and distinguished professional service.*" Dr. Mittal is also the recipient of the 1997 John A Wagnon Technical Achievement Award of the International Microelectronics and Packaging Society (IMAPS). He was, furthermore, invited as a Project Specialist in the area of materials science by the International Advisory Panel and the Chinese Review Commission (under the auspices of the World Bank) to visit the Shanghai Jiao Tong University, Shanghai, PRC in 1985. The worldwide colloid and surface science community honored him by inaugurating the Kash Mittal Award in the surfactant field in his honor for his great efforts and significant contributions to surface and colloid science in 2002.

Maria Curie–Skłodowska University in Lublin, Poland, awarded him the title of Doctor *Honoris Causa* in 2003. Also, three books dedicated to him have already been published in his honor, and this marks the 4th book. Moreover, many issues of several journals have been published in his honor.

Dr. Mittal recently moved to New Delhi, India, and has been active on many fronts. He is a sought-after speaker and travels all over India to give lectures. In 2025, he founded the Adhesion Society of India (ASI) to promote research in Adhesion and bring the Indian adhesion research community together. He also established five Awards at different academic institutions in India where he had his education, and he calls it "Give Back Program".

Originally, this Festschrift was planned to be published in honor of Dr. Mittal's 75th birthday. A Symposium "The Wide World of Adhesion" was organized during the annual meeting of the American Chemical Society (ACS) in the Fall of 2020, in San Francisco, to honor him. Unfortunately, that was the year when the COVID pandemic hit the world, and ACS was forced to change the conference from in-person to a virtual format. With no in-person contact possible, the symposium had to be postponed to the 2022 ACS Fall Meeting in Chicago. The symposium was a grand success, attended by several researchers from many countries who saluted him for the significant contributions he had made not just to Adhesion Science but to other fields as well. This Festschrift, delayed by 5 years, is now in honor of Dr. Mittal's 80th birthday. We thank all researchers who attended the symposium, as well as those who could not make it to the Symposium but contributed to this Festschrift. We wholeheartedly salute Dr. Mittal for his magnificent contributions and extend our sincere wishes for his good health and further contributions in the years to come.

Anil N. Netravali
Frank M. Etzler

Appendix 1

List of Books Edited by Dr. K.L. Mittal

1. K. L. Mittal, Editor, *Adsorption at Interfaces*, ACS Symposium Series No. 8, American Chemical Society, Washington, D.C., 1975
2. K. L. Mittal, Editor, *Colloidal Dispersions and Micellar Behavior*, ACS Symposium Series No. 9, American Chemical Society, Washington, D.C., 1975

3. K. L. Mittal, Editor, Micellization, *Solubilization, and Microemulsions, Vol. 1*, Plenum Press, New York, 1977. Part of this book has been translated into Russian.
4. K. L. Mittal, Editor, *Micellization, Solubilization, and Microemulsions, Vol. 2*, Plenum Press, New York, 1977. Part of this book has been translated into Russian.
5. K. L. Mittal, Editor, *Adhesion Measurement of Thin Films, Thick Films and Bulk Coatings*, American Society for Testing and Materials, Philadelphia, PA, 1978. Also based on this book he was the recipient of the 1990 Charles B. Dudley Award of the ASTM.
6. K. L. Mittal, Editor, *Solution Chemistry of Surfactants, Vol. 1*, Plenum Press, New York, 1979.
7. K. L. Mittal, Editor, *Solution Chemistry of Surfactants, Vol. 2*, Plenum Press, New York,1979.
8. K. L. Mittal, Editor, *Surface Contamination: Genesis. Detection, and Control, Vol. 1*, Plenum Press, New York, 1979.
9. K. L. Mittal, Editor, *Surface Contamination: Genesis, Detection, and Control, Vol. 2*, Plenum Press, New York, 1979.
10. K. L. Mittal and E. J. Fendler, Editors, *Solution Behavior of Surfactants: Theoretical and Applied Aspects, Vol. 1*, Plenum Press, New York, 1982.
11. K. L. Mittal and E. J. Fendler, Editors, *Solution Behavior of Surfactants: Theoretical and Applied Aspects, Vol. 2*, Plenum Press, New York, 1982.
12. K. L. Mittal, Editor, *Physicochemical Aspects of Polymer Surfaces, Vol. 1*, Plenum Press, New York, 1983.
13. K. L. Mittal, Editor, *Physicochemical Aspects of Polymer Surfaces, Vol. 2*, Plenum Press, New York, 1983.
14. K. L. Mittal, Editor, *Adhesion Aspects of Polymeric Coatings, Plenum Press*, New York, 1983.
15. K. L. Mittal and B. Lindman, Editors, *Surfactants in Solution (Lund Proceedings) Vol. 1*, Plenum Press, New York, 1984.
16. K. L. Mittal and B. Lindman, Editors, *Surfactants in Solution (Lund Proceedings) Vol. 2*, Plenum Press, New York, 1984.
17. K. L. Mittal and B. Lindman, Editors, *Surfactants in Solution (Lund Proceedings) Vol. 3*, Plenum Press, New York, 1984.
18. K. L. Mittal, Editor, *Adhesive Joints: Formation, Characteristics and Testing*, Plenum Press, New York, 1984.
19. K. L. Mittal, Editor, *Polyimides: Synthesis, Characterization and Applications, Vol. 1*, Plenum Press, New York, 1984.
20. K. L. Mittal, Editor, *Polyimides: Synthesis, Characterization and Applications, Vol. 2*, Plenum Press, New York, 1984.
21. K. L. Mittal and P. Bothorel, Editors, *Surfactants in Solution (Bordeaux Proceedings), Vol. 4*, Plenum Press, New York, 1986.
22. K. L. Mittal and P. Bothorel, Editors, *Surfactants in Solution (Bordeaux Proceedings), Vol. 5*, Plenum Press, New York, 1986.

23. K. L. Mittal and P. Bothorel, Editors, *Surfactants in Solution (Bordeaux Proceedings), Vol. 6*, Plenum Press, New York, 1986.
24. K. L. Mittal, Editor, *Treatise on Clean Surface Technology, Vol. 1*, Plenum Press, New York, 1987.
25. K. L. Mittal, Editor, *Surface and Colloid Science in Computer Technology*, Plenum Press, New York, 1987.
26. G. G. Fuller and K. L. Mittal, Editors, *Opportunities and Research Needs in Adhesion Science and Technology*, Proceedings of The NSF Workshop on Adhesion Science and Technology held in Lake Tahoe, CA, October 14-16, 1987, published in 1988.
27. K. L. Mittal, Editor, *Particles on Surfaces 1: Detection, Adhesion and Removal*, Plenum Press, New York, 1988.
28. K. L. Mittal, Editor, *Particles in Gases and Liquids 1: Detection, Characterization and Control*, Plenum Press, New York, 1989.
29. K. L. Mittal, Editor, *Surfactants in Solution (New Delhi Proceedings), Vol. 7*, Plenum Press, New York, 1989.
30. K. L. Mittal, Editor, *Surfactants in Solution (New Delhi Proceedings), Vol. 8*, Plenum Press, New York, 1989.
31. K. L. Mittal, Editor, *Surfactants in Solution (New Delhi Proceedings) Vol. 9*, Plenum Press, New York, 1989.
32. K. L. Mittal, Editor, *Surfactants in Solution (New Delhi Proceedings) Vol. 10*, Plenum Press, New York, 1989.
33. K. L. Mittal, Editor, *Particles on Surfaces 2: Detection, Adhesion and Removal*, Plenum Press, New York, 1989.
34. K. L. Mittal, Editor, *Polymers in Information Storage Technology*, Plenum Press, New York, 1989.
35. K. L. Mittal and J. R. Susko, Editors, *Metallized Plastics I: Fundamental and Applied Aspects*, Plenum Press, New York, 1989.
36. K. L. Mittal, Editor, *Particles in Gases and Liquids 2: Detection, Characterization and Control*, Plenum Press, New York, 1990.
37. K. L. Mittal and H. R. Anderson, Jr., Editors, *Acid-Base Interactions: Relevance to Adhesion Science and Technology*, VSP, Zeist, The Netherlands, 1991.
38. K. L. Mittal, Editor, *Particles on Surfaces 3: Detection, Adhesion and Removal*, Plenum Press, New York, 1991.
39. K. L. Mittal, Editor, *Metallized Plastics 2: Fundamental and Applied Aspects*, Plenum Press, New York, 1991.
40. K. L. Mittal and D. O. Shah, Editors, *Surfactants in Solution (Gainesville 1990 Proceedings), Vol. 11*, Plenum Press, New York, 1991.
41. K. L. Mittal, Editor, *Silanes and Other Coupling Agents*, VSP, Zeist, The Netherlands, 1992.
42. K. L. Mittal, Editor, *Metallized Plastics 3: Fundamental and Applied Aspects*, Plenum Press, New York, 1992.
43. K. L. Mittal, Editor, *Particles in Gases and Liquids 3: Detection, Characterization and Control*, Plenum Press, New York, 1993.

44. K. L. Mittal, Editor, *Contact Angle, Wettability and Adhesion*, VSP, Zeist, The Netherlands, 1993.
45. A. Pizzi and K. L. Mittal, Editors, *Handbook of Adhesive Technology*, Marcel Dekker, New York, 1994.
46. M. Strobel, C. S. Lyons and K. L. Mittal, Editors, *Plasma Surface Modification of Polymers: Relevance to Adhesion*, VSP, Utrecht, The Netherlands, 1994.
47. K. L. Mittal, Editor, *Particles on Surfaces: Detection, Adhesion and Removal*, Marcel Dekker, New York, 1995.
48. D. S. Rimai, L. P. DeMejo and K. L. Mittal, Editors, *Fundamentals of Adhesion and Interfaces*, VSP, Utrecht, The Netherlands, 1995.
49. K. L. Mittal, Editor, *Adhesion Measurement of Films and Coatings*, VSP, Utrecht, The Netherlands, 1995.
50. K. L. Mittal, Editor, *Polymer Surface Modification: Relevance to Adhesion*, VSP, Utrecht, The Netherlands, 1996.
51. M. K. Ghosh and K. L. Mittal, Editors, *Polyimides: Fundamentals and Applications*, Marcel Dekker, New York, 1996.
52. A. K. Chattopadhyay and K. L. Mittal, Editors, *Surfactants in Solution (Caracas 1994 Proceedings)*, Surfactant Science Series, No. 64, Marcel Dekker, New York, 1996.
53. K. L. Mittal and K.-W. Lee, Editors, *Polymer Surfaces and Interfaces: Characterization, Modification, and Application*, VSP, Utrecht, The Netherlands, 1997.
54. K. L. Mittal, Editor, *Metallized Plastics: Fundamentals and Applications*, Marcel Dekker, New York, 1998.
55. K. L. Mittal, Editor, *Metallized Plastics 5&6: Fundamental and Applied Aspects*, VSP, Utrecht, The Netherlands, 1998.
56. K. L. Mittal and A. Pizzi, Editors, *Adhesion Promotion Techniques: Technological Applications*, Marcel Dekker, New York, 1999.
57. P. Kumar and K. L. Mittal, Editors, *Handbook of Microemulsion Science and Technology*, Marcel Dekker, New York, 1999.
58. K. L. Mittal, Editor, *Particles on Surfaces 5&6: Detection, Adhesion and Removal*, VSP, Utrecht, The Netherlands, 1999.
59. K. L. Mittal and P. Kumar, Editors, *Emulsions, Foams and Thin Films*, Marcel Dekker, New York, 2000.
60. J. Drelich, J. S. Laskowski and K. L. Mittal, Editors, *Apparent and Microscopic Contact Angles*, VSP, Utrecht, The Netherlands, 2000.
61. K. L. Mittal, Editor, *Acid-Base Interactions: Relevance to Adhesion Science and Technology, Vol. 2*, VSP, Utrecht, The Netherlands, 2000.
62. K. L. Mittal, Editor, *Silanes and Other Coupling Agents, Vol. 2*, VSP, Utrecht, The Netherlands, 2000.
63. K. L. Mittal, Editor, *Polymer Surface Modification: Relevance to Adhesion, Vol. 2*, VSP, Utrecht, The Netherlands, 2000.
64. K. L. Mittal, Editor, *Adhesion Measurement of Films and Coatings, Vol. 2*, VSP, Utrecht, The Netherlands, 200l.

65. K. L. Mittal, Editor, *Adhesion Aspects of Thin Films, Vol. 1*, VSP, Utrecht, The Netherlands, 2001.
66. K. L. Mittal, Editor, *Polyimides and Other High Temperature Polymers: Synthesis, Characterization and Applications, Vol. 1*, VSP, Utrecht, The Netherlands, 2001.
67. K. L. Mittal, Editor, *Metallized Plastics 7: Fundamental and Applied Aspects*, VSP, Utrecht, The Netherlands, 200I.
68. K. L. Mittal, Editor, *Adhesive Joints: Formation, Characteristics and Testing, Vol. 2*, VSP, Utrecht, The Netherlands, 2002.
69. K. L. Mittal, *Particles on Surfaces 7: Detection, Adhesion and Removal*, VSP, Utrecht, The Netherlands, 2002.
70. K. L. Mittal, Editor, *Contact Angle, Wettability and Adhesion, Vol. 2*, VSP, Utrecht, The Netherlands, 2002.
71. K. L. Mittal and D. O. Shah, Editors, *Adsorption and Aggregation of Surfactants in Solution*, (Gainesville 2000 Proceedings) Surfactant Science Series, No. 109, Marcel Dekker, New York, 2003.
72. K. L. Mittal, Editor, *Adhesion Aspects of Polymeric Coatings, Vol. 2*, VSP, Utrecht, The Netherlands, 2003.
73. A. Pizzi and K. L. Mittal, Editors, *Handbook of Adhesive Technology, 2nd Edition*, CRC Press, Boca Raton, FL, 2003.
74. K. L. Mittal, Editor, *Surface Contamination and Cleaning, Vol. 1*, VSP, Utrecht, The Netherlands, 2003.
75. K. L. Mittal, Editor, *Polyimides and Other High Temperature Polymers: Synthesis, Characterization and Applications, Vol. 2*, VSP, Utrecht, The Netherlands 2003.
76. K. L. Mittal, Editor, *Particles on Surfaces 8: Detection, Adhesion and Removal*, VSP, Utrecht, The Netherlands, 2003.
77. K. L. Mittal, Editor, *Contact Angle, Wettability and Adhesion, Vol. 3*, VSP, Utrecht, The Netherlands, 2003.
78. K. L. Mittal, Editor, *Polymer Surface Modification: Relevance to Adhesion, Vol. 3*, VSP Utrecht, The Netherlands, 2004.
79. K. L. Mittal, Editor, *Silanes and Other Coupling Agents, Vol. 3*, VSP, Utrecht, The Netherlands, 2004.
80. K. L. Mittal, Editor, *Polyimides and Other High Temperature Polymers: Synthesis, Characterization, and Applications, Vol. 3*, VSP/Brill, Leiden, The Netherlands, 2005.
81. K. L. Mittal, Editor, *Adhesion Aspects of Thin Films, Vol. 2*, VSP/Brill, Leiden, The Netherlands, 2005.
82. J. Drelich and K.L. Mittal, Editors, *Atomic Force Microscopy in Adhesion Studies*, VSP/Brill, Leiden, The Netherlands, 2005.
83. K. L. Mittal, Editor, *Contact Angle, Wettability and Adhesion, Vol. 4*, VSP/Brill, Leiden, The Netherlands, 2006.
84. K. L. Mittal, Editor, *Particles on Surfaces 9: Detection, Adhesion, and Removal*, VSP/Brill, Leiden, The Netherlands, 2006.

85. K. L. Mittal, Editor, *Polymer Surface Modification: Relevance to Adhesion, Vol. 4*, VSP/Brill, Leiden, The Netherlands, 2007.
86. K. L. Mittal Editor, *Silanes and Other Coupling Agents, Vol. 4*, VSP/Brill, Leiden, The Netherlands, 2007.
87. K. L. Mittal, Editor, *Adhesion Aspects of Thin Films, Vol. 3*, VSP/Brill, Leiden, The Netherlands, 2007.
88. K.L. Mittal, Editor, *Polyimides and Other High Temperature Polymers: Synthesis, Characterization and Applications, Vol. 4*, VSP/Brill, Leiden, The Netherlands, 2007.
89. R. Kohli and K. L. Mittal, Editors, *Developments in Surface Contamination and Cleaning: Fundamentals and Applied Aspects*, William Andrew Publishing, Norwich, NY, 2008.
90. G. Biresaw and K. L. Mittal, Editors, *Surfactants in Tribology*, CRC Press, Boca Raton, FL, 2008.
91. K. L. Mittal, Editor, *Contact Angle, Wettability and Adhesion, Vol. 5*, VSP/Bril, Leiden, The Netherlands, 2008.
92. R. Gomatam and K. L. Mittal, Editors, *Electrically Conductive Adhesives*, VSP/Brill, Leiden, The Netherlands, 2008.
93. K. L. Mittal, Editor, *Polymer Surface Modification: Relevance to Adhesion, Vol. 5*, VSP/Brill, Leiden, The Netherlands, 2009.
94. K. L. Mittal, Editor, *Silanes and Other Coupling Agents, Vol. 5*, VSP/Brill, Leiden, The Netherlands, 2009.
95. K. L. Mittal, Editor, *Polyimides and Other High Temperature Polymers: Synthesis, Characterization and Applications, Vol. 5*, VSP/Brill, Leiden, The Netherlands, 2009.
96. A. Carré and K. L. Mittal, Editors, *Superhydrophobic Surfaces*, VSP/Brill, Leiden, The Netherlands, 2009.
97. K. L. Mittal, Editor, *Contact Angle, Wettability and Adhesion, Vol. 6*, VSP/Brill, Leiden, The Netherlands, 2009.
98. J. P. Matinlinna and K. L. Mittal, Editors, *Adhesion Aspects in Dentistry*, VSP/Brill, Leiden, The Netherlands, 2009.
99. K. L. Mittal and A. Pizzi, Editors, *Handbook of Sealant Technology*, CRC Press, Boca Raton, FL, 2009.
100. R. Kohli and K. L. Mittal, Editors, *Developments in Surface Contamination and Cleaning, Vol Two: Particle Deposition, Control and Removal*, Elsevier, Oxford, UK, 2010.
101. A. Pizzi and K. L. Mittal, Editors, *Wood Adhesives*, VSP/Brill, Leiden, The Netherlands, 2010
102. G. Biresaw and K. L. Mittal, Editors, *Surfactants in Tribology, Vol. 2*, CRC Press, Boca Raton, FL, 2011.
103. R. Kohli and K. L. Mittal, Editors, *Developments in Surface Contamination and Cleaning, Vol. Three, Methods for Removal of Particle Contaminants*, Elsevier, Oxford, UK, 2011.

104. A. Carré and K. L. Mittal, Editors, *Surface and Interfacial Aspects of Cell Adhesion*, VSP/Brill, Leiden, The Netherlands, 2011.
105. S. H. Kim, M. T. Dugger and K. L. Mittal, Editors, *Adhesion Aspects in MEMS/NEMS*, VSP/Brill, Leiden, The Netherlands, 2011.
106. R. Kohli and K. L. Mittal, Editors, *Developments in Surface Contamination and Cleaning, Vol. Four, Detection, Characterization and Analysis of Contaminants*, Elsevier, Oxford, UK, 2012.
107. G. Biresaw and K. L. Mittal, Editors, *Surfactants in Tribology, Vol. 3*, CRC Press, Boca Raton, FL, 2013.
108. R. Kohli and K. L. Mittal, Editors, *Developments in Surface Contamination and Cleaning, Vol. Five, Contaminant Removal and Monitoring*, Elsevier, Oxford, UK, 2013.
109. K. L. Mittal, Editor, *Advances in Contact Angle, Wettability and Adhesion, Vol. 1*, Wiley-Scrivener, Beverly, MA, 2013.
110. M. Thomas and K. L. Mittal, Editors, *Atmospheric Pressure Plasma Treatment of Polymers: Relevance to Adhesion*, Wiley-Scrivener, Beverly, MA, 2013.
111. S. Kumar and K. L. Mittal, Editors, *Advances in Modeling and Design of Adhesively Bonded Systems*, Wiley-Scrivener, Beverly, MA, 2013.
112. R. Kohli and K. L Mittal, Editors, *Developments in Surface Contamination and Cleaning, Volume Six, Methods of Cleaning and Cleanliness Verification*, Elsevier, Oxford, UK, 2013.
113. K. L. Mittal and Tanweer Ahsan, Editors, *Adhesion in Microelectronics*, Wiley-Scrivener, Beverly, MA, 2014.
114. K. L. Mittal and Thomas Bahners, Editors, *Laser Surface Modification and Adhesion*, Wiley-Scrivener, Beverly, MA, 2015.
115. G. Biresaw and K. L. Mittal, Editors, *Surfactants in Tribology, Vol 4*, CRC Press, Boca Raton, FL, 2015.
116. K. L. Mittal and Ravi Jaiswal, Editors, *Particle Adhesion and Removal*, Wiley-Scrivener, Beverly, MA, 2015.
117. R. Kohli and K. L. Mittal, Editors, *Developments in Surface Contamination and Cleaning, Volume 7, Cleanliness Validation and Verification*, Elsevier, Oxford, UK, 2015.
118. R. Kohli and K. L. Mittal, Editors, *Developments in Surface Contamination and Cleaning, Volume 8, Cleaning Techniques*, Elsevier, Oxford, UK, 2015.
119. K. L. Mittal, Editor, *Progress in Adhesion and Adhesives*, Wiley-Scrivener, Beverly, MA, 2015.
120. K. L. Mittal, Editor, *Advances in Contact Angle, Wettability and Adhesion, Volume 2*, Wiley-Scrivener, Beverly, MA, 2016.
121. R. Kohli and K. L. Mittal, Editors, *Developments in Surface Contamination and Cleaning: Fundamentals and Applied Aspects, Second Edition*, Elsevier, Oxford, UK, 2016.
122. A. N. Netravali and K. L. Mittal, Editors, *Interface/Interphase in Polymer Nanocomposites*, Wiley-Scrivener, Beverly, MA (2017).

123. R. Kohli and K. L. Mittal, Editors, *Developments in Surface Contamination and Cleaning, Volume Nine, Methods for Surface Cleaning*, Elsevier, Oxford, UK (2017).
124. R. Kohli and K. L. Mittal, Editors, *Developments in Surface Contamination and Cleaning, Volume Ten, Types of Contamination and Contamination Resources*, Elsevier, Oxford, UK (2017).
125. K. L. Mittal, Editor, *Progress in Adhesion and Adhesives, Volume Two*, Wiley-Scrivener, Beverly, MA (2017)
126. K. L. Mittal and F. M. Etzler, Editors, *Adhesion in Pharmaceutical, Biomedical and Dental Fields*, Wiley-Scrivener, Beverly, MA (2017)
127. K. L. Mittal and Thomas Bahners, *Textile Finishing: Recent Developments and Future Trends*, Wiley-Scrivener, Beverly, MA (2017).
128. G. Biresaw and K. L. Mittal, Editors, *Surfactants in Tribology, Volume 5*, CRC Press, Boca Raton, FL (2017).
129. A. Pizzi and K. L. Mittal, Editors, *Handbook of Adhesive Technology, Third Edition*, CRC Press, Boca Raton, FL (2018).
130. K. L. Mittal, Editor, *Advances in Contact Angle, Wettability and Adhesion, Volume 3*, Wiley-Scrivener, Beverly, MA (2018).
131. K. L. Mittal and W.-S. Lei, Editors, *Laser Technology: Applications in Adhesion and Related Areas*, Wiley-Scrivener, Beverly, MA (2018).
132. K. L. Mittal, Editor, *Progress in Adhesion and Adhesives, Volume 3*, Wiley-Scrivener, Beverly, MA (2018).
133. R. Kohli and K. L. Mital, Editors, *Developments in Surface Contamination and Cleaning. Volume 11: Applications of Cleaning Techniques*, Elsevier, Oxford, UK (2019).
134. R. Kohli and K. L. Mittal, Editors, *Developments in Surface Contamination and Cleaning. Volume 12: Methods for Assessment and Verification of Cleanliness of Surfaces and Characterization of Surface Contaminants*, Elsevier, Oxford, UK (2019).
135. K. L. Mittal, Editor, *Progress in Adhesion and Adhesives, Volume 4*, Wiley-Scrivener, Beverly, MA (2019).
136. G. Biresaw and K. L. Mittal, Editors, *Surfactants in Tribology. Volume 6*, CRC Press, Boca Raton, FL (2020).
137. K. L. Mittal, Editor, *Advances in Contact Angle, Wettability and Adhesion, Volume 4*, Wiley-Scrivener, Beverly, MA (2020)
138. K. L. Mittal and S. K. Panigrahi, Editors, *Structural Adhesive Joints: Design, Analysis and Testing*. Wiley-Scrivener, Beverly, MA (2020).
139. K. L. Mittal, I. S. Bakshi and J. K. Narang, Editors, *Bioadhesives in Drug Delivery*, Wiley-Scrivener, Beverly, MA (2020).
140. K. L. Mittal and C.-H. Choi, Editors, *Ice Adhesion: Mechanism, Measurement and Mitigation*, Wiley-Scrivener, Beverly, MA (2020).
141. K. L. Mittal, Editor, *Progress in Adhesion and Adhesives, Volume 5*, Wiley-Scrivener, Beverly, MA (2020).

142. K. L. Mittal and H. S. Bui, Editors, *Surface Science and Adhesion in Cosmetics*, Wiley-Scrivener, Beverly, MA (2021).
143. K. L. Mittal, Editor, *Progress in Adhesion and Adhesives, Volume 6*, Wiley-Scrivener, Beverly, MA (2021).
144. R. Kohli and K. L. Mittal, Editors, *Surfactants in Precision Cleaning: Removal of Contaminants at Micro and Nanoscale*, Elsevier, Oxford, UK (2022).
145. M. Dunky and K. L. Mittal, Editors, *Biobased Adhesives: Sources, Characteristics and Applications*, Wiley-Scrivener, Beverly, MA (2023).
146. K. L. Mittal and S. K. Panigrahi, Editors, *Structural Adhesives: Properties, Characterization and Applications*, Wiley-Scrivener, Beverly, MA (2023).
147. K. L. Mittal and S. Neogi, Editors, *Adhesives for Biomedical Applications*, Wiley-Scrivener, Beverly, MA (2023).
148. K. L. Mittal and A. N. Netravali, Editors, *Polymer Surface Modification to Enhance Adhesion: Techniques and Applications Wiley-Scrivener*, MA (2024).
149. K. L. Mittal, Editor, *Progress in Adhesion and Adhesives, Volume 7*, Wiley-Scrivener, Beverly, MA (2024).
150. K. L. Mittal, Editor, *Progress in Adhesion and Adhesives, Volume 8*, Wiley-Scrivener, Beverly, MA (2024).
151. K. L. Mittal, Editor, *Progress in Adhesion and Adhesives, Volume 9*, Wiley-Scrivener, Beverly, MA (2025).
152. K. L. Mittal, Editor, *Progress in Adhesion and Adhesives, Volume 10*, Wiley-Scrivener, Beverly, MA (2026).
153. K.L. Mittal and D.A. Aronovich, Editors, *Cyanoacrylate Adhesives: Synthesis, Properties and Applications*, Wiley-Scrivener, Beverly, MA, in press.
154. K. L. Mittal, Editor, *Progress in Adhesion and Adhesives, Volume 11*, Wiley-Scrivener, Beverly, MA, in preparation.

Appendix 2

List of publications by Dr. K.L. Mittal

1. H. Singh and K. L. Mittal, "Negative Catalyst," J. Chem. Ed., 46, 185-186 (1969).
2. J. O'M. Bockris, K. L. Mittal and R. K. Sen, "Mechanism of Activation in Electron Transfer in Solution," Nature Physical Sciences, 234, 118-119 (1971).
3. K. L. Mittal, "Conceptual Clarification of the Terms Used to Describe Emulsion Behavior," J. Soc. Cosmetic Chem., 22, 815-824 (1971).
4. J. McHardy, K L. Mittal and J. O'M. Bockris, "Electrocatalysis of Oxygen Reduction by the Tungsten Bronzes," Final Report, U.S. Army Mobility Equipment Command, Ft. Belvoir, Virginia, Contract No. DA44-009-AMC-469 (T), 73 pages, December 1971. AD- 747 766.
5. R. D. Vold and K. L. Mittal, "The Effect of Lauryl Alcohol on the Stability of Oil-in-Water Emulsions," J. Colloid Interface Sci., 38, 451-456 (1972).

6. S. H. Kim, K. L. Mittal and J. O'M. Bockris, "Ellipsometric Studies of Passivation," Final Report, Office of Saline Water, Washington, D.C., Project No. 14-30-2801, 19 pages, April 1972.
7. R. D. Vold and K. L. Mittal, "Difference in Ultracentrifugal Stability of Various Oil-in- Water Emulsions," J. Soc. Cosmet. Chem., 23, 171-188 (1972).
8. R. D. Vold and K. L. Mittal, "Determination of Sodium Dodecyl Sulfate in the Presence of Lauryl Alcohol," Anal. Chem., 44, 849-850 (1972).
9. R. D. Vold and K. L. Mittal, "Effect of Age on the Ultracentrifugal Stability of Liquid Petrolatum-Water Emulsions," J. Pharm. Sci., 61, 769-772 (1972).
10. K. L. Mittal, "Determination of CMC of Polysorbate 20 in Aqueous Solution by Surface Tension Method," J. Pharm. Sci., 61, 1334-1335 (1972).
11. K. L. Mittal and R. D. Vold, "Effect of the Initial Concentration of Emulsifying Agents on the Ultracentrifugal Stability of Oil-in-Water Emulsions," J. Am. Oil Chem. Soc., 49, 527- 532 (1972).
12. J. O'M. Bockris, R. K. Sen and K. L. Mittal, "On Quantum Electrochemical Kinetics," J. Res. Inst. of Catalysis (Hokkaido), 20, No.3, 153-184 (1972).
13. R. D. Vold and K. L. Mittal, "Effect of Temperature on the Stability of Nujol-Water Emulsions," J. Colloid Interface Sci., 42, 436-440 (1973).
14. D. F. O'Kane and K. L. Mittal, "Plasma Cleaning of Metal Surfaces," J. Vac. Sei. Technol., 11, 567-569 (1974).
15. K. L. Mittal, "Adherate: A New and General Term," J. Adhesion, 6, 377-378 (1974).
16. K. L. Mittal. "Ultracentrifugal Technique in the Study of Emulsions," in: *Colloidal Dispersions and Micellar Behavior,* K. L. Mittal, Editor, ACS Symposium Series No. 9, pp. 76-96, American Chemical Society. Washington, D. C., 1975.
17. K. L. Mittal, "A Critical Appraisal of the Methods for Measuring Adhesion of Electrodeposited Coatings," in: *Properties of Electrodeposits: Their Measurement and Significance.* R. Sard. H. Leidheiser, Jr., and F. Ogburn, Editors, Chap. 17, pp. 273-306, Electrochemical Society. Princeton, NJ, 1975.
18. A.N. Saxena and K. L. Mittal, "Optical and Dielectric Constants of Hafnium and its Anodic Oxide Films," J. Appl. Phys., 46, 2788-2789 (1975).
19. K. L. Mittal, "Surface Chemical Criteria for Maximum Adhesion and Their Verification Against the Experimentally Measured Adhesive Strength Values," in: *Adhesion Science and Technology*, L. H. Lee, Editor, Part A, pp. 129-168, Plenum Press, New York, 1975.
20. K. L. Mittal, "Adhesion Aspects of Metallization of Organic Polymer Surfaces," J. Vac. Sci. Technol., 13, 19-25 (1976). Adapted in Italian in *Galvanotecnica* 28(7), 113-115 (1977).
21. K. L. Mittal, "Adhesion Measurement of Thin Films," Electrocomponent Sei. Technol., 3. 21-42 (1976).
22. K. L. Mittal and D. F. O'Kane, "Vapor Deposited Silanes and Other Coupling Agents," J. Adhesion. 8, 93-97 (1976).

23. K. L. Mittal, "Reply to the Letter by W. A. Dukes," J. Adhesion, 8, 101-102 (1976).
24. R. P. Misra, S. Bhatia and K. L. Mittal. "A Study of Surface Degradation of Polymeric High Voltage Insulating Materials due to UV Exposure," in: *1976 Annual Report, Conference on Electrical Insulation and Dielectric Phenomena*, pp. 325-331, National Academy of Sciences, Washington, D.C., 1978.
25. K. L. Mittal and P. Mukerjee. "The Wide World of Micelles," in: *Micellization. Solubilization, and Microemulsions*, K. L. Mittal, Editor, Vol. 1, pp. 1-21, Plenum Press, New York (1977).
26. K. L. Mittal. "The Role of the Interface in Adhesion Phenomena," Polym. Eng. Sci.. 17. 467-473 (1977).
27. S. Bhatia, R. P. Misra and K. L. Mittal, "Surface Degradation of Polymeric High Voltage Insulator Materials due to Environmental Stresses," Electrochemical Society Meeting, Extended Abstracts, Vol. 77-1, pp. 58-60, May 1977.
28. G. DiGiacomo and K. L. Mittal, "Reaction of Lead Film in Perfluoro-n-Hexane (Low Conductivity) in the Presence of Added HCI and CH_3 COOH," Electrochemical Society Meeting. Extended Abstracts, Vol. 77-2, pp. 294-296, October 1977.
29. S. P. Bhatia, R. P. Misra and K. L. Mittal, "Effect of UV Exposure on Leakage Current of High Voltage Insulating Materials," in *1977 Annual Report, Conference on Electrical Insulation and Dielectric Phenomena*, pp. 360-364, National Academy of Sciences, Washington, D.C., 1979.
30. K. L. Mittal and H. R. Anderson, Jr., "Interfacial Chemistry and Adhesion: Recent Developments and Prospects," Science-Ciencia, 6, No. 1, 19-20 (1978).
31. K. L. Mittal, "Adhesion Measurement: Recent Developments, Unsolved Problems, and Prospects," in: *Adhesion Measurement of Thin Films, Thick Films, and Bulk Coatings*, STP640, K. L. Mittal, Editor, pp. 5-17, American Society for Testing and Materials, Philadelphia, PA, 1978.Translated in Italian in *Galvanotecnica*, pp. 139-145 (July/August 1979).
32. J. Ahn, K. L. Mittal and R. H. MacQueen, "Hardness and Adhesion of Filmed Structures as Determined by the Scratch Technique," in: *Adhesion Measurement of Thin Films, Thick Films, and Bulk Coatings*, STP640, K. L. Mittal, Editor, pp. 134-157, American Society for Testing and Materials. Philadelphia, PA, 1978.
33. R. D. Vold, K. L. Mittal and A. U. Hahn, "Ultracentrifugal Stability of Emulsions," in: *Surface and Colloid Science*, E. Matijevic, Editor, Vol. 10, pp. 45-97, Plenum Press, New York, 1978.
34. S. Bhattacharya and K. L. Mittal, "Mechanics of Removing Glass Particulates from a Solid Surface," Surface Technol., 7, 413-425 (1978).
35. K. L. Mittal, "Surface Contamination: An Overview," in: *Surface Contamination: Genesis. Detection and Control*, K. L. Mittal, Editor, Vol. 1, pp. 3-45, Plenum Press, New York, 1979.

36. K. L. Mittal, "Factors Affecting Adhesion of Lithographic Materials," in *Durability of Macromolecular Materials*, R. K. Eby, Editor, pp. 371-391, ACS Symposium Series No. 95, American Chemical Society, Washington, D.C., 1979. Also Reprinted in Solid State Technology, pp. 89-95, 100 (May 1979).
37. A.U. Hahn and K. L. Mittal, "Mechanism of Demulsification of Oil-in-Water Emulsion in the Centrifuge," Colloid Polym. Sci., 257, 959-967 (1979).
38. K. L. Mittal, "Surface Contamination: Its Genesis, Detection and Control," in: Proceedings (pp. 4-9) of the Nordic Association for Contamination Control Meeting held in Abo (Turku), Finland, May 21-23. 1979.
39. K. L. Mittal, "Progress in Adhesion Science," Adhesives Age, 36-37 (January 1980).
40. K. L. Mittal and R. O. Lussow, "Adhesion Measurement of Polymeric Films and Coatings with Special Reference to Photoresist Materials," in *Adhesion and Adsorption of Polymers*, L. H. Lee, Editor, Part B, pp. 503-520, Plenum Press, New York, 1980. Preprint published in ACS Organic Coatings and Plastics Preprints 40, 527-532 (April 1979).
41. K. L. Mittal, "Interfacial Chemistry and Adhesion: Recent Developments and Prospects," Pure Appl. Chem., 52, 1295-1307 (1980).
42. N. K. Eib, K. L. Mittal and A. Friedrichs, "Wettability Studies of Plasma Deposited Hexamethyldisilazane," J. Appl. Polym. Sci., 25, 2435-2438 (1980).
43. K. L. Mittal, "Clean Surface Technology," Adhesives Age, pp. 18-19 (January 1981).
44. K. L. Mittal, "Adhesion Measurement of Polymeric Coatings," Electrochemical Society Meeting Extended Abstracts Vol. 81-1, pp. 326-327 (1981).
45. K. L. Mittal, "Adhesion Aspects of Polymeric Coatings: An Overview," Electrochemical Society Meeting Extended Abstracts, Vol. 81-1, pp. 263-265 (1981).
46. K. L. Mittal, "Microscopic Aspects of Adhesion: Recent Trends and Prospects," in *Microscopic Aspects of Adhesion and Lubrication*, J. M. Georges, Editor, pp. 153-165, Elsevier, Amsterdam, 1982.
47. T. Furukawa, K. L. Mittal and H. R. Anderson, Jr. "Fundamental Interfacial Studies Using Inelastic Electron Tunneling Spectroscopy and Their Relevance to Polymer Adhesion," in: Proceedings of the IUPAC Macro 82 held at the University of Massachusetts, Amherst, July 12-16, 1982, p. 697.
48. T. Furukawa, N. K. Eib, K. L. Mittal and H. R. Anderson, Jr., "Inelastic Electron Tunneling Spectroscopic Study of the Silane Coupling Agents: 1. Aminophenyltrimethoxysilane Adsorbed on Plasma Grown Aluminum Oxide and Effects of High Humidity," Surface Interface Anal., 4, 240-244 (1982).
49. T. Furukawa, N. K. Eib, K. L. Mittal and H. R. Anderson, Jr., "Inelastic Electron Tunneling Spectroscopic Study of Silane Coupling Agents. II. γ -Aminopropyltriethoxysilane Adsorbed from Vapor on Aluminum Oxide Surface and Effects of High Humidity," J. Colloid Interface Sci., 96, 322-338 (1983).

50. M. EI-Shabasy, B. Szikora, G. Peto, I. Szabo and K. L. Mittal, "Investigation of Multilayer Systems by the Scratch Method," Thin Solid Films, 109, 127-136 (1983).
51. D. Suryanarayana and K. L. Mittal, "Effect of pH of Silane Solution on the Adhesion of Polyimide to Silica Substrate," J. Appl. Polym. Sci., 29, 2039-2043 (1984).
52. N. K. Eib, T. Furukawa and K. L. Mittal, "Digital Low Noise DC Power Supply for Tunneling Spectrometers," Rev. Sci. Instrum., 55, 1161-1163 (1984).
53. H. R. Anderson, Jr., A. I. Baise, L. P. Buchwalter, P. M. Cotts, J. Greenblatt, R. H. Lacombe, H. Lee and K. L. Mittal, "Adhesion Aspects of Organic Insulators in Semiconductor Technologies: I. Polyimides," in: *Proceedings of the International Adhesion Conference held at the University of Nottingham, Nottingham*, U.K., September 12-14, 1984.
54. D. Suryanarayana and K. L. Mittal, "Effect of Polyimide Thickness on Its Adhesion to Silicon Nitride Substrate with and without Adhesion Promoter," J. Appl. Polym. Sci., 30. 3107-3111 (1985).
55. K. L. Mittal, "Selected Bibliography on Adhesion Measurement of Films and Coatings. In: *"Testing of Metallic and Inorganic Coatings"*, STP 947, W.B. Harding and G.A. DiBari, Editors, pp. 343-362, American Society for Testing and Materials, Philadelphia, PA, 1987.
56. K. L. Mittal, "Opportunities for Joint Indo-US Research in Materials Relative to Microelectronics," in: *Proceedings of the NSF Workshop on Indian Scientific Strengths: Selected Opportunities for Indo-US Cooperation* held at SRI International, Arlington, VA, March 12-13, 1987, pp. 137-153.
57. K. L. Mittal, "Selected Bibliography on Adhesion Measurement of Films and Coatings," J. Adhesion Sci. Technol. 1, 247-259 (1987).
58. K. L. Mittal, "Interfacial Phenomena in Microelectronics: Research Needs," in *Proceedings of the NSF Workshop on Interfacial Phenomena in the New and Emerging Technologies* held at the University of Colorado, Boulder, CO, May 29-31, 1986, pp. 2-1 to 2-10, published in 1987.
59. K. L. Mittal, "A Multidisciplinary Approach to Surface Contamination Control" (interview), Microcontamination, 16-24 (May 1988).
60. G. G. Fuller and K. L. Mittal, "Workshop on Adhesion Science and Technology: Executive Summary" in: *Opportunities and Research Needs in Adhesion Science and Technology*, G. G. Fuller and K. L. Mittal, Editors. Proceedings of The NSF Workshop on Adhesion Science and Technology held in Lake Tahoe, CA, October 14-16, 1987, published in September 1988, pp. VII- XII.
61. K. L. Mittal, "Surface Cleaning: Recent Developments and Needs," Proc. 34th Annual Technical Conference of the Society of Vacuum Coaters, pp. 13-17 (1991).
62. K. L. Mittal, "Adhesion Measurement of Thin Films: A Critical Assessment," Proc. Fifth International Web Coating Conference, pp. 201-212 (1991).

63. K. L. Mittal, "Surface Modification Techniques: A Summary Report," Proc. Surface Processing Workshop, NREL/CP-412-5007, pp. 9-1 to 9-4 (Sept. 1992).
64. K. L. Mittal, "Surface Contamination Control Primer" (interview), CleanRooms, 10-14 (April 1993).
65. F. J. Boerio, G. D. Davis, J. E. deVries, C. E. Miller, K. L. Mittal, R. L. Opila and H. K. Yasuda, "Polymer-Metal (Oxide) Interfaces" Crit. Rev. Surface Chem., 3(1), 81-89 (1993).
66. K. L. Mittal, "Surface Contamination Concepts and Concerns." Precision Cleaning, 17-22 (Jan. 1995).
67. K. L. Mittal, "Adhesion Measurement of Films and Coatings: A Commentary." in *Adhesion Measurement of Films and Coatings*, K. L. Mittal, Editor, pp. 1-13, VSP, Utrecht, The Netherlands, 1995.
68. R. Kohli and K. L. Mittal, "Nanoscale Removal of Contaminant Particles and Films," Vacuum Technology & Coating, 2. No. 8, 50-57 (August 2001).
69. K. L. Mittal, "Adhesion Measurement of Films and Coatings: A Commentary". Vacuum Technology & Coating, 2, No. 12, 44-48 (December 2001).
70. Tito E. Garza-Lopez, Erol Sancaktar and K. L. Mittal, "Excimer Lasers: How They Work and Ablate", Vacuum Technology & Coating, 3, No. 5, 52-59 (May 2002).
71. William Birch, Alain Carré, and Kashmiri L. Mittal, "Wettability Techniques to Monitor the Cleanliness of Surfaces", in: *Developments in Surface Contamination and Cleaning: Fundamental and Applied Aspects*, R. Kohli and K. L. Mittal (Eds), Chap. 13, pp. 693-723, William Andrew Publishing, Norwich, NY (2008).
72. K. L. Mittal and Frank M. Etzler, "Is the World Basic? Lessons from Surface Science," ANNALES (Universitatis Mariae Curie-Sklodowska, Lublin-Polonia) LXIII, Sectio AA(Chemia), 1-16 (2008).
73. K. L. Mittal and Frank M. Etzler, "Is the World Basic? Lessons from Surface Science", in: *Contact Angle, Wettability and Adhesion*, Vol. 6, K. L. Mittal (Ed.) pp. 112-123, VSP/Brill, Leiden (2009).
74. A. N. Netravali and K. L. Mittal, Fiber Surface Treatment: Relevance to Interfacial Characteristics," in: *Wiley Encyclopedia of Composites*, 2nd edition, L. Nicolais, A. Borzacchiello and S. M. Lee (Eds.) pp. 980-998, John Wiley & Sons, Hoboken, New Jersey (2012).
75. D. Williams and K. L. Mittal, "Wettability Techniques to Monitor the Cleanliness of Surfaces," in: *Developments in Surface Contamination and Cleaning: Fundamental and Applied Aspects*, Second Edition, R. Kohli and K. L. Mittal (Eds.) Chap. 10, Elsevier, Oxford, UK (2016).
76. Wei-Sheng Lei, Kash Mittal and Ajay Kumar, "Adhesion Measurement of Thin Films and Coatings: Relevance to Biomedical Applications", in: *Medical Coatings and Deposition Technologies*, D. Glocker and S. Ranade (Eds.). pp. 631-670, Wiley-Scrivener, Beverly, MA (2016).

77. Wei-Sheng Lei, Kash Mittal and Zhishui Yu, "Adhesion Measurement of Coatings on Biodevices/Implants: A Critical Review", Rev. Adhesion Adhesives 4, 367-397 (2016).
78. Thomas Bahners and Kash Mittal, " Striving for Self-Cleaning Textiles - Critical Thoughts on Current Literature," in: *Textile Finishing: Recent Developments and Future Trends*, K. L. Mittal and Thomas Bahners (Eds.) Chap. 3, pp. 129-169, Wiley-Scrivener, Beverly, MA (2017).
79. Wei-Sheng Lei and Kash Mittal, "Laser Surface Modification for Adhesion Enhancement," in: *Laser Technology: Applications in Adhesion and Related Areas*, K. L. Mittal and Wei-Sheng Lei (Eds.), Chap. 4, pp. 123-174, Wiley-Scrivener, Beverly, MA (2018).
80. H. S. Panda, Rigved Samant, K. L. Mittal and S. K. Panigrahi, "Durability Aspects of Structural Adhesive Joints," in: *Structural Adhesive Joints: Design. Analysis and Testing*, K. L. Mittal and S. K. Panigrahi (Eds), pp. 97 - 134, Wiley-Scrivener, Beverly, MA (2020).

A Personal Tribute to Dr. Kash Mittal for His Excellent Help and Collaboration with Me and My Colleagues

Emil Chibowski

Professor Volunteer, Department of Interfacial Phenomena, Institute of Chemical Sciences, Faculty of Chemistry, Maria Curie-Skłodowska University, Lublin, Poland

Introduction

The celebration of the 80th birthday of Dr. Kash Mittal is a wonderful occasion for our research group to express our profound gratitude for his excellent and long-time cooperation in the fields of adhesion, wetting, spreading, contact angle, surfactants in solution, and others. He often helped us attend the events by waiving the registration fees for the conferences and symposia organized by him. This enabled presenting the results of our studies in person.

We are also indebted to him for his fastidious review and for suggestions made to improve our papers sent for publication in the *Journal of Adhesion Science and Technology* under his editorship. Chapters in books edited by him, and for the invitations to participate in the Symposia: *Surfactants in Solution; Contact Angle, Wettability, and Adhesion; Acid-Base Interactions: Relevance to Adhesion*, and others. We were also asked to review papers sent by other authors for publication in the *Journal of Adhesion Science and Technology*, for which I was an *Editorial Board* member for many years.

It is not our intention to describe in this tribute Dr. Kash Mittal's contribution to Surface and Colloid Chemistry, because he is a world-renowned person. We wish to depict the events that were of great importance for our research group and very much helped us to present results of our investigations, as well as to attend many international symposia and conferences.

Email: e.chibowski@gmail.com

Anil N. Netravali and Frank M. Etzler (eds.) The Wide World of Adhesion: A Festschrift in Honor of Dr. Kash Mittal on his 80th Birthday, (21–48) © 2026 Scrivener Publishing LLC

The Beginning of Our Cooperation

It happened in 1981 when Dr. Kash Mittal invited us to attend the Symposium on "Physicochemical Aspects of Polymer Surfaces" organized by him in New York City (August 24 -28, 1981). This Symposium was held as a part of the American Chemical Society Meeting. The Division of Colloid and Surface Chemistry was the principal sponsor. At that time, Dr. K. L. Mittal was employed at the IBM Corporation, and he was the Chairman of this Symposium. Unfortunately, we were not able to participate in person in this Symposium, but we contributed a paper. The Proceedings of the Symposium were published in 1983, Volumes 1 and 2. The paper: E. Chibowski, B. Jańczuk, and W. Wójcik, Properties of n-Alkane Films in the System: Teflon/n-Alkane-Water. *In: Physicochemical Aspects of Polymer Surfaces. K.L. Mittal (Ed.) Springer, Boston, MA*, was published in Vol. 2, pages 217-227. https://doi.org/10.1007/978-1-4615-7584-9_13. The original cover of the book and the first page of our paper from that book are shown in Figure 1.

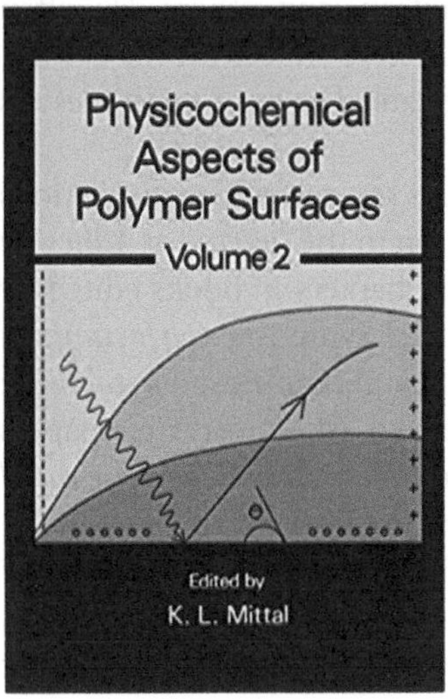

E. Chibowski, B. Jańczuk and W. Wójcik

Department of Physical Chemistry, Institute of Chemistry
Maria Curie-Sklodowska University
Lublin, Poland

Although Teflon possesses a very low surface free energy, n-alkane thick liquid films may exist in Teflon/n-alkane-water system. We examined properties of the films by measurements of contact angle, adhesion force of air bubbles and zeta potential for Teflon/n-alkane in water. Series of hydrocarbons ranging from n-hexane to n-hexadecane were tested. The results obtained show that equilibrium values of the measured parameters change nonlineary with chain length of the alkanes. The changes of the parameters correlate between themselves and with melting temperature of the alkanes. For Teflon/n-hexane in water system, the film pressure values were calculated from the experimental relationship between specific film volume and zeta potential. In the extreme case, about 4 statistical monolayers of n-hexane were present on the Teflon surface. If air bubbles are attached and then removed from Teflon surface wetted completely with n-hexane, they take away some amounts of n-hexane. This results in an increase in the adhesion force and decrease in contact angle, both of which reach equilibrium values. About 13 air bubbles are needed to attain the equilibrium, and 13-14 air bubbles remove the film almost completely. Comparison of the results from zeta potential and contact angle measurements leads to the conclusion that the most stable film exists at 1.5 to 2.5 statistical monolayers.

217

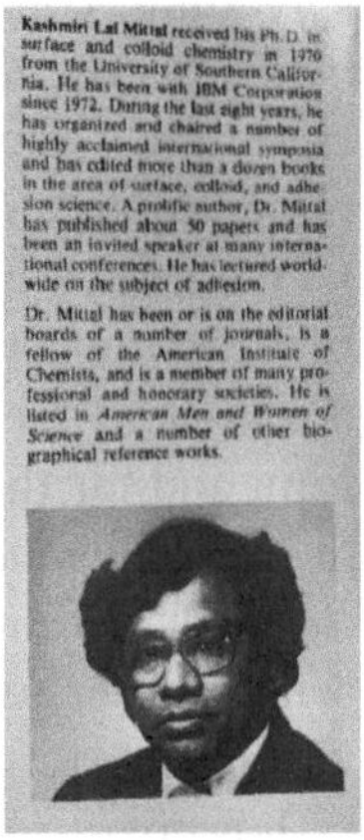

Kashmiri Lal Mittal received his Ph.D. in surface and colloid chemistry in 1970 from the University of Southern California. He has been with IBM Corporation since 1972. During the last eight years, he has organized and chaired a number of highly acclaimed international symposia and has edited more than a dozen books in the area of surface, colloid, and adhesion science. A prolific author, Dr. Mittal has published about 50 papers and has been an invited speaker at many international conferences. He has lectured worldwide on the subject of adhesion.

Dr. Mittal has been or is on the editorial boards of a number of journals, is a fellow of the American Institute of Chemists, and is a member of many professional and honorary societies. He is listed in *American Men and Women of Science* and a number of other biographical reference works.

Figure 1 Cover of the book *Physicochemical Aspects of Polymer Surfaces* edited by Kash Mittal (1983) and the first page of our first paper published in this book.

Dr. Kash Mittal Granted the Title Doctor *Honoris Causa* of UMCS

Having profound gratitude to Dr. Mittal, in 2003, this author applied to the authorities of Maria Curie-Skłodowska University (UMCS) with the request to grant him the doctor *honoris causa* degree. My petition was approved by the Senate of our university, and it was a great satisfaction for me. Then on April 30, 2003, Dr. Mittal was awarded this valuable degree, and he personally collected it on October 24, 2003. The original UMCS Senate Resolution dealing with this nomination is shown in Figure 2 (in Polish). As mentioned above, the ceremony took place on October 24, 2003, a day after the official opening of the Academic Year 2003/2004 at UMCS. The Senate session with Dr. Kash Mittal presence is shown in Figure 3.

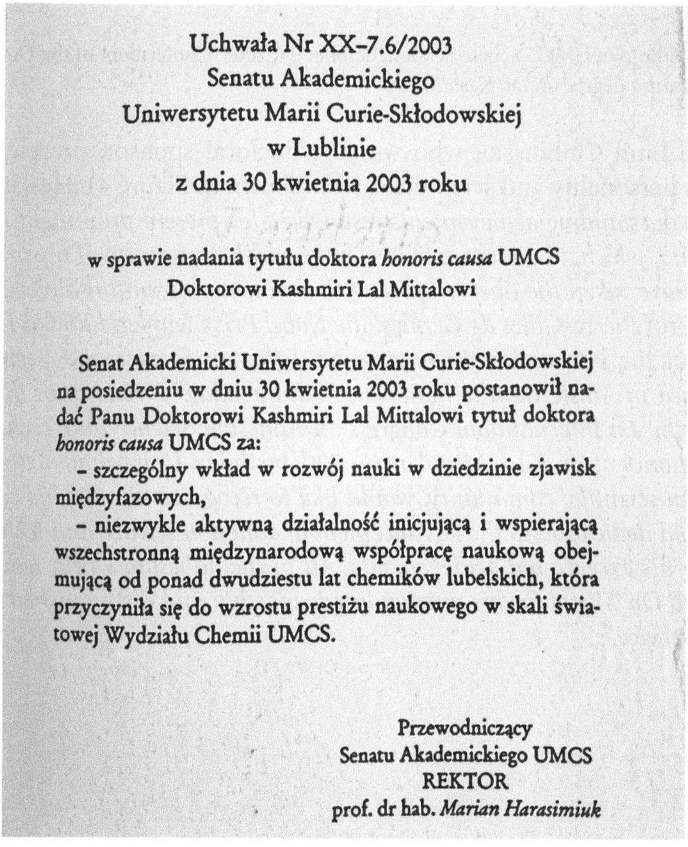

Uchwała Nr XX–7.6/2003
Senatu Akademickiego
Uniwersytetu Marii Curie-Skłodowskiej
w Lublinie
z dnia 30 kwietnia 2003 roku

w sprawie nadania tytułu doktora *honoris causa* UMCS
Doktorowi Kashmiri Lal Mittalowi

Senat Akademicki Uniwersytetu Marii Curie-Skłodowskiej na posiedzeniu w dniu 30 kwietnia 2003 roku postanowił nadać Panu Doktorowi Kashmiri Lal Mittalowi tytuł doktora *honoris causa* UMCS za:

- szczególny wkład w rozwój nauki w dziedzinie zjawisk międzyfazowych,

- niezwykle aktywną działalność inicjującą i wspierającą wszechstronną międzynarodową współpracę naukową obejmującą od ponad dwudziestu lat chemików lubelskich, która przyczyniła się do wzrostu prestiżu naukowego w skali światowej Wydziału Chemii UMCS.

Przewodniczący
Senatu Akademickiego UMCS
REKTOR
prof. dr hab. *Marian Harasimiuk*

Figure 2 A copy of the original UMCS Senate Resolution regarding Dr. K.L. Mittal's nomination (in Polish).

Figure 3 Photo of UMCS Senate Session, October 24, 2003. Conferment of the Doctor *Honoris Causa* degree on Dr. Kash Mittal.

Prof. Emil Chibowski, who was the doctoral sponsor, presented Dr. Mittal's personality and scientific career (shown in Figure 4). He said, "*Dr. Mittal's outstanding achievements resulted in his present prominent rank in academia, which, to a large extent, is not typical. As a motto of my presentation, I have taken the liberty of using the dedication from the silver plaque, which* Prof *Pierre-Gilles de Gennes, the Nobel Prize winner, handed Dr. K.L. Mittal at the International Congress on Adhesion Science and Technology organized in Amsterdam to honor his 50th birthday." It reads: On the occasion of the 1st International Congress on Adhesion Science and Technology, held in honor of Dr. Kash Mittal on his 50th birthday, I, on behalf of the global adhesion scientific community, would like to recognize him for his commitment and dedication to the advancement of adhesion science and technology and for his exceptional achievements. All in the adhesion world would like to thank Dr. Mittal for his outstanding leadership and distinguished professional service."*

Figure 4 Prof. Emil Chibowski presents Dr. Kash Mittal's personality and scientific career.

Then Dr. Kash Mittal, in his *lectio doctoris*, said: "I am deeply thankful to the Maria Curie-Skłodowska University for bestowing upon me the title of Doctor *honoris causa*. . . . I am particularly delighted today for two reasons: first, I feel most gratified that my contributions have been acknowledged, and second, I have had a long association with many faculty members of this university. So it is a momentous occasion and a special day for me".

And continuing, he stated: "The topic of my presentation 'The Wide World of Surfaces and Interfaces' is very apropos as the Maria Curie-Sklodowska University has had a long tradition of carrying out world-class research, both theoretical and experimental,...and many faculty members have made sterling contributions over a long period".

Next, he explained why surfaces were fascinating, citing: "Surfaces, surfaces everywhere, their study is fun, nobody can deny the importance of bulk, but still the surface is number one", and: "Only after knowing the surface of things one can throw himself into the search for what is underneath. But the surface of things is endless" (Italio Calvino, Palomar, trans. and cited by M. Morra). As for his personal scientific interest, Dr. Mittal said: "I have been involved with surfaces and interfaces throughout my career", starting with his M.Sc. Thesis ("Ion Exchange Capacity of Some Clays"), then Ph.D. Thesis (Factors Influencing the Ultracentrifugal Stability of Oil-in-Water Emulsions) and postdoctoral work (corrosion and passivation). In 1972, he found the subject of adhesion to be a wonderful one. So he says, "since 1972, my romance with adhesion science has continued unabated, and I plan to keep this love affair going even more intensely in times to come", and it appears to be true till now.

The moment of handing the doctor *honoris causa* diploma to Dr. Kash Mittal by the Rector UMCS Prof. Marian Harasimiuk can be seen in Figure 5. A few more pictures taken during the ceremony of the doctor *honoris causa* celebration are displayed in Figures 6–9.

Figure 5 Rector UMCS is presenting the Doctor Honoris Causa diploma to Dr. Kash Mittal in 2003. On the left is the Dean of the Faculty of Chemistry UMCS Prof. Andrzej Dąbrowski.

Figure 6 Dr. Kash Mittal with a bunch of flowers presented to him during the ceremony.

Figure 7 Cover of the booklet prepared on the occasion of the Doctor *Honoris Causa* award for Dr. Kash Mittal.

Figure 8 In the banquet, Rector UMCS Prof. Harasimiuk proposes a toast to Doctor HC Kash Mittal.

Figure 9 Dr. Mittal and the reviewers who evaluated his achievements – to the left Prof. W. Zwierzykowski (Technical University, Gdańsk) and to the right Prof. W. Wójcik (UMCS, Lublin).

Cooperation with the Journal of Adhesion Science and Technology (JAST)

Dr. Kash Mittal was, for many years (until 2012), the Editor-in-Chief of this Journal. A lot of top-quality papers have been published by internationally recognized authors, which have been cited numerous times. Also, the members of the Editorial Advisory Board were of worldwide reputation. I had the great pleasure of being one of these members. As an example, the International Advisory Board in the year 2010 is presented in Figure 10. Typically, three reviewers evaluated each manuscript submitted for publication in the Journal and also each paper was carefully examined by Dr. Kash Mittal personally, and the example is given in Figure 11. Our research group from the Department of Interfacial Phenomena, Faculty of Chemistry UMCS in Lublin, Poland also published many papers in this Journal during our over 40- year cooperation. Some of the papers were published with our colleagues from Granada University, Spain. They are listed below.

Journal of Adhesion Science and Technology

Aims

The *Journal of Adhesion Science and Technology* provides a forum for the basic aspects, theories and mechanisms of adhesion and particularly welcomes contributions dealing with applications of adhesion principles and theories in all areas of technology.

The progress of the understanding of mechanisms of bonding and bond failure is a result of the advent of surface-sensitive analytical tools and their application to the study of adhesion problems. This, combined with the advancement of new theories and views on adhesion or adhesion failure, has led to the need for this journal covering both principles and theories and their application in several new areas.

Scope

The scope of the *Journal of Adhesion Science and Technology* encompasses, but is not restricted to, the following areas within the purview of adhesion: theories of adhesion; surface energetics; fracture mechanics; development and applications of surface-sensitive methods to study adhesion phenomena; nondestructive methods for bond failure studies; techniques for the measurement of adhesion of thin films, thick films and coatings; surface modification of polymers to improve their adhesion; mechanisms and studies of coupling agents; adhesive joints, wood bonding; adhesion aspects of composite materials; biomedical materials; structural adhesives; paint and coatings (e.g. in relation to corrosion control); adhesion aspects in microelectronics; adhesion of lithographic materials; metallized plastics; rubber and elastomer adhesion; adhesion of aerospace materials.

The journal publishes review articles pertaining to mechanisms or theories of adhesion and on adhesion aspects in certain areas of technological importance. In addition, original research papers or short communications on any of the above mentioned aspects are accepted. The *Journal of Adhesion Science and Technology* also publishes a diary on international events related to any aspect of adhesion science and plans to highlight institutions which have a concentrated activity dealing with adhesion. Short reports on current research and "Letters to the Editor" can also be submitted.

ISSN 0169-4243 print/ISSN 1568-5616 online

http://dx.doi.org/10.1080/01694243.2012.731244
http://www.tandfonline.com

Figure 10 The Editorial Advisory Board of the *Journal of Adhesion Science and Technology* in 2010.

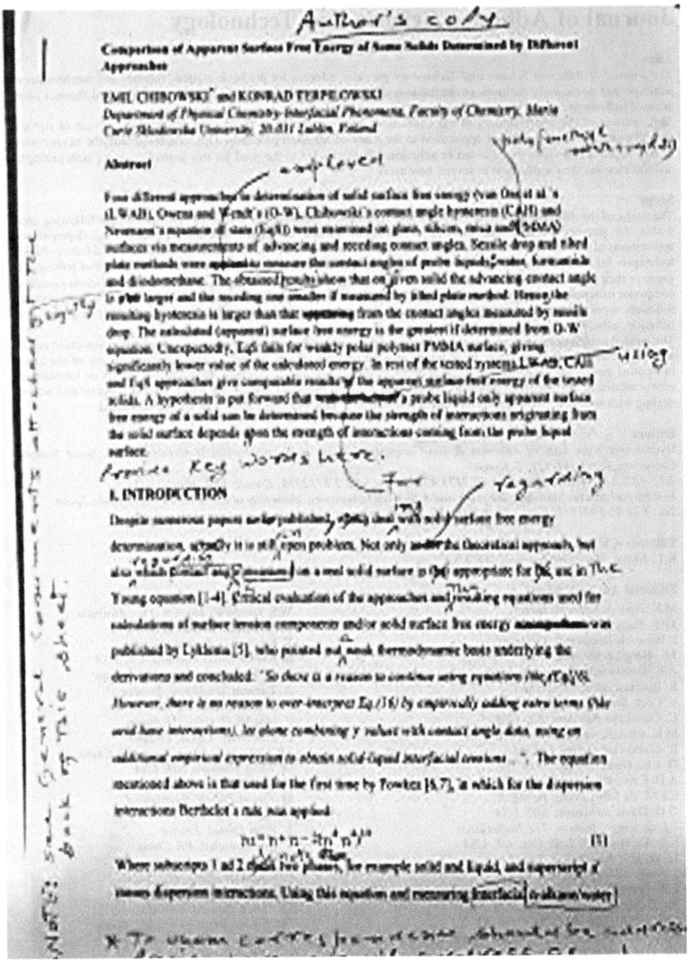
Comparison of Apparent Surface Free Energy of Some Solids Determined by Different Approaches

EMIL CHIBOWSKI and KONRAD TERPILOWSKI

Abstract

1. INTRODUCTION

Figure 11 An example of the questions, comments, and corrections written by Dr. Kash Mittal on a manuscript. This is only shown to illustrate the detailed evaluation and not to read it.

Papers Published in the Journal of Adhesion Science and Technology (JAST)

1. E. Chibowski and L. Hołysz, Surface free energy of synthetic chalcocite from zeta potential and contact angle techniques. *J. Adhesion Sci. Technol.* 3, 575-585 (1989).
2. E. Chibowski, Solid surface free energy components determination by the thin-layer wicking technique. *J. Adhesion Sci. Technol.* 6, 1069-1090 (1992).

3. F. Gonzalez-Caballero, R. Perea-Carpio and E. Chibowski, On the adhesion between particles and air bubbles in the celestite/n-alkyloammonium chloride system. *J. Adhesion Sci. Technol.* 7, 1001-1011 (1993).
4. E. Chibowski and F. Gonzalez-Caballero, Interpretation of contact angle hysteresis. *J. Adhesion Sci. Technol.* 7, 1195-1209 (1993).
5. L. Hołysz, E. Chibowski, W. Wójcik, K.W. Busch and M.A.Busch, Parameters determining deposition of calcium carbonate into a glass capillary. *J. Adhesion Sci. Technol.*, 8, 181-193 (1994).
6. J.M. Bruque, M.L. Gonzalez-Martin, J.A. Mendez-Sierra and R. Perea-Carpio, Adhesion of Air Bubbles to a Fluorite Surface in the Presence of Oleate Species. *J. Adhesion Sci. Technol.* 8, 289-300 (1994).
7. B. Jańczuk, J.M. Bruque, M.L. Gonzalez-Martin and R. Perea-Carpio, Influence of n-Alkyl ammonium Chlorides on the Adhesion of Air Bubbles to the Fluorite Surface, *J. Adhesion Sci. Technol.* 8, 1017-1025 (1994).
8. E. Chibowski, L. Hołysz, B.Dzikowska and R. Ogonowski, Studies on impurity ion and radiofrequency electric field effect on carbonate deposition. *J. Adhesion Sci. Technol.* 10, 421-430 (1996).
9. J.D.G. Duran, A. Ontiveros, A.V. Delgado, F. Gonzalez-Caballero and E. Chibowski, A study on the adhesion of calcium carbonate to glass. Energy balance in the deposition process. *J. Adhesion Sci. Technol.* 10, 847-868 (1996).
10. A. Ontiveros, J.D.G. Duran, A.V. Delgado, F. Gonzalez-Caballero and E. Chibowski, Adhesion of colloidal ZnS on silicon. Effect of ionic strength and radio frequency electric field. *J. Adhesion Sci. Technol.*, 10, 999-1019 (1996).
11. E. Chibowski and L. Hołysz, On the use of Washburn equation for wetting contact angle determination. *J. Adhesion Sci. Technol.*, 11, 1289-1301 (1997).
12. F. Durando, F.M. Gama, E. Chibowski and M. Mota, Characterization of cellulose surface free energy. *J. Adhesion Sci. Technol.*, 12, 1081-1090 (1998).
13. E. Chibowski, L. Hołysz and M. Lubomska, Effect of external radiofrequency electric field on the surface free energy components of calcium carbonate in the presence of cationic and anionic surfactant. *J. Adhesion. Sci. Technol.*, 13, 1103-1117 (1999).
14. W. Wójcik, B. Jańczuk, and R. Ogonowski, Interparticle Interaction between Coal Grains through an Alkane, *J. Adhesion Sci. Technol.* 14, 1021-1033 (2000).
15. W. Wójcik, B. Jańczuk, and R. Ogonowski, Adhesion of a coal grain to a coal grain in an alcohol, *J. Adhesion Sci Technol.*, 14, 1665-1676 (2000).
16. W. Wójcik, B. Jańczuk, and R. Ogonowski, Determination of the attachment and detachment forces in a coal grain/liquid/coal grain system by detachment experiments, *J. Adhesion Sci. Technol.* 15, 1399-1401 (2001).
17. M. Espinosa-Jimenez, R. Padilla-Weigand, A. Ontiveros-Ortega, R. Perea-Carpio, M.M. Ramos-Tejeda and E. Chibowski, Investigation of the polyamide 6,6 dyeing process with acid blue 45 dye. Part I. thermodynamics of acid blue 45 adsorption. *J. Adhesion Sci. Technol.*, 16, 285-301 (2002).

18. M. Espinosa-Jimenez, R. Padilla-Weigand, A. Ontiveros-Ortega, R. Perea-Carpio, M.M. Ramos-Tejeda, and E. Chibowski, Investigation of the polyamide 6,6 dyeing process with acid blue 45 dye. Part II. Surface free energy, zeta potential and dye/polyamide interactions. *J. Adhesion Sci. Technol.*, 16, 303-316 (2002).
19. E. Chibowski, A. Ontiveros-Ortega, and R. Perea-Carpio, On the interpretation of contact angle hysteresis. *J. Adhesion Sci. Technol.*, 16, 1367-1404 (2002).
20. H. Radelczuk, L. Hołysz and E. Chibowski, Comparison of the Lifshitz-van der Waals/acid-base and contact angle hysteresis approaches to determination of the solid surface free energy.*J. Adhesion Sci. Technol.*, 16, 1547-1568 (2002).
21. W. Wójcik, B. Jańczuk, and R. Ogonowski,The effect of long chain alkanes on adhesion between Silica Particles, *J. Adhesion Sci. Technol.* 17, 277-289 (2003).
22. W. Wójcik, B. Jańczuk and M.Krasowska, Adhesion between silica particles in an alcohol medium, *J. Adhesion Sci. Technol.*17, 1945-195 (2003).
23. E. Chibowski, L. Hołysz and K. Terpiłowski, Effect of magnetic field on deposition and adhesion of calcium carbonate particles on different substrates. *J. Adhesion Sci. Technol.*, 17, 2005-2021 (2003).
24. A. Szcześ, L. Hołysz, and E. Chibowski, Influence of magnetic field on the properties of freshly precipitated calcium phosphate and its adhesion to glass surface, *J. Adhesion Sci. Technol.* 20, 345-358 (2006).
25. T. Białopiotrowicz, Influence of erroneous data on the results of calculations from acid-base surface free energy theories. I. Simulations for a small input data set, *J. Adhesion Sci. Technol.* 21, 1539-1556 (2007).
26. T. Białopiotrowicz, Influence of erroneous data on the results of calculations from acid-base surface free energy theories. II. Why are negative values of square roots obtained?, *J. Adhesion Sci. Technol.* 21, 1557-1573 (2007).
27. R. Ogonowski, B. Jańczuk and W. Wójcik, Interaction of silica particles through a liquid, *J. Adhesion Sci. Technol.* 22, 111-120 (2008).
28. K. Szymczyk and B. Jańczuk, Adhesion of aqueous solution of binary surfactant mixtures to polytetrafluoroethylene, *J. Adhesion Sci. Technol.*22, 1145-1157 (2008).
29. T. Białopiotrowicz, Influence of erroneous data on the results of calculations from acid-base surface free energy theories. III. Solution of three-equation set in the case of homoscedastic error, *J. Adhesion Sci. Technol.* 23, 799-813 (2009).
30. T. Białopiotrowicz, Influence of erroneous data on the results of calculations from acid–base surface free energy theories. IV. Solution of an overdetermined set in the case of homoscedastic error, *J. Adhesion Sci. Technol.* 23, 815-825 (2009).
31. E. Chibowski, Apparent surface free energy of superhydrophobic surfaces, *J. Adhesion Sci. Technol.* 25, 1323-1336 (2011).

32. K. Szymczyk and B. Jańczuk, Wettability of polymeric solids by aqueous solutions of anionic and nonionic surfactant mixtures, *J. Adhesion Sci. Technol.* 25, 2641–2657 (2011).

Chapters We Published in the Books Edited by Dr. Kash Mittal

Our scientific cooperation with Dr. Mittal, among others, is reflected in the above listed 32 published papers. However, maybe our chapters published in the books edited by Dr. Mittal, currently over 150, even better depict the cooperation. This is because many of them reflect the results presented during various symposia and/or conferences organized by him. Below are listed the chapters published in the books.

1. E. Chibowski, B. Jańczuk and W. Wójcik, Properties of n-alkane films in the system : Teflon/n-alkane-water, in: *Physicochemical Aspects of Polymer Surface*, Vol.2, K.L. Mittal (Ed.), pp. 217- 227, Plenum Press (1983).
2. E. Chibowski, Solid surface free energy components by thin-layer wicking technique, in: *Contact Angle, Wettability and Adhesion*, Festschrift in honour of Professor Robert J. Good, K.L. Mittal (Ed.), pp. 641-662, VSP, Utrecht, The Netherlands (1993).
3. E. Chibowski, L. Hołysz, A. Zdziennicka, and F. Gonzalez-Caballero, Is contact angle hysteresis due to a film present behind the drop, in: *Surfactants in Solution*, A.K. Chattopadhyay and K.L. Mittal (Eds.), pp. 31-53, Marcel Dekker, New York (1996).
4. E. Chibowski and A. Sołtys, The effect of radiofrequency electric field on the properties of dispersed systems: relevance to adhesion, in: Festschrift in honor of Dr. K.L. Mittal on the occasion of his 50th birthday, W.J van Ooij and H.R. Anderson, Jr. (Eds.), pp. 837-855, VSP Zeist, The Netherlands (1998).
5. E. Chibowski and L. Hołysz, Effect of PMMA precoverage on acid-base interactions of carbonate rock, in: *Acid-Base Interactions: Relevance to Adhesion*, Vol, 2, K.L. Mittal (Ed.), pp. 615-624, CRC Press, Boca Raton, FL (2000).
6. E. Chibowski, Thin layer wicking – methods for the acid-base free energy interactions determination, in: *Acid-Base Interactions; Relevance to Adhesion*, Vol.2, K.L. Mittal (Ed.), pp. 419-437, CRC Press, Boca Raton, FL (2000).
7. E. Chibowski, Contact angle hysteresis due to a film present behind the drop, in: *Contact Angle, Wettability and Adhesion*, K.L. Mittal (Ed.), Vol. 2, pp. 265-288, CRC Press, Boca Raton, FL (2002).
8. L. Hołysz, E. Chibowski and K. Terpiłowski, Influence of ambient humidity on the apparent surface free energy of poly(methyl methacrylate) (PMMA), in: *Contact Angle, Wettability and Adhesion*, K.L. Mittal (Ed.), Vol. 5, pp. 95–111, CRC Press, Boca Raton, FL (2008).

9. E. Chibowski and K. Terpiłowski, Comparison of apparent surface free energy of some solids determined by different approaches, in: *Contact Angle, Wettability and Adhesion,* K.L. Mittal (Ed.), Vol. 6, pp. 283-299, CRC Press, Boca Raton, FL (2009).
10. E. Chibowski, K. Terpiłowski and L. Hołysz, Effect of relative humidity on contact angle and its hysteresis on phospholipid DPPC bilayer deposited on glass, in: *Advances in Contact Angle, Wettability and Adhesion,* K.L. Mittal (Ed.), Vol.1, pp. 329-346, Wiley- Scrivener, Beverly, MA (2013).
11. E. Chibowski, M. Jurak and L. Hołysz, Wettability of solid supported lipid layers, in: *Surfactant Science and Technology,* (Kash Mittal Festschrift), pp. 119-145, Laurence S. Romsted (Ed.), CRC Press, Boca Raton, FL (2014).
12. K. Terpiłowski, Surface free energy of superhydrophobic materials obtained by deposition of polymeric particles on glass, in: *Advances in Contact Angle, Wettability and Adhesion,* K.L. Mittal (Ed.), Vol. 2, pp. 379-395, Wiley-Scrivener, Beverly, MA (2015).
13. E. Chibowski, L. Holysz and A. Szcześ, Wettability of powders, in: *Adhesion Science in Pharmaceutical Biomedical and Dental Fields,* K.L. Mittal and F. Etzler (Eds.), pp. 23-44, Wiley-Scrivener, Beverly, MA (2017).
14. K. Terpiłowski, D. Rymuszka, O. Goncharuk and L. Yakovenko, Equilibrium contact angle and determination of apparent surface free energy using hysteresis approach on rough surface, in: *Advances in Contact Angle, Wettability and Adhesion,* K.L. Mittal (Ed.), Vol.3, pp. 331-348, Wiley-Scrivener, Beverly, MA (2018).
15. K. Terpiłowski, M. Bielska, K. Prochaska, and E. Chibowski, Wettability of ultrafiltration membranes, in: *Advances in Contact Angle, Wettability and Adhesion,* K.L. Mittal (Ed.), Vol.4, pp. 57-72, Wiley-Scrivener, Beverly, MA (2019).

First International Congress on Adhesion Science and Technology on the Occasion of the 50th Birthday of Dr. K.L. Mittal

Among many symposia and conferences that took place during our long-time cooperation with Dr. Kash Mittal, this Congress needs a special mention. It was held in the Grand Hotel Krasnapolsky in Amsterdam, October 16-20, 1995, where 235 papers from 38 countries were presented to celebrate the 50th Birthday of Dr. Mittal. Among those present were many luminaries, such as Nobel Prize winner Prof. Pierre-Gilles de Gennes, who handed Dr. Mittal a special plaque with the inscription, which is presented in Figure 12. During the Congress, two papers were presented by this author. Then in the Festschrift the invited papers were published in seven Parts, i.e. Part 1: Fundamental aspects of adhesion and general topics; Part 2: Contact angle, wettability and surface energetics; Part 3: Surface modification: Relevance to adhesion; Part 4: Adhesives and adhesive joints; Part 5: Adhesion aspects of polymeric coatings, and polymer-polymer interphase; Part 6: Metal-polymer and metal-ceramic adhesion; and Part 7: General papers. In the final section of it, the titles, authors, and affiliations of other papers that were included in the technical program of the conference were listed. The cover of the Festschrift and the titles of Part 7 papers are presented in Figure 13. On the occasion of this celebration, Dr. S. Bandyopadhay from UNSW Australia wrote a poem, *"Fun at the Adhesion Conference in Amsterdam, 1995,"* which was also published in the Festschrift. It is worth citing a few lines of it:

Figure 12 Prof. Pierre-Gilles de Gennes, Nobel laureate in Physics, 1991, presenting the plaque to Dr. Kash Mittal during the Congress in Amsterdam, October 16, 1995, and below text on the plague: *On the occasion of the 1st International Congress on Adhesion Science and Technology, held in honor of Dr. Kash Mittal on his 50th birthday, I, on behalf of the global adhesion scientific community, would like to recognize him for his commitment and dedication to the advancement of adhesion science and technoogy and for his exceptional achievements. All in the adhesion world would like to thank Dr. Mittal for his outstanding leadership and distinguished profesional service. Professor P.-G. de Gennes.*

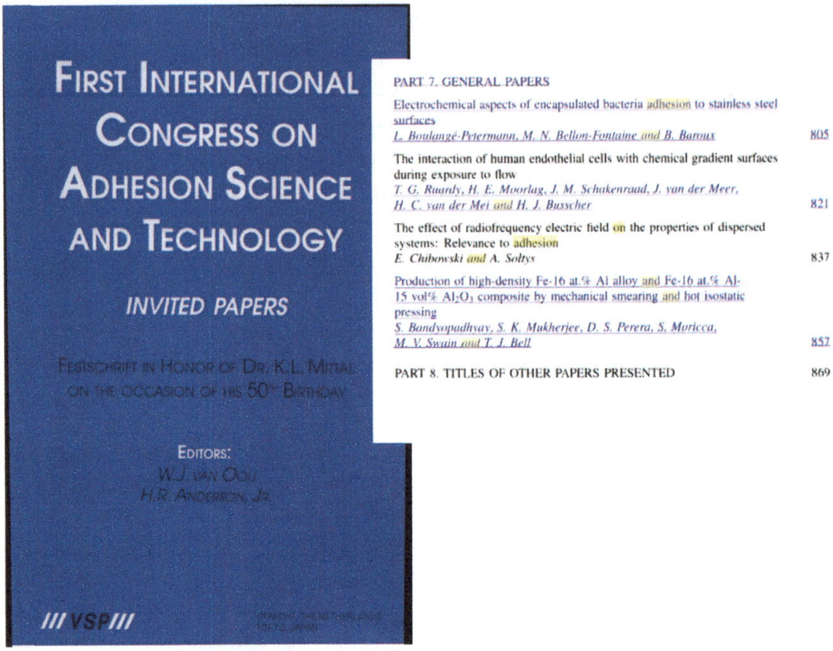

Figure 13 Cover of the Festschrift in honor of Dr. Kash Mittal contains invited papers presented during First International Congress on Adhesion Science and Technology and Part 7. General Papers, containing our paper.

"I am Dr. Bando, Teach polymers in the school,
It is very easy, As can see any fool. . . .
. . .
When I come to this conference, On science of adhesion
I find it very educative, From the session one
. . .
Mittal spoke about 300 ways, To determine adhesion
Kendall says oh, no, no "Fracture mechanics is the only one"
. . .
Nothing I can do about it, (But) I have learnt a great deal
. . .
Also thank the beautiful girls, Who work always there
To make the conference a very good success
No efforts did they spare."

And indeed, it was true, as can be seen in Figure 14.

Figure 14 First International Congress on Adhesion Science and Technology, Amsterdam 1995. Dr. Kash Mittal with ladies of the organizing staff.

The International Symposia in Which We Participated In-Person

Obviously, we were very proud to attend the event in Amsterdam. But, thanks to Dr. Mittal's help, we attended many other scientific events, and most often the "Surfactants in Solution" (SIS) meetings, which Dr. Mittal initiated in 1976. The symposia and the presented papers are listed below.

1. F. Gonzalez-Caballero, M. Bolivar and E. Chibowski, Nitrofurantoin surface free energy changes due to adsorption of some aminoacids. 9th International Symposium on Surfactants in Solution, Varna (Bulgaria), June 10-15, 1992.
2. E. Chibowski and L. Hołysz, Surfactant (collector) coverage, surface free energy changes and flotability of minerals. 10th International Symposium on Surfactants in Solution, Caracas (Venezuela), June 26-30, 1994.
3. E. Chibowski, Is contact angle hysteresis due to a film present behind the drop? 10th International Symposium on Surfactants in Solution, Caracas, (Venezuela), June 26-30, 1994.
4. A. Ontiveros, J.D.G. Duran, E. Chibowski and F. Gonzalez-Caballero, Adhesion of colloidal ZnS on silicon. Effect of pH, ionic strength and radiofrequency electric field. 1st International Congress on Adhesion Science and Technology, in honor of Dr. Kash Mittal on his 50th Birthday, October 16-20, 1995, Amsterdam (The Netherlands).
5. E. Chibowski and A. Sołtys, The effect of radiofrequency electric field on the properties of dispersed systems relevant to adhesion. 1st International Congress on Adhesion Science and Technology, in honor of Dr. Kash Mittal on his 50th Birthday, Amsterdam (The Netherlands), October 16-20, 1995.
6. E. Chibowski and R. Ogonowski, Thin layer wicking - a method for determination of the gradient hydrophobicity of a solid surface due to

surfactant adsorption. 11th International Symposium on Surfactants in Solution, Jerusalem (Israel), June 9-13, 1996.

7. E. Chibowski, Thin-layer wicking - A method for the acid-base free energy interaction determination, Second International Symposium on Acid-Base Interactions: Relevance to Adhesion, Newark, NJ, USA, October 19-21, 1998.
8. E. Chibowski, Is the contact angle hysteresis due to a film present behind the drop – 5 years later, Second International Symposium on Contact Angle Wettability and Adhesion, Newark, NJ, June 21-23, 2000, (invited speaker).
9. E. Chibowski, A.E. Wiącek, K. Wilk and L. Syper, Investigation of new dicephalic saccharide-derived surfactants as stabilizers for oil-in-water emulsion, 12th International Symposium on Surfactant in Solution, Gainesville, Florida, June 11-16, 2000.
10. H. Radelczuk, L. Hołysz and E. Chibowski, Comparison of the acid-base components and contact angle hysteresis approaches to the solid surface free energy determination, Third International Symposium on Acid-Base Interactions: Relevance to Adhesion, Newark, NJ, USA, June 13–15, 2001.
11. E. Chibowski and A.Wiącek, Use of α-lactalbumin and β-casein as emulsifier for oil-in-water emulsion, 14th International Symposium on Surfactants in Solution, Barcelona, Spain, June 9-14, 2002.
12. E. Chibowski, Electrokinetic properties of paraffin wax suspension in surfactant solutions,15th International Symposium on Surfactants in Solution, Fortaleza, Brazil, June 6-11, 2004 (invited lecture).
13. E. Chibowski, M. Jurak and L. Hołysz, Topography and surface free energy of DPPC layers deposited on glass, mica or PMMA support, 16th International Symposium on Surfactants in Solution, 2006, Seoul, Korea, June 4-9, 2006.
14. A.E. Wiącek, E. Chibowski and L. Hołysz, Electrokinetic properties of silica gel in the presence of lecithin (DPPC), 16th International Symposium on Surfactants in Solution, 2006, Seoul, Korea, June 4-9, 2006.
15. E. Chibowski, K. Terpiłowski and L. Hołysz, Influence of ambient humidity on the apparent surface free energy of smooth solid surface, Fifth International Symposium on Contact Angle, Wettability and Adhesion, Toronto, Canada, June 21-23, 2006.
16. E. Chibowski, K. Terpiłowski and L. Holysz, Superhydrophobic effect due to deposition of nano- and/or microparticles on a solid surface, Fifth International Symposium on Contact Angle, Wettability and Adhesion, Toronto, Canada, (invited), June 21-23, 2006.
17. M. Jurak and E. Chibowski, Properties of solid supported lipid layers and their changes after an enzyme action, 17th International Symposium in Surfactants in Solution, Berlin, August 17-23, 2008.
18. A. Szcześ and E. Chibowski, Synthesis of calcium carbonate in water-in-oil emulsions, 17th International Symposium on Surfactants in Solution, Berlin, August 17-22, 2008.
19. A.E. Wiącek, M. Mirosław, L. Hołysz and E. Chibowski, Comparison of n-tetradecane/electrolyte emulsions properties stabilized by DPPC and

DPPC liposomes in the electrolyte solution, 17th International Symposium in Surfactants in Solution, 17-23 August, 2008, Berlin.

20. K. Terpiłowski, L. Hołysz and E. Chibowski, Comparison of apparent surface free energy of some solids determined by different approaches, Sixth International Symposium of Contact Angle Wettability and Adhesion, University of Maine, Orono, Maine USA, 14-16 July, 2008.
21. E. Chibowski, L. Hołysz, A. E. Wiącek, M. Jurak and A. Szcześ, Effects of phospholipid adsorption on some dispersed system properties, 18th International Symposium on Surfactants in Solution, Melbourne (Australia), 14-19 November 2010.
22. E. Chibowski, and M. Jurak, Surface free energy of solid supported phospholipid layers without and with enzymatic action, 18th International Symposium on Surfactants in Solutions, Melbourne (Australia), 14-19 November, 2010.
23. K. Terpiłowski, M. Bielska, K. Prochaska and E. Chibowski, Apparent surface free energy of ultrafiltration membranes, Seventh International Symposium on Contact Angle, Wettability and Adhesion, Danbury, CT, USA, June 23-25, 2010.
24. E. Chibowski, K. Terpiłowski and L. Hołysz, Effect of relative humidity on contact angle and its hysteresis on phospholipid DPPC bilayer deposited on glass, Eight International Symposium on Contact Angle, Wettability and Adhesion, Quebec City, Canada. June 13-15, 2012.
25. K. Terpiłowski, D. Rymuszka, L. Hołysz, E. Goncharuk, I. Sulym and E. Chibowski, Wettability of modified silica layers deposited on glass support, Ninth International Symposium on Contact Angle, Wettability and Adhesion, Bethlehem, PA, USA, June 16-18, 2014.
26. L. Hołysz, M. Jurak and E. Chibowski, Effect of phospholipid DPPC and phospholipase A2 enzyme on surface properties of silica materials, 20th International Symposium Surfactants in Solution, Coimbra, Portugal, 22-27June, 2014.
27. M. Jurak, A. Szczes and E. Chibowski, Investigation of phospholipid DOPC bilayer and liposome enzymatic hydrolysis in the presence of cholesterol, 20th International Symposium Surfactants in Solution, Coimbra, Portugal, 22-27June, 2014.
28. E.Chibowski, Y. Yan and A.Szcześ, Modification of titanium alloy surface with phospholipid in aspect of biocompability, 21st International Symposium on Surfactants in Solution, Jinan, China, June 5-11, 2016.
29. K. Terpiłowski, M. Tomczyńska-Mleko, S. Mleko and E. Chibowski, The surface properties of biopolymers obtained with the presence of gluten, Tenth International Symposium on Contact Angle, Wettability and Adhesion, Hoboken, NJ, USA, July 13-15, 2016.
30. E. Chibowski and A. Szcześ, Magnetic field effects on anionic and cationic surfactant solutions, 22nd International Symposium on Surfactants in Solution, Norman, OK, USA, June 3-8, 2018.

31. A. Gozdecka, K. Przykaza, M. Jurak, A.E. Wiącek and E. Chibowski, Physicochemical characteristic of biocompatible hybrid films, 22nd International Symposium on Surfactants in Solution, Norman, OK, USA, June 3-8, 2018.
32. K. Terpiłowski, S. Peréz Huertas and Y. Bolbukh, Surface properties of plasma activated chitosan foils, 22nd International Symposium on Surfactants in Solution, Norman, OK, USA, June 3-8, 2018.
33. K. Szymczyk, M. Szaniawska and K. Terpilowski, Temperature-dependent aggregation of aqueous solutions of binary mixtures of Kolliphors – volumetric and viscometric approach, 22nd International Symposium on Surfactants in Solution, Norman, OK, USA. June 3-8, 2018.
34. K. Szymczyk, Interaction in the flavonoid-surfactant systems, 23rd International Symposium on Surfactants in Solution, Lublin, Poland, September 11-16, 2022.
35. K. Terpiłowski, A. Gal, L. Hołysz, and E. Chibowski, Effect of static magnetic field on surface tension of surfactant solutions, 23rd International Symposium on Surfactants in Solution, Lublin, Poland, September 11-16, 2022.
36. K. Terpiłowski, K. Grządka, M. Chodkowski, M. Nastaj, and S. Perez-Huertas, Essential oil emulsions stabilized with polymers as insect repellants, 23rd International Symposium on Surfactants in Solution, Lublin, Poland, September 11-16, 2022.
37. B. Jańczuk, A. Zdziennicka, J. Krawczyk and K. Szymczyk, Composition of the mixed surfactants monolayer and prediction of their solutions surface tension, 23rd International Symposium on Surfactants in Solution, Lublin, Poland, September 11-16, 2022.
38. B. Jańczuk, K. Szymczyk, A. Zdziennicka, and J. Krawczyk, Usefulness of the Szyszkowski equation for the prediction of the surface tension of multicomponent surfactant systems, 23rd International Symposium on Surfactants in Solution, Lublin, Poland, September 11-16, 2022.
39. A. Zdziennicka, B. Jańczuk, A. Rekiel, J. Krawczyk and K. Szymczyk, Surface tension of the nonionic surfactant+biosurfactant mixtures, 23rd International Symposium on Surfactants in Solution, Lublin, Poland, September 11-16, 2022.
40. K. Szymczyk, K. Terpiłowski, M. Nastaj, and B.G. Sołowiej, Studies on the properties of mixtures of polysorbates and sorbitol in context of their use in the preparation of processed cheeses, 23rd International Symposium on Surfactants in Solution, Lublin, Poland, September 11-16, 2022.
41. J. Krawczyk, A. Zdziennicka, K. Szymczyk, and B. Jańczuk, Prediction of polar and apolar polymers wetting by aqueous solutions of sugar surfactants, 23rd International Symposium on Surfactants in Solution, Lublin, Poland, September 11-16, 2022.
42. J. Krawczyk, K. Szymczyk, A. Zdziennicka, and B. Jańczuk, Critical surface tension of different polymers wetting and their surface tension in the systems

including sugar surfactants, 23rd International Symposium on Surfactants in Solution, Lublin, Poland, September 11-16, 2022.
43. A. Zdziennicka, A. Rekiel, B. Jańczuk, K. Szymczyk, and J. Krawczyk, Description and prediction of contact angle of the TX-165+biosurfactant mixture on different solids.
44. M. Jurak, K. Szafran, A.E. Wiącek, and K. Przykaza, Langmuir-Blodgett films as a drug delivery platform, 23rd International Symposium on Surfactants in Solution, Lublin, Poland, September 11–16, 2022.
45. K. Przykaza, M. Jurak, and A.E. Wiącek, Effect of naproxen on model lipid membrane formed on the water –chitosan subphase, 23rd International Symposium on Surfactants in Solution, Lublin, Poland, September 11-16, 2022.
46. K. Terpilowski, N. Guzenko, M. Galaburda, and O. Goncharuk, Hybrid hydrogel sorbents based on natural polysaccharides for extraction of dyes from aqueous solutions, 24th International Symposium on Surfactants in Solution, Goa, India, June 16-21, 2024.
47. R. Sarsembekova, K. Szymczyk, M. Krzan, P. Brewczak, S. Aidarova, B. Malinowska, and Konrad Terpilowski, Determination of volumetric properties of non-ionic surfactants used in industry for the preparation of cleaning products, 24th International Symposium on Surfactants in Solution, Goa, India, June 16-21, 2024.
48. B. Malinowska and K. Terpiłowski, Flotation separation of ABS/PS flakes mixture from used refrigeration equipment, 24th International Symposium on Surfactants in Solution, Goa, India, June 16-21, 2024.
49. B. Malinowska and K. Terpiłowski The effect of ethoxylated alcohols on the sedimentation rate of calcium carbonate. Potential for use in flotation separation of plastics, 24th International Symposium on Surfactants in Solution, Goa, India, June 16-21, 2024.

Appendix

This tribute was presented during the Symposium on "*The Wide World of Adhesion*", which was held at the *American Chemical Society* Fall Meeting in Chicago, IL, August 21-25, 2022. The Symposium consisted of invited papers from international scientists to honor the contributions of Dr. Kash Mittal to various aspects of adhesion science and technology and to coincide with the celebration of his 75th birthday, which was in 2020. In fact, it was postponed by 2 years because of the COVID-19 pandemic and no possibility for the celebration in person. The topics at the symposium indeed dealt with a "wide world of adhesion", i.e. contact angle, wettability and adhesion; superhydrophobicity; surface free energy characterization; polymer surface modification; adhesion of thin films, thick films and coatings; adhesion measurement techniques; adhesive joints: formation, characteristics and testing; adhesives–biobased and nanoreinforced; particle adhesion and removal; and metallized plastics. The Symposium organizers were Frank Etzler and Anil Netravali. Many very interesting presentations were given in person during the Symposium. Dr. Kash Mittal presented his "*Half-a-Century Romance with Adhesion Science*," and Dr. F. Etzler's talk was about "*Considerations for the Evaluation of Surface Free Energies from Contact Angle Data*". Dr. Konrad Terpiłowski, with co-authors: M. Gradecki, M. Chodkowski, E. Pakhlov, from our University in Lublin, Poland, presented "*Plasma-activated glass surface hydrophobization by hexamethyldisilazane*". In the picture in Figure 15, Kash Mittal, Frank Etzler, and this author are just waiting for the beginning of the next session.

Several weeks later, September 11-16, 2022, at Maria Curie-Skłodowska University in Lublin, Poland, we organized the 23rd Symposium of the series "*Surfactants in Solution (SIS)*". This Symposium was also postponed by two years because of the COVID-19 pandemic. In Figure 16 is shown the picture taken in the UMCS Rector's Office during the Symposium.

Figure 15 Waiting for the next session of the Symposium on "The Wide World of Adhesion" at the ACS Fall Meeting, 2022 in Chicago. From left: Frank Etzler, Kash Mittal and Emil Chibowski.

Figure 16 In the UMCS Rector's office during the 23rd SIS in Lublin, September 2022. From left: v-Director of the Institute of Chemical Sciences UMCS Prof. K. Szymczyk, Local Scientific Organizer Prof. K. Terpiłowski, Rector UMCS Prof. R. Dobrowolski, General Chairman Dr. Kash Mittal and Local Chairman Prof. Emil Chibowski.

Then, from June 16-21, 2024, the 24th SIS Symposium was held in Goa, India. Our University represented Prof. Konrad Terpiłowski, who was also a Member of the International Advisory Committee. In Figure 17 are shown Dr. Kash Mittal, Prof. Konrad Terpiłowski, and Prof. Suman Kundu, just before the lunch serving.

Figure 17 Waiting for lunch at the 24th Symposium "Surfactants in Solution" in Goa, India, June 2024. From left: Dr. Kash Mittal, Prof. Suman Kundu, Director BITS Pilani Goa Campus, and Prof. Konrad Terpiłowski.

Moreover, the 25th SIS Symposium will be held in Madrid.

It is worth mentioning that the *Surfactants in Solution* Symposia were initiated in 1976 in Albany, New York, by Dr. Kash Mittal. The next two meetings were held in Knoxville and Potsdam, USA, before the symposium became truly international. Thanks to the help of Dr. Kash Mittal, this author could participate in-person (shown in italics) in 13 of these 24 Symposia.

Albany, USA (1976)
Knoxville, USA (1978)
Potsdam, USA (1980)
Lund, Sweden (1982)
Bordeaux, France (1984)
New Delhi, India (1986)
Ottawa, Canada (1988)
Gainesville, USA (1990)
Varna, Bulgaria(1992)
Caracas, Venezuela (1994)
Jerusalem, Israel (1996)
Stockholm, Sweden (1998)

Gainesville, USA (2000)
Barcelona, Spain (2002)
Fortaleza, Brazil (2004)
Seoul, S. Korea (2006)
Berlin, Germany (2008)
Melbourne, Australia (2010)
Edmonton, Canada (2012)
Coimbra, Portugal (2014)
Jinan, China (2016)
Norman, OK,USA (2018)
Lublin, Poland (2022)
Goa, India (2024)

Acknowledgements

Half-a-Century Romance with Adhesion Science

Kash Mittal

Former Editor-in-Chief: Journal of Adhesion Science and Technology
Former Editor-in-Chief: Reviews of Adhesion and Adhesives

This write-up reflects my half-century romance with adhesion science. I thought it would be better to first describe my scientific activity and experience with interesting anecdotes. So, I have divided the text into pre-and adhesion eras and more recent involvements.

Pre-Adhesion Era

I must say that I was always fascinated by surfaces and interfaces, as an interface is the seat of action. Even my M.Sc. thesis research was concerned with "Ion Exchange Capacity of Some Clays". When I applied for a Ph.D. research program in 1966 in the United States, I was accepted at various universities, but I opted for the University of Southern California, Los Angeles, to carry out research on "Factors Influencing the Ultracentrifugal Stability of Oil-in-Water Emulsions" (colloid chemistry). My professor of quantum chemistry at IIT-Delhi (where I obtained my M.Sc. degree) asked me what my Ph.D. research area would be. I answered very eagerly and enthusiastically: colloid chemistry, and he retorted, "Don't you have anything better to do?" Some people have the naïve belief that the thinking of those undertaking research in colloidal chemistry was also cloudy, as many colloidal systems were cloudy in appearance. My very first publication was "Conceptual Clarification of the Terms Used to Describe Emulsion Behavior," published in The Journal of the Society of Cosmetic Chemists in 1970. My post-doctoral research also dealt with interfacial phenomena: First, dealing with S/L colloidal dispersions at Penn State University,

Email: raarareviews@gmail.com

Anil N. Netravali and Frank M. Etzler (eds.) The Wide World of Adhesion: A Festschrift in Honor of Dr. Kash Mittal on his 80th Birthday, (49–70) © 2026 Scrivener Publishing LLC

followed by passivation of cobalt in the Electrochemistry Laboratory of Prof. John O'Mara Bockris at the University of Pennsylvania. At UPenn, I coauthored a paper that was published in Nature Physical Sciences in 1971. When I was in India, I heard that once one has been published in Nature, one has reached the pinnacle and does not need to do anything more.

Now let me turn to the adhesion era.

Adhesion Era at IBM (1972-1993)

My first foray into the world of adhesion was in 1972 when I joined IBM Research in San Jose, California. I must thank Dr. H.R. Anderson, Jr., who impressed upon me that adhesion was ubiquitous in the industrial world and was very important to IBM technologies. I became engrossed and spent time every weekend in the IBM library, and made copies of many, many published papers. I must admit that the more I read about adhesion, the more confused I got. Apropos, when I left IBM in 1994, I had almost 200 3-ring binders replete with adhesion literature.

Now I will describe my activities in the broad realm of adhesion in a chronological order.

1972-1974

My first assignment at IBM was to enhance the adhesion of parylene to different substrates. Surface cleaning is a sine *qua non* for adhesion: if it ain't clean, it will not stick.

Surface contamination is a *bete noire* when it comes to adhesion. So perforce we had to devise an efficient way to clean a Ni surface, and D.F. O'Kane and yours truly published a paper on "Plasma Cleaning of Metal Surfaces" in the Journal of Vacuum Science & Technology in 1974. (my very first publication from IBM).

In 1974, I coined the term "adherate" to denote anything that goes on the adherend (substrate) in analogy to surface chemistry. In surface chemistry, we have "adsorbate" and "adsorbent". I believe there is a need to adopt the lovely term "adherate". This neologism can be a convenient term to describe what goes on top of an adherend. In any case, it rhymes.

1974-1976

During this period, I was at IBM with the organization called Product Assurance, and our department was affectionately called "cook and look" department. There was some element of experimental adhesion research. At this time I also developed interest in particle adhesion. Moreover, in 1975 I was fascinated by the topic of metallized plastics and published a paper in the Journal of Vacuum Science Technology in 1976, which was also translated into Italian.

1976-1985

This period was very fertile and encompassed many significant activities, and we worked on a number of different topics. I personally became very infatuated with wettability and its relation to bond strength. During this period, I published a paper entitled "The Role of the Interface in Adhesion Phenomena," which is one of my most cited papers. Also, during this period, I got very enamored by the topic of adhesion measurement of a coating. In the literature, many divergent and discordant views were described to express what was actually measured and what it meant.

Based on my analysis, personal opinion, and other factors, I published three reviews and two original research papers. The review papers have been widely cited, and thus my purpose was served. As a matter of fact, I published a list of 350 adhesion measurement techniques in an alphabetical order. Some of these are humorous and entertaining. I also stressed that there were many myths about adhesion measurement and delineated certain myths and facts. Here, I would like to share with the researchers my main findings and recommendations (See pp. 51-61).

First of all, what is the difference between a film and a coating: very profound—a film is a thin coating, and a coating is a thick film. Below, I have summarized what I had proposed during 1975-1977. Adhesion of a coating can be expressed in three different forms.

Different Forms of Adhesion

1. **Fundamental adhesion**, also known as basic adhesion, chemical adhesion, specific adhesion, primary adhesion, theoretical adhesion.

- It is the summation of all interfacial intermolecular interactions, including covalent bonds, dipole–dipole (Debye), dipole–induced dipole (Keesom), induced dipole–induced dipole (London) interactions, and hydrogen bonds.
- It can be calculated but cannot be measured for a coating.
- Once two materials in contact are fixed, it is fixed.

2. **Thermodynamic adhesion**, also known as wetting adhesion, surface energetic adhesion, or adsorption adhesion for a liquid coating.

$$W_A = \gamma_L \,(1+\mathrm{Cos}\theta)$$

where W_A is the work of adhesion
γ is the surface tension of the liquid coating.
Θ is the contact angle.

3. **Practical adhesion**, also known as adhesion strength, experimental adhesion, measured adhesion, bond strength, mechanical adhesion, or adherence.

WHAT IS MEASURED?
ANSWER: Practical adhesion (defined as the force or the work required to remove the coating from the substrate).
PRACTICAL ADHESION = f (FUNDAMENTAL ADHESION, MANY OTHER FACTORS)

Other factors include:

- Stress in the coating
- Mode of applying the load
- Mode of failure
- Technique variables
- Inelastic deformation of the coating and substrate
- Thickness of coating
- Mechanical properties of coating
- Viscoelastic properties of coating
- Glass transition temperature of coating
- Hardness of coating

- Environment (temp., RH)
- Mood and sex of the operator
- Time of the day

Also, it was emphasized that along with practical adhesion, the locus of failure must also be ascertained. It can be an "interfacial failure" or "adhesional failure" or simply "adhesion failure". Also, it was suggested not to use the term "adhesive failure" as it might be misconstrued that some adhesive failed, as in the case of an adhesive joint, which actually is "cohesive failure". The problem is that the word "adhesive" is used both as a noun and as an adjective. So, it can be used as a material as well as related to the adhesion phenomenon. The diagram below shows various loci of failure: #3 represents "interfacial failure", #2 and #4 represent "interphasial failure", and #1 and #5 represent cohesive failure. Apropos, an interface is a two-dimensional plane or a sharp boundary, whereas an interphase has some thickness.

LOCI OF FAILURE
3 Interfacial failure, adhesional failure, or simply adhesion failure
2 and 4. Interphasial failure
1 and 5. Cohesive failure

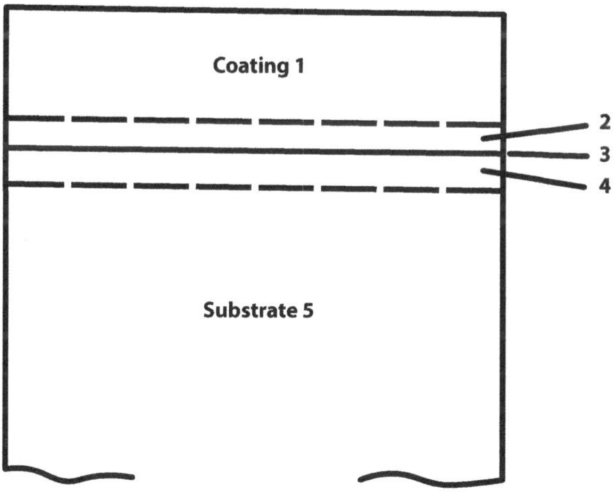

Criteria for an ideal practical adhesion test

1. Quantitative
2. Reproducible

3. Not very time consuming
4. Easily adaptable to routine testing
5. Relatively simple to perform
6. Applicable to all coating–substrate combinations
7. Independent of operator's experience
8. Applicable over a wide range of sample sizes
9. Applicable to products and processes
10. Applicable over a wide range of coating thicknesses
11. Minimum test variables
    - Nondestructive
    - Independent of bulk properties of adherend and coating

Also, the following caveats were proposed:

- From the observed values of practical adhesion, one cannot deduce the values of fundamental adhesion unless contributions due to non-adhesion factors are known; at most, practical adhesion values are related to fundamental adhesion.
- For a given coating–substrate combination, the measured values of practical adhesion may differ with technique; and even using the same technique, different operators may obtain different values.
- Efforts to improve fundamental adhesion may not directly reflect in measured practical adhesion unless contributions due to extraneous (non-adhesion) factors are kept constant.

It was emphasized that the following points should be kept in mind.

- Use the term "practical adhesion."
- Specify all experimental parameters to facilitate reproduction of data.
- Practically any adhesion measurement technique is good for comparative purposes.
- The best technique is the one that simulates actual usage stress conditions as closely as possible.

Also, proper and modern units should be used:

- In terms of force/area – N/m^2 or Pascal (Pa) – I have seen in the old literature some strange units, e.g., ton/square foot
- In terms of force/length – N/m – some old unit: ton/yard

Ways to measure adhesion:

- Experimentally, practical adhesion can be measured in three ways:
- In terms of the force, defining practical adhesion as the force per unit area exerted when two materials are separated.
- In terms of work or energy, defining practical adhesion as the work done in separating or detaching the two bonded materials from one another.
- In terms of the time it takes in an accelerated environment (e.g., pressure cooker) for delamination to take place. The longer the time, the better the practical adhesion.

Some myths and facts:

- Myth: Fundamental adhesion of a coating can be measured.
  Fact: Fundamental adhesion of a coating cannot be measured; it can only be calculated.
- Myth: One can predict practical adhesion from fundamental adhesion.
  Fact: Only if one knows the contributions of all "other factors."
- Myth: One can convert practical adhesion determined using one technique into values for other techniques.
  Fact: No, because the values of "other factors" differ across techniques.
- Myth: All "other factors" affect all measurement techniques similarly.
  Fact: No. "Other factors" are different for different techniques.
- Myth: All operators should get the same values of practical adhesion for a given coating–substrate combination.
  Fact: Not necessarily true unless the environment and all test variables are kept constant.
- Myth: For a given coating–substrate system, one should obtain the same value at all locations.
  Fact: Not necessarily so unless temperature and RH are kept the same.
- Myth: One can always increase practical adhesion by enhancing fundamental adhesion.
  Fact: Not always true.

- Myth: By carrying out an accelerated test, one can derive the value at normal conditions.
  Fact: Not easy; it requires extensive calculation and many assumptions.

For a given coating-substrate combination, below are shown practical adhesion results obtained by three different techniques, showing very different trends, whereas the fundamental adhesion is the same in all three cases.

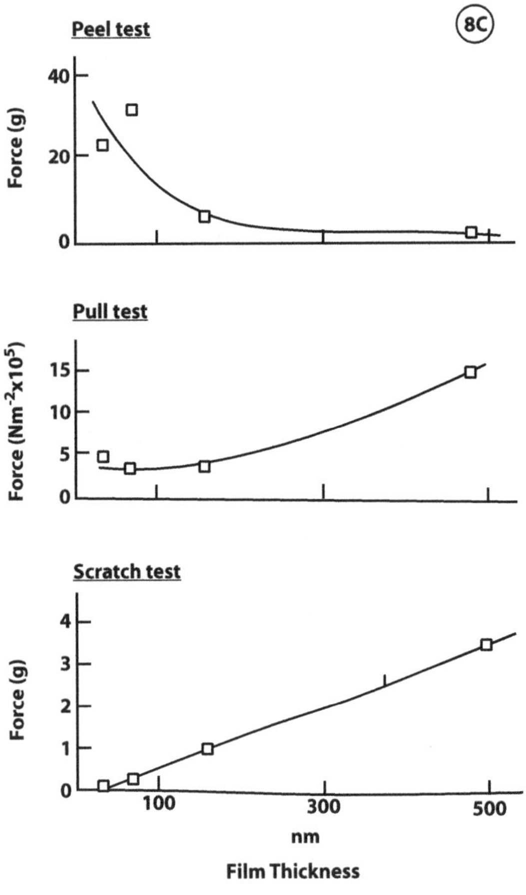

Summary and conclusions

1. There are very few cases where a true interface exists; in most adhering systems, there is an interfacial region called the "interphase."

2. Separation either at the interface or in the interphase should be a measure of practical adhesion.
3. Conditions at the interface or in the interphase modify the locus of separation and the separation strength.
4. Cohesive failure simply signifies that the practical adhesion is higher than the cohesive strength of the fractured phase.
5. Some so-called adhesion measurement tests simply measure the cohesive strength of either the adherend or the coating.
6. Some tests are qualitative in nature, so no numerical values are obtained.
7. Some tests may modify the interface or interphase, thereby altering practical adhesion.
8. When reporting practical adhesion values, one should explicitly specify the test conditions and the pertinent parameters of both the adherend and the coating.
9. If separation occurs in patches, interpretation becomes difficult.
10. Efforts to improve fundamental adhesion may not directly reflect in practical adhesion unless contributions due to "extraneous factors" are kept constant.
11. When reporting practical adhesion data, one should also examine and report the locus of failure.
12. Modern surface analytical instruments are very helpful in examining failure surfaces; one should not and cannot rely on visual observation alone.
13. For comparative or relative purposes, any technique may be used.
14. Accelerated tests (e.g., pressure cooker test) are useful for screening, relative ranking, and for optimizing material and process conditions.

Also during this period, we at IBM became interested in the use of silanes to promote adhesion of parylene to different substrates. Silanes are commonly applied from the solution phase, but we also explored vapor-phase and plasma-polymerization modes.

In retrospect, the year 1976 turned out to be of mega importance for me. As I always had an insatiable desire to learn and a penchant to teach, I looked around for courses being offered in the field of adhesion science and technology in the United States. To the best of my knowledge, no course was being offered in the United States on adhesion science and technology, so I thought it was a golden opportunity to fill this lacuna. By the way, in 1976 courses were offered in the area of polymer science, and adhesives were treated as a stepchild in a polymer course. Of course, when one uses an adhesive, adhesion is important. However, there are multifarious instances where one does not use any adhesive, yet the phenomenon of adhesion is of cardinal importance. So, I decided to develop two versions (a 2-day and a 3-day) of the course called "Adhesion Science and Technology." I started offering it, and it was enthusiastically received and became an instant success. It has been offered for about 50 years now, and more than 5,000 participants all over the world have benefited from it.

Here I would like to share some interesting foils from this course.

There are agents in nature able to make the particles of joints stick together by very strong attraction, and it is the business of experimental philosophy to find them out.
–Sir Isaac Newton

It is not possible for two things to be fairly united without a third, for they need a bond between them which shall join them both, that as the first is to the middle, so is the middle to the last, and the last and the first become the middle, of necessity, all will come to be the same, and being the same, with one another, all will be a unity.
–Plato

Various Views About Adhesion

- Minimalist's View ⇒ Fragmentation into Many Areas
- Maximalist's View ⇒ Encompasses Adhering of Anything to Anything
- Negativist's View ⇒ Plenty of Adhesionists or Stickologists (Kitchen Science)
- Positivist's View ⇒ Needs Inter-, Multi-, or Trans-disciplinary Approach to Adhesion

Some Terms and Definitions

- Adherend or Substrate
- Adherate (thin films, thick films, coatings, adhesives, etc…)
- Adherate-Adherend Combination or Adhering System (one interface, two bulk phases)– Adhesion strength
- Adhesive Joint or Adhint (two interfaces, three bulk phases)– Adhesive Joint Strength or Adhint Strength

- Heterohesion
- Homohesion, Autohesion, or Self -Adhesion
- Abhesion (Lack of Adhesion, Anti-adhesion)
- Hesion (Who knows?)

Note: Adherate signifies anything adhering to an adherend

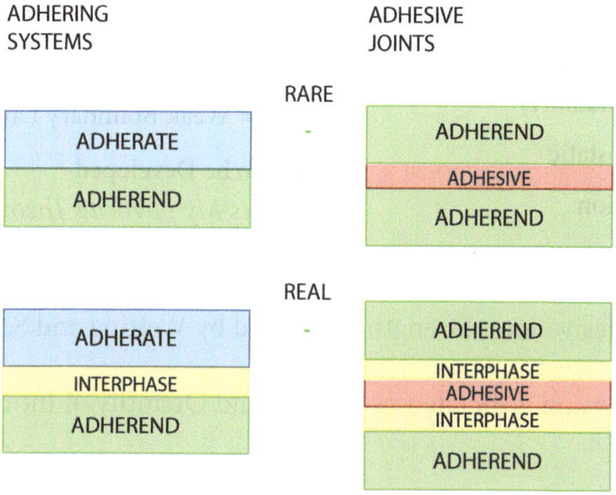

The Following Two Aspects are Important

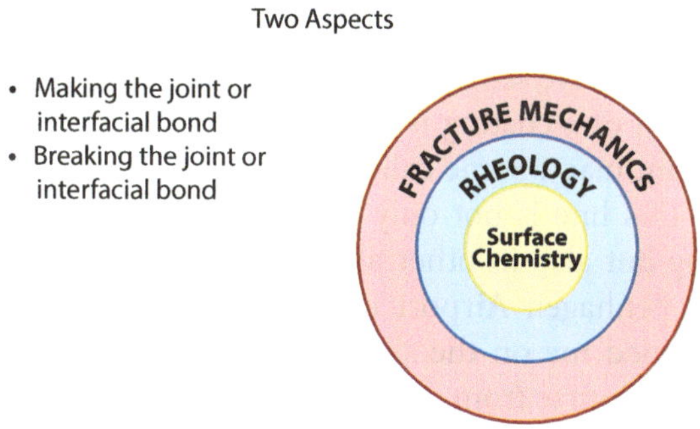

The topic of adhesion is veritably inter-, trans-, and multi-disciplinary subject. Along with surface chemistry which constitutes the core, both rheology and fracture mechanics are of cardinal importance.

The Most Acceptable Theory of Adhesion (To add some humor)

$$\Phi = \alpha\varphi_M + \beta\varphi_A + \gamma\varphi_{SE} + \delta\varphi_E + \varepsilon\varphi_D + \eta\varphi_{AB} + \theta\varphi_C + \rho\varphi_R + \sigma\varphi_{WBL} \cdots\cdots \omega\varphi_X$$

Where: $(\alpha^2 + \beta^2 + \gamma^2 + \delta^2 + \varepsilon^2 + \eta^2 + \theta^2 + \rho^2 + \sigma^2 \ldots \omega^2)^{1/2} = 1$

M = Mechanical

A = Adsorption

SE = Surface Energy (Contact Angle, Wettability)

E = Electrostatic

D = Diffusion

AB = Acid-Base

C = Chemical

R = Rheological

WBL = Weak Boundary Layer

X = To be Developed
(This is My Favorite Theory)

- Adhesive Joint Strength is dictated by Wetting and Setting and
- Adhesion is dictated by Quality and Quantity of Intimacy

Notes:

1. The behavior of an adhesive joint depends on adhesive wetting and adhesive setting (curing, crosslinking). For good joint strength, one needs perfect adhesive wetting and proper adhesive setting.
2. Adhesion of a coating is dictated by the quality and quantity of intimacy. Some call it the "universal theory of adhesion". This line is not only applicable to the adhesion of a coating but also in other aspects of life. One day, I was at the Copenhagen Airport in Denmark, and this guy came and tapped me on the shoulder. He said he had taken an adhesion course from me in Amsterdam eight years ago and added that he had forgotten everything I had taught him except the line "quality and quantity of intimacy."

"I have been telling my wife that if we do not want delamination (divorce), we should increase both the quality and quantity of intimacy, and she fully concurred," he said. "Dr. Mittal, I assure you that we are still intact."

Some Suggestions/Recommendations

RFT (Read from Top)

- Metallized Plastic---------------Metal/Polymer
- Polymer Coating on Metal-----Polymer/Metal

Do Not Use

- Adhesive Failure
- Adhesive Strength

Note: The use of these two terms causes great confusion as the word "adhesive" is used both as noun as well as adjective.

Use

- Interfacial Failure or Adhesional Failure
- Adhesion Strength
- Joint Strength

In 1978 I edited the book *Adhesion Measurement of Thin Films, Thick Films and Bulk Coatings*, the first book on this topic published by ASTM International. The book was very enthusiastically received by the community, and because of its high impact I received the prestigious 1990 Charles B. Dudley Award of the ASTM.

During 1980–1985 we at IBM were very interested in devising ways to promote adhesion of polyimide (a high-temperature polymer), as this material was of high interest in microelectronics.

One of the complex problems is how to study, on a molecular level, the deleterious effect of water on the adhesion of a coating or the bond strength of an adhesively bonded joint. ESCA (XPS) and other sophisticated techniques are not helpful in this regard, so a relatively uncommon technique called Inelastic Electron Tunneling Spectroscopy (IETS) had been explored for this purpose. Around 1982, when we were working with IETS at IBM, I had a postdoc who was an electromechanical genius but eschewed writing. One day he came to me and said, "Kash, I have made our IETS unit so automatic that I can stay in my apartment and press a button, and out comes the manuscript ready for submission." I retorted that we should at least read it before sending it to a journal. By the way, when we were working with IETS, to the best of my knowledge, there were only seven units in the world, and we published three papers. If this postdoc had stayed with me two more years, we would have been *numero uno* in the world.

1985-1993

In 1985 I moved from IBM East Fishkill (NY) to IBM Corporate Technical Institutes (CTI), the so-called IBM internal university that provided technical education to IBM employees worldwide. As part of CTI, there was the Manufacturing Technology Institute (MTI), which was responsible for delivering materials science education, and I joined this entity, which was a niche for me. Later CTI was transformed into Skill Dynamics… an IBM Company. My job at Skill Dynamics was even better than a United Nations job. In 1987 I founded the new journal christened *Journal of Adhesion Science and Technology* (JAST). The journal was an instant and huge success, and I was its skipper for about 25 years.

In May 1992 I was (along with two or three other adhesionists) interviewed by a correspondent from *The New York Times*, and the write-up was published in the May 31, 1992 issue. However, in 1993 the whole IBM went through very tough times, and there were massive layoffs. As a result, Skill Dynamics was closed down, and my tenure at IBM ended.

Life After IBM (1994 to Present)

Since leaving IBM in early 1994, I have been busy teaching courses on adhesion science and technology worldwide. Everyone is born with a purpose in life, and my life mission has been to proselytize as many individuals as I could to the religion of adhesion. In my childhood in India, I heard that if you cannot become viceroy (during British Empire), become a professor. Teaching is the noblest profession. Anyway, imparting education has been in my DNA.

The year 1995 reached the zenith of my career and was the most gratifying. In October 1995 I was recognized by the worldwide adhesion community by organizing the 1st International Congress on Adhesion Science and Technology in my honor on my 50th birthday in Amsterdam, The Netherlands. Two hundred thirty-five papers were presented from 38 countries. A number of luminaries graced the occasion. Prof. Pierre-Gilles de Gennes (Nobel Laureate in Physics, 1991) was the opening speaker. He was followed by Prof. Gabor Somorjai of the University of California, Berkeley, and a number of other dignitaries. My presentation was entitled "A Quarter Century Romance with Adhesion Science." On this occasion I came up with four ditties, listed below, regarding adhesion science. I hope you will like and enjoy these.

People Climb Mountains
Because They are There
Scientists Study Surfaces
Because They are Everywhere

Surfaces, Surfaces Everywhere
Their Study is Fun
Nobody can Deny the Importance of Bulk
But Still the Surface is Number One

Adhesion, O Adhesion
How Wonderful You are
From High-tech to Low-tech
You are Concern to All

Too Good Adhesion is No Good
Like Salt in the Food
The Trick is to Obtain the Right Degree
Using Scientific Concepts to Avoid Misery

Many attendees enthusiastically came up with poems to share their excitement on this occasion, which are included in the Festschrift volume published after this symposium. On this occasion Professor de Gennes presented a plaque to me (see picture shown), and the text of the plaque is also reproduced below. I will always cherish this plaque and this wonderful memento from Prof. de Gennes.

1st International Congress on

ADHESION SCIENCE & TECHNOLOGY

IN HONOR OF DR. KASH MITTAL ON HIS 50TH BIRTHDAY IN 1995

On the occasion of the 1st International Congress on Adhesion Science & Technology, held in honor of Dr. Kash Mittal on his 50th birthday, I, on behalf of the global adhesion scientific community, would like to recognize him for his commitment and dedication to the advancement of adhesion science and technology and for his exceptional achievements. All in the adhesion world would like to thank Dr. Mittal for his outstanding leadership and distinguished professional service.

Professor P. G. de Gennes

Text of the plaque presented to Dr. K. L. Mittal

2003 was a very special year for me, as in 2003 I was honored by the Maria Curie-Sklodowska University, Lublin, Poland, by awarding me the title of doctor *honoris causa*. At the time of acceptance I gave a talk on "The Wide World of Surface Science" and presented a table (shown below) demonstrating the importance and relevance of surface science from A to Z.

Table Examples of products, processes and technologies where surfaces/interfaces are or crucial importance.

A: Abhesion, Adhesion, Adsorption, Aerosols, Agriculture, Anesthesiology, Audiotapes, Autophobicity	**N:** Nanocomposites, Nanodevices, Nanomaterials, Nanotechnology, NEMS
B: Bioadhesion, Bifouling, Biomaterials, Biomembranes, Biotechnology, Blood clotting	**O:** Oil recovery, Oil spills, Olfaction, Ostwald ripening
C: Capillarity, Catalysis, Chromatography, Cleaning, Coagulation, Coalescence, Coatings, Colloids, Composites, Corrosion, Cosmetics	**P:** Painting, Paper sizing, Paving, Pharmaceuticals, Photography, Photolithography, Polymer blends, Printing, Prosthetics
D: Decontamination, Detergency, Dewetting, Dispersions, Drug delivery	**Q:** Quarrying, Quenching, Quicksand
E: Electrocapillarity, Electronic components, Electrophoresis, Electrowetting, Emulsions	**R:** Reinforced composites, Rusting
F: Flotation, Fluidics, Foams, Foods, Fouling, Friction	**S:** Self-assembled monolayers, Semiconductor devices, Sintering, Soap bubbles, Sol-gels, Sols
G: Gels, Glass fiber composites, Glazing, Glueing, Grinding	**T:** Tarnishing, Tertiary oil recovery, Thin film devices, Tribology
H: Haemodialysis, Hard disks, Heterocoagulation, Hydroplaning, Hydrosols, Hygienic products	**U:** Ultrasonic welding, Urology

(Continued)

Table Example of products, processes and technologies where surfaces/interfaces are or crucial importance. (*Continued*)

I: Information storage devices, Ink jet, Integrated circuits fabrication	**V:** Vehicle skidding, Veneer laminates, Videography, Videotapes
J: Jelly, Jet breakup, Joggling, Joining, Junctions (semiconductors)	**W:** Washing, Waterproofing, Wear, Weathering, Welding, Wettability, Wicking, Wine tears, Wound healing
K: Karyology, Kinesiology	**X:** Xerography
L: LB films, Liposomes, Liquid crystals, Lithography, Lubrication	**Y:** Yeast, Yellowing, Yogurt
M: Marangoni effect, Marine products, Membranes, MEMS, Microelectronics, Molecular beam epitaxy	**Z:** Zygology

In 2012 I initiated a new journal with the rubric *Reviews of Adhesion and Adhesives (RAA)*, but left it in July 2022. In 2013 I launched a new series of books called *Progress in Adhesion and Adhesives*. Already ten volumes have appeared under this banner, and more are planned at regular intervals.

I was covered in the Indian Express, North American Edition, p 22, May 4, 2007. I was recognized for publication of my 100 edited books by organizing the International Symposium in my honor on Recent Advances in Adhesion Science and Technology held at the American Chemical Society meeting in Boston, MA, August 22-26, 2010.

I feel blessed and honored that four books and 5 special journal issues have been compiled by professional colleagues and dedicated to me.

I must say I had a second love affair, i.e., with surfactants. Actually, it was a hangover from my Ph.D. days when I used a lot of surfactants. In 1976 I organized a symposium with the appellation "Micellization, Solubilization and Microemulsions" in Albany, NY. On popular demand it became a biennial ritual and has been attended by researchers from all corners of the world. Starting in 1982 with the meeting in Lund, Sweden, it became universally known by the rubric "Surfactants in Solution (SIS)," and the SIS 2026 meeting was a milestone event marking the 50th anniversary and was held in Madrid, Spain. I had two more intense interests: surface contamination and cleaning, and polyimides.

Over the last 50 years, I have found and heard many gems, pearls, and witty statements relative to adhesion and adhesives, and these are listed below for your entertainment.

Well, this is my journey. I hope you will enjoy reading it and will find it interesting. Let me conclude by saying that initially I was fascinated by the topic of adhesion, but now I am stuck to it:

Viva la Adhesion!

Pearls, Gems, and Witty Statements regarding Adhesion/Adhesives

1. Those working in adhesion and adhesives can—indeed, should—be scientifically called *adhesionists*, or colloquially and humorously, *stickologists*. Likewise, the science of adhesion can be labeled *stickology* or *adhesiology*.

2. Wolfgang Pauli: "God created matter; surfaces were invented by the devil."

3. "Only after knowing the surface of things can one throw himself into the search for what is underneath; but the surface of things is endless."
—Italo Calvino, *Palomar* (translated and cited by M. Morra)

4. Water is God's cruelest liquid when it comes to adhesion—and soapy water is even worse.

5. Any child can peel a hot potato; try peeling a cold one. (A lesson in the useful effects of heat and moisture.)

6. Thin is in: a thin coating generally sticks better. However, in some cases, thin can be sin.

7. The difference between a coating and a film is profound: a film is a thin coating, and a coating is a thick film.

8. What we really need is *social* adhesion. Fortunately, a glass of good wine can work wonders to promote social adhesion.

9. A barnacle is a very smart creature; erudite scientists can learn a great deal from this lowly organism. It is an equal-opportunity adherer—it sticks to every surface—and it does so under water, a remarkable feat.

10. The gecko is the consummate adhesionist and performs tricks that learned scientists have yet to master. If you are ever seeking an adhesion specialist, hire a gecko.

11. Scientists working in fracture mechanics are, well… cracked.

12. Our adhesive can join any two broken things—except broken hearts.

13. Never work in adhesion or in passivation: one is too simple, the other is too complex. If one glue does not work, try another.

14. There is a clear and sharp distinction between *adhesion* and *adhesive*: adhesion is a phenomenon, while an adhesive is a material. Adhesion is ubiquitous and plays an important role in countless industries.

15. If there were no adhesion, there would be no civilization. Think about that.

16. Do not screw it—glue it.

17. When all else fails, use bloody long nails.

18. Smart countries do not fight in the rain.

19. Nature does not waste time producing frivolous or useless things.

20. Nature is a great teacher.

21. Most surface scientists think of Thomas Young as the father of scientific research on contact angles and wetting. He published his work in 1805, but an even earlier recognition of wetting phenomena came from Galileo in *Bodies That Stay atop Water, or Move in It* (1612). Galileo may well be considered the grandfather of the field.

Another momentous milestone came in 1997, when Barthlott and Neinhuis published their groundbreaking study on the sacred lotus (*Nelumbo nucifera*), discovering extreme water repellency (superhydrophobicity) and the self-cleaning mechanism now famously known as the *Lotus Effect*.

22. It is remarkable that five Nobel Laureates—Albert Einstein, Erwin Schrödinger, Niels Bohr, Walter Lippmann, and Pierre-Gilles de Gennes—showed interest in the field of adhesion. Lippmann and de Gennes, in particular, made highly significant contributions to wettability and adhesion.

23. In the June 1, 2015 issue of *C&EN* of the American Chemical Society published a thought-provoking article titled "Sleeping Beauties Wake Up." Notably, two of these "sleeping beauties" were seminal papers in wetting science:

- A. B. D. Cassie and S. Baxter (1944)
- R. N. Wenzel (1936)

  These classic papers only began receiving widespread attention in 2002 and 2003 respectively. Today, no discussion of wetting is complete without them.

24. There are little gremlins holding hands together—that is why things stick.

25. As was once said of a well-known adhesion scientist:
Half of the time he was right, and the other half he was absolutely wrong, but he himself never knew which half he was talking about.

26. Mechanical interlocking or hooking theory is the oldest theory of adhesion.

Part 1

CONTACT ANGLE, WETTABILITY AND ADHESION

Part I

1

Application of Tadmor Equations to the Description of Wettability of Solid Surfaces

Salvador Pérez-Huertas[1,2] and Konrad Terpiłowski[3]*

[1]*Department of Chemical Engineering, University of Granada, Granada, Spain*
[2]*Department of Organic Chemistry, University of Córdoba, Córdoba, Spain*
[3]*Department of Interfacial Phenomena, Maria Curie Skłodowska University, Lublin, Poland*

Abstract

Wettability and adhesion are critical properties that dictate the performance of coatings, adhesives, and surface treatments across various industries. Traditionally, Young's equation has been used to study the contact angles of droplets on ideal, homogeneous, and smooth solid surfaces, serving as the basis for determining surface free energy (SFE). However, real surfaces exhibit roughness and chemical heterogeneity, resulting in contact angle hysteresis and discrepancies in SFE calculations. In response, advanced models such as those proposed by van Oss-Chaudhury-Good, Owens–Wendt, and Neumann have been developed, yet they frequently neglect the full impact of hysteresis. Tadmor's equilibrium contact angle model addresses these limitations by incorporating both advancing and receding contact angles. The refined approach based on Tadmor's equations offers enhanced accuracy in predicting adhesion performance and helps in the development of next-generation adhesive systems. This chapter reviews interesting studies dealing with the application of Tadmor's equations in determining equilibrium contact angles for surface free energy calculations and explores their implications for adhesion science.

Keywords: Wettability, surface free energy, tadmor, contact angle, hysteresis, coatings

**Corresponding author*: konrad.terpilowski@mail.umcs.pl

Anil N. Netravali and Frank M. Etzler (eds.) The Wide World of Adhesion: A Festschrift in Honor of Dr. Kash Mittal on his 80th Birthday, (73–96) © 2026 Scrivener Publishing LLC

1.1 Introduction

Surface free energy is one of the quantities describing properties of solids. It represents a different energetic state of molecules on the surface of a solid. Precise estimation of surface free energy value is of significant importance in many industrial areas, including the deposition of protective coatings, painting, passivation or the adhesion of different types of surfaces. Surface free energy can be calculated from the contact angles of probe liquids measured on the solid. The relation between the contact angle and surface free energy is given by Young [1]. According to him, the surface free energy value (γ_{SV}) depends on the contact angle of the experimental liquid (θ), its surface tension (γ_{LV}), and the interfacial energy (γ_{SL}) between the liquid and the solid surface:

$$\gamma_{SV} = \gamma_{SL} + \gamma_{LV} \cos\theta \tag{1.1}$$

Nevertheless, of the described parameters, only two can be experimentally measured, i.e., liquid surface tension and contact angle. The contact angle should be measured under the equilibrium conditions between the drop deposited on the surface and the atmosphere within the measuring chamber. However, such equilibrium conditions cannot be fully achieved experimentally. When a liquid droplet is placed in the measuring chamber, equilibrium is established not only between the droplet and the surrounding atmosphere but also between the droplet and the thin liquid film adsorbed on the surface. Thus, given the challenges associated with contact angle measurement, the term "apparent surface free energy" is often used in accordance with Young's definition [1]. Accordingly, the optimal method for measuring contact angles remains a topic of discussion in the literature. Current research efforts are directed towards enhancing the technical capabilities of the measurement apparatus and improving the observational resolution at the three-phase contact point [2–5].

Theoretical approaches are employed for the estimation of apparent surface free energy. Thus, the calculated energy value depends on both the type of liquid used for measuring the contact angle and the approach used for its calculation [6]. From a practical point of view, the most popular are those proposed by van Oss, Chaudhury and Good [7–9], Owens and Wendt [10], as well as based on equation of state given by Neumann [11–13]. An increasing number of reports in the literature have focused on the calculation of apparent surface free energy based on the contact angles hysteresis. This approach, developed by Chibowski [14–16], is used to determine apparent surface free energy regardless of the surface character, including its hydrophilic [17, 18], hydrophobic [19], and even superhydrophobic [20, 21] properties. The significance of this approach is proved by the number of citations and the growing number of publications based on its theoretical foundations [15]. Nevertheless, the importance and limitations of the methods used to determine apparent surface free energy are still being debated [22–25]. Recently, even the methodology of contact angle measurements and nomenclature used in the literature have been the subject of debate [4, 26–29], and it is even difficult to establish a predominant trend based on available reports. For instance, Marmur and colleagues [25] focussed on nomenclature related to the description of wettability. The paper presents a systematized nomenclature for a plethora of terms related to contact angles based on the accumulative knowledge of the surface chemistry community over recent decades. It is generally accepted that the highest metastable contact angle that can be measured is known as the advancing angle, while the angle measured after liquid volume reduction or formed due to the surface slope on which the droplet is placed is referred to as the receding angle. The term "apparent surface free energy" is used to describe the quantity calculated based on these angles.

As mentioned earlier, according to Young's theory [1], the contact angle is defined for a perfectly flat surface, which is practically impossible to achieve on most surfaces due to their inhomogeneity. Marmur [30] studied the problem concerning the equilibrium contact angle measurements. The author proposed a contact angle measurement method based on a capillary bridge formed between two solids, where the angle values are determined by comparing the measured force with a numerical computation of the expected forces. However, the complexity of the proposed procedure makes its practical application in the laboratory challenging. Another approach for the calculation of the equilibrium contact angle, which is more suitable for laboratory conditions, was proposed by Tadmor [31–35]. The contact angle observed on real surfaces is an average, as the angle can vary along the three-phase contact line when the drop is observed. This variation arises from both topographic and energetic reasons. According to Tadmor, the advancing contact angle represents the maximum value on a given surface, while the receding angle is the minimum. The equilibrium contact angle must be situated between these two values. By associating the line energy (along the three-phase contact line composed of local interfacial energies) with various values of contact angles on the same surface and assuming that heterogeneities on a given surface are of a similar nature, Tadmor proposed the following equations [31]:

$$\theta_0 = \arccos\left(\frac{\Gamma_a \cos\theta_a + \Gamma_r \cos\theta_r}{\Gamma_a + \Gamma_r}\right) \tag{1.2}$$

where

$$\Gamma_a \equiv \left(\frac{\sin^3\theta_a}{\left(2 - 3\cos\theta_a + \cos^3\theta_a\right)}\right)^{1/3} \tag{1.3}$$

and

$$\Gamma_r \equiv \left(\frac{\sin^3\theta_r}{\left(2 - 3\cos\theta_r + \cos^3\theta_r\right)}\right)^{1/3} \tag{1.4}$$

where θ_0 represents the equilibrium angle, θ_a is the advancing angle, θ_r is the receding angle, Γ_a is the coefficient related to the advancing angle, and Γ_r is the coefficient related to the receding angle.

It is important to note that although the Tadmor equilibrium contact angle lies between the advancing and receding angles, it is not their average. The greater the difference between the advancing and equilibrium angles, the larger the value of contact angle hysteresis is. Calculating the contact angle based on this approach cannot be identified with the physical measurement of the equilibrium contact angle described in Young's work. However, it can be considered a reasonable approximation of the contact angle described by Young [1]. This chapter reviews interesting studies dealing with the application of Tadmor equations in determining equilibrium contact angles for surface free energy calculations and explores their implications for adhesion science. Additionally, it discusses the influence of plasma activation on surface properties, the role of surface roughness, and the experimental validation of the Tadmor approach for different materials, including glass, polymers, and metallic substrates.

1.2 Surface Preparation and Plasma Activation

The activation of solid surfaces with a plasma is valuable to alter the surface properties. Plasma-activated surfaces usually become more polar. Action of plasma on the surface enables its cleaning [36], removing surface contaminants, or introducing new functional groups, thereby altering its wetting property [37]. The effect of plasma treatment depends on various parameters, including the gas used for plasma formation [38], treatment time [39], distance from the plasma source [40] and plasma generator power [41, 42]. Plasma action results in changes in surface layer properties and can significantly enhance the adhesion of different types of surface coatings. Literature shows that plasma is one of the most frequently applied techniques for enhancing the adhesion of coatings to solid substrates. Studies have shown that the adhesion of hydrophobic and superhydrophobic coatings to the glass is often inadequate. Figure 1.1 illustrates the damaged coating resulting from the deposition of polypropylene nanoparticles dispersed in a PMMA poly(methyl methacrylate)/chloroform solution. Despite exhibiting superhydrophobic property ($\theta > 150°$), the coating showed poor mechanical durability, as the droplet removed the layer from the surface during contact angles measurements.

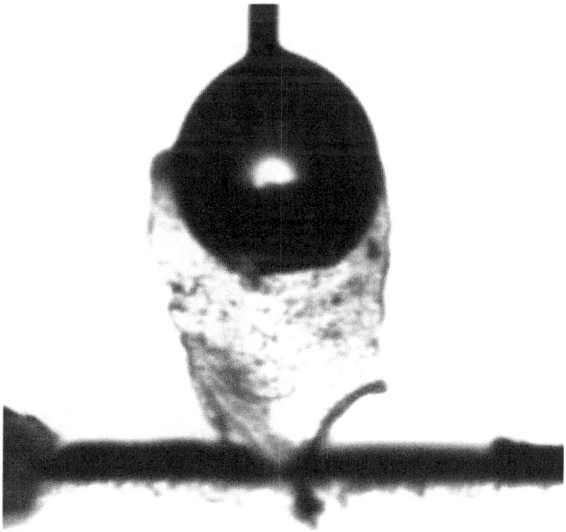

Figure 1.1 An example of insufficient adhesion of a superhydrophobic film to the substrate surface [43].

In response, some authors have used plasma-based deposition techniques on substrates before coating them in order to enhance the mechanical durability of the superhydrophobic coatings [43, 44]. For instance, Terpiłowski *et al.* [45] used air plasma treatment to enhance the adhesion of the coatings to the glass substrate. The coatings were prepared with hexamethyldisilazane (HMDS) and polydimethylsiloxane (PDMS) modified silica, dispersed in a polystyrene/chloroform solution, and the resulting mixture was applied to the substrate by spin coating. Contact angle measurements showed that the wettability of the surfaces strongly depended on the degree of surface modification with trimethylsilyl groups. Advancing contact angles ranged from 59.7° to 155° for samples with low (0.14) and high (1.0) degree of surface coverage, respectively. Contact angle hysteresis on these surfaces was approximately 30°. The apparent surface free energy decreased from 51.3 mJ/m^2 for low hydrophobicity samples to 1.0 mJ/m^2 for the most hydrophobic samples. The apparent surface free energy, calculated using equilibrium contact angles, was about 13 mJ/m^2 higher compared to calculations that did not incorporate equilibrium angles from Tadmor's approach. For surfaces with lower contact angle hysteresis, this difference was reduced to approximately 5 mJ/m^2. Furthermore, the authors highlighted the role of hierarchical roughness in achieving such superhydrophobicity. Figure 1.2 shows the samples with medium (0.53) and high (1.0) degree of surface coverage.

In order to better describe the surface roughness, subregions were separated from the larger representation using computer software. The surface with low surface coverage (a) exhibited craters with a subregion size of 57.4 μm, leading to wetting property consistent with the Wenzel model. In contrast, Figure 1.2 (b) showed a subregion size of 15.9 μm, characterized by

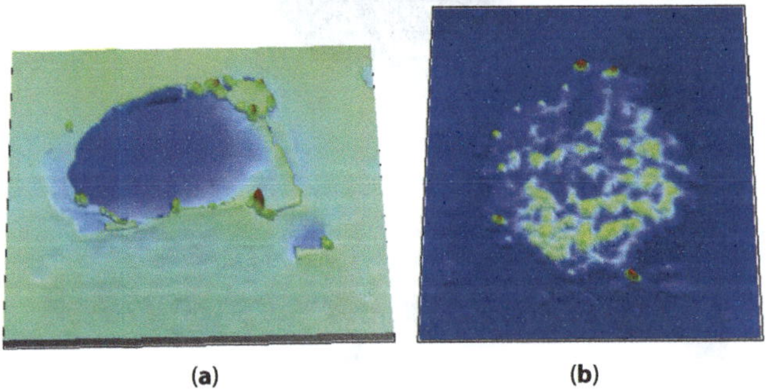

Figure 1.2 Subregion for the samples with medium (0.53 TMS (trimethylsilyl)) (a) and high (1 TMS) (b) degree of surface coverage [45].

non-homogeneous microroughness with additional nanoroughness on the top of the hills. The authors [45] concluded that using Tadmor equilibrium contact angles for surface free energy calculations offered a more accurate characterization of the wetting property of rough hydrophobic surfaces. Terpiłowski and Rymuszka also investigated the influence of plasma activation on the wetting property of glass plates using air, nitrogen, oxygen and argon gases [46]. The findings revealed that all plasma gases reduced the surface roughness of glass, with argon plasma producing the most significant effect. Due to a significant increase in surface hydrophilicity, measuring water contact angles was not feasible, but the kinetics of water spreading was explored. As follows from Figure 1.3, surface modification depended on the gas used, and the surface was most hydrophilic when nitrogen or air was used for plasma generation.

XPS spectra revealed the incorporation of functional groups after the surface activation, leading to changes in the surface composition, such as an increase in the number of SiO groups. The authors concluded that the introduction of characteristic functional groups on the treated surface was more effective in enhancing adhesion property than surface roughness modification. Another technique for enhancing the weak adhesion of hydrophobic coatings to substrates is the sol-gel method. Incorporating silica particles into sol-gel solutions and depositing them on solid surfaces may enhance the coating stability. For instance, silica powder can be used as a structuring agent for the design of hydrophobic rough surfaces [47]. In another study, stable hydrophobic coatings were prepared with the sol-gel

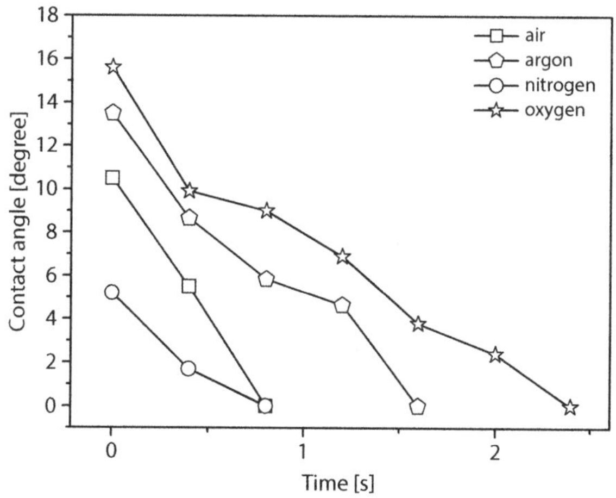

Figure 1.3 Kinetics of water drops spreading on the plasma activated glass [46].

method, incorporating silica powder into a sol-gel binder and followed by an exposure to hexamethyldisilazane (HMDS). A detailed description of sol-gel solution preparation can be found in [47]. As can be seen in Figure 1.4 (a), the samples treated with HMDS (denoted as SN) became hydrophobic. For all surfaces, the contact angles hysteresis was minimal, typically within a few degrees. However, as shown in Figure 1.4 (b), significant differences in the total value of apparent surface free energy were observed when equilibrium contact angles were used in calculations. The prepared samples of hydrophobized films on the glass surface were denoted according to the binder (silicate – ST or siloxane – SN) and silica powder used, e.g. SN – MP4 corresponds to the film composed of siloxane bound silica micropowder with the primary particles of 4 μm or ST – NP9 – film composed of silicate bound silica nanopowder with the primary particles of 9 nm.

As seen, the application of equilibrium contact angle for the calculation of apparent surface free energy is justified even for hydrophobic surfaces characterised by low contact angle hysteresis. SEM images revealed that the adsorption of HDMS vapours on the surface of sol-gel films changes their structure completely (Figure 1.5).

Moreover, surface topography confirmed the formation of hierarchical roughness structures. The obtained sol-gel films exhibited hierarchical structure with hills and grooves of varying sizes, leading to diverse wetting behaviours. Some regions were completely wetted, others partially, while certain areas trapped air and remained unwetted (Figure 1.6).

According to the authors, neither the Wenzel nor the Cassie-Baxter model adequately describes the wetting behavior of these rough surfaces. Instead, the mixed wettability model proposed by Bormashenko [27]

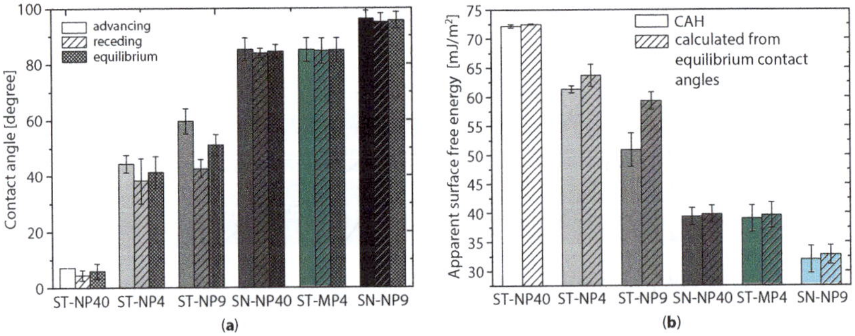

Figure 1.4 (a) Contact angles on the films obtained with contribution of silica. (b) Apparent surface free energy for the film obtained with contribution of silicas [47].

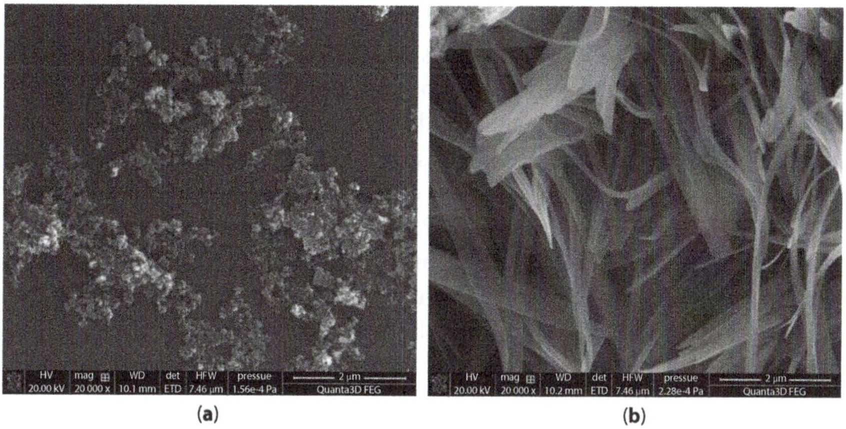

Figure 1.5 SEM mapping magnification 20000x for the surface obtained with contribution of silica particles with a diameter of 9 nm. (a) Sol-gel film, (b) sol-gel film after the contact with HDMS [47].

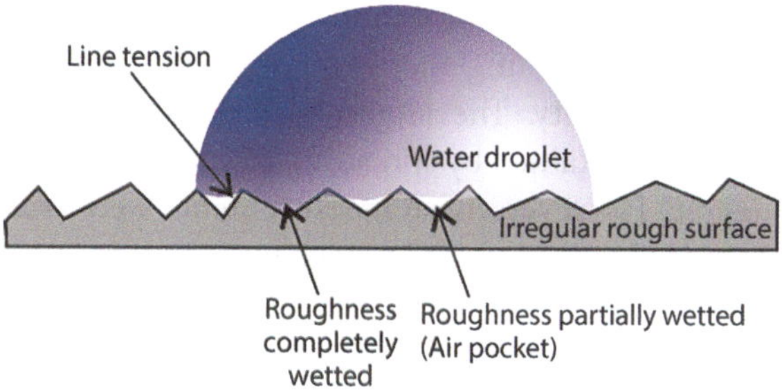

Figure 1.6 Wettability model on an irregular rough surface [47].

appears more suitable for explaining the observed phenomena. The authors [47] concluded that the surface free energy values derived from equilibrium contact angles offered a more accurate representation of the energetic property of these surfaces, providing a more comprehensive understanding of their wettability. In another study, Hołysz *et al.* [48] investigated the wettability property of polystyrene (PS) films filled with silanized colloidal silica particles deposited on glass. The authors analyzed the effects of different silica surface coverage levels on water contact angles, contact angle hysteresis, and surface free energy. To obtain coatings with specific properties, a solution of polystyrene (PS) in chloroform (0.1 g/100 ml) was prepared and used as a glue to obtain the surface roughness. The specific surface area of the initial silica was 300 m^2/g and was hydrophobicized by adding dimethylodichlorosilane. Silica particles with varying degrees of silane coverage (1-0.16) were synthesized. The coatings were applied using three methods: (i) dispersing silica particles onto a PS-coated surface, (ii) spreading a PS-silica dispersion, and (iii) employing a spin coater. In the first method, 0.5 ml PS solution was spread onto a glass plate (2.5x2.7 cm), followed by the deposition of 0.5 mg of silica. In the second method, a dispersion of 0.2 g of silica in 3 ml of PS solution was prepared, from which 0.5 ml was spread on a glass plate. In the last method, 0.5 ml of the same dispersion was poured onto the centre of a rotating glass plate (20 mm diameter) at 1080 rpm, obtaining a homogeneous coating in 15 s. Before the measurements, all samples were kept at room temperature for 24 h. In the case of samples prepared using the third method, some samples were placed in the vacuum drier at a pressure of 117 mbar. The wettability property of the silica was determined by measuring the contact angles on pressed silica pellets. Contact angle measurements revealed that when the degree of silica silanization was less than 0.93, the surfaces were completely wetted by water drops. For surfaces obtained by spreading silica dispersion, the largest advancing contact angles were obtained for silica of 0.93 silanization degree, confirming that both surface chemistry and topography influence wettability. The largest water contact angle (157°) was recorded for surfaces obtained by spreading dispersion of silica with a 0.68 silanization degree. However, these measurements were characterized by a significant standard deviation and a large hysteresis of contact angles. In contrast, surfaces obtained *via* spin coating and dried under vacuum had lower advancing contact angles (~120°) but significantly lower hysteresis values, indicating a more stable wetting behavior. Figure 1.7 illustrates the roughness profiles of surfaces prepared by spin coating and dried under vacuum.

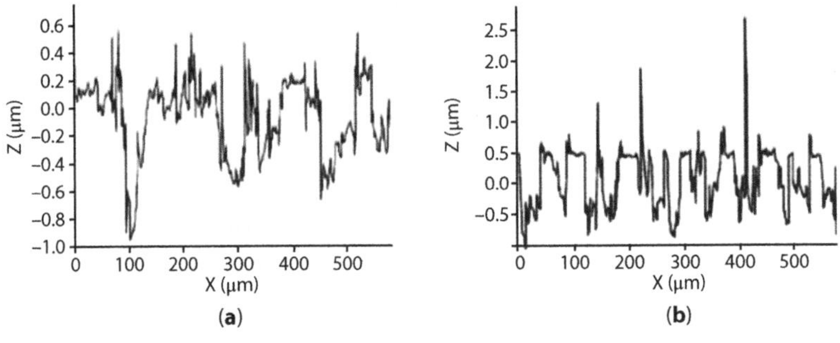

Figure 1.7 Lateral profiles of roughness of surfaces prepared by the spin coating technique and dried in vacuum (a) degree of silanization 0.93, (b) degree of silanization 0.42 [48].

Differences in roughness scales were evident: one surface exhibited dominant roughness at the micrometric scale, while the other showed nanometric roughness. A hierarchical roughness structure helped retain water on the surface, whereas roughness confined to the microscale allowed water penetration into surface cavities, leading to complete wetting. Wettability of these surfaces was also described using contact angles calculated from the Tadmor model (Table 1.1).

When the calculated equilibrium angles were used for the determination of surface free energy, larger values of SFE were obtained compared to those calculated from advancing and receding contact angles. However, the overall trend remained consistent: the lower the hysteresis values, the smaller the difference between the apparent surface free energy values calculated from contact angle hysteresis and equilibrium angles. As seen, the use of Tadmor equilibrium contact angles proved consistent for surface free energy assessment for surfaces exhibiting both high and low contact angle hysteresis.

On the other hand, several studies have also sought to experimentally validate the applicability of the Tadmor equilibrium contact angle for wettability description. For instance, Cho *et al.* [49] studied the wetting behavior of hierarchical superhydrophobic surfaces by combining experimental measurements and theoretical models. In particular, the authors explored the relationship between surface roughness, wettability, and wetting transitions using contact angle hysteresis, small-angle X-ray scattering (SAXS), and theoretical models such as Cassie, and Tadmor's approaches. The authors highlighted that Tadmor approach was particularly useful in describing wettability on rough surfaces, where local surface defects act as pinning points. The results also indicated that surfaces with smaller nanofeatures exhibit more significant deviation between Cassie- and

Table 1.1 Advancing, receding and equilibrium contact angles as well as hysteresis for the surfaces obtained by the *spin coating* technique and dried in vacuum [48].

Degree of silanization	Advancing angle θ_a	Receding angle θ_r	Contact angle hysteresis	Equilibrium angle θ_0
1.0	145.6	135.6	10	138.2
0.93	141.6	130.7	10.9	135.3
0.68	121.5	105.2	16.3	112.3

Tadmor-predicted contact angles. By comparing contact angles, SAXS-based air entrapment analysis, and theoretical models, the study validates Tadmor's prediction that equilibrium contact angles are significantly influenced by pinning and adhesion effects. The findings also suggest that the petal state, a partially wetted intermediate state, aligns well with Tadmor's model. Recently, Caruana and coworkers [50] also provided experimental validation for the Tadmor equilibrium contact angle model in the context of Indirect Evaporative Cooling (IEC) plate wettability. In this study, the authors investigated the wettability property of coated and uncoated plates used in IEC systems. In particular, the authors experimentally analyzed the static and dynamic contact angles of water droplets on three different surfaces, i.e., uncoated aluminum alloy (AL), standard epoxy coating (STD), and a hydrophilic lacquer (HPHI), and the equilibrium contact angles predicted using Tadmor equation were compared with experimentally obtained static contact angles. For measuring advancing and receding contact angles, the authors utilized high-speed imaging and sessile drop analysis. The results showed that dynamic contact angles obtained from drop impact experiments closely followed the trends predicted by Tadmor's equilibrium contact angle model, with a relatively small error between the predicted and measured values. The difference between the model-predicted and experimentally measured equilibrium contact angles on vertical surfaces ranged from 9° to 12°. Using dynamic angles from high-speed videos, the prediction was more accurate for STD (5° error) but remained similar to vertical surface results for HPHI and AL (11° and 9° errors, respectively). The authors concluded that Tadmor's model was a reliable approach for assessing wettability in IEC applications, confirming its applicability to non-ideal surfaces. Another interesting study was recently conducted by Li *et al.* [51]. In this study, the authors developed a thermodynamic framework for analyzing equilibrium contact angles and contact angle hysteresis in the Wenzel state. The authors introduced a theoretical model that calculates free energy changes associated with contact line motion, allowing them to predict advancing and receding contact angles under different surface conditions. Using numerical simulations, the authors confirmed that contact angle hysteresis is strongly influenced by the local interactions between the drop boundary and the microstructures of the surface, supporting Tadmor's premise that equilibrium contact angles must account for line energy contributions.

In a further study, Terpilowski *et al.* [52] investigated the macro- and micro-wettability of hydrophobic siloxane films with hierarchical surface roughness, synthesized using silica microparticles (MPs) and sol-gel dip-coating method. The hydrophobic films were deposited on glass surface that were preliminarily activated by plasma. Silica was dispersed over the surfaces using a sol-gel solution, with concentrations of the filler in the solution set at 55, 110, and 165 mg/ml. Profilometric studies revealed that plasma activation of the glass substrate contributed to the formation of a more homogeneous surface coating. On surfaces where the glass supports were subjected to a plasma treatment, the measured water contact angles exceeded 150°, indicating superhydrophobic behaviour. These surfaces also exhibited low contact angle hysteresis. Contact angles were measured using the tilted plate method, with 6 μl droplets. Interestingly, the droplets exhibited behaviour similar to that observed on superhydrophobic surfaces; however, upon removal, they underwent deformation, a characteristic typically associated with hydrophilic surfaces. In order to confirm the presence of hydrophilic domains, droplets were deposited and then detached from the surface. As shown in Figure 1.8, after detachment, a smaller droplet remained on the surface, which can be attributed to the presence of small hydrophilic domains on the surface.

Thus, it can be said that macroscopically, this surface appeared predominantly superhydrophobic but at the microscopic level, it exhibited hybrid superhydrophobic/hydrophilic property. In order to elucidate this phenomenon, the authors conducted scanning electron microscopy measurements in environmental mode (a setting that allows to control of the humidity in the chamber) (Figure 1.9). The sample was placed in the microscope chamber, and the relative humidity was gradually increased

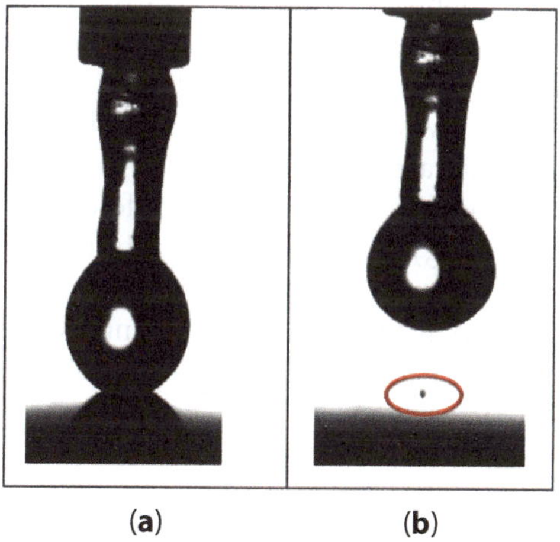

Figure 1.8 Drop deposition on the surface [52]. (a) Placing a water droplet on the surface, (b) a microdroplet of water remaining after removing a macroscopic droplet.

(a) (b)

Figure 1.9 SEM mapping from the sample containing 165 mg/ml filler in the sol-gel solution deposited on the air plasma activated glass (a) 6500x, (b) 20000x [52].

from 20% to 80%. This induced the formation of small droplets on the surface, as shown in Figure 1.9.

Figure 1.9 shows the presence of hydrophilic sites on the surface, indicating that while the surface is macroscopically superhydrophobic, it contains microscale hydrophilic domains. If a larger magnification is applied to the triple-phase line, it becomes evident that this line is not uniform (Figure 1.9 (b)). Points of engagement (Figure 1.9) can be observed where the surface exhibits more hydrophilic property. According to the authors, the surface was characterized as a mosaic structure, combining both superhydrophobic and hydrophilic patches. This type of structure is observed in nature, particularly in the desert beetle (*Stenocara*), which uses this unique surface morphology to collect water from fog. Finally, when the apparent surface free energy was calculated using equilibrium contact angles, the values were approximately 2 mJ/m^2 higher than those calculated using advancing and receding contact angles. This difference is significant since the total surface free energy did not exceed 10 mJ/m^2. The authors reported that the surface free energy calculated using the equilibrium contact angles was potentially closer to that predicted with Young´s equation, confirming the validity of this approach for more accurate wettability characterization on non-ideal surfaces.

Plasma technology is widely available for enhancing adhesion of biopolymers to polymer surfaces. In another interesting study, nitrogen plasma was applied to polyetheretherketone (PEEK) surface to improve the adhesion of chitosan, a biopolymer known for its antibacterial property [53]. Wetting property of both activated- and non-activated surfaces was discussed in terms of apparent surface free energy changes calculated using different theoretical approaches. In the case of chitosan deposited on untreated PEEK, thermal analysis revealed that the biopolymer detached from the surface and decomposes at 86.2°C. However, when the surface was activated with the nitrogen, no distinct decomposition peak corresponding to the chitosan was observed. Instead, the entire system decomposed at ca. 250°C, indicating strong adhesion between chitosan and the activated PEEK surface. The morphological effects of chitosan surface deposited on nitrogen-plasma-activated PEEK were further examined using SEM (Figure 1.10).

The images revealed that on untreated PEEK, chitosan formed distinct epitaxial structures due to the polymer semicrystalline nature [54]. XRD analysis confirmed that chitosan crystallized in an orthorhombic system on the PEEK surface. Wettability of the described surfaces was studied using three test liquids, i.e., water, diiodomethane and formamide. The liquids were applied to determine acid-base and Lifshitz-van der Waals interactions and total value of apparent surface free energy [7–9]. On the plates activated with nitrogen plasma, formamide completely spread over the PEEK surface. This is associated with interaction of nitrogen groups introduced on the surface due to plasma action with formamide molecules. Hysteresis of liquid contact angles did not exceed 10° on most surfaces. The study utilized equilibrium contact angles derived from Tadmor approach to calculate Lifshitz-van der Waals interactions and electron acceptor and donor parameters. The van Oss, Chaudhury and Good [7–9], Owens and

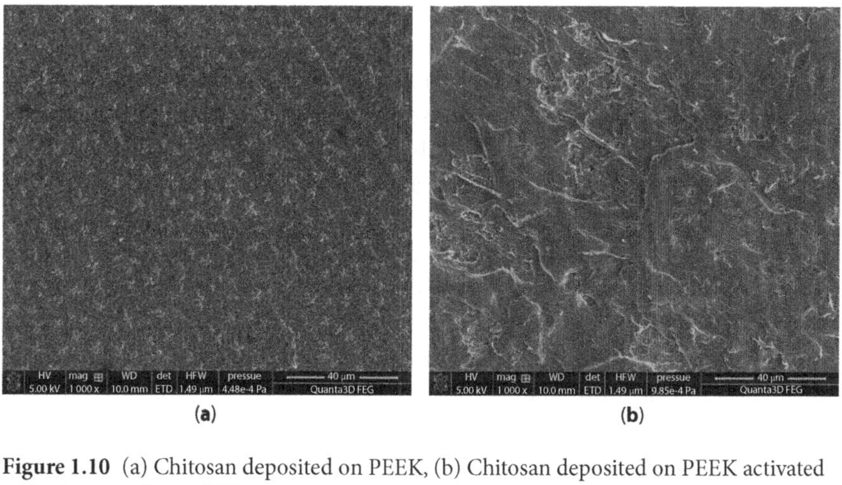

(a) (b)

Figure 1.10 (a) Chitosan deposited on PEEK, (b) Chitosan deposited on PEEK activated with nitrogen plasma [53].

Wendt [10] and Neumann [11–13] approaches for apparent surface free energy estimation usually utilize only advancing contact angles. As was previously mentioned, incorporating equilibrium angles into calculations allows to take into account the effect of receding angle, providing a more comprehensive assessment of the wetting property. Notably, when using equilibrium angles, the total apparent surface free energy values were approximately 3 mJ/m^2 higher than those obtained using advancing and receding contact angles alone. Given the potential applications of chitosan coatings in antibacterial surfaces for the medical industry, this difference is significant. Therefore, using equilibrium contact angles for wettability description enhances the description of surface properties even in the case of surfaces characterized by small hysteresis of contact angles.

On the other hand, Pérez-Huertas and coworkers conducted a series of studies to explore whether the plasma activation of several substrates can affect the structure of different types of gel-type biopolymer films formed on them [55–57]. Another aim was to assess whether these bio-based surfaces could be studied by advancing and receding contact angle measurements and if they could be applied for further calculations of equilibrium angles from Tadmor equation. The study involved the preparation of a series of gelatine gels, varying the gelatine concentration from 10% to 20% [55]. The gels (1mm thick films) were deposited on different glass surfaces pre-treated with air, argon and oxygen plasma. Wetting property was investigated for the gels after they were detached from the treated and untreated glass plates. Wettability was studied by measuring advancing and receding contact angles of water, diiodomethane and formamide. The results showed that plasma activation of the support resulted in an increase in gelatine-gel roughness. Moreover, the total value of apparent surface free energy increased with rising gelatine concentration. Analysis of surface free energy components revealed that plasma activation of the support caused a significant increase in the value of the electron-donor parameter. In the case of gels with the lowest gelatine concentration, the electron donor parameter increased from 16.5 mJ/m^2 to 52 mJ/m^2, 44 mJ/m^2 and 47 mJ/m^2 in the case of gels obtained on oxygen-, argon-, and air-plasma treated plates, respectively. As known from earlier studies [46], activation of glass surface with plasma leads to the introduction of new functional groups. However, this study, for the first time, confirmed that the formation *in situ* of gel films on plasma-activated supports can induce significant modifications in the gel surface properties upon interaction with plasma-excited surfaces. This effect was evidenced by changes in roughness, total surface free energy, and its dispersion and polar components. Furthermore, the authors found that surface free energy values calculated using the CAH approach were consistently higher than those derived from van Oss–Chaudhury–Good LWAB (Lifshitz-van der Waals/ acid-base) approach. These differences were attributed to the use of the Tadmor equilibrium contact angles for the CAH approach, since in the LWAB approach, only the advancing contact angle was used. As seen, incorporating equilibrium contact angles allows for a more comprehensive characterization of surface energetics, even of hydrophilic gel-type surfaces. This innovative approach to characterize the wetting property of the biopolymers has significant implications for applications in food, biomedical, and engineering fields. In another study, Nucia and colleagues studied the potential changes induced in the surface properties of gluten films deposited on plasma-activated supports [58]. Gluten films were prepared from eight different wheat cultivars. Gluten wetting changes were assessed by the calculation of apparent surface free energy using CAH and LWAB approaches, its dispersion component γ_s^{LW} and other parameters (γ_s^+ and γ_s^-). It was found that gluten surfaces become more hydrophilic after surface activation with plasma due to the incorporation of different kinds of functional polar groups, as evidenced by the IR-ATR spectra. The value of water contact angle changes more than that of other liquids due to the significant polar interactions of water. The apparent surface free energy values increased for plasma-treated gluten films, primarily due to a significant rise in the polar component of the surface free energy. For example, the polar component of apparent surface free energy for the Zura cultivar increased fourfold after plasma activation (Table 1.2). Furthermore, apparent surface free energy calculated using the CAH method was consistently higher than that obtained from the Owens-Wendt approach, which can be attributed to the way of energy calculation. In the former approach, surface free energy is calculated for individual liquids separately and then averaged. In the latter approach, a three-equation system is used, considering the contributions of individual liquids simultaneously.

Where CAH_W and CAH_D are the apparent surface free energies calculated from the hysteresis approach using water and diiodomethane, respectively; O-W represents the apparent surface free energy calculated from the Owens-Wendt approach; and γ_s^d and γ_s^p are the dispersion and polar components of apparent surface free energy, respectively. Comparing the apparent surface free energy values obtained from these two approaches, it can be seen that higher values were obtained when the energy is calculated using the equilibrium contact angles (Table 1.2). Thus, the use of equilibrium contact angles further refined the wettability analysis, providing a more accurate description of the surface properties.

Table 1.2 Apparent surface free energy and its components for some cultivar samples deposited on untreated and air-plasma-treated glass support [58].

Cultivar	CAH_W	CAH_D	O-W	γ_s^d	γ_s^p
Kandela	51.7	45.0	46.1	42.2	3.8
Kandela (air)	67.8	44.9	56.1	42.1	13.9
Katoda	45.5	38.9	31.9	31.2	0.7
Katoda (air)	58.7	39.4	35.9	35.8	4.7
Zura	50.7	43.3	42.1	39.7	2.4
Zura (air)	60.0	43.3	49.0	40.6	8.2

1.3 Impact of Hysteresis on Contact Angle Measurement

Contact angle hysteresis occurs on all solid surfaces, including those that are presumed to be flat, homogeneous, and inert. This presents a challenge, as the idealized conditions described by Young's equation are virtually impossible to achieve in practical applications. There can be distinguished five main reasons for differences between the advancing and receding angles: surface roughness, molecular reorientation, liquid penetration inside the solid surface, effect of drop volume change, and chemical heterogeneity on the microscale [59]. Nevertheless, Decker *et al.* [60] reported that there is no direct relation between the amount of roughness of the droplet contact line and the amount of contact angle hysteresis. Even for surfaces exhibiting a high advancing contact angle and low contact angle hysteresis, the measured value can deviate significantly from the equilibrium contact angle. This discrepancy increases with the magnitude of the contact angle hysteresis. Therefore, it is recommended in all cases to use the equilibrium contact angle for calculating the surface free energy—an approach that becomes increasingly important as the hysteresis grows. Relying solely on advancing contact angles to determine surface free energy is erroneous in such instances. While this method does not resolve all challenges associated with contact angle measurements and surface free energy calculations, it represents a necessary step toward improving the reliability and applicability of current experimental results. For instance, the study by Molina *et al.* [61] on the influence of plasma treatment on polyester fibres to enhance graphene adhesion raises potential controversies regarding the determination of apparent surface free energy. The paper reports apparent surface free energy values calculated using Wu's model based on advancing contact angles of water, glycerol and glycol [62]. Interestingly, in most cases, the total apparent surface free energy does not exceed 100 mJ/m^2 which constitutes a calculation error, as the apparent surface free energy calculated from these liquids cannot exceed the value of 72.8 mJ/m^2, which is the surface tension of water. Taking this into consideration, it becomes evident that the study of wettability has not been thoroughly explored, highlighting the need for further investigations to refine measurement methodologies and improve accuracy in surface free energy estimations, thereby improving the overall understanding of wettability. In any case, as has been seen, numerous authors agree that using Tadmor equilibrium contact angles provides a more accurate estimation of apparent surface free energy, yielding values comparable to those predicted by Young's equation for ideal surfaces.

1.4 Conclusions

- Tadmor's model offers a refined approach for determining equilibrium contact angles, particularly for surfaces exhibiting contact angle hysteresis.
- Integrating Tadmor's method with existing models, such as those proposed by Neumann and Chibowski, provides a more realistic assessment of wetting property on non-ideal surfaces.
- Plasma activation is an effective technique for introducing functional groups onto the substrate surfaces in order to increase the adhesion of coatings of different nature.
- Adsorption of PDMS during silica synthesis allows the production of hydrophobic particles, which can be applied to obtaining functional superhydrophobic coatings.
- The greater the contact angle hysteresis, the larger the difference between the advancing and the equilibrium contact angles.
- Apparent surface free energy calculated using the equilibrium contact angles is usually larger than that calculated from the advancing angles.

- Application of the equilibrium contact angle is justified for the calculation of apparent surface free energy, even on hydrophobic surfaces with small contact angle hysteresis.
- Hysteresis is a key reason for applying the Tadmor equilibrium contact angles in apparent surface free energy determination. This approach is particularly relevant when using the acid-base approach or Neumann equation of state for apparent surface free energy calculation, as it ensures the consideration of receding contact angles in the analysis.

AI disclosure statement: The author confirms that no artificial intelligence (AI)-generated content or AI-assisted writing tools were used in the preparation of this chapter (including summary and the abstract). All text, analysis, and conclusions are the result of the author's original work.

References

1. T. Young, An essay on the cohesion of fluids, *Philos. Trans. R. Soc. Lond.*, 95, 65-87 (1805).
2. M. W. Yang and S. Y. Lin, A method for correcting the contact angle from the θ/2 method, *Colloids Surf. A*, 220, 199-210 (2003).
3. C. Della Volpe, D. Maniglio, M. Morra and S. Siboni, The determination of a 'stable-equilibrium' contact angle on heterogeneous and rough surfaces, *Colloids Surf. A*, 206, 47-67 (2002).
4. F.J. Montes Ruiz-Cabello, M.A. Rodríguez-Valverde, and M.A. Cabrerizo-Vílchez, Equilibrium contact angle or the most-stable contact angle?, *Adv. Colloid Interface Sci.*, 206, 320-327 (2014).
5. M. A. Rodríguez-Valverde, F. J. M. Ruiz-Cabello, P. M. Gea-Jódar, H. Kamusewitz, and M. A. Cabrerizo-Vílchez, A new model to estimate the Young contact angle from contact angle hysteresis measurements, *Colloids Surf. A*, 365, 21-27 (2010).
6. G. Kumar and K. Narayan Prabhu, Review of non-reactive and reactive wetting of liquids on surfaces, *Adv. Colloid Interface Sci.*, 133, 61-89 (2007).
7. C.J van Oss, R.J Good, and M.K Chaudhury, The role of van der Waals forces and hydrogen bonds in "hydrophobic interactions" between biopolymers and low energy surfaces, *J. Colloid Interface Sci.*, 111, 378-390 (1986).
8. C.J van Oss, M.K Chaudhury, and R.J. Good, Interfacial Lifshitz-van der Waals and polar interactions in macroscopic systems, *Chem. Rev.*, 88, 927-941 (1988).
9. C.J. van Oss, R.F. Giese, and W. Wu, On the predominant electron-donicity of polar solid surfaces, *J. Adhesion*, 63, 71-88 (1997).

10. D.K. Owens and R. Wendt, Estimation of the surface free energy of polymers, *J. Appl. Polym. Sci.*, 13, 1741-1747 (1969).
11. A. Neumann, R. Good, C. Hope, and M. Sejpal, An equation of state approach to determine surface tensions of low-energy solids from contact angles, *J. Colloid Interface Sci.*, 49, 291-304 (1974).
12. D. Li and A. Neumann, A re-formulation of the equation of state for interfacial tensions, *J. Colloid Interface Sci.*, 137, 304-307 (1990).
13. D. Li and A. Neumann, Equation of state for interfacial tensions of solid-liquid systems, *Adv. Colloid Interface Sci.*, 39, 299-345 (1992).
14. E. Chibowski, Contact angle hysteresis due to a film present behind the drop, in: *Contact Angle, Wettability and Adhesion*, Vol2, K.L. Mittal (Ed.) pp. 265-288, Wiley-Scrivener, Beverly, MA (2002).
15. E. Chibowski and R. Perea-Carpio, Problems of contact angle and solid surface free energy determination, *Adv. Colloid Interf. Sci.*, 98, 245-264 (2002).
16. E. Chibowski, Surface free energy of a solid from contact angle hysteresis, *Adv. Colloid Interface Sci.*, 103, 149-172 (2003).
17. M. Jurak, Surface Gibbs energy interaction of phospholipid/cholesterol monolayers deposited on mica with probe liquids, *Chem. Phys. Lipids,* 183, 60-67 (2014).
18. S.J. Pogorzelski, P. Rochowski, and J. Szurkowski, *Pinus sylvestris* L. needle surface wettability parameters as indicators of atmospheric environment pollution impacts: Novel contact angle hysteresis methodology, *Appl. Surf. Sci.*, 292, 857-866 (2014).
19. I.S. Bayer, C.M. Megaridis, J. Zhang, D. Gamota, and A. Biswas, Analysis and surface energy estimation of various model polymeric surfaces using contact angle hysteresis, *J. Adhesion Sci. Technol.*, 21, 1439-1467 (2007).
20. E. Chibowski, L. Hołysz, K. Terpiłowski, and M. Jurak, Investigation of super-hydrophobic effect of PMMA layers with different fillers deposited on glass support, *Colloids Surf. A*, 291, 181-190 (2006).
21. G.A. Georgiev, S. Baluschev, P. Eftimov, M. Bacheva, and K. Landfester, Addressing the apparent controversies between the contact angle-based models for estimation of surface free energy: A critical review, *Colloids and Interfaces*, 8, 62 (2024).
22. P. G. Rouxhet, Contac angle and surface energy of solids: relevance and limitations, in: *Advances in Contact Angle, Wettability and Adhesion*, Vol. 1, K.L. Mittal (Ed.), pp. 379-395, Wiley-Scrivener, Beverly, MA (2013).
23. F. M. Etzler, Characterization of surface free energies and surface chemistry of solids, in: *Advances in Contact Angle, Wettability and Adhesion*, Vol. 3, K.L. Mittal (Ed.), pp. 219-264, CRC Press, Boca Raton, Fl (2003).
24. B. Jańczuk and J.M. Bruque, M.L. González-Martín, J. Moreno del Pozo, A. Zdziennicka, F. Quintana-Gragera, The usefulness of the equation of state for interfacial tension estimation in some liquid-liquid and solid-liquid systems, *J. Colloid Interface Sci.,* 181, 108-117 (1996).

25. A. Marmur, C. Della Volpe, S. Siboni, A. Amirfazli, and J. W. Drelich, Contact angles and wettability: towards common and accurate terminology, *Surface Innovations*, 5, 3-8 (2017).
26. E. Bormashenko, Wetting of real solid surfaces: New glance on well-known problems, *Colloid Polym. Sci.*, 291, 339-342 (2013).
27. E. Bormashenko, General equation describing wetting of rough surfaces, *J. Colloid Interface Sci.*, 360, 317-319 (2011).
28. H. Y. Erbil, The debate on the dependence of apparent contact angles on drop contact area or three-phase contact line: A review, *Surf. Sci. Rep.*, 69, 325-365 (2014).
29. L. Makkonen, On the methods to determine surface energies, *Langmuir*, 16, 7669-7672 (2000).
30. A. Marmur, Equilibrium contact angles: theory and measurement, *Colloids Surfaces A*, 116, 55-61 (1996).
31. R. Tadmor, Line energy and the relation between advancing, receding, and Young contact angles, *Langmuir*, 20, 7659-7664 (2004).
32. R. Tadmor and P.S. Yadav, As-placed contact angles for sessile drops, *J. Colloid Interface Sci.*, 317, 241-246 (2008).
33. R. Tadmor, Line energy, line tension and drop size, *Surf. Sci.*, 602, L108-L111 (2008).
34. R. Tadmor, P. Bahadur, A. Leh, H. E. N'guessan, R.Jaini, and L. Dang, Measurement of lateral adhesion forces at the interface between a liquid drop and a substrate, *Phys. Rev. Lett.*, 103, 266101 (2009).
35. R. Tadmor, Approaches in wetting phenomena, *Soft Matter*, 7, 1577-1580 (2011).
36. B. Dong, M. S. Driver, I. Emesh, R. Shaviv, and Jeffry A. Kelber, Surface chemistry and fundamental limitations on the plasma cleaning of metals, *Appl. Surf. Sci.*, 384, 294-297 (2016).
37. S. P. Huertas, K. Terpiłowski, M. Tomczyńska-Mleko, and S. Mleko, Ł. Szajnecki, Time-based changes in surface properties of poly(ethylene terephthalate) activated with air and argon-plasma treatments, *Colloids Surfaces A*, 558, 322-329 (2018).
38. S. Cheruthazhekatt, M. Černák, P. Slavíček, and J. Havel, Gas plasmas and plasma modified materials in medicine, *J. Appl. Biomedicine*, 8, 55-66 (2010).
39. L. Bárdos and H. Baránková, Cold atmospheric plasma: Sources, processes, and applications, *Thin Solid Films*, 518, 6705-6713 (2010).
40. Y. Wang, A. Chen, S. Li, D. Ke, X. Wang, D. Zhang, Y. Jiang, and M. Jin, Influence of distance between sample surface and focal point on spectral intensity of nanosecond laser-induced silicon plasma in air, *AIP Advances*, 7, 095204 (2017).
41. A. Van Deynse, P. Cools, C. Leys, N. De Geyter, R. Morent, Surface activation of polyethylene with an argon atmospheric pressure plasma jet: Influence of applied power and flow rate, *Appl. Surf. Sci.*, 328, 269-278 (2015).

42. S. Perni, M.G. Kong, and P. Prokopovich, Cold atmospheric pressure gas plasma enhances the wear performance of ultra-high molecular weight polyethylene, *Acta Biomater.*, 8, 1357-1365 (2012).
43. K. Terpiłowski, M. Wiśniewska, and V. I. Zarko, Influence of solution pH, supporting electrolyte presence and solid content on the stability of aqueous nanosilica suspension, *J. Ind. Eng. Chem.*, 30, 71-76 (2015).
44. E. Chibowski, L. Hołysz, K. Terpiłowski, and M. Jurak, Investigation of super-hydrophobic effect of PMMA layers with different fillers deposited on glass support, *Colloids Surf. A*, 80, 395-403 (2006).
45. K. Terpiłowski, D. Rymuszka, O. V. Goncharuk, I. Ya. Sulym, and V. M. Gun'ko, Wettability of modified silica layers deposited on glass support activated by plasma, *Appl. Surf. Sci.* 353, 843-850 (2015).
46. K. Terpiłowski and D. Rymuszka, Surface properties of glass plates activated by air, oxygen, nitrogen and argon plasma, *Glass Phys. Chem.* 42, 535-541 (2016).
47. K. Terpiłowski, D. Rymuszka, O. Goncharuk, and L. Yakovenko, Equilibrium contact angle and determination of apparent surface free energy using hysteresis approach on rough surfaces, in: *Advances in Contact Angle, Wettability and Adhesion*, Vol.3, K.L. Mittal (Ed.), pp. 331-348, Wiley-Scrivener, Beverly, MA (2018).
48. L. Holysz, K. Terpilowski, V. Zarko, and E. Chibowski, Superhydrophobic polystyrene layers filled with silica on glass, *Surf. Innovations* 1, 52-59 (2013).
49. K.L. Cho, A.H.F. Wu, I.I. Liaw, D. Cookson, and R. Lamb, Wetting transitions on hierarchical surfaces, *J. Phys. Chem.*, 116, 26810-26815 (2012).
50. R. Caruana, S. De Antonellis, L. Marocco, P. Liberati, and M. Guilizzoni, Experimental characterization of the wettability of coated and uncoated plates for indirect evaporative cooling systems, *Fluids,* 8, 122 (2023).
51. Y. Li, J. Liu, J. Dong, Y. Du, J. Han, and Y. Niu, Theoretical analysis of contact angle and contact angle hysteresis of Wenzel drops on superhydrophobic surfaces, *Nanomaterials*, 14, 1978 (2024).
52. K. Terpiłowski, O. Goncharuk, and V. M. Gun'ko, Macro and micro wettability of hydrophobic siloxane films with hierarchical surface roughness, *Smart Mater. Struct.* 27, 075002 (2018).
53. K. Terpiłowski, A.E. Wiącek, and M. Jurak, Influence of nitrogen plasma treatment on the wettability of polyetheretherketone and deposited chitosan layers, *Adv. Polym. Technol.*, 37, 1557-1569 (2017).
54. R.N. Wenzel, Resistance of solid surfaces to wetting by water, *Ind. Eng. Chem.*, 28, 988-994 (1936).
55. S. Pérez-Huertas, K. Terpiłowski, M. Tomczyńska-Mleko, S. Mleko, E. García-Gareta, E. Pakhlov, and F. González-Caballero, Tailoring the surface and rheological properties of gelatine-based hydrogel films *via* indirect cold plasma treatment for engineering applications, *J. Food Eng.*, 363, 111781 (2024).

56. S. Pérez-Huertas, K. Terpiłowski, M. Tomczyńska-Mleko, K. Nishinari, and S. Mleko, Surface and rheological properties of egg white albumin/gelatin dispersions gelled on cold plasma-activated glass, *Food Hydrocolloids*, 96, 224-230 (2019).
57. S. Pérez-Huertas, K. Terpiłowski, M. Tomczyńska-Mleko, and S. Mleko, Surface modification of albumin/gelatin films gelled on low-temperature plasma-treated polyethylene terephthalate plates, *Plasma Process. Polym.*, 17, 1900171 (2020).
58. A. Nucia, M. Tomczyńska-Mleko, S. Okoń, K. Kowalczyk, K. Terpiłowski, S. Pérez-Huertas, and S. Mleko, Surface properties of gluten deposited on cold plasma-activated glass, *Food Hydrocoll.*, 118, 106778 (2021).
59. H.Y. Erbil, G. McHale, and S.M. Rowan, M.I. Newton, Determination of the receding contact angle of sessile drops on polymer surfaces by evaporation, *Langmuir*, 15, 7378-7385 (1999).
60. E.L. Decker, B. Frank, Y. Suo, and S. Garoff, Physics of contact angle measurement, *Colloids Surf. A*, 156, 177-189 (1999).
61. J. Molina, J. Fernández, M. Fernandes, A.P. Souto, M.F. Esteves, J. Bonastre, and F. Cases, Plasma treatment of polyester fabrics to increase the adhesion of reduced graphene oxide, *Synth. Metals*, 202, 110-122 (2015).
62. S. Wu, Surface tension of solids: An equation of state analysis, *J. Colloid Interface Sci.*, 17, 605-609 (1979).

2

Spreading and Adhesion - The Two Sides of Wettability and Their Role in Multiphase Flow through Porous Media of Petroleum Reservoirs

Dandina N. Rao* **and Bikash D. Saikia**

Craft & Hawkins Department of Petroleum Engineering, Louisiana State University, Baton Rouge, Louisiana, USA

Abstract

There exists some confusion about spreading and adhesion phenomena and how they relate to wettability of petroleum reservoir rocks. This stems from the fact that most of our attempts to understand the interfacial interactions occurring in the solid-liquid-liquid systems, such as the rock-oil-brine systems encountered in petroleum reservoirs, have been based in the surface science studies involving solid-liquid-vapor systems. We have explored the differences between these systems in an earlier publication. Here we explore the differences between fluid-fluid spreading - as characterized by the spreading coefficient – versus fluid-solid spreading related to the Zisman-type correlation between the liquid-liquid interfacial tension and the receding contact angle in solid-liquid-liquid systems of differing wettabilities. The influence of these interfacial phenomena on multiphase flow through the porous media of the petroleum reservoir rocks is explored here with further experimental evidence.

Keywords: Spreading, critical spreading tension, spreading coefficient, Zisman-type spreading, interfacial tension, wettability, gravity drainage

**Corresponding author*: dnrao@lsu.edu

Anil N. Netravali and Frank M. Etzler (eds.) The Wide World of Adhesion: A Festschrift in Honor of Dr. Kash Mittal on his 80th Birthday, (97–114) © 2026 Scrivener Publishing LLC

2.1 Background: Surface Science and Zisman's Seminal Contribution

From the production of semiconductor chips, fuel cells, pharmaceutical products, biomaterials, weather-resistant paints to stealth fighter jets, the fruits of advancement of knowledge in surface science and technology have been considerable in our lives. This has been made possible by tireless work of scientists who have dedicated their lives to fundamental understanding as well as ingenious application of the surface properties of materials. William Albert Zisman was one such remarkable scientist, who and his colleagues at the Naval Research Laboratory, Washington D.C. carried out some of the early pioneering work in the field of wettability and adhesion. Not only did the technologies developed from their work immensely benefited the US navy in its war efforts, but also contributed greatly to the advancement of knowledge in surface science. The wealth of data generated from their work has provided the essential foundation for many theories advanced in the field of wettability. Spreading of liquids on high and low energy surfaces, contact potential measurement and the concept of critical surface tension were some of his works that were seminal in those times influencing later developments. Figure 2.1 shows the contact angle (θ) formed by a wetting liquid drop on a solid surface and the interfacial tension components that balance the drop on the surface determining the wettability of the surface.

Zisman theory and its graphical representation in the Zisman plot of characterizing the wettability of a surface were particularly elegant. Zisman plot is used for rapid wettability quantification of a solid surface, through a parameter called critical surface tension (γ_c), by simply measuring the liquid contact angle. Figure 2.2 shows a typical Zisman plot. The cosine of the measured contact angle plotted against the surface tension of the contacted

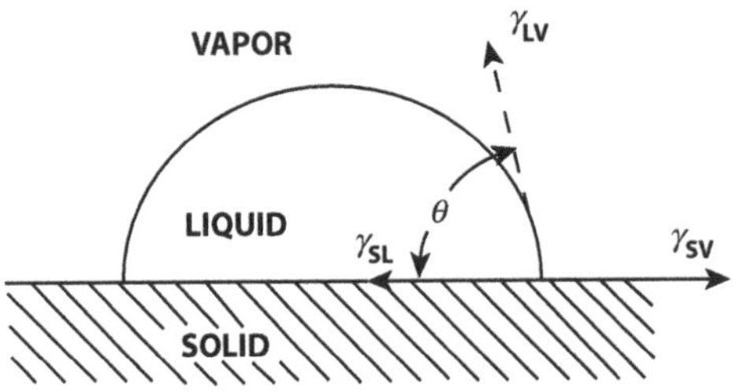

Figure 2.1 Contact angle (θ) and interfacial tension components for a liquid drop in a solid-liquid-vapor (S-L-V) system [1].

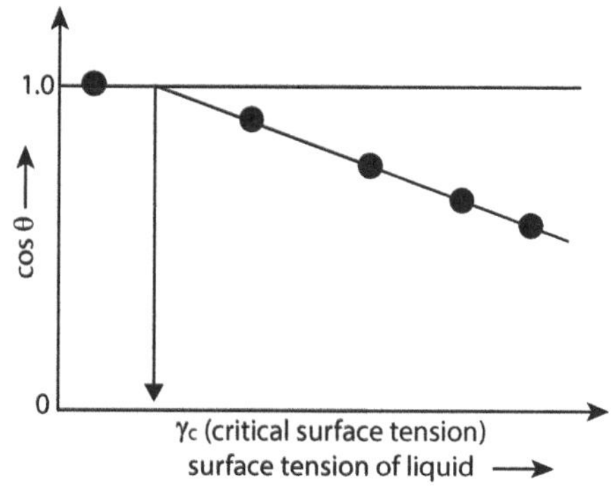

Figure 2.2 A typical "Zisman plot" for solid-liquid-vapor (S-L-V) system.

liquid wetting the solid surface yields the critical surface tension (γ_c), when extrapolated to zero contact angle (or cos θ=1).

The above concept of critical surface tension that Zisman forwarded in 1964 is still used widely for quick wettability quantification. At this point, it is worth noting that the concept of critical surface tension that Zisman so cogently introduced was for Solid-Liquid-Vapor (S-L-V) systems. In petroleum recovery operations, however, one for the most part encounters Solid-Liquid-Liquid (S-L-L) systems, rather than S-L-V systems, with petroleum reservoir rock as the solid phase and liquid phases being the crude oil and brine. Logically speaking, the concept of critical surface tension that Zisman proposed for S-L-V systems should likely be applicable to S-L-L systems as well. Unfortunately, the data to support this extrapolation has been lacking in the literature to our best knowledge. Fortuitously, Rao and co-workers [2, 3] have identified the same fundamental Zisman-type spreading phenomenon in rock-oil-brine (S-L-L) systems while working on two completely different aspects (effect of temperature and effect of salinity) with two different sets of crude oil systems, namely Lindbergh heavy oil from onshore Canada and Yates light oil in Permian basin in the US. Interestingly, these chance encounters supporting extrapolation of Zisman-type spreading behavior to S-L-L systems also happened years apart and that too at the hands of different researchers engaged in applied petroleum research. Thus, there apparently exists a couple of anecdotal evidences substantiating the applicability of Zisman's concept to S-L-L systems. Are these two occurrences just coincidences or is this Zisman-type spreading behavior generally applicable to rock-oil-brine systems also? If it is the latter, then the serious consequences of such spreading behavior might compel us to re-evaluate our current practices of petroleum recovery process. Here we attempt to seek an answer to the above question, namely: Is Zisman-type spreading behavior applicable to rock-oil-brine systems of petroleum reservoirs?

2.2 Evidence of Zisman-Type Spreading in Solid-Liquid-Liquid (S-L-L) Systems

The following paragraphs briefly elaborate the two cases of Zisman-type spreading mentioned earlier in case of Solid-Liquid-Liquid (S-L-L) systems from the Lindbergh heavy oil system in Alberta (Canada) and the Yate's light oil system in Texas (US), respectively.

The Lindbergh heavy oil field is located in the Province of Alberta, Canada. While investigating how temperature affected petroleum recovery from this field, the first case of Zisman-type spreading in S-L-L system was observed by Rao [2]. Figure 2.3 shows the picture of the oil spreading over the solid substrate (quartz) by completely displacing the water film at a certain threshold value of interfacial tension. There were some important distinctions, however, between the spreading in the S-L-V and S-L-L systems. Unlike quick spontaneous spreading in Zisman's work with S-L-V systems, the spreading here was, however, quite gradual. As the interfacial tension value approached ~12 mN/m, the receding contact angle rose to over 90°, clearly indicating the propensity for the oil to spread. Similarly, on further reduction of the interfacial tension value to 9.6 mN/m, the receding contact angle rose further to 122°, and when the interfacial tension further dropped to 6.8 mN/m, the oil was seen to completely spread on the

Figure 2.3 Zisman-type spreading with oil spreading over a solid substrate displacing the water film for a solid-liquid-liquid (S-L-L) system, seen with Lindbergh heavy oil and Quartz substrate [2].

solid surface. This gradual spreading on the solid (quartz) surface can be explained by the fact that in an S-L-L system, the spreading liquid has to displace another liquid from the solid surface in order to be able to spread. In an S-L-V system, on the other hand, the liquid phase needs to displace a lighter and less viscous vapor phase from the surface, which can occur spontaneously below the critical surface tension.

A plot, similar to Zisman's, is made in Figure 2.4 where the cosine of the receding contact angle is plotted against the interfacial tension. In case of S-L-L systems, receding angle more likely depicts the spreading behavior [2]. As can be seen in the Zisman-type plot in Figure 2.4, the least squares fitted data for data points above 20 mN/m with contact angles ranging from 0-30°, when extrapolated to zero contact angle (cos θ=1), intersected at an interfacial tension value of 12 mN/m, which can be denoted as the critical spreading tension which is quite similar in its meaning to Zisman's critical surface tension. The results of Test No. 2 shown in the same Figure 2.4 similarly indicate the interfacial tension value when the oil completely spreads over the quartz substrate, as 6.8 mN/m, which is less than the critical spreading tension of 12 mN/m. Thus, these data apparently corroborate Zisman-type spreading in the domain of S-L-L systems. There appears to be an indication that a Zisman-type relationship may exist in S-L-L systems also, where wettability or spreading preference of one liquid over another on a specific solid substrate is governed by the liquid-liquid interfacial tension between the liquids.

Yate's reservoir (the second case) is a light oil field located in the Permian basin, Texas, USA. While examining the effect of salinity on wettability, the

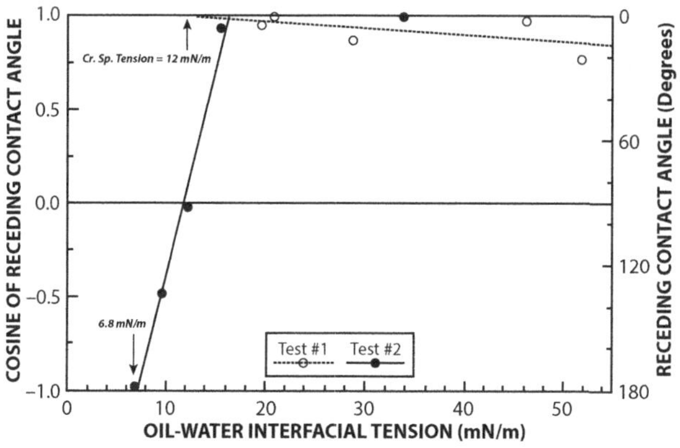

Figure 2.4 Zisman plot for solid-liquid-liquid (S-L-L) system with Lindbergh heavy oil and Quartz substrate [2] showing critical spreading tension of 12 mN/m.

second fortuitous encounter showing Zisman-type spreading in S-L-L systems occurred [3]. Figure 2.5 shows the effect of brine salinity on oil-water interfacial tension. Initial interfacial tension value at 100% concentrated brine was 27.3 mN/m. As the reservoir brine was diluted, the interfacial tension decreased to a minimum (moving to the left of the x-axis) and then increased on further dilution. The minimum value was 10.2 mN/m for a 50:50 volume mixture. The advancing and receding contact angles were also measured for the various mixtures. It was seen that as soon as the mixture with the lowest interfacial tension of 10.2 mN/m was tested, it immediately spread over the solid substrate like a pancake, attaining large receding angles, reaching an equilibrium value of 140°. To confirm the Zisman-type spreading, a mixture of 60% brine and 40% de-ionized water with an interfacial tension of 12.4 mN/m was tested next. To our surprise, the drop did not spread unlike the previous drop, reaching a receding angle of just 19°, with advancing angle still large at 150°. To further test the phenomenon, an intermediate mixture of 55% brine and 45% deionized water with an interfacial tension of 10.9 mN/m was tested and it too had oil spreading instantaneously as soon as it was placed on the surface, but the spreading stopped when receding contact angle of 90° was reached. These observations are in line with the observation made in the case of Lindberg heavy oil system, further corroborating the existence of Zisman-type spreading phenomenon in an S-L-L system.

The Zisman-type plot for the Yate's oil system is shown in Figure 2.6. The trends seen in this plot are unmistakably of the Zisman-type behavior with critical spreading tension of ~10.9 mN/m.

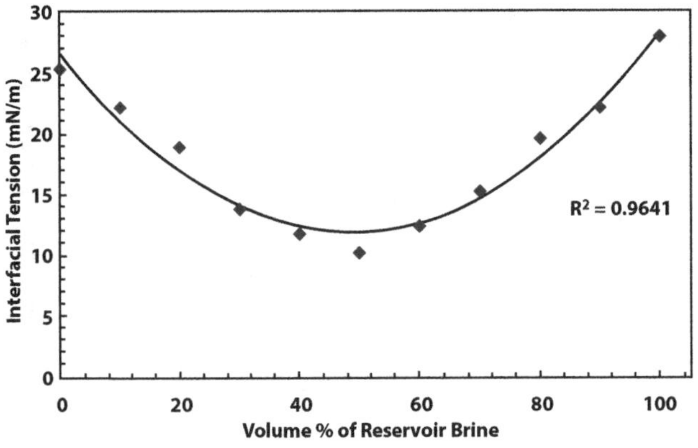

Figure 2.5 Effect of brine dilution on interfacial tension between reservoir brine and Yate's light oil [3].

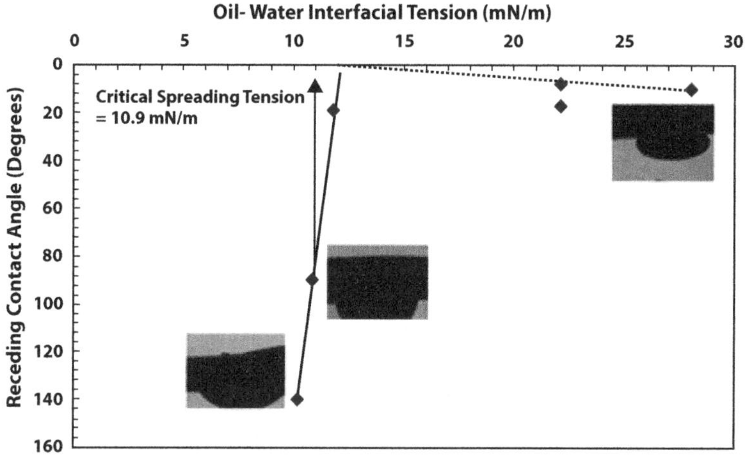

Figure 2.6 Zisman-type plot for solid-liquid-liquid (S-L-L) system with Yate's light oil and dolomite substrate [3].

2.3 The Need for Exploration of Zisman-Type Spreading in S-L-L Systems

Examples laid out in the previous section clearly provide a few chance instances of possible extension of fundamental Zisman-type spreading phenomenon to S-L-L systems, albeit requiring further investigation. It is indeed fascinating to watch the spreading phenomenon that Zisman so brilliantly predicted for S-L-V systems extend to S-L-L systems. Exciting as it is, it also calls us to action to explore further and establish this concept solidly in the domain of surface science involving S-L-L systems, including understanding the very molecular level forces that govern such phenomenon. Understanding of this fundamental phenomenon would undoubtedly enrich the applied side of the process by better informing us about the origin of the macroscopic manifestation. Rao has presented such a need in many forums including Fall 2022 ACS presentation. Rapid developments in the state-of-the-art instrumentation capabilities including high-resolution camera, microscopy, associated computer algorithms and software, etc. have better prepared us to answer these fundamental questions in much more detail than we could in the past. Another point worth mentioning at this stage is that the slew of data generated by Zisman has been in the macroscopic space and many inferences bank on plausible existing theories and principles. With the availability of nano-to-micro level instrumentation prowess nowadays, it is time we delve deeper to pinpoint the molecular level interactions that enable such macroscopic manifestations.

2.4 Consequences of Zisman-Type Spreading in Rock-Oil-Brine (S-L-L) Systems

Extension of Zisman-type spreading to rock-oil-brine (S-L-L) systems would certainly establish the universality of the phenomenon encompassing interactions between solids with fluids, be these liquids or gases. This would further provide credence to the significance of his insightful discovery. Furthermore, the determination of solid-vapor and solid-liquid interfacial tensions is important in many areas of pure and applied science and use of liquid-vapor interfacial tension to indirectly determine the solid-fluid interfacial tension is very attractive. However, liquid-vapor interfacial tension measurements are prone to errors due to unavoidable vibrations. Fortunately, such vibrations are not so much of a nuisance in case of liquid-liquid interfacial tension measurements, because of damping offered by the surrounding liquid phase. Extension of Zisman spreading to solid-liquid-liquid systems would facilitate confident use of liquid-liquid interfacial tensions in this regard. This could even have further consequences for theories regarding the origin of total surface tension of a surface such as surface tension component approach [4, 5] or that based on deriving functional relationship between L-L, S-L and S-V interfacial tension based on contact angle data patterns [6]. Even in practical terms, this will have an impact across many engineering disciplines. For example in petroleum production, the spreading of oil and water on rock surfaces has profound implications in the recovery of oil. If Zisman type spreading were to truly occur in a rock-oil-water (S-L-L) system, as observed in these few instances, meaning if oil were to spread and replace the water film on the rock surface and strongly adhere to the exposed surface, then the oil would have tendency to stick to the rock, resisting production. Thus, presence of Zisman-type spreading in a rock-oil-brine system could fundamentally change surfactant flooding, as we know it, in petroleum reservoirs. Surfactant flooding is one of the principal chemical enhanced oil recovery (EOR) methods to recover additional oil from petroleum reservoirs. Surfactant flooding attempts to either lower the interfacial tension between oil and water phases or change the wettability of rock surface to favorable water-wettedness to get additional oil from oil reservoirs [7]. Mechanism of wettability alteration using surfactant itself has not been fully understood and many approaches such as ion pairing to desorb polar organic components and micellar solubilization of adsorbed organic components have been proposed which stress on desorption of the oil phase in some form [8]. The existence of Zisman-type spreading could upend the very effects that we endeavor to create in oil reservoirs by completely dislodging the water layer from the rock surface. In such a scenario, surfactant flooding would have to be re-evaluated, so as not to initiate Zisman-type spreading. To keep things in perspective, the majority of the current oil reservoirs across different basins in US are depleted and past the primary stage, needing enhanced oil recovery. If one of the primary processes in our efforts at enhanced oil recovery from these depleted reservoirs is impacted, it can truly have significant consequences for the oil industry. With ever increasing demand for energy and in the absence of a reliable, energy-intensive alternative to oil, we may have to continue to rely on oil at least in the near to medium-term. Thus, broadly speaking, it also would have a cascading effect on our energy security.

2.5 Fluid-Fluid Spreading Phenomenon in S-L-L Systems

Other roles of interfacial tension presented here are its influence on the distribution of oil, water and gas within the pores of the reservoir rock matrix, the Zisman-type preferential spreading phenomenon, and oil-rock adhesion characteristics. The imbalance of the three fluid-fluid interfacial tensions at oil-water, water-gas and gas-oil interfaces determines the manner in which these fluids arrange themselves with respect to one another. This distribution affects their relative flow behavior through porous media.

2.5.1 Spreading Coefficient S_o and Oil, Water and Gas Distributions in Porous Media

The manner in which the three fluids arrange themselves in a porous solid substrate depends not only on the spreading coefficient, but also on the wettability of the solid surface. Figure 2.7 displays these dependencies along with the defining equation of spreading coefficient of oil.

If the solid surface is water-wet, a positive spreading coefficient would result in the spreading of oil between gas and water, while a negative spreading coefficient would result in oil lenses floating on the gas-water interface. In the case of an oil-wet rock surface, the positive spreading coefficient

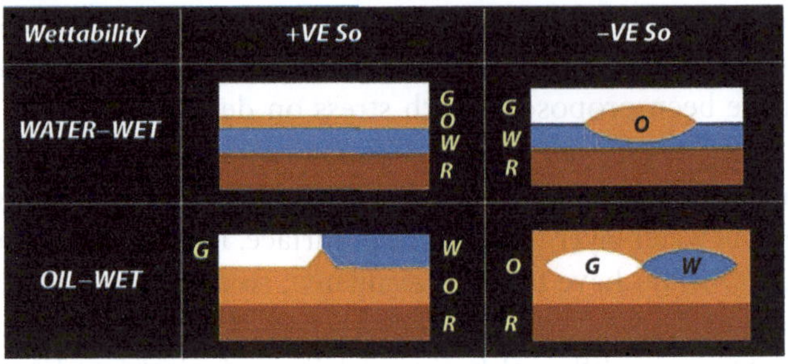

Figure 2.7 Spreading coefficient and oil-water-gas distribution in O/W/G/R (Rock) systems.

would disable gas-water contact by spreading between them and a negative spreading coefficient would require that both gas and water phases flow as discrete globules entrained in the oil phase. The reason for such distributions evolves from the combination of the interfacial force balance at the three fluids contact and the definition of the spreading coefficient as shown in Figure 2.8 with reference to a water-wet case.

At equilibrium, the interfacial force balance along the water surface at the point of three-phase contact yields:

$$\sigma_{og} \text{Cos}(a) + \sigma_{ow} = \sigma_{wg} \tag{2.1}$$

The defining equation for the oil spreading coefficient is:

$$S_o = \sigma_{wg} - (\sigma_{ow} + \sigma_{og}) \tag{2.2}$$

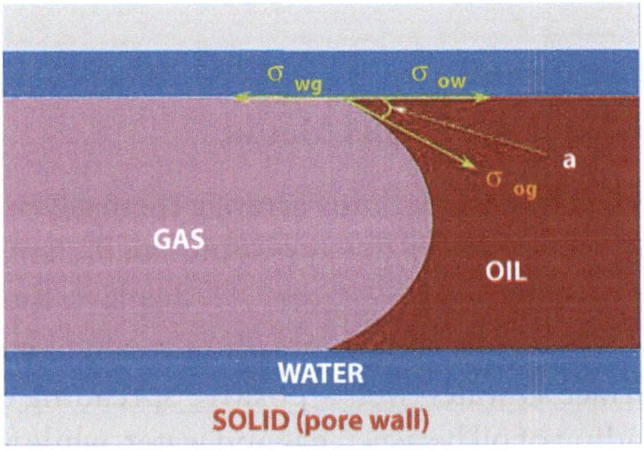

Figure 2.8 Spreading behavior in a solid-liquid-liquid system (water-wet rock).

Combining Equations (2.1) and (2.2):

$$\text{Cos (a)} = 1 + (S_o/.\sigma_{og}) \tag{2.3}$$

Cos (a) lies between +1 and -1 and hence S_o can be zero or negative at equilibrium distribution. Otherwise, if S_o is positive, three-phase contact is not possible and the oil spreads between water and gas, as mentioned before.

Similarly, for the oil-wet case shown in Figure 2.9, the following derivation applies:

At equilibrium,

$$\sigma_{wg}\,\text{Cos (b)} + \sigma_{ow} = \sigma_{og} \tag{2.4}$$

The definition of water spreading coefficient is,

$$S_w = \sigma_{og} - (\sigma_{ow} + \sigma_{wg}) \tag{2.5}$$

Combining Equations (2.4) and (2.5), we get

$$\text{Cos (b)} = 1 + (S_w/\sigma_{wg}) \tag{2.6}$$

Since Cos (b) lies between +1 and -1, S_w can be zero or negative for three-phase contact to occur. If S_w is positive, three-phase contact is impossible and water spreads between oil and gas. However, since $S_w = -(S_o + 2\sigma_{ow})$, for positive S_w, S_o is negative and larger than $2\sigma_{ow}$. Similar fluids distribution maps have been presented by Grattoni *et al.* [9].

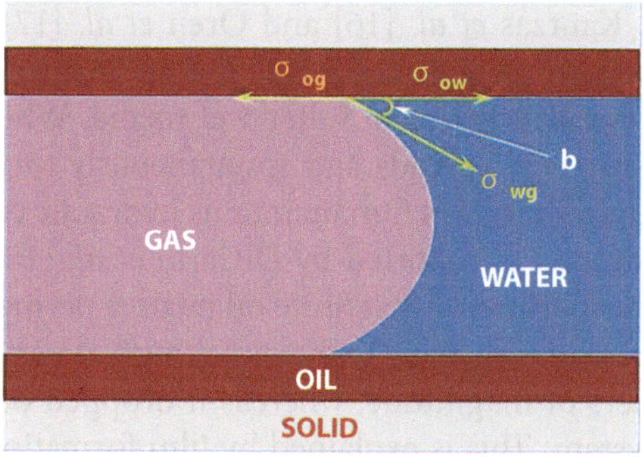

Figure 2.9 Spreading behavior in a solid-liquid-liquid system (oil-wet rock).

Significant impact of this spreading coefficient on the dynamics of oil-water-gas flow in porous media of different wettability states is illustrated in the work of Oren and Pinczewski [10]. They used micromodels to investigate these effects and their results highlight the interplay of spreading and wettability in oil recovery. While positive spreading coefficient was beneficial in terms of higher oil recovery in a water-wet system, the oil-wet system showed an opposite trend. Dzulkarnain [11] examined the effect of fluid-fluid spreading behavior on gravity drainage experiments using water-wet and oil-wet sand packs. These results [11] are explored in the following subsections.

2.5.2 Gravity Drainage in Water-Wet Sand

Figure 2.10 shows the gravity drainage recovery profiles with water-wet sands (sand pack) in both spreading (Soltrol as the oil phase) and non-spreading system (Decane as the oil phase). Recovery profiles are illustrated for three scenarios – Free gravity drainage (FGD), Controlled (or Forced) gravity drainage (CGD) in secondary mode, and Controlled (or Forced) gravity drainage (CGD) in tertiary mode. As can be seen, in case of the water-wet sands, the recoveries have been consistently higher for spreading systems vis-à-vis non-spreading systems under both free and controlled gravity drainages in different modes. For example, FGD in spreading system (Soltrol) recovered 64.8% of the Original Oil In Place (OOIP) compared to 57% for the non-spreading counterpart (decane). Average recoveries for all injection modes (secondary and tertiary) in water-wet sands was 61.4% for spreading systems against 50.5% for the non-spreading systems.

These results are similar to findings from [12–15]. They have also reported higher recoveries for gravity drainage in spreading systems with water-wet sands. Kantzas *et al.* [16] and Oren *et al.* [17], assign spreading-film phenomenon as the reason behind these high recoveries. Oil phase is the intermediate phase in water-wet media. When the spreading coefficient is positive, oil spreads and spontaneously forms a film in the presence of gas and water. This film maintains hydraulic connectivity even at very low saturations as reported by DiCarlo *et al.* [18]. DiCarlo *et al.* observed in their experiments that finite oil relative permeability persisted for the spreading system (hexane and octane) even at very low saturations spanning six orders of magnitude, whereas it dropped off to zero for the non-spreading system. This is explained by film formation in the spreading system that provided continuous path for the bypassed oil blobs to get

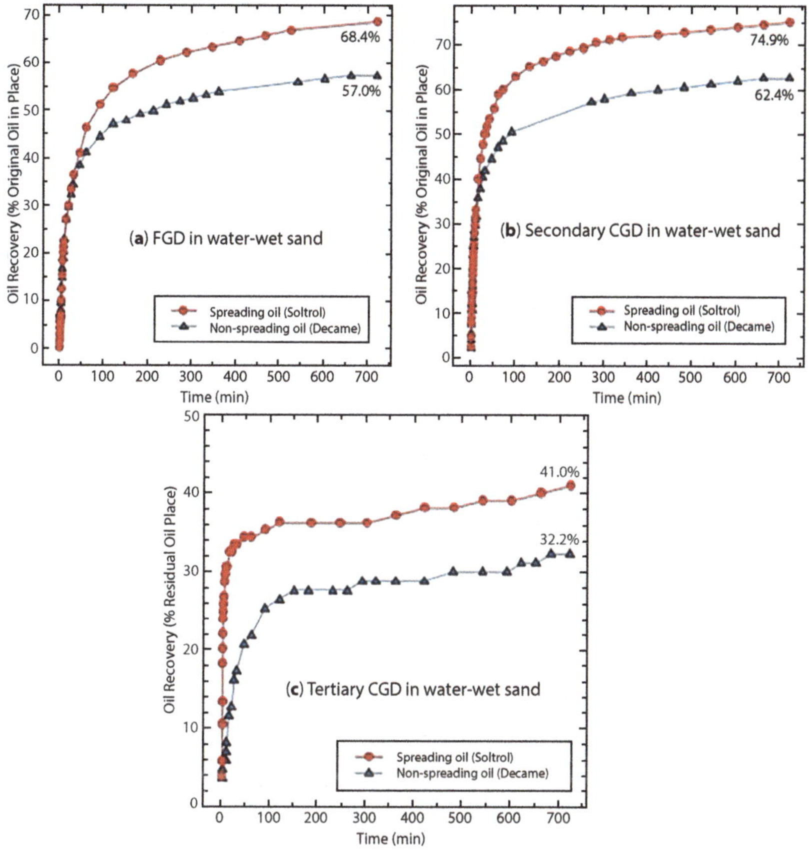

Figure 2.10 Recovery profiles of gravity drainage experiments in spreading and non-spreading systems with water-wet sands (a) free gravity drainage (FGD) (b) secondary controlled gravity drainage (CGD) and (c) tertiary controlled gravity drainage [11].

re-connected and get drained. Whereas in the non-spreading system, the oil did not spread and the conductive film for the residual oil flow was non-existent.

2.5.3 Gravity Drainage in Oil-Wet Sand

Figure 2.11 shows gravity drainage recovery profiles in both spreading (Soltrol) and non-spreading (decane) in case of oil-wet sands (sand packs) for the same modes as in case of water-wet sands.

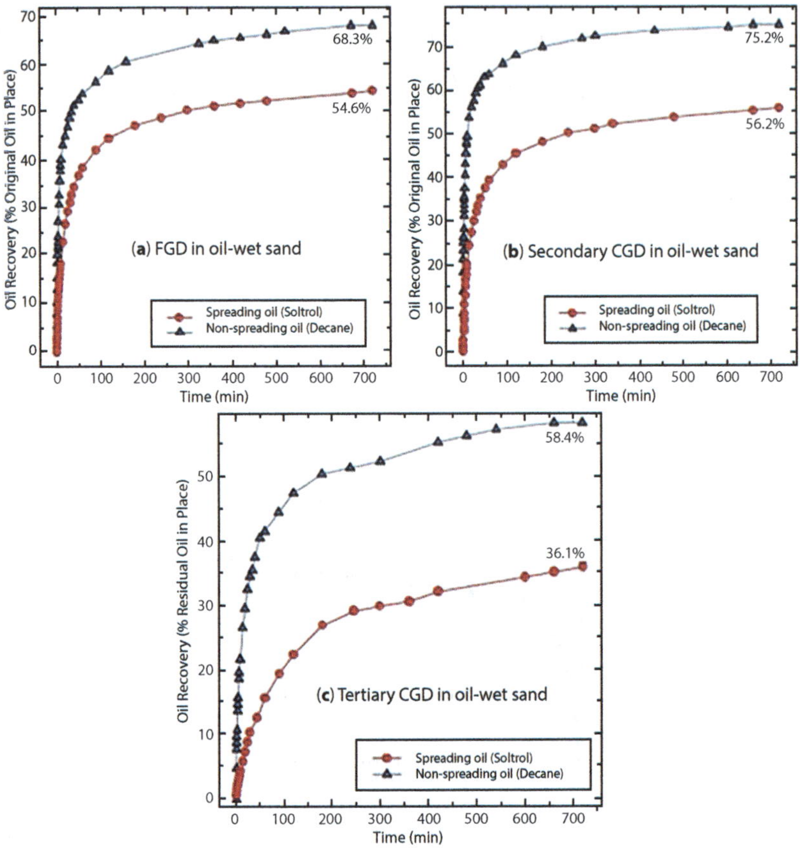

Figure 2.11 Recovery profiles of gravity drainage experiments in spreading and non-spreading systems with oil-wet sands (a) free gravity drainage (FGD) (b) secondary controlled gravity drainage (CGD) and (c) tertiary controlled gravity drainage [11].

In case of the oil-wet sands, it is the non-spreading system that performed better in terms of recoveries. For instance, non-spreading (decane) system fetched recovery of 68.3% OOIP as against 54.6% for the spreading (Soltrol) case. This same behavior was consistent across all modes. Average oil recovery in case of the non-spreading system was 67.3% and in case of spreading system it was 49%. Similar, yet somewhat contradictory, findings were reported by Oren and Pinczewski [10] and Vizika and Lombard [19]. Oren and Pinczewski [10] had found highest recovery in oil-wet porous media irrespective of their spreading or non-spreading behavior. They carried out 3-phase micromodel flow experiments and reported highest recovery for oil-wet media, followed by water-wet media with spreading system, followed by water-wet media with non-spreading system. They explained that in oil-wet media, sign of the spreading coefficient, i.e., formation of films, was not important since oil could drain through the continuous wetting phase in the absence of spreading oil films. According to them, the drainage rate in case of oil-wet media was in fact faster because the wetting films were thicker, leading to higher conductive flow path for oil. In their experimental study, Vizika and Lombard [19] found the lowest oil recovery for the oil-wet case, when compared to water-wet and fractional-wet sands. There was no difference in recoveries for spreading or non-spreading systems in case of oil-wet sands, signifying spreading behavior has no impact on recovery. They attributed the presence of non-wetting water phase blocking gas passage in larger pores to displace oil as the reason behind this low recovery. While comparing the work of Oren and Pinczewski [10] and Vizika and Lombard [19], it is important to note that their studies were performed at different scales - pore scale vs core scale, respectively. Also, the saturation histories were different for the two cases with former conducted at residual oil saturation, while the latter conducted at residual water saturation. These differences in the experimental conditions thus do not allow a direct comparison between their findings.

Results from our laboratory [11] generally agreed well with Oren and Pinczewski [10]. For example in the case of tertiary floods, recovery was highest for non-spreading oil-wet system (58.4%), followed by spreading water-wet system (41%), and lowest for non-spreading water-wet system (32.2%). Similar behavior was observed in secondary floods. However, in our experiments there was stark contrast in the recovery from spreading and non-spreading systems for oil-wet media, unlike that found by Oren and Pinczewski.

2.6 Summary and Conclusions

We have attempted here to clarify the distinction between the roles played by spreading and adhesion phenomena on wettability of oil-rock-brine systems encountered in petroleum reservoirs.

Based on our observations of two very different rock-oil-brine systems, it seems reasonable to conclude that Zisman-type fluid-solid spreading behavior, correlated by a relationship between the water receding contact angle and the oil-water interfacial tension at the respective representative pressure and temperature conditions, is applicable to S-L-L systems also. Further research is indeed essential to address the questions: (1) whether or not the wetting water film on the rock surface is completely displaced by the oil (and consequently making the rock surface oil-wet) during this Zisman-type spreading behavior? If not, (2) does the oil adhere to the rock surface due to some components in the oil interacting with the rock surface through the wetting water film?

The fluid-fluid spreading behavior that is governed entirely by the three interfacial tensions in the three-fluid-phase (oil-brine-gas) containing systems in addition to a solid phase (reservoir rock), plays an important role in the flow through porous media. This is because it affects the three-phase flow behavior involving the mechanisms of film flow, gravity drainage and displacement of oil irrespective of the wettability of the solid phase. The results presented here clearly bring out the need to distinguish between Zisman-type solid-fluid spreading which would alter wettability of the solid surface involved in the system and the fluid-fluid spreading which influences the flow through porous media without affecting the wettability of the solid phase involved in the system.

Thus, while reservoir wettability can be confidently characterized by the water advancing contact angle which reflects the adhesion aspect of wettability of rock-oil-brine systems, the fluid-fluid spreading behavior governed by the three individual fluid-fluid interfacial tensions and the Zisman-type solid-fluid spreading behavior constitute the other side of the three-phase flow behavior.

AI disclosure statement: The author confirms that no artificial intelligence (AI)-generated content or AI-assisted writing tools were used in the preparation of this chapter (including summaries and the abstract). All text, analysis, and conclusions are the result of the author's original work.

References

1. W.A. Zisman, Relation of equilibrium contact angle to liquid and solid constitution, in: *Advances in Chemistry Series*, Vol. 43, pp. 1-51, American Chemical Society, Washington DC (1964).
2. D.N. Rao, Wettability effects in thermal recovery operations, *SPE Reservoir Evaluation & Eng.*, 2, 420–430 (1999).
3. C.S. Vijapurapu and D.N. Rao, Effect of brine dilution and surfactant concentration on spreading and wettability, in: *Proceedings of SPE International Symposium on Oilfield Chemistry*, (2003).
4. F.M. Fowkes, Attractive forces at interfaces, *Ind. Eng. Chem.*, 56 (12), 40-52(1964).
5. D.K. Owens and R.C. Wendt, Estimation of surface free energy of polymers, *J. Appl. Polym. Sci.* 13,1741-1747(1969).

6. D. Y. Kwok and A. W. Neumann, Contact angle measurement and contact angle interpretation, *Adv. Colloid Interface Sci.,* 81, 167-249(1999).
7. A. Bashir, A.S. Haddad and R. Rafati, A review of fluid displacement mechanisms in surfactant-based chemical enhanced oil recovery processes: analyses of key influencing factors, *Petroleum Sci.*, 19, 1211-1235(2022).
8. R. Gupta and K.K. Mohanty, Wettability alteration mechanisms for oil recovery from fractured carbonate rocks, *Transport Porous Media,* 87, 635–652 (2011).
9. C.A. Grattoni, X.D. Jing, and R.A. Dawe, Dimensionless groups for three-phase gravity drainage flow in porous media, *J. Petrol. Sci. Eng.*, 29, 53–65(2001).
10. P.E. Oren and W. V. Pinczewski, The effect of wettability and spreading on the recovery of waterflood residual oil by immiscible gasflooding", *SPE Formation Evaluation,* 9, 149–156(1994).
11. I. Dzulkarnain, Investigation of flow mechanisms in gas-assisted gravity drainage process, Ph.D. thesis, Louisiana State University, Baton Rouge, USA(2018).
12. A. Kantzas, I. Chatzis, and F.A.L. Dullien, Enhanced oil recovery by inert gas injection, in: Proceedings of SPE Enhanced Oil Recovery Symposium, pp. 653-662(1988).
13. O. Vizika, Effect of the spreading coefficient on the efficiency of oil recovery with gravity drainage, in: Proceeding of 205th American Chemical Society national meeting, Symposium on Enhanced Oil Recovery (1993)
14. F. Kalaydjian, O. Vizika, J. Moulu, and P. K. Munkerud, The role of wettability and spreading in gas injection processes under secondary conditions, Geol. *Soc. Spec. Publ.,* 84, 63–71(1995).
15. H. Maeda and K. Okatsu, EOR using thin oil film drainage mechanism in water wet oil reservoir, in: Proceedings of SPE Asia Pacific Oil and Gas Conference and Exhibition (2008).
16. A. Kantzas, A., I. Chatzis, and F.A. L. Dullien, Mechanisms of capillary displacement of residual oil by gravity-assisted inert gas injection, in: Proceedings of SPE Rocky Mountain Regional Meeting, pp. 297-307(1988).
17. P.E. Oren, J. Billiotte and W.V. Pinczewski, Mobilization of waterflood residual oil by gas injection for water-wet conditions, *SPE Formatioin Evaluation,* 7, 70–78(1992).
18. D.A. DiCarlo, A. Sahni and M.J. Blunt, Three-phase relative permeability of water-wet, oil-wet, and mixed-wet sandpacks, *SPE Journal,* 5, 82–91(2000).
19. O. Vizika and J.M. Lombard, Wettability and spreading: two key parameters in oil recovery with three-phase gravity drainage. *SPE Reservoir Eng.*, 11, 54–60(1996).

3

Application of Integrated Analytical Approach (IAA) to the Investigation of Release and Permeation Processes of Active Substances from Ointments Using the Contact Angle Methodology

D. Rossi[1*], S. Banella[2], E. Vettorato[1], E. Franceschinis[1], G. Lazzari[1], S. Dall'Acqua[1] and N. Realdon[2]

[1]*Pharmaceutical Technical Laboratory, Department of Pharmaceutical and Pharmacological Sciences, Univesity of Padua, Padua, Italy*
[2]*Department of Diagnostics and Public Health, University of Verona, Verona, Italy*

Abstract

The Integrated Analytical Approach (IAA) is applied to complex systems investigations by integrating different analytical techniques within the same field of study. The IAA has undergone an evolution thanks to the integration of the data obtained from analytical techniques in different fields of investigation. In our work, the Integrated Analytical Approach was applied to the study of the release property of semi-solid formulations used for skin application, such as Lipogel loaded with carnitine and a commercial ointment called Gel-ALPHA containing a specific substance.

Our research has highlighted the integration of the surface tensiometry data, obtained with the Solid-like methodology using the Rossi number, and the concentration of active substances released during *ex vivo* permeation process. The study has demonstrated how the different amounts of the released and permeated drug can generate variations in the surface chemistry and surface tensiometry of acceptor solution over time.

In this context, the surface tensiometry demonstrated the importance of the non-invasive, rapid, and economical approach to the characterization of release and *ex vivo* permeation processes. The IAA underlined the importance of

**Corresponding author*: davide.rossi@unipd.it

Anil N. Netravali and Frank M. Etzler (eds.) The Wide World of Adhesion: A Festschrift in Honor of Dr. Kash Mittal on his 80th Birthday, (115–144) © 2026 Scrivener Publishing LLC

multidisciplinarity in the study of complex systems, focusing on the observations of the phenomena that govern their nature in pharmaceutical technology field.

***Keywords*:** Integrated analytical approach, surface tensiometry, drug release, drug permeation, Rossi number, semi-solid formulations, solid-like methodology

3.1 Introduction

The Integrated Analytical Approach (IAA) can be defined as a method of combining multiple data sources, or information, to gain a more comprehensive understanding of a particular phenomenon or problem. This approach involves integrating data from various disciplines, perspectives, or domains, to provide a holistic view of the issue at hand. IAA has found application in various fields of investigation and has demonstrated wide applicability and the importance of a multidisciplinary approach in complex systems investigation, thanks to the multiple information it can provide.

For example, the IAA is applied in studies relating to the identification and quantification of steroid hormones in the plasma and urine of calves treated orally with "hormone-like" substances by combining different analytical techniques (e.g., Western Blot, LC-APCI- Q-MS/MS analysis) [1]. Similarly, in the chemical-geological [2] and chemical-mineralogical fields [3], the combination of different analytical techniques in the same field of investigation proved to be very important for a more in-depth knowledge of the studied phenomena.

In the most recent definition of Integrated Analytical Approach on the same material sample s and at the same t (IAA_t^s), the data integration is obtained directly from the same material sample and at the same time [4], in such a way as to be able to evaluate its chemical, rheological and surface energy properties simultaneously (Figure 3.1).

The advantage that the IAA_t^s offers is that of obtaining the highest possible quantity of information allowing to have the broadest possible view of the object of investigation.

Recent examples of the application of IAA_t^s [5, 6] have considered the integration of biological, geochemical, and surface tensiometry data for Lourdes water samples. The correlation analysis between the data from the various analytical techniques made it possible to identify the water samples from the Sacred Fountain of Massabielle in Lourdes, France.

Recently, a new method was developed in surface tensiometry called Solid-like methodology (SLM) [7]. The SLM is an innovative method able

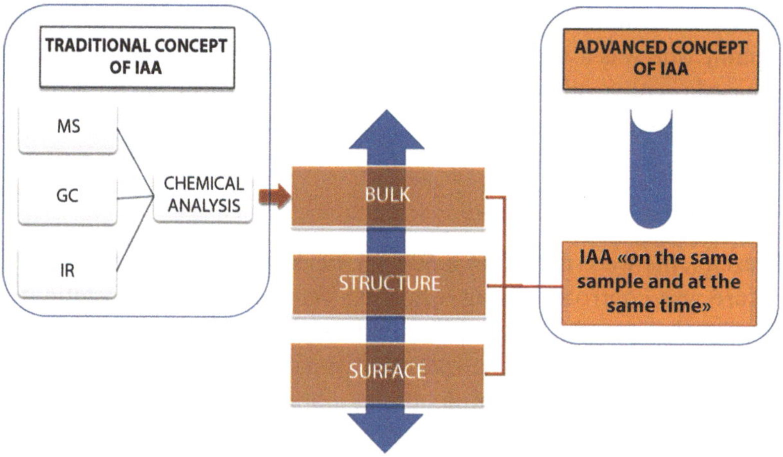

Figure 3.1 Integrated analytical approach (IAA) for the same samples, and at the same time (t).

to determine the Surface Tension (ST: mN/m), Dispersion Component (DC; mN/m), and Polar Component (PC; mN/m) of a liquid without the influence of the interfacial friction forces and the roughness factor (r) using a perfluoropolyether liquid film (PFPEf) of Fomblin HC/25®PFPE (PFPE) as a solid substrate for the characterization of liquid systems, and a drop of the same perfluoropolyether (PFPEd) for the determination of Surface Free Energy (SFE: mJ/m^2), Dispersion Component (DC; mJ/m^2), Polar Component (PC: mJ/m^2) of solid systems [7].

The main aim of our work was the evaluation of the discriminating capability of the surface tensiometry technique to evaluate the drug release and *ex vivo* permeation capability [8–10] of different formulations through an integrated *chemical-surface* approach by applying the SLM, and integrating the chemical data from High Performance Liquid Chromatography – Evaporative Light Scattering Detection (HPLC-ELSD) with the surface tensiometry data.

3.2 Experimental

3.2.1 Formulations

The study was focused on the characterization of two ointments, one with high lipophilicity and one with high hydrophilicity [11]. In particular, the hydrophilic ointment was a commercial formulation named "Gel-ALPHA" loaded with an active substance *X*, so named for copyright reasons, and the lipophilic one was a Lipogel loaded with carnitine. The Lipogel ointment was based on olive oil, bees wax, cholesterol, cetostearilic alcohol, and carnitine, all purchased from A.C.E.F. (Italy).

3.2.1.1 *Lipogel Preparation*

The excipients were melted in a 600 mL beaker in a water bath. The temperature was controlled using a heating plate with a thermostatic probe at 70°C. Subsequently, the preparation was placed under constant stirring using a paddle stirrer with a metal mesh rotor set at 60 rpm to achieve the gelling process. The ointment thus obtained was transferred into a container with a screw cap, sealed with Parafilm and stored at room temperature, and protected from light. Table 3.1 shows the composition of the Lipogel.

Table 3.1 Composition of the Lipogel.

Component	**Content (wt %)**	**Content (g)**
Bees wax	6	6
Cholesterol	3	3
Cetostearilic alcohol	2	2
Olive oil	*q.s 100	89

*q.s = quantum sufficit

Table 3.2 Composition of the Lipogel with water polar phase.

Component	**Content (wt%)**	**Content (g)**
Bees wax	6	6
Cholesterol	3	3
Cetostearilic alcohol	2	2
Water	10	10
Olive oil	*q.s. 100	79

*q.s = quantum sufficit

Table 3.2 shows the composition of the Lipogel containing 6% bees wax with 10% water.

The quantum sufficit (or "as much as suffices" or quantum satis) is a term used in the practice of pharmacy and means sufficient quantity (also called "just enough") used on medical prescriptions to indicate the amount determined by the formulator [12].

3.2.1.1.1 Lipogel with 4% w/w Carnitine

The Lipogel containing 4% w/w of carinitine was prepared by adding its dispersion at 50° C to the base Lipogel.

The composition of the Lipogel loaded with 4% wt/wt of carnitine is reported in Table 3.3.

3.2.1.2 Drug Release

The release assay was performed seven days after the preparation of Lipogel loaded with carnitine. The Perspex® cells were equipped with a container for the preparation under examination, on whose surface was positioned a cellulose nitrate membrane (diameter 25 mm, pores 0.45 μm, Prat Dumas, France) previously hydrated in milli-Q water for 24 hours (Figure 3.2). The upper flange of the cell was positioned above the lower one and separated by an O-ring on the membrane. Finally, the system was completed by screwing the flanges at three points to ensure the closure of each cell that was placed in the vessels of a dissolution apparatus (Figure 3.2) The Perspex® cells containing the preparation were set up together with

Table 3.3 Composition of Lipogel with 4% wt/wt carnitine.

Component	Content (wt %)	Content (g)
Bees wax	6	6
Cholesterol	3	3
Cetostearilic alcohol	2	2
Carnitine	4	4
Water milli-Q	10	10
Olive oil	*q.s.100	75

*q.s = quantum sufficit

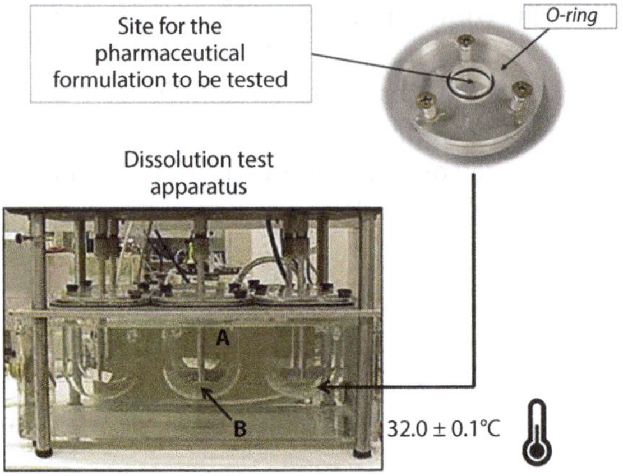

Figure 3.2 Example of Perspex® cell used in the release test already assembled and a dissolution apparatus with (A) vessel containing 400 ml of milli-Q water and (B) Perspex® cells installation site.

controls without the active ingredient. The release assay was performed by filling each vessel with 400 mL milli-Q water and the test was carried out at 32.0 ± 0.1°C and a rotation speed of the rotating blades was 60 rpm for the entire duration. Aliquots of 2 mL were taken every 15 minutes during the first hour and every 30 minutes for the following hours, for a total of 6 hours. At each withdrawal, the volume of the acceptor phase was restored with an equal amount of fresh ultrapure water. At the end of the release assay, the samples were stored at +4 °C until the execution of the HPLC analysis and the surface tensiometry analysis by SLM.

The analysis of carnitine was carried out by HPLC with an Agilent 1260 model coupled to an ELSD detector (SEDEX LT60). The HPLC-ELSD is equipped with a Phenomenex Kinetek C-18 4.6 x 150 mm column. The analysis was performed using acetonitrile (ACN) and water + 1% formic acid (HCOOH) with gradient elution (5%:95% ACN: H_2O + 1% HCOOH up to 100% ACN) in the first 15 minutes, and continuing in isocratic conditions up to 30 minutes.

The flow rate was 1.1 mL/min, and the ELSD detector was set to an amplificator factor of 12 to adjust the strength of the signal according to the sample. The temperature was 60°C, and the pressure of N_2 was equal to 2.3 bar. The wavelength used for the detection of carnitine was 250 nm, and the retention time approximately 1.41 minutes. The quantitative analysis was carried out using a calibration line obtained from the analysis of five samples with known concentrations of carnitine between 10-600 μg/mL.

3.2.1.3 Drug Permeation

The formulations tested were the commercial ointment Gel ALPHA loaded with the active substance *X*. Five Franz diffusion cell systems (B, D, F, H, and L) were used for the *ex vivo* study of the barrier effect at the skin permeation of active substance using ear tissues samples taken from a living pig (PermeGear, SES, GmbH, Germany) (Figure 3.3a). The Franz cells were equilibrated with a thermostated system at 38.0 ± 0.5°C to maintain the temperature of the acceptor phase at 32 ± 1°C and thawed the porcine skin for 30 minutes (Figure 3.3b).

400 μL samples were withdrawn from the acceptor phase of each cell at intervals of 30 minutes, 1 hour, and 3 hours. After each sampling, the volume of the acceptor phase was restored with fresh water solution of ethanol at 30% v/v. The collected samples were stored in Eppendorf® tubes at +4°C and protected from light up to quantitative analysis by HPLC and surface tensiometry analysis by SLM.

The barrier effect was evaluated against the skin permeation of a given xenobiotic from the active substance *X* in a 1% w/v aqueous solution. The control was performed by applying one layer of stringy vaseline to the porcine skin; a lipophilic and occluding base already proposed as a negative control [10] against the passage of active substance *X*.

Five membranes of about 20×20 mm were prepaed by cutting the central area of the skin of the dorsal side of the porcine ear. Each cell was filled with 5 mL of 30% v/v ethanol water solution (acceptor phase). 100 mg of each was applied and spread over the surfaces of the four membranes prepared to create a thick film as homogeneous as possible. Each membrane thus treated plus four untreated membranes were positioned in the center

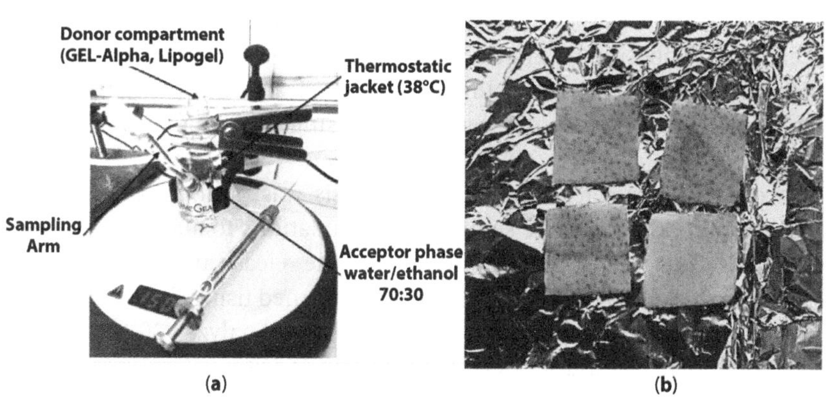

Figure 3.3 (a) Franz cell system. (b) Porcine skin.

of the opening of each Franz cell, ensuring complete contact between the acceptor phase and the dermis. Finally, the cells were placed in each compartment, and the system was closed with forceps. The acceptor phases of the cells were subjected to a constant magnetic stirring at 1500 rpm, and after 10 minutes of equilibration of the assembled system, 500 μL of active substance solution was added to the compartments of all cells. The temperature was kept at 31.0 ± 0.3 °C for the entire duration of the experiment. The quantification of the active substance *X* content in the samples was carried out with the analytical method adopted for the release study.

3.2.2 Surface Tensiometry Analysis

The term "surface tensiometry" was defined for the first time as "the analytical technique capable of characterizing surface free energy (SFE) and surface tension (ST) using the contact angle method" in 2021 [4]. The new "surface tensiometry" term avoids the equivalence with surface tension. It substitutes the term "tensiometric" often associated with the analytical technique for the measurement of liquid contact angles and their surface tension measured using the pendant drop method. The new definition of surface tensiometry substitutes also the term "surface tensiometric parameter" which often indicates SFE and ST of natural and artificial complex systems [13].

In general, surface tensiometry uses a tensiometer (optical goniometer) to characterize ST and SFE of complex systems in a non-invasive and rapid way, avoiding the destructive approaches represented by chemical and rheological analytical techniques. For example, surface tensiometry is important to measure the ST of liquids using the sessile drop method [14] and to investigate surface and interfacial properties of cationic and anionic surfactants [15]. Surface tensiometry was also used in the study of intercellular adhesion by Tissue Surface Tensiometry (TST) to quantify the intercellular cohesive energy under physiological conditions [16] and to differentiate tissues from different organs using the Langmuir monolayer technique [13].

In our work, surface tensiometry analysis was performed seven days after the preparation of the semi-solid formulations to allow their stabilization. The surface tensiometry analysis of the release and *ex vivo* permeation from Lipogel and Gel ALPHA was performed using the contact angle method (CA: deg) [17, 18] using a Drop Shape Analyzer 30 tensiometer (Kruss GmbH, Hamburg, Germany) equipped with a camera aiming at a light source capable of displaying a real-time image of the CA formed when a test liquid comes in contact with the substrate (Figure 3.4).

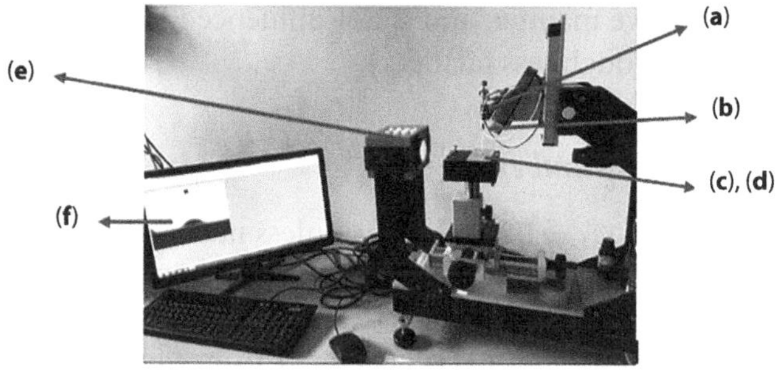

Figure 3.4 Tensiometer (a) syringe support, (b) syringe, (c) sample holder support, (d) substrate to be analysed, (e) light source, (f) calculator and dedicated software.

3.2.2.1 Solid-Like Methodology (SLM)

The SLM is applied to determine the surface tensiometry characteristics of liquid systems after depositing 0.5 ml (PFPE) on an inert support (Figure 3.5) [7].

PFPE is a substance belonging to the class of perfluoropolyethers having hydrophobic, lipophobic and "self-repellent" characteristics [19]. Thanks to its particular characteristics, PFPE is able to form a PFPE film (PFPEf) on the surface of an inert support [7]. The SLM method is based on the possibility of determining the CA of an unknown liquid when it comes in contact with the PFPEf. The PFPEf is therefore considered a solid substrate compatible with the tensiometric study of unknown liquids [7]. Considering the particular physico-chemical characteristics of PFPEf, with SLM it is possible to carry out a surface tensiometry analysis of a liquid in

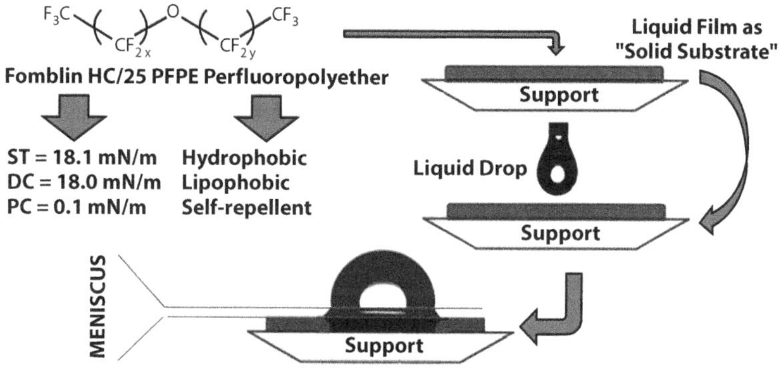

Figure 3.5 The solid-like methodology (SLM).

a rapid, non-invasive manner, and is not influenced by interfacial friction forces and surface roughness factor (r).

3.2.2.2 *The Rossi Number* (χ^{L}_{PFPEf})

The Rossi number (χ^{L}_{PFPEf}) is a dimensionless number for the characterization of unknown liquids [7]. The χ^{L}_{PFPEf} is a physical quantity expressed as a number without any unit and is defined as a pure number capable of determining the deformation (D) generated from the contact between a liquid and a liquid film of a particular substance called perfluoropolyether (PFPEf).

Thus, when a drop of an unknown liquid contacts the surface of the PFPEf, a characteristic meniscus (M) is formed at the interface between the liquid and PFPEf. The size of this meniscus depends on the surface tension characteristics of the unknown liquid [7]. The surface tensiometry parameter that describes the size of the meniscus of PFPEf and its variation is called the Rossi number (Equation 3.1).

$$\chi^{L}_{PFPEf} = CA_{M} - CA_{nM} \tag{3.1}$$

Where χ^{L}_{PFPEf} (χ=Rossi number symbol) is the Rossi number of liquid (L) in contact with a perfluoropolyether liquid film (PFPEf), CA_M is the CA of the liquid (L) measured by placing the baseline at the PFPEf surface, and CA_{nM} is the CA measured by placing the baseline at the interface between the unknown liquid and the meniscus of PFPEf (Figure 3.6).

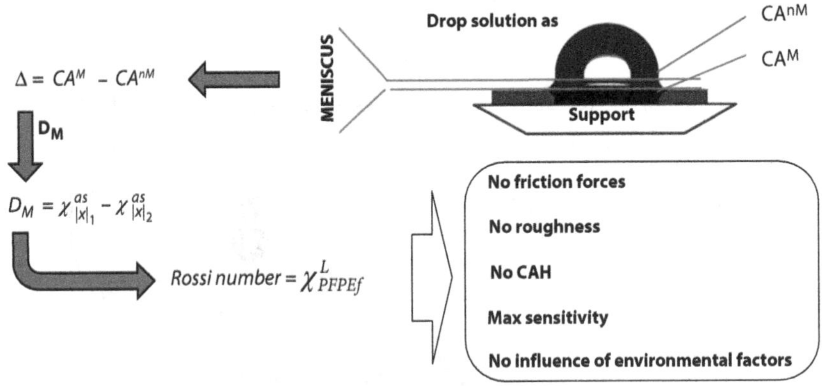

Figure 3.6 From solid-like methodology to Rossi number where CAH is Contact Angle Hysteresis, D_M is Meniscus Deformation, $|x_1|$ is solute concentration 1, $|x_2|$ is solute concentration 2.

The CAs of the released droplets were measured after 0.5s from contact with PFPEf, and after periodically withdrawing the volume for up to 6 hours. The droplets were left on the PFPEf.

3.2.2.2.1 The Rossi Number (χ^{L}_{PFPEf}) and Surface Chemistry of Complex Systems

Generally, surface chemistry focuses on the adsorption, migration, assembly, activation, reaction, and desorption of atoms and molecules at the surfaces of liquids and solids complex systems [20]. The surface chemistry of complex systems can be characterized using scanning tunnelling microscopy (STM), atomic force microscopy (AFM) and scanning probe microscopy (SPM) that provide tools for imaging, spectroscopy and manipulation at the single-atom level [20].

In particular, water-solid interfaces can be studied by surface-sensitive imaging and spectroscopic techniques such as vibrational spectroscopy, high resolution electron energy loss spectroscopy (HREELS), low-energy electron diffraction (LEED), X-ray absorption spectroscopy (XAS), X-ray photoelectron spectroscopy (XPS) and combined XAS/XPS [19].

The link between surface chemistry and surface tension is based on the imbalance between the intermolecular attractive forces and the cohesive forces due to the interactions between the molecules in all directions. On this basis, the scientific significance of χ^{L}_{PFPEf} is its capability to express both surface chemistry and surface tensiometry of a liquid, which can be simultaneously evaluated in a rapid and non-invasive way using the CA method without the influence of friction forces, roughness, contact angle hysteresis and environmental factors (Figure 3.7).

Figure 3.7 shows the capability of χ^{L}_{PFPEf} in the surface chemistry characterization of liquids using the CA method. Thus, the Rossi number leads to the simultaneous evaluation of liquids surface chemistry (SC) and surface tension (ST) and their variations over time (Figure 3.7). In surface tensiometry, χ^{L}_{PFPEf} represents a new tool for the determination of ST of liquids avoiding the interfacial phenomena that occur at liquid/solid (l/s) and liquid/liquid (l/l) interfaces. The interfacial phenomena generated at l/s and l/l interfaces are represented by the physicochemical interactions over time occurring in the two phases that produce changes in their surface chemistry and surface tensiometry. For example, Figure 3.8 reports the differences between crude oil/water (o/w), perfluoropolyether/water (PFPE/w), and perfluoropolyether/crude oil (PFPE/o) interfaces in terms of ST (γ) variations over time.

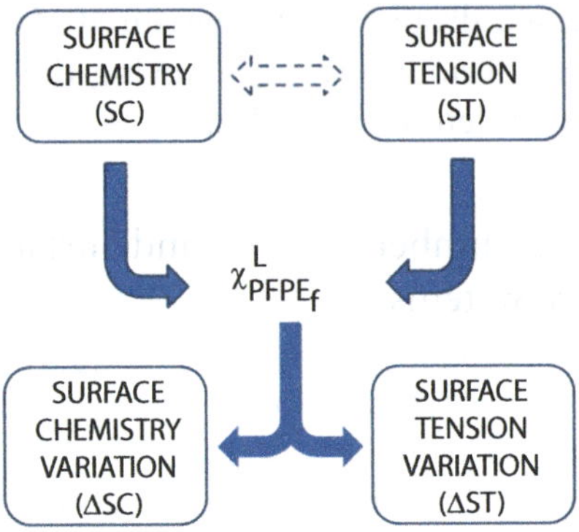

Figure 3.7 Simultaneous evaluation of liquid surface chemistry (SC) and surface tension (ST) and their variations (ΔSC, ΔST) over time using the Rossi number.

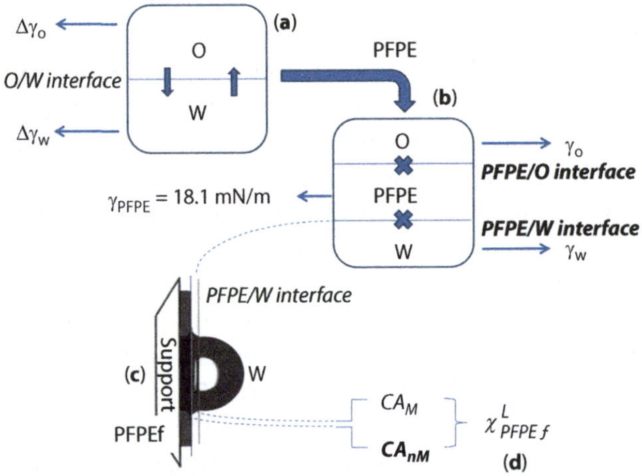

Figure 3.8 Scientific significance of Rossi number where (a) represents the physicochemical interactions (↓↑) at crude oil/water (O/W) interface and the variations in surface tension of crude oil ($\Delta\gamma_o$) and water ($\Delta\gamma_w$) over time, (b) shows perfluoropolyether/crude oil (PFPE/O) and perfluoropolyether/water (PFPE/W) interfaces generated in absence (⊗) of interfacial interactions and without variations in surface tensions of perfluoropolyether (γ_{PFPE}), crude oil (γ_o) and water (γ_w) over time, (c) shows the generation of perfluoropolyether/water (PFPE/W) interface after contact between a liquid film of PFPE (PFPEf) and a drop of water (W) by SLM methodology, and (d) reports the determination of Rossi number (χ^L_{PFPEf}) on the basis of the variation in CAs measured at PFPEf/W interface.

The absence of physicochemical interactions at the PFPEf/W interface leads to the determination of χ^{L}_{PFPEf} by the measurement of water CA (CA_{nM}). As a consequence, the χ^{L}_{PFPEf} expresses the real surface chemistry and surface tension of the liquid studied. The variation of water CA_{nM} measured at the interface with PFPEf depends exclusively on the variation of its surface chemistry and surface tension. Thus, the absence of physicochemical interactions at the interface between PFPEf and the liquid analyzed represents the main factor to determine a pure value of ST unaffected by friction forces, roughness, contact angle hysteresis and environmental factors.

The application of SLM and the determination of Rossi number can help interdisciplinary researchers evaluate surface chemistry and surface tensiometry characteristics of liquids in a rapid and non-invasive way within the concept of complex systems. Furthermore, many natural and artificial liquids that have undergone modifications in their surface chemistry over time can be assessed easily using surface tensiometry and the CA method. As SLM can be applied universally to all kinds of liquids, possible fields of interest could be the evaluation of the surface activity of surfactants in the pharmaceutical sector [21] and the evaluation of disperse systems used in the food industry, cosmetics, health care and environmentally-friendly materials [22].

In the context of this research work, the SLM and the χ^{L}_{PFPEf} applied to the study of the surface tension variations of the solution (acceptor phase) during the release test were performed on Lipogel charged at 4% w/w of carnitine, and a commercial ointment Gel-ALPHA charged at 4% active substance X.

3.2.2.2.2 The Rossi Number (χ^{L}_{PFPEf}) and the State of a Complex System

A complex system can defined in various ways. For example, a system is complex because it is "formed by many interacting elements that give rise to emergent phenomena" or "a complex system is a distributed set of entities with many interconnections (usually networked), where each entity self-operates locally with its neighbouring entities, and exhibiting globally emergent behaviour" [23]. Another definition reports that "complex systems are formed by many agents interacting with each other in nonlinear ways, thus yielding complex phenomena that cannot emerge from the simple sum of the agents". Anyway, a system's and subsystem's complexity depends mainly on their hierarchical organization within a spontaneous order [23].

More recently, Estrada defined a complex system as "a system characterized by a bidirectional non-separability between the identities of the parts and the identity of the whole, where the whole is the constituents of the system and its identity is the nature of the interactions occurring between them" [23]. When more entities are added to the system, its quality changes with increased complexity because more interactions occur with other parts. The interactions (called Morinian interactions) are a source of "organization" in a general system that increases in number and richness when it becomes a complex system [23].

The increase in order of the system due to the interactions between the entities has been compensated by the rise in internal disorder. The Estrada definition of a complex system can be simplified considering that a system is complex if its behaviour crucially depends on the details of the system [24]. On this basis, the system state at time t can be represented by the vector X (t) having N components. The time evolution of X (t) uses the equation of motion (3.2).

$$G\big(X(t)\big)=\frac{dX}{dt} \tag{3.2}$$

Where X is the system state in time t, and G(X(t)) represents the motion equation of X in time t.

Considering the vector X of a system within two moments t_0 (G_0 (X)) and t1 (G_1 (X)), the two systems X_0 and X_1 can be described using the equation of motion (Equation 3.3).

$$G_1(X)-G_0(X)=O\,(\varepsilon) \tag{3.3}$$

Where $G_1(X)$ is the equation of motion of complex system at t1, $G_0(X)$ is the equation of motion of complex system at t_0, O is the variation in the equation of motion leading to a great variation in the behaviour of the complex system, and ε is the order value [24].

Similarly to the complex system's state variation $\frac{dX}{dt}$, the motion equation of a liquid also changes over time because its surface chemistry (ΔSC) and surface tensiometry (ΔST) change due to external modifications ($\frac{d}{dt}$). As a consequence, χ^{L}_{PFPEf} led to the simultaneous evaluation of liquid surface chemistry (SC) and surface tension (ST) and their variations over time (Figure 3.7).

In this context, Equation 3.4 indicates the value of the physical deformation of the PFPEf meniscus expressed by the variation of χ^{L}_{PFPEf} of two different solutions having different concentrations of solute [7].

$$\chi^{s}_{x1} - \chi^{s}_{x2} = D_M \tag{3.4}$$

Where D_M is the physical deformation of the meniscus, χ^{s}_{x1} is the Rossi number value related to solution x1 and χ^{s}_{x2} is the Rossi number value related to solution x2.

In the case of a solution that changes over time, Equation 3.4 can be rewritten as Equation 3.5.

$$\chi^{s}_{t1} - \chi^{sol}_{t0} = D_M(\varepsilon) \tag{3.5}$$

Where D_M is the physical deformation of the meniscus, χ^{s}_{t1} is the Rossi number of the solution (s) at t_1 after adding the solute s, χ^{sol}_{t0} is the Rossi number of the solvent only at t_0 and ε is the order value. In this case $\chi^{s}_{t2} > \chi^{sol}_{t0}$ and D_M depends only on the concentration of a single solute ($[x^1]$).

On these bases, we introduce here a theory (called "surface state" theory) by which there could exist a close correlation between χ^{L}_{PFPEf} and G(X(t)) because the complex system represented by the Rossi number can also be represented by the vector X (t) through the equation of motion. As a consequence, Equation 3.6 shows the equivalence between Equations 3.3 and 3.5.

$$O(\varepsilon) \equiv D_M(\varepsilon) \tag{3.6}$$

Where O is the variation of the equation of motion, DM is the physical deformation of the meniscus that reflects the complex system variation, and ε is the order value. This theory is one of the possible physical interpretations of Rossi number that hypothesizes its link with the variation of motion's equation of a complex system over time.

In the end, O (ε) and D_M (ε) both appear to be linked to interdisciplinary and transdisciplinary approaches [25]. In this context, the Integrated Analytical Approach "on the same material sample and at the same t" assumes particular importance [4].

3.2.2.3 *Lipogel Water Test (LWT)*

In this work, a particular test called the "Lipogel Water Test" (LWT) was developed, from which it was possible to evaluate the presence of an excess of oil on the surface of the Lipogel preparation. Consequently, the LWT was developed to select the more suitable Lipogel formulation for the release test. The LWT test led to determining the CA of a drop of water deposited on the surface of the formulation and its spreading over time. The CAs were measured by placing the baseline on the Lipogel surface (CA_{ls}) and then at the interface between the oil meniscus and the water drop (CA_{om}). The amount of oil on the surface can be determined using Equation 3.7.

$$[SO] = CA_{ls} - CA_{om} \tag{3.7}$$

where [SO] is the amount of surface oil, CA_{ls} is the contact angle measured by placing the baseline on the Lipogel surface, and CA_{om} is the contact angle measured by placing the baseline at the interface between the meniscus of oil formed upon contact with water.

3.3 Results and Discussion

3.3.1 Drug Release

To evaluate the consistency of the preparation, Figure 3.9 shows the frames of the profiles obtained by depositing the drops of water on the surfaces of three Lipogels with different contents of beeswax, water, and carnitine.

The frames demonstrate the capability of olive oil on the surface of the Lipogels to generate a characteristic meniscus when in contact with a water drop. This phenomenon occurs because the preparations contain the same percentage of gelling agent (beeswax).

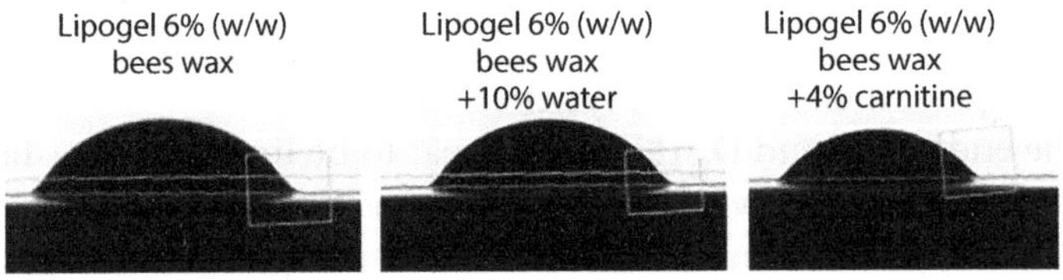

Figure 3.9 Shapes of water drops left on the surfaces of Lipogel 6% (w/w) bees wax, Lipogel 6% (w/w) bees wax with 10% water, and Lipogel 6% (w/w) bees wax with 4% carnitine.

The size of the meniscus changes in accord with the composition of the Lipogel. Comparing the Lipogel at 6% w/w of beeswax with the Lipogel containing 6% w/w beeswax with added 10% w/w of water, the meniscus of the latter appears to have a lower size than the Lipogel containing 6% w/w of beeswax only. The presence of 10% water leads to a visible contraction of the meniscus after the contact with the test liquid. Finally, in the case of the 6% w/w beeswax Lipogel loaded with 4% w/w carnitine, the meniscus size becomes negligible. The 4% carnitine added to the 6% beeswax Lipogel helped to give the preparation greater consistency.

The results obtained therefore demonstrated that the contact angle generated at the interface between the surface of Lipogel and the water droplet is strongly influenced by the presence of water and carnitine's aqueous solution. Figure 3.10 shows a comparison of the dimensions of the surface oil menisci measured over time during spreading of the water drop on the surfaces of the Lipogel samples considered.

Figure 3.10 demonstrates that measuring the spreading of a water droplet over time can discriminate between 6% w/w beeswax 6% w/w Lipogel, beeswax Lipogel with water, and the beeswax 6% w/w Lipogel with an aqueous solution of carnitine. The formation of menisci having different dimensions at the interfaces between water and the three Lipogels confirms the high sensitivity of the surface tensiometry in discriminating between preparations having different consistencies.

In this case, the decrease in the meniscus size appears to be linked to the increase of water in the preparation with and without carnitine.

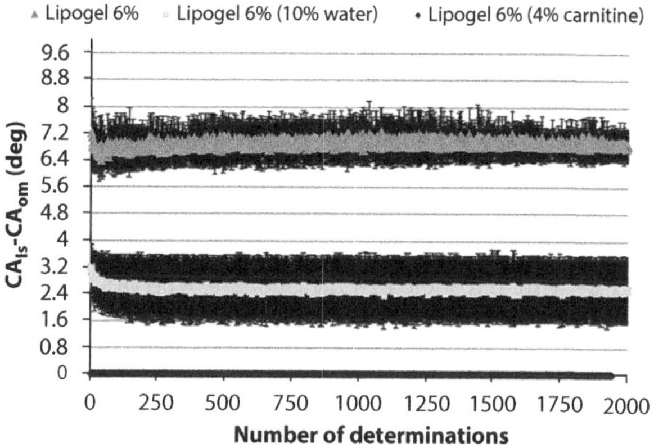

Figure 3.10 Comparison between Lipogel 6% (w/w) bees wax, Lipogel 6% (w/w) bees wax with 10% water, and Lipogel 6% (w/w) bees wax with 4% carnitine, where CA_{ls} is the contact angle measured on Lipogel surface and CA_{om} measured at oil/water interface.

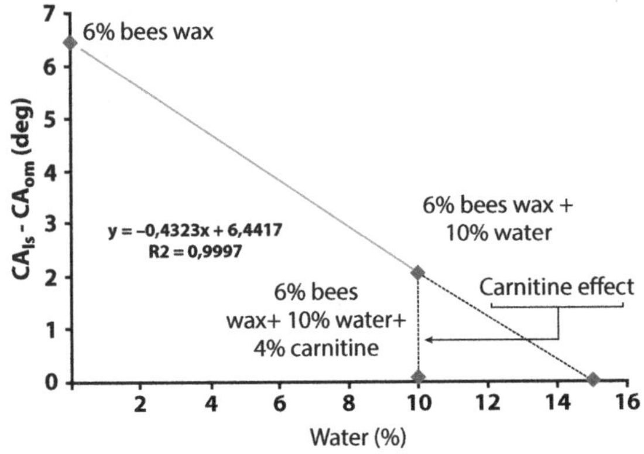

Figure 3.11 Correlation between the water content and the meniscus size where CA_{ls} is the CA measured by placing the baseline on the Lipogel surface and CA_{om} is the CA measured by placing the baseline at the interface between the meniscus of oil and the water.

Figure 3.11 shows the surface tensiometry quantitative correlation trend between the variation in the meniscus size and the amount of water with and without carnitine added to the preparation.

Figure 3.11 demonstrates the influence of the presence of water in the formulation (Lipogel at 6% w/w beeswax and 10% w/w water) and the presence of carnitine (Lipogel at 6% w/w beeswax, 10% w/w water and 4% w/w carnitine). The measurement of the size of the oil meniscus on the surface was shown to be a non-invasive analytical tool capable of determining both the consistency of a Lipogel and the effect of adding water and carnitine on the surface characteristics. Based on these results, we selected Lipogel 6% beeswax with 4% carnitine added for release investigations because of its optimal consistency.

3.3.1.1 Surface Tensiometry Evaluation

The results derived from the SLM analysis of the carnitine solution (acceptor phase) after 6-hour release from Lipogel allowed obtaining information on the surface tension variations of the system over time.

The SLM was applied to the system constituted by carnitine's solutions sampled from a cell chosen with random criteria (cell A). Figures 3.12a, 3.12b, 3.12c, 3.12d, and 3.12e show respectively the χ^{L}_{PFPEf} trends relating to the selected series in the timeframes 0-1h, 0-1.5h, 2h-3.5h, 3.5h-4h, and 4.5h-5h.

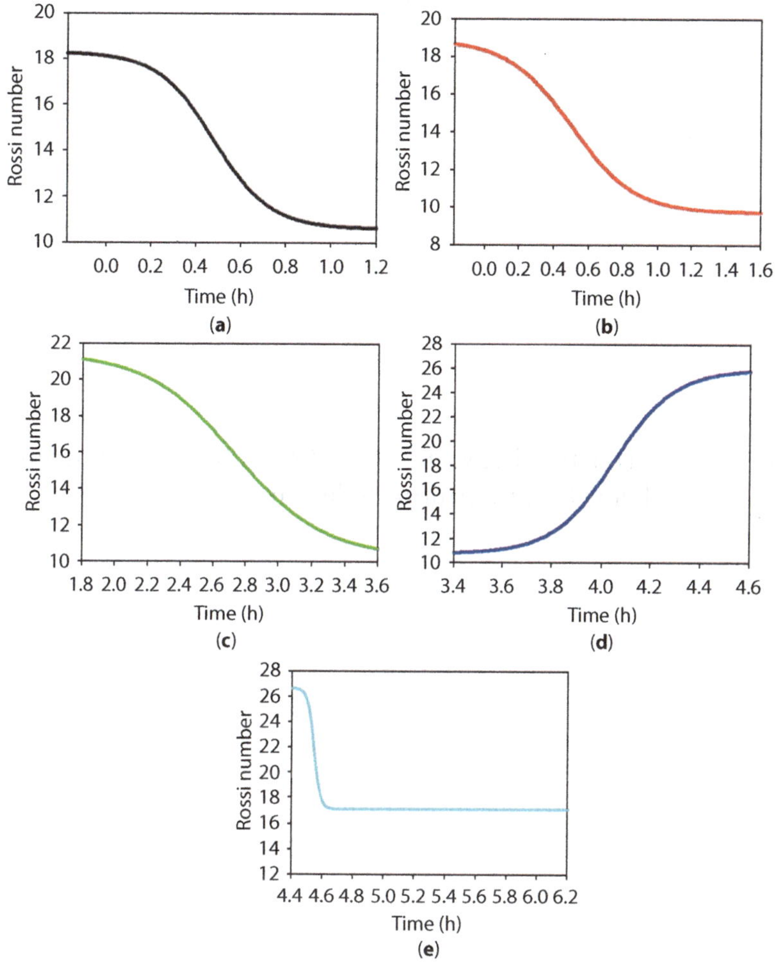

Figure 3.12 Rossi number trends for times (a) 0-1h, (b) 0-1.5h, (c) 2h-3.5h, (d) 3.5h-4.5h, (e) 4.5h-5h.

The behavior of χ^{L}_{PFPEf} (Rossi number) measured in the range 0 to 1h of carnitine release (Figure 3.12a) demonstrated a sigmoidal correlation with the time. The sigmoid 4-parameter (4P) correlation equation was applied because it is a mathematical model which gives a shape similar to the S-shaped dose response commonly used in bioassay data [26]. In our case, the dose is represented by the amount of the drug released, and the response is the χ^{L}_{PFPEf}. The sigmoid 4-parameter (4P) equation showed a correlation degree between χ^{L}_{PFPEf} and time of $R^2 = 1$. As the same time, the ehaviour of the χ^L_{PFPEf} measured in the range 0-1h (Figure 3.12b) demonstrated a sigmoidal correlation with the time and a correlation degree between the χ^L_{PFPEf} and time of R^2= 0.9966.

As regards the behaviour of χ^L_{PFPEf} measured in the range 2h-3.5h (Figure 3.12c), the application of a sigmoid 4-parameter correlation equation demonstrated a correlation degree between χ^L_{PFPEf} and time of R^2 = 1. In the end, the behaviour of χ^L_{PFPEf} measured in the range 3.5h-4.5h (Figure 3.10d) shows a correlation degree between χ^L_{PFPEf} and time of R^2= 1.

The final behaviour of the carnitine release process from Lipogel in the range 4.5h-6h (Figure 3.12e) demonstrates a the decrease of the correlation degree between χ^L_{PFPEf} and time (R^2=0.8739).

3.3.1.1.1 Correlation between χ^L_{PFPEf} (Rossi Number) and Surface Chemistry of Carnitine Solution

The surface of a complex system having defined components and structure is characterized by typical chemical properties. As a consequence, surface chemistry research plays a crucial role in the energy of complex systems [27].

Taking into account the relationships between the chemistry of a complex system and its surface and energy characteristics, the most compatible hypothesis to explain the behaviour of χ^L_{PFPEf} (Rossi number) over time (Figures 3.12a-3.12e) regards the tendency of each system to reach its equilibrium after an external energy intervention.

According to this hypothesis, the system undergoes an energy increase in the first 1.5h (χ^L_{PFPEf} decreases), followed by an energy decrease from 1.5h to 2h (χ^L_{PFPEf} increases). The phenomenon observed continues from 2h to 3.5h, 3.5h to 4.5h, and from 4.5h to 5h. After 5h, the system appears to trend to the overall equilibrium with minimal energy variations.

The correlations between χ^L_{PFPEf} and time demonstrated the differences in the decrease and increase behaviors of χ^L_{PFPEf} in relation to the differences in the energy increase and decrease in the system because of the surface chemistry variations of the solution as a consequence of the carnitine release.

Figure 3.13a shows the overall trend of the χ^L_{PFPEf} (Rossi number) variation up to 5h and the relative degree of linear correlation over time, and Figure 3.13b reports its maximum and minimum values.

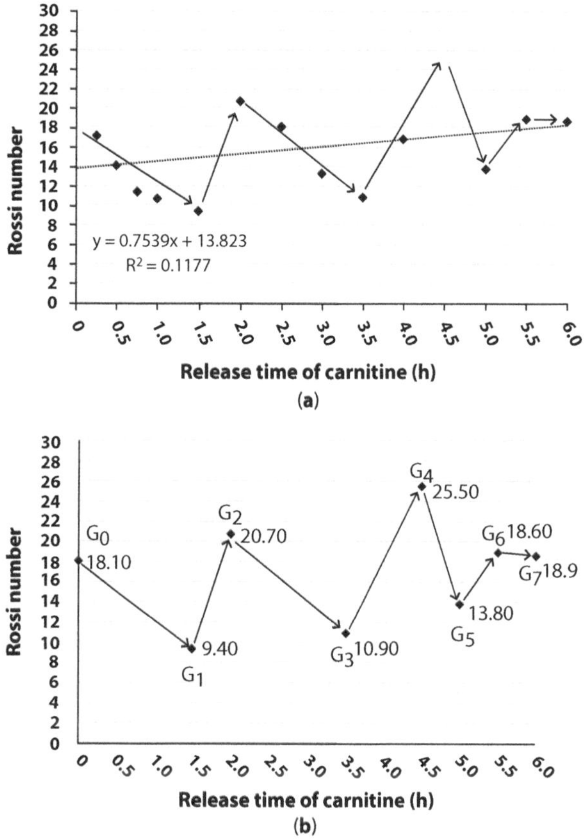

Figure 3.13 (a) Rossi number variation after carnitine release from Lipogel and (b) maximum and minimum values of Rossi number corresponding to the variation of surface tensiometry of the system (G) over time.

Figure 3.13 shows the relationship between χ^{L}_{PFPEf} (Rossi number), surface chemistry, and surface tensiometry variations of carnitine solution over time.

In accord with the "surface state" theory proposed here, Figure 3.13a shows that the variations of motion's equation $G_{2h}(X)-G_{0}(X)$ and $G_{4.5h}(X)-G_{2h}(X)$ could correspond respectively to $\chi^{s}_{2h}-\chi^{sol}_{t0}$ and $\chi^{s}_{4.5h}-\chi^{s}_{2h}$ variations. $G_{3.5h}(X)-G_{1.5h}(X)$ and $G_{5.0h}(X)-G_{3.5h}(X)$ could correspond respectively to $\chi^{s}_{3.5h}-\chi^{s}_{1.5h}$ and $\chi^{s}_{5.0h}-\chi^{s}_{3.5h}$ variations. The link between the variation of motion's equation and the progressive increase of D_M hypothesizes an increase in the stability of the acceptor solution system over time and a progressive decrease in its energy. Differently, Figure 3.13b highlights the link between the changes in the dimension of the meniscus measured using χ^{L}_{PFPEf} and the variation of the equation of motion (O) following each step of the release process. In the first phase of carnitine release, the variation of system's motion equation $G_{1.5h}(X)-G_{0h}(X)$ corresponds to a decrease of meniscus size $\chi^{sol}_{0h}-\chi^{s}_{t1.5h}$. Thus, the variation of G reflects the modification of the acceptor solution system due to the addition of carnitine released from Lipogel measured using χ^{L}_{PFPEf}. Successively, after 2h the variation of system's motion equation $G_{2h}(X)-G_{1.5h}(X)$ corresponds to an increase of meniscus size $\chi^{sol}_{2h}-\chi^{s}_{t1.5h}$ in accord with what observed in Figure 3.13a in the range 0-2 hours.

In the end, O (ε) after 2h, 3.5h, 4.5h, 5h, 5.5h and 6h of carnitine release corresponds respectively to an increase of meniscus size $\chi^{s}_{2h}-\chi^{s}_{t1.5h}$, a decrease of the meniscus size $\chi^{s}_{2h}-\chi^{s}_{t3.5h}$, an increase of the meniscus size $\chi^{s}_{4.5h}-\chi^{s}_{t3.5h}$, a decrease of the meniscus size $\chi^{s}_{4.5h}-\chi^{s}_{t5.0h}$, an increase of the meniscus size $\chi^{s}_{5.5h}-\chi^{s}_{t5.0h}$ and at a slightly decrease of the meniscus size $\chi^{s}_{5.5h}-\chi^{s}_{t6.0h}$ taking the system near the equilibrium.

Based on the results obtained and on the surface state hypothesis, Figures 3.13a and 3.13b show the variations in the surface chemistry of the acceptor solution after carnitine release over time and demonstrates that it is possible to provide a dynamic profile of the energy of a liquid system *via* surface tensiometry using the SLM. The accuracy with which the system tries to reach the final equilibrium demonstrates the high sensitivity of the χ^{L}_{PFPEf} in detecting fastly the energy variations caused by the release process of the carnitine over time.

In Figure 3.14 is shown the comparison between the behaviours of carnitine amount released and χ^{L}_{PFPEf} (Rossi number).

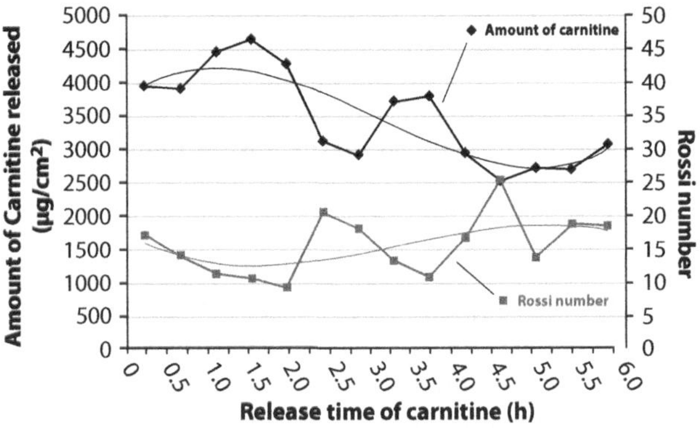

Figure 3.14 Comparison between carnitine amount released and Rossi number values measured over time.

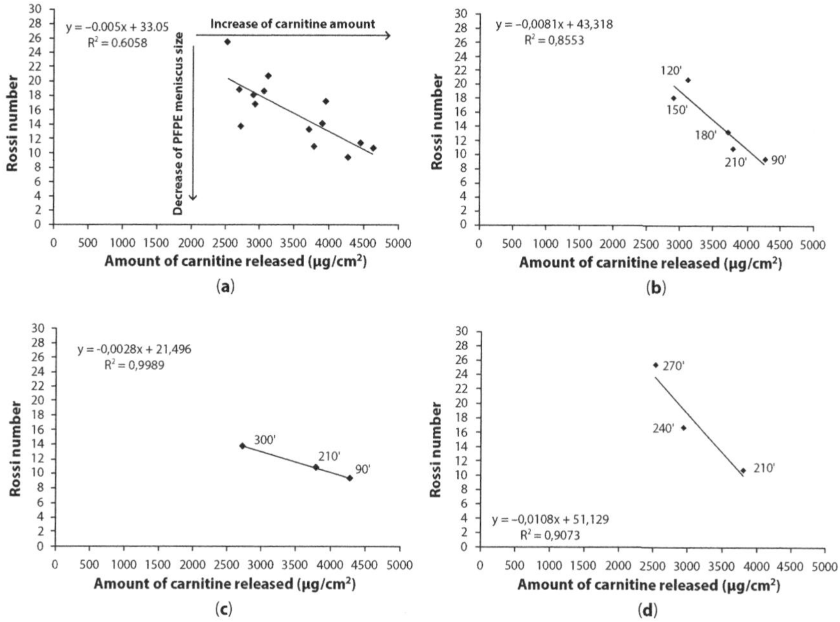

Figure 3.15 Rossi number trends vs carnitine amount released (a) overall, (b) 1.5h-3.5h, (c) 1.5-5h, (d) 3.5h-4.5h.

Figure 3.14 shows the comparison between the amount of carnitine released and the χ^{L}_{PFPEf}, and demonstrates a strong link between the decrease of the dimension of PFPEf meniscus and the increase of carnitine amount released over time. In Figures 3.15a, 3.15b, 3.15c, and 3.15d is reported the overall effect of the amount of carnitine released on the PFPEf meniscus size.

Figure 3.15a shows the overall release process of carnitine, and the decrease of PFPEf meniscus to be correlated to the increase of the amount of the drug. Figures 3.15b, 3.15c, and 3.15d demonstrate the relationships between χ^{L}_{PFPEf} (Rossi number) and the amount of carnitine released over time.

The results obtained appear in accord with the hypothesis formulated here by which the SLM test has great sensitivity in the detection of the variation in the quantity of carnitine released with time.

3.3.2 Drug Permeation Evaluation

The following figures show comparisons between the quantitative value (µg/mL) expressed as the amount of the active substance X [10], and the

Rossi number χ^{L}_{PFPEf}. To better appreciate the inverse proportionality between χ^{L}_{PFPEf} and quantitative data, trend correlation analysis between the two data sets is reported for each Franz cell considered B (Figures 3.16a, 3.16b), D (Figures 3.17a, 3.17b), F (Figures 3.18a, 3.18b), H (Figures 3.19a, 3.19b), and L (Figures 3.20a, 3.20b).

Figures 3.16a and 3.16b, Figures 3.17a and 3.17b, Figures 3.18a and 3.18b, Figures 3.19a and 3.19b, Figures 3.20a and 3.20b demonstrate that χ^{L}_{PFPEf} can evaluate the modifications of the energy level of the active substance *X* solution system during its permeation. The PFPEf meniscus decreases after 6 h of permeation due to the increase in the amount of

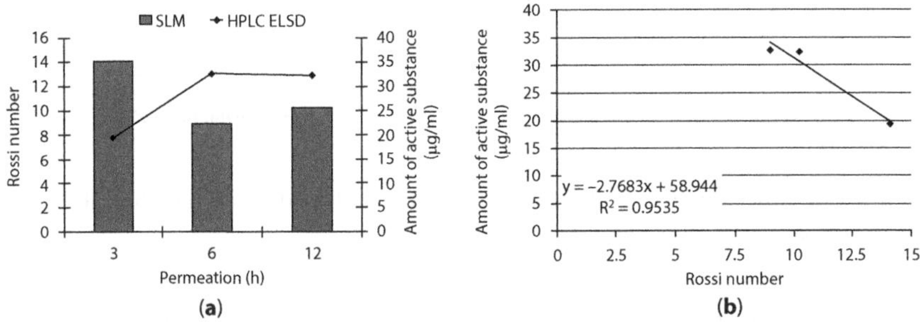

Figure 3.16 (a) Comparison between Rossi number and the amount of active substance X detected using HPLC with Evaporative Light Scattering Detector (ELSD) in relation to permeation process in Franz cell B, (b) Correlation trend between Rossi number and amount of substance X in relation to permeation process in Franz cell B.

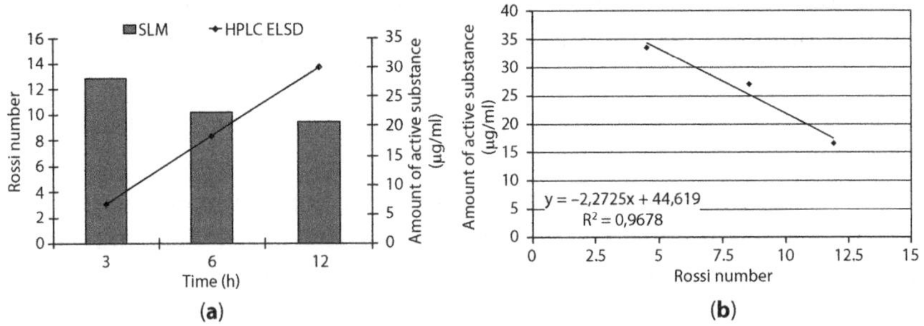

Figure 3.17 (a) Comparison between Rossi number and the amount of substance X detected using HPLC with Evaporative Light Scattering Detector (ELSD) in relation to permeation process in Franz cell D, (b) Correlation trend between Rossi number and amount of active substance X in relation to permeation process in Franz cell D.

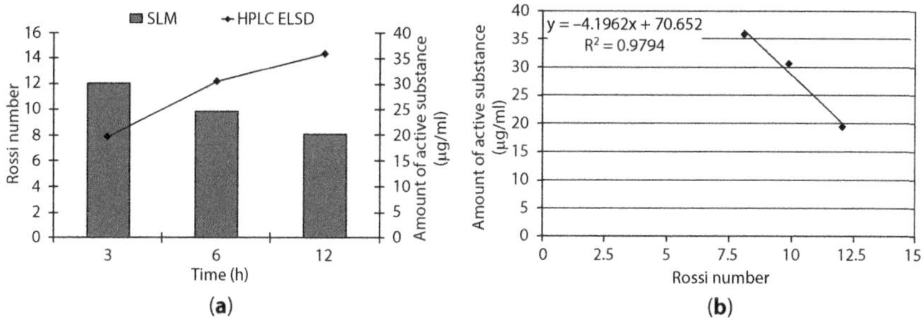

Figure 3.18 (a) Comparison between Rossi number and amount of substance X detected using HPLC with Evaporative Light Scattering Detector (ELSD) in relation to permeation process in Franz cell F, (b) Correlation trend between Rossi number and amount of active substance X in relation to permeation process in Franz cell F.

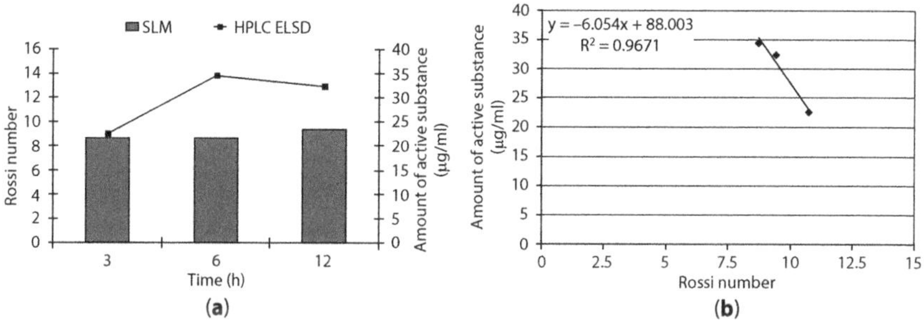

Figure 3.19 (a) Comparison between Rossi number and amount of substance X detected using HPLC with Evaporative Light Scattering Detector (ELSD) in relation to permeation process in Franz cell H, (b) Correlation trend between Rossi number and amount of active substance X in relation to permeation process in Franz cell H.

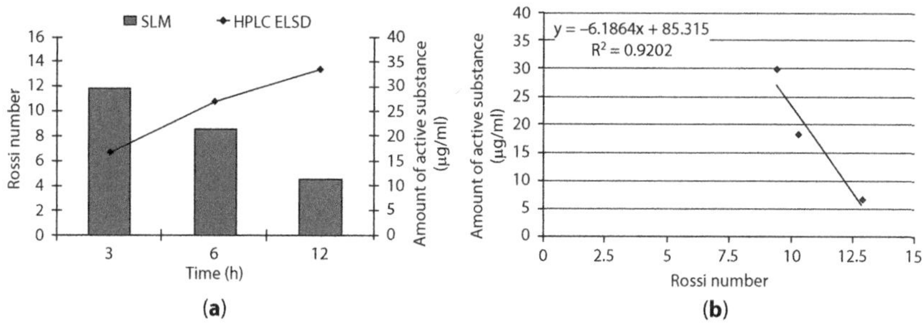

Figure 3.20 (a) Comparison between Rossi number and amount of substance X detected using HPLC with Evaporative Light Scattering Detector (ELSD) in relation to permeation process in Franz cell L, (b) Correlation trend between Rossi number and amount of active substance X in relation to permeation process in Franz cell L.

active substance *X* passing through the biological membrane that is detectable in the acceptor phase.

Subsequently, in the case of B and H Franz cells, the meniscus tends to increase after 12 h corresponding to the decrease in the amount of permeated active substance *X*. Conversely, as for the other cells, after 12 h the meniscus undergoes a contraction because the active substance *X* in the donor compartment has increased. The resuts show that inverse proportionality between the size of the meniscus and the active substance *X* concentration detected in the donor compartment by HPLC analysis. The application of the χ^{L}_{PFPEf} to the study of active substance *X* permeation through porcine epidermis from Gel Alpha confirmed a high degree of correlation between χ^{L}_{PFPEf} and the quantitative data. Overall, the results shown in this correlation study further demonstrated that the B and H cells show different performances in the permeation of the active substance *X*. In particular, SLM is capable of verifying the reproducibility of the permeation process between the Franz cells in terms of carnitine release from Lipogel. The degree of correlation between the calculated χ^{L}_{PFPEf} and active substance *X* amount in the context of IAA^{s}_{t} was satisfactory, demonstrating a high sensitivity of the SLM. The sensitivity demonstrated by the SLM method in describing the active substance *X* permeation process through the porcine epidermis, together with the high degree of correlation obtained between quantitative data and surface tensiometry paved the way for the subsequent experimentation that envisaged the application of the SLM to the samples obtained from the carnitine-loaded Lipogel release process.

Specifically, the surface tensiometry approach has proven to be able to describe the dynamics of caffeine release and active substance *X* permeation in a rapid and reproducible manner. The study of permeation of active substance *X* by χ^{L}_{PFPEf} confirms the importance of increasing the number of determinations over time per Franz cell to measure the correlation degree between chemical and surface tensiometry data.

3.4 Summary and Conclusions

The characterization of the release and permeation performances of lipophilic and hydrophilic semi-solid formulations for skin application using the IAA^{s}_{t} has made possible to obtain greater discriminating capability compared to that of individual analytical techniques used separately. The new concept of Integrated Analytical Approach confirms its wide applicability in the correlation studies between surface chemistry and surface tensiometry.

These correlations confirm the high sensitivity of SLM in describing in a non-invasive way the transcutaneous permeation process of a single molecule such as substance X, and the realase of carnitine from Lipogel. χ^{L}_{PFPEf} demonstrated its capability to detect in a rapid and non-invasive way the variations of carnitine and substance X amounts in relation to the variations of the PFPE meniscus size.

Thanks to the correlation analysis performed here, the IAA approach ensured an extensive applicability to multiple research areas. The correlation study reported here confirms the possibility of considering the variation of the liquid film meniscus size as a general quality parameter capable of evaluating the release and permeation processes in a non-invasive way and including the matrix effect.

The link between surface chemistry and surface tensiometry of complex systems led to the hypothesis that the variations of caffeine and carnitine concentrations could be linked to the increase or decrease of the energy of the system that occur during the release and permeation processes which can be characterized by SLM and measured using χ^{L}_{PFPEf}. This seems to be supported by the theory of "surface state" proposed here, which hypothesizes a direct link between the variations in the complex system state X and the variation in χ^{L}_{PFPEf}.

Our results encourage us to continue the research in the field of surface tensiometry with attention to the interface phenomenon between polyperfluoromethylisopropyl ethers and natural and artificial complex systems. In the end, the wide applicability of χ^{L}_{PFPEf} and the non-invasinevess of the surface tensiometry approach suggest here the possibility of introducing surface tensiometry in the evaluation of *in vitro* permeation processes by introducing the SLM as a new tool for permeation process characterization and Franz cells performances evaluation.

List of Abbreviations

CA	Contact Angle (deg)
CS	Complex System
DC	Dispersion Component
HPLC-ELSD	High Performance Liquid Chromatography – Evaporative Light Scattering Detector
IAA^{s}_{t}	Integrated Analytical Approach
PC	Polar Component
PFPE	Polyperfluoromethylisopropyl Ether (Perfluoropolyether) Fomblin HC/25®PFPE

PFPEd	Polyperfluoromethylisopropyl Ether (Perfluoropolyether) Fomblin HC/25®PFPE drop
PFPEf	Polyperfluoromethylisopropyl Ether (Perfluoropolyether) Fomblin HC/25®PFPE liquid film
SCA	Static Contact Angle
SFE	Surface Free Energy
SLM	Solid-like Methodology
SO	Surface Oil
ST	Surface Tension
χ^{L}_{PFPEf}	Rossi number

AI disclosure statement: The author confirms that no artificial intelligence (AI)-generated content or AI-assisted writing tools were used in the preparation of this chapter (including summaries and the abstract). All text, analysis, and conclusions are the result of the author's original work.

References

1. R Draisci, C. Montesissa, B. Santamaria, C. D'Ambrosio, G. Ferretti, R. Merlanti, C. Ferranti, M. De Liguoro, C. Cartoni, E. Pistarino, L. Ferrara, M. Tiso, A. Scaloni and M.E. Cosulich, Integrated analytical approach in veal calves administered the anabolic androgenic steroids boldenone and boldione: Urine and plasma kinetic profile and changes in plasma protein expression. *Proteomics,* 7, 3184–3193 (2007).
2. M.A.W. Marks, T. Wenzel, M. J. Whitehouse, M. Loose, T. Zack, M. Barth, L. Worgard, V. Krasz, G. N. Eby, H. Stosnach and G. Markl, The volatile inventory (F, Cl, Br, S, C) of magmatic apatite: An integrated analytical approach. *Chemical Geology* 291, 241–255 (2012).
3. V. Comite, J.S. Pozo-Antonio, C. Cardell, L. Randazzo, M. F. La Russa and P. Fermo, A multi-analytical approach for the characterization of black crusts on the Facade of an historical cathedral. *Microchem J* 158, 105121 (2020).
4. D. Rossi and N. Realdon, Characterization of formulations for face skin treatments in the beauty farm field: Perspectives in surface tensiometry of cosmetic products, in: *Surface Science and Adhesion in Cosmetics*, K.L. Mittal and H.S. Bui (Eds.), pp. 309–352, Wiley-Scrivener Beverly, MA (2021).
5. D. Rossi, S. Rossi, D. Dobrzyński, A. Dolmella and N. Realdon, Assessment of spring waters from Lourdes (France) by contact angle method. Surfaces *Interfaces*, 19, 1-9 (2020).
6. D. Rossi, D. Dobrzyński, I. Moro, M. Dal Zotto, E. Moschin and N. Realdon, Contact angle keying method for an integrated analytical approach to

characterising waters in the Lourdes area (Pyrenees, France). *Int J. Water* 14 (1), 29-68 (2020).
7. D. Rossi, S. Rossi and N. Realdon, Determination of surface tension components of aqueous solutions using Fomblin HC/25 Perfluoropolyether liquid film as a solid substrate, in: *Advances in Contact Angle, Wettability and Adhesion*, Vol. 4, K.L.Mittal (Ed.), pp. 145-178, Wiley-Scrivener, Beverly, MA (2019).
8. P. Padmaa, A.J. Preethy, M.S. Chitral and P. Christopher, Release kinetiks – Concepts and applications. *Int. J. Pharmacy Res. Technol.*, 8, 12-20 (2019).
9. V.P. Shah, T.L. Cecil, S.V. Srinivasan and R.L. Williams, Progressively reducing regulatory burden. *AAPS J,* 16, 621-624 (2014).
10. A. Casiraghi, F. Ranzini, U.M. Musazzi, S. Franzè, M. Meloni and P. Minghetti, *In vitro* method to evaluate the barrier properties of medical devices for cutaneous use. *Regulatory Toxicol Pharmacol,* 90, 42-50 (2017).
11. D. Rossi, E. Vettorato and N. Realdon, Surface tensiometry evaluation of *ex-vivo* permeation process of active substance using Solid-like method, in Proceedings of the 25th International Conference on Multidisciplinary Studies, p. 452 (2021).
12. D. L. Cowen and D. F. Kent, Medical and pharmaceutical practice in 1854. *Pharmacy in History,* 39, 91–100 (1997).
13. A. Preetha, R. Banerjee and N. Huilgol, Dynamic surface tensiometry of tissues using Langmuir films. *Colloids Surfaces B: Biointerfaces,* 40, 35-43 (2005).
14. M. Ijavi, R.W. Style, L. Emmanouilidis, A. Kumar, S. M. Meier, A.L. Torzynski, F. H. T. Allain, Y. Barral, M. O. Steinmetzc and E. R. Dufresne, Surface tensiometry of phase separated protein and polymer droplets by the sessile drop method. *Soft Matter,* 17, 1655-1662 (2021).
15. M.K. Omer, I.H. Ali and S.H. Kareem, Surface tensiometry study of the interfacial phenomena in the mixed surfactant solutions of sodium di-2-ethylhexylsulfosuccinate (AOT) and hexadecyl benzylammonium chloride (HDBAC). *Chemical Papers,* 76, 7189–7193 (2022).
16. R.A. Foty, Measurement of intercellular cohesion by Tissue Surface Tensiometry, in: *Tissue Morphogenesis. Methods in Molecular Biology,* Vol. 1189, C. Nelson (Ed.), pp. 237–254, Humana Press, New York (2015).
17. F.M. Etzler, Characterization of surface free energies and surface chemistry of solids, in: *Contact Angle, Wettability and Adhesion*, Vol. 3, K.L. Mittal (Ed.), pp. 219-264, CRC Press, Boca Raton, FL. (2003).
18. F.M. Etzler, Determination of the surface free energy of solids, *Rev. Adhesion Adhesives*, 1, 3-45 (2013).
19. G. Pantini, F. Brunetta and F. Guidolin, Perfluoropolyethers: status and new developments. *Cosmetics&Toiletries,* 106, 71-80 (1991).
20. Xin Li, Zhen Xu, Donglei Bu, Jinming Cai, Huamei Chen, Qi Chen, Ting Chen, Fang Cheng, Lifeng Chi, Wenjie Dong, Zhenchao Dong, Shixuan Du, Qitang Fan, Xing Fan, Qiang Fu, Song Gao, Jing Guo, Weijun Guo, Yang He,

Shimin Hou, Ying Jiang, Huihui Kong, Baojun Li, Dengyuan Li, Jie Li, Qing Li, Ruoning Li, Shuying Li, Yuxuan Lin, Mengxi Liu, Peinian Liu, Yanyan Liu, Jingtao Lü, Chuanxu Ma, Haoyang Pan, JinLiang Pan, Minghu Pan, Xiaohui Qiu, Ziyong Shen, Shijing Tan, Bing Wang, Dong Wang, Li Wang, Lili Wang, Tao Wang, Xiang Wang, Xingyue Wang, Xueyan Wang, Yansong Wang, Yu Wang, Kai Wu, Wei Xu, Na Xue, Linghao Yan, Fan Yang, Zhiyong Yang, Chi Zhang, Xue Zhang, Yang Zhang, Yao Zhang, Xiong Zhou, Junfa Zhu, Yajie Zhang, Feixue Gao, Yongfeng Wang, Recent progress on surface chemistry I: Assembly and reaction. *Chin Chem Lett,* 35, 110055 (2024).

21. E. O. Olorunsola and M.O. Adedokun, Surface activity as basis for pharmaceutical applications of hydrocolloids: A review. *J. Appl. Pharm. Sci.,* 4 (10), 110-116 (2014).
22. C. Courrégelongue and D. Baigl, Functionalization of emulsion interfaces: Surface chemistry made liquid. *Chem. Eur. J.,* 1-15 (2024).
23. E. Estrada, What is a complex system, after all? *Foundations of Sci.,* 29, 1143–1170 (2024).
24. G. Parisi, Complex systems: A physicist's viewpoint, *Physica. A.,* 263, 557-564 (1999).
25. O. Descalzi, S. Curilef, L. Velazquez and V. Muñoz, Complex systems and inter/transdisciplinary research: A review. *Chaos,* 34 (1), 010401 (2024).
26. G.Y. Lee, J.A. Myers, S.M. Perez, K.P. Singh and H.N. Green, Cisplatin combined with laser activated nanotherapy as an adjuvant therapy for head and neck cancer. *Int. J. Oncology.,* 57, 1169-1178 (2020).
27. Xin L., Zhen X., Donglei B., Jinming C., Huamei C., Qi C., Ting C., Fang C., Lifeng C., Wenjie D., Zhenchao D., Shixuan D., Qitang F., Xing F., Qiang F., Song G., Jing G., Weijun G., Yang H., Shimin H., Ying J., Huihui K., Baojun L., Dengyuan L., Jie L., Qing L., Ruoning L., Shuying L., Yuxuan Li., Mengxi L., Peinian L., Yanyan L., Jingtao L., Chuanxu M., Haoyang P., JinLiang P., Minghu P., Xiaohui Q., Ziyong S., Qiang S., Shijing T., Bing W., Dong W., Li W., Lili W., Tao W., Xiang W., Xingyue W., Xueyan W., Yansong W., Yu W., Kai W., Wei X., Na X., Linghao Y., Fan Y., Zhiyong Y., Chi Z., Xue Z., Yang Z., Yao Z., Xiong Z., Junfa Z., Yajie Z., Feixue G. and Yongfeng W., Recent progress on surface chemistry II: Property and characterization. *Chin. Chem Lett.,* 36 (1), 110100 (2025).

4

Coffee-Stain Effect – Evaporation of Wetting and Nonwetting Liquid Drops: A Critical Review

Andrew Terhemen Tyowua[1,2*], Augusta Ijeoma Ezekwuaku[1] and Msugh Targema[1]

[1]*Applied Colloid Science and Cosmeceutical Group, Department of Chemistry, Benue State University, Makurdi, Benue State, Nigeria*
[2]*Microencapsulation and Micromanipulation Group, School of Chemical Engineering, University of Birmingham, Edgbaston, Birmingham, UK*

Abstract
The coffee-stain effect is a consequence of the pinning of a drop edge on a substrate and the rapid evaporation of solvent molecules at the pinned site, in which the drop leaves circular patterns of suspended particulate matter upon drying. The effect has both drawbacks and applications. For example, traces of solutes that would have been impossible to detect concentrate themselves *via* the coffee-stain effect mechanism and become detectable in analytical and biochemical assays. Contrarily, the coffee-stain effect is a problem in inkjet printing and drop coating, where uniform film formation of the dispersed particulate matter is desired. Here, we examine the fundamental aspects of coffee-stain formation, methods of prevention, drawbacks and applications, highlighting areas for future research focus.

Keywords: Contact angle hysteresis, capillary flow, Marangoni flow, liquid marble, adhesion, coffee-ring

4.1 Introduction

When a drop of coffee is allowed to dry completely on a tabletop or on the floor (*i.e.*, on a solid substrate), a circular residual stain is observed.

**Corresponding author*: a.t.tyowua@bham.ac.uk

Anil N. Netravali and Frank M. Etzler (eds.) The Wide World of Adhesion: A Festschrift in Honor of Dr. Kash Mittal on his 80th Birthday, (145–172) © 2026 Scrivener Publishing LLC

This phenomenon is the so-called "coffee-stain" or "coffee-ring" effect. Although its mechanism of formation was initially sketched by Robert Brown in his second paper [1], the coffee-stain effect was first thoroughly studied in 1997 by Professor Sidney Nagel and his co-workers [2] at the University of Chicago, USA. This phenomenon is not peculiar to coffee, it can also be observed when particle suspensions dry completely on solid surfaces. In everyday life, coffee-stain effect occurs after wine, fruit juice, yoghurt, soup (or sauce), blood splatter, sugar and salt solutions dry on a solid substrate. The coffee-stain effect has both advantages and disadvantages. Here, we discuss the mechanism of coffee-stain formation, methods of suppressing it as well as its advantages and applications, highlighting areas for future research focus.

4.2 Evaporation of Wetting Liquid Drops: Mechanism of Coffee-Stain Formation

The coffee-stain effect occurs because of two factors: (i) pinning of the liquid drop edge to the solid substrate and (ii) rapid evaporation of the liquid (solvent) molecules at the pinned drop edge compared with other parts of the liquid drop [3]. The edge of the liquid drop is pinned to the substrate when it wets it such that the contact angle (θ, Figure 4.1(a)) it makes with the substrate is less than 90°. The pinning is amplified by contact angle hysteresis (*i.e.*, the difference between the advancing and receding contact angles) [4]. Because the pinned drop edge is much thinner than the bulk liquid drop, evaporation occurs there much faster than at any other part of the liquid drop. Consequently, a strong outward radial (capillary) flow of liquid to the drop edge is induced to replenish the evaporated liquid molecules. As the liquid flows to the drop edge, it sweeps the dispersed particles to the pinned drop edge, depositing them there. The process begins slowly with a "rush-hour" effect (*i.e.*, rapid capillary flow and particle deposition) occurring at the final stage of the drying process where $\theta \sim 0°$ [5, 6]. The outward radial capillary flow is countered by the circulatory Marangoni flow which tends to redisperse the deposited particles into the drop. If the volume fraction of particles in the drop is $<< 1$, the scaled evaporation of the solvent at the drop edge is $>> 1$ and the circulatory Marangoni flow does not redisperse the deposited particles into the drop, and the deposited particles form circular residue patterns (*i.e.*, the so-called coffee-stain) at the end of solvent evaporation [7]. Three types or patterns of particle deposits are commonly observed as shown in Figure 4.1(b,c):

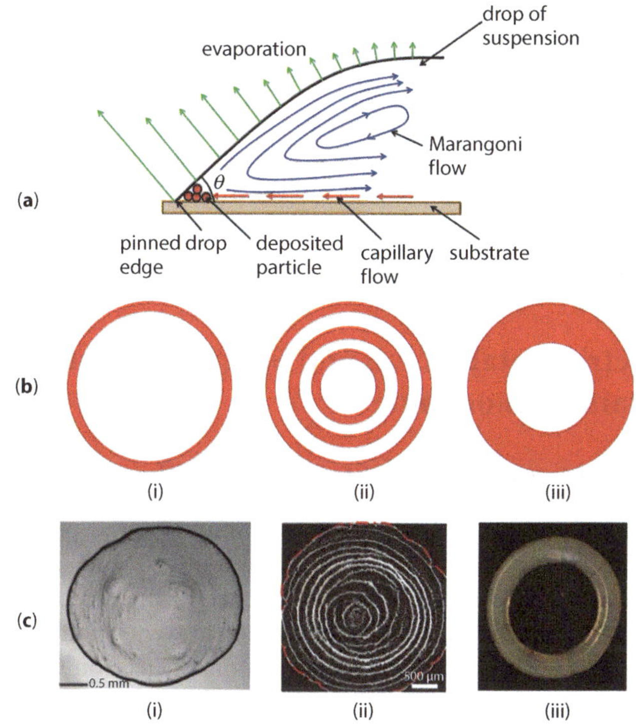

Figure 4.1 Schematic illustration of (a) the cross section of a drop of a particle suspension resting on a planar substrate with contact angle θ, showing the pinned drop edge, outward radial capillary flow that carries suspended particles to the drop edge, and the circulatory Marangoni flow that tends to disperse the particles; (b) the three main types of particle deposits from dried drops of particle suspensions: (i) single ring, (ii) multiple concentric rings, and (iii) broadband ring and (c) the images of the various types of particle deposits: (i) an image of a single ring from a dried water drop containing spherical polystyrene particles, reproduced with permission from ref. [14], copyright 2011 Springer Nature, (ii) an image of multiple concentric rings from a drop of an aqueous suspension of red fluorescent polymer microspheres, reproduced with permission from ref. [15], copyright 2020 Proceedings of the National Academy of Sciences of the USA, and (iii) an image of a broadband particle deposit from a dried aqueous suspension of Al_2O_3 particles, reproduced with permission from ref. [16], copyright 2010 Springer Nature.

single rings [3, 8], multiple concentric rings [9, 10] and broad bands [11, 12]. Complex patterns of particle deposit, like concentric crystal band, annular band, fractal-like lace and doughnut-like shape have also been observed with various dried suspensions and salt solutions on both hydrophilic and hydrophobic substrates [12]. More complex patterns like oval splatter, crown-shaped, fern-like island, crown needle-like, bump-like, dendritic branch and star-like deposits have been reported from dried suspensions of drugs like Budesonide, Clonazepam, Ciprofloxacin and Methotrexate, containing 5% NaCl and 95% drug [13]. Single rings are formed when the drop edge is strongly pinned to the solid substrate such that it remains there throughout the evaporation timeframe. If, however, the drop edge is weakly pinned to the substrate, it may "jump" (*i.e.*, pin between certain points) or "move slowly" as the drop evaporates and concentric rings or broadband residue patterns are formed, respectively.

4.3 Detrimental Effects and Suppression of Coffee-Stain Phenomenon

4.3.1 Detrimental Effects of Coffee-Stain Phenomenon

The coffee-stain effect is detrimental to some industrial processes if left uncontrolled. For example, in the electronics industry, inkjet printing of 2-D crystals is considered as a major advance in the creation of printed electronics. However, the coffee-stain effect prevents uniform distribution of crystals during industrial-scale printing of electronic components, thereby leading to faulty devices. In painting, the coffee-stain effect causes the pigments in watercolor to accumulate at the rim of a painting, thereby reducing its artistic appeal. Furthermore, the coffee-stain effect prevents uniform deposition of thin films of particles or active materials on surfaces. The coffee-stain effect is also problematic in material jetting, an advanced additive manufacturing method that is useful for printing electronics like flexible circuit boards, capacitors, transistors, and inductors [17]. These electronics are created by dried stack drops of particle suspensions. Therefore, if it is not controlled, the coffee-stain effect can cause these electronics to lack uniform topography, especially when uniform topography is targeted. The coffee-stain effect is also a nuisance in the coating industry where the formation of a uniform film is required. In analytical chemistry and biochemistry, the coffee-stain effect is problematic to those who use the matrix-assisted laser desorption/ionization time-of-flight mass spectrometry to analyze (or detect) ionized large organic and biomolecules (analytes) from a dried salt solution (matrix). Due to this effect, the analyte is deposited as a ring rather than a uniform film, making it difficult to pinpoint where the analyte is sufficiently deposited, thus limiting the success of an analysis [18]. In the textile industry, the coffee-stain effect affects the uniformity of pattern design on fabrics if left unchecked. The problematic nature of the coffee-stain effect is not limited to only the areas mentioned here, but from the foregoing, it can be generalized that the effect is a bottleneck whenever uniform deposition of particle residue from a dying suspension is desirable, thereby necessitating the need to suppress it.

4.3.2 Suppression of the Coffee-Stain Phenomenon

Various methods for suppressing the coffee-stain effect have emerged over the years. A critical analysis of these methods [19] reveals that majority of them target the causative factors of the phenomenon: (i) pinning of the drop edge to the solid substrate and (ii) rapid evaporation of solvent molecules at the pinned drop edge, which induces an outward radial capillary flow that sweeps particles to the drop edge where they remain after the drop dries, provided they are not redispersed by the Marangoni flow. Some of the approaches for suppressing the coffee-stain effect are discussed in the subsequent sections.

4.3.2.1 Evaporation of Nonwetting Liquid Drops: Suppression of Coffee-Stain Effect by Depinning the Drop Edge

Hydrophilic surfaces favor the occurrence of the coffee-stain phenomenon because these are wettable and the drop edge is easily pinned to it and as a result, contact angle hysteresis is also relatively high. The pinning force F_p increases with contact angle hysteresis as given by the relationship between F_p, surface tension γ of the liquid drop and the advancing θ_A and the receding θ_R contact angles in Equation (4.1).

$$F_p = \gamma(\cos\theta_A - \cos\theta_R) \tag{4.1}$$

Increasing the hydrophobicity of the surface decreases wettability and disfavors pinning of the drop edge and lowers contact angle hysteresis as well as the pinning force, all of which disfavor the coffee-stain phenomenon *via* a smooth receding of the drop edge as evaporation progresses, in line with theoretical predictions [20–22]. The hydrophobicity of a surface is increased by patterning it either through attaching non-polar organic moieties to the surface (*i.e.*, chemical hydrophobization), by physically roughening the surface (*i.e.*, physical hydrophobization) [23] or by doing both [24]. Once hydrophobized, a liquid drop exists in two different states: the Wenzel, and the Cassie-Baxter (Figure 4.2). In the former, the drop penetrates the spaces between the projected regions ("pillars") on the surface [25] while it does not in the latter; rather, the drop sits on top of the

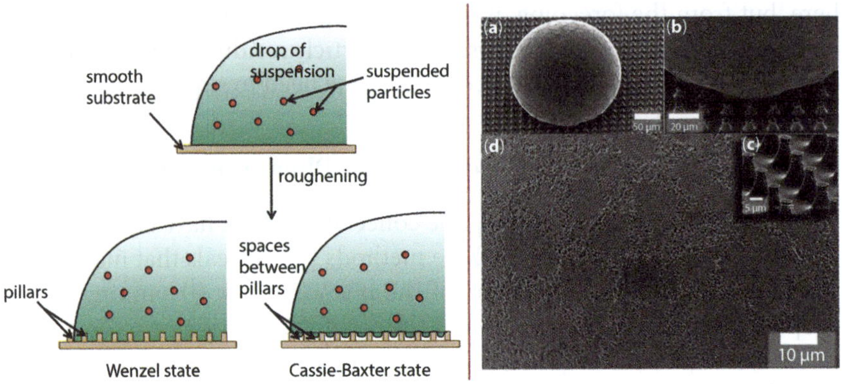

Figure 4.2 (left) Schematic illustration of a drop of a suspension on a smooth planar substrate and on a rough substrate where the drop is either in the Wenzel or Cassie-Baxter wetting state. (right) An image of a ball-like particle deposit on a rough superhydrophobic surface (a), showing how it sits on top of the micropillars of the rough surface (b), which are magnified in (c), as well as a detailed arrangement of individual particles on its surface (d). Reproduced with permission from ref. [29], copyright 2012 Proceedings of the National Academy of Sciences of the USA.

projected regions [26]. A transition from the Wenzel state to the Cassie-Baxter state, and vice versa, has been reported for a drop of an aqueous suspension drying on a superhydrophobic surface [27, 28]. Because the drop edge is not pinned in the Cassie-Baxter state, it leaves a ball-like (Figure 4.2) [29] or disc-shaped [27, 30] residue after evaporation instead of a coffee-stain. Contrarily, the drop edge is pinned in the Wenzel state unless the surface is nanotextured and superhydrophobic [31]. The edges of a drop in the Wenzel state on a superhydrophobic nanotextured surface are not pinned, thus it dries without forming a coffee-stain [30]. If the pinned edge of a liquid drop in the Wenzel state gels, an inward circulatory flow that brings the deposited particles back into the liquid drop is induced. This redistributes the particles and prevents coffee-stain formation [32]. Also, a liquid drop containing particles with low surface activity (*e.g.*, $CuSO_4$) forms spot-like deposit when it dries on a surface with low contact angle hysteresis [33].

The opposite of placing a liquid drop on a hydrophobic or superhydrophobic surface is coating it with hydrophobic or superhydrophobic powdered particles. For particle suspensions, this normally leads to the so-called particle suspension marble [34]. The hydrophobic particle coating prevents the drop (marble) edge from pinning when on a hydrophilic surface, thereby forming a spot-like deposit rather than a particle ring after evaporation (Figure 4.3) [34].

There are two drawbacks associated with preventing the coffee-stain effect by depinning the drop edge using hydrophobic or superhydrophobic surfaces. Firstly, making a substrate hydrophobic or superhydrophobic is difficult and time-consuming. Secondly, when the liquid drop contains surface-active or strongly adsorbing molecules like in the case of biological liquids (*e.g.*, blood, urine, and aqueous suspensions of microorganisms), the molecules adsorb on the substrate and promote pinning of the drop edge, making it difficult to suppress coffee-ring formation. In this regard, coating the liquid with hydrophobic or superhydrophobic particles before allowing it to dry will be a viable option [34], but this is yet to be explored.

Electrowetting, where wetting of a liquid drop is induced and controlled by an applied electric field [35, 36], has also been used to suppress coffee-stain formation by preventing the pinning of the drop edge [37–40]. In electrowetting, the liquid suspension is deposited on a substrate (thickness τ and dielectric constant ε) and then placed on an electrode in a vacuum

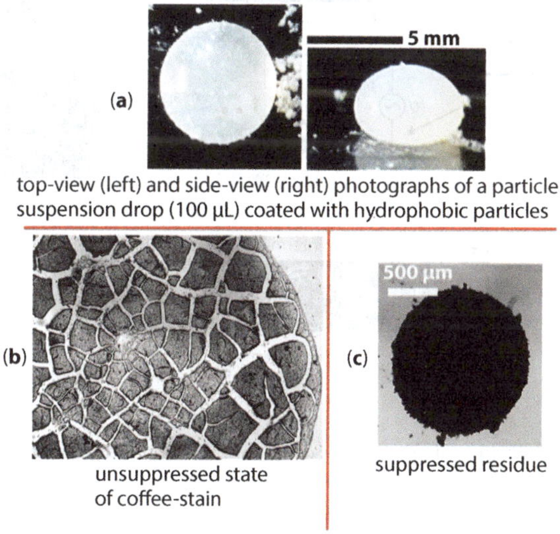

Figure 4.3 (a) Top-view and side-view photographs of a drop of an aqueous suspension of hydrophilic silica particles coated with hydrophobic silica particles, (b) an optical microscope image of an unsuppressed state of a coffee-stain from a drop of an aqueous suspension containing hydrophilic silica particles after drying on a hydrophilic glass slide, and (c) an optical microscope image of the suppressed residue from the marble of the corresponding suspension (in (b)) after drying on the same glass slide, in which the marble was stabilized by hydrophobic silica particles. Reproduced with permission from ref. [34], copyright 2021 Elsevier.

of permittivity ε_0. An electric force F_e (Equation 4.2) that counters the F_p (Equation 4.1) is set-up when potential U is applied between the drop and the electrode.

$$F_e = \frac{\varepsilon\varepsilon_0 U}{2\tau} \tag{4.2}$$

The drop edge is pulled outwardly when $F_e > F_p$ and if an alternating current is used, the drop edge moves periodically away from the pinning site in response to the sinusoidal nature of the current. This allows the drop edge to recede freely as it dries, thus promoting uniform or spot-like deposition of the particles [38] as illustrated in Figure 4.4(a). The periodic movement of the drop edge also sets up an internal circulatory flow in

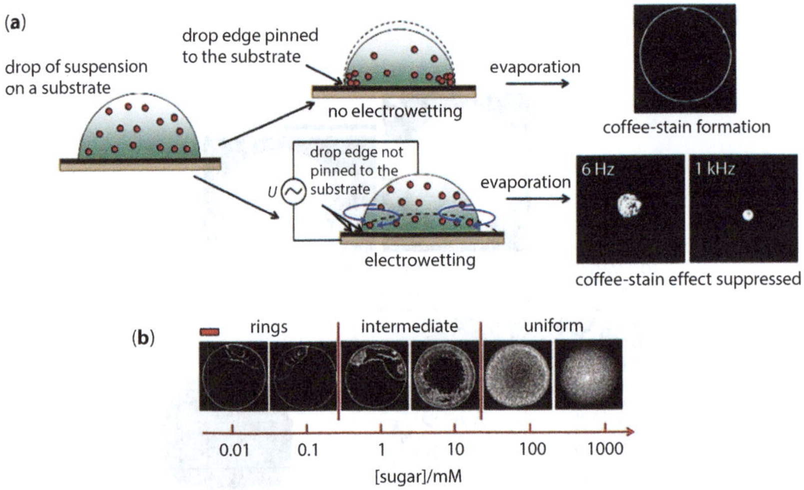

Figure 4.4 (a) Illustration of an aqueous drop containing polystyrene particles evaporating in the absence and presence of electrowetting. In the absence of electrowetting, the drop edge is pinned to the substrate and evaporation leads to coffee-stain formation, but when subjected to electrowetting, the drop gives a spot-like residue after evaporation. The diameter of the spot-like residue decreases as the frequency of the applied electric field increases. Reproduced with permission from ref. [37], copyright 2011 Royal Society of Chemistry. (b) Change in particle deposit from dried drops of polystyrene particle suspension as the concentration of added sugar increases. Coffee-rings form at relatively low sugar concentrations and transit through an intermediate state, at moderate concentrations, to uniform particle deposit at relatively high concentrations. Scale bar = 500 μm. Reproduced under Creative Commons CC BY from ref. [11], Copyright 2018 Authors.

the drop that disturbs the outward radial capillary flow and redisperses particles transported to the drop edge, promoting uniform film formation or spot-like deposition of particles. Such spot-like deposits are important in matrix-assisted laser desorption/ionization mass spectrometry because they increase the signal strength of the analyte species [41]. The prevention of coffee-stain formation through electrowetting can also be achieved with direct current as reported by Zhang *et al.* [39]. Generally, electrowetting is effective even when the liquid drop contains a high percentage of the solute, but its efficacy reduces when the solute is an adsorbing species. Solute adsorption enhances drop edge pinning, making drop edge depinning and the internal circulatory flow less effective.

Another technique that prevents the coffee-stain effect by depinning the drop edge is by adding a relatively high amount of sugar to the particle suspension. During evaporation, the added sugar is precipitated from the drop periphery, preventing the drop edge from pinning and hence the coffee-stain effect. By varying the sugar concentration in the suspension, the minimum required for obtaining a uniform particle film rather than a coffee-ring was identified, with an intermediate between the ring and the uniform film state, as shown in Figure 4.4(b) [11]. However, the success of this method might be limited if the sugar molecules react with the suspended particles, changing their surface chemistry. Also, if the resultant film is meant to be used for an assay, molecules of the added sugar might interfere with the result of the experiment.

4.3.2.2 Suppression of Coffee-Stain Effect by Preventing Rapid Evaporation at the Drop Edge

The formation of homogeneous particle deposit from a dried drop of a particle suspension can be achieved by applying a film of nonvolatile oil on the substrate [15, 42]. When this is done, the formation of the deposit is a function of the wetting state of the drop, the movement of the drop edge, and the flow of liquid during evaporation. By varying the thickness and viscosity of the oil film as well as the hydrophilicity of the substrate, a uniform particle deposit or supra-particle structures can be obtained [42]. *Via* this method, the non-volatile oil at the drop edge prevents rapid evaporation there and by extension the outward radial capillary flow responsible for carrying particles to the drop edge and hence preventing ring formation. Rather, evaporation occurs at the center of the drop, creating an upward flow of liquid that carries the suspended particles which are trapped at the liquid-air interface once the drop dries. Adding surfactants to the

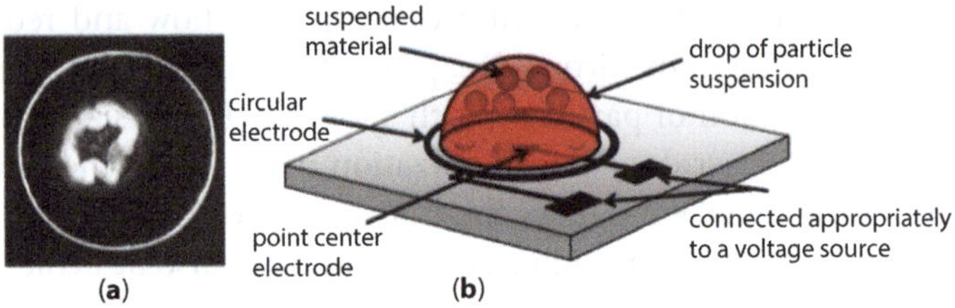

Figure 4.5 (a) An optical microscope image of a coffee-eye formed by a dried drop of polystyrene aqueous suspension. Reproduced with permission from ref. [43]; copyright 2015, Royal Society of Chemistry. (b) Schematic illustration of the experimental set-up for suppressing coffee-ring by radial electroosmotic flow induced by an applied external electric field *via* a circular and a center point microelectrode [44].

suspension decreases the oil-liquid interfacial tension and enables greater control over the movement of the drop edge on the oil-coated substrate and hence the nature of the final deposit [15]. Coupled with the absence of evaporation at the drop edge and the upward flow of liquid in the drop, either a uniform particle film or a "coffee-eye" (*i.e.*, a thin ring with a central spot-like particle deposit, Figure 4.5(a)) is obtained [15]. Nonetheless, as surfactant molecules can adsorb on the suspended particles and alter their surface chemistry, this technique might not be applicable to some real-life applications. Also, the technique cannot be generalized as only one oil type (silicone oil) was explored.

4.3.2.3 Suppression of Coffee-Stain Effect by Preventing Evaporation-Driven Capillary Flow of Liquid to the Drop Edge

To reiterate, there is a flow of liquid from the bulk to the edge of a particle suspension drop resting on a planar substrate whilst it evaporates, known as evaporation-driven capillary flow. As the liquid flows to the drop edge to replenish liquid molecules lost by evaporation, the suspended particles in the drop are also carried along. If these particles are not redispersed by other flow fields in the drop, they remain there until the drop dries, forming a ring. Minimizing, disturbing, or stopping the capillary flow also minimizes, disturbs or stops the flow of particles to the drop edge, making it difficult for the particles to assemble into a ring after the drop dries. This has been achieved using various means, including electrowetting (Sec. 3.2.1), electroosmosis, mechanical vibration, surface acoustic waves, and induction of Marangoni flow.

For a particle suspension containing an electrolyte, there exists an electric double layer at the drop edge where it rests on the substrate. Therefore, the outward capillary flow of liquid to the drop edge generates a streaming potential that influences the transportation and accumulation of particles at the drop periphery. This electric double layer can be harnessed to generate electroosmotic flow by subjecting the interior of the drop to an external electric field. This has been demonstrated by using a microelectrode which consists of a circular and a point central electrode (Figure 4.5(b)) [44]. The drop is placed in the circular electrode on the point central electrode so that the circular electrode surrounds the edge of the drop while the point electrode is in the center of the drop (Figure 4.5(b)). An electric field E passes across the drop when a potential is applied across the electrodes and an inward radial electroosmotic flow (velocity V_e) is generated around the base of the drop by the circular electrode. The V_e can be written in terms of the dielectric constant of the liquid ε_r, the permittivity of free space ε_0, the zeta potential of the substrate ζ and the viscosity of the suspension η as

$$V_e = -\frac{\varepsilon_r \varepsilon_0 \zeta E}{\eta} \tag{4.3}$$

By varying the applied potential and changing the polarities of the electrodes, the strength and direction of the radial flow are respectively controlled, and by extension the nature of the particle deposit. When the inward radial flow balances the outward capillary flow, a uniform distribution of particle deposit occurs. However, when the inward radial electroosmotic flow is greater than the capillary flow, the particles are deposited centrally as a spot [44]. If the substrate is charged, deviation from coffee-stain formation is also enhanced by increase in particle deposition velocity at the final stage of the evaporation process due to electrokinetic effect [45]. The success of this technique strongly depends on the precise deposition of the drop on the electrodes and an added electrolyte is needed to make the drop responsive to the applied external electric field. This technique can be used to concentrate samples of electroactive substances like protein and DNA for biochemical or biomedical analysis. Although this is yet to be investigated, the applied electric field might induce the Joule heating effect, leading to an increase in temperature of the liquid drop as well as disturbances in concentration, electric and flow fields [46]. If the drop contains substances that can be destroyed by heat, *e.g.*, biological substances, they will be destroyed or their innate properties will be altered, and by extension, the outcome of any assay involving them. If the drop does not contain such substances, the Joule heating effect can be an added advantage by promoting uniform particle film formation *via* the thermal Marangoni flow; however, this should be investigated experimentally.

The coffee-stain effect is also a problem in convection deposition [47] and spray coating [48]. By vibrating the substrate laterally in the deposition direction, a uniform layer of particle deposit can be obtained after deposition and evaporation of the deposited suspension [49]. The particle layer forms faster with the vibration than in the absence of it [49]. This technique has also been applied in spray coating to obtain a uniform layer of particles for thin electrodes and organic solar-cell applications [50]. In this technique, the coffee-stain effect is defeated because the lateral movement of the substate deforms the substrate-drop interface laterally and creates a variation in pressure in the thin film that influences the liquid flow profile. It is the liquid flow profile that prevents the accumulation of particles at the substrate-drop interface as well as the formation of a particle ring. Whether or not the success of this technique is limited to convection deposition and spray coating remains to be investigated.

Surface acoustic waves [51] can also be used to disrupt coffee-stain formation in an evaporating drop of a particle suspension [52]. Here, the surface acoustic wave generates a standing pressure and capillary waves in the drying drop, while the acoustic force F_{ac} transports the suspended particles to the nodes of the standing waves (Figure 4.6). A value for F_{ac} (in the x-direction) can be estimated from [52]:

$$F_{ac} = 4\pi d^3 k E_{ac} \alpha \sin(2kx) \tag{4.4}$$

where d represents the diameter of a suspended particle, k the wavenumber, E_{ac} the time-average acoustic energy density and α the acoustic contrast factor, which depends on the compressibility and density of the particles as well as those of the liquid. The particles are confined in the nodes when the F_{ac} is greater than the drag force that is associated with the capillary flow and thus preventing them from being swept to the drop-substrate interface and hence coffee-stain formation. This technique gives either a particle film or a spot-like deposit depending on the frequency of the acoustic wave as proposed theoretically and verified experimentally by Mhatre *et al.* [53]. Compared with other techniques, this technique does not require modification of the substrate (although an elastic substrate is recommended) and can be successfully applied to a wide variety of particle suspensions, irrespective of their nature, as well as to biological liquids like blood. However, it is not effective if the suspended particles are small such that the acoustic

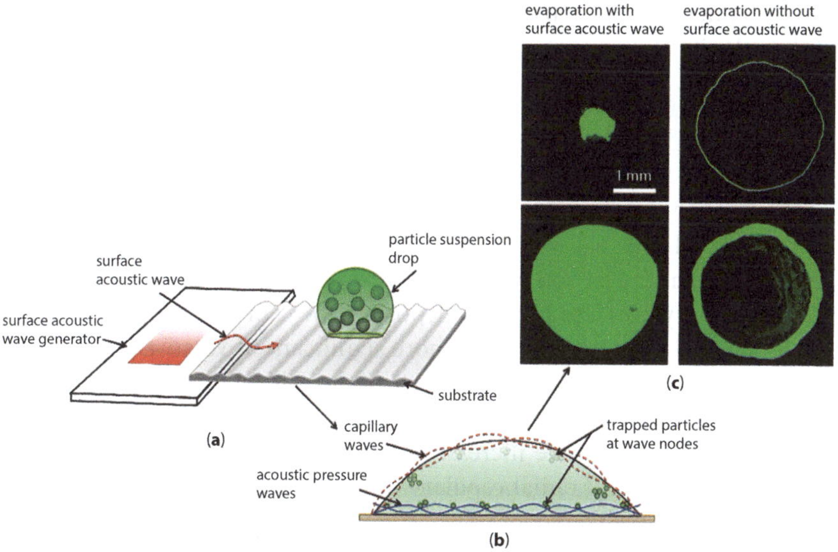

Figure 4.6 Coffee-stain suppression by surface acoustic waves. (a) Schematic illustration of a liquid drop containing suspended particles under the influence of surface acoustic waves. (b) The surface acoustic waves induce standing capillary and acoustic pressure waves, and the suspended particles are trapped at their nodes. (c) Fluorescent images of residues left by a dried water drop containing polystyrene particles in the presence or absence of surface acoustic waves, showing a spot-like deposit or a uniform particle film in the former and coffee-rings in the latter. Reproduced under Creative Commons Attribution 3.0 Unported Licence from ref. [52], copyright 2015 Royal Society of Chemistry.

force is less than the drag force. In this case, acoustic streaming flow can be employed [54, 55]. Also, if the suspended materials in the drop are sensitive to heat, the acousto-thermal effect, associated with surface acoustic waves, will affect their properties after the drop dries [56], bringing the universality of the technique to question. During evaporation, the drop cools, with molecules at the liquid-air interface cooling more rapidly than those in the bulk of the drop, thereby creating a temperature gradient. The temperature gradient in turn creates a surface tension gradient which induces Marangoni flow, *i.e.*, the flow of liquid molecules from a region of low surface tension to that of high surface tension [57, 58]. Because the flow is induced by temperature, it is also known as thermal Marangoni flow. Thermal Marangoni flow is influenced by the relative conductivities of the evaporating liquid drop and the substrate. If the ratio of the latter to the former is < 1.45, the edge of the drop is cooler than its bulk and an inward Marangoni flow to the center of the drop, opposite to the outward radial capillary flow, is observed. The opposite is true if the ratio is > 1.45. If the velocity V_{Ma}

$$V_{Ma} \approx \frac{\partial \gamma}{\partial T}\frac{\Delta T}{\eta} \tag{4.5}$$

of the inward thermal Marangoni flow (where γ is surface tension, T is temperature and η is viscosity of the drop) is greater than the outward radial capillary flow, the particles deposit at the center after the drop dries [19]. Thermal Marangoni flow is more significant in volatile nonaqueous solvents than in water [59]. Irrespective of the solvent, thermal Marangoni flow can be increased by increasing the temperature of the substrate [43, 60]. Heating the substrate increases both the inward thermal Marangoni flow and the outward radial capillary flow, leading to both a particle ring and a central spot-like deposit (*i.e.*, coffee-eye), after the drop dries [43]. The inward thermal Marangoni flow can also be induced by heating the drop surface using a laser beam [61]. Depending on the heating time, laser exposure size and position as well drop size, uniform disc, spot-like and coffee-eye residues can be observed [61]. Because thermal Marangoni flow is more significant in volatile nonaqueous solvents, this method has limited applicability in aqueous systems. Also, increasing the temperature of the liquid drop by any means can affect the physical or chemical nature of the suspended material, especially if the material is sensitive to heat.

Surface tension gradient, and hence Marangoni flow, across an evaporating liquid drop is also created by adding a solvent (*e.g.*, an alcohol [62], a ketone [63] or a long chain alkane like dodecane [64]) or surfactant [65] to the drop, which generally decreases the surface tension of the liquid drop. Due to the solvent or surfactant, a concentration gradient develops between the liquid-air interface and the bulk liquid as evaporation progresses, inducing surface tension gradient and thus Marangoni flow. Contrary to the outward evaporation-driven radial capillary flow, Marangoni flow is circulatory in nature and can be used to control the nature of particle film left after a liquid drop containing suspended particles dries [59]. As liquid molecules are pulled from the liquid-air interface, where the surface tension is relatively low, back into the bulk of the drop, where surface tension is relatively high, *via* the Marangoni flow mechanism, particles that have been swept to the drop edge and the liquid-air interface are redispersed. Depending on the extent of the Marangoni flow, this generally leads to a uniform particle film or a spot-like particle deposit after the liquid drop dries up. Biosurfactants have also been used to hinder coffee-stain formation in biological systems *via* the same Marangoni flow mechanism [66]. From the foregoing, the suppression of the coffee-stain effect through the Marangoni flow mechanism requires either heating the substrate or the liquid drop itself, addition of a solvent or surfactant. These treatments can alter the chemistry of the sample, especially if they are biological in nature, limiting it to only certain applications like inkjet printing.

4.3.2.4 Suppression of Coffee-Stain Effect by Other Miscellaneous Methods

The coffee-stain effect can also be prevented by using other methods that hinder the effect by various mechanisms other than those discussed in the preceding sections. These methods include tuning the particle-particle and particle-substrate interactions, tuning the rheological property of the suspension, humidity adjustment, use of a porous substrate, and use of multiple drops.

When the attractive (cohesive) interaction between the particles is promoted, particle aggregates (heavier) form and settle on the substrate under the influence of gravity compared with the individual particles which are easily swept to the drop edge by the outward radial capillary flow. This gives rise to a uniform particle deposit rather than a coffee-stain [67–73]. The attractive interaction between the particles is promoted by suppressing the electrostatic repulsive interaction between them by either salt addition [71] or pH adjustment [74]. The nature of the film formed is governed by the relative competition between the rate of aggregate formation and the rate of evaporation: a uniform deposit is formed when the rate of particle aggregation is faster than the rate of evaporation while a coffee-stain is formed when the opposite happens [67]. The nature of the deposit formed is also influenced by the electrostatic and van der Waals interactions between the particles and the substrate as well as the Marangoni flow. The sum of the electrostatic and van der Waals interactions is known as the Derjaguin-Landau-Verwey-Overbeek (DLVO) interaction [75]. A uniform deposit is formed when the DLVO interaction between the particles and the substrate is attractive as the particles are prevented from being carried to the drop edge. Meanwhile, a coffee-ring is formed when the interaction is repulsive because the particles are easily carried to the drop edge while a spot-like deposit results from the action of Marangoni flow [74]. The relative competition between these factors and the resultant deposit is summarized in a 2-D phase diagram by Bhardwaj *et al.* [74]. Also, the adsorption of particles at the liquid-air interface can suppress the coffee-stain effect because once adsorbed, the particles deform the interface, promoting attractive capillary interaction between the adsorbed particles. Since the particles are not carried to the drop edge, they form a homogeneous residue when the drop dries. This effect is more prominent with ellipsoidal particles as well as their mixtures with spherical particles [14] and particles that are partially wetted by the liquid-air interface, *e.g.*, partially hydrophobic particles. The wettability of the particles can be tuned *in-situ* by adding surfactants [76] or *ex-situ* by coating them with a suitable agent. Photosensitive surfactants have an added advantage of being switched between hydrophilic and hydrophobic such that the wettability of the particles they adsorb on can be altered accordingly [76].

If the viscosity of a drop of suspension can suppress the outward radial capillary flow, the coffee-stain effect can also be suppressed. However, high viscosity (of the order of 10^7 Pa s for water) which is undesirable in the printing industry is needed for this to happen, making this method less viable [19]. In laponite suspension, such viscosity is achieved during evaporation where a sol-to-gel transition occurs. The viscosity opposes the sweeping of particles to the drop edge by the outward capillary flow, thereby preventing the coffee-stain effect [77].

If the ring residue of a dried suspension is placed in a humid environment and dried again repeatedly, the resultant residue is a disc-like particle deposit [78]. This happens because the residue is re-fluidized to the initial spherical-cap during which there is an upward flow that carries the initially deposited particles to the center of the drop, forming a uniform deposit after the drop dries [79]. Although no force field or modification of the suspended particles or solvent is required here, this method is only applicable if the particle residues can be re-fluidized in the presence of the vapor of the solvent. This technique is more suitable for producing polymer-organic light emitting diode display devices [79].

Evaporation of a suspension on a porous substrate, *e.g.*, paper [80] can also prevent coffee-stain formation with the nature of film formed depending on the porosity of the substrate as well as the depth of the pores. In the Wenzel wetting state, where the solvent imbibes the pores of the substrate (Figure 4.2), the rate of solvent imbibition as well as rate of transport of particles to the drop edge govern the resultant deposition pattern. The coffee-stain effect is suppressed when the rate of the former is faster than the rate of the latter. In the Cassie-Baxter wetting state, where the drop does not imbibe into the pores (Figure 4.2), the resultant film pattern is dictated by the rate of particle transportation to the drop edge and the rate of evaporation. The coffee-stain effect is prevented when the rate of evaporation is faster than the rate at which the particles are carried to the drop edge by the outward radial capillary flow [81]. Like hydrophobic and superhydrophobic surfaces, porous substrates are difficult to create. Repeated drying of different drops of a suspension at the same spot also gives a uniform particle deposit rather than a coffee-stain [82].

4.4 Applications of Coffee-Stain Phenomenon

4.4.1 Coffee-Stain Phenomenon in Detection of a Material

Generally, evaporation of a suspension increases the concentration of the suspended material. In analytical chemistry, biochemistry and biomedical science, this enables easy detection of the material. The deposition of the material as a ring makes it readily detectable by analytical equipment. This is the basis for the self-ordered ring (SOR) method used in biochemical analysis to detect substances that are present in trace amounts in human urine and blood serum. For example, the SOR method has been used to enhance the fluorescence detection of traces of berberine in human urine [83], tetracycline [84] and quinidine sulfate [85] in serum samples after oral administration. The SOR method has also been used to develop a biosensor for rapid detection of aptamer, which enables traces of α-thrombin in whole blood (at 4× dilution) and human serum to be detected [86], and diagnosis of malaria by detecting traces of the biomarker *Plasmodium falciparum* histidine-rich protein II [87, 88]. Droplet coating deposition Raman spectroscopy, which involves drying drops (~ 2 μL) of dilute analyte solutions on substrates, has several advantages in proteomics [89–92], *e.g.*, obtaining Raman spectra from very dilute protein solutions, is also based on the SOR method. Surface-enhanced Raman spectroscopy (SERS) [93], a sensitive technique that enhances the Raman scattering of traces of molecules on a nanostructured material support, also uses the SOR method to concentrate traces of the analyte on the surface of the material support. SERS finds applications in chemistry, nanotechnology, catalysis, food science, biomedicine, biology, and biochemical and environmental analysis. Coffee-ring-enhanced SERS has been used to detect polycyclic aromatic hydrocarbons in water samples [94] and the biomarker glycated hemoglobin that enables the long-term monitoring of blood glucose level [95]. In medicine, specialized personnel and equipment are traditionally needed to track changes in drug quality to ensure optimal efficacy, but this is capital-intensive and unaffordable in developing countries. By analyzing the residue patterns and coffee-rings left by dried drops of drug suspensions alone or in combination with NaCl, impurities and changes in the drug concentration can be readily detected with high accuracy in a cost-effective manner [96].

4.4.2 Coffee-Stain Phenomenon in Forensic Science and Disease Diagnosis

Drops of biological fluid samples like blood [97–101] and protein solution [102] produce distinctive characteristic patterns, which can be easily visualized, when they dry and this can be utilized in forensic science for crime investigation and in medicine as an efficient, robust and relatively cheap means of disease diagnosis [100]. For instance, the pattern produced by blood drops in a crime scene can be analyzed forensically to reveal the angle and the velocity with which they were deposited or splattered as well as how they were produced [103]. In medicine, patterns of dried blood can be used to diagnose non-communicable and infectious diseases [104]. When human blood is mixed with polystyrene particles, the patterns formed upon drying enable differentiation of a sickle cell hemoglobin from a normal one [105]. Similarly, the residue pattern formed by a dried drop of Tris-ethylenediaminetetraacetic acid buffer containing fluorescent polystyrene microsphere was used for the qualitative and quantitative analysis of nucleic acid [106]. The polystyrene microspheres were functionalized with a single-stranded oligonucleotide probe which hybridizes with the added complementary target DNA, thereby forming particle aggregates. These particle aggregates promoted the formation of the coffee-ring where the ratio of the ring thickness to the area of the residue increases with the concentration of the added target DNA [106]. This is similar to the detection of M13 bacteriophage using superparamagnetic particles functionalized with anti-M13 monoclonal [107]. In the presence of M13 bacteriophage, the particles form aggregates which are concentrated as a spot-like deposit by Marangoni flow upon drop evaporation. When M13 bacteriophage was absent, particle aggregation did not occur and a drop containing the particles dried to form a coffee-ring [107]. From the foregoing, the coffee-stain effect can be used to develop simple, robust and cost-effective means of diseases diagnosis, especially in developing countries where the lack of modern equipment often hinders efficient diagnosis of common diseases.

4.4.3 Material Separation Using the Coffee-Stain Effect

When a liquid drop containing a mixture of suspended materials dries, the materials are arranged from the drop edge to the center according to their size. For example, a drop of an aqueous suspension containing a mixture of polystyrene particles (diameter ~ 40 nm, ~ 1 μm and ~ 2 μm) formed three distinct rings with the outer ring containing the smallest particles (diameter ~ 40 nm), followed by the ring containing particles of diameter ~ 1 μm while the innermost ring was formed by the largest particles (diameter ~ 2 μm) as shown in Figure 4.7(a) [108]. The universality of this concept was demonstrated using an aqueous mixture containing biological materials like proteins, microorganisms, and mammalian cells [108]. In this case, the outermost ring was formed by protein, followed by a ring of microorganisms and then an innermost ring containing mammalian cells. Once the materials are carried to the drop edge by capillary flow, they are separated into their various sizes based on the geometry of the liquid meniscus which rather acts as a wedge (Figure 4.7(b)). Once this happens, rapid evaporation at the drop edge and a set of pinning and depinning events, from the competition between surface tension and adhesion forces, further enhance the separation of the materials [108]. The separation can also be achieved or enhanced by heating the substrate to bring about thermal Marangoni flow [109]. This concept can be used to separate a mixture of nano and microparticles as well as a mixture of biological materials.

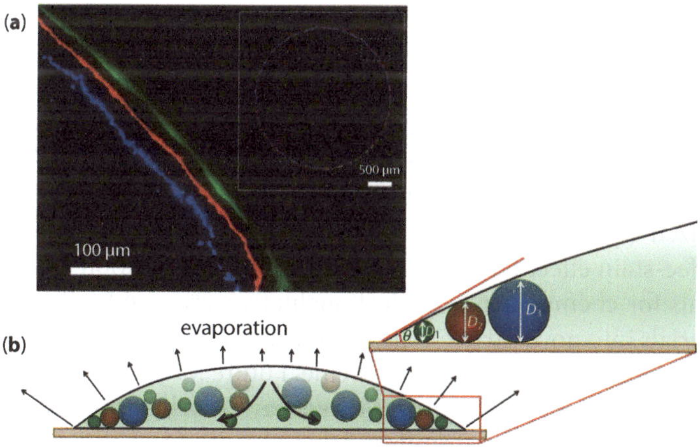

Figure 4.7 (a) Optical fluorescence image showing the separation of three different polystyrene particles of diameter ~ 40 nm (green), ~ 1 μm (red) and ~ 2 μm (blue) into three distinct coffee-rings after an aqueous drop containing a mixture of the particles dries. (b) Schematic illustration of how the particles arrange themselves at the drop edge during evaporation, leading to their separation. The arrangement is based on the geometry of the drop meniscus as it rests on the substrate with a contact angle θ. Reproduced with permission from ref. [108], copyright 2011 American Chemical Society.

4.4.4 Materials Self-Assemblage by Coffee-Stain Effect

The deposition of suspended materials after a drop of a suspension dries has been used to self-assemble nano and microparticles into 2-D [110] and 3-D structures [110–112]. These structures have various applications, including photonics [111, 113, 114], gas sensing [115], electronics [111, 116, 117] and medicine [118].

4.5 Summary

Relevant literature regarding the mechanism of coffee-stain formation is reviewed, including related consequences and advantages of the phenomenon as well as methods of suppressing it. The coffee-stain effect occurs because the edge of a drying drop, containing suspended materials, is pinned on the substrate causing fast evaporation of solvent molecules there. Consequently, the evaporated molecules are replenished by the outward radial capillary flow which carries suspended particles to the pinned drop edge. These particles form a ring-like residue, the so-called coffee-stain, when the drop dries. The coffee-stain effect is detrimental to some experimental procedures and industrial processes, therefore its formation is often prevented or suppressed by preventing or suppressing processes that favor its formation, leading to numerous coffee-stain suppression techniques. These techniques have varying advantages and disadvantages. For example, they cannot be applied to all samples or situations. Additionally, some of them require relatively expensive specialized equipment. Therefore, future research should focus on techniques that do not require expensive specialized equipment and have wide applicability. So far, applications of the coffee-stain effect fall into two categories: concentration of suspended materials for chemical or biological qualitative assay and self-assemblage of materials into novel structures for applications in photonics, electronics, environmental monitoring and medicine. Future research should focus on using the phenomenon for quantitative assay as well as the assemblage of materials for applications in other fields like the food and the cosmetic industries.

AI disclosure statement: The author confirms that no artificial intelligence (AI)-generated content or AI-assisted writing tools were used in the preparation of this chapter (including summaries and the abstract). All text, analysis, and conclusions are the result of the author's original work.

References

1. R. Brown, Additional remarks on active molecules. *Philos. Mag.*, 6, 161-166 (1829).
2. R.D. Deegan, O. Bakajin, T.F. Dupont, G. Huber, S.R. Nagel and T.A. Witten, Capillary flow as the cause of ring stains from dried liquid drops. *Nature*, 389, 827-829 (1997).
3. R.D. Deegan, O. Bakajin, T.F. Dupont, G. Huber, S.R. Nagel and T.A. Witten, Contact line deposits in an evaporating drop. *Phys. Rev. E.*, 62, 756-765 (2000).
4. A.T. Tyowua and S.G. Yiase, Contact angle hysteresis - advantages and disadvantages: A critical review. *Rev. Adhesion Adhesives*, 8, 47-67 (2020).
5. Á.G. Marín, H. Gelderblom, D. Lohse and J.H. Snoeijer, Rush-hour in evaporating coffee drops. *Phys. Fluids*, 23, 091111 (2011).
6. G. Huber, Rush hour in a drop of coffee. *Physics*, 4, 65 (2011).
7. C.N. Kaplan and L. Mahadevan, Evaporation-driven ring and film deposition from colloidal droplets. *J. Fluid Mech.*, 781, R2-13 (2015).
8. Á.G. Marín, H. Gelderblom, D. Lohse and J.H. Snoeijer, Order-to-disorder transition in ring-shaped colloidal stains. *Phys. Rev. Lett.*, 107, 085502 (2011).
9. L. Shmuylovich, A.Q. Shen and H.A. Stone, Surface morphology of drying latex films: Multiple ring formation. *Langmuir*, 18, 3441-3445 (2002).
10. S. Maheshwari, L. Zhang, Y. Zhu and H.-C. Chang, Coupling between precipitation and contact-line dynamics: Multiring stains and stick-slip motion. *Phys. Rev. Lett.*, 100, 044503 (2008).
11. S.F. Shimobayashi, M. Tsudome and T. Kurimura, Suppression of the coffee-ring effect by sugar-assisted depinning of contact line. *Sci. Rep.*, 8, 17769 (2018).
12. P. Takhistov and H.-C. Chang, Complex stain morphologies. *Ind. Eng. Chem. Res.*, 41, 6256-6269 (2002).
13. Y.J.P. Carreón, O. Díaz-Hernández, G.J. Escalera Santos, I. Cipriano-Urbano, F.J. Solorio-Ordaz, J. González-Gutiérrez and R. Zenit, Texture analysis of dried droplets for the quality control of medicines. *Sensors*, 21, 4048 (2021).
14. P.J. Yunker, T. Still, M.A. Lohr and A.G. Yodh, Suppression of the coffee-ring effect by shape-dependent capillary interactions. *Nature*, 476, 308-311 (2011).
15. Y. Li, C. Diddens, T. Segers, H. Wijshoff, M. Versluis and D. Lohse, Evaporating droplets on oil-wetted surfaces: Suppression of the coffee-stain effect. *Proc. Natl. Acad. Sci. USA*, 117, 16756-16763 (2020).
16. K. Sefiane, On the formation of regular patterns from drying droplets and their potential use for bio-medical applications. *J. Bionic Eng.*, 7, S82-S93 (2010).
17. F. Zhang, E. Saleh, J. Vaithilingam, Y. Li, C.J. Tuck, R.J.M. Hague, R.D. Wildman and Y. He, Reactive material jetting of polyimide insulators for complex circuit board design. *Addit. Manuf.*, 25, 477-484 (2019).

18. H.B. Eral, D. van den Ende and F. Mugele, Say goodbye to coffee stains. *Physics World*, 25, 33-37 (2012).
19. D. Mampallil and H.B. Eral, A review on suppression and utilization of the coffee-ring effect. *Adv. Colloid Interface Sci.*, 252, 38-54 (2018).
20. J. Freed-Brown, Evaporative deposition in receding drops. *Soft Matter*, 10, 9506-9510 (2014).
21. L. Frastia, A.J. Archer and U. Thiele, Dynamical model for the formation of patterned deposits at receding contact lines. *Phys. Rev. Lett.*, 106, 077801 (2011).
22. X. Man and M. Doi, Ring to mountain transition in deposition pattern of drying droplets. *Phys. Rev. Lett.*, 116, 066101 (2016).
23. D. Tian, Y. Song and L. Jiang, Patterning of controllable surface wettability for printing techniques. *Chem. Soc. Rev.*, 42, 5184-5209 (2013).
24. C.R. Crick and I.P. Parkin, Preparation and characterisation of super-hydrophobic surfaces. *Chem. Eur. J.*, 16, 3568-3588 (2010).
25. R.N. Wenzel, Resistance of solid surfaces to wetting by water. *Ind. Eng. Chem.*, 28, 988-994 (1936).
26. A.B.D. Cassie and S. Baxter, Wettability of porous surfaces. *Trans. Faraday Soc.*, 40, 546-551 (1944).
27. M. Dicuangco, S. Dash, J.A. Weibel and S.V. Garimella, Effect of superhydrophobic surface morphology on evaporative deposition patterns. *Appl. Phys. Lett.*, 104, 201604 (2014).
28. C. Semprebon, P. Forsberg, C. Priest and M. Brinkmann, Pinning and wicking in regular pillar arrays. *Soft Matter*, 10, 5739-5748 (2014).
29. Á.G. Marín, H. Gelderblom, A. Susarrey-Arce, A.v. Houselt, L. Lefferts, J.G.E. Gardeniers, D. Lohse and J.H. Snoeijer, Building microscopic soccer balls with evaporating colloidal fakir drops. *Proc Natl Acad. Sci. USA*, 109, 16455-16458 (2012).
30. V.D. Ta, A. Dunn, T.J. Wasley, J. Li, R.W. Kay, J. Stringer, P.J. Smith, E. Esenturk, C. Connaughton and J.D. Shephard, Laser textured superhydrophobic surfaces and their applications for homogeneous spot deposition. *Appl. Surf. Sci.*, 365, 153-159 (2016).
31. X. Dai, B.B. Stogin, S. Yang and T.-S. Wong, Slippery Wenzel state. *ACS Nano*, 9, 9260-9267 (2015).
32. L. Cui, J. Zhang, X. Zhang, Y. Li, Z. Wang, H. Gao, T. Wang, S. Zhu, H. Yu and B. Yang, Avoiding coffee ring structure based on hydrophobic silicon pillar arrays during single-drop evaporation. *Soft Matter*, 8, 10448-10456 (2012).
33. Y.-F. Li, Y.-J. Sheng and H.-K. Tsao, Evaporation stains: Suppressing the coffee-ring effect by contact angle hysteresis. *Langmuir*, 29, 7802-7811 (2013).
34. A.T. Tyowua and A.I. Ezekwuaku, Overcoming coffee-stain effect by particle suspension marble evaporation. *Colloids Surf. A Physicochem. Eng. Asp*, 629, 127386 (2021).
35. A.T. Tyowua, *Liquid Marbles: Formation, Characterisation and Applications*, CRC Press, Boca Raton, Florida (2019).

36. Y.-P. Zhao and Y. Wang, Fundamentals and applications of electrowetting. *Rev. Adhesion Adhesives*, 1, 114-174 (2013).
37. H.B. Eral, D.M. Augustine, M.H.G. Duits and F. Mugele, Suppressing the coffee stain effect: How to control colloidal self-assembly in evaporating drops using electrowetting. *Soft Matter*, 7, 4954-4958 (2011).
38. D. Mampallil, H.B. Eral, D. van den Ende and F. Mugele, Control of evaporating complex fluids through electrowetting. *Soft Matter*, 8, 10614-10617 (2012).
39. J. Zhang, M.K. Borg, K. Ritos and J.M. Reese, Electrowetting controls the deposit patterns of evaporated salt water nanodroplets. *Langmuir*, 32, 1542-1549 (2016).
40. A.W. Wray, D.T. Papageorgiou, R.V. Craster, K. Sefiane and O.K. Matar, Electrostatic suppression of the "coffee stain effect". *Langmuir*, 30, 5849-5858 (2014).
41. O. Kudina, B. Eral and F. Mugele, e-MALDI: An electrowetting-enhanced drop drying method for MALDI mass spectrometry. *Anal. Chem.*, 88, 4669-4675 (2016).
42. A. Gao, J. Liu, L. Ye, C. Schönecker, M. Kappl, H.-J. Butt and W. Steffen, Control of droplet evaporation on oil-coated surfaces for the synthesis of asymmetric supraparticles. *Langmuir*, 35, 14042-14048 (2019).
43. Y. Li, C. Lv, Z. Li, D. Quéré and Q. Zheng, From coffee rings to coffee eyes. *Soft Matter*, 11, 4669-4673 (2015).
44. S.J. Kim, K.H. Kang, J.-G. Lee, I.S. Kang and B.J. Yoon, Control of particle-deposition pattern in a sessile droplet by using radial electroosmotic flow. *Anal. Chem.*, 78, 5192-5197 (2006).
45. S. Das, S. Chakraborty and S.K. Mitra, Ring stains in the presence of electrokinetic interactions. *Phys. Rev. E*, 85, 046311 (2012).
46. X. Xuan, Joule heating in electrokinetic flow. *Electrophoresis*, 29, 33-43 (2008).
47. K. Joshi, Behavior of thin colloidal films and fundamental studies of convective deposition. PhD thesis, Lehigh University, USA (2017).
48. A. Reale, L. La Notte, L. Salamandra, G. Polino, G. Susanna, T.M. Brown, F. Brunetti and A. Di Carlo, Spray coating for polymer solar cells: An up-to-date overview. *Energy Technol.*, 3, 385-406 (2015).
49. T. Muangnapoh, A.L. Weldon and J.F. Gilchrist, Enhanced colloidal monolayer assembly *via* vibration-assisted convective deposition. *Appl. Phys. Lett.*, 103, 181603 (2013).
50. F. Zabihi and M. Eslamian, Substrate vibration-assisted spray coating (SVASC): Significant improvement in nano-structure, uniformity, and conductivity of PEDOT:PSS thin films for organic solar cells. *J. Coat. Technol. Res.*, 12, 711-719 (2015).
51. P. Hess, Surface acoustic waves in materials science. *Physics Today*, 55, 42-48 (2002).

52. D. Mampallil, J. Reboud, R. Wilson, D. Wylie, D.R. Klug and J.M. Cooper, Acoustic suppression of the coffee-ring effect. *Soft Matter*, 11, 7207-7213 (2015).
53. S. Mhatre, A. Zigelman, L. Abezgauz and O. Manor, Influence of a propagating megahertz surface acoustic wave on the pattern deposition of solute mass off an evaporating solution. *Langmuir*, 32, 9611-9618 (2016).
54. J. Wu, Acoustic streaming and its applications. *Fluids*, 3, 108 (2018).
55. R.J. Shilton, M. Travagliati, F. Beltram and M. Cecchini, Nanoliter-droplet acoustic streaming *via* ultra high frequency surface acoustic waves. *Adv. Mater.*, 26, 4941-4946 (2014).
56. T. Zheng, C. Wang, Q. Hu and S. Wei, The role of electric field in microfluidic heating induced by standing surface acoustic waves. *Appl. Phys. Lett.*, 112, 233702 (2018).
57. R. Tadmor, Marangoni flow revisited. *J. Colloid Interface Sci.*, 332, 451-454 (2009).
58. X. Xu and J. Luo, Marangoni flow in an evaporating water droplet. *Appl. Phys. Lett.*, 91, 124102 (2007).
59. H. Hu and R.G. Larson, Marangoni effect reverses coffee-ring depositions. *J. Phys. Chem. B*, 110, 7090-7094 (2006).
60. M. Parsa, S. Harmand, K. Sefiane, M. Bigerelle and R. Deltombe, Effect of substrate temperature on pattern formation of nanoparticles from volatile drops. *Langmuir*, 31, 3354-3367 (2015).
61. V.D. Ta, R.M. Carter, E. Esenturk, C. Connaughton, T.J. Wasley, J. Li, R.W. Kay, J. Stringer, P.J. Smith and J.D. Shephard, Dynamically controlled deposition of colloidal nanoparticle suspension in evaporating drops using laser radiation. *Soft Matter*, 12, 4530-4536 (2016).
62. J. Park and J. Moon, Control of colloidal particle deposit patterns within picoliter droplets ejected by inkjet printing. *Langmuir*, 22, 3506-3513 (2006).
63. B.-J. de Gans and U.S. Schubert, Inkjet printing of well-defined polymer dots and arrays. *Langmuir*, 20, 7789-7793 (2004).
64. J.A. Lim, W.H. Lee, H.S. Lee, J.H. Lee, Y.D. Park and K. Cho, Self-organization of inkjet-printed triisopropylsilylethynyl pentacene *via* evaporation-induced flows in a drying droplet. *Adv. Funct. Mater.*, 18, 229-234 (2008).
65. H. Yildirim Erbil, Control of stain geometry by drop evaporation of surfactant containing dispersions. *Adv. Colloid Interface Sci.*, 222, 275-290 (2015).
66. W. Sempels, R. De Dier, H. Mizuno, J. Hofkens and J. Vermant, Autoproduction of biosurfactants reverses the coffee ring effect in a bacterial system. *Nat. Commun.*, 4, 1757 (2013).
67. S. Choi, A.P. Pisano and T.I. Zohdi, An analysis of evaporative self-assembly of micro particles in printed picoliter suspension droplets. *Thin Solid Films*, 537, 180-189 (2013).
68. A. Crivoi, X. Zhong and F. Duan, Crossover from the coffee-ring effect to the uniform deposit caused by irreversible cluster-cluster aggregation. *Phys. Rev. E*, 92, 032302 (2015).

69. B. Fleury, G. Dantelle, S. Darbe, J.P. Boilot and T. Gacoin, Transparent coatings made from spray deposited colloidal suspensions. *Langmuir*, 28, 7639-7645 (2012).
70. Y. Tang, W. He, S. Wang, Z. Tao and L. Cheng, The superiority of silver nano-ellipsoids synthesized *via* a new approach in suppressing the coffee-ring effect during drying and film formation processes. *Nanotechnology*, 25, 125602 (2014).
71. V.R. Dugyala and M.G. Basavaraj, Evaporation of sessile drops containing colloidal rods: Coffee-ring and order–disorder transition. *J. Phys. Chem. B*, 119, 3860-3867 (2015).
72. P. Sun, R. Ma, K. Wang, M. Zhong, J. Wei, D. Wu, T. Sasaki and H. Zhu, Suppression of the coffee-ring effect by self-assembling graphene oxide and monolayer titania. *Nanotechnology*, 24, 075601 (2013).
73. A. Crivoi and F. Duan, Amplifying and attenuating the coffee-ring effect in drying sessile nanofluid droplets. *Phys. Rev. E*, 87, 042303 (2013).
74. R. Bhardwaj, X. Fang, P. Somasundaran and D. Attinger, Self-assembly of colloidal particles from evaporating droplets: Role of DLVO interactions and proposition of a phase diagram. *Langmuir*, 26, 7833-7842 (2010).
75. Z. Adamczyk and P. Weroński, Application of the DLVO theory for particle deposition problems. *Adv. Colloid Interface Sci.*, 83, 137-226 (1999).
76. M. Anyfantakis and D. Baigl, Dynamic photocontrol of the coffee-ring effect with optically tunable particle stickiness. *Angew. Chem. Int. Ed.*, 53, 14077-14081 (2014).
77. E.L. Talbot, L. Yang, A. Berson and C.D. Bain, Control of the particle distribution in inkjet printing through an evaporation-driven sol–gel transition. *ACS Appl. Mater. Interfaces*, 6, 9572-9583 (2014).
78. T. Kajiya, W. Kobayashi, T. Okuzono and M. Doi, Controlling profiles of polymer dots by switching between evaporation and condensation. *Langmuir*, 26, 10429-10432 (2010).
79. A.D. Eales and A.F. Routh, Elimination of coffee-ring formation by humidity cycling: A numerical study. *Langmuir*, 32, 505-511 (2016).
80. A. Nilghaz, L. Zhang and W. Shen, Coffee stains on paper. *Chem. Eng. Sci.*, 129, 34-41 (2015).
81. M. Pack, H. Hu, D.-O. Kim, X. Yang and Y. Sun, Colloidal drop deposition on porous substrates: Competition among particle motion, evaporation, and infiltration. *Langmuir*, 31, 7953-7961 (2015).
82. N.C. Schirmer, S. Ströhle, M.K. Tiwari and D. Poulikakos, On the principles of printing sub-micrometer 3D structures from dielectric-liquid-based colloids. *Adv. Funct. Mater.*, 21, 388-395 (2011).
83. Y. Liu, C.Z. Huang and Y.F. Li, Fluorescence assay based on preconcentration by a self-ordered ring using berberine as a model analyte. *Anal. Chem.*, 74, 5564-5568 (2002).
84. C.Z. Huang, Y. Liu and Y.F. Li, Microscopic determination of tetracycline based on aluminum-sensitized fluorescence of a self-ordered ring formed by

a sessile droplet on glass slide support. *J. Pharm. Biomed. Anal.*, 34, 103-114 (2004).

85. C. Yang and C. Huang, Fluorescent microscopic determination of quinidine sulfate in serum samples with self-ordered ring technique by capillary flow effect. *Chin. J. Anal. Chem.*, 34, 183-187 (2006).
86. J.T. Wen, C.-M. Ho and P.B. Lillehoj, Coffee ring aptasensor for rapid protein detection. *Langmuir*, 29, 8440-8446 (2013).
87. C.P. Gulka, J.D. Swartz, J.R. Trantum, K.M. Davis, C.M. Peak, A.J. Denton, F.R. Haselton and D.W. Wright, Coffee rings as low-resource diagnostics: Detection of the malaria biomarker *Plasmodium falciparum* histidine-rich protein-II using a surface-coupled ring of Ni(II)NTA gold-plated polystyrene particles. *ACS Appl. Mater. Interfaces*, 6, 6257-6263 (2014).
88. J.R. Trantum, D.W. Wright and F.R. Haselton, Biomarker-mediated disruption of coffee-ring formation as a low resource diagnostic indicator. *Langmuir*, 28, 2187-2193 (2012).
89. J. Filik and N. Stone, Drop coating deposition Raman spectroscopy of protein mixtures. *Analyst*, 132, 544-550 (2007).
90. D. Zhang, Y. Xie, M.F. Mrozek, C. Ortiz, V.J. Davisson and D. Ben-Amotz, Raman detection of proteomic analytes. *Anal. Chem.*, 75, 5703-5709 (2003).
91. C. Ortiz, D. Zhang, Y. Xie, A.E. Ribbe and D. Ben-Amotz, Validation of the drop coating deposition Raman method for protein analysis. *Anal. Biochem.*, 353, 157-166 (2006).
92. R.A. Halvorson, W. Leng and P.J. Vikesland, Differentiation of microcystin, nodularin, and their component amino acids by drop-coating deposition Raman spectroscopy. *Anal. Chem.*, 83, 9273-9280 (2011).
93. X.X. Han, R.S. Rodriguez, C.L. Haynes, Y. Ozaki and B. Zhao, Surface-enhanced Raman spectroscopy. *Nat. Rev. Methods Primers*, 1, 87 (2021).
94. J. Xu, J. Du, C. Jing, Y. Zhang and J. Cui, Facile detection of polycyclic aromatic hydrocarbons by a surface-enhanced Raman scattering sensor based on the Au coffee ring effect. *ACS Appl. Mater. Interfaces*, 6, 6891-6897 (2014).
95. I. Barman, N.C. Dingari, J.W. Kang, G.L. Horowitz, R.R. Dasari and M.S. Feld, Raman spectroscopy-based sensitive and specific detection of glycated hemoglobin. *Anal. Chem.*, 84, 2474-2482 (2012).
96. Y.J.P. Carreón, O. Díaz-Hernández, G.J. Escalera Santos, I. Cipriano-Urbano, F.J. Solorio-Ordaz, J. González-Gutiérrez and R. Zenit, Texture analysis of dried droplets for the quality control of medicines. *Sensors*, 21, 4048 (2021).
97. D. Brutin, B. Sobac, B. Loquet and J. Sampol, Pattern formation in drying drops of blood. *J. Fluid Mech.*, 667, 85-95 (2010).
98. J. Choi, W. Kim and H.-Y. Kim, Crack density in bloodstains. *Soft Matter*, 16, 5571-5576 (2020).
99. P. Nandakishore and L. Goehring, Crack patterns over uneven substrates. *Soft Matter*, 12, 2253-2263 (2016).
100. R. Chen, L. Zhang, D. Zang and W. Shen, Blood drop patterns: Formation and applications. *Adv. Colloid Interface Sci.*, 231, 1-14 (2016).

101. M. Mukhopadhyay, R. Ray, M. Ayushman, P. Sood, M. Bhattacharyya, D. Sarkar and S. DasGupta, Interfacial energy driven distinctive pattern formation during the drying of blood droplets. *J. Colloid Interface Sci.*, 573, 307-316 (2020).
102. E. Golbraikh, E.G. Rapis and S.S. Moiseev, On the crack pattern formation in a freely drying protein film. *Tech. Phys.*, 48, 1333-1337 (2003).
103. D. Attinger, C. Moore, A. Donaldson, A. Jafari and H.A. Stone, Fluid dynamics topics in bloodstain pattern analysis: Comparative review and research opportunities. *Forensic Sci. Int.*, 231, 375-396 (2013).
104. J.E. George and D. Paul, Dried blood patterns for diagnosis of non-communicable and infectious diseases, in: V. Borse, P. Chandra and R. Srivastava (Eds.) *BioSensing, Theranostics, and Medical Devices: From Laboratory to Point-of-Care Testing*, pp. 299-316, Springer, Singapore (2022).
105. S. Devineau, M. Anyfantakis, L. Marichal, L. Kiger, M. Morel, S. Rudiuk and D. Baigl, Protein adsorption and reorganization on nanoparticles probed by the coffee-ring effect: application to single point mutation detection. *J. Am. Chem. Soc.*, 138, 11623-11632 (2016).
106. Y. Li, Z. Zhao, M.L. Lam, W. Liu, P.P. Yeung, C.-C. Chieng and T.-H. Chen, Hybridization-induced suppression of coffee ring effect for nucleic acid detection. *Sens. Actuators B Chem.*, 206, 56-64 (2015).
107. J.R. Trantum, M.L. Baglia, Z.E. Eagleton, R.L. Mernaugh and F.R. Haselton, Biosensor design based on Marangoni flow in an evaporating drop. *Lab Chip*, 14, 315-324 (2014).
108. T.-S. Wong, T.-H. Chen, X. Shen and C.-M. Ho, Nanochromatography driven by the coffee ring effect. *Anal. Chem.*, 83, 1871-1873 (2011).
109. H. Jeong, J. van Tiem, Y.B. Gianchandani and J. Park, Nanoparticle separation using Marangoni flow in evaporating droplets, in: Proceedings of Solid-State Sensors, Actuators and Microsystems Workshop, Hilton Head Island, South Carolina (2014).
110. I.B. Dogru, C. Kosak Soz, D.A. Press, R. Melikov, E. Begar, D. Conkar, E.N. Firat Karalar, E. Yilgor, I. Yilgor and S. Nizamoglu, 3D coffee stains. *Mater. Chem. Front.*, 1, 2360-2367 (2017).
111. S. Choi, S. Stassi, A.P. Pisano and T.I. Zohdi, Coffee-ring effect-based three dimensional patterning of micro/nanoparticle assembly with a single droplet. *Langmuir*, 26, 11690-11698 (2010).
112. X. Mu and D.G. Gray, Droplets of cellulose nanocrystal suspensions on drying give iridescent 3-D "coffee-stain" rings. *Cellulose*, 22, 1103-1107 (2015).
113. D.J. Norris, E.G. Arlinghaus, L. Meng, R. Heiny and L.E. Scriven, Opaline photonic crystals: How does self-assembly work? *Adv. Mater.*, 16, 1393-1399 (2004).
114. J. Zhang, Z. Sun and B. Yang, Self-assembly of photonic crystals from polymer colloids. *Curr. Opin. Colloid Interface Sci.*, 14, 103-114 (2009).

115. L. Malfatti, Y. Tokudome, K. Okada, S. Yagi, M. Takahashi and P. Innocenzi, Coffee stain-driven self-assembly of mesoporous rings. *Microporous Mesoporous Mater.*, 163, 356-362 (2012).
116. Y. Li, L. Lan, P. Xiao, S. Sun, Z. Lin, W. Song, E. Song, P. Gao, W. Wu and J. Peng, Coffee-ring defined short channels for inkjet-printed metal oxide thin-film transistors. *ACS Appl. Mater. Interfaces*, 8, 19643-19648 (2016).
117. Z. Zhang, X. Zhang, Z. Xin, M. Deng, Y. Wen and Y. Song, Controlled inkjetting of a conductive pattern of silver nanoparticles based on the coffee-ring effect. *Adv. Mater.*, 25, 6714-6718 (2013).
118. S. Khapli, I. Rianasari, S. Sharma, T. Blanton and R. Jagannathan, Fabrication of hierarchically structured porous films of metal oxides and carbonates through coffee ring effect. *Mater. Today Proc.*, 3, 362-368 (2016).

Part 2
ADHESION ASPECTS

5

Controlling the Way Surfaces Interact and Adhere in Cosmetics. From Polymers to Biophysical and Dermatological Applications

Gustavo S. Luengo

L'Oréal Research and Innovation, Aulnay-sous-Bois, France

Abstract

Evaluating surface adhesion gives a glimpse of the fascinating world of molecular interactions. This chapter provides a brief review of my research in the field of surface forces and interactions, based on the results of fundamental studies exploring the relationship between adhesion and friction using thin polymer films, as well as of more applied studies involving cosmetics, dermatology, and tactile perception.

Hair and skin are biological surfaces that are the target of many personal care products. The performance of these products is strongly linked to the way their ingredients attach to or interact with these surfaces. This review focuses on several aspects of these interactions, in particular the structure and surface properties of hair fibers and their interaction with hair care ingredients, skin surface morphology and the importance of corneocyte adhesion for skin barrier function, the influence of surface patterning on the adhesion and sensorial perception of biomimetic coatings, and the crucial role of surface science in the eco-design of sustainable cosmetic formulations.

One name, that of Kash Mittal, has been with me as my knowledge of the fascinating field of adhesion research has expanded. The broad impact of Dr Mittal in this field has made it possible to connect researchers from many different areas, making adhesion a truly interdisciplinary topic.

Keywords: Hair, skin, cosmetics, surface forces

Email: gluengo@rd.loreal.com

Anil N. Netravali and Frank M. Etzler (eds.) The Wide World of Adhesion: A Festschrift in Honor of Dr. Kash Mittal on his 80th Birthday, (175–194) © 2026 Scrivener Publishing LLC

5.1 Introduction

In August 2022, I was invited to "The Wide World of Adhesion" symposium in honor of Kash Mittal held at the American Chemical Society (ACS) meeting in Chicago. This was a great opportunity to celebrate many years of research in the field of adhesion in the company of a person who had contributed so much of the core literature that has surrounded me during my professional life. It should also be mentioned that the international series of Symposia on Surfactants in Solution (SIS) initiated by him has been running since 1976! A previous ACS symposium in his honor in 2010, together with a request to contribute to a special issue of the Journal of Adhesion Science and Technology (JAST) in 2010 addressing "Surface and Adhesion Aspects of Hair and Skin", as well as other published books [1], are examples of events that I have followed personally.

Through his books, myself and other L'OREAL researchers have had the opportunity to work more closely with Kash Mittal, most notably during an invited seminar in 2017 in Paris. Through this collaboration, I had the chance to contribute to the successful book "Surface Science and Adhesion in Cosmetics", which was recently published [2, 3]. Since then, few days have passed when I did not have some kind of book contribution involving Kash Mittal to finish, accompanied by the consequent difficulties of publication in the current industrial world.

Before beginning this chapter, I would like to take this opportunity to also mention two of my main mentors in the field. The first is Professor Ramon G. Rubio at the Complutense University, Madrid, Spain, who directed my thesis in 1993 and from whom I learned essential skills and the principles of physical chemistry and polymer science [4, 5]. The second is Prof. Jacob Israelachvili at the University of California Santa Barbara, USA, who regrettably passed away in 2018, with whom I was exposed to the physics of intermolecular forces as the basis for the surface science of soft matter, using the famous Surface Forces Apparatus technique [6, 7].

Tribology and adhesion have followed me during my career as an industrial researcher, first as a lubricant chemist at the REPSOL Oil company in Madrid from 1991-1993, and then since 2000 in my current company position at L'OREAL.

5.2 Hair Care and Surface Forces

5.2.1 The Hair Substrate

In a first approximation, a hair fiber can be thought of as cylindrical in shape with a mean diameter between 50 and 100 μm (see for example [8, 9]). Hair fiber is a strong structure composed mainly of keratin, and it is this protein structure that gives hair its high Young's modulus. Much work has been dedicated to understanding the internal structure of hair fibers, including a first of its kind study conducted by our group that assessed the structure during hair formation along the follicle [10]. In this work we used Atomic Force Microscopy (AFM) to measure the Young's modulus of the internal part of a longitudinally sectioned hair follicle. This approach allowed us to follow the progressive hardening of the hair fiber during its formation and identify different stages of protein structuration before the fiber reached its differentiated state (Figure 5.1a).

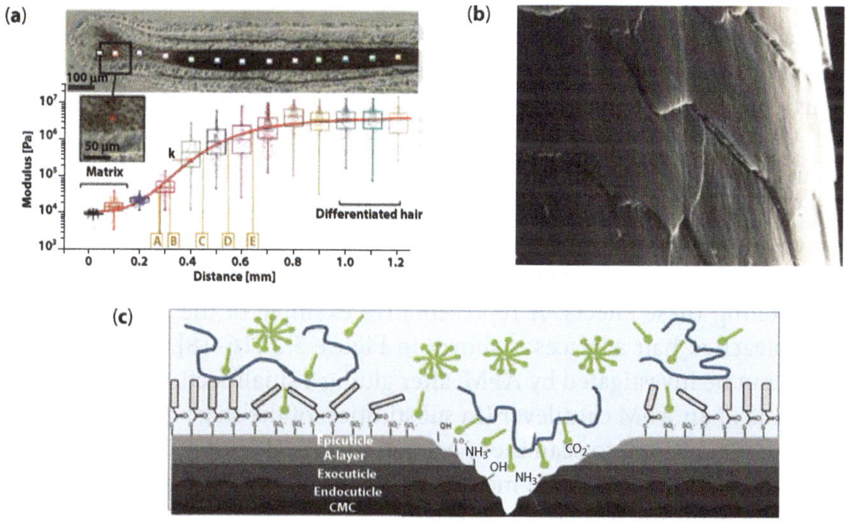

Figure 5.1 (a) top: Optical image of a longitudinal section of the hair follicle indicating the places where indentations were performed using Atomic Force Microscopy (AFM); bottom: Young's modulus measurements showing an overall increase during hair formation [10] (b) SEM image of the surface of hair. (c) Diagram illustrating the model structure of the lipid layer of 18-MEA, the most abundant lipid that covers the hair, showing the thioester linkage with the protein underneath. This lipid is lost progressively toward the tip, due to weathering and/or some hair treatments (see text). CMC represents the cell membrane complex that lies between the two cuticle cells.

Once hair is formed, its surface properties are essential. A rough estimate based on a typical hair length of 20 cm and the number of hairs on a human head (about 120,000–150,000) results in a rather large surface area of about 6 m^2. The outermost layer (the cuticle), which has an overlapping scale-like structure, is largely responsible for the surface attributes of hair (Figure 5.1b). Each cuticle scale is around 50 μm in width and ~0.5 μm thick, and these scales are oriented in one direction (i.e., from root to tip). The surface properties of hair are greatly impacted by variability in this macro-topographical organization [11].

The physicochemical and tribological properties of hair are thought to be significantly influenced by both morphological and chemical characteristics. Recent studies using, for example, X-ray photoelectron spectroscopy (XPS) have helped to characterize the different chemical species present at the hair surface, including highly crosslinked proteins (~75%) and fatty acids (~25%), with 18-methyleicosanoic acid (18-MEA) being the most abundant lipid (Figure 5.1c) [2]. Protein residues can be exposed, and this organization can have implications for the use of hair care products. In particular, sulfonate groups are especially important when considering bleaching hair. Any generic cosmetic water-based formulations containing surfactants or polymers have to interact with the hair surface and, through intermolecular forces, adsorb, adhere and form a coating that will result in particular cosmetic effects (e.g., lubrication or conditioning) [12, 13].

Another situation in which surface forces play an essential role is hair manageability [14, 15]. The three-dimensional aspect of hair in static or dynamic situations can influence visual and tactile perceptions (e.g., the smoothness of movement of hairs passing each other can influence general hair mobility). The intensity of fiber-fiber interactions is a key parameter controlling these effects. A representative example of the forces between two bleached hair surfaces is shown in Figure 5.2 [16–18]. These interactions can be investigated by AFM, after gluing a small section of hair to the extreme of an AFM cantilever (in substitution of the sharp tip (probe) that is generally present to scan the specimen surface (Figure 5.2a). The overall repulsive force can then be measured in several conditions (Figure 5.2b), allowing determination of the surface charge.

In the aforementioned experimental setup, the hairs can also be made to shear against one another at various weights, speeds, and relative sliding angle directions (Figure 5.2c). The so-called "differential friction effect"—represented by a higher friction coefficient when sliding takes place against the scales—occurs during shearing and is determined by the shape of the hair fiber surface and the presence of cuticle scales. Figure 5.2d shows the difference in friction coefficients when shearing two bleached hair fibers

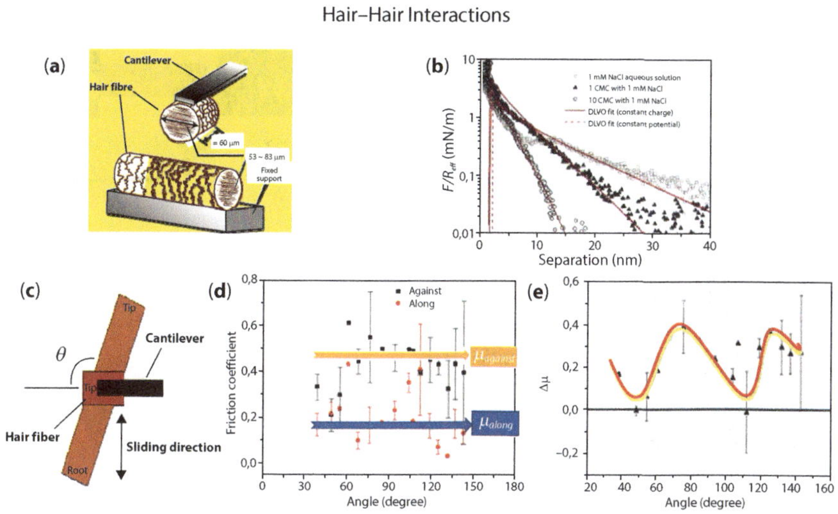

Figure 5.2 (a) Atomic force microscopy (AFM) setup used for measurement of the forces between two fibers. (b) Example of the forces between two bleached hair fibers measured in 1mM NaCl aqueous solution and other conditions. (c) Top-down view of the AFM setup for friction measurements, showing the angle between fibers. (d) Friction coefficient between two hair fibers at different angles and as a function of the shearing direction (against or along the scales). (e) Oscillation of the friction force increment ($\Delta\mu$) between two hair fibers as a function of the relative contact [17, 18].

against or along the scales of the cuticle. In addition, the strong angular dependence (Figure 5.2e) makes it evident that the probability for interlocking is periodic in nature, depending on the angle between the fibers, and influences the intensity of this effect.

5.2.2 Shampoo-Surface Effects

Simultaneously with advancing our knowledge of hair surface interactions, it also improves our understanding of the mechanisms of how common ingredients interact and deposit on hair. This is the case for shampoos and conditioners, topics to which much of our work has been dedicated. The basic mechanisms of adsorption have been known for some time and rely on the precipitation of coacervates formed between cationic polyelectrolytes and anionic surfactants (Figure 5.3a) [19–22]. Dilution with increasing amounts of water results in precipitation, but the precise intermolecular interactions between the different compounds allows them to remain adhered to the hair keratin substrate. Most current efforts toward improving the sustainability of formulations require a transition from classic

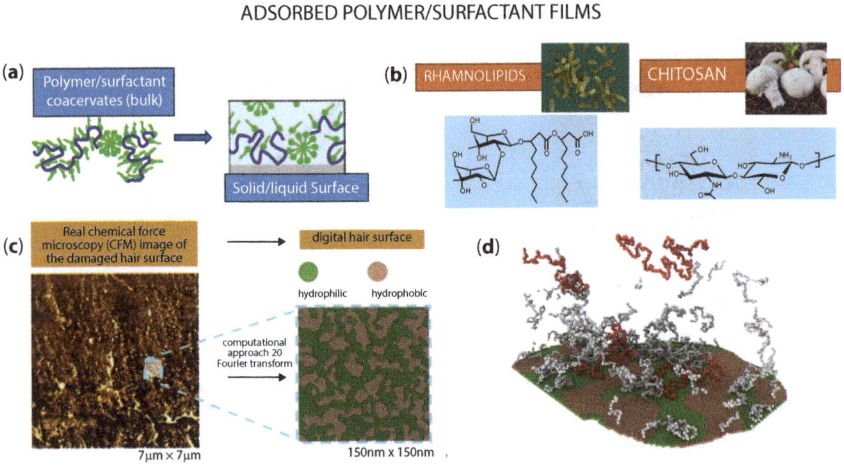

Figure 5.3 (a) Scheme to illustrate the deposition of coacervates of polymer/surfactant onto the surface of the hair upon dilution (e.g. dilution of a shampoo with water during application). (b) Chemical structures of two examples of bio-based ingredients: rhamnolipids from bacteria and chitosan from mushrooms. (c) Image of the surface of the hair obtained by Chemical Force Microscopy showing hydrophobic and hydrophilic zones. The image data can be used to generate digital models, which can then be used to perform Molecular Dynamic (MD) simulations. (d) 3D MD simulation of cationic polymers interacting with the digital model hair surface [25].

petrochemical-based compounds to more bio-based and biodegradable ingredients [23]. This has provided a renewed incentive to increase our understanding of the physicochemical factors that underly the long-lasting adsorption of product ingredients on hair. Examples of these new, more sustainable, potential ingredients include polysaccharides and biosurfactants, for which extracts can be obtained from bacteria or, in the case of chitosan, mushrooms (see molecular structures in Figure 5.3b). Several physicochemical adsorption studies using these ingredients have improved our understanding in this area, for example, such studies have investigated the influence of di-rhamnolipid (two rhamnose sugar moieties) versus mono-rhamnolipids on coating thickness, and have compared deposits of chitosan with other classical polyelectrolytes such as polydiallyldimethylammonium chloride (PDADMAC) [24].

Experimental testing is not sufficient to cope with the diversity of structures of all the ingredients and physicochemical parameters at play. Numerical approaches such as molecular dynamics (MD) are also essential tools, but they need to be adapted so that they can be applied to the hair surface. In our studies [25, 26], we used data collected from chemical force microscopy analysis of real hair samples to generate realistic simulated digital models of hair surfaces (see Figure 5.3c). When chain molecules representing the polymers adsorb, they do so onto structured surfaces rather than onto classical homogeneous ones (Figure 5.3d). This approach has been crucial to better understand what is needed in terms of polymer structure to maximize the adsorption and durability of hair coatings.

5.3 Skin Care and Adhesion

5.3.1 Skin Surface Morphology

Human skin is a living substrate with a complicated structure, making it difficult for cosmetic films to adhere to it. The chemistry and physical characteristics of human skin, including the face and lips, have been thoroughly researched and described in previous publications [27–31]. As stated previously, the surface free energy, surface roughness, and elasticity of the skin are significant characteristics that strongly influence cosmetic film adhesion after product application. Additionally, other elements including skin hair, skin curvature, movement of the skin substrate, sweat, and sebum affect how well and how long a product adheres to the skin. To address the adhesion property of skin, it is again important to differentiate between the mechanical and the chemical aspects of this substrate.

One of the main elements of the skin that contributes to adhesion is the outermost layer called the stratum corneum (SC). It is composed of ca. 15 to 20 layers of dead cells called corneocytes. Flat and pseudo-hexagonal in shape, these cells cover the outermost surface of the skin creating a barrier against external aggressors (see Figure 5.4a). The SC is the hardest of all skin layers and plays a key role in determining the skin's mechanical properties [32]. In addition, the exposed surface of this layer has a complex morphology, from the fine microrelief (directional grooves) patterns found on the surface of young skin to the extreme wrinkles visible in aged individuals. Finally, sweat and sebum are also part of the surface ecosystem, and together with the structural lipids present at the surface of the corneocytes, these components give the skin a mostly hydrophobic character. Several studies have addressed how these characteristics impact cosmetic products, such as how make-up could be formulated to better adhere to this complicated surface, taking into account the fact that make-up products have to form coatings that can support considerable stress (i.e., withstand the movement associated with facial gestures).

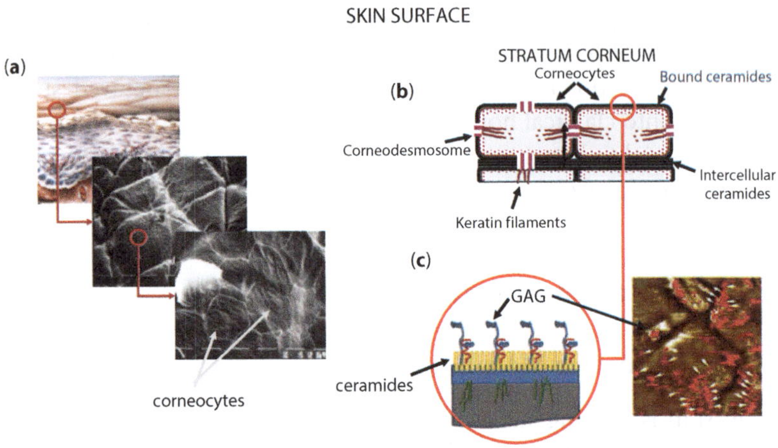

Figure 5.4 (a) Sketch and two scanning electron microscopy images of the skin surface at different levels of magnification, showing the flat pentagonal cells called corneocytes. (b) "Brick and mortar" model of the stratum corneum, the outermost layer of skin, showing the main different components of interest for adhesion, such as the corneodemosomes. (c) A close-up illustration of the corneocyte envelope showing, in addition to ceramides, the presence of glycosaminoglycans (GAG, branched polysaccharides with adhesive characteristics), as well as a top-down Atomic Force Microscopy (AFM) image of a corneocyte with the presence of these components shown as red adhesive spots. Adapted from [34, 35].

However, there is another aspect of adhesion that determines how the SC can be such an efficient skin barrier and resist mechanical disruption. Figure 5.4b shows how corneocytes are connected to each other through specific adhesion junctions (corneodesmosomes) [33, 34]. These cells are then connected and embedded in intercellular lipids (mostly ceramides) following the well-known "brick and mortar" model. The glycosylation of proteins on the surface of corneocytes is also believed to play an important role in cellular adhesion in the SC of human skin. Several of our studies have used AFM to provide insights into the distribution of these glycoproteins and their adhesive characteristics. Figure 5.4c shows a sketch representing the localization of these proteins on the surface of the corneocytes, together with an example of a typical AFM image that allowed their localization to be determined.

5.3.2 Intercellular Adhesion

Inter-corneocyte adhesion is difficult to measure directly. Nevertheless, the relatively large surface of the corneocytes (~30-50μm diameter) allows estimates to be made using an approach involving small glass microcantilevers that pull individual corneocytes away from their surrounding neighbors in the SC (Figure 5.5a). Using this system, we have examined both peripheral adhesion among corneocytes within an SC layer, and interplanar adhesion between corneocytes in different SC layers. Figure 5.5b shows a slightly different setup where this method is used to separate two corneocytes previously extracted from the SC using an adhesive tape (tape stripping). Figure 5.5c shows microscopic optical images taken before and after the detachment of two corneocytes. Notice the clean separation between the two previously adhered corneocytes.

Analyses of detachment forces measured using this approach showed that the strength of both peripheral and interplanar corneocyte adhesion increases significantly with increasing SC depth. The corneocyte population at the SC surface showed significant adhesion heterogeneity, with detachment forces for corneocytes placed next to one another varying by more than one order of magnitude. The measured detachment forces in the upper-middle layers of the SC showed that peripheral adhesion was stronger than interplanar adhesion in this SC region. These studies also confirmed that stronger peripheral SC corneocyte adhesion promotes the development of a more effective skin barrier, which can then withstand significant mechanical pressure and is more resistant to damage.

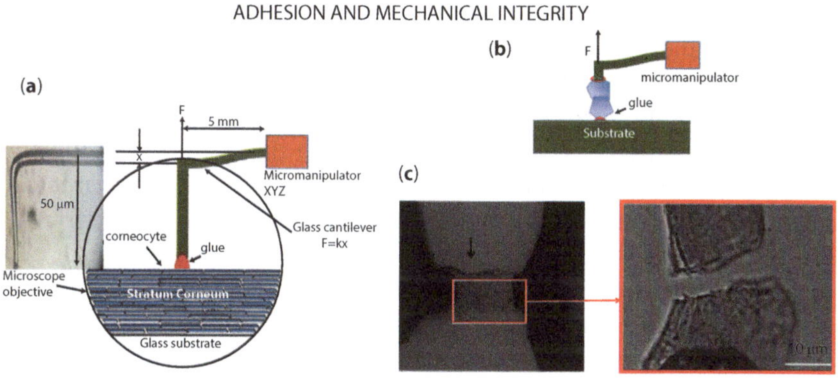

Figure 5.5 Studies investigating the interaction of corneocytes. (a) Experimental setup used to evaluate corneocyte detachment from the stratum corneum. A glass microcantilever is glued to the surface of a single corneocyte and then pulled using a micromanipulator. An optical microscope is used to allow precise positioning of the cantilever on top of the cell. (b) Example of the optical microscopy images used to evaluate the peripherical separation of corneocytes. One corneocyte is glued to the substrate and the other to the cantilever. (c) A representative microscopic image showing the complete detachment of two previously adhered corneocytes; adapted from [36].

5.3.3 Skin Tribology and Tactile Perception

In addition to providing the desired cosmetic effect (hair manageability, skin hydration, skin tone, anti-aging, etc.), cosmetic products also have to provide consumers with an acceptable sensorial experience [37, 38]. Consumer perceptions on opening the cosmetic container or touching a cream or a gel, as well as the simple self-assessment of skin softness after application, all depend on the tribological (i.e., adhesion, friction) properties of a product on contact with human skin. These effects can be static such as the feeling of comfort on the skin, or dynamic such as the softness felt during the tactile application of a product. Most studies evaluating the tribological properties of cosmetic products rely on expert or consumer panels that can describe the overall sensorial experience of a formulation. Predicting the potential sensorial properties of a product is an important step during both the design and the formulation phases of product development. In the particular case of touch, many different tribological approaches have been used to evaluate *in vitro* the properties of ingredients and formulations [39]. Complex correlations can then be used to help determine the basic physicochemical properties needed to obtain the desired sensorial effects (e.g., softness) for a particular family of products.

The mechanisms underlying the sense of touch are complex. Apart from the neurological aspects, the skin has developed specific mechanoreceptors that are sensitive to vibrations, such as those generated during an active touch of the skin [40–42]. In recent years, we have focused on improving our understanding of how tactile perception evolves with age by using a combined psychophysical approach in experiments based on tactile discrimination of the surfaces of different artificial linear textures imprinted on a polymer surface (Figure 5.6a) [43, 44]. In these studies, a cohort of young and elderly individuals were asked to discriminate between tactile surfaces based on the wavelength of the textured surface. Tactile friction was also measured during the tests, allowing us to correlate touch perception with friction forces. As shown in Figure 5.6b, the percentage of valid responses (i.e., individuals who correctly determined whether a test stimulus texture differed from a reference sample with a wrinkle wavelength of 100 μm) was plotted against the texture wavelength of the test stimulus. The results showed that as the wavelength of the stimulus texture approached that of the reference sample, the task became more challenging for both groups, as evidenced by a decrease in the number of correct responses. The gap between the young and elderly participants was particularly large when discriminating between stimuli textures with wrinkled surface wavelengths of ~20 or 40 μm. Further studies are needed in this area to determine the relative influence of age-associated decreases in skin elasticity versus mechanoreceptor density on these findings [45–47].

Surface texture is a key factor in determining the adhesion property, and more generally the tribological or physicochemical attributes, of natural surfaces (e.g., rose petals, lotus, etc.). Altering surface textures is a key strategy in multiple industries, for example, changes in packaging can be used to alter consumer touch perceptions. In cosmetics, surface wrinkling, or more generally surface patterning, can also be explored as a way of modifying the perception of the skin and skin cosmetic coatings. Numerous senses, such as shine, color, and touch, are influenced by the surface topography (microrelief) of the skin. A whole different set of perceived qualities can be produced by imprinted patterns, which can alter the original microrelief. The tribological effects of imprinted geometrical structures on the SC have been assessed in our studies, with the results showing a great

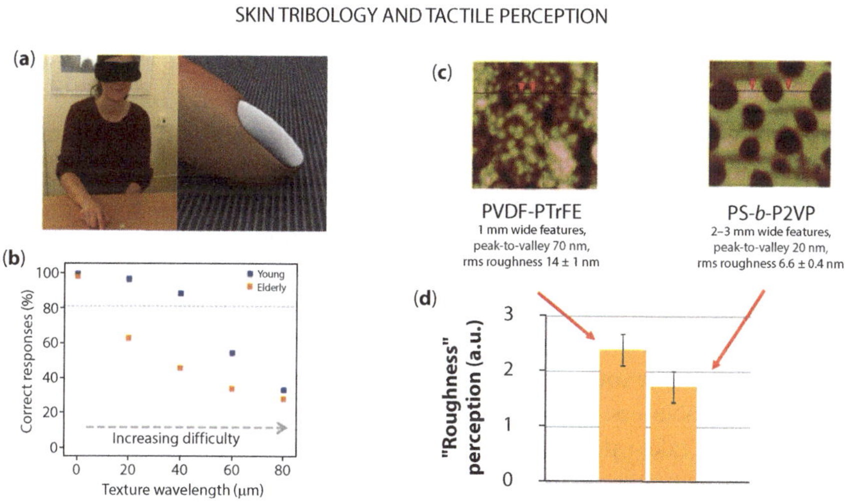

Figure 5.6 Examples of studies related to tribological behavior and the impact of sensorial tactile perception. (a) *In vivo* tactile tests to evaluate and discriminate surfaces of different calibrated roughness. Participants were exposed to binary combinations of randomized textured surfaces and asked to determine if they were similar or different. (b) Percentage of correct responses as a function of the wavelength of the undulated surfaces used and as a function of the age of the participant. The results show a decrease in tactile acuity with age, as well as with an increase in roughness wavelength. (c) Textured (patterned) co-polymer coatings (PDF-PTrFE and PS-b-P2VP, see text) obtained by solvent casting and exhibiting rough (peak to valley feature height: 70nm) and a flatter patterned (peak to flattened valley depth 20nm) surfaces. (d) Tactile sensorial evaluation of a small panel of volunteers assessing the roughness of the PVDF-PTrFE polymer surface. Adapted from [49].

diversity of friction coefficient responses and induced vibration stimuli associated with surface texture changes [48].

Apart from mechanical imprinting, it is well known that certain polymers can also develop surface patterns upon annealing. In Figure 5.6c we show examples of rough and hollow patterns obtained with a random copolymer of poly(vinylidene difluoride) and polytrifluoroethylene (PVDF-PTrFE) and a block copolymer of polystyrene-b-poly-2-vinylpyridine (PS-b-P2VP) [48, 49]. In addition to the diversity of friction responses observed with different sample textures studied, as would be expected due to the different topographical characteristics of the surface patterns, it is important to note that some of the patterns were also perceived differently when touched (Figure 5.6d).

5.3.4 Surface Science to Help Sustainability

Improving the sustainability of cosmetic formulations by switching from petrochemical-based ingredients to more sustainable alternatives is a challenging process, particularly when it comes to replacing the polymers used in cosmetic products with naturally occurring and biodegradable macromolecules [50]. A complete switch to using only natural polymers in cosmetic products would require some concessions, such as accepting the greater structural variability of alternative materials caused in part by inherent natural diversity. In addition, the properties of many natural polymers still need to be optimized for specific applications of interest to cosmetic manufacturers. These issues persist despite numerous attempts to create polymer structures with a low environmental impact.

The sustainable sourcing of materials is a key aspect in the eco-design of functional polymers. To be truly eco-designed, the product development process must take into account the renewable plant origin and sustainability of the raw materials used for polymer synthesis, as well as ensure that the potential products are environmentally friendly and preserve biodiversity, before any development in the laboratory [51, 52].

Surface science is essential for developing eco-designed formulations that provide the same level of performance as those containing existing ingredients. For example, polymer deposits are essential for aesthetic applications. Even with petrochemical polymers, the physiochemistry underlying the creation of films by evaporating solvents, a process often used in cosmetics, is still far from being fully understood. Cosmetic films must perform several tasks, including protection, wrinkle filling, and pigment stabilization. Regardless of the application, the films must also adhere well to the target surface (i.e. hair, skin, or nails), while still being easy to remove when needed (i.e., ease of make-up removal). Optimizing and controlling these properties is crucial for improving the performance of cosmetic products containing natural ingredients. In general, these properties depend on the structure of the polymer ingredients, which can be examined by microscopic analyses and then altered to change the microstructure of the films and thus their properties (i.e., roughness).

In addition to incorporating Green Chemistry principles when creating new polymers for cutting-edge cosmetic formulations, it is crucial to consider their environmental impact after consumers have used them, just like with other ingredients. After use, cosmetics are frequently released with wastewater and end up in the water supply. A major step in reducing the environmental impact of cosmetic products can be achieved by focusing on two easily manageable eco-design levers: biodegradability and the gray water footprint. In cosmetics and many other industries, combining high biodegradability and adequate functional efficiency is one of the major challenges for the future of polymer eco-design [23].

Another property of interest for components, and notably for polymers, is non-persistence in the environment. The persistence of a formulation depends on the ability of the chemical components to completely breakdown over time, through physical, chemical, and/or biological processes, preventing the buildup of the parent component or any potential byproducts in the environment. The technical guidelines that accompany chemical regulations include criteria and methods for evaluating the persistence of compounds [51].

5.4 Summary

In this chapter, I have explored the significance of surface contacts and adhesion in the cosmetics industry, with a special focus on hair and skin. This review provides a concise summary of findings from my several years of research on surface forces, ranging from fundamental investigations conducted with polymer films to applied research in the fields of cosmetics, dermatology, and tactile perception.

In the first section dedicated to hair care, the structure of hair was discussed, with particular emphasis placed on the role of the cuticle in defining hair surface characteristics and the influence of 18-MEA, an essential lipid. The section described research that was mainly conducted using surface analysis techniques, such as AFM, to quantify the forces and friction that are present between hair fibers and explain how surface forces influence the manageability of hair on the scalp. Shampoo components and the methods that I have employed to quantitatively evaluate the deposition of coacervates from bulk solutions onto the hair surface were also discussed. Particular attention was paid to the transition toward bio-based and biodegradable alternatives, as well as the use of molecular dynamic simulations to help better understand the process of polymer adsorption and predict cosmetic performance.

The next section focusing on skin care provided an overview of the structure of the surface of human skin, with a particular emphasis on the SC and the role it plays in adhesion. The "brick and mortar" model of the skin barrier was described, as well as the role that corneodesmosomes play in skin barrier function. Key findings from the research that was conducted using microcantilevers to study corneocyte adhesion were presented, highlighting the strength of these interactions, as well as their heterogeneity.

In the Tribology and Tactile Perception section I examined the significance of sensorial experiences in the cosmetics industry, as well as the role that tribological properties, such as friction, play in the relationship between these properties and tactile perception. I also explored the importance and function of surface patterning in influencing touch perception and provided details about research that has been conducted using textured surfaces to investigate age-related loss of tactile acuity and how, by creating different polymer surface patterns, we can modify the tactile perception of skin.

I also emphasized the importance of sustainability with a particular focus on the necessity of using sustainable ingredients in cosmetics and the difficulties associated with substituting petrochemical-based polymers for natural alternatives. The importance of biodegradability and limiting the environmental impact of cosmetic products is also discussed, as well as the role that surface science plays in enhancing the performance of formulations that are built based on environmentally-friendly principles.

In summary, this chapter underscores the important role that surface science and adhesion research play in improving our understanding of the way in which cosmetic products interact with hair and skin, and highlights current research that is being conducted to develop cosmetic formulations that are both environmentally-friendly and highly effective. Adhesion and tribology are essential properties that need to be taken into account during the sustainable transformation of the cosmetics industry. Finally, I would like to conclude with my sincere recognition of the tremendous contribution that Kash Mittal's work has made to the field of adhesion research over the years.

5.5 Acknowledgments

I would like to thank my collaborators at L'OREAL for the many years of support on hair and skin research, as well as all the external collaborators who have worked with me. I would also like to thank Drs Emma Pilling and Marielle Romet (Santé Active Edition - Synergy Pharm) for language-editing services.

AI disclosure statement: The author confirms that no artificial intelligence (AI)-generated content or AI-assisted writing tools were used in the preparation of this chapter (including summaries and the abstract). All text, analysis, and conclusions are the result of the author's original work.

References

1. W. Gutowski and H. Dodiuk (Eds.), *Recent Advances in Adhesion Science and Technology in honor of Dr. Kash Mittal*, CRC Press, Boca Raton, FL, USA (2014).
2. G.S. Luengo and A.J. Greaves, Advances in the chemical structure of the hair surface, surface forces and interactions, in: *Surface Science and Adhesion in Cosmetics*, K.L. Mittal and H.S. Bui (Eds.) pp. 185–214, Wiley-Scrivener, Beverly, MA, USA (2021).
3. G.S. Luengo, E. Guzman, L. Fernández-Peña, F. Leonforte, F. Ortega and R.G. Rubio, Interaction of polyelectrolytes and surfactants on hair surfaces. Deposits and their characterization, in: *Surface Science and Adhesion in Cosmetics*, K.L. Mittal and H.S. Bui (Eds.) pp. 401–449, Wiley-Scrivener, Beverly, MA, USA (2021).
4. J. Aracil, G. Luengo, B.S. Almeida, M.M. Telo da Gama, R.G. Rubio and M. Diaz Pena, Surface properties of mixtures of molecular fluids: An experimental and theoretical study of carbon disulfide + dichloromethane and carbon disulfide + carbon tetrachloride, *Journal of Physical Chemistry*, **93**, 3210 - 3218 (1989).
5. M.A. Monedero, G.S. Luengo, S. Moreno, F. Ortega, R.G. Rubio, M.G. Prolongo and R.M. Masegosa, Calorimetric and dielectric study of a blend containing a conductive polymer: Poly(3-octylthiophene)+poly(ethylene-co-vinylacetate), *Polymer*, **40**, 5833–5842 (1999).
6. J. Israelachvili, S. Giasson, T. Kuhl, C. Drummond, A. Berman, G. Luengo, J.M. Pan, M. Heuberger, W. Ducker and N. Alcantar, Some fundamental differences in the adhesion and friction of rough versus smooth surfaces, in: *Thinning Films and Tribological Interfaces, Proceedings of the 26th Leeds-Lyon*

Symposium on Tribology, Volume 38, D. Dowson, M. Priest, *et al.* (Eds.) pp. 3–12, Elsevier (2000).

7. M. Heuberger, G. Luengo and J.N. Israelachvili, Tribology of shearing polymer surfaces. 1. Mica sliding on polymer (PnBMA), *Journal of Physical Chemistry B*, **103**, 10127–10135 (1999).
8. C. Bouillon and J. Wilkinson (Eds.), *The Science of Hair Care*, CRC Press, Boca Raton, FL, USA (2005).
9. J.A. Swift, Human hair cuticle: Biologically conspired to the owner's advantage, *J. Cosmet. Sci.*, **50**, 23–47 (1999).
10. T. Bornschlogl, L. Bildstein, S. Thibaut, R. Santoprete, F. Fiat, G.S. Luengo, J. Doucet, B.A. Bernard and N. Baghdadli, Keratin network modifications lead to the mechanical stiffening of the hair follicle fiber, *Proc Natl Acad Sci U S A*, **113**, 5940–5945 (2016).
11. J.A. Swift and J.R. Smith, Atomic force microscopy of human hair, *Scanning*, **22**, 310–318 (2000).
12. B. Bhushan, Nanoscale characterization of human hair and hair conditioners, *Prog. Mater. Sci.*, **53**, 585–710 (2008).
13. R.Y. Lochhead, *Shampoo and Conditioner Science*, pp. 75–116, Alluredbooks, Carol Stream, IL, USA (2012).
14. H. Bryant, F. Dixon, A. Ellington and C. Porter, Hair straightening, in: *Cosmetic Dermatology: Products and Procedures*, Z.D. Draelos (Ed.) pp. 262–269, John Wiley & Sons, Chichester, UK (2015).
15. J. Hicks, D.W. Metry, J. Barrish and M. Levy, Uncombable hair (cheveux incoiffables, pili trianguli et canaliculi) syndrome: Brief review and role of scanning electron microscopy in diagnosis, *Ultrastructural Pathology*, **25**, 99–103 (2001).
16. G.S. Luengo, H. Mizuno and M.W. Rutland, Hair-hair contact dynamics and interactions studied with atomic force microscopy, in: *Measuring the Skin*, P. Humbert, F. Fanian, *et al.* (Eds.) pp. 1–11, Springer International Publishing, Cham, Switzerland (2015).
17. H. Mizuno, G.S. Luengo and M.W. Rutland, Interactions between crossed hair fibers at the nanoscale, *Langmuir*, **26**, 18909–18915 (2010).
18. H. Mizuno, G.S. Luengo and M.W. Rutland, New insight on the friction of natural fibers. Effect of sliding angle and anisotropic surface topography, *Langmuir*, **29**, 5857–5862 (2013).
19. Y. Akiyama, Y. Matsue, A. Kasagawa, T. Mori and S. Nishijima, Adsorption behavior of coacervates on the human hair surface, *Chemistry Letters*, **41**, 1351–1353 (2012).
20. G.S. Luengo, F. Leonforte and N. Baghdadli, From polymers at interfaces to shampoos in cosmetics, *L'actualité chimique*, **424**, 18–19 (2017).
21. V. Andre, R. Norenberg, P. Hossel and A. Pfau, The role of polymer-surfactant interactions in the adsorption process of hair-conditioning polymers, *Macromolecular Symposia*, **145**, 169–179 (1999).

22. R.Y. Lochhead, Shampoo and conditioner science, in: Practical Modern Hair Science. T. Evans, and R.R. Wickett (Eds.) pp. 75–116, Alluredbooks, Carol Stream, IL, USA (2012).
23. G.S. Luengo, F. Leonforte, A. Greaves, R.G. Rubio and E. Guzman, Physicochemical challenges on the self-assembly of natural and bio-based ingredients on hair surfaces: Towards sustainable haircare formulations, *Green Chemistry*, **25**, 7863–7882 (2023).
24. L. Fernandez-Pena, E. Guzman, F. Leonforte, A. Serrano-Pueyo, K. Regulski, L. Tournier-Couturier, F. Ortega, R.G. Rubio and G.S. Luengo, Effect of molecular structure of eco-friendly glycolipid biosurfactants on the adsorption of hair-care conditioning polymers, *Colloids Surfaces B. Biointerfaces*, **185**, 110578 (2020).
25. I. Adroher-Benítez, T.I. Morozova, G. Catalini, N.A. García, J.-L. Barrat, G.S. Luengo and F. Léonforte, Effect of polymer architecture on the adsorption and coating stability on heterogeneous biomimetic surfaces, *Macromolecules*, **56**, 10285–10295 (2023).
26. T.I. Morozova, N.A. Garcia, J.L. Barrat, G.S. Luengo and F. Leonforte, Adsorption and desorption of polymers on bioinspired chemically structured substrates, *ACS Appl Mater Interfaces*, **13**, 30086–30097 (2021).
27. P.G. Agache, F. Fanian, P. Humbert and H.I. Maibach (Eds.), *Agache's Measuring the Skin*, Springer, Berlin (2020).
28. K. Levi, A. Kwan, A.S. Rhines, M. Gorcea, D.J. Moore and R.H. Dauskardt, Emollient molecule effects on the drying stresses in human stratum corneum, *British Journal of Dermatology*, **163**, 695–703 (2010).
29. H. Joodaki and M.B. Panzer, Skin mechanical properties and modeling: A review, *Proceedings of the Institution of Mechanical Engineers, Part H: Journal of Engineering in Medicine*, **232**, 323–343 (2018).
30. J.L. Leveque and P. Corcuff, The surface of the skin — the microrelief, in: *Noninvasive Methods for the Quantification of Skin Functions*, P.J. Frosch and A.M. Kligman (Eds.) pp. 3–24, Springer-Verlag Berlin, Heidelberg, Germany, (1993).
31. R.K. Sivamani and H.I. Maibach, Tribology of skin, *Proceedings of the Institution of Mechanical Engineers, Part J: Journal of Engineering Tribology*, **220**, 729–737 (2006).
32. R. Santoprete and B. Querleux, Cellular-scale mechanical model of the human stratum corneum, in: *Computational Biophysics of the Skin*, B. Querleux (Ed.) pp. 161–184, Pan Stanford Publishing, New York (2015).
33. M. Donovan, R. Duke, L. Simonetti, N. Cavusoglu, P.M. Rudd and D. Bernard, N-glycans are stratum corneum biomarkers of aging skin, *J Invest Dermatol*, **143**, 492–494 e410 (2023).
34. C. Rankl, R. Zhu, G.S. Luengo, M. Donovan, N. Baghdadli and P. Hinterdorfer, Detection of corneodesmosin on the surface of stratum corneum using atomic force microscopy, *Expl. Dermatol.*, **19**, 1014–1019 (2010).

35. J. Danzberger, M. Donovan, C. Rankl, R. Zhu, S. Vicic, C. Baltenneck, R. Enea, P. Hinterdorfer and G.S. Luengo, Glycan distribution and density in native skin's stratum corneum, *Skin Res. Technol.*, **24**, 450–458 (2018).
36. S. Guo, Y. Domanov, M. Donovan, B. Ducos, Y. Pomeau, C. Gourier, E. Perez and G.S. Luengo, Anisotropic cellular forces support mechanical integrity of the stratum corneum barrier, *J. Mech. Behav. Biomed. Mater.*, **92**, 11–23 (2019).
37. X. Zeng and E. van der Heide, Perception of skin friction, *Household and Personal Care Today*, **10**, 14–20 (2015).
38. A. Giboreau, De l'analyse sensorielle au jugement perceptif : L'exemple du toucher, (Habilitation Thesis) University Claude Bernard - Lyon 1, Lyon, France (2009).
39. G.S. Luengo, A. Galliano and C. Dubief, Aqueous lubrication in cosmetics, in: *Aqueous Lubrication*, N.D. Spencer (Ed.) pp. 103–144, World Scientific Publishing, Singapore, (2014).
40. S. Ding and B. Bhushan, Tactile perception of skin and skin cream by friction induced vibrations, *J Colloid Interface Sci*, **481**, 131–143 (2016).
41. J. Scheibert, S. Leurent, A. Prevost and G. Debregeas, The role of fingerprints in the coding of tactile information probed with a biomimetic sensor, *Science*, **323**, 1503–1506 (2009).
42. K.O. Johnson, The roles and functions of cutaneous mechanoreceptors, *Curr. Opin. Neurobiol.*, **11**, 455–461 (2001).
43. S. Cozzolino, P. Gutfreund, A. Vorobiev, R.J.L. Welbourn, A. Greaves, F. Zuttion, M.W. Rutland and G.S. Luengo, Adsorption hierarchy of surfactants and polymers to a damaged hair model: Effect of composition, order and polymer size, *Phys Chem Chem Phys*, **27**, 1089–1099 (2025).
44. L. Skedung, C. El Rawadi, M. Arvidsson, C. Farcet, G.S. Luengo, L. Breton and M.W. Rutland, Mechanisms of tactile sensory deterioration amongst the elderly, *Sci. Rep.*, **8**, 5303 (2018).
45. L. Samain-Aupic, M. Dione, E. Ribot-Ciscar, R. Ackerley and J.M. Aimonetti, Relations between tactile sensitivity of the finger, arm, and cheek skin over the lifespan showing decline only on the finger, *Frontiers in Aging Neuroscience*, **16**, 1387136 (2024).
46. M. Dione, R.H. Watkins, J.M. Aimonetti, R. Jourdain and R. Ackerley, Effects of skin moisturization on various aspects of touch showing differences with age and skin site, *Sci. Rep.*, **13**, 17977 (2023).
47. L. Samain-Aupic, L. Gilbert, N. Andre, R. Ackerley, E. Ribot-Ciscar and J.M. Aimonetti, Applying cosmetic oil with added aromatic compounds improves tactile sensitivity and skin properties, *Sci. Rep.*, **13**, 10550 (2023).
48. R. Jin, I.C. Liao, C. Cazeneuve, J.C. Chang, M. Ruths and G.S. Luengo, Effects of imprinted 3d surface patterning on localized changes in the tribology of human stratum corneum, *Langmuir*, **35**, 15573–15584 (2019).
49. R. Jin, L. Skedung, C. Cazeneuve, J.C. Chang, M.W. Rutland, M. Ruths and G.S. Luengo, Bioinspired self-assembled 3d patterned polymer textures as

skin coatings models: Tribology and tactile behavior, *Biotribology*, **24**, 100151 (2020).

50. M. Philippe, J. L'Haridon, J. Portal, S. Chodorowski and G.S. Luengo, Eco-friendly polymers for cosmetic formulations, *L'actualité chimique*, 456-457-**458**, 101–107 (2020).
51. J. L'Haridon, L. Patouillard, J. Pedneault, A.M. Boulay, F. Witte, M. Vargas-Gonzalez, P. Bonningue, I. Rollat, T. Blanchard, G. Goncalves, A. Hervio and L. Gilbert, Spot: A strategic life-cycle-assessment-based methodology and tool for cosmetic product eco-design, *Sustainability (Switzerland)*, **15**, (19) 14321 (2023).
52. J. L'Haridon, P. Martz, J.C. Cheneble, J.F. Campion and L. Colombe, Ecodesign of cosmetic formulae: Methodology and application, *Int. J. Cosmet. Sci.*, **40**, 165–177 (2018).

6

Role of Surface Topography in Engineering Interfacial Adhesion: Fundamentals, Advancements and Prospects

Anubha Jaiswal[1] and Ravi P. Jaiswal[2]*

[1]*Department of Physics, G. B. College Ramgarh, Veer Kunwar Singh University, Ara, Bihar, India*
[2]*Department of Chemical Engineering & Technology, Indian Institute of Technology (BHU), Varanasi, Uttar Pradesh, India*

Abstract
Surface topography plays a pivotal role in regulating interfacial adhesion across a wide range of industrial and practical applications. This review presents a comprehensive overview of the fundamental mechanisms by which surface textures—ranging from the micro- to nanoscale—affect adhesion at both solid–solid and solid–liquid interfaces, along with methods for their characterization. A variety of surface modification techniques, including grit blasting, focused ion beam milling, laser surface texturing, nanoimprint lithography, chemical and plasma etching, and chemical mechanical polishing, are discussed in terms of their ability to engineer precise topographical features for targeted functionalities. The article also reviews recent scientific advancements in surface topography engineering aimed at altering adhesion in specific contexts, such as bioinspired and biomimetic surfaces, anti-fouling and antibacterial coatings, and superhydrophilic and superhydrophobic surfaces. Finally, key challenges are identified, and future research directions are proposed to guide continued progress in this evolving field.

Keywords: Adhesion, surface topography, interfacial forces, surface free energy, topographical modification methods, applications of topographical engineering

**Corresponding author*: ravi.che@iitbhu.ac.in

Anil N. Netravali and Frank M. Etzler (eds.) The Wide World of Adhesion: A Festschrift in Honor of Dr. Kash Mittal on his 80th Birthday, (195–230) © 2026 Scrivener Publishing LLC

6.1 Introduction

Surface topography—defined by micro- to nanometer-scale textures—has long been recognized as a critical factor influencing the mechanical and physical properties of interfacial contact [1, 2]. A wide range of interfacial phenomena, including adhesion, friction, heat transfer, corrosion resistance, aerodynamic behavior, and biocompatibility, are significantly affected by both the scale and structural features of surface textures. In recent decades, extensive research has focused on leveraging engineered surface topographies to enhance functional performance across various applications [3, 4]. For instance, microstructured surfaces inspired by gecko feet have been employed in wall-climbing robots and anti-slip materials to achieve strong, reversible adhesion [5]. Similarly, tailored surface textures have been designed to promote nucleation sites for bubble formation in boilers and advanced cooling systems, as well as to enhance convective heat transfer in heat exchangers [6]. In biomedical contexts, ultra-smooth implant surfaces help minimize the initiation of corrosion [7], while increased surface roughness or micro-grooved patterns are used to improve coating adhesion on pipelines and other structural substrates. Similar surface topographies have also been applied in the automotive and aerospace industries, including wind turbine blades and high-performance swimwear, to reduce aerodynamic drag by promoting the formation of turbulent boundary layers [8, 9].

Surface topography plays a vital role in governing adhesion at both solid–solid and solid–liquid interfaces. At solid–solid interfaces, the introduction of nanoscale roughness can reduce the real contact area, thereby lowering adhesion forces [10, 11]. This principle is utilized in microelectromechanical systems (MEMS), where it helps prevent stiction and subsequent device failure [12]. Similarly, nano-patterned surfaces are employed in medical devices —such as catheters and surgical instruments—to inhibit bacterial adhesion and colonization by physically disrupting microbial membranes [13]. Conversely, micro-scale asperities can enhance adhesion through mechanical interlocking. This mechanism is exploited in composite-to-metal bonding to improve adhesion between polymer resins and metal substrates, especially in aerospace and automotive engineering—for instance, the bonding of carbon fiber composites with aluminum in aircraft fuselages [14]. At solid–liquid interfaces, surface roughness significantly influences wettability by amplifying a surface's intrinsic hydrophilic or hydrophobic behavior, as described by the Wenzel and Cassie–Baxter models [15]. This phenomenon is harnessed to create superhydrophilic and superhydrophobic surfaces, which are widely used in anti-fog coatings, satellite optics, building facades, solar panels, and other self-cleaning or dust-repellent technologies. Additionally, surface topography-controlled wetting and liquid spreading are utilized in applications such as printed electronics, textile printing, and point-of-care medical diagnostics [16].

Surface topography plays a multifaceted role in governing interfacial adhesion by modulating key factors such as mechanical interlocking, real contact area, capillary forces, and intermolecular interactions—including van der Waals forces and chemical bonding. As such, a comprehensive understanding of how surface texture influences these mechanisms is crucial for the design and optimization of both established and emerging technologies across a wide range of fields.

6.2 Basics of Surface Topography

Surface topography is generally characterized by three primary components: form, waviness, and roughness, as illustrated in Figure 6.1 [17]. *Form* refers to the overall shape or macro-contour of an object. *Waviness* denotes medium-scale surface deviations, typically resulting from factors such as vibrations, tool deflection, or imbalance during machining. *Roughness* describes the finely spaced surface irregularities—such as peaks and valleys—that occur at the smallest scale. The nature of surface roughness is primarily determined by the manufacturing process employed. The surface roughness can span a wide range of scales: from the micronscale

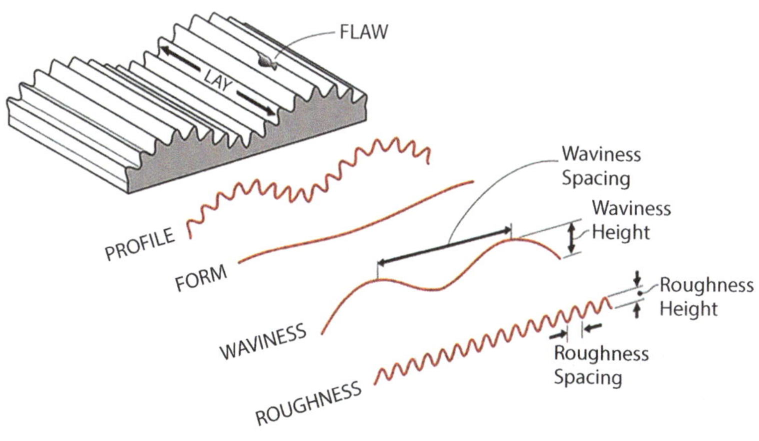

Figure 6.1 Different elements of a surface topography [17].

(1–100 μm), which is advantageous for enhancing mechanical interlocking, to the nanometer scale (typically <1 μm), which is effective in increasing surface free energy and influencing interfacial phenomena. In some cases, hierarchical structures—which combine both micro- and nanoscale features—are deliberately engineered to synergistically leverage the benefits of multiple length scales.

Surface roughness is commonly quantified using a range of standardized parameters, with the most widely used being Arithmetic Average Roughness (Ra), Root Mean Square Roughness (Rq), and *Ten-Point Mean Roughness* (*Rz*). The *Ra* is defined as the average of the absolute values of the surface height deviations from the mean line over a given length, *L*

$$Ra = \frac{1}{L}\int_0^L |y(x)|\, dx \tag{6.1}$$

The Rq is defined as the square root of the average of the squared deviations from the mean line within a sampling length.

$$Rq = \sqrt{\frac{1}{L}\int_0^L y^2(x)dx} \tag{6.2}$$

The *Rz* parameter is defined as the average vertical distance between the five highest peaks and the five deepest valleys within a a specified sampling length. While no single roughness parameter provides a complete description of a surface's texture, each one offers distinct insights into specific aspects of surface roughness.

6.2.1 Fabrication Methods for Topographically Engineered Surfaces

The methods to alter surface topography of solid substrates can be broadly classified into physical and chemical methods.

6.2.1.1 Physical Methods

6.2.1.1.1 Grit Blasting Method

Grit or sand blasting is a widely used mechanical surface treatment technique that involves the high-velocity impact of abrasive particles onto a substrate to clean, texture, or roughen its surface [18]. It is typically employed as a pre-treatment step prior to coating, painting, or bonding applications to enhance interfacial adhesion. Owing to its versatility, grit blasting is applicable to a wide range of materials, including metals, ceramics, polymers, and composites.

A variety of abrasive media can be used depending on the target material and desired surface finish, including glass beads, silicon carbide, alumina, steel grit, and plastic particles (for treating softer substrates). Key process parameters influencing the outcome include the type and size of abrasive particles, blasting pressure (supplied by a compressed air system), nozzle distance and angle, substrate hardness, and blasting duration. By adjusting these parameters, a broad range of surface roughness values (Ra ~0.5 to 10 μm) can be achieved. While grit blasting is effective in generating micro-scale surface roughness, it is not suitable for applications requiring nanoscale surface modification. Despite being a relatively simple, cost-effective, and scalable process, it can produce significant dust and may lead to surface damage—particularly on softer or more delicate materials.

6.2.1.1.2 Focused Ion Beam Milling

Focused Ion Beam (FIB) systems utilize a finely focused beam of ions—typically gallium (Ga^+)—which is accelerated through an electric field and directed onto the target surface. These high-energy ions remove material *via* sputtering, wherein momentum transfer causes the ejection of surface atoms, allowing for nanometer-scale precision in material modification. FIB technology can achieve milling resolutions below 10 nm, making it particularly well-suited for localized surface texturing at both the micro- and nanoscales. The depth and lateral dimensions of the milled features can be precisely controlled by adjusting parameters such as beam current and energy, ion dose, dwell time, and incidence angle. Over time, FIB systems have evolved into versatile tools for three-dimensional nanofabrication and advanced surface engineering [19]. FIB-induced surface topography typically results from a combination of direct sputtering and ion beam instabilities. While FIB offers exceptional spatial resolution and is compatible with a broad range of materials—including metals, semiconductors, and polymers—it is not well-suited for large-area processing due to its relatively low throughput.

6.2.1.1.3 Laser Surface Texturing

Laser surface texturing (LST) is a non-contact, highly precise, and scalable technique for the controlled modification of surface topography [20, 21]. It utilizes focused laser pulses to induce localized ablation or melting followed by re-solidification of the substrate, enabling the creation of micro- and nanoscale surface features with high spatial resolution. The photothermal interaction between the laser and the target material is influenced by several key parameters, including laser wavelength and pulse duration (commonly in the nanosecond, picosecond, or femtosecond range), the thermal and optical properties of the substrate, and the surrounding environment (e.g., air, vacuum, or inert gas). By systematically tuning these parameters, a wide variety of surface morphologies—such as grooves, ridges, dimples, and hierarchical structures—can be engineered. LST is valued for its speed, non-contact nature, and minimal contamination risk, and is applicable to a broad range of materials. However, it may induce thermal damage or surface oxidation in thermally sensitive substrates. Additionally, LST is generally limited in its ability to fabricate high-aspect-ratio structures, which may restrict its use in certain advanced applications.

6.2.1.1.4 Nanoimprint Lithography

Nanoimprint Lithography (NIL) represents a significant advancement in nanoscale patterning, offering a precise, repeatable, and cost-effective method for fabricating features below 100 nanometers [22, 23]. As traditional photolithography approaches its resolution and economic limitations at these scales, NIL has emerged as a compelling alternative due to its simplicity, high throughput, low cost, and exceptional resolution capability. The NIL process is fundamentally mechanical in nature. It involves pressing a pre-patterned mold into a deformable resist layer to physically replicate nanoscale features. These molds—typically fabricated from hard materials such as silicon or quartz—are created using high-resolution patterning techniques like electron beam lithography or focused ion beam (FIB) milling. A polymer-based resist, either a thermoplastic (e.g., PMMA) or a UV-curable resin (e.g., acrylate) is spin-coated onto the target substrate.

In thermal NIL, the mold is imprinted into the resist under elevated temperature and pressure conditions. In contrast, UV-NIL operates at room temperature, where ultraviolet light is used to cure the resist during imprinting. Following the imprinting process, the mold is carefully demolded, leaving behind a high-fidelity replica of the pattern in the resist layer. This pattern can then be transferred into the underlying substrate using anisotropic etching techniques, such as reactive ion etching (RIE).

By optimizing key parameters—such as mold geometry, resist formulation, imprinting temperature and pressure, and exposure time, NIL enables precise engineering of surface topography across micro- and nanoscales.

Despite its many advantages, one of the main challenges associated with NIL is the fabrication of high-resolution master molds, a process that remains both capital- and time-intensive.

6.2.1.1.5 Plasma Treatment

Plasma is an ionized gas comprising electrons, ions, neutral species, radicals, and photons. When interacting with solid surfaces, plasma facilitates both energy and mass transfer, leading to modifications in surface structure and chemistry. The primary mechanisms responsible for altering surface topography include physical sputtering and ion bombardment—typically induced by inert plasmas such as argon—which cause localized erosion and morphological changes, as well as chemical etching, driven by reactive plasmas [24, 25]. The extent and nature of surface modification are governed by several key parameters, including ion energy density, plasma composition (e.g., O_2, CF_4, Ar), treatment duration, and the inherent properties of the substrate material. For instance, oxygen plasma is effective in roughening polymeric and organic surfaces (e.g., Poly(methyl methacrylate), polyimide), often resulting in nanoporous or fibrous textures. Plasmas based on air, nitrogen, or hydrogen are frequently employed to enhance surface hydrophilicity through topographical and chemical modifications. In contrast, fluorocarbon-based plasmas (e.g., Ar/CF_4, SF_6, or CF_4/SF_6 mixtures) can generate hierarchical surface structures on materials such as silicon, glass, and semiconductors. Notably, plasma treatment can also lead to surface smoothing under specific conditions—either through ion-induced reflow of surface features or through thin-film deposition mechanisms. Despite the versatility and fine controllability of plasma-based techniques for tailoring surface topography at both micro- and nanoscales, achieving consistent and reproducible roughness remains a significant challenge, particularly in large-area or industrial-scale applications.

6.2.1.2 Chemical Methods

6.2.1.2.1 Etching

Chemical etching is a well-established surface modification technique that enables precise control over surface topography through the selective removal of material *via* controlled chemical reactions. This method can be applied to tailor surfaces at both micro- and nanoscales for a wide range of functional applications [26, 27]. Based on the physical state of the etchant, chemical etching is generally categorized as either wet etching (liquid-phase) or dry etching (vapor-phase). Isotropic etching, which removes

Table 6.1 Commonly used etchants for various materials.

Metals	**Etchants**
Aluminum (Al)	Phosphoric acid + Acetic acid + Nitric acid (PAN solution)
Copper (Cu)	Ferric chloride ($FeCl_3$), Ammonium persulfate $(NH_4)_2S_2O_8$, Nitric acid
Chromium (Cr)	Ceric ammonium nitrate in acetic acid
Nickel (Ni)	Nitric acid, Nitric acid–Acetic acid mixture
Gold (Au)	Aqua regia (3:1 $HCl:HNO_3$)
Stainless Steel	$HCl + HNO_3$ (aqua regia)
Titanium (Ti)	Hydrofluoric acid (HF), Nitric acid
Semiconductors	
Silicon	Potassium hydroxide (Anisotropic etching), Tetramethylammonium hydroxide (TMAH), $HF + HNO_3$ (Isotropic etching)
Silicon Dioxide (SiO_2)	Dilute HF or buffered HF (BHF)
Silicon Nitride (Si_3N_4)	Hot phosphoric acid (H_3PO_4)
Gallium Arsenide (GaAs)	$H_2SO_4:H_2O_2:H_2O$, or $NH_4OH:H_2O_2:H_2O$
Germanium (Ge)	H_2O_2 + Acetic acid, Nitric acid
Indium Phosphide (InP)	$HCl:H_2O_2:H_2O$
Ceramics and Oxides	
Alumina (Al_2O_3)	Hot phosphoric acid, NaOH
Glass	HF or buffered HF (BHF)
Indium Tin Oxide (ITO)	$HCl + HNO_3$
Titania (TiO_2)	HF or Oxalic acid
Zirconia (ZrO_2)	$HCl + H_2SO_4$, HF
Polymers	
Polystyrene (PS)	Chromic acid (CrO_3 in H_2SO_4), Ozone
Polydimethylsiloxane (PDMS)	Piranha solution ($H_2SO_4 + H_2O_2$), O_2 plasma
Polyethylene (PE)	Chromic acid, Nitric acid
Polyimide (PI)	KOH, Oxygen plasma
Poly(methyl methacrylate) (PMMA)	Acetone, O_2 plasma

material uniformly in all directions, is commonly employed for general surface roughening. In contrast, anisotropic etching, where the etch rate depends on crystallographic orientation, is used to fabricate well-defined, directionally selective nanostructures. For example, hydrofluoric acid (HF) is widely used to produce nanostructures on a glass surface. When followed by fluorosilane treatment, this process yields hierarchical topographies that exhibit superhydrophobic behavior, with water contact angles exceeding 150°. A mixture of HF and nitric acid (HNO_3) is typically used for isotropic etching of silicon wafers, offering a more uniform surface modification. Conversely, potassium hydroxide (KOH) selectively etches (100)-oriented silicon wafers in an anisotropic manner, producing pyramidal surface features due to the differing etch rates between the (100) and (111) crystal planes. These anisotropic microstructures are particularly advantageous for improving light trapping and absorption in photovoltaic devices.

Reactive ion etching (RIE) is a highly versatile technique extensively employed for the precise modification of surface topography at both micro- and nanoscales [28]. By combining the effects of chemical reactivity with physical ion bombardment, RIE enables highly anisotropic etching with excellent control over feature dimensions and surface characteristics. For instance, oxygen plasma RIE is commonly used to roughen polymeric materials such as poly{methyl methacrylate}(PMMA) and polyimide (PI) through mechanisms involving polymer chain scission and oxidative degradation [24]. Likewise, fluorine-based gases—such as sulfur hexafluoride (SF_6) and carbon tetrafluoride (CF_4)—are employed to generate nanostructured surface textures on substrates like silicon and silicon dioxide. In optical applications, fluorocarbon plasmas (e.g., CF_4/O_2 mixtures) are used to texture glass surfaces, thereby improving their anti-reflective property. Meanwhile, chlorine-based chemistries (such as Cl_2 and BCl_3) are particularly effective for etching conductive materials, enabling precise topographical control in microelectronic and MEMS (Micro-Electro-Mechanical Systems) device fabrication.

A summary of commonly used etchants for various materials is provided in Table 6.1, highlighting the diversity of chemical systems available for surface engineering through etching.

6.2.1.2.2 Chemical Mechanical Polishing

Chemical Mechanical Polishing (CMP) and lapping are widely adopted surface finishing techniques that enable precise control over surface topography by integrating mechanical abrasion with chemical activity [29].

Originally developed for planarization in semiconductor manufacturing, these techniques have evolved into critical tools for tailoring surface roughness at both micro- and nanoscales. The CMP is highly effective for achieving atomic-scale smoothness through uniform material removal. The process involves a rotating polishing pad, a chemically active slurry containing nanometer-scale abrasive particles (such as silica, alumina, or ceria), and the application of mechanical pressure. Material removal is facilitated by a synergistic interaction between the abrasive action of the slurry particles and chemical reactions between the slurry constituents and the substrate surface.

In contrast to CMP, lapping is a predominantly mechanical process designed for bulk material removal and surface roughness control, particularly on hard substrates such as metals, glass, and ceramics. During lapping, the workpiece is pressed against a flat lapping plate in the presence of a hard abrasive slurry, typically composed of materials such as diamond, silicon carbide, or alumina. This technique is commonly used to establish a consistent baseline surface topography or to achieve flatness and parallelism across larger components. Both CMP and lapping are highly versatile and scalable, making them essential for tailoring surface characteristics to meet the specific requirements of diverse applications across a broad spectrum of materials and industries.

6.3 Fundamentals of Adhesion Involving Solid Surfaces

Every solid surface has an associated free energy which has its origin in the creation of the surface. A molecule located in the bulk of a material interacts with the neighbouring molecules lying in an imaginary sphere around it, however a molecule situated at the surface interacts with neighbouring molecules lying in an imaginary hemisphere. When a bulk molecule is separated from its neighbouring molecule and brought to the surface, more number of physical bonds are broken than the newly formed bonds. Therefore, the additional work done in this process is stored in the surface molecules, giving rise to the surface free energy. At macroscale, if an interface of two solid surfaces (1 and 2), with contact area A, is separated in a medium (3) to create two individual surfaces as shown in Figure 6.2, the work done, aka work of adhesion per unit contact area, (W_{123}^{Ad}) is obtained using Dupre̍ equation as

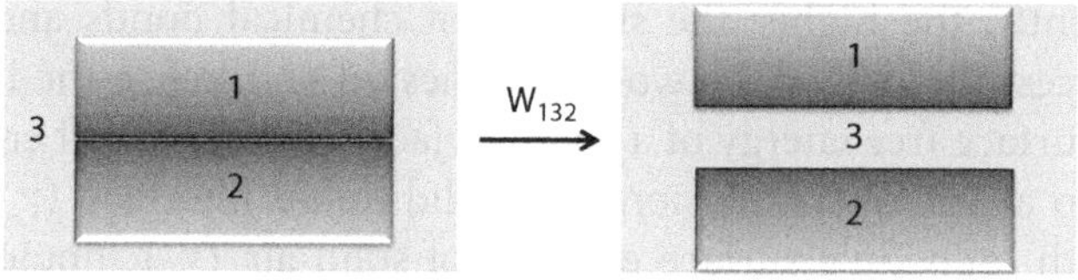

Figure 6.2 A schematic exhibiting the creation of the solid-liquid interfaces by separating two pieces (1 & 2) of solid in a medium (3).

$$W_{132}{}^{Ad} = \gamma_{13} + \gamma_{23} - \gamma_{12} \tag{6.3}$$

Where γ_1, γ_2 and γ_{12} are surface or interfacial energy per unit area for the individual surfaces and the interface, respectively. If both solids are made of the same material, the work of adhesion, W_{131}, is obtained as

$$W_{131}{}^{Ad} = 2.\gamma_{13} \tag{6.4}$$

The work of adhesion at solid–solid and solid–liquid interfaces corresponds to the energy required to overcome intermolecular forces, including chemical bonds (CB) and physical interactions such as hydrogen bonding (Hyd), van der Waals forces (vdW), Coulombic electrostatic forces (ES), and capillary forces (Cap) [30]. Among these, chemical and hydrogen bonds are the strongest and operate over the short ranges (below 0.4 nm). The van der Waals is the weakest and is also short ranged (1-10 nm) in nature. However, the capillary and electrostatic forces act over the medium and long ranges, respectively. In general, the surface free energy of a solid (1) in a given medium (3) is described as:

$$\gamma_{13} = \tfrac{1}{2} W_{131}{}^{Ad} = \tfrac{1}{2}\left(W_{131}{}^{CB} + W_{131}{}^{Hyd} + W_{131}{}^{vdW} + W_{131}{}^{Cap} + W_{131}{}^{ES}\right) \tag{6.5}$$

Generally in a non-humid environment at solid-solid contact, the chemical bonds and van der Waals forces are the dominating forces, leading to the surface free energy of a solid approximated as:

$$\gamma_{13} = \tfrac{1}{2} W_{131}{}^{Ad} = \tfrac{1}{2}\left(W_{131}{}^{CB} + W_{131}{}^{vdW}\right) \tag{6.6}$$

Apparently, the higher the strength of chemical bonds and van der Waals forces, the higher the work of adhesion and hence the higher the value of surface free energy of a solid surface. The interfacial energy, also referred to as interfacial tension, of a solid-liquid interface (γ_{SL}) can be linked with individual surface energies of solid-air (γ_S), liquid-air (γ_L) interfaces through Young's Equation as:

$$\gamma_S = \gamma_L . Cos\, \theta + \gamma_{SL} \tag{6.7}$$

Table 6.2 summarizes the surface free energy (or surface tension in case of liquids) of several common materials used in various applications. It is evident that metals have very high surface free energy due to strong metallic bonds, followed by ionic compounds. Polymers show very low surface free energy due to relatively weak covalent bonds. Their surface free energy

Table 6.2 Surface free energy (or surface tension for liquids) of common materials.

Material	Surface free energy (or surface tension for liquids) (mJ/m²)
Metals	**> 500** [31, 32]
Iron	2030
Copper	1100
Aluminium	840
Zinc	753
Tin	526
Ionic Compounds	**100-500**
Calcium fluoride	450
Sodium chloride	400
Lithium fluoride	340
Barium fluoride	280
Calcium carbonate	230

(Continued)

Table 6.2 Surface free energy (or surface tension for liquids) of common materials. (*Continued*)

Material	**Surface free energy (or surface tension for liquids) (mJ/m²)**
Potassium chloride	110
Glass	83.4
Liquids	**20-80**
Water	72.8
Glycerol	64
Formamide	58
Ethylene glycol	48
Heptane, Dodecane, Methanol, Ethanol, Octanol Acetone, Ethyl acetate, Benzene	20-30
Polymers	20-50 [31]
Polyethersulfone	50
Styrene Butadiene Rubber (SBR)	48
Polycarbonate	46
Polyamide, Polyurethane	43
Poly(ethylene terephthalate) (PET)	41
Poly(vinyl chloride) (PVC)	39
Polystyrene (PS)	38
Poly(vinyl alcohol) (PVA)	37
Epoxy, Poly(methyl methacrylate) (PMMA), Polyamide	36
Polyethylene (PE)	31
Polypropylene (PP)	29
Poly(vinyl fluoride) (PVF)	25
Silicone	24
Polytetrafluoroethylene (PTFE)	18

is even lower than that of water, therefore, they demonstrate hydrophobic behaviour due to negative spreading coefficient (defined by $\gamma_S - \gamma_L - \gamma_{SL}$).

6.3.1 Chemical Bonds

Chemical bonds are fundamental in determining the surface free energy of solid materials, directly influencing their interfacial behavior and adhesion characteristics. At the atomic level, the primary types of chemical bonding involved at a solid-solid interface include metallic, ionic, and covalent bonds [30]. Metallic bonding is particularly important in the formation of joints between metals or metallic particles, such as in soldering, sintering, and welding processes. This type of bonding is non-directional and arises from a delocalized electron cloud shared among metal atoms, contributing to cohesive strength across metallic interfaces. Ionic bonding occurs in systems where there is a substantial difference in electronegativity between atoms at the interface, such as at metal–electrolyte and electrochemical interfaces. These bonds are characterized by electrostatic attraction between oppositely charged ions and are generally less directional than covalent bonds. Covalent bonding involves the sharing of electrons between atoms across an interface and is highly directional in nature. It plays a critical role in systems where strong, localized interfacial adhesion is required. Examples include semiconductor interfaces (e.g., Si/SiO_2), crosslinked polymer networks, and composite materials incorporating functionalized fillers. Covalent bonds at interfaces are typically introduced through surface functionalization or plasma treatments. For instance, in epoxy–aluminum interfaces, surface hydroxyl groups on aluminum can chemically react with epoxide groups to form covalent bonds, thereby significantly enhancing adhesion strength.

6.3.2 van der Waals (vdW) Force

The vdW force is electrodynamic in nature as it originates from the electromagnetic interactions between the adhering bodies. It has three components: dipole-dipole interaction (orientation or Keesom interaction), dipole-induced dipole interaction (induction or Debye interaction) and induced dipole-induced dipole interaction (dispersion or London force) [30, 33]. Each of these components varies with the inverse sixth power of the separation distance between atoms. The net vdW interaction energy between two atoms or molecules can be given as

$$W_{atom-atom}{}^{vdW} = -\frac{C_{vdW}}{r^6} \quad (6.8)$$

Where C_{vdW} is the vdW constant. The third component (London dispersion) is the most dominating component over the other two, except for highly polar molecules. Since the dispersion component is always present irrespective of the nature of the interacting materials, it makes the vdW interaction ubiquitous. Notably, the London theory of dispersion does not account for the 'many body interactions', i.e. the interaction associated with the electromagnetic field that emanates from the atom of one body and reaches to the atoms of another body by 'reflection' from neighboring atoms of the medium. This limits London dispersion theory to be applicable to the gaseous medium only, and cannot be applied for the condensed medium, e.g. solvent medium, in strict sense.

When two atoms are at appreciable distance apart (with intermolecular separation (r) >50nm), the time required for electromagnetic wave, emanating from the induced dipoles, to travel between interacting bodies can become comparable with the time period of the dipole fluctuation. This makes the induced dipole-induced dipole interaction less favorable because of unfavorable alignments of dipoles with respect to the traveling electromagnetic field. This effect is known 'retardation effect'. Only the dispersion component of the vdW suffers through retardation, it is referred as retarded vdW force (or Casimir force). Casimir demonstrated that the retarded dispersion interaction between two molecules becomes proportional to $1/r^7$ than $1/r^6$ (decays even faster) at larger separation distance (r >100nm).

At macroscale, the determination of vdW force between two solid bodies becomes extremely challenging and computationally expensive. In 1956, Lifshitz proposed a general macroscopic approach to determine vdW forces between two semi-infinite bodies, considering the retardation and the non-pairwise additivity of vdW interactions [30, 33]. The Lifshitz theory was based on the quantum electrodynamics and required knowledge of dielectric constants and refractive indices of the interacting materials over a wide range of electromagnetic spectrum. However, such information is available only for limited number of materials.

In 1937, Hamaker had proposed a simple microscopic approach based on 'pairwise additivity' to determine the vdW forces between macroscopic bodies, without accounting for the retardation and many-body interactions [34, 35]. Hamaker's expressions for the vdW potential ($W_{1-3-1}{}^{vdW}$)

and vdW force ($F_{1-3-1}{}^{vdW}$) between two parallel plates (1) (of surface area A) separated by distance D in a medium (3) are given by

$$W_{1-3-1}{}^{vdW} = -\frac{A.H_{1-3-1}}{12\pi D^2} \; and \; F_{1-3-1}{}^{vdW} = \frac{A.H_{1-3-1}}{6\pi D^3} \tag{6.9}$$

Here, H_{1-3-1} is Hamaker constant which is a lumped constant given as $\pi^2.C^{vdW}\rho_1.\rho_1$, where C^{vdW} is the vdW constant as defined in Eq. 6.8, and ρ_1 is the number atomic density of the adhering plates.

Hamaker constant is a characteristic of material properties of the interacting bodies and the intervening media. The assumption made in Hamaker's methods restricts it to be applicable only in the gaseous or less condensed medium. However, Hamaker's approach is well accepted and widely used to study vdW adhesion between solid bodies because it is simple and incorporates only a single measurable force constant (i.e. Hamaker constant). The higher the magnitude of Hamaker constant, the stronger the vdW attraction.

$$H_{1-3-2} \approx (\sqrt{H_{1-0-1}} - \sqrt{H_{3-0-3}})(\sqrt{H_{2-0-2}} - \sqrt{H_{3-0-3}}) \tag{6.10}$$

An estimate of the Hamaker constant (H_{1-3-2}) between two dissimilar adhering bodies (1&2) and in a medium (3) can be obtained using the above equation where, H_{i-0-i} is the self Hamaker constant of a material, i, in vacuum (or air). It is apparent from Eq. 6.10 that the magnitude of H can be negative (i.e. repulsive vdW) only if $H_{1-0-1} > H_{3-0-3} > H_{2-0-2}$ or $H_{1-0-1} < H_{3-0-3} < H_{2-0-2}$. As the value of Hamaker constant for a solid is always greater than liquid, the vdW force between two solids in any medium (air or liquid) is always attractive due to a positive value for H. Moreover, the vdW force between solids in liquid is always lower than that in air (or vacuum) due to screening effect of the intervening medium. The value of H for solids is typically of the order of 10^{-19} J in the gaseous media and lies in the range of 10^{-21} - 10^{-20} J in the condensed media. Further, the Hamaker constant for metals (gold, silver, aluminium, etc.) is generally higher (around 20–40 × 10^{-20} J in air) than that of the inorganic materials such as silica, alumina, titania, and quartz (5–15 × 10^{-20} J in air). This is because metals have high electron densities and free electrons, leading to strong fluctuations in electromagnetic fields (stronger London dispersion forces) and causing extremely high polarizability. Conversely, the inorganic materials, such as metal oxide, ceramic, have moderate Hamaker constant due to lack of free electrons, despite high dielectric constants and refractive indices. However, polymeric and organic materials, such as polystyrene, polyethylene and Teflon, have low values of Hamaker constant in the range of 3–8 × 10^{-20} J (in air) due to very low polarizability and lack of free or delocalized electrons.

6.3.3 Capillary Forces

When two solid surfaces are in contact in a humid environment, the moisture can condense on at the solid-solid contact and form a liquid bridge, It leads to the generation of liquid-air interface (aka meniscus) connecting the two adhering solids, as shown in Figure 6.3. The formation of liquid meniscus leads to the generation of capillary pressure force (F_P^{Cap}), described by the following Young-Laplace equation for a columnar water bridge as [36]:

$$F_P^{Cap} = \Delta P.A = A.\gamma_L\left(\frac{1}{r_n} + \frac{1}{r_p}\right) \tag{6.11}$$

where ΔP is the pressure difference across the meniscus (the Laplace pressure), A is liquid-solid interfacial area (i.e. wetted contact), γ_L is the surface tension of the liquid condensate, and r_n and r_p are the two principal radii of curvature (ROC) of the liquid bridge between the surfaces. The capillary pressure force pulls the two solid surfaces together for a concave meniscus.

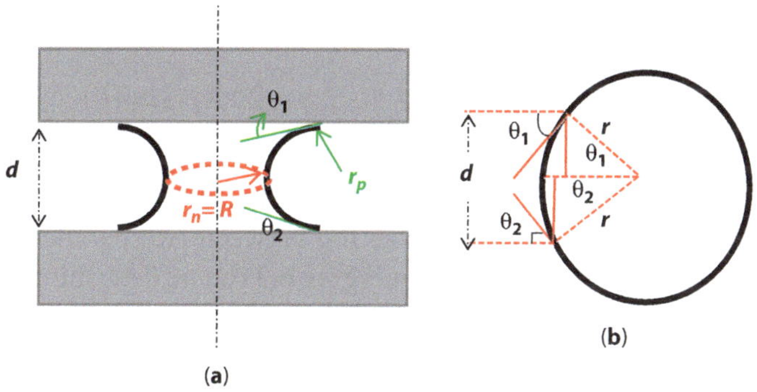

Figure 6.3 (a) Two parallel solid plates linked by a columnar water bridge; (b) A zoomed-in view of the radius of curvature (ROC) r_p.

The ROC, r_p, of the film meniscus can be related to the film thickness, *d*, using geometry shown in Figure 6.3b:

$$d = r_p(cos(\theta_1) + cos(\theta_2)) \tag{6.12}$$

where θ_1 and θ_2 are the contact angles of the liquid against the two plates. Finally, the capillary pressure force for this system can be determined as:

$$F_P^{Cap} = A.\gamma_L\left(\frac{(cos(\theta_1) + cos(\theta_2))}{d} + \frac{1}{r_n(= -R)}\right) \tag{6.13}$$

If both plates shown in Figure 6.3a are made of the same material with the liquid contact angle $\theta_1 = \theta_2 = \theta$, the capillary pressure force between them can be determined as:

$$F_P^{Cap} = \pi R^2.\gamma_L\left(\frac{2.cos(\theta_1)}{d} - \frac{1}{R}\right) \tag{6.14}$$

If the radius, r_n of the liquid column is much larger than the column height (i.e. $R >> d$),

$$F_P^{Cap} \sim \frac{2\pi R^2 \gamma_L cos(\theta_1)}{d} \tag{6.15}$$

In addition, the unbalanced surface tension force, Fst, acting at solid-liquid boundary pulls the solid plates in Figure 6.3 towards each other and can be obtained as:

$$F_{st} = \gamma_L \cos(\pi/2 - \theta_1)\, l_{wetted} = 2\pi R \gamma_L sin(\theta_1) \tag{6.16}$$

Where l_{wetted} is the wetted perimeter on the particle, given by $2\pi R$. Therefore, the total capillary force acting between two parallel solid surfaces of the same materials shown in Figure 6.3 can be determined by combining the capillary pressure force with the surface tension force as:

$$F_{tot}^{Cap} = \pi R^2 \gamma_L\left(\frac{2\cos\theta_1}{d} - \frac{1}{R}\right) + 2\pi R \gamma_L sin(\theta_1) \tag{6.17}$$

6.3.4 Effects of Surface Topography on Interfacial Adhesion

The topography of a solid body can alter surface free energy, and hence interfacial adhesion, by various mechanisms described below:

6.3.4.1 Mechanical Interlocking

Microscale surface topographies can enhance the potential for mechanical interlocking, particularly when a compliant material conforms to a rigid substrate. This interlocking effect provides physical anchoring at the interface, which increases resistance to separation under shear or tensile loading. The extent of interlocking is influenced by several factors, including the scale and geometry of the surface features, the viscosity and mobility of the contacting materials, as well as the applied pressure and contact time during bonding. For example, in fiber-reinforced composites, roughened fiber surfaces promote improved adhesion by allowing the resin to infiltrate and anchor within surface irregularities [37].

6.3.4.2 Increased Effective Separation Between Solid Bodies

The incorporation of surface topography increases the effective separation between interacting solid bodies in comparison to atomically smooth surfaces, as illustrated in Figure 6.4. In other words, surface roughness displaces the adhering bodies further apart, which can significantly hinder the formation of chemical bonds—interactions that are inherently short-ranged. Additionally, the strength of physical interactions, such as van der Waals forces and capillary forces, also diminishes with increasing interfacial separation, given their inverse dependence on distance [38]. Furthermore, surface irregularities may trap contaminants such as moisture, oxides, or organic residues, which can further inhibit both chemical and physical bonding mechanisms at the interface.

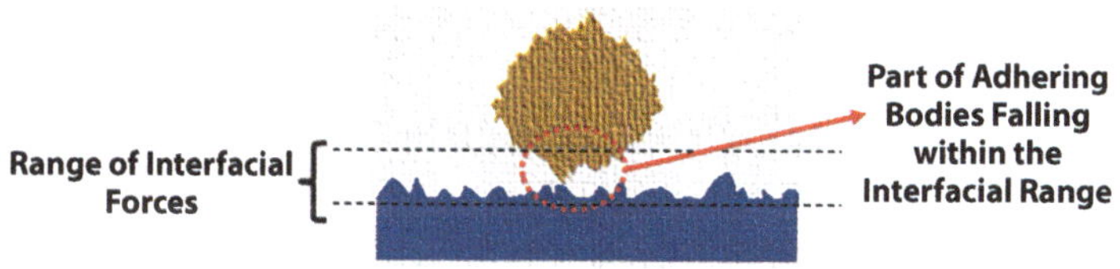

Figure 6.4 A schematic depicting a rough particle adhering to a rough substrate.

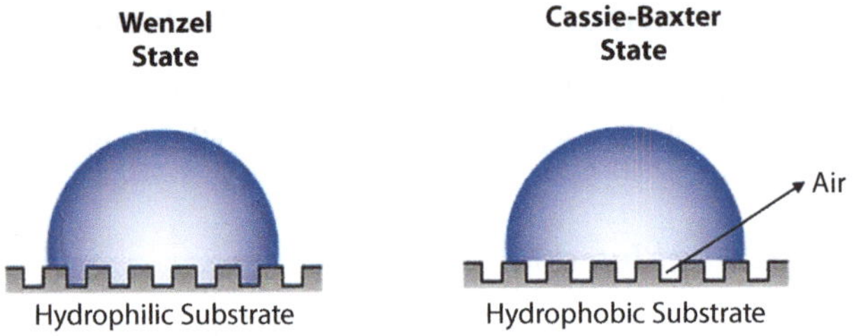

Figure 6.5 Schematics of Wenzel and Cassie-Baxter states of water droplets sitting on the textured hydrophilic and hydrophobic substrates.

6.3.4.3 Reduced Interfacial Mass and Contact Area

The introduction of surface topography can also reduce the interfacial mass and effective contact area between adhering bodies within the range of interfacial forces, as illustrated in Figure 6.4. This reduction diminishes vdW interactions, which are proportional to the interfacial mass present within the vdW domain. Furthermore, the decreased solid–solid contact area limits the formation of large capillary bridges in humid environments, thereby mitigating capillary forces that would otherwise contribute to adhesion [38].

6.3.4.4 Incorporation of Air Pockets

Surface topography can also introduce air pockets between adhering solids, thereby altering the wettability of the substrate [39]. If the substrate is hydrophilic, these air pockets may be displaced by liquid infiltration, resulting in enhanced wetting—a phenomenon described by the Wenzel state (as illustrated in Figure 6.5). Conversely, on hydrophobic substrates, the air pockets tend to remain trapped within the surface features, which amplifies the apparent hydrophobicity of the surface. This condition is known as the Cassie–Baxter state. Consequently, surface topography serves as a powerful tool to manipulate the wettability of a material, and in doing so, can significantly influence interfacial capillary forces.

6.4 Experimental Measurements

6.4.1 Surface Topography Characterization

Sub-micrometer surface topography can be characterized using Field Emission Scanning Electron Microscopy (FESEM), which provides high-resolution imaging of surface morphology. For nanometer-scale features such as surface roughness, Atomic Force Microscopy (AFM) is commonly employed, operating in either contact mode or non-contact (tapping) mode. AFM offers exceptional vertical and lateral resolutions, making it ideal for nanoscale surface analysis [40]. However, one of its limitations is the relatively small lateral scan range, typically restricted to a maximum of approximately 100 μm × 100 μm. For characterizing surface topography over larger areas, stylus-based surface profilometry is generally preferred, as it enables scanning of millimeter-scale regions with adequate vertical resolution.

6.4.2 Interfacial Adhesion Forces Measurement

For macroscale systems, the peel test is commonly employed to evaluate the adhesion of flexible films or tapes bonded to rigid substrates. In this method, the bonded flexible material is peeled away from the substrate at a controlled angle and speed, and the peel force per unit width (typically in N/m) is recorded. Although this test is simple and widely used, it is highly sensitive to factors such as material stiffness and test geometry. As a result, reproducibility can be compromised if standardized measurement procedures are not strictly followed.

For microscale and nanoscale systems, Atomic Force Microscopy (AFM) is a powerful and precise technique for measuring interfacial adhesion forces between solid surfaces. In a typical setup, a solid particle—such as a nanoparticle—is affixed to the tip of an AFM cantilever, as illustrated in Figure 6.6a. The second substrate (e.g., a silicon wafer) is mounted on the piezoelectric stage of the AFM. During the experiment, the nanoparticle attached to the AFM probe is gradually brought into contact with the substrate, and the deflection of the cantilever is continuously monitored, as shown in Figure 6.6b. In this force–distance curve, points a, b, c, and d correspond to different stages as the particle approaches the substrate. Once contact is established, the AFM system begins to retract the cantilever. At point e, sufficient force has been applied to detach the particle from the substrate. The pull-off force (typically in nN) recorded at this point reflects the net adhesion force between the two solid surfaces. However, due to variations in surface topography, the local interaction between the nanoparticle and the substrate can vary significantly with each contact. As illustrated in Figure 6.6c, surface asperities may align differently on each approach, affecting the degree of mechanical interlocking or intimate contact between surfaces. A better topographical match generally leads to a higher pull-off force. Consequently, repeated measurements often yield a

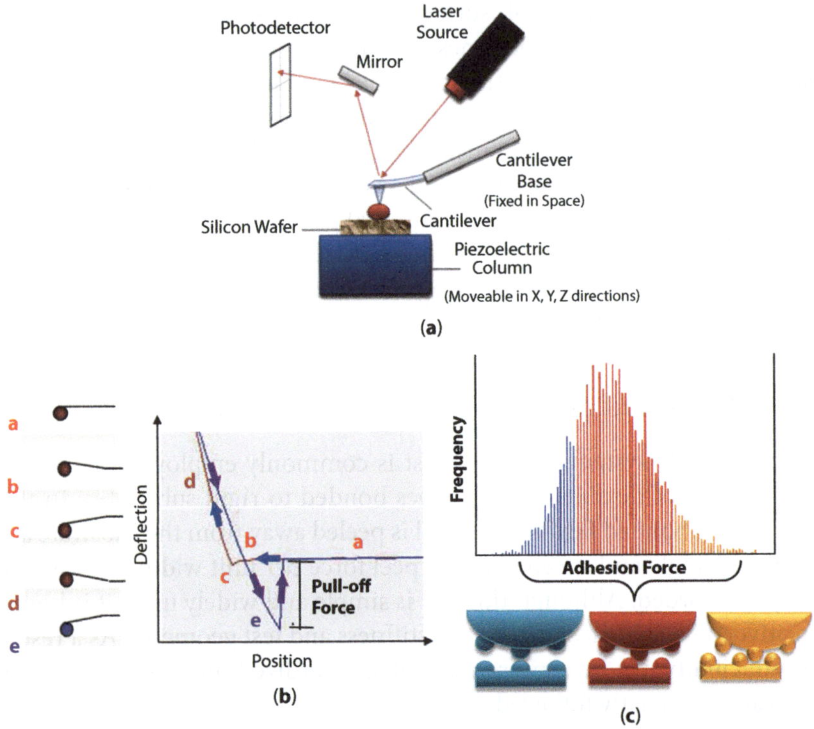

Figure 6.6 (a) A schematic depicting functioning of AFM; (b) Typical AFM force curve; (c) Distribution of measured pull-off forces.

distribution of pull-off forces (i.e. net adhesion forces), reflecting the stochastic nature of morphological interactions at the nanoscale.

6.4.3 Surface Free Energy Measurement

Surface free energy is a fundamental thermodynamic property that characterizes the energetic state of a solid surface. Unlike liquids where surface free energy, referred to as surface tension, can be directly measured, the surface free energy of solids cannot be measured directly. Instead, it is typically inferred through indirect methods such as contact angle measurements using a few probe liquids, thermodynamic models, or analyses of interfacial interactions. As noted above, surface free energy arises from various intermolecular interactions, including chemical bonds and physical forces such as vdW, ES, and capillary forces. These interactions are commonly classified into two components: **dispersion** (e.g., London vdW dispersion) and **polar** (e.g., hydrogen bonding, dipole-dipole, dipole-induced dipole, and ion-dipole interactions). Van Oss, Chaudhury and Good [41] further subdivided the polar component into **acidic** and **basic** interactions. The *acidic interaction* of a solid surface refers to its tendency to accept electron density from the contacting medium, whereas the *basic interaction* reflects the surface's ability to donate electron density to the medium. Based on various theoretical frameworks that incorporate intermolecular interactions, several methods are employed to determine the surface free energy of a solid.

6.4.3.1 One-Component Model (Zisman Method)

Zisman theory [41, 42] is a one-parameter model used to estimate the surface free energy of solid surfaces. It characterizes surface free energy using a single value: the surface tension of the liquid that completely wets the solid, i.e., the liquid that produces a contact angle of 0°. According to Zisman, the surface free energy of a solid is equivalent to the surface tension of this fully wetting liquid. This theory is particularly effective for non-polar surfaces, such as polyethylene and polypropylene. While it can be applied to other polymeric surfaces, its accuracy diminishes significantly when the surface exhibits even a modest degree of polar interactions. In such cases, the Zisman method yields misleading results.

6.4.3.2 Two-Component Model (Owens-Wendt Method)

The Owens–Wendt model—also referred to as the Owens, Wendt, Rabel, and Kaelble (OWRK) model [44]—decomposes the surface free energy of a solid into two primary components: dispersion (non-polar, denoted as *d*) and polar (denoted as *p*). This theory proposes the surface free energy of solid in contact with a liquid to be defined as:

$$\gamma_{SL} = \gamma_S + \gamma_L - 2\sqrt{\gamma_S{}^d \gamma_L{}^d} - 2\sqrt{\gamma_S{}^p \gamma_L{}^p} \tag{6.18}$$

The above equation used with Young Equation (6.7) leads to the following:

$\gamma_L(1 + Cos\,\theta) = 2(\sqrt{\gamma_S{}^d \gamma_L{}^d} + \sqrt{\gamma_S{}^p \gamma_L{}^p})$ which is re-written as

$$\frac{\boldsymbol{\gamma_L(1 + Cos\,\theta)}}{2\sqrt{\gamma_L{}^d}} = \sqrt{\gamma_S{}^p}\sqrt{\frac{\gamma_L{}^p}{\gamma_L{}^d}} + \sqrt{\gamma_S{}^d} \tag{6.19}$$

Where γ_{SL}: Interfacial tension between liquid and solid, γ_S: Surface energy of the solid, $\gamma_S{}^d$: Dispersion component of the surface free energy of the solid, $\gamma_S{}^p$: Polar component of the surface free energy of the solid, γ_L: Surface tension of the test liquid, $\gamma_L{}^d$: Dispersion component of the surface tension of the test liquid, $\gamma_L{}^p$: Polar component of the surface tension of the test liquid, θ: Equilibrium contact angle at solid-liquid interface.

The values of $\gamma_S{}^d$ and $\gamma_S{}^p$ are determined, as the intercept and slope respectively, from a linear plot between $\frac{\gamma_L(1+\mathrm{Cos}\,\theta)}{2\sqrt{\gamma_L{}^d}}$ (Y axis) and $\sqrt{\frac{\gamma_L{}^p}{\gamma_L{}^d}}$ (X axis). A minimum of two probe liquids is required to solve for the two unknowns ($\gamma_S{}^d$ and $\gamma_S{}^p$). Generally, water and diiodomethane are used as two probe liquids with their values for $\gamma_L, \gamma_L{}^d$ and $\gamma_L{}^p$ are given as 72.8, 21.8 and 51 mJ/m^2, respectively for water; and 50.8, 50.8 and 0 mJ/m^2, respectively for diiodomethane. Finally, the surface free energy for solid is determined as:

$$\gamma_S = \gamma_S{}^d + \gamma_S{}^p \tag{6.20}$$

6.4.3.3 *Three-Component Model (van Oss–Chaudhury–Good Method)*

Van Oss, Chaudhury, and Good [45] extended the two-component approach by further decomposing the polar part of surface free energy into **Lewis acid** (γ^+) and **Lewis base** (γ^-) components. This led to the development of a three-component model for determining the surface free energy of a solid. In this framework, the total surface free energy of a solid in contact with a liquid is expressed as:

$$\gamma_{SL} = \gamma_S + \gamma_L - 2\sqrt{\gamma_S{}^d\gamma_L{}^d} - 2\sqrt{\gamma_S{}^+\gamma_L{}^-} - 2\sqrt{\gamma_S{}^-\gamma_L{}^+} \tag{6.21}$$

The above equation, along with Young equation, yields the following:

$$\gamma_L(1+Cos\,\theta) = 2\left(\sqrt{\gamma_S{}^d\gamma_L{}^d} + \sqrt{\gamma_S{}^+\gamma_L{}^-} + \sqrt{\gamma_S{}^-\gamma_L{}^+}\right) \tag{6.22}$$

Where $\gamma_S{}^+, \gamma_S{}^-, \gamma_L{}^+, \gamma_L{}^-$ are the Lewis acid and base components for solid and liquid respectively. Apparently, there are three unknowns

$(\gamma_S{}^d, \gamma_S{}^+, \gamma_S{}^-)$ and hence it requires a minimum three probe liquids to determine surface free energy of a solid-air interface. Generally, water, formamide, and diiodomethane are used as three probe liquids with $\gamma_L, \gamma_L{}^d, \gamma_L{}^+$ and $\gamma_L{}^-$ as 72.8, 21.8, 25.5, 25.5 mJ/m^2 respectively for water, 58.0, 39.0, 2.28, 39.6 mJ/m^2 respectively for formamide, and 50.8, 50.8, 0.0, 0.0 mJ/m^2 respectively for diiodomethane. Finally, the surface free energy for solid is determined as:

$$\gamma_S = \gamma_S{}^d + \gamma_S{}^+ + \gamma_S{}^- \tag{6.23}$$

There are many other methods to determine surface free energy of solids using contact angles of probe liquids [46, 47].

6.4.3.4 *Inverse Gas Chromatography Method*

Inverse Gas Chromatography (IGC) has emerged as a powerful alternative, particularly for powders, porous materials, fibers, and polymers, to measure surface free energy [48]. IGC measures how vapor-phase probe molecules interact with a solid surface, offering quantitative insights for the determination of surface free energy of the solid materials. Unlike conventional gas chromatography, where a known stationary phase is used to analyze unknown analytes, the role is revered in the IGC. Here, the unknown solid sample acts as the stationary phase, and well-characterized probe molecules (one non-polar and two polar) are injected in the gas phase. Specifically, finely powdered solid or packed fibers/polymers are loaded into the IGC column and preconditioned under a dry carrier gas (e.g. helium or nitrogen) to remove adsorbed moisture or contaminants. Infinitely diluted vapor-phase probes are injected into the IGC column. Generally, n-alkanes (e.g., hexane, heptane, octane) are used as non-polar probe gas, and dichloromethane, ethyl acetate, acetone, or chloroform are used as polar probes. A flame ionization detector (FID) for hydrocarbons or thermal conductivity detector (TCD) for both organic and inorganic gases is used in the IGC. Their strength and nature of interaction with the solid surface is inferred by analyzing the retention times of these probe molecules, leading to the determination of surface free energy.

6.5 Applications of Topographical Modifications to Engineer Interfacial Adhesion

6.5.1 Adhesion-Promoting Surfaces

Numerous bio-inspired and biomimetic surfaces have been developed to enhance interfacial adhesion and friction across various applications, including robotics, packaging, biomedical devices, and aerospace systems. For example, Shi *et al.* [49] fabricated tree-frog-inspired nanopillars using polycaprolactone *via* a molding process and demonstrated a significant improvement in both adhesion and friction compared to conventional hexagonal micropillar arrays. Similarly, Xue *et al.* [50] created hierarchical micropillar structures composed of polydimethylsiloxane (PDMS) embedded with polystyrene (PS) nanopillars, showing that the hierarchical architecture improved adhesion and friction by altering the stress distribution at the contact interface. Yun *et al.* [51] introduced periodic, micrometer-scale square dot patterns on steel surface to enhance adhesion to epoxy layers. The improved interfacial bonding was attributed to increased cohesive fracture area and topographical interlocking. In another study, Balkundhi *et al.* [52] modified the surface morphology of aluminum sheets through a combination of physical and chemical treatments to improve adhesion with carbon fibers. They reported significant adhesion enhancement at the aluminum/carbon fiber interface using sulfuric acid etching and electric discharge-based surface texturing. Overall, the literature demonstrates that surface topography can be strategically modified—ranging from the nanoscale to the microscale—to enable single or multilayered textures on deformable materials, thereby promoting improved interfacial adhesion.

6.5.2 Anti-Fouling Surfaces

While surface texturing has been shown—particularly in the previous section—to enhance adhesion through deformation on polymeric surfaces or mechanical interlocking on rigid surfaces, it can also serve the opposite function: mitigating adhesion by reducing interfacial forces between contaminants and the underlying substrate. Wang *et al.* [53] demonstrated dust-mitigation behavior on aluminum surfaces modified with multiscale micro-/nano-rough structures. They reported a fourfold reduction in dust deposition, attributed to a decreased contact area between dust particles and the textured substrate. Similarly, Lee *et al.* [54] fabricated large-area nanostructures with tailored geometries on polycarbonate substrates and observed a significant reduction in dust accumulation. This was explained by the aggregation of particles due to strong interparticle forces, which facilitated gravity-driven removal from the surface. Jaiswal [55] compared nanopillar and nanowell surface topographies and found that nanopillars were more effective in suppressing capillary condensation, thereby reducing interfacial adhesion and promoting dust-resistant characteristic of a substrate. Ling *et al.* [56] investigated the role of surface topography on ice adhesion to metallic substrates and identified two competing effects:

a) Increased ice adhesion due to mechanical interlocking offered by surface asperities, and
b) Reduced ice adhesion facilitated by the formation of micro-cracks and trapped air pockets.

They reported a 93% decrease in ice adhesion for an untreated fine stainless steel mesh compared to polished stainless steel. These recent advancements underscore the potential of using biomimetic designs and engineered nanostructures to develop anti-fouling, dust-repellent, and ice-phobic surfaces, offering functional benefits across environmental, aerospace, and optical applications.

6.5.3 Anti-Bacterial Surfaces

Anti-bacterial surfaces are critically important in several applications including surgical tools, medical implants and devices, and antimicrobial coatings. Physical surface modification strategies aimed at imparting anti-bacterial property generally operate through two primary mechanisms:

i) Anti-biofouling characteristic, which prevents bacterial adhesion and colonization, and
ii) Contact killing, achieved through bactericidal nanotopographies that physically disrupt bacterial membranes.

Both low and high surface free energy substrates have been fabricated in the literature to exhibit anti-biofouling traits. Low surface free energy substrates such as fluoropolymers, silicones, zwitterionic polymers have been shown to prevent bacterial adhesion and colonization due to weak van der Waals forces and absence of hydration layers. Conversely, high surface free energy surfaces are known to repel hydrophobic substances, such as bacterial lipids, thereby impeding adhesion. Cox *et al.* [57] demonstrated that highly hydrophilic surfaces were more effective in reducing biofouling than hydrophobic surfaces.

Whereas earlier research primarily focused on minimizing bacterial adhesion to surfaces [58], more recent efforts have shifted toward developing bactericidal surfaces using engineered nanostructures. Sharp nanofeatures, with optimized shapes and spacing in the range of ~100–200 nm, have been shown to mechanically rupture bacterial membranes, thereby enabling contact killing of both Gram-negative and Gram-positive bacteria. Various nanotopographies have been employed not only to enhance the efficacy of known antimicrobial materials—such as copper and silver [59]—but also to impart inherent bactericidal property to otherwise inert substrates [60, 61]. For instance, Zhao *et al.* [62] evaluated the mechano-bactericidal performance of five different nanostructured surfaces and systematically optimized the shape, size, and spacing of the nanofeatures. Their results showed that nanopillar topographies exhibited superior bactericidal activity compared to nanocone structures, and that careful optimization of surface nanotopography allowed for enhanced control over the bactericidal response against *E. coli* cells. Although the bactericidal potential of nanostructures is now well established, the current research frontier lies in the development of surfaces that are both antibacterial and cytocompatible, ensuring efficacy against pathogens while maintaining compatibility with human cells.

6.5.4 Superhydrophilic and Superhydrophobic Surfaces

Surface topography plays a critical role in achieving and enhancing superhydrophilic behavior. By precisely engineering micro- and nanostructures, surfaces can attain near-zero water contact angles and exhibit rapid water spreading, which is essential for applications such as anti-fogging, self-cleaning, and water–oil separation. Additionally, superhydrophilic surfaces are known to enhance biocompatibility and cell adhesion, making them particularly valuable in medical implants and diagnostic devices. For instance, Mundo *et al.* [63] investigated the effect of surface nanotexturing on the durability of superhydrophilic coatings. They deposited a thin layer of silica onto commercial transparent polycarbonate substrates, both with and without oxygen plasma-induced nanotexturing. While silica deposition alone yielded temporary superhydrophilicity and antifogging effects—lasting approximately four months— nanotexturing of the substrate extended the superhydrophilic and antifogging performance to over one year, highlighting the critical role of topography in long-term stability. Similarly, Zhang *et al.* [64] modified the surface topography of poly(acrylic acid) (PAA)-grafted polyetheretherketone (PEEK) to fabricate superhydrophilic PEEK surfaces. This modification significantly improved cellular affinity, underscoring the utility of topographically engineered surfaces in orthopedic implant applications.

Surface topography can also be engineered to enhance hydrophobicity, enabling the fabrication of superhydrophobic surfaces, which are highly effective for anti-fouling, anti-fogging, anti-corrosion, self-cleaning, and oil–water separation applications [65, 66]. By introducing micro- and nanoscale roughness, surface features can trap air and reduce the solid–liquid contact area, thereby amplifying water repellency according to the Cassie–Baxter model. For instance, Saffar *et al.* [67] developed a superhydrophobic PTFE–SiO_2-coated copper substrate by incorporating hierarchical surface roughness, demonstrating effective performance in anti-icing and self-cleaning applications. Majhy *et al.* [68] demonstrated that a hydrophobic coating of PDMS (polydimethylsiloxane) could be transformed into a superhydrophobic surface by depositing carbon soot—generated from a candle flame—onto a partially cured PDMS layer (dissolved in *n*-hexane), followed by soot removal using a high-speed water jet. The resulting nano-textured PDMS surface exhibited significantly enhanced water repellency. In another study, Amiriafshar *et al.* [69] investigated the hydrophobicity of stainless-steel substrates with varying surface roughness after undergoing **zinc** electrodeposition and subsequent stearic acid coating. They found that the initial surface roughness critically influenced the hydrophobicity, adhesion durability, and corrosion resistance of the treated surfaces. Given their potential to deliver corrosion protection, drag reduction, and anti-icing functionality, numerous research efforts have focused on rendering metallic surfaces superhydrophobic for diverse engineering and industrial applications [70].

In addition, the surface topography and roughness become crucial in a wide range of other applications, such as fiber-reinforced composites [71], thin film coatings [72], semiconductor device fabrication [73] and regulating biological response [74].

6.6 Challenges and Future Perspectives

Despite significant advancements in understanding and utilizing surface topography to engineer interfacial adhesion, several challenges remain that constrain the full potential of this approach in both industrial and biomedical applications. These challenges include: (a) limited predictability of topographical effects due to the difficulty in quantitatively assessing their influence on adhesion; (b) insufficient durability and long-term stability of engineered surface features under operational conditions; and (c) issues related to the reproducibility, scalability, and economic viability of fabrication processes. Future research efforts should focus on these challenges through the convergence of advanced fabrication technologies, data-driven surface design, and interdisciplinary integration.

6.7 Summary

This review has comprehensively discussed the fundamental principles and recent advancements in utilizing surface topography to engineer and optimize interfacial adhesion across a broad spectrum of materials and applications. Surface topography—from micro- to nanoscale range—plays a pivotal role in governing interfacial phenomena by influencing mechanical interlocking, real contact area, and interfacial forces such as van der Waals interactions, capillary forces, and chemical bonding. Key surface roughness parameters (e.g., Ra, Rq, Rz) were discussed alongside a variety of physical and chemical surface modification techniques, including grit blasting, focused ion beam milling, laser surface texturing, nanoimprint lithography, plasma treatment, chemical and plasma etching, and chemical mechanical polishing. These methods offer precise control over surface morphology at micro- and nanoscales, directly affecting adhesion performance. The article also reviewed experimental techniques for characterizing surface topography and quantifying interfacial adhesion forces and energy. Recent research efforts aimed at topographically engineering surfaces for applications such as biomimetic surfaces, anti-fouling, antibacterial performance, and superhydrophilic or superhydrophobic behavior were highlighted. Overall, the review underscored the immense potential of surface topography as a versatile tool for tailoring interfacial adhesion in next-generation functional surfaces. Despite current challenges, continued advancements in fabrication technologies, mathematical model-driven surface design and interdisciplinary research are expected to unlock new opportunities across both established industries and emerging technologies.

AI disclosure statement: The author acknowledges the use of AI-based tools for minor language editing or grammatical correction. However, all intellectual contributions, analyses, and conclusions presented in this chapter are entirely the author's own.

References

1. R. P. Jaiswal, G. Kumar, C. Kilroy and S. P. Beaudoin, Modeling and validation of the van der Waals force during the adhesion of nanoscale objects to rough surfaces, *Langmuir*, 25, 10612–10623 (2009).
2. G. Kumar, S. Smith, R. P. Jaiswal and S. Beaudoin, Scaling of van der Waals and electrostatic adhesion interactions from the micro- to the nano-scale, *J. Adheion. Sci. Technol.* 22, 407-428(2008).
3. J. Song, H. Huang, X. Wang and W. Shi, Status and prospects of surface texturing: design, manufacturing and applications, *Surf. Sci. Technol.*, 1, 21 (2023).
4. A. A. Al-Amiery, M. A. Fayad, H. A. A. Wahhab, W. K. Al-Azzawi, J. K. Mohammed and H. Sh. Majdi, Interfacial engineering for advanced functional materials: Surfaces, interfaces, and applications, *Results Eng.* 22, 102125 (2024).
5. J. Zhu, Y. Zhu and P. Zhang, Review of advancements in wall climbing robot techniques, *Franklin Open*, 8, 100148 (2024).
6. T. Liu and M. S. Mauter, Heat transfer innovations and their application in thermal desalination processes, *Joule*, 6, 1199-1229 (2022).
7. A. W. Hashmi, H. S. Mali, A. Meena, K. K. Saxena, S. Ahmad, M. K. Agrawal, B. Sagbas, A. P. V. Puerta and M. I. Khan, A comprehensive review on surface post-treatments for freeform surfaces of bio-implants, *J. Mater. Res. Technol.*, 23, 4866-4908 (2023).
8. S. G. Natarajan and R. J. B. Ralphin, Performance enhancement of futuristic airplanes by nature inspired biomimetic fish scale arrays—A design approach, *Biomim. Intell. Robot.*, 2, 2, 100045 (2022).
9. B. Dean and B. Bhushan, Shark-skin surfaces for fluid-drag reduction in turbulent flow: A review, *Philos. Trans. R. Soc. A.*, 368, 4775–4806 (2010)
10. R. P. Jaiswal and S. P. Beaudoin, Nanoparticle adhesion models: Applications in particulate contaminant removal from extreme ultraviolet lithography photomasks, *J. Adhesion Sci. Technol.*, 25, 781-797 (2011).
11. C. Kilroy, R. P. Jaiswal and S. P. Beaudoin, Adhesion of contaminant particles to advanced photomask Materials, *IEEE Trans. Semicond. Manuf.*, 25, 37-44 (2012).
12. Y. Zhao, Stiction and anti-stiction in MEMS and NEMS, *Acta Mechanica Sinica*, 19, 1–10 (2003).
13. S. Rajaramon, H, David, A. Sajeevan, K. Shanmugam, H. Sriramulu, R. Dandela and A. P. Solomon, Multi-functional approach in the design of smart surfaces to mitigate bacterial infections: a review, *Front. Cell. Infect. Microbiol.*, 13, 1-15 (2023).
14. A. Pramanik, A.K. Basak, Y. Dong, P.K. Sarker, M.S. Uddin, G. Littlefair, A.R. Dixit and S. Chattopadhyaya, Joining of carbon fibre reinforced polymer (CFRP) composites and aluminium alloys – A review, *Composites Part A*, 101, 1-29 (2017).

15. T. Josyula, L. K., Malla, T. M. Thomas, S. S. Kalichetty, P. S. Mahapatra, and A. Pattamatta, Fundamentals and applications of surface wetting, *Langmuir*, 40, 8293–8326 (2024).
16. P. Ihalainen, A. Määttänen, J. Järnström, D. Tobjörk, R. Österbacka, and J. Peltonen, Influence of surface properties of coated papers on printed electronics, *Ind. Eng. Chem. Res.*, 51, 6025–6036 (2012).
17. ZYGO: Roughness Measurements - Let's Take a Closer Look, https://www.zygo.com/insights/blog-posts/roughness-measurements (Accessed on June 30, 2025).
18. A. Rudawska, I. Danczak, M. Müller and P. Valasek, The effect of sandblasting on surface properties for adhesion, *Int. J. Adhesion Adhesives*, 70, 176-190 (2016).
19. R. G. Mote, Fabrication and structure manipulation at nanoscale—A focused ion beam (FIB) approach, *Comprehensive Mater. Process.*, 10, 205-218 (2024)
20. J. Zhou, J. Wu S. Tang and Y. Li, Review of laser texturing technology for surface protection and functional regulation of aluminum alloys: Wettability, anti-icing, corrosion resistance, and wear resistance, *Coatings*, 15, 567-578 (2025).
21. A. F. Obilor, M. Pacella, A. Wilson and V. V. Silberschmidt, Micro-texturing of polymer surfaces using lasers: A review, *J. Adv. Manuf. Technol.*, 120, 103–135 (2022).
22. S. Barcelo and Z. Li, Nanoimprint lithography for nanodevice fabrication *Nano Convergence* 3: 21, 1-9 (2016).
23. A. Sharma, A. Jaiswal and R. P. Jaiswal, Fabrication of a tailorable polystyrene nanoscale mesh of honeycomb morphology using nano sphere lithography, *Thin Solid Films*, 763, 139567 (2022).
24. R. A. Faase, N. M. Keeling, J. S. Plaut, C. Leycam, G. A. Munares, M. T. Hinds, J. E. Baio and P. L. Jurney, Temporal changes in the surface chemistry and topography of reactive ion plasma-treated poly(vinyl alcohol) alter endothelialization potential, *ACS Appl. Mater. Interfaces*, 16, 389–400 (2024).
25. A. Prakash, K. Manohara, S. Kumar and M. Katiyar, Effect of plasma processing on the morphological and electrical properties of polyethylene terephthalate (PET) substrate, *Results Surf. Inter.*, 20, 100559 (2025).
26. L. Wu , P. Chen , L. Deng , P. Zhang, B. Yu and L. Qian, Effects of crystal planes on topography evolution of silicon surface during nanoscratch-induced selective etching, *Mater. Sci. Semicond. Process.*, 124, 105606 (2021).
27. N. U. Taskin and F. Ordu, Effect of etching duration on roughness and wettability of different carbon steel substrates, *Mater. Chem. Phys.*, 257, 123746 (2021).
28. C. Weigel, U. Brokmann, M. Hofmann, A. Behrens, E. Rädlein, M. Hoffmann, S. Strehle and S. Sinzinger, Perspectives of reactive ion etching of silicate glasses for optical microsystems, *J. Opt. Microsyst.*, 1, 4, 040901 (2021).
29. X. Zhu, J. Ding, Z. Mo, X. Jiang, J. Sun, H. Fu, Y. Gui, B. Ban, L. Wang and J. Chen, Evaluation of chemical mechanical polishing characteristics using

mixed abrasive slurry: A study on polishing behavior and material removal mechanism, *Appl. Surf. Sci.*, 679, 161157 (2025).
30. J. N. Israelachvili, *Intermolecular and Surface Forces*, Academic Press, New York (2011).
31. https://www.innotape.de/en/blog/surface-energy (Accessed on June 30,2025).
32. https://www.jbc-tech.com/blog/posts/bonding-low-surface-energy-lse-materials/ (Accessed June 30, 2025).
33. R. P. Jaiswal, Adhesion between particles and nano-structured films, (Doctoral Dissertation), Purdue University Press, West Lafayette, IN (2008).
34. R. P. Jaiswal and S. P. Beaudoin, Approximate scheme for calculating van der Waals interactions between finite cylindrical volume elements, *Langmuir*, 28, 8359–8370 (2012).
35. R. P. Jaiswal and S. P. Beaudoin, London-van der Waals force field of a chemically patterned surface to enable selective adhesion, *Langmuir*, 35, 86–94 (2019).
36. S. P. Beaudoin, P. Jaiswal, A. Harrison, J. Laster, K., Smith, M. Sweat and M. Thomas, Fundamental forces in particle adhesion, in: *Particle Adhesion and Removal*, K. L. Mittal and R. Jaiswal (Eds.) Wiley-Scrivener, Beverly, MA (2015).
37. C. H. Lee, A. Khalina and S. H. Lee, Importance of interfacial adhesion condition on characterization of plant-fiber-reinforced polymer composites: A review, *Polymers,* 13, 438 (2021).
38. K. L. Mittal and R. P. Jaiswal (Eds), *Particle Adhesion and Removal*, Wiley-Scrivener, Beverly, MA (2015).
39. X. Wang, C. Fu, C. Zhang, Z. Qiu and B. Wang, A comprehensive review of wetting transition mechanism on the surfaces of microstructures from theory and testing methods, *Materials,* 15, 4747 (2022).
40. T.A. Ruzova and B. Haddadi, Surface roughness and its measurement methods - Analytical review, *Results Surf. Interf.*, 19, 100441 (2025).
41. C. J. van. Oss, M. K. Chaudhury and R. J. Good, Interfacial Lifshitz-van der Waals and polar interactions in macroscopic systems, *Chem. Rev.*, 88, 927–941 (1988).
42. H. Fox and W. Zisman, The spreading of liquids on low energy surfaces. I. polytetrafluoroethylene, *J. Colloid Sci.*, 5, 514-531 (1950).
43. G. McHale, N. Afify, S. Armstrong, G. G. Wells and R. Ledesma-Aguilar, The liquid Young's law on SLIPS: Liquid–liquid interfacial tensions and Zisman plots, *Langmuir*, 38, 10032–10042 (2022).
44. D. K. Owens and R. C. Wendt, Estimation of the surface free energy of polymers, *J. Appl. Polym. Sci.*, 13, 1741-1747 (1969).
45. C. J. Van Oss, R. J. Good and M. K. Chaudhury, The role of van der Waals forces and hydrogen bonds in "hydrophobic interactions" between biopolymers and low energy surfaces, *J. Colloid Interface Sci.*, 111, 378-390 (1986).

46. F. M. Etzler, Characterization of surface free energies and surface chemistry of solids, in: *Contact Angle, Wettability and Adhesion*, Vol. 3, K. L. Mittal (Ed.) 219-264, CRC Press, Boca Raton, FL (2003).
47. F. M. Etzler, Determination of the surface free energy of solids: A critical review. *Rev Adhesion Adhesives* 1, 3-45 (2013).
48. S. Mohammadi-Jam and K.E. Waters, Inverse gas chromatography applications: A review, *Adv. Colloid Interface Sci.*, 212, 21-44 (2014)
49. Z. Shi, D. Tan, Q. Liu, F. Meng, B. Zhu, and L. Xue, Tree frog-inspired nanopillar arrays for enhancement of adhesion and friction, *Biointerphases* 16, 021001 (2021).
50. L. Xue, B. Sanz, A. Luo, K. T. Turner, X. Wang, D. Tan, R. Zhang, H. Du, M. Steinhart, C. Mijangos, M. Guttmann, M. Kappl and A. d. Campo, Hybrid surface patterns mimicking the design of the adhesive toe pad of tree frog, *ACS Nano*, 11 , 9711–9719 (2017).
51. I. H. Yun, W-S Kim, K. H. Kim; J-M Jung, J-J Lee and H-T Jung, Highly enhanced interfacial adhesion properties of steel-polymer composites by dot-shaped surface patterning, *J. Appl. Phys.* 109, 074302 (2011).
52. M. Balkundhi, S. S. Baloor and G. Bolar, Assessment of surface treatment methods for strengthening the interfacial adhesion in CARALL fiber metal laminates, *Sci. Rep.* 14, 30909 (2024).
53. X. Wang, W. Wang, H. Shao, S. Chao, H. Zhang, C. Tang, X. Li, Y. Zhu, J. Zhang, X. Zhang and Y. Lu, Lunar dust-mitigation behavior of aluminum surfaces with multiscale roughness prepared by a composite etching method, *ACS Appl. Mater. Interfaces*, 14, 34020–34028 (2022).
54. S. S. Lee, L. Micklow, A. Tunell, K. C. Chien, S. Mohanty, N. Cates, S. Furst and C-H Chang, Engineering large-area antidust surfaces by harnessing interparticle forces, *ACS Appl. Mater. Interfaces*, 15, 13678–13688 (2023).
55. A. Jaiswal, Modulation of the capillary force profile at the solid–solid interface through topographical modifications, *Langmuir*, 39, 13178–13188 (2023).
56. E. J. Y. Ling, V. Uong, J-S Renault-Crispo, A-M Kietzig and P. Servio, Reducing ice adhesion on nonsmooth metallic surfaces: wettability and topography effects, *ACS Appl. Mater. Interfaces*, 8, 8789–8800 (2016).
57. H. J. Cox, I. Cooper, H. F. Kaspar, M. A. Packer and J. P. S. Badyal, Antibiofouling functional surfaces for marine aquaculture, *Colloids Surfaces A 639, 128313 (2022)*.
58. S. Wu, B. Zhang, Y. Liu, X. Suo and H. Li, Influence of surface topography on bacterial adhesion: A review, *Biointerphases*, 13, 060801 (2018).
59. D. W. Müller, C. Pauly, K. Brix, R. Kautenburger, and F. Mücklich, Modifying the antibacterial performance of Cu surfaces by topographic patterning in the micro- and nanometer scale, *Biomater. Adv.*, 169, 214184 (2025).
60. I. Georgakopoulos-Soares, E. L. Papazoglou, P. Karmiris-Obratański, N. E. Karkalos and A. P. Markopoulos, Surface antibacterial properties enhanced

through engineered textures and surface roughness: A review, *Colloids Surfaces:B* , 231, 113584 (2023).
61. J. Hasan and K. Chatterjee, Recent advances in engineering topography mediated antibacterial surfaces, *Nanoscale*, 7, 15568-15575 (2015).
62. Z. Zhao, Y. Matsushita, N. Ogawa, G. Inoue, T. Shimizu, S. Shingubara, and T. Ito, Bactericidal performance of nanostructures resin surfaces: Nanopillars versus nanocones, *ACS Appl. Nano Mater.*, 7, 14040–14049 (2024).
63. R. D. Mundo, R. d'Agostino and F. Palumbo, Long-lasting antifog plasma modification of transparent plastics, *ACS Appl. Mater. Interfaces*, 6, 17059–17066 (2014).
64. B. Zhang, J. Leng, Z. Ouyang, Z. Yang, Q. Zhang, Q. Li, D. Li and H. Zhao, Superhydrophilic and topography-regulatable surface grafting on PEEK to improve cellular affinity, *Biomater Adv.* 146, 213310 (2023).
65. S. Parvate P. Dixit and S. Chattopadhyay, Superhydrophobic surfaces: Insights from theory and experiment, *J. Phys. Chem. B*, 124, 1323–1360 (2020).
66. F. Sotoudeh, S. M. Mousavi, N. Karimi, B. J. Lee, J. Abolfazli-Esfahani and M. K.D. Manshadi, Natural and synthetic superhydrophobic surfaces: A review of the fundamentals, structures, and applications, *Alex. Eng. J.*, 68, 587-609 (2023).
67. M. A. Saffar, A. Eshaghi and M. R. Dehnavi, Fabrication of superhydrophobic, self-cleaning and anti-icing ZnO/PTFE-SiO_2 nano-composite thin film, *Mater. Chem. Phys*, 259, 124085 (2021).
68. B. Majhy, R. Iqbal and A. K. Sen, Facile fabrication and mechanistic understanding of a transparent reversible superhydrophobic – superhydrophilic surface, *Sci. Reports.*, 8, 18018 (2018).
69. M. Amiriafshar, M. Rafieazad, X. Duan and A. Nasiri, Fabrication and coating adhesion study of superhydrophobic stainless steel surfaces: The effect of substrate surface roughness, *Surfaces Interfaces*, 20, 100526 (2020).
70. K. Ellinas, P. Dimitrakellis, P. Sarkiris and E. Gogolides, A review of fabrication methods, properties and applications of superhydrophobic metals, *Processes*, 9, 666 (2021).
71. M. A. Sankar, V. Rajkumar and S. A. Kumar, Surface topography of additively manufactured carbon fiber-reinforced polymer composites, *Adv. Additive Manuf.*, 2, 321–332 (2025).
72. P. Kang, S. Huang, L. Feng, K. Ehmann and J. Cao, Hierarchical surface textures for improved coating durability using double-sided incremental forming, *J. Manuf. Sci. Eng.*, 147, 081004 (2025).
73. S. Kim, N. Saka and J-H Chun, The role of pad topography in chemical-mechanical polishing, *IEEE Trans. Semicond. Manuf.*, 27, 431-442 (2014).
74. M. A. Tokmedash, C. Kim, A. P. Chavda, A. Li, J. Robins and J. Min, Engineering multifunctional surface topography to regulate multiple biological responses, *Biomaterials*, 319, 123136 (2025).

7

Adhesion Aspects in Color Cosmetics

Hy Si Bui

L'Oréal Research & Innovation, Clark, NJ, USA

Abstract

The complex surface topographies of lips, facial skin, eyelashes and nails make uniform cosmetic film deposition a significant challenge. Topical products strive to modify these surfaces for protection, added functionality, or enhanced appearance, often by forming a film through solvent evaporation. However, achieving consistent film thickness, homogeneity, and adhesion is difficult given the intricate reliefs of these substrates. This challenge is directly linked to consumer demand for long-lasting beauty benefits, as seen in products like non-transfer lipstick, all-day foundation, and durable nail polish, or semi-permanent mascara. Creating such long-lasting, transfer-resistant color cosmetics is particularly complex. The resulting film is not a simple polymer layer, but rather a complex matrix of ingredients including pigments, fillers, thickeners, and surfactants, all impacting adhesion and flexibility. This cosmetic film must resist transfer, withstand environmental factors like water, sebum, and sweat, and avoid undesirable sensory experiences. Balancing this performance with ease of removal is a considerable hurdle for cosmetic chemists, especially considering the dynamic nature of human skin. This chapter focuses on the crucial role of adhesion in cosmetic science, examining how various products adhere to the dynamic substrates of skin, eyelashes, lips, and nails. It delves into the key factors that influence this adhesion, impacting product performance and longevity. Finally, it explores future innovations in developing high-performance cosmetics with enhanced adhesion, while prioritizing sustainability and eco-friendliness.

***Keywords*:** Adhesion, long-wear, color cosmetics, surface free energy, film-formers, pigments, spreading, wetting

Email: hy.bui@loreal.com

Anil N. Netravali and Frank M. Etzler (eds.) The Wide World of Adhesion: A Festschrift in Honor of Dr. Kash Mittal on his 80th Birthday, (231–298) © 2026 Scrivener Publishing LLC

7.1 Introduction

Adhesion plays a crucial role in cosmetics, directly impacting both product performance and consumer satisfaction. Modern consumers seek long-lasting benefits across a range of beauty products, from non-transfer lipstick and all-day foundation to voluminous and long-lasting mascara, and chip-resistant nail polish (Figure 7.1). These desires translate to specific needs like food-proof lip color, long-lasting foundation that withstands sweat and activities, mascara that holds curl and volume without flaking, and nail polish that maintains its shine and color through daily wear and tear. However, formulating products to meet these diverse needs presents significant challenges. Creating long-lasting formulation often requires strong adhesion, which can potentially compromise comfort and ease of removal. For example, a highly adhesive lipstick might feel dry or sticky, while a tenacious foundation could be difficult to cleanse. Balancing these competing demands—long-lasting wear, comfortable texture, easy removal, and compatibility with different substrates (lips, skin, lashes, nails)—requires ongoing innovation in formulation science, polymer science, polymer physics, surfactant technology and a deep understanding of the interactions between cosmetic ingredients and the target surfaces. The pursuit for the perfect balance between enduring adhesion and desirable cosmetic attributes remains a driving force in the beauty industry. Therefore, such consumer demand is fueling innovation in long-wear color cosmetics, particularly for products that offer extended wear alongside diverse color options, textures, and finishes. Achieving these desired effects hinges critically on adhesion science. Long-lasting, waterproof, and smudge-proof claims rely on advanced adhesion technologies to ensure the

Figure 7.1 The consumer desire for long-lasting beauty products.

product adheres securely to the skin and maintains its intended appearance throughout the day. Simultaneously, the growing importance of sustainability is impacting formulation choices, pushing brands to develop eco-friendly ingredients, formulations, and packaging without compromising performance [1, 2]. This dual focus on long-wear performance and sustainability presents a significant challenge, demanding innovative ingredients and formulations that are both highly effective and environmentally responsive [3, 4].

Dedicated to Dr. Kash Mittal on his 80th birthday, this chapter honors his contributions to the field of adhesion science. It delves into the intricacies of adhesion within color cosmetics, exploring the factors influencing product performance on various substrates. The chapter reviews relevant film technologies and their applications in long-wear color cosmetics and concludes by examining future directions in the development of more sustainable and long-lasting products.

7.2 Principles of Adhesion in Cosmetics

Cosmetic adhesion, essential for long-lasting makeup, describes a product's ability to effectively adhere to dynamic biological substrates like skin, hair, lips and nails. While initially detailed by Gardner for medical and dental adhesives [5], seven key adhesion mechanisms offer valuable insights into cosmetic science. Five of these theories can directly apply to cosmetic adhesion, including mechanical interlocking with the substrate texture, electrostatic attraction between charged particles, adsorption of the cosmetic product onto the substrate, thermodynamic adhesion governed by surface and interfacial energies. In addition, tack, a key characteristic of elastomer-based adhesives, is the mechanism enabling bonding to a substrate under light pressure. This is exemplified in pressure-sensitive adhesives (PSAs). Similarly, tack plays a significant role in how cosmetic products adhere to skin and hair. A thorough understanding of these mechanisms empowers cosmetic scientists to develop products with optimized wear and enhanced performance characteristics.

7.2.1 Mechanical Interlocking

Mechanical interlocking plays a significant role in cosmetic adhesion, relying heavily on the surface texture of the application area. When a liquid or creamy product spreads on a porous surface, it penetrates the pores and crevices. Upon drying or setting, the product solidifies, creating a physical, interlocked bond with the substrate. This mechanism is particularly relevant for products like foundation, where the formulation fills in fine lines and pores on the skin, creating a smoother surface and enhancing adhesion. With lipstick, the product conforms to the lip's natural texture, including its micro-crevices, contributing to longer wear. Even mascara benefits from this mechanism, as the formulation can grip individual lashes, adding volume and preventing flaking. Similarly, nail polish adheres to the nail plate, utilizing the microscopic ridges and irregularities on the nail's surface to enhance grip and prevent chipping.

7.2.2 Electrostatic Attraction

Electrostatic attraction is a key principle in developing cosmetic products for improved hair and skin conditioning, manageability, and appearance. This principle leverages the interaction between charged ingredients in cosmetic products and oppositely charged sites on the hair or skin. When a charged product comes into contact with its target substrate, an electrical double layer forms at the interface, creating an attractive force. This is particularly relevant for positively charged polymers, which readily interact with negatively charged hair and skin surfaces. In hair styling, for instance, these polymers bind to hair shafts, providing hold and style retention. Similarly, some skincare products use cationic surfactants or polymers to enhance adhesion to the skin, forming a barrier that helps retain moisture. Furthermore, cationic polymers can interact with anionic surfactants during shampooing, forming and depositing complex coacervates that condition the hair [6, 7].

7.2.3 Adsorption

The adsorption theory of adhesion is central to the performance of many long-wear cosmetic products. It describes the process where molecules adhere to a surface through relatively weak but numerous intermolecular forces like van der Waals forces, hydrogen bonds, and dipole-dipole interactions. This mechanism involves the penetration of a liquid adhesive material into the pores and crevices of the substrate, creating a strong mechanical interlocking effect augmented by these secondary chemical bonds. Polymers with polar or ionic groups are particularly adept at forming these interactions due to their stronger intermolecular forces. The overall adhesion strength correlates with the bond energy: covalent bonds are strongest, followed by hydrogen bonds, and then the weaker van der Waals forces. While mechanical interlocking provides initial grip, adsorption through these secondary bonds significantly enhances adhesion and longevity.

This is evident in various long-wear products. Long-wearing foundations utilize silicone resins and silicone acrylates which enhance the adsorption of pigments strongly to the skin, creating a durable, smudge-resistant film. Similarly, liquid lipsticks employ film-forming polymers that adsorb to the lip surface, locking in color. Long-wear mascaras rely on polymers that strongly adsorb to lashes, providing water and smudge resistance. Setting sprays create a thin film over makeup through adsorption, increasing its longevity. Finally, even long-lasting nail polish, while primarily relying on mechanical interlocking, benefits from the adsorption of polymers to the nail plate. In all these cases, adsorption works alongside other adhesion mechanisms to ensure a strong and lasting bond between the cosmetic and the substrate, with the specific formulation dictating the relative importance of each mechanism.

7.2.4 Thermodynamic Adhesion (Wetting Theory)

The thermodynamic theory of adhesion plays a key role in understanding how long-wear cosmetic products, like foundations, lipsticks, mascaras, and nail polishes adhere to the skin, lips, eyelashes and fingernails and resist removal throughout the day. Adhesion is driven by the interactions at the interface between the cosmetic product and the surface to which it is applied (skin, hair, nails). The role of interface in adhesion phenomena discussed by Mittal [8] can be applicable directly to long-wear cosmetics as below:

- **Surface free energy:** Every surface (skin, cosmetic product) possesses a certain amount of surface energy, representing the excess energy at the surface compared to the bulk material. High surface free energy means the surface is ready to interact with other substances to lower that energy. Therefore, balancing the surface energies of the cosmetic and the skin is critical for optimizing the adhesion.
- **Wetting:** For a liquid cosmetic to adhere well, it needs to wet the surface effectively. Good wetting occurs when the surface free energy of the skin (or hair) is higher than the surface tension of the cosmetic product. This allows the product to spread and make close contact with the surface to form a strong bond.

- **Interfacial Tension:** This is the energy at the boundary between two different substances, like the cosmetic and the skin. Lower interfacial tension indicates a stronger attraction between the two materials. Thus, formulating cosmetic products with ingredients that have favorable interactions with the skin surface will reduce interfacial tension.
- **Work of Adhesion (W_a):** This represents the energy required to separate the cosmetic from the skin. A high work of adhesion indicates strong adhesion and thus better long-wear property. The work of adhesion is related to the surface free energies of the cosmetic (γ_c) and the skin (γ_s), and the interfacial free energy between them (γ_{cs}):

$$\mathbf{W}_a = \gamma_s + \gamma_c - \gamma_{cs}$$

Therefore, the interfacial energy between product and substrate (γ_{cs}) is a critical factor in determining adhesive strength. Minimizing γ_{cs} maximizes the work of adhesion ($\mathbf{W}_a$), resulting in stronger adhesion. This principle, highlighting the crucial role of wetting and interface in adhesion phenomena, is thoroughly discussed by Mittal [8].

7.2.5 Mechanism of Elastomer-Based Adhesive

Pressure-sensitive adhesives (PSAs) adhere by spreading and wetting the substrate under light pressure. The initial stickiness, or tack, crucial for forming adhesive bond, is often enhanced in elastomer-based PSAs by adding tackifier resins. While tack facilitates immediate bonding, Gardner suggests that sustained adhesion relies on additional factors like surface tension, wettability, and mechanical interlocking, especially as tack often decreases after application. Therefore, tack can be viewed as one mechanism contributing to overall adhesion in elastomer-based and other adhesives exhibiting this initial high-tack behavior [5]. In color cosmetics, the PSA principle has been utilized in developing long-wear, transfer-resistant makeup products for the last three decades. For instance, high molecular weight poly(dimethylsiloxane) (PDMS) or thermoplastic silicone elastomers like polysilicone-polyamide, tackified with silicone resin MQ, create a strong initial tack during solvent evaporation in products like lipstick and foundation [9, 10]. This tack diminishes significantly after film formation on the skin. Similarly, long-lasting gloss liquid lipsticks often incorporate a tackifier such as a low molecular weight but high T_g hydrocarbon resin with styrene-based triblock copolymers in the formulations [11, 12]. This tackifier enhances adhesion to the lips, resulting in a longer-lasting gloss compared to formulations without a tackifier.

7.3 Factors Affecting Cosmetic Adhesion -The Role of Substrate

Achieving lasting cosmetic benefits faces significant challenges due to the complex nature of skin, lips, hair, eyelashes, and nail substrates. Figure 7.2 shows the challenges of cosmetic product adhesion to various surfaces. It highlights key surface characteristics that influence how cosmetics interact with skin, hair, lips and nails. These characteristics include surface roughness, the presence of sebum and sweat, elasticity, sensitivity and breathability, and the renewable nature of these surfaces and their low surface free energy.

Figure 7.2 The challenges of cosmetic product adhesion to various surfaces.

7.3.1 Skin

Human skin's dynamic nature makes it challenging for adhesion [9, 10, 13, 14]. Its uneven, elastic surface constantly renews itself, producing sweat, oil, and hair at varying rates. The stratum corneum's low surface energy hinders adhesion. Overly strong adhesion risks skin damage upon removal, and certain ingredients can cause irritation. While cleansing briefly improves adhesion, skin's constant changes—oil and sweat production, hair growth, and cell turnover every two weeks—make consistent, long-lasting adhesion difficult.

This difficulty is compounded by skin's multi-layered structure and elasticity, requiring adhering products to flex and stretch. Its rough texture further complicates adhesion. Additionally, skin secretions, dirt, dead cells, and bacteria create a contaminated surface.

Table 7.1 Effect of skin surface on the adhesion of skin products.

Property	Description	Impact on adhesion
Skin Type	Oily, dry, normal, combination.	Influences adhesion and sensory feel. Sebum (oil) negatively impacts time of foundation wear.
Surface Free Energy (SFE) [9, 10, 18–22]	Skin's hydrophobic surface has lower surface free energy (SFE=38.5 - 46 mJ/m^2) than typical cosmetic products. Cleansing can lower SFE.	Affects wetting, spreading, and adhesion. Silicone-based foundations spread easily due to low SFE.
Young's Modulus (E_y) [10, 23–29]	Measures skin's stiffness. Varies based on location, hydration, age. Lower for younger individuals. E_Y= 10^3 -10^5 Pa (cheek, nose and forehead).	Film formation from product should have a lower modulus than skin for flexibility and comfort.
Surface Roughness [10, 30]	Varies with age and location on the face. Aging increases roughness.	Can impact product application and evenness.
Sebum and sweat [16, 17]	Sebum excretion rate = 0.5-2.5 μg/cm^2/min.	Can delaminate or plasticize makeup films, affecting color and transfer resistance.

Several factors influence the effectiveness and longevity of cosmetic products on skin [9, 10, 14, 15] (Table 7.1):

- **Skin Properties:** Skin type, hydration, oil content, and pH all affect adhesion and long-wear performance. Dry skin, for example, can hinder adhesion due to its rough texture and reduced elasticity.
- **Skin Surface:** The skin's roughness, composition (lipid content, hydration), and condition (health, pH) are crucial for adhesion. A rougher skin generally promotes better adhesion through mechanical interlocking, while dry skin can impede it. A slightly acidic pH (around 5.5) can be beneficial to adhesion of certain materials.
- **Additional Factors:** Skin microrelief, hair, curvature, movement, and the presence of sebum and sweat all influence product adhesion. Hair acts as a barrier, reducing contact, while facial movements necessitate flexible films.

Understanding these skin properties is crucial for developing effective facial products with good adhesion and lasting power. Formulations must be tailored to the specific characteristics of the target skin type.

7.3.2 Lips

The human lip is a complex structure composed of muscles, connective tissues, and a unique type of skin [31, 32]. Its reddish color, particularly prominent in the vermilion (or red lip), arises from underlying blood vessels visible through the thin, translucent epithelium – a key distinction from the rest of the face.

The lips are divided into three distinct zones (Figure 7.3):

a. the cutaneous lip, like regular facial skin; the vermilion,
b. a transitional zone between the skin and the oral mucosa characterized by a thinner stratum corneum and incomplete keratinization; and
c. the labial oral mucosa, the inner moist lining of the mouth.

The vermilion zone (lips) is constantly exposed to environmental factors and has unique characteristics. These include a specialized keratin pattern, less melanin (pigment), and a tighter connection between its surface layer and the underlying tissue due to a higher density of finger-like projections

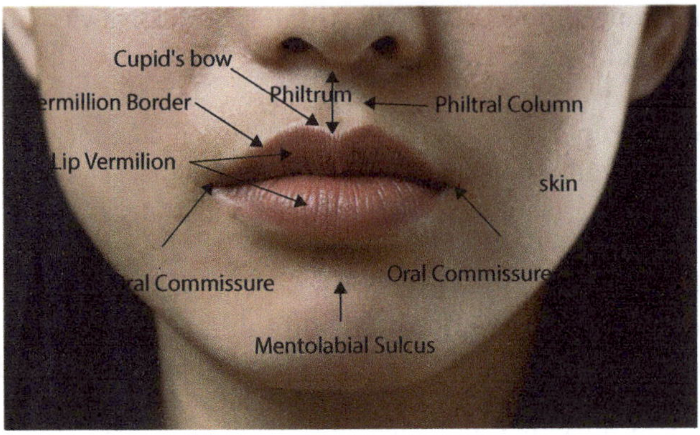

Figure 7.3 Anatomy of the lip [31].

called rete pegs, compared to the skin surrounding the lips [31]. While the vermilion lacks hair follicles and sweat glands, contributing to its susceptibility to dryness and chapping, particularly in certain weather conditions, the presence of numerous nerve endings throughout the lips contributes to their high sensitivity, vital for functions like speech, eating, and expressing emotions. The oral mucosa, constantly moistened by saliva, plays a role in lip hydration, though this is not sufficient to prevent dryness in the exposed vermilion.

Key biophysical properties of the lips, as reported in the literature [31–39], are summarized in Table 7.2.

The extent of lipstick adhesion, and therefore its wear, is heavily influenced by the complex structure of the lips. Several factors contribute to this intricate relationship:

- **Surface Texture:** The microscopic texture of the lips, including fine lines, wrinkles, and unique lip prints, plays a dual role. While these crevices can enhance adhesion through mechanical interlocking, excessive roughness can lead to uneven, patchy application. Dry, flaky lips further exacerbate this issue, preventing the formation of a smooth, continuous lipstick film and resulting in flaking and reduced wear.
- **Lip Physiology:** The outermost layer of the lip, the stratum corneum, is crucial for adhesion. Its hydration level and lipid content directly impact how well lipstick adheres. A well-hydrated stratum corneum provides a smoother, more receptive surface. Conversely, natural moisture loss and

Table 7.2 Biophysical properties of lips.

Biophysical property	Description	Relevance to lip products
Stratum Corneum (SC) Thickness	Significantly thinner in the vermilion (~1/3 thickness of facial skin stratum corneum, ~19.5 μm)	Impacts product penetration and requires formulations specifically designed for the delicate vermilion.
Young's Modulus E (Stiffness) [23]	E= 33.7 kPa (Lower lip)	Product deposited should be softer than the lips for comfort.
Hydration/Water Content	Lip skin is comparatively well-hydrated. Vermilion experiences higher water content but also higher trans epidermal water loss (TEWL). Water content and TEWL change with age, with younger individuals experiencing more dryness	Requires products that balance hydration with barrier protection, especially in the vermilion. Age-specific formulations may be beneficial.
Trans Epidermal Water Loss (TEWL)	Higher in the vermilion due to impaired barrier function (~51 $g/m^2/h$). Influenced by environmental factors	Products should focus on reducing TEWL and reinforcing the skin barrier in the vermilion.
Natural Moisturizing Factors (NMFs)	Present in lip skin, but lower levels in the vermilion compared to other skin sites. Shows relatively high variability	Incorporating NMFs in lip products can support hydration and barrier function in the vermilion.
Surface pH [38]	pH= 6.2-7.6	Lip product pH should be compatible with the natural lip pH.

sebum production can hinder adhesion, causing dryness and slippage.

- **Chemical Composition:** The chemical interplay between the lips and lipstick is also key. Proteins and lipids on the lip surface interact with the lipstick's ingredients, influencing the adhesion forces. The slightly acidic pH of the lips can further affect the adherence of certain lipstick components.
- **Lip Dynamics:** The constant movement of the lips during talking, eating, and facial expressions creates a challenging environment for lipstick. The lipstick film must be flexible and resistant to cracking or flaking under this stress. Furthermore, exposure to saliva and the moist environment of the mouth contributes to wear and tear, particularly on the inner lip.

7.3.3 Eyelashes

The properties of eyelashes (Figure 7.4) were reviewed [40–47] and are summarized in Table 7.3. From these properties, several eyelash characteristics might influence how well mascara can adhere. Thickness or density, which varies along the lash (thicker at the root, thinner at the tip) and between individuals, affects rigidity. Thicker lashes are inherently stiffer and harder to curl, requiring mascara formulations with higher rigidity for comparable curling effects across different eyelash types. While length is a factor, its influence on adhesion is less significant compared to that of thickness. Longer lashes provide more surface area for adhesion but can

Figure 7.4 Human eyelash.

Table 7.3 Eyelash properties.

Property	Description	Ethnic variations	Comments
Quantity (Lower Lid) [41, 42]	3-4 rows, totaling 70-75 lashes.	May vary slightly.	
Quantity (Upper Lid) [41, 42]	5-6 rows, totaling 90-160 lashes.	May vary slightly.	
Growth Rate [42, 44]	50-140 μm/day.	Similar across ethnicities.	Influenced by several factors including prostaglandin analogs, which are commonly used in medical treatments designed to enhance eyelash growth
Length [42, 44]	6-12 mm (upper lid), 6-8 mm (lower lid).	Can vary, but maximum length generally similar.	Never exceeds 12mm on upper lid regardless of ethnicity.
Thickness [42, 43]	Varies along the lash and between individual lashes; thicker at the root, thinner at the tip.	Varies, with Asian lashes generally thicker.	Influences rigidity and curl.
Curvature/Curl [42, 43]	Present in all individuals; initiated at the bulb and continues to the tip.	Varies, with Caucasian lashes typically more curled.	Influenced by asymmetric cell types within the follicle.

(Continued)

Table 7.3 Eyelash properties. (*Continued*)

Property	Description	Ethnic variations	Comments
Pigmentation [45, 46]	Primarily determined by melanogenesis; retained longer in eyelashes compared to scalp hair (gray eyelashes are less common).	May vary slightly.	Influenced by the quantity of melanocytes in the follicle.
Life Cycle [41]	Anagen (growth), catagen (degradation), telogen (resting); complete cycle: 4-11 months (shorter than scalp hair).	Similar across ethnicities.	

also be more prone to smudging if the mascara film is not flexible enough. Curvature, another ethnically variable characteristic, impacts the final curl achieved. Straighter lashes, more common in Asian populations, present a greater challenge for curling mascaras. The modulus or rigidity of the eyelash, influenced by its protein structure, determines how it bends and flexes, impacting the film's ability to conform without cracking. Finally, the blinking rate (15 - 20 blinks/min) introduces a mechanical challenge [47]. Frequent blinking generates friction between upper and lower lashes, increasing the risk of flaking and transfer if the mascara film is not robust enough.

7.3.4 Nail

The human nail unit is a complex structure with several key components contributing to its functionality as a protective barrier, a versatile scraping tool, and a sensory organ [48–50] (Figure 7.5). The nail plate itself is a hardened, keratinized structure comprised of three tightly bound layers: the dorsal (outermost), intermediate, and ventral (innermost) layers. The intermediate layer, rich in cystine and disulfide bonds, forms the main

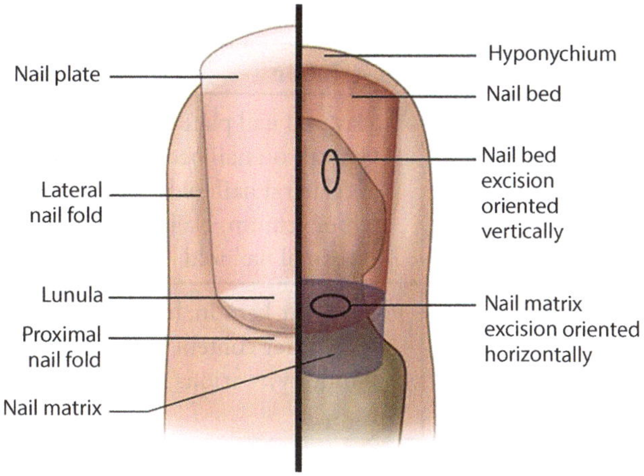

Figure 7.5 Nail anatomy [48].

barrier. Keratin, an α-type fibrous protein, is the primary structural component, organized in transversely aligned filaments in the intermediate layer and in a less-ordered fashion in the dorsal and ventral layers. This structure contributes to the nail's anisotropic mechanical properties. The nail plate sits on the nail bed, a vascularized epithelial layer contributing to nail plate growth and providing sensory input. The nail matrix, located beneath the proximal nail fold, is the site of nail plate production. The proximal and lateral nail folds frame and protect the nail matrix, with the cuticle (a thickened keratinous rim) sealing the proximal border. Finally, the hyponychium thickens the epidermis beneath the free distal edge, providing additional protection. Nails have a relatively low surface energy (around 34 mJ/m^2) and are considered hydrophilic due to the water-binding capacity of keratin, despite a low lipid content [51]. This hydrophilicity is coupled with low water permeability due to the tight keratin structure. The nail plate's pH is typically around 5, slightly acidic, with toenails having a slightly higher pH than fingernails [52–54]. The nail plate is harder than the surrounding skin and exhibits significant elasticity, which is greater along the growth axis than transversely. These physical properties of human nails are summarized in Table 7.4.

The structure of the fingernail significantly influences the adhesion and longevity of nail polish. The nail plate itself, composed of tightly packed keratin layers, presents a relatively smooth, but slightly porous surface [50]. This porosity can aid initial adhesion by allowing the nail polish to penetrate and anchor slightly into the porous layers. However, excessive moisture within the nail plate can weaken this bond, leading to chipping

Table 7.4 Physical properties of human nails.

Property	Description
Structure [50]	Tri-layered nail plate (dorsal, intermediate, ventral) on nail bed; framed by proximal and lateral nail folds, cuticle, and hyponychium. Nail matrix beneath proximal nail fold.
Composition [50]	Primarily α-keratin; low lipid content (0.1-1%); water content 7-12% (varies with humidity); various minerals (e.g., Na, K, Ca, Mg).
Surface Free Energy [51]	~34 mJ/m^2 (relatively low)
Hydrophilicity/Lipophilicity	Hydrophilic (due to keratin) but low water permeability (due to tight structure); low lipid content contributes to moderate hydrophobicity.
pH [52–54]	~5 (slightly acidic), toenails slightly higher pH than fingernails.
Rigidity [55–57]	Hard; anisotropic elasticity (greater along growth axis). Modulus of Elasticity: ~4 GPa (parallel to growth), ~2 GPa (perpendicular to growth)
Growth rate [50, 58]	Fingernail: 3 mm/month Toenail: 1 mm/month

and peeling. The free edge of the nail, being thinner and more exposed, is particularly susceptible to wear and tear, impacting the longevity of the nail polish at the tips. The surface texture of the nail plate, including ridges, grooves, or damage from previous manicures, also plays a role. These imperfections can create uneven surfaces that prevent the nail polish from forming a smooth, continuous film, increasing the risk of chipping and peeling. Additionally, the cuticle region, where the nail plate meets the skin, is a critical area for maintaining proper adhesion. Proper cuticle preparation is essential to ensure a clean, even surface for the nail polish to adhere to, preventing lifting and premature wear. Finally, the presence of natural oils and moisture on the nail plate can interfere with nail polish adhesion, highlighting the importance of thorough cleaning and dehydration prior to application.

7.4 Factors Affecting Cosmetic Adhesion – Cosmetic Ingredients

Generally, long-lasting, non-transfer color cosmetics with desirable sensory properties rely on three key components: film-formers, pigments, and fillers [10, 14, 15, 62].

Film-formers enhance transfer resistance, pigments determine color longevity and vibrancy, while fillers contribute to the overall feel and texture on the skin as seen in Figure 7.6.

7.4.1 Film-Forming Technologies in Long-Lasting Color Cosmetics

Long-lasting, transfer-resistant color cosmetics achieve their durability through specialized formulations and film-forming ingredients. These products are typically available as anhydrous liquids, solids (like lipsticks and eyeshadows), or water-in-oil or water-in-silicone emulsions (common in long-wear foundations). The key to their longevity lies in the use of polymers, which act as film-formers.

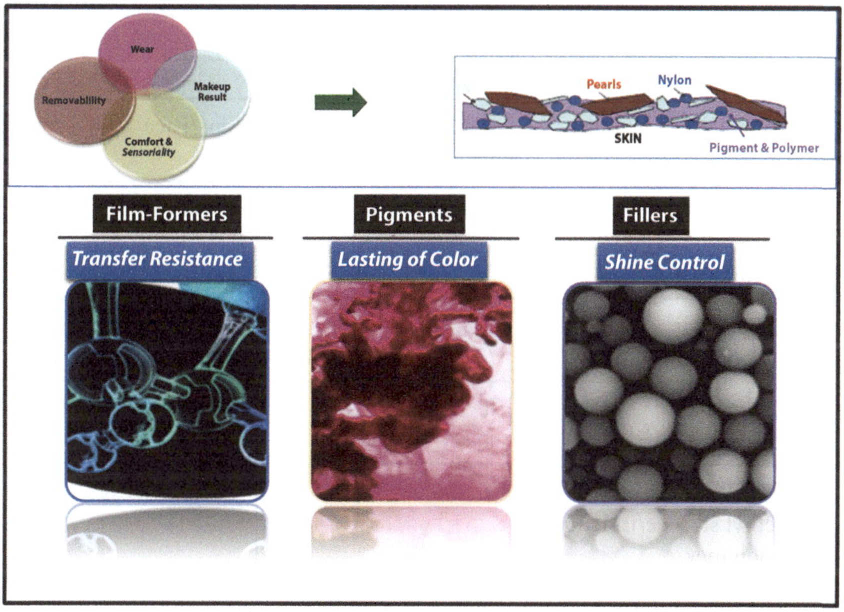

Figure 7.6 The key components of long-wear and non-transfer color cosmetic formulations.

These polymers, dissolved in volatile solvents, create a thin, adhering film upon application as the solvent evaporates. The film properties, including elasticity, flexibility, and thickness, are precisely controlled by adjusting the polymer type with selected glass transition temperature T_g, molecular weight, concentration, and other ingredients in the formulation. While film formation can sometimes cause skin stress due to drying and shrinkage, ongoing advancements in polymer technology strive to minimize it. The resulting cosmetic film typically offers all-day wear yet is readily removable with common makeup removers or cleansers.

Silicone-based polymers, such as silicone acrylates, trimethylsiloxysilicate (MQ) resins, and high molecular weight polydimethylsiloxane (PDMS), are frequently used due to their excellent film-forming property, and water resistance. Other synthetic polymers like water dispersible (latexes) or oil-dispersible acrylate copolymers, and nitrocellulose or UV-cured acrylate oligomers for nail polish, also play significant role.

Table 7.5 Common film technologies used in color cosmetics.

Film-forming material	Property	Application areas
MQ (Trimethylsiloxysilicate)	Rigid Particle (200°C $<T_g<$ 370°C)	Lips/Face/Eyes
Silicone Acrylate	Soft or Rigid (20°C $< T_g <$ 100°C)	Lips/Face/Eyes
T-Propyl Silsesquioxane	Soft ($T_g <$ 25°C)	Lips/Face/Eyes
Water-dispersible Acrylate Copolymer (Latex) (or polyurethane)	Soft or Rigid (-40°C $< T_g <$ 100°C)	Eye lashes/Nails
Oil-dispersible Acrylate Copolymer	Soft or Rigid (25°C $< T_g <$ 100°C)	Lips/Face/Eyes/Nails
Hydrogenated Polycyclopentadiene	$T_g \sim$ 56 °C	Lips/Face/Eyes/Nails
Hydrogenated styrene/methyl styrene/indene copolymer	$T_{softening} \sim$ 100 °C	Lips/Face/Eyes
Rosin Ester	30°C $< T_g <$ 60°C	Eyes/Lips
Nitrocellulose	Rigid ($T_g >$100°C)	Nails

While natural film-formers exist, they currently face challenges in performance, stability, and sensory feel compared to synthetic alternatives, limiting use in long-wear cosmetics. Table 7.5 summarizes some common materials used for film formation on lips, skin, eyelash and nails.

7.4.1.1 *Silicone-Based Materials*

Previous research has extensively documented the use of silicone chemistry in cosmetics, the mechanism of adhesion for long-wear products, and the various materials employed for film formation on skin [9, 10, 14, 15, 38, 40, 59–62], further supporting the importance of these technologies in long-lasting cosmetics.

- Trimethylsiloxysilicate (MQ) (Figure 7.7a) is a common silicone resin used in long-wear lipsticks, foundations, and mascaras. It acts as a tackifier for high molecular weight PDMS or other silicone elastomers [9, 10, 59]. This highly branched, cube-shaped molecule, with a molecular weight of approximately 10,000 Da and a glass transition temperature (T_g) between 200°C and 370°C, significantly influences the performance of long-wear cosmetics. The interaction between MQ resin and PDMS through hydrogen bonding creates a physical crosslinked network, contributing to the material tackiness. The observed increase in the dried film tack, directly correlating with higher MQ content, suggests improved adhesion to the skin or lips. Similarly, tack or adhesion is also greater when the MQ resin possesses a higher hydroxyl density. Moreover, MQ can also function as a rigid filler. For instance, when dispersed in a thermoplastic elastomer like polysiloxane polyamide copolymer (PSPA), MQ acts both as a rigid filler and a physical crosslinker *via* hydrogen bonding, enhancing the final film viscoelasticity and mechanical properties.
- Polypropylsilsesquioxane or T-propyl resin, a more flexible silicone with a ladder-like structure (Figure 7.7b) and a softening temperature of 16°C. Its lower T_g eliminates the need for plasticizers, even acting as a plasticizer for higher T_g polymers. Its inherent stickiness promotes skin adhesion and compatibility with other oils. T-propyl resin functions as a liquid binder, enhancing pigment and filler adhesion. T-propyl resins are also used in other long-wear cosmetics

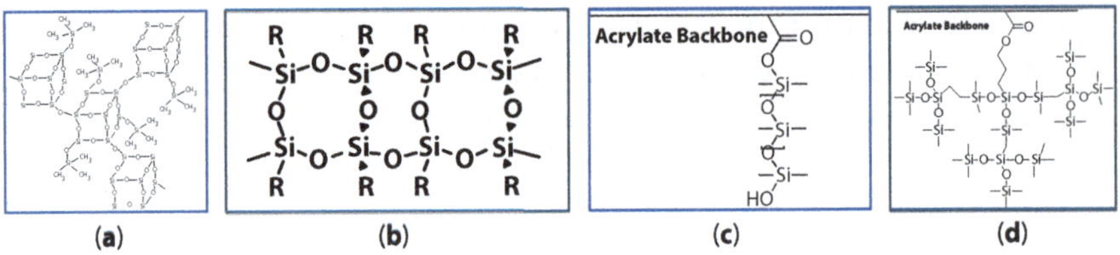

Figure 7.7 Typical silicone-based film-formers. (a) MQ resin, (b) T-propylsilsesquioxane, (c) acrylate/dimethicone copolymer, and (d) silicone acrylate dendrimer.

besides foundation, such as lipsticks, eyeshadows, and eyeliners [9, 10, 14, 59].

- Another long-wear makeup film-former is silicone acrylate, a hybrid polymer combining the adhesion of acrylates with the texture of silicones. During solvent evaporation, phase separation occurs, with acrylate segments anchoring to the skin and silicone segments migrating to the surface, creating a non-tacky, water- and oil-repellent film. Silicone acrylates offer benefits like comfort, skin protection, skin tightening, sebum resistance, and long-lasting wear. One example is the acrylate/dimethicone copolymer (Figure 7.7c), available as ShinEtsu KP-545 and KP-550. With a T_g of around 23°C - 40°C, this copolymer can be plasticized by T-propyl silsesquioxane [9, 10, 59].
- Another type of silicone acrylate copolymer is the acrylate polytrimethylsiloxymethacrylate, characterized by its dendritic silicone structure (Figure 7.7d). This copolymer typically forms a brittle film with a high glass transition temperature (T_g) of 70°C, thus necessitating plasticization. However, by carefully balancing the ratio of the silicone to the acrylate components, the T_g can be adjusted; a higher silicone content will result in a softer material. In cosmetic formulations, the silicone dendrimer functions as a binder, imparting a smooth, water-resistant, and transfer-resistant finish. Concurrently, fillers enhance the product's matte appearance and contribute to long-lasting color [10, 62].

7.4.1.2 Hydrocarbon-Based Materials

Several non-silicone film-formers contribute to long-wear color cosmetics, each with its own strengths and weaknesses. These include hydrogenated polycyclopentadiene, hydrogenated styrene/methyl styrene/indene copolymer, rosin esters, water-dispersed or non-aqueous dispersed acrylate copolymers and polyurethanes. While generally offering good adhesion to skin, hair, lips and eyelashes, along with high water resistance, these materials are often sensitive to oils.

- Hydrogenated polycyclopentadiene (KOBOGUARD® 5400) enhances adhesion, water resistance, rub resistance, and film stability, while also imparting gloss and aiding pigment suspension. Its glass transition temperature (T_g) is 54°C, and its high refractive index falls between 1.55 and 1.57 that contributes to high gloss. This polymer is used in many cosmetic formulations to enhance the adhesion to skin, eyelashes, hair, lips and nails [63].
- Hydrogenated styrene/methyl styrene/indene copolymer (Regalite™ R1100 CG), derived from petrochemicals, acts as a film-former or a tackifier for thermoplastic elastomer, contributing to adhesion and transfer resistance. It offers excellent stability, solubility, and compatibility with non-polar cosmetic ingredients, producing clear, glossy films suitable for eye makeup, face makeup, lipstick, and gloss [11, 12, 64].
- Rosin esters, known for their hydrophobicity, flexibility, cohesion, adhesion, and shine, are also effective film-formers in long-wear cosmetics [40, 65–68]. Pentaerythritol rosin ester, with its low critical surface tension and high water contact angle (76°), readily wets and adheres to keratin substrates [69], making it a suitable candidate for semi-permanent mascara formulations [67, 68]. Other rosin-based film-formers used in mascaras include tall oil glycerides, glyceryl rosinate, and their hydrogenated versions, often in combination [68].
- Aqueous dispersions of acrylate copolymers (latex) or polyurethanes are commonly employed in mascara formulations to enhance adhesion and provide long-lasting wear [40, 70, 71].
- Similarly, a non-aqueous dispersion (NAD) of acrylate copolymer, capable of dispersing in volatile oils such as isododecane, has been developed and extensively utilized across various cosmetic applications, including long-wear lipsticks, mascaras, eyeshadows, and skin-tightening products [72–76]. This poly(meth)acrylate copolymer exhibits exceptional

adhesion to keratin substrate, attributed to a carefully balanced combination of its glass transition temperature (T_g), tackiness, and Young's modulus. The copolymer composition has been engineered to incorporate diverse functionalities, such as polar moieties and glass transition modulating units. By precisely tuning the polymer composition, it is possible to influence key properties like glass transition temperature, Young's modulus, flexibility, surface free energy, oil resistance, cohesion and adhesion characteristics [72, 73].

7.4.2 Surface Treated Pigments in Long-Wear Color Cosmetics

Long-wear cosmetics rely on a diverse range of pigments (inorganic and organic) to achieve both vibrant, lasting colors and specific effects. For enhancing their wear performance, treating pigments with surface modifiers is essential. These treatments improve pigment dispersion, compatibility with other formulation components, and overall wear by altering how the pigment interacts with its surroundings [10, 15, 62]. For example, silicone-based ingredients like polymethyldisiloxane (PDMS, or methicone) improve water repellency and spreadability. For better adhesion and stability, stearoyl glutamic acid is often used, frequently in combination with

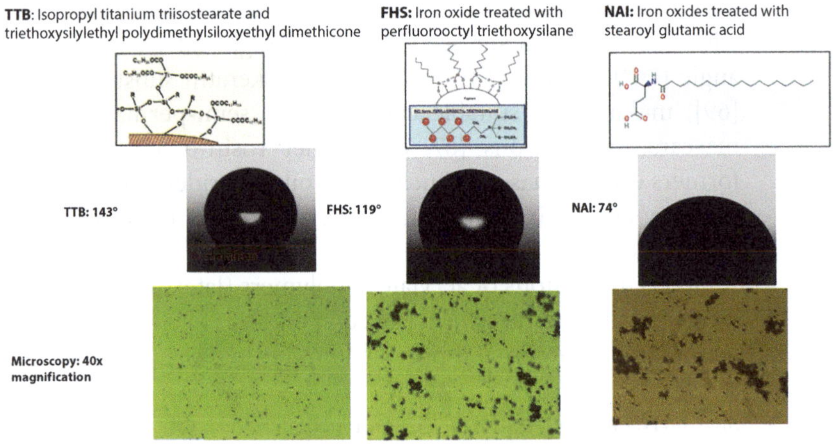

Figure 7.8 Pigment surface-treated with TTB demonstrates the highest water resistance, indicated by a water contact angle of 143°, and good pigment dispersion in silicone fluid (as seen in the micrograph). In contrast, NAI-treated pigment shows the lowest water resistance, with a water contact angle of 74°, and exhibits greater pigment aggregation [10].

aluminum hydroxide. Alkoxysilanes enhance compatibility with other silicone-based ingredients in the formulations. Organic titanates, such as isopropyl titanium triisostearate (ITT), boost adhesion and durability. Finally, for superior water and oil resistance, perfluorinated treatment is employed. The selection of both the pigment and its surface treatment depends on the desired properties of the final product, pigment type, and other ingredients, with compatibility and optimal dispersion being critical factors as shown in Figure 7.8. Ultimately, these modifications contribute to transfer resistance, smudge-proof property, and fade resistance, essential for achieving the long-lasting wear expected of these cosmetics.

7.4.3 Fillers in Long-Wear Color Cosmetics

Fillers in long-wear cosmetics contribute significantly to both sensory feel and oil or water (or both) absorption, improving the product's performance and wear [10, 15, 62].

Silica, a common filler, offers both oil and water absorption alongside optical effects, creating a smoother appearance. Talc acts as a bulking agent and also absorbs oil and water. Perlite and kaolin both provide oil and water absorption, with Perlite giving a soft-focus (blurring or smoothing) effect and kaolin also acting as a bulking agent. Nylon 12 contributes to a soft-focus finish and acts as a bulking agent, absorbing both oil and water. Poly (methyl methacrylate) (PMMA) particles also absorb oil and water and provide a soft-focus effect. Boron nitride, another soft-focus filler, absorbs oil and water while potentially enhancing spreadability. Figure 7.9 illustrates common fillers utilized in long-wear color cosmetics to absorb oil, enhance the soft-focus effect, and improve sensory attributes. Finally, Aluminum starch octenylsuccinate (ASO) modifies texture and sensorial perception, absorbing oil and potentially adding a smooth, velvety feel.

Figure 7.9 Typical functional fillers used in long-wear color cosmetics [62].

Newer fillers on the market include cellulose-based Tegofeel C10 (Evonik) for sensory modification, Amaze Nordic barley (Nouryon) which combines sensory modification with absorption, EcoFil B110 (Eckart) for soft-focus and opacity, and Spherilex 10PC (Evonik) for soft-focus and absorbency. Formulators select fillers based on the desired texture, oil control needs, and other properties they want to achieve in the final product [62].

7.5 Techniques to Measure Cosmetic Product Adhesion and Performance

Several scientific techniques can be employed to evaluate the performance of long-wear color cosmetics, focusing on bulk viscoelasticity, surface properties, adhesion and cohesion of film formation from products. By assessing these critical attributes, a thorough understanding of the physical properties and performance of long-wear formulations is achieved, as summarized in Table 7.6.

Table 7.6 Summary of test methodologies for evaluating long-lasting film/products.

Test method	Property evaluated
Contact angle measurement [10, 14, 15, 38, 51, 73, 77–86]	▪ Wetting and spreading of sebum, sweat, and water on films ▪ Resistance of film to sebum, oil and water ▪ Surface free energy of skin, hair, nails and deposited films, residual film on skin
Tape test/peel test [38, 73, 80, 81, 89]	▪ Adhesion and cohesion of film to selected substrate
Laser Shock Adhesion Test [90]	Adhesion strength from degree of delamination of film from substrate
Texture Analyzer [10, 38, 61, 73, 85, 91, 92]	▪ Stickiness or tack from wet to dried film ▪ Adhesion and cohesion properties of materials

(*Continued*)

Table 7.6 Summary of test methodologies for evaluating long-lasting film/products. (*Continued*)

Test method	Property evaluated
Mechanical rubbing test [38, 40]	Durability of films/wear of film under various conditions
Colorimetry (lightness (L*)) [40, 79, 87, 95]	▪ Color transfer of pigmented films before and after exposure to sebum or water ▪ Wear resistance of film after rubbing
Stretching test/fragmentation test [15, 38, 73]	Integrity and cohesiveness of films under stretching condition
Dynamic Mechanical Analysis (DMA) [40, 73, 88]	Young's modulus, elastic modulus and glass transition temperature (T_g) of cosmetic film
Instron tensile testing [38, 73]	Young's modulus and flexibility/durability of film
DSC [38, 73]	Thermal property of film: glass transition temperature (T_g) and melting temperature (T_m)
SEM/TEM [40, 73–75, 83]	▪ Microstructures of substrates ▪ Nano and microstructures of film formed
Atomic Force Microscopy (AFM) [73, 101]	Nano-mechanical and nano-adhesion property of films
Rheology [15, 40, 103, 104]	Viscoelasticity and shear flow of polymer solutions or products related to wetting, spreading, tack, stability and application

7.5.1 Surface Analysis

7.5.1.1 Contact Angle (CA) Measurement

Adhesion in cosmetic products is fundamentally governed by wettability and the physical interactions occurring between the product and biological substrates like skin, nail or hair. To comprehend this relationship and correlate surface properties, including surface free energy (SFE) and surface roughness, with adhesion performance, contact angle (CA) measurement

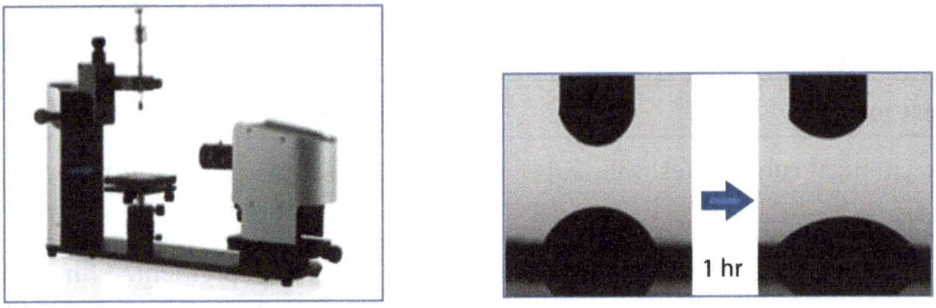

Figure 7.10 Sebum resistance in a long-wear foundation film was assessed by measuring the sebum contact angle over a one-hour period.

proves to be a valuable technique. Specifically, CA measurements are crucial for characterizing the inherent surface properties of skin, assessing the impact of cosmetic thin films or products applied to it, and determining the surface free energy of the cosmetic products themselves or the films they create [10, 14, 15, 38, 51, 73, 77–86]. The fundamental theories and methods for determining the surface free energy of solids have been extensively reviewed and discussed by Etzler [77, 78]. Furthermore, analyzing the contact angles of various liquids, such as oil, water, sweat or sebum, on cosmetic film substrates provides significant insights into their wetting behavior, spreading characteristics, and the film resistance to these substances. For example, a noticeable change in the sebum contact angle over time on a foundation film can indicate its inadequate sebum resistance [10, 15, 79] (Figure 7.10). Another practical application involves evaluating the water and oil resistance of lipstick formulations that incorporate film-formers [14, 38, 73], again utilizing contact angle measurements.

7.5.2 Adhesion and Cohesion

7.5.2.1 Peel Test

To assess the adhesion of cosmetic products to skin, several methods are employed. One common technique is the peel test, which can be conducted directly on skin or on synthetic substrates like bio-skin or *in vitro* nail using Scotch® tape [38, 80, 89]. Within the peel test, two primary approaches are used: one involves pressing the tape onto the dried product surface and then visually rating the amount of color or product transferred to the tape after peeling [38, 73]; the other utilizes instruments such as an Instron machine or a texture analyzer to precisely record adhesion strength from peel test and work of adhesion from pull-off test [80, 89]. In addition, a rotational rheometer can determine the adhesion of the adhesive film by measuring the normal force (tension mode) or maximum stress required to separate 2 parallel plates having material between them from a stationary position during a controlled pull-away at a fixed rotation speed [81].

Another simple and quick method, commonly used in the coating industry and applicable for cosmetic films, is the cross-hatch adhesion test. In this test, the film is scratched with a specific pattern, and a Scotch® tape is applied over the scratched area and manually pulled off. The adhesion strength is then evaluated based on the amount of product remaining on the surface [38, 80, 89].

7.5.2.2 Laser Shock Adhesion Test

Beyond the peel test, another noteworthy method for evaluating film adhesion is the Laser Shock Adhesion Test (LASAT). In LASAT, a laser shock is applied to the film surface, generating a pressure load that subsequently delaminates the film from its substrate, as detailed by Richard [90].

7.5.2.3 Tack Test

Tack, a critical adhesion mechanism for pressure-sensitive adhesives (PSAs), is equally vital for the performance of long-wear color cosmetics. It refers to the ability of a material to form a bond with a substrate under light pressure and short contact time. In cosmetic applications, tack testing evaluates a formulation stick to a substrate after solvent evaporation and film formation [10, 38, 61, 73, 85]. These tests also assess the duration of tack during product application, which directly influences the sensory experience of consumers.

Various methods are employed to evaluate tack, ideally yielding a force versus distance failure curve that accounts for variables such as contact time, contact area, temperature, pull-off rate, and applied load. Among these, probe tack tests are commonly performed using a Texture Analyzer from Stable Micro System equipped with a cylindrical or spherical probe [91]. This method measures the force needed to separate the probe from a wet formulation or dried film at a constant speed after a fixed contact time [91, 92]. The resulting force-displacement curve provides key insights: the maximum force indicates tack or adhesion strength, while the area under the curve reflects the sample cohesiveness. For wet samples, rotational or extensional rheometers offer an effective alternative for tack measurement [93, 94]. This is due to complex nature of tack, stemming from an interplay of adhesive viscosity, elasticity, and surface energy. For instance, a rotational rheometer can assess the tackiness and cohesive/adhesive properties of materials like hair wax, toothpaste, or honey through a pull-away test that monitors normal force (tension) [94]. This involves measuring the force required to separate two parallel plates with a defined volume of material between them, starting from a stationary position with no initial pressure. In this context, the peak in negative normal force (tension) signifies 'tack,' the area under the force-time curve represents adhesion or cohesion strength, and the time taken for the peak force to decay by 90% offers a comparative measure of the failure rate. By utilizing these diverse methodologies, cosmetic formulators can effectively optimize the tack and adhesion of long-wear color cosmetics, thereby ensuring both desirable application properties and prolonged product performance.

7.5.2.4 Color Transfer Resistant Test/Rubbing Test

Testing for color transfer resistance is a crucial aspect of evaluating long-wear cosmetic products. Initially, this involves a simple tapping test where a cotton pad is pressed onto the dried film (typically after 30 minutes or 24 hours of drying) to observe any color transfer [15, 38, 73]. For products like lipstick, real-world scenarios such as kissing a tea or coffee cup are simulated to assess transfer to objects [38, 60, 87]. During the COVID-19 pandemic, evaluating color transfer to face masks also became a common method [87, 95, 96].

To mimic the product's durability under challenging conditions, such as high heat and humidity, rubbing test is performed. In this method, a drop of artificial sebum, sweat, oil or water is applied to the cosmetic film on a substrate [15, 38, 73]. A Kimwipe or cotton pad then presses and rubs the film a specified number of times. The assessment involves two key observations: the amount of color transferred to the wipe and the extent of damage to the film on the substrate. For example, as illustrated in Figure 7.11,

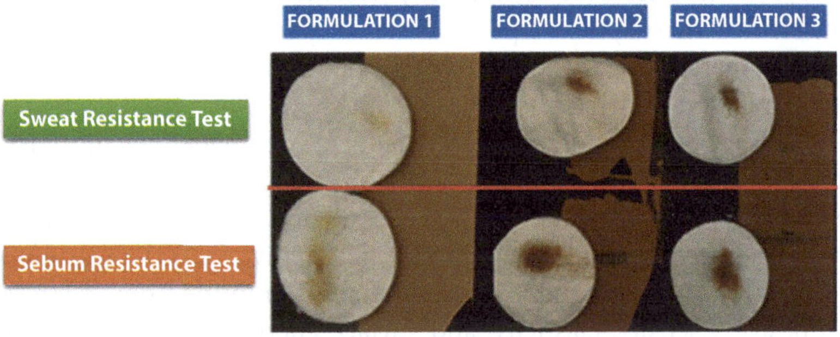

Figure 7.11 The color transfer resistance of the foundation films to sweat and sebum was evaluated by rubbing them with a cotton pad.

the color transfer from foundation films to cotton pads after rubbing with sweat and sebum can be observed across different formulations.

Beyond visual assessment, colorimeters offer a quantitative approach to evaluate color transfer [40, 79, 95]. By measuring the color difference (ΔL^*) of samples before and after rubbing, the degree of transfer can be precisely quantified. The L^* value, which represents brightness on a scale from 0 (black) to 100 (white), is used for this purpose. A high ΔL^* indicates significant color transfer, while a low ΔL^* signifies excellent transfer resistance or "transfer-proof" when $\Delta L^* = 0$ (no color transfer) qualities.

7.5.2.5 Mechanical Property Assessment

To determine the cohesiveness or mechanical properties of the film, the dynamic mechanical analysis (DMA) or tensile test using Instron can be performed on a thin film [40, 73, 88]. Another quick way to evaluate the cohesiveness is a stretching test, in which the sample is deposited on a rubber band or artificial skin and is stretched [15, 38, 73]. The cohesiveness of the film is visually assessed and graded.

7.5.3 Microscopic and Nanoscale Analysis

7.5.3.1 Scanning Electron Microscopy (SEM)

Scanning Electron Microscopy (SEM) is a powerful technique for visualizing the integrity and formation of films on substrates [40, 73–75, 83]. For example, the images below (Figure 7.12) show mascara applied to a hair

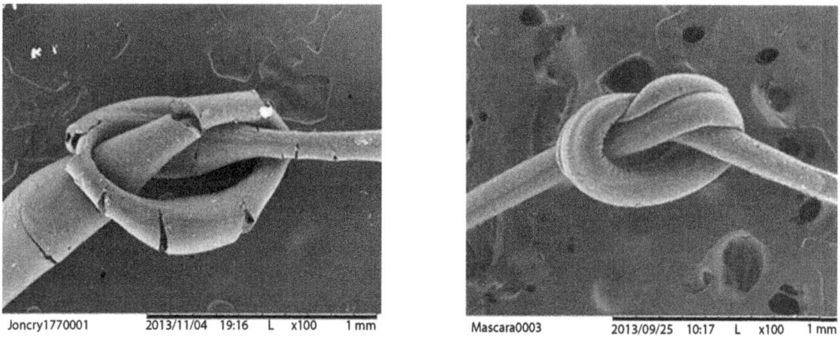

(a) (b)

Figure 7.12 SEM images of film formation from mascara deposited on hair fiber: (a) without plasticizer and (b) with plasticizer.

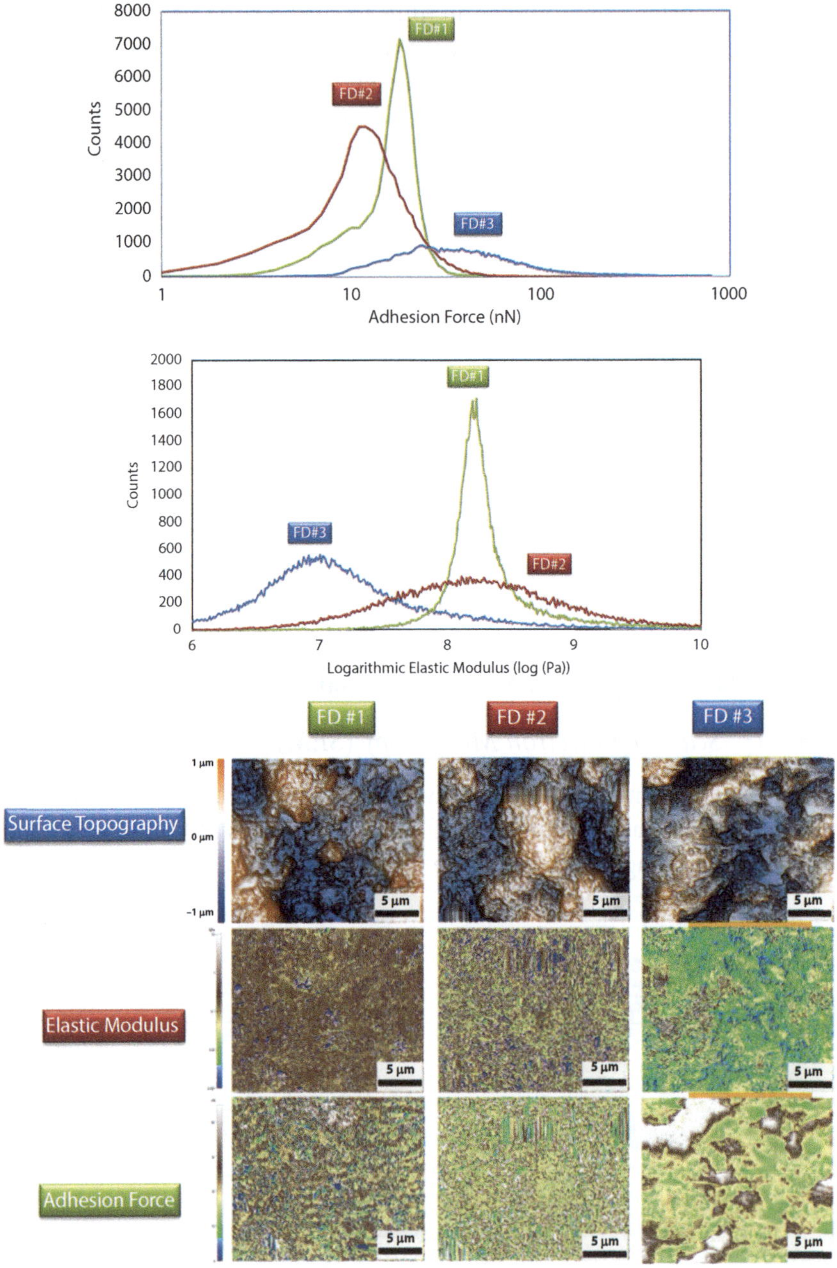

Figure 7.13 Elastic modulus and adhesion force determined by AFM for 3 long-wear foundations (scan size: 25µm x 25 µm).

fiber [40]. This technique allows for detailed examination of the film surface morphology, revealing characteristics such as smoothness, uniformity, cracks, or imperfections. It also helps to understand how the film interacts with the substrate, including adhesion and coverage.

7.5.3.2 Atomic Force Microscopy (AFM)

Atomic Force Microscopy (AFM) has emerged as a widely applied technique in the hair and skin domains, holding significant importance for the cosmetics and pharmaceutical industries [73, 83, 97–102]. Beyond its capability for rapid imaging of surface topography, AFM enables the measurement of various nanomechanical properties, including elastic modulus, adhesion force, and frictional force, alongside the acquisition of associated qualitative data. For examining the structures and nanomechanical properties of films, AFM has proven to be a valuable tool for determining surface roughness, elastic modulus, and adhesion force of cosmetic films, as extensively discussed in various references [73, 102]. Figure 7.13 is example of elastic modulus and adhesion force determined by AFM for 3 long-wear foundations.

7.6 Factors Affecting Cosmetic Adhesion – Cosmetic Formulation

7.6.1 Long-Wear and Transfer-Resistant Liquid Foundation

In today's fast-paced world, consumers demand high-performance foundations that can keep up with their active lifestyles. The desire for long-wear formulation stems from a need for makeup that remains flawless throughout the day, regardless of environmental challenges. Water resistance is a key benefit, ensuring that foundation stays put even during exposure to water or perspiration. This is crucial for maintaining coverage and preventing streaks or patchiness. Similarly, oil and sweat-proof properties are essential for preventing makeup breakdown and transfer. These properties help control shine and keep the complexion looking fresh, even when the skin's natural oils or perspiration come into play.

Furthermore, consumers seek foundations that can withstand heat and humidity. Heat resistance is vital for preventing melting, creasing, and the dreaded "makeup meltdown" in warmer climates. This ensures that the foundation maintains its coverage and even finish, even in challenging conditions. Increasingly, consumers are also prioritizing anti-pollution

Figure 7.14 Benefits of face long-wear makeup.

properties in their foundation choices. These properties protect the skin from environmental aggressors and help prevent makeup from absorbing pollutants that can lead to a dull, lackluster appearance. Ultimately, these long-lasting benefits are paramount for consumers seeking a foundation that can truly withstand the demands of their day-to-day lives, providing a confident and flawless look from morning to night (Figure 7.14).

7.6.1.1 Long-Wear and Non-Transfer Foundation Compositions

Long-wear and transfer-resistant foundations are complex formulations, often structured as water-in-oil (W/O) or water-in-silicone (W/Si) emulsions. As detailed by Li and Bui [10], these typically include pigments, fillers, film-formers, emulsifiers, oils (hydrocarbons or silicones), and water. Silicone film-formers like MQ resins or T-propyl resins act as binders for treated pigments, enhancing longevity and transfer resistance [9, 10, 15, 62]. Various fillers, such as magnesium oxide and magnesium stearate, help absorb and manage sebum and sweat [10, 62]. These fillers can even solidify sebum's fatty acid components, effectively reducing shine. Furthermore, silicone-treated pigments are frequently used to optimize pigment dispersion within the oil phase of these long-wear foundations. Table 7.7 provides a more detailed breakdown of the key components in typical long-wear, transfer-resistant foundation formulations.

Table 7.7 Main components in long-wear and transfer-resistant liquid foundation.

Component	Function	Key considerations
Volatile Fluids	Major components act as carriers. Water and volatile silicones/ isododecane. Influence emulsion characteristics, stability, and user experience (spreadability, feel).	Choice affects performance and sensation perceived by consumers. Cyclopentasiloxane (D5) is common but faces environmental restrictions.
Treated Pigments	Provide color and influence wear performance.	Surface treatments (silicone, silane, amino acid, organo-titanate) enhance dispersion, compatibility, and performance (water resistance).
Film-Formers	Create a film on the skin for long-wear and pigment adhesion.	Type and combination influence flexibility, tackiness, and adhesion. MQ resins are common; T-Propyl resin adds flexibility.
Emulsifiers	Stabilize water-in-silicone emulsions.	Influence emulsion stability and compatibility with other ingredients. Silicone emulsifiers are commonly used.
Fillers	Adjust viscosity, deposition, blendability, skin feel, tack, and sebum resistance, optical effect.	Type and particle morphology influence sensory properties and performance. Used for soft-focus effects, oil absorption.

MQ resin in cyclopentasiloxane (D5) is a common binder for color pigments in long-wear, transfer-resistant liquid foundations, enhancing their staying power. The adhesion strength of these foundations depends on several factors, including the concentration of MQ resin, the dispersion medium (e.g., PDMS or hydrocarbon oils), and interactions with other ingredients, especially silicone emulsifiers. Film flexibility is influenced by the non-volatile solvent, dispersion matrix, and solid content. Water-in-silicone (W/Si) emulsions with volatile silicone oils like cyclopentasiloxane and silicone elastomers are often used to enhance comfort and provide a silky feel. Oil-absorbent fillers, such as silica, nylon, or methyl silsesquioxane, further enhance sebum and sweat absorption. The MQ resin binds these fillers and pigments, creating a durable, long-lasting color deposit on the skin.

The interplay between MQ resin and silicone emulsifiers significantly impacts the foundation's performance [10, 61]. Balancing their compatibility is essential for optimizing properties like tack, flexibility, pigment adhesion, and sebum resistance, as detailed in Table 7.8. Careful emulsifier selection is necessary to ensure compatibility with the film-former while minimizing interaction with sebum, which can compromise long-wear performance [10]. The data underscores the complexity of these formulations and the importance of optimizing ingredient ratios and chemical structures to achieve the desired results.

Table 7.8 MQ resin/silicone polyether interactions [10].

Property	Trend observed	Impact on performance
Tackiness	Increases with MQ content up to 50 wt%, then decreases.	Influences adhesion of product, potential for transfer and sensation.
Flexibility	Decreases with increasing MQ content. Adequate silicone polyether improves flexibility.	Prevents cracking upon application and facial movement.
Pigment Adhesion (Rub-off Test)	Higher MQ content enhances rub-off resistance.	Enhances long-wear properties and resistance to transfer of color.
Sebum Resistance	Emulsifier choice significantly impacts sebum resistance. Alkyl functional groups in emulsifiers decrease sebum resistance.	Crucial for maintaining even color in the presence of sebum.

7.6.1.2 *Enhancing a Long-Wear Foundation Performance Using Setting Spray*

Setting sprays offer a modern solution for extending the wear time of face makeup [105, 106]. These sprays form a protective layer over makeup, acting as a shield against degradation. Typically formulated with polymers like acrylics dissolved in a solvent (often alcohol), and fillers, they create a water-resistant barrier. Polymers such as Poly(vinyl pyrrolidone) (PVP), Poly(vinyl pyrrolidone)/Poly(vinyl alcohol) (PVP/PVA) copolymer, Acrylates Copolymer, or Polyurethane-14 are key ingredients that help prevent smudging and fading, effectively locking makeup in place. Beyond longevity, modern setting sprays offer additional benefits, including mattifying or glow finishes, with some even incorporating Sun Protection Factor (SPF). This versatility allows for flexible application – before, during, or after makeup application – depending on the desired effect.

While generally suitable for all skin types, choosing the right formulation is crucial. Hydrating formulations benefit dry skin, matte finish formulations control shine on oily skin, and fragrance-free options cater to sensitive skin. Table 7.9 explores the relationship between common setting spray ingredients and their adhesion property, which is key to achieving long-lasting wear.

Generally, several factors influence how well a foundation adheres to the skin. The interfacial interaction between ingredients from the liquid foundation and skin substrate plays a significant role in the adhesion. Table 7.10 summarizes the factors that can have impact on the performance of a long-wear foundation on facial skin.

In summary, long-lasting liquid foundations are desirable for their extended wear, full coverage, and natural finish. Current high-performance, transfer-resistant formulations rely heavily on silicone technology—resins, film-formers, emulsifiers, elastomers, and oils—to achieve both strong adhesion and a comfortable feel. However, the increasing demand for sustainable cosmetics presents a challenge: matching the efficacy of silicones with eco-friendly alternatives. This challenge is amplified by the need to balance robust adhesion with skin comfort and the growing "clean beauty" trend. Formulating with natural ingredients adds further complexity, as naturally-derived adhesives may not perform as well as their synthetic counterparts. Therefore, innovation in adhesion, i.e., focused formulations using sustainable materials, is crucial for the future of long-lasting liquid foundations.

Table 7.9 Some common ingredients used in setting spray products.

Ingredient category	Example ingredients
Film-Formers/Fixatives	▪ PVP (Polyvinyl pyrrolidone): Provides good hold, water-soluble. ▪ Acrylates Copolymer (various types): e.g., Acrylates/Ethylhexyl Acrylate Copolymer, VP/Eicosene Copolymer - offer varying degrees of flexibility, hold, and water resistance. ▪ Polyurethane-14: Creates a strong, durable film, excellent for long-wear. ▪ AMP-Acrylates Copolymer: Often used in hairspray but can also be found in some setting sprays for extra hold.
Silicones	Dimethicone, Cyclopentasiloxane, Cyclohexasiloxane
Humectants	Glycerin, Hyaluronic Acid, Butylene Glycol, Propanediol
Powders	Silica, Starch, Talc (Primarily in mattifying setting sprays)
Alcohol	Denatured Alcohol, Ethanol

7.6.2 Long-Wear and Transfer-Resistant/Transfer-Proof Lipsticks

Modern consumers demand more than just longevity from their long-wear lipsticks. They seek a comprehensive combination of comfort, lasting color, diverse finishes (from matte to high shine), and even lip care benefits. This desire for truly enduring, even "food proof" lipstick stems from the need for convenience and confidence. Busy lifestyles require makeup that seamlessly transitions from day to night, eliminating the need for constant touch-ups. A lipstick that withstands meals and drinks provides peace of mind, ensuring a polished, confident look that lasts all day, regardless of activity (Figure 7.15).

Achieving this performance hinges on adhesion. A long-wear lipstick's ability to adhere firmly to the lips is fundamental to its longevity and transfer resistance. Strong adhesion ensures the pigments and ingredients bind securely, forming a durable film that can endure eating, drinking, and talking. This prevents smudging, fading, and transfer, maintaining a flawless look. Formulating such high-performance lipsticks requires a delicate

Table 7.10 Factors affecting adhesion of face makeup products on facial skin.

Factor	Description	Impact on adhesion
Skin Condition	Skin type (oily, dry, combination), texture (smooth, rough), and cleanliness significantly affect how well makeup adheres.	Clean, properly moisturized skin provides the best base for makeup adhesion. Primers can help create a smooth, even surface.
Individual Skin Chemistry	Natural variations in skin pH, oil production, and skin cell turnover can affect how certain formulations interact with individual skin.	Makeup performance and adhesion can vary between individuals due to these inherent differences.
Makeup Formulation	The ingredients, including oils, waxes, silicones, polymers, and pigments, determine the product's texture, coverage, and ability to adhere to the skin.	Formulations designed for specific skin types and desired finishes (matte, dewy, etc.) offer varying levels of adhesion.
Application Technique	Using brushes, sponges, fingers, or airbrushing, as well as layering and blending methods, impacts how evenly and smoothly the makeup is applied.	Proper application techniques maximize product adherence and minimize streaking or patchiness.
Environmental Factors	Humidity, temperature, and exposure to water, sweat, or friction can compromise makeup's staying power.	Setting sprays or powders can improve makeup longevity and resistance to environmental factors.
Product Interactions	The compatibility of different products layered together (e.g., moisturizer, primer, foundation, concealer, powder) influences overall adhesion and appearance.	Layering compatible products enhances adhesion, while incompatible products may lead to peeling or separation.

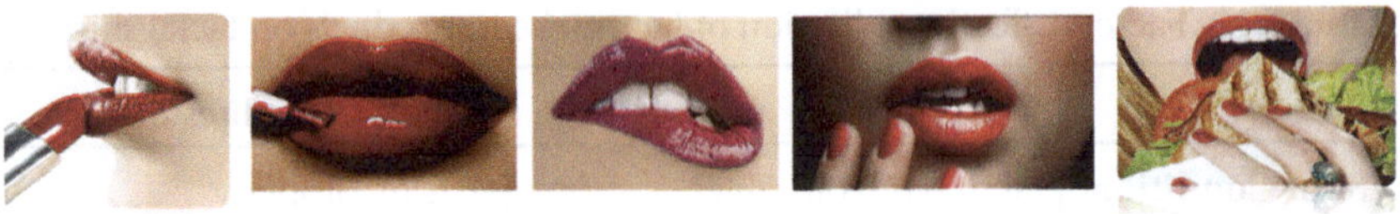

Figure 7.15 The quest for long lasting lipstick: shine/non-transfer/long- wear and food-proof.

balance of film-formers, pigments, binders, oils, and fillers, all rigorously tested.

Silicone MQ resins revolutionized long-wear lipstick formulations in the late 1990s [9, 60]. Initial formulations were in the liquid form containing MQ resin, using volatile solvents like cyclomethicone (D5) and low viscosity of PDMS that evaporate to leave a pigmented film. Later, a key advancement involved combining MQ resins with dimethicone gum, or with thermoplastic elastomers and a volatile solvent as either in D5 or isododecane (IDD) [9, 10, 59, 60]. This significantly improved film durability, adhesion to the lips, and resistance to abrasion and stretching. In addition, formulating these lipsticks involves careful consideration of pigment chemistry, as organic lakes have higher oil absorption than inorganic pigments. Surface treatments, like dimethicone and silane, can improve pigment loading and reduce dryness [62]. However, the resulting films from both these approaches tend to be dry and matte, leading to less comfortable wear. To address this, topcoats are often used to improve comfort, shine, and protect the color. These topcoats, available in liquid or solid form, typically contain high amounts of non-volatile hydrocarbons or high refractive index silicone oils like phenyldimethicone fluid [9]. The incompatibility of these oils with the base coat further extends wear.

Despite these advancements, the demand for single-step long-wear lipsticks that offer both extended wear and shine persists. Attempts to formulate these single-step products using the same MQ technology have faced challenges. Incorporating emollient oils for comfort and shine can compromise the film's integrity, leading to transfer and reduced wear. Balancing adhesion, cohesion, transfer resistance, comfort, and shine in a single-step formulation had remained a significant challenge in cosmetic science for a long time [9].

Table 7.11 compares the performances of one-step and two-step long-wear lipsticks.

Recently, a breakthrough in long-wear liquid lipstick technology was reported that could deliver lasting color, shine, and transfer resistance in a single application [107, 108]. This innovative formulation utilizes a

Table 7.11 Performances of one-step and two-step long-wear lipsticks.

Property	1-step long-wear, transfer-resistant lipstick	2-step long-wear, transfer-resistant lipstick
Application	Single application	Two steps: color-based coat followed by sealant/gloss topcoat
Film Formation	Forms thin film after solvent evaporates	Color coat forms a base; sealant locks in color and provides shine/finish
Transfer Resistance	Moderate to high, depending on the formulation	Typically, more than 1-step due to the sealant layer topcoat
Longevity	Moderate to high (4h-12 h)	Usually longer lasting than 1-step (8h-24 h)
Comfort	Can sometimes feel dry, depending on formulation	Often more comfortable due to the moisturizing/protective sealant layer
Finish	Varies (matte, satin, shimmer)	Varies depending on the topcoat (matte, glossy, etc.)
Removal	May require dedicated makeup remover	Can be easier or more difficult to remove depending on the formulation

phase-separation mechanism, creating a unique unstable dispersion. Upon application and solvent evaporation, the formulation self-stratifies into a bilayer film on the lips. This bilayer structure is key to its performance. The bottom layer, rich in pigments, silicone resins, and film-formers, adheres strongly to the lips, providing long-lasting color that resists transfer even on contact with food or tissues. The top layer, composed primarily of silicone fluids, forms a smooth, shiny surface that enhances comfort. The longevity of the shine is attributed to the balanced viscoelastic properties of the silicone polymers in this upper layer. Furthermore, long-chain silicone polymers likely become anchored at the interface between the two layers during film formation, promoting interlayer adhesion and contributing to the "self-healing" shine effect. This ingenious approach overcomes the long-standing challenge of combining long wear, comfort, and shine in a single-step lipstick formulation (Figure 7.16).

Shiseido developed "Water Sensing Technology" for long-lasting, non-transfer lipstick. This technology utilizes water-compatible oily ingredients that attract color pigments. Upon application, these ingredients react with lip moisture forming a pigment network that adheres closely, preventing color transfer (Figure 7.17). This approach avoids the stiff feel of traditional long-lasting lipsticks by maintaining coating fluidity. The resulting lipstick offers a smooth application, lasting color and gloss, and reduced friction, addressing consumer demand for comfortable, high-performance lip makeup. This innovation is featured in the Maquillage Dramatic Essence Rouge [109] (Figure 7.18).

As previously discussed, a lipstick's long-wear performance relies heavily on the adhesion between its formulation and the lip surface. Table 7.12 outlines the factors influencing this adhesion.

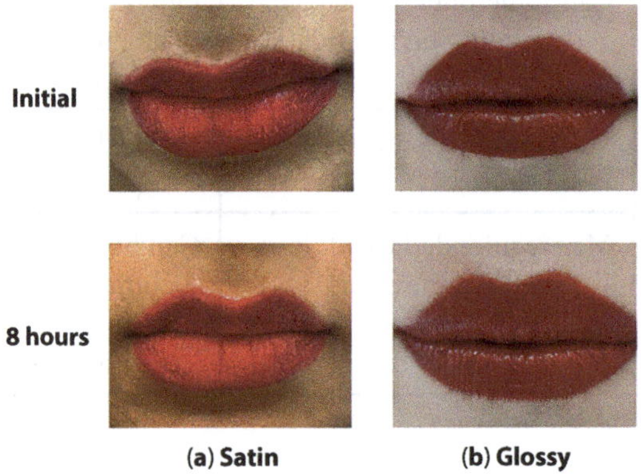

Figure 7.16 Visual evaluation of the wear of long-wear and shine single-step lipstick [107].

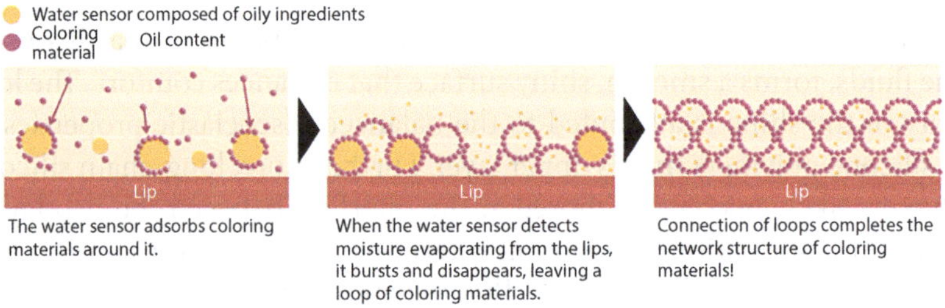

Figure 7.17 Water sensing technology lipstick [109].

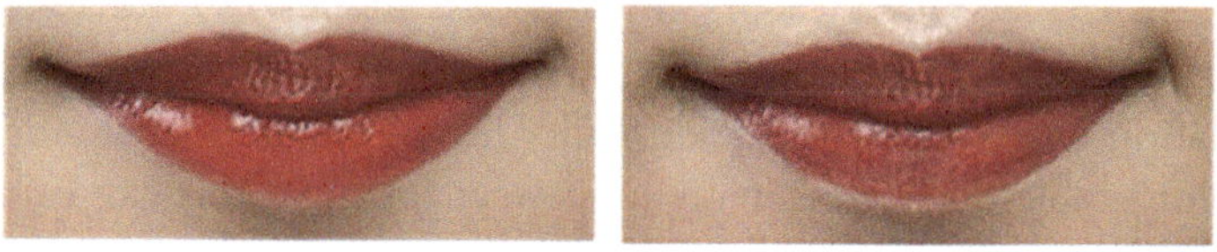

Figure 7.18 Water sensing technology lipstick application [109].

Table 7.12 Factors affecting lipstick adhesion to lips.

Factor	Description	Impact on adhesion
Lip Condition	Dryness, chapped lips, or excess oils can affect how well the lipstick adheres.	Reduced adhesion on dry or overly oily lips; Optimal adhesion on smooth, hydrated lips.
Individual Lip Chemistry	Natural variations in lip pH and sebum production can influence how different lipstick formulations interact and adhere.	Lipstick performance and adhesion can vary from person to person due to individual lip characteristics.
Lipstick Formulation	Ingredients like film-formers, polymers, waxes, and oils influence the lipstick's ability to form a continuous film and adhere to the lips.	Specific ingredients and their ratios directly impact adhesion and film formation.
Pigment Properties	Pigment surface treatments and particle size can impact how well they disperse and bind within the lipstick formulation, ultimately influencing adhesion to the lips.	Treated pigments often enhance adhesion and dispersion, leading to a more even and long-lasting application.
External Factors	Eating, drinking, talking, and environmental conditions like humidity and temperature can affect the lipstick's staying power and transfer resistance.	Reduced adhesion and increased transfer potential on exposure to external factors.

7.6.2.1 Summary and Future Trends

Developing long-lasting and non-transfer lipsticks presents a complex formulation challenge. The key difficulty lies in balancing extended wear and comfortable lip feel, as many long-wear formulations can cause dryness and tightness. Strong adhesion plays a vital role in ensuring that the lipstick stays in place without compromising comfort. Creating a lipstick resistant to transfer during eating and drinking, which involves exposure to both water and oil, adds another layer of complexity. The lips, with their thin, sensitive skin, require gentle, moisturizing ingredients that also promote good adhesion. Moreover, the constant movement of the lips and their high trans-epidermal water loss (TEWL) make it even more challenging to maintain long-lasting color and adhesion. Therefore, a thorough understanding of lip surface properties, including its barrier function and unique composition, is crucial for developing effective lip care and comfortable, high-performance, long-wear lipsticks with excellent adhesion.

The growing focus on environmental sustainability is significantly shaping the future of lipsticks. Eco-friendly formulations incorporating biodegradable compounds and refillable packaging are expected to become increasingly popular. Technological innovation is driving advancements in multiple areas, including adhesion technology. New formulations offer longer wear, richer color payoff, and enhanced moisturizing benefits, significantly improving the user experience while maintaining superior adhesion. Sleek, eco-friendly packaging with features like magnetic closures further attracts environmentally-conscious consumers. Advances in pigment technology are also expanding the possibilities for unique, customizable shades without compromising adhesion performance. Virtual try-on technology using augmented reality (AR) and AI-assisted formulation are revolutionizing the landscape of long-wear, non-transfer lipsticks. These technological improvements, including those focused on adhesion, are not only boosting consumer demand but also opening doors to new and emerging markets.

7.6.3 Long-Wear Eye Makeup

Long-wear eye makeup products are essential for maintaining a flawless look throughout the day, offering smudge-proof, crease-proof, and fade-resistant formulations that withstand various conditions. These products, including eyeshadows, eyeliners, and mascaras, utilize advanced film-forming polymers and pigments that adhere firmly to the skin and lashes, ensuring vibrant color and definition for extended periods (Table 7.13).

Table 7.13 Comparison of long-wear mascara, eyeliner and eyeshadow products.

Feature	Long-wear mascara	Long-wear eyeliner	Long-wear eyeshadow
Desired Properties	Volume, length, curl, smudge-proof, waterproof	Precise application, intense color, smudge-proof, long-lasting	Intense color, crease-proof, long-lasting
Key Ingredients	Film-forming polymers (e.g., acrylates, silicones), waxes, pigments (iron oxides, carbon black)	Film-forming polymers, pigments (iron oxides, carbon black), sometimes oils or waxes for creamy texture	Pigments (micas, iron oxides, etc.), binders (zinc stearate, silicones), fillers (talc, mica, sericite), film-forming polymers
Form	Liquid or cream in a tube with a wand applicator	Liquid, pencil, gel, or powder	Powder (pressed or loose), cream, stick
Application	Applied to lashes with a wand	Applied to eyelids with a brush, pencil, or applicator	Applied to eyelids with a brush or fingers
Challenges	Clumping, flaking, smudging	Smudging, fading, transfer	Creasing, fading, peeling
Removal	Makeup remover, sometimes waterproof remover needed	Makeup remover	Makeup remover

Such adhesion is crucial for long-wear performance, as it allows the makeup to form a durable bond with the skin or lashes, resisting transfer, smudging, and fading. This longevity eliminates the need for frequent touch-ups, providing convenience and confidence for busy individuals or special occasions. Long-wear formulations cater to diverse needs, offering matte, shimmer, and satin finishes, while some also incorporate skincare

Figure 7.19 Long wear eye makeup.

benefits to nourish the delicate eye area. Whether it is a dramatic or a subtle everyday look, long-wear eye makeup provides enduring beauty that lasts (Figure 7.19).

7.6.3.1 Long-Wear Mascara

In the cosmetics, the long-wear mascara can be classified in 4 types: smudge- proof, waterproof, one-day wear, and semi-permanent as shown in Table 7.14. One-day wear and semi-permanent mascaras are available in both O/W emulsion and anhydrous formulations. Silicone-based resins or acrylate copolymers (latices) are frequently used in these formulations to achieve the desired long-wear properties.

a. Typical Composition of Long-Wear Mascara

The mascara's composition and physical properties significantly influence adhesion (Table 7.15). Film-formers are the cornerstone of long-wear performance, acting as binders for pigments and other ingredients and creating the film that adheres to lashes. Silicone resins (like MQ resins), rosin esters, and acrylate latexes are common choices [40]. Rosin esters, derived from natural sources offer a balance of hydrophobicity and flexibility but can be sensitive to sebum. Latex-based film-formers provide good film formation and hold, but achieving both long wear and easy removal is challenging. The volatile solvent (water or isododecane) carries the ingredients and

Table 7.14 Various types of long-wear mascaras [40].

Mascara type	Key characteristics	Typical ingredients
Waterproof	Requires oil-based makeup remover, long-lasting (6-8 hours).	Isododecane, waxes, film-formers (oil-soluble polymers), surfactants (for W/O emulsions), thickeners, pigments
Smudge-proof/ Smudge-Resistant	Water-resistant, provides length and volume, easily removed with warm water.	Water, film-formers (higher levels of acrylate copolymers), waxes (low amounts), surfactants, thickeners, pigments
Long-wear (1-day)	Lasts up to 24-48 hours, often anhydrous.	Volatile solvents, film-forming polymers (silicone resins, rosin esters, acrylate latex)
Semi-Permanent (3-7 days)	Multi-day wear, typically contains silicone film-formers, hydrocarbon resins, or oxidative dyes.	Film-formers (silicone resins, rosin esters, latex), volatile solvents, pigments, (typically no wax)

evaporates after application, influencing drying speed and film formation. Pigments provide color, and their dispersion and treatment affect how well they adhere within the film and resist transfer. Waxes are common in traditional mascaras for thickening and volumizing, but high wax content can interfere with film formation and flexibility, particularly in smudge-proof formulations. Thickeners and gellants adjust viscosity and influence application, playing a crucial role in smudge-proof mascaras where wax content is typically lower. The critical pigment volume concentration *(CPVC)* is a crucial factor [110–114]. Exceeding the CPVC results in a brittle film with poor adhesion. Formulations designed for multi-day wear typically contain high levels of film-formers and minimize wax content to stay below the CPVC and ensure a flexible, continuous film [40].

Table 7.15 Effect of long-wear mascara composition on adhesion to eyelashes.

Component	Description	Impact on adhesion
Film-Forming Polymers	These create a continuous film that coats the lashes, providing hold and contributing to long-lasting wear. Examples include MQ resin, acrylates copolymers, PVP.	Essential for strong adhesion; contribute to smudge-proof and water-resistant properties. Flexibility of the polymer influences the feel and durability of the film.
Waxes	Provide structure, thickness, and help the mascara adhere to the lashes. Examples include beeswax, carnauba wax, paraffin wax.	Contribute to buildable volume and hold; can influence the stiffness and flexibility of the mascara film, affecting adhesion and wear.
Oils and Emollients	Can improve the spreadability and flexibility of the mascara, but excessive amounts can compromise adhesion and lead to smudging.	Used in small quantities to enhance application and prevent dryness; excessive amounts can negatively impact adhesion and long-wear.
Pigments	Provide color and can influence the texture and adhesion of the mascara depending on their properties and surface treatments.	Treated pigments generally enhance dispersion and adhesion. Pigment concentration affects the intensity of color and can also influence the film properties.
Thickeners	Help create the desired consistency and can influence how the mascara clings to the lashes. Examples include clays, gums, starches.	Contribute to the mascara's texture and can affect its ability to coat the lashes evenly, impacting adhesion and preventing clumping.

Table 7.16 Factors affecting adhesion of mascara.

Factor	Description	Impact on adhesion
Lash Condition	Cleanliness, oiliness, and dryness of the lashes impact how well the mascara adheres. Existing product buildup can also interfere with adhesion.	Clean, dry lashes provide optimal adhesion. Oily or product-laden lashes hinder adhesion.
Individual Lash Characteristics	Natural variations in lash thickness, curl, and length can influence how different mascara formulations perform and adhere.	Mascara performance and adhesion can vary depending on individual lash characteristics.
Mascara Formulations	The composition of the mascara, including polymers, waxes, oils, and film-forming agents, directly affects its ability to coat and cling to the lashes.	Formulation with appropriate ratios of film-formers, waxes, and polymers enhances adhesion.
Brush Type and Design	The shape, size, and bristle material of the mascara wand dictate how effectively the mascara is distributed and how well it coats each lash.	Different brush designs cater to specific needs (volume, length, separation) and can affect how evenly and securely the mascara applies.
Application Technique	Layering, wiggling the wand, and avoiding pumping the tube can impact how evenly and smoothly the mascara is applied, influencing adhesion and clumping.	Proper technique promotes even coating and minimizes clumping, leading to better adhesion.
Environmental Factors	Humidity, temperature, and exposure to water or other substances can affect the mascara's staying power and smudge resistance.	High humidity or exposure to water can compromise adhesion and lead to smudging.

The factors such as eyelash structure and condition, mascara formulation and brush design would have big impacts on the long-wear performance of mascara (Table 7.16).

b. Recent development of long-wear mascara and ease of removability
For a long-wearing mascara alternative, tubing mascara offers a unique solution [115, 116]. Its innovative formulation utilizes film-forming polymers, such as acrylates/ethylhexyl acrylate copolymer, acrylate copolymer, ammonium acrylate copolymer, and vinylpyrrolidone/eicosene copolymer, to create individual tubes that encase each lash. Upon application, these polymers adhere tightly to the keratin, forming a strong, flexible, and water-resistant film. This "tubing" action not only lengthens a lash but also provides exceptional smudge-proof and long-lasting wear. Unlike traditional mascaras, tubing mascara is removed easily with warm water; the tubes slide off intact, leaving no residue. This gentle removal process makes it ideal for sensitive eyes. The combination of long-lasting wear, easy removal, and a focus on lash definition makes tubing mascara a compelling alternative in the evolving mascara landscape.

In summary, the current mascara market prioritizes long-lasting wear, and advancements in adhesion play a crucial role in achieving this. While waterproof and smudge-proof formulations have been the focus, the true innovation lies in enhancing the mascara's bond with the lashes. This focus on adhesion is evident in formulations that incorporate advanced polymers and film-forming agents, creating a flexible coating that grips each lash without clumping or flaking. This strong adhesion ensures that the mascara stays in place throughout the day, resisting smudging, even in challenging conditions. This emphasis on adhesion, alongside the trends of natural ingredients, multi-functional benefits, and sustainable packaging, demonstrates the industry's commitment to providing high-performing, long-wearing mascaras that meet the demands of today's consumer.

7.6.3.2 Long-Wear Eyeshadow

Today's consumers expect eyeshadows to deliver long-lasting, vibrant color without smudging or creasing (Figure 7.20). This long-wear performance is achieved through a complex formulation of ingredients. Micas, often treated with coatings like hybrid silane/polyhydroxystearic acid for improved adhesion and dispersion, provide a smooth, blendable texture. Zinc stearate binds pigments, adding depth and intensity to the color [62].

Figure 7.20 Eyeshadow.

Advanced film-forming polymers, such as hydrogenated cyclopentadiene, hydrogenated styrene/isoprene copolymer, T-propyl silsesquioxane or MQ resin, create a water-resistant barrier that locks in color, enhances skin adhesion and prevents transfer [40]. Fillers like talc, mica, and sericite contribute to the overall feel and application. Highly concentrated pigments, from classic iron oxides to vibrant organic dyes, deliver rich color payoff. This carefully balanced combination of ingredients ensures that long-wear eyeshadows provide a flawless base for any eye look, minimizing touch-ups and maintaining a fresh, vibrant appearance throughout the day. These long-wear formulations are often available in various forms including pressed powders, creams, liquids, and sticks, each offering unique application benefits with its special texture.

Here are the most common types and the best type of eyeshadow depends on personal preference, desired look, and skin type [117–119]:

- **Powder Eyeshadows:** This is the most common and versatile type, coming in pressed or loose forms. Powder eyeshadows offer a wide range of finishes, from matte to shimmer to metallic.
- **Cream Eyeshadows:** These eyeshadows have a creamy, blendable texture, often providing a light to medium coverage with a natural finish. They can also be used as a base for powder eyeshadows.
- **Liquid Eyeshadows:** Liquid eyeshadows typically offer intense pigmentation and a long-lasting finish. They can range from metallic and shimmery to matte.

- **Stick Eyeshadows:** These eyeshadows come in a convenient stick form, making them easy to apply and blend. Their creamy or shimmery texture makes them ideal for effortless eye makeup application.
- **Loose Pigment Eyeshadows:** Highly pigmented loose powders offer intense color payoff but can be messy to work with.
- **Baked Eyeshadows:** Baked eyeshadows are cooked in an oven, giving them a unique texture and finish. They can be used wet or dry, often delivering a shimmery or metallic effect.

Different eyeshadow forms exhibit varying levels of adhesion to the eyelid due to variations in formulation and manufacturing processes, as detailed in Table 7.17.

The long-lasting performance of eyeshadows depends on the eyelid-eyeshadow interfacial adhesion, its formulation/format and application technique. The factors affecting adhesion of eyeshadow to eyelids are summarized in Table 7.18.

Table 7.17 Adhesion characteristics of eyeshadow types.

Eyeshadow type	Adhesion characteristics
Powder	Can be prone to fallout and creasing, especially on oily lids. Adhesion varies depending on the formulation and ingredients.
Cream	Typically adheres well, especially on dry lids. Can crease if the formulation is too emollient or if eyelids are oily.
Liquid	Generally, liquid form offers good adhesion and long-lasting wear. Can be prone to creasing if not properly set.
Stick/Crayon	Adhesion varies greatly depending on the formulation. Can be convenient for quick application but may require blending and setting.
Loose Pigment	Requires a mixing medium or sticky base for optimal adhesion.
Baked	Adhesion is similar to powder eyeshadows. Can be used wet for increased color intensity and adhesion.

Table 7.18 Factors affecting adhesion of eyeshadow.

Factor	Impact on adhesion
Skin Type	Oily eyelids tend to cause eyeshadow creasing and fading. Dry eyelids might need a primer for better adhesion.
Eyelid Primer	Primers create a smooth base, absorb excess oil, and improve eyeshadow adhesion.
Eyeshadow Formulation	Cream, powder, liquid, and other formulations have different adhesion property. Ingredients like silicones, polymers, and binders affect how well the eyeshadow sticks.
Application Technique	Using brushes, fingers, or applicators influences how evenly eyeshadow is applied, impacting adhesion. Layering and blending techniques also matter.
Environmental Factors	Humidity, temperature, and sweat can all impact eyeshadow wear.
Eyelid Anatomy	Deep-set or hooded eyes might experience more creasing due to the eyelid's structure.

In summary, advancements in eyeshadow technology prioritize adhesion and long-lasting wear, aligning with consumer demand. New formulations, including silicone-enhanced powders and hybrid gel-creams, improve adhesion, reducing flaking and creasing for up to 12 hours of wear. This focus on adhesion ensures vibrant, long-lasting color, a key factor driving consumer preference and innovation in the eyeshadow market. While other trends like sustainability, inclusivity, and personalized shades are important, the emphasis on improved adhesion through advanced formulations directly addresses the desire for high-performing, crease-proof eyeshadow.

7.6.4 Nail Polish

Nail polish serves to both beautify and protect nails. An ideal formulation offers smooth application, strong adhesion, quick drying, vibrant color, high gloss, good coverage, uniform finish, flexibility, appropriate hardness, water resistance, stability, safety, and easy removal without staining.

(a) (b) (c)

Figure 7.21 (a) Conventional nail polish, (b) Water-based nail polish, and (c) UV-cured gel nail polish.

Achieving these properties requires a carefully balanced formulation that withstands daily wear. Adhesion of nail polish depends on the nail's surface, the polish's composition, brush type and application technique.

Long-lasting nail polish comes in three main forms: conventional, water-based, and UV-cured gel (Figure 7.21). Conventional nail polishes rely on nitrocellulose, plasticizers, resins, solvents, and pigments [120, 121]. Color development is essential, with preference given to specific red and black dye lakes. Water-based nail polishes, driven by consumer demand for natural products, use acrylate copolymers (latex) but face challenges matching the durability and chip resistance of conventional nail polishes, as well as pigment staining. UV-cured gel nail polish provides a long-lasting, high-gloss finish. The process involves applying and curing multiple layers—a base coat, color coat, and topcoat—under a UV lamp [122, 123]. This technology offers a durable alternative to traditional nail polish and acrylic nail extensions. The UV light triggers a polymerization reaction: photoinitiators within the gel react with monomers (like acrylates) to create a strong polymer network. This cross-linking ensures good adhesion to the natural nail and enhances chip resistance. Key benefits of UV gel nail polish over traditional formulations include faster drying, extended wear, a broader color palette, and potential benefits for nail health. The typical compositions of these three nail polishes and their performance are summarized in Table 7.19.

Nail polish longevity and adhesion are directly linked to its chemical composition, whether it is a solvent-based, water-based, or UV gel formulation. Each type interacts uniquely with the nail plate, resulting in different wear times. Adhesion mechanisms generally involve mechanical interlocking, adsorption, and/or chemical bonding [81]. UV gel nail polishes typically exhibit the longest wear due to the formation of chemical bonds between the molecules from the formulation and the keratin in the nail plate.

Table 7.19 Comparison of 3 different types of nail polishes.

Feature	Conventional nail polish	Water-based nail polish	UV nail polish (Gel polish)
Film-Former	Nitrocellulose	Acrylate copolymers	Acrylate/methacrylate oligomers
Solvent	Organic solvents (e.g., ethyl acetate, butyl acetate)	Water	Minimal or no solvents
Pigments	Wide range, including dyes and pigments	Limited range, staining can be an issue	Wide range, typically highly pigmented
Wear Time	Moderate (up to 7 days claimed, but often less)	Generally lower than conventional	Longest lasting (up to 2 weeks or more)
Durability	Prone to chipping	Less durable than conventional, chipping a common issue	More resistant to chipping and wear
Removal	Requires nail polish remover (acetone-based)	Can be difficult to remove without damaging nails. Hot water removal is sometimes viable	Requires soaking in acetone or specialized removal systems
Sustainability	Concerns about VOCs from solvents	More sustainable due to lack of organic solvents	Concerns about UV exposure, disposal of products

The adhesion of nail product on nails depends on several key factors [124, 125]:

1. **Cleanliness of the Nail Plate:** Contaminants like dirt and oil can block adhesion. Thorough cleansing is crucial.
2. **Nail Hydration:** Moisture on the nail surface interferes with molecular bonding. Dehydrators are essential to temporarily remove this moisture for better product bonding.
3. **Nail pH:** A slightly alkaline nail surface promotes better adhesion with acidic components in some products. pH balancers or preparations can adjust the nail's pH.
4. **Buffing:** Follow product instructions regarding buffing. Some products require it, while others forbid it to prevent nail damage.
5. **Primer/Bonder Use:** These products create a chemical bond between the natural nail and the nail polish, significantly improving adhesion and preventing lifting.
6. **Product Compatibility:** Using a base coat, nail polish, and topcoat from the same brand system ensures the products are designed to work together optimally.
7. **Proper Application:** Applying thin, even coats of nail polish and allowing each layer to dry properly promotes better adhesion.
8. **Aggressive Filing:** Overfiling thins the nail plate, weakening it and potentially compromising adhesion. This should be avoided.
9. **Cuticle Removal:** Any cuticle left on the nail plate will prevent proper adhesion, highlighting the importance of thorough cuticle removal.

Essentially, clean, dry, properly prepped nails with a balanced pH are key for strong nail polish adhesion. Using the right products, following manufacturer instructions, and avoiding harsh filing are equally important for achieving long-lasting manicures.

7.6.4.1 Summary and Perspective

The adhesion across all types of nail polishes—solvent-based, water-based, and UV gel—is unequivocally of paramount importance, fundamentally determining a product's effectiveness and consumer satisfaction. Without robust adhesion, the nail polish film cannot remain securely attached to the nail plate, leading to premature chipping, peeling, or flaking, thus compromising both aesthetic appeal and the protective barrier it provides. Strong adhesion ensures that the nail polish withstands daily wear and tear, extending its longevity and maintaining a flawless finish. Ultimately, optimal adhesion is the indispensable factor transforming a liquid formulation into a durable, long-lasting cosmetic layer.

This critical bond is significantly influenced by the physical and surface properties of the natural nail plate itself [50, 51, 54, 55]. Composed of keratin, lipids, and water, the nail's varying surface energy, presence of oils, moisture content, and micro-irregularities dictate how effectively a polish can wet and spread. Moreover, the nail's porosity and dynamic swelling/shrinkage due to water absorption can stress the adhesive bond [55–57]. Therefore, enhancing adhesion involves a multi-faceted approach, beginning with optimal nail plate preparation—thorough cleaning and dehydration—to maximizing surface receptivity. This is complemented by the use of specialized base coats that act as intermediary bonding layers and the strategic incorporation of adhesion promoters within the nail polish formulation, designed to create stronger chemical or physical links with the nail and improve film compatibility.

Despite these strategies, achieving consistently strong adhesion presents distinct challenges across different nail polish chemistries.

For **traditional solvent-based nail polishes**, issues arise from the inherent brittleness of their dried films, making them prone to chipping, and their sensitivity to inadequate nail preparation or environmental degradation.

Water-based nail polishes face significant hurdles due to the water-soluble nature of some components and the complex film-forming mechanism, making them susceptible to lifting or peeling with water exposure if film formation (influenced by factors like minimum film forming temperature (MFFT), surfactants, and critical pigment volume concentration (CPVC)) is not perfectly controlled.

UV gel nail polishes, while often offering superior durability, demand meticulous nail preparation and precise UV/LED curing to avoid undercured, poorly adhering films; polymerization shrinkage and product incompatibility also pose risks to the adhesive bond.

The future of nail polish, driven by the demand for sustainability and "cleaner" beauty, is increasingly exploring biomaterials and natural film-formers. This shift, however, introduces new adhesion challenges. Bio-based materials like cellulose derivatives or polysaccharides can yield inherently more brittle films prone to chipping or possess increased hydrophilicity, making the nail polish susceptible to swelling and lifting with water exposure. Achieving robust and long-lasting adhesion with these materials necessitates overcoming their natural limitations in flexibility and water resistance. Even in the realm of bio-based UV gel nail polishes [126, 127], critical challenges persist: achieving efficient and complete polymerization with novel bio-based monomers, balancing flexibility and hardness, and managing polymerization shrinkage effectively remain complex scientific endeavors. Thus, the continuous pursuit for enhanced adhesion remains central to all advancements in nail polish technology, ensuring optimal wear, chip resistance, and consumer satisfaction, regardless of the underlying chemical foundation.

7.7 Challenges and Future Directions for Long-Wear Color Cosmetics

The remarkable staying power of the long-wear and non-transfer color cosmetic products is primarily attributed to silicone resin MQ (trimethylsiloxysilicate) and other silicone film-formers. These film-formers, particularly MQ resin, work in synergy with other silicone components such as emulsifiers, gums, fluids, and thermoplastic elastomers.

The superior performance of silicone MQ resin compared to other silicone or hydrocarbon film-formers in terms of long wear and transfer resistance explains its prevalence in long-wear color cosmetics, including lipsticks, foundations, mascaras, and eyeshadows. However, the beauty industry's shift towards sustainability has cast spotlight on MQ resin or silicone film-former base, prompting the search for high-performing, eco-friendly alternatives.

To look for silicone alternatives in terms of adhesion property, other natural polymers such as pullulan or chitosan can be employed in long-wear color cosmetic products such as face and eye makeup. These materials are derived from renewable resources and are often biodegradable. For example, starch-based adhesives, which are biodegradable and derived from abundant agricultural sources, have shown potential in various applications [128]. Pullulan, a versatile polysaccharide derived from starch fermentation, boasts a unique set of properties that make it a star ingredient in various cosmetics [129]. Its low viscosity, excellent adhesion, and film-forming prowess allow for the creation of luxurious textures, while its moisture-retaining property contributes to skin hydration and plumpness. However, one drawback of pullulan film is its poor water resistance, a limitation addressable through crosslinking [130]. Chitosan, a natural polymer derived from chitin, has shown promise as a sustainable film-former. Its wide applications in hair and skin care are already reported [131]. Its adhesion property stems from the numerous hydroxyl and amine groups that can interact with various surfaces. The effectiveness of chitosan as an adhesive depends on two key factors: acetylation degree and molecular weight. Lower acetylation degree results in a higher distribution of positively charged amine groups ($R\text{-}NH_3^+$), which enhance adhesion through electrostatic interactions. Higher molecular weight chitosan possesses a greater number of positive charges, leading to stronger adhesion forces. However, like pullulan, its water resistance requires enhancement by crosslinking or hydrophobization [132, 133]. In addition, the complex heterogeneity of chitosan also poses challenges for achieving consistent adhesion performance.

The development of bio-based synthetic polymers for adhesives marks significant progress towards sustainability in the cosmetics industry [134, 135]. These polymers, derived from renewable resources, aim to replicate the properties of traditional synthetic polymers, particularly silicone film-formers. However, challenges remain in achieving comparable performance. Natural alternatives often fall short in terms of durability, comfort, and longevity. Maintaining consistent quality and performance with natural polymers like chitosan also presents a significant hurdle for large-scale production. Despite these obstacles, the pursuit of bio-based and natural alternatives to silicone remains crucial for creating more sustainable color cosmetics. Ongoing research and innovation are essential to developing high-performing, eco-friendly products that satisfy consumer demand for both efficacy and environmental responsibility. Further advancements in bio-inspired and stimuli-responsive adhesives [101, 135–137] offer promising avenues for enhancing long-wear color cosmetics without relying on silicone resins or film- formers.

7.8 Summary

This chapter delves into the multifaceted world of adhesion in color cosmetics, exploring the intricate interplay between product formulations and the dynamic nature of human substrates – skin, eyelashes, lips, and nails. The quest for long-lasting, transfer-resistant color cosmetics presents a significant challenge, demanding a deep understanding of surface chemistry, polymer science, and materials properties.

In conclusion, achieving optimal adhesion in color cosmetics requires a holistic approach that considers the unique properties of both the product and the substrate. Balancing long-lasting wear with comfort, transfer resistance with ease of removal, and performance with sustainability demands continuous innovation in formulation science and a thorough understanding of the complex interactions that govern adhesion at the cosmetic-substrate interface. The ongoing pursuit of novel materials and technologies promises exciting advancements in the field, paving the way for even more sophisticated and sustainable color cosmetic products.

Acknowledgements

The author would like to thank Ronni Weinkauf, Connie Avramis, L'OREAL Research and Innovation, French Reading Committee and L'OREAL communication team to support this work.

AI disclosure statement: The author acknowledges the use of AI-based tools for minor language editing or grammatical correction. However, all intellectual contributions, analyses, and conclusions presented in this chapter are entirely the author's own.

References

1. A. M. Martins and J. M. Marto, A sustainable life cycle for cosmetics: From design and development to post-use phase, Sustainable Chemistry and Pharmacy, 101178, 1-28 (2023)
2. N. Lionetti and M. Deola. Formulating Forum: Long-lasting Color with Treatment Efficacy, Cosmetics & Toiletries, (2024). https://www.cosmeticsandtoiletries.com/magazine/article/22888058/cosmetics-toiletries-magazine-formulating-forum-longlasting-color-with-treatment-efficacy
3. https://magazine.elkem.com/personal-care-and-consumer-goods/sustainable-classic-ingredients-for-the-next-generation-of-beauty/
4. https://www.pristinemarketinsights.com/makeup-market-report
5. D. J. Gardner, Theories and mechanisms of adhesion in the pharmaceutical, biomedical and dental fields, in: *Adhesion in Pharmaceutical, Biomedical and Dental Fields*, K. L. Mittal and F. M. Etzler (Eds.), pp. 3–21, Wiley-Scrivener, Beverly, MA (2017).
6. K. Aramaki, Y. Shiozaki, S. Kosono, and N. Ikeda, Coacervation in cationic polyelectrolyte solutions with anionic amino acid surfactants, J. Oleo Sci. 69, 1411-1416 (2020)

7. Y. Kakizawa and M. Miyake, Creation of new functions by combination of surfactant and polymer- complex coacervation with oppositely charged polymer and surfactant for shampoo and body wash, J. Oleo Sci. 68, 525-539 (2019)
8. K. L. Mittal, The role of the interface in adhesion phenomena, Polym. Eng. Sci. 17, 467-473 (1977)
9. H. S. Bui and D. Coleman-Nally, Film-forming technology and skin adhesion in long-wear cosmetics, in: *Adhesion in Pharmaceutical, Biomedical and Dental Fields*, K. L. Mittal and F. M. Etzler (Eds.), pp. 141–166, Wiley-Scrivener, Beverly, MA (2017).
10. Z. Li and H. S. Bui, Factors affecting cosmetics adhesion to facial skin, in: *Surface Science and Adhesion in Cosmetics*, K. L. Mittal and H. S. Bui (Eds), pp. 543-584, Wiley-Scrivener, Beverly, MA (2021)
11. H. S. Bui, S. Lu, F. Pradier, R. Mercado, and X. Blin, Cosmetic compositions containing block copolymers, tackifiers and shine enhancing agents, US Patent 8,557,230, assigned to L'OREAL (2013).
12. H. S. Bui, S. Lu, F. Pradier, R. Mercado, and X. Blin, Cosmetic compositions containing block copolymers, tackifiers and a solvent mixture, US Patent 8,673,283, assigned to L'OREAL (2014).
13. J. Renvoise, D. Burlot, G. Marin, and C. Derail, Adherence performances of pressure sensitive adhesives on a model viscoelastic synthetic film: A tool for the understanding of adhesion on the human skin, International Journal of Pharmaceutics, 368, 83-88 (2009)
14. G. Luengo, H. Bui and J. Portal, Formation and performance of cosmetic films in cosmetics, in: *Handbook of Cosmetic Science and Technology*, F. Dreher, E. Jungman, K. Sakamoto, and H. I. Maibach (Eds.) 5th Edition, pp. 167-181, CRC Press, Boca Raton, FL (2023).
15. H. S. Bui, M. Hasebe, and J. Ebanks, Evaluation of sebum resistance for long-wear face make-up products using contact angle measurements, in: *Advances in Contact Angle, Wettability and Adhesion* Vol. 4, K. L. Mittal (Ed.), pp. 193-221, Wiley-Scrivener, Beverly, MA (2020)
16. C. Piérard-Franchimont, M. Lesuisse, J. Courtois, C. Ritacco, and G. E. Piérard, Sebum production, in: *Textbook of Aging Skin*, M. Farage, K. Miller, and H. Maibach (Eds), pp. 739-751, Springer, Berlin (2017)
17. G. Honari and H. Maibach, Skin structure and function, in: *Applied Dermatotoxicology*, H. Maibach and G. Honari (Eds), pp. 1-10, Academic Press (2014).
18. A. Elkhyat, S. Mar-Mary and P. Humbert, Skin wettability and friction, in: *Handbook of Cosmetic Science and Technology*, A. O. Barel (Ed.), pp. 427-436, Informa Heathcare Inc, New York (2009).
19. A. Mavon, H. Zahouani, D. Redoules, P. Agache, Y. Gall and P. Humbert, Sebum and stratum corneum lipids increase surface free energy as determined from contact angle measurements: A study on two anatomical sites, *Colloids Surfaces B*, 8, 147-155 (1997).

20. A. Mavon, D. Redoules, P. Humbert, P. Agache and Y. Gall, Changes in sebum levels and skin surface free energy components following skin surface washing, *Colloids Surfaces B,* 10, 243–250 (1998).
21. J. Krawczyk, Surface free energy of the human skin and its critical surface tension of wetting in the skin/surfactant aqueous solution/air system, *Skin Res.Technol.,* 21, 214–223 (2015).
22. C. Pailler-Mattei, S. Nicoli, F. Pirot, R. Vargiol and H. Zahouani, A new approach to describe the skin surface physical properties *in vivo, Colloids Surfaces B,* 68, 200–206 (2009).
23. V. Luboz, E. Promayon and Y. Payan, Measurements of the linear elastic properties of the face soft tissues using an aspiration device: Towards patient specific characterization, *Annals Biomed. Eng.,* 42, 2369–2378 (2014).
24. C. Pailler-Mattei, S. Bec and H. Zahouani, *In vivo* measurements of the elastic mechanical properties of human skin by indentation test, *Medical Eng. Phys.,* 30, 599–606 (2008).
25. H. Zahouani, C. Pailler-Mattei,, B. Sohm, R. Vargio and R. Debret, Characterization of the mechanical properties of a dermal equivalent, *Skin Res. Technol.,*15, 68-76 (2009).
26. S.Diridollou, M. Berson, V. Vabre, D. Black, B. Karlsson, F. Auriol,, J. M. Gregoire, C. Yvon, L. Vaillant, Y. Gall and F. Patat, An *in vivo* method for measuring the mechanical properties of the skin using ultrasound, *Ultrasound Med Biol.,* 24, 215–224 (1998).
27. D.L. Bader and P. Bowker, Mechanical characteristics of skin and underlying tissues *in vivo, Biomaterials,* 4, 305–308 (1983).
28. J. Jachowicz, R. McMullen and D. Prettypaul, Indentometric analysis of *in vivo* skin and comparison with artificial skin models, *Skin Res.Technol.,* 13, 299–309 (2007).
29. P.G. Agache, C. Monneur, J. L. Leveque and J. De Rigal, Mechanical properties and Young's modulus of human skin *in vivo, Archives Dermatological Res.,* 269, 221–232 (1980).
30. V. Korn, C. Surber and G. Imanidis, Skin surface topography and texture analysis of sun-exposed body sites in view of sunscreen application, *Skin Pharmacol Physiol,* 29, 291-299 (2016).
31. J. Shang, X. Feng, Y. Chen, Z. Gu, and Y. Liu, Human lip vermilion: Physiology and age-related changes. *J Cosmetic Dermatol.* 23, 2676-2680 (2024)
32. Lip Anatomy: Surface Guide for Aesthetics Practitioners, https://www.harleyacademy.com/aesthetic-medicine-articles/lip-anatomy-surface-guide-for-aesthetics-practitioners/
33. H. Kobayashi and H. Tagami, Functional properties of the surface of the vermilion border of the lips are distinct from those of the facial skin. Br J Dermatol., 150, 563-567 (2004)
34. E. Chirikhina, A. Chirikhin, P. Xiao, S. Dewsbury-Ennis, and F. Bianconi, *In vivo* assessment of water content, trans-epidermial water loss and thickness in

human facial skin. *Applied Sciences*. 10, 6139 (2020). https://doi.org/10.3390/app10176139

35. J. Kim, H. Yeo, T. Kim, E. T. Jeong, J. M. Lim, and S. G. Park. Relationship between lip skin biophysical and biochemical characteristics with corneocyte unevenness ratio as a new parameter to assess the severity of lip scaling, Int J Cosmet Sci. 43, 275-282 (2021)
36. E. Tamura, J. Ishikawa, K. Sugata, K. Tsukahara, H. Yasumori, and T. Yamamoto, Age-related differences in the functional properties of lips compared with skin, Skin Res Technol. 24, 472-478 (2018)
37. H. Kim, M. Lee, S. Y. Park, Y. M Kim, J. Han, and E. Kim, Age-related changes in lip morphological and physiological characteristics in Korean women, Skin Res Technol. 25, 277-282 (2019)
38. R. Barresi and I. C. Liao, Lip biophysical properties and characterization methods for long-wear lipsticks, in: *Surface Science and Adhesion in Cosmetics*, K. L. Mittal and H. S. Bui (Eds), pp. 3-34, Wiley-Scrivener, Beverly, MA (2021)
39. S. Baliga, S. Muglikar, and R. Kale, Salivary pH: A diagnostic biomarker, J Indian Soc Periodontol. 17, 461-465 (2013)
40. C. Pang and H.S. Bui, Adhesion aspect in semi-permanent mascara in: *Surface Science and Adhesion in Cosmetics*, K. L. Mittal and H. S. Bui (Eds), pp. 585-634, Wiley-Scrivener, Beverly, MA (2021)
41. S. Thibaut, E. De Becker, L. Caisey, D. Baras, S. Karatas, O. Jammayrac, P.J. Pisella and B.A. Bernard, Human eyelash characterization, British J. Dermatol., 162, 304-310 (2010)
42. J.I. Na, O.S. Kwon, B.J. Kim, W.S. Park, J.K Oh, K.H Kim, K.H. Cho and H.C Eun, Ethnic characteristics of eyelashes: a comparative analysis in Asian and Caucasian females, British J. Dermatol., 155,1170-1176 (2006)
43. M. Kikuchi, K. Matsuda, Y. Ishihara, T. Yanai, K. Yasuta, K. Hosokawa and T. Uemura, A study of normal eyelashes in Japanese individuals, Global Dermatol., 2, 74-77 (2015)
44. S. Aumond and E. Bitton, The eyelash follicle features and anomalies: A review. J. Optometry, 11, 211-222 (2018)
45. R. Paus, I. Burgoa, C. I. Platt, T. Griffiths, E. Poblet, and A. Izeta, Biology of the eyelash hair follicle: an enigma in plain sight, British J. Dermatol., 174, 741 – 752 (2016)
46. D. A. Glaser, D. Jones, J. Carruthers, A. Campo, S. Moench, G. Tardie, J. Largent, and C. Caulkins, Epidemiologic analysis of change in eyelash characteristics with increasing age in a population of healthy women. Dermatologic Surgery, 40, 1208-1213 (2014)
47. A. Mandel, S. Helokunnas, E. Pihko, and R. Hari, Neuromagnetic brain responses to other person's eye blinks seen on video, Eur. J. Neurosci., 40, 2576 – 2580 (2014)
48. https://theimperialroom.us/blogs/news/basic-nail-anatomy-of-a-nail
49. D. de Berker, Nail anatomy, J. Clin Dermatol., 5, 509-515 (2013)

50. S. Baswan, GB. Kasting, SK. Li, R. Wickett, B. Adams, S. Eurich, and R. Schamper, Understanding the formidable nail barrier: A review of the nail microstructure, composition, and diseases, Mycoses, 60, 284-295 (2017).
51. S. Murdan, C. Poojary, D.R. Patel, J. Fernandes, A. Haman, P. S. Saundh, and Z. Sheikh, *In vivo* measurement of the surface energy of human fingernail plates, Int J Cosmet Sci. 34, 257-262 (2012)
52. S. Murdan, G. Milcovich, and G. S. Goriparthi, The pH of the human nail plate, in: *Agache's Measuring the Skin*, P. Humbert, F. Fanian, H. Maibach, and P. Agache (Eds.), pp. 883-889, Springer, Cham. (2017)
53. S. Murdan, G. Milcovich, and G. S. Goriparthi. An assessment of the human nail plate pH. Skin Pharmacol Physiol. 24, 175-181 (2011).
54. M. Batory, P. Namieciński, and H. Rotsztejn, Evaluation of structural damage and pH of nail plates of hands after applying different methods of decorating, Int J Dermatol, 58, 311-318 (2019)
55. H. P. Baden, The physical properties of nail, J. Investigative Dermatology, 55, 115-122 (1970)
56. L. Farran, A.R. Ennos, M. Starkie, and S.J. Eichhorn, Tensile and shear properties of fingernails as a function of a changing humidity environment, J. Biomechanics, 42, 1230-1235 (2009)
57. H. Tohmyoh and M. Abukawa, Nanoindentation study of human fingernail for determining its structural elasticity. *Skin Res Technol.*; 29, e13456 (2023) doi:10.1111/srt.13456
58. S. Yaemsiri, N. Hou, M. Slining, and K. He, Growth rate of human fingernails and toenails in healthy American young adults. Journal of the European Academy of Dermatology and Venereology, 24, 420-423 (2010)
59. A. B. Pawar and B. Falk, Use of advanced silicone materials in long-lasting cosmetics, in: *Surface Science and Adhesion in Cosmetics*, K. L. Mittal and H. S. Bui (Eds), pp. 151-182, Wiley-Scrivener, Beverly, MA (2021)
60. R. Lochhead and M. Lochhead, Two Decades of Transfer-resistant Lipstick https://www.cosmeticsandtoiletries.com/research/chemistry/Two-Decades-of-Transfer-resistant-Lipstick-290207561.html
61. Z. Li, B. Maxon, K. Nguyen, M. Lee, M. Gu, and P. Pretzer, A general formulation strategy toward long-wear color cosmetics with sebum resistance, *J Cosmet Sci.*,68, 91-98 (2017)
62. H. Lam, Factors enhancing adhesion of color cosmetic products to skin: The role of pigments and fillers, in: *Surface Science and Adhesion in Cosmetics*, K. L. Mittal and H. S. Bui (Eds), pp. 487-541, Wiley-Scrivener, Beverly, MA (2021)
63. https://www.koboproductsinc.com/downloads/kobo-koboguard5400.pdf
64. https://www.ulprospector.com/additives/en/datasheet/264685/regalite-s1100-hydrocarbon-resin
65. W. F. Buth, Film forming lipstick, US Patent 3088876, assigned to Kolmar Laboratories, Inc., a corporation of Delaware (1963)

66. A. L. Howell and B. C. Mei, Cosmetic composition for impacting wear resistance and shine, US Patent 6303105, assigned to Avon Products, Inc. (2001)
67. J.H. Dempsey, A.M. Fabula, T.E. Rabe, J.M. Lubbers and M. Ye, Development of a semi-permanent mascara technology, Int. J. Cosmet. Sci., 34, 29-35 (2012)
68. J. H. Dempsey and T. E. Rabe, Semi-permanent mascara compositions, US Patent 10,034,829, assigned to Noxell Corporation (2018)
69. J. Comyn, Surface characterization of pentaerythritol rosin ester, Int. J. Adhesion and Adhesives, 15, 9-14 (1995)
70. C. Li, H.S. Bui, M. Kanji, B.T. Bavouzet, K. Patel, and D. Patel, Mascara containing an aqueous dispersion of polyurethane and an acrylic film former, US Patent 8920787, assigned to L'Oreal (2014)
71. B. Atis, H. S. Bui and M. Kanji, Mascara compositions containing at least two acrylate film formers. US Patent 8846018, assigned to L'Oreal (2014).
72. J. Portal, X. Schultze, S. Taupin, and M. Vicic, Dispersion of polymer particles in a non-aqueous medium and cosmetic use thereof, US Patent 10,071,046, assigned to L'OREAL (2018).
73. J. Portal, X. Schultze, S. Taupin, M. Arnaud-Roux, J. Bonnard, G. Naudin, M. Hely, H. Bui and N. Biderman, Adhesion aspect and film-forming properties of hydrocarbon polymers-based lipsticks, in: *Surface Science and Adhesion in Cosmetics*, K. L. Mittal and H. S. Bui (Eds), pp. 451-486, Wiley-Scrivener, Beverly, MA (2021)
74. AL. S. Bernard, Y. Deng, H. S. Bui, L. Daubersies, and R. Debeaud, Compositions and methods for improving the appearance of the skin, US Patent 11382855, assigned to L'OREAL (2022).
75. AL. S. Bernard, Y. Deng, H. S. Bui, L. Daubersies, R. Debeaud, and A. Farran, Compositions and methods for improving the appearance of the skin, US Patent 10864157, assigned to L'OREAL (2020).
76. L. Daubersies, C. Lison, S. Douezan, H. S. Bui, and R. El-Khouri, Composition comprising polymer particles, a hydrocarbon-based oil and a hydrocarbon-based block copolymer, and process using the same, US Patent 10576025, assigned to L'OREAL (2020).
77. F. M. Etzler, Characterization of surface free energies and surface chemistry of solids in: *Contact Angle, Wettability and Adhesion*, Vol. 3, K. L. Mittal (Ed.) pp. 219-264, CRC Press, Boca Raton, FL (2003)
78. F. M. Etzler, Determination of surface free energies of solids: A critical review. Rev. *Adhesion Adhesives* 1, 3-45 (2013)
79. J. V. Badami and H.S. Bui, Quantification of the color transfer from long-wear face foundation products: The relevance of wettability, in: *Surface Science and Adhesion in Cosmetics*, K. L. Mittal and H. S. Bui (Eds), pp. 379-400, Wiley-Scrivener, Beverly, MA (2021)
80. S. Murdan, A. Bari, S. Ahmed, B. Hossin, and L. Kerai, Nail lacquer films' surface energies and *in vitro* water-resistance and adhesion do not predict

their *in vivo* residence, *British Journal of Pharmacy* 2, (2017) doi: https://doi.org/10.5920/bjpharm.2017.03
81. F. Laubé, A. Poupon, P. Zinck, C. Müller-Goymann, S. Reichl, and V. Nardello-Rataj, Physicochemical investigations of native nails and synthetic models for a better understanding of surface adhesion of nail lacquers, European Journal of Pharmaceutical Sciences, 131, 208-217 (2019)
82. D. Rossi and N. Realdon, Surface tensiometry approach to characterize cosmetic products in the beauty sector, in: *Surface Science and Adhesion in Cosmetics*, K. L. Mittal and H. S. Bui (Eds), pp. 311-352, Wiley-Scrivener, Beverly, MA (2021)
83. A. Siokowska, B. Kaczmarek, M. Michalska, K. Lewandowska, and S. Grabska, Preparation and characterization of collagen/chitosan/hyaluronic acid thin films for application in hair care cosmetics, *Pure Applied Chemistry*, 89, 1829 – 1839 (2017)
84. R. Hagens, T. Mann, V. Schreiner, H.G. Barlag, H. Wenck, K. P. Wittern, and W. Mei, Contact angle measurement - a reliable supportive method for screening water-resistance of ultraviolet-protecting products *in vivo*. Int J Cosmet Sci. 29, 283-291 (2007)
85. F. Eudier, D. Hirel, M. Grisel, C. Picard, and G. Savary, Prediction of residual film perception of cosmetic products using an instrumental method and non-biological surfaces: The example of stickiness after skin application. Colloids Surfaces B Biointerfaces.,174, 181-188 (2019)
86. E. Faucheux, C. Picard, M. Grisel, and G. Savary, Residual film formation after emulsion application: Understanding the role and fate of excipients on skin surface. Int J Pharm.,30, 585: 119453 (2020) https://doi.org/10.1016/j.ijpharm.2020.119453
87. C. Messaraa and M. Mangan, Lasting Impression: A Consumer-centric Approach to Measure Transfer-proof Lipstick March 1, 2021 https://www.cosmeticsandtoiletries.com/testing/efficacyclaims/Lasting-Impression-A-Consumer-centric-Approach-to-Measure-Transfer-proof-Lipstick-573894481.html
88. B. Yu, SY. Kang, A. Akthakul, N. Ramadurai, M. Pilkenton, A. Patel, A. Nashat, D. G. Anderson, F. H. Sakamoto, B. A. Gilchrest, R. R. Anderson, and R. Langer, An elastic second skin. *Nat Mater.*;15, 911-918 (2016)
89. S. Murdan, L. Kerai, and B. Hossin, To what extent do *in vitro* tests correctly predict the *in vivo* residence of nail lacquers on the nail plate, J. Drug Delivery Science and Technology, 25, 23-28 (2015)
90. C. Richard, Lipstick adhesion measurement, in: *Surface Science and Adhesion in Cosmetics*, K. L. Mittal and H. S. Bui (Eds), pp. 635-662, Wiley-Scrivener, Beverly, MA (2021).
91. https://www.stablemicrosystems.com/MeasureStickiness.html
92. A. E. O'Connor and N. Willenbacher, The effect of molecular weight and temperature on tack properties of model polyisobutylenes, Int. J. Adhesion and Adhesives, 24, 335-346 (2004)

93. TA Instruments, Pressure-sensitive adhesive tack using the ARES rheometer (1999) http://www.tainstruments.com/pdf/literature/L2084_Tack_Testing_ARES.pdf
94. Malvern Instruments, Application note: Assessing tackiness and adhesion using a pull away test on a rotational rheometer (2015) https://cdn.technology-networks.com/TN/Resources/PDF/AN150527AssessingTackinessPullAway.pdf
95. The best transfer proof foundations for face masks https://www.anniesnoms.com/2021/02/04/the-best-transfer-proof-foundations-for-face-masks/
96. www.popsugar.com/beauty
The Best Transfer-Proof Makeup Products For Face Masks | PS Beauty
97. J. Pereda, C. M. Khatib, S. Kezic, M. O. Christensen, S. Yang, J. P. Thyssen, C. Chu, C. Riethmüller, H. Liao, I. Akhtar, B. Ungar, E. Guttman-Yassky, M. Hædersdal, and E. Hwu, A review of atomic-force microscopy in skin barrier function assessment, J. Investigative Dermatology, 144, 2136-2144 (2024)
98. R. L. Mcmullen and G. Zhang, Investigation of the internal structure of human hair with atomic force microscopy. J. Cosmet. Sci.,71 ,117–131 (2020)
99. S. Breakspear, B. Noecker, and C. Popescu, AFM for hair surface characterisation in: *Surface Science and Adhesion in Cosmetics*, K. L. Mittal and H. S. Bui (Eds), pp. 215-244, Wiley-Scrivener, Beverly, MA (2021).
100. D. Sychev, S. Schubotz, Q. A. Besford, A. Fery, and G. K. Auernhammer, Critical analysis of adhesion work measurements from AFM-based techniques for soft contact, J Colloid and Interface Sci., 642, 216-226 (2023)
101. R. Jin, L. Skedung, C. Cazeneuve, J. C. Chang, M. Rutland, M. Ruths, and G. Luengo, Bioinspired self-assembled 3D patterned polymer textures as skin coatings models: Tribology and tactile behavior. *Biotribology*, 24, 100151 (2020) https://doi.org/10.1016/j.biotri.2020.100151
102. N. Biderman and H. S. Bui, Atomic force microscopy (AFM) as a surface characterization tool for hair, skin, and cosmetic deposition, in: *Surface Science and Adhesion in Cosmetics*, K. L. Mittal and H. S. Bui (Eds), pp. 245-278, Wiley-Scrivener, Beverly, MA (2021).
103. L. N. Jimenez, C. D. V. Martínez Narváez, C. Xu, S. Bacchi, and V. Sharma, Rheological properties influence tackiness, application and performance of nail polish/lacquer formulations, in: *Surface Science and Adhesion in Cosmetics*, K. L. Mittal and H. S. Bui (Eds), pp. 109-150, Wiley-Scrivener, Beverly, MA (2021)
104. L. N. Jimenez, C. D. V. Martínez Narváez, C. Xu, S. Bacchi, and V. Sharma. The rheologically-complex fluid beauty of nail lacquer formulations. Soft Matter, 17, 5197-5213 (2021)
105. H. J. Shin, M. J. Kim, and S. H. Jung, Spray type cosmetic composition having makeup-maintaining ability and spraying power, US Patent 11,759,413 assigned to LG H&H Ltd (2023).

106. S. Coste, Makeup-fixing cosmetic composition, US Patent 11,026,878 assigned to Laboratoire GARANCIA, Paris (2021)
107. T. Huang, J. Ebanks, H. Bui, C. Li, and V. Roulier, One-step long-lasting color/shine lip gloss *via* liquid-liquid phase separation, podium presentation, (IFSCC), London, UK, September 19-22, (2022)
108. X. Zhou, T. Huang, J. Ebanks, M. Hasebe, H. S. Bui, C. Li, R. J. El-Khouri, and R. Ahmad, Lip compositions capable of forming a multilayer structure after application to lips, US Patent 11712411 assigned to L'OREAL (2023).
109. Shiseido develops "Water Sensing Technology," an innovative technology to achieve color lasting and secondary adhesion-less effects - Lipstick that realizes both smooth and light wearing feel and long-lasting color (2023) https://corp.shiseido.com/en/newsimg/3692_t2d68_en.pdf
110. G. D. Rio and A. Rudin, Latex particle size and CPVC, Prog. Organic Coatings, 28, 259-270 (1996)
111. R. F. G. Brown, C. Carr, and M. E. Taylor, Effect of pigment volume concentration and latex particle size on pigment distribution, Prog. Organic Coatings, 30, 185-194 (1997)
112. G. Bierwagen, R. Fishman, T. Storsved and J. Johnson, Recent studies of particle packing in organic coatings, Prog. Organic Coatings, 35, 1-9 (1999)
113. D. Y. Perera, Effect of pigmentation on organic coating characteristics, Prog. Organic Coatings, 50, 247-262 (2004)
114. M. T. Rodriguez, J. J. Gracenea, A. H. Kudama and J. J. Suay, The influence of pigment volume concentration (PVC) on the properties of an epoxy coating Part I. Thermal and mechanical properties, Prog. Organic Coatings, 50, 62-67 (2004)
115. https://reads.alibaba.com/mascara-unlocking-the-secret-to-captivating-lashes/
116. https://www .prnewswire.com/news-releases/blinc-just-invented-the-perfect-mascara-301247882.html
117. https://www.allure.com/story/best-cream-eye-shadows#different
118. https://www.de-lanci.com/blogs/makeup-blog/guide-to-different-types-of-eyeshadows?
119. https://luxurybysofia.com/blogs/tutorials/a-guide-to-the-6-different-types-of-eyeshadow?
120. F. C. Pagano, A Review of Nail Polish: The Industrial Cosmetic (2013) https://www.cosmeticsandtoiletries.com/formulas-products/color-cosmetics/article/21836978/a-review-of-nail-polish-the-industrial-cosmetic
121. F.C.Pagano,Technology-drivenTrendsinNailPolishColorandTexture(2014) https://www.cosmeticsandtoiletries.com/formulas-products/color-cosmetics/article/21836270/technology-driven-trends-in-nail-polish-color-and-texture
122. L. Pavlovic, The science of UV cure nail gels: a short review, Radtech report (2018). Microsoft Word - UV Nail Gel Review Final.docx

123. I. Mieriņa, Z. Grigale-Sorocina ,and I Birks. The chemistry of behind the UV-curable nail polishes. Polymers (Basel), 17, 1166 (2025)
124. H. Schipper, All About Adhesion: How Nail Products Stick to the Nails, May 17, 2018 https://www.nailsmag.com/378664/all-about-adhesion-how-nail-products-stick-to-the-nails
125. Top Causes of Lifting or Chipping in Gel Manicures, (2024) https://www.vibrantsalonandspa.com/post/top-causes-of-lifting-or-chipping-in-gel-manicures
126. F. Zareanshahraki and V. Mannari," Green" UV-LED gel nail polishes from bio-based materials, Int J Cosmet Sci, 40: 555-564 (2018)
127. J. Khun, Advancing bio-based alternatives to UV gel nail products, The Aggie Transcript at UC Davis, July 6 2024
128. J. Watcharakitti , E. E. Win, J. Nimnuan , and S. M. Smith, Modified starch-based adhesives: a review. Polymers (Basel). 14, 2023 (2022)
129. Nagase. Pullulan (cosmetic grade). https://www.nagase.com/products/texturizer/pullulan_cosmetic_grade.
130. C-T. Chen, K-I. Chen, H-H. Chiang, Y-K. Chen, and K-C. Cheng, Improvement on physical properties of pullulan films by novel cross-linking strategy, J Food Sci., 82, 108-117 (2017)
131. K. Kulka and A. Sionkowska, Chitosan based materials in cosmetic applications: a review. Molecules, 28, 1817 (2023) https://doi.org/10.3390/molecules28041817
132. M. Abdelmoula, H. B. Hlima, F. Michalet, G. Bourduche, J.-Y. Chavant, A. Gravier, C. Delattre, M. Grédiac, J.-D. Mathias, S. Abdelkafi, P. Michaud, and H. deBaynast, Chitosan-based adhesive: optimization of tensile shear strength in dry and wet conditions, Polysaccharides, 2, 110-120 (2021)
133. J. Silvestre, C. Delattre, P. Michaud, and H. de Baynast, Optimization of chitosan properties with the aim of a water-resistant adhesive development. Polymers (Basel).,13, 4031 (2021) https://doi.org/10.3390/polym13224031
134. S. Magalhães, L. Alves, B. Medronho, A. C. Fonseca, A. Romano, J. F. J. Coelho, and M. Norgren. Brief overview on bio-based adhesives and sealants, Polymers (Basel). 11:1685 (2019) https://doi.org/10.3390/polym11101685
135. L. A. Heinrich, Future opportunities for bio-based adhesives—advantages beyond renewability. Green Chem., 21, 1866-1888 (2019).
136. https://www.learnbiomimicry.com/blog/Best-biomimicry-products
137. W. Li, R. Zhou, Y. Ouyang, Q. Guan, Y. Shen, E. Saiz, M. Li, and X. Hou, Harnessing biomimicry for controlled adhesion on material surfaces. *Small.*, 20, 2401859 (2024) https://doi.org/10.1002/smll.202401859

8

Investigation on the Adhesion of Electrospun Polyacrylonitrile (PAN) Nanofiber Mats to Polypropylene (PP) Nonwoven Substrate

Zaynab Daneshzand[1], Somaye Akbari[1]*, Parastoo Saeedi[1] and Mohsen Mohseni[2]

[1]Textile Engineering Department, Amirkabir University of Technology (Tehran Polytechnic), Tehran, Iran
[2]Polymer Department, Amirkabir University of Technology (Tehran Polytechnic), Tehran, Iran

Abstract
The adhesion of electrospun nanofibrous mats to supporting substrates is a critical area of research, particularly given the diverse applications of these materials. One of the primary challenges associated with nanofibers is their inadequate mechanical properties, which can be mitigated by employing nonwoven fabrics as a supportive layer. This study investigates the adhesion characteristics of electrospun polyacrylonitrile (PAN) nanofiber mats to polypropylene (PP) nonwoven substrate as a supporting layer. Peel tests were performed to assess the adhesion strength over three distinct electrospinning time intervals and different fiber diameters on two types of PP nonwoven densities, as indicated by weight per square meter (g/m^2) used as supporting layers. Our findings reveal that the adhesion between the electrospun fibers and the substrate is influenced by multiple factors. The results have shown that for a substrate with a higher weight per square meter (30 g/m^2), increasing the fiber diameter by enhancing mechanical entanglement can lead to a significant increase in adhesion force, sometimes up to two times. Conversely, in a lower weight per square meter (20 g/m^2) substrate, by reducing the fiber diameter by expanding the contact surface and enveloping more fibers around the substrate fibers can also improve adhesion, offering practical insights

**Corresponding author*: akbari_s@aut.ac.ir.

Anil N. Netravali and Frank M. Etzler (eds.) The Wide World of Adhesion: A Festschrift in Honor of Dr. Kash Mittal on his 80th Birthday, (299–312) © 2026 Scrivener Publishing LLC

for substrate engineering. These insights could help in developing innovative strategies for substrate design in future applications of nanofibrous materials.

Keywords: Adhesion, polyacrylonitrile, nanofiber, nonwoven substrate, electrospun fibers, electrospun layers

8.1 Introduction

Nanofibers, a material class with unparalleled potential, have sparked the curiosity of researchers and industries alike. Their unique features, such as high specific area, high porosity, easy size control, easy surface functionalization, and high surface area-to-volume ratio are key to solving many challenges [1, 2]. There are various methods for producing this class of materials. In general, nanomaterials synthesis follows two general methods: top-down and bottom-up. The bulk material is transformed into nanofibers using physical, biological, or chemical treatments or is produced based on molecular assemblies [3]. More precisely, and from another perspective, the methods by which nanofibers are classified into electrospinning and non-electrospinning. Among the electrospinning methods, melt electrospinning [4–9], near-field electrospinning [10–12], coaxial electrospinning [13–19], solution blow electrospinning [20–23], centrifugal electrospinning [24–28], and microfluidic electrospinning [29, 30] can be mentioned Non-electrospinning methods include: template synthesis [31, 32] and phase separation [33, 34].

Electrospinning, a versatile method that allows to produce nanofibers from polymeric materials, offers a range of advantages. These include relative ease, low cost, high speed, vast materials selection, and, most importantly, the ability to control fiber diameter, microstructure, and arrangement [35–38]. During the electrospinning process, an electric field was introduced by using two conductors with opposite charges that are shaped like fibers, applied to the solution or molten substance [39]. When a polymer solution is charged with a high voltage, the electrostatic field between the electrodes plays an important role and immediately applies an electrical pressure to the solution droplet. The surface tension deforms the droplet into a conical shape called a Taylor cone. When the solution jet is formed, it is immediately stretched into a long and thin filament due to the tensile force. In other words, in the electrospinning process, electrostatic forces are the key players that complement or replace the conventional mechanical forces (e.g., hydrostatic, pneumatic) used to form the jet and reduce the fiber size. After a short cross section, it experiences a rapid solvent evaporation process. It is worth noting that there are also melt spinning devices, but the most common is solvent electrospinning. The resulting fibers can range from nanometers to several micrometers, highlighting the complexity and diversity of the electrospinning process. Finally, the fibers are collected on a collector with random orientation [40, 41]. Fiber properties, such as diameter, morphology, and how the fibers are arranged together, are affected by many variables and are very important to understand because they are the factors that determine the properties of the fibers that play a role in the study of adhesion behavior. Variables include process factors (voltage, distance, and feed rate), solution properties (viscosity, polymer molecular weight, polymer solution concentration, pH, spatial arrangement of polymers, surface tension, and solvent type), and environmental factors such as temperature and humidity [42].

There are also problems and obstacles in the fabrication of electrospun fiber. For example, nanofibers are produced in a mat form with poor mechanical strength [43, 44]. Various approaches have been implemented to address the mechanical strength issue, including blending polymers, utilizing epoxy composites reinforced with nanomaterials, applying dip-coating, employing self-reinforcing methods, and using thermal lamination [1]. Each method poses environmental problems or is not sufficiently efficient. One suitable method that addresses these issues involves the use of a supportive layer, which not only enhances the mechanical strength of the nanofibers but also mitigates some of the environmental concerns associated with other methods.

Various materials, such as aluminum foil or textiles, are used as supports [45]. One of these efficient substrates, which one can confidently rely on, are nonwoven fabrics. Nonwoven, with their versatile applications in medical, industrial, and daily use, are indispensable in many areas of our lives. Their difference from woven or knitted fabrics is the elimination of one of the yarn production steps. In nonwoven production, fibers are placed together and connected, usually randomly arranged [46]. The most common polymers in nonwoven fabrics are polypropylene (PP), polyamides, and polyester. They are from the family of thermoplastic polymers with high molecular weight and broad molecular weight distribution [46, 47]. The spun-bond method involves three steps that are precisely similar to spinning synthetic fibers. In addition, the stabilization of the obtained layer is carried out through mechanical, chemical, or thermal bonding methods [48]. In the needle punching method, the creation of mechanical friction between the fibers is a crucial step, controlling significantly the fabric's structure and properties [49]. The two key aspects must also be considered. 1) The effect of non-conductive textiles placed between the needle and the collector. 2) The adhesion of the electrospun nanofibers to the substrate on which they are placed [50].

Adhesion is the manifestation of the attractive forces that exist between all atoms [51]. The term adhesion is ambiguous, meaning both the formation of surface bonds and the mechanical load required to rupture an assembly. One of the main difficulties in studying adhesion mechanisms is that the subject lies at the border of several scientific fields, including macromolecular science, surface and interface physics, materials science, and rheology. In the study of adhesion, different concepts are used, which depend on the individual's field of expertise, and, therefore, the interpretation of the observed phenomena can vary considerably. This diversity of approaches is emphasized by the fact that many theoretical models of adhesion have been proposed. These models, while often complementary, can also be contradictory, adding a layer of complexity and richness to our understanding of adhesion [52]. Understanding the adhesion process is of paramount importance. To this end, scientists have developed categorization of adhesion mechanisms or theories, based on the fundamental behavior of the materials to be bonded [53].

The mechanical interlocking theory, proposed by McBain and Hopkins in 1925 [54], considers the mechanical locking of two surfaces in the cavities and pores of the solid surfaces as the main factor determining the strength of two adhering materials [52]. For relatively rough surfaces, increasing the surface area may lead to a proportional increase in adhesion as long as the roughness of the contact between the surfaces is not reduced. A classic example of the mechanical interlocking theory of adhesion is where one phase is "keyed" to another. The acceptance of the theories has varied over time and with different investigations. The mechanical interlocking theory has historically proven valuable because it has focused attention on surface roughness and its effect on adhesion. Roughening of interfaces, on a scale ranging from hundreds of micrometers to nanometers, is important in the more effective use of bonding methods and developing new materials. In examining the effect of roughness on adhesion, the concepts of adsorption, diffusion, and mechanical theories not only overlap but also integrate seamlessly to provide a comprehensive model of experimental observations. Furthermore, scientific theories are intellectual models. Mental constructs used to rationalize observations are no more accurate than the phenomena from which they are derived [54].

In this study, we examine how fiber diameter and the density of nonwoven substrates influence the mechanical engagement by measuring the adhesion of polyacrylonitrile (PAN) nanofibers to polypropylene (PP) substrates. To do this, PAN nanofibers with three different diameters were electrospun onto the PP nonwoven substrates with two different weight per square meter (g/m^2). The results were evaluated by measuring the adhesion energy between the electrospun layers and nonwoven substrates as well as by conducting air permeability tests.

8.2 Materials and Methods

The polypropylene nonwoven fabrics were purchased from Bafineh Yazd Company, Iran weighing 20±3 and 30±1 g/m^2. Polyacrylonitrile powder (PAN) (Mw = 70 kDa) was obtained from Isfahan Polyacryl Company, Isfahan, Iran. N, N-dimethylformamide (DMF, 99.5%) was purchased from Neutron Pharmaceutical Chemistry Company, Iran.

8.2.1 Electrospinning Procedure

PAN was dissolved in DMF to obtain 11%, 12% and 13% (w/v) solutions. In order to produce nanofibers, various electrospinning parameters were varied to obtain bead-free fibers and stable spinning conditions, according to Table 8.1. The nanofiber layer production process was carried out on two substrates of 20 and 30 g/m^2 with dimensions of 25 cm x 25 cm. Finally, electrospun layers were detached from the collector for further tests. To facilitate testing and to accurately compare the results of multiple samples of polyacrylonitrile nanofibers, the samples were coded from left

Table 8.1 Electrospinning parameters for fabrication of PAN fibers on PP substrate. Feeding rate and the distance between the needle and the collector were 1 ml/h and 100 mm respectively.

Voltage (kV)	Substrate weight (g/m^2)	Concentration of the polymer solution (mol/l)	Sample code
10	20	11	P20-11
12	20	12	P20-12
10	20	13	P20-13
10	30	11	P30-11
12	30	12	P30-12
10	30	13	P30-13

to right according to the pattern of weight per square meter and concentration of the electrospinning solution in Table 8.1. The duration of electrospinning influences both the thickness of the layer and the coverage of the electrospun fibers on the substrate. In this study, electrospun layers were produced over time intervals of 15, 30, 60, and 120 minutes.

8.2.2 SEM Analysis

The surface morphology of PAN nanofibers was studied using SEM (30XL, Philips, Netherlands). Electro-spun nanofiber was coated with gold up to 10 nm to minimize charge and produce a conductive surface. Then, using ImageJ software (V-1.52), 100 fibers were selected randomly for SEM imaging, and their diameters were measured and reported as mean and standard deviation (SD).

8.2.3 Peel Test

The peel test was performed using an Instron machine according to ASTM D1876 standard with a force of 100 N. The electrospun samples were cut into 150 mm x 18 mm sections, and a layer of adhesive was applied. The adhesive tape was used to the prevent tearing the nanofiber layer to fit into the clamp. Then, the unattached end of the test sample was clamped in the clamps of the test machine, while the adhesive layer was attached between nanofibrous layer and the nonwoven substrate with an angle of 180 degrees between the two ends. The load was applied at a constant rate of 10 mm/min until the entire length of the sample was detached. At least five different samples were tested for each two-layer system. The outcomes of the peel test are shown as force-extension graphs, which illustrate the amount of work associated with the adhesion force.

8.2.4 Air Permeability Analysis

The amount of air passing through the samples was evaluated using a 1998SDL device, Shirley Development, Ltd (SDL), made in UK, according to ASTM D737 standard. The pressure of the device was set to 100 Pa. The amount of air passing through the sample in milliliters per 5 square centimeters was measured, and the samples were compared with each other on this basis.

8.3 Results and Discussion

Adhesion is a crucial concept in various fields and is essential for the effectiveness of the substrate that supports the nanofiber layer in producing this key product. Therefore, identifying the relationship between the factors that affect the adhesion of the electrospun layers to nonwoven substrates is a crucial step in this process. The relationship between the images obtained from the examining electron microscope and the analysis of air permeability and peel test results is explored.

The results shown in Figure 8.1 reveal that by increasing polymer solution concentration, the diameter of the nanofibers increased due to the increasing viscosity of the solution by improving the entanglement between chains [17]. An analysis of the air permeability test results in Table 8.2

(a) P20-11 (b) P20-12 (c) P20-13

Figure 8.1 SEM images of PAN nanofibers (a) 11% (b) 12% (c) 13% concentration of PAN.

Table 8.2 Average diameter and air permeability of each sample.

sample	Concentration of the polymer solution (mol/l)	Average diameter (nm)	Air permeability (ml/S cm^2)
P20-11	11	240±37	120±37
P20-12	12	304±47	87±15
P20-13	13	455±70	223±47
P30-11	11	274±60	130±1
P30-12	12	298±51	132±19
P30-13	13	333±58	156±15

indicates that for two identical substrates, a greater fiber diameter reveals greater air permeability due to more space between fibers [4].

The force-extension curves obtained from the peel test are crucial in evaluating the performance of nanofibers. Padhye *et al.* [55] reported the work of adhesion by reporting the average value of the area under the force-extension diagram per unit area. Figure 8.2 shows the adhesion energy between electrospun nanofibers with various fiber diameters on two substrates. As the time of the electrospinning procedure impacts the coverage of electrospun fibers on the nonwoven layer, four types of electrospun fibers were fabricated as illustrated in Figures 8.2 a, b, c and d.

The results confirm that adhesion energy is higher for higher g/m^2 substrate. For instance, a weight of 30 g/m^2 reveals a higher adhesion than the substrate with a weight of 20 g/m^2. During the beginning of electrospinning (15 minutes of electrospinning), a thin layer of nanofibers covered the substrate surface. By increasing the quantity of fibers per unit area from 20 g/m^2 to 30 g/m^2 substrate, an additional interface with the nanofibers

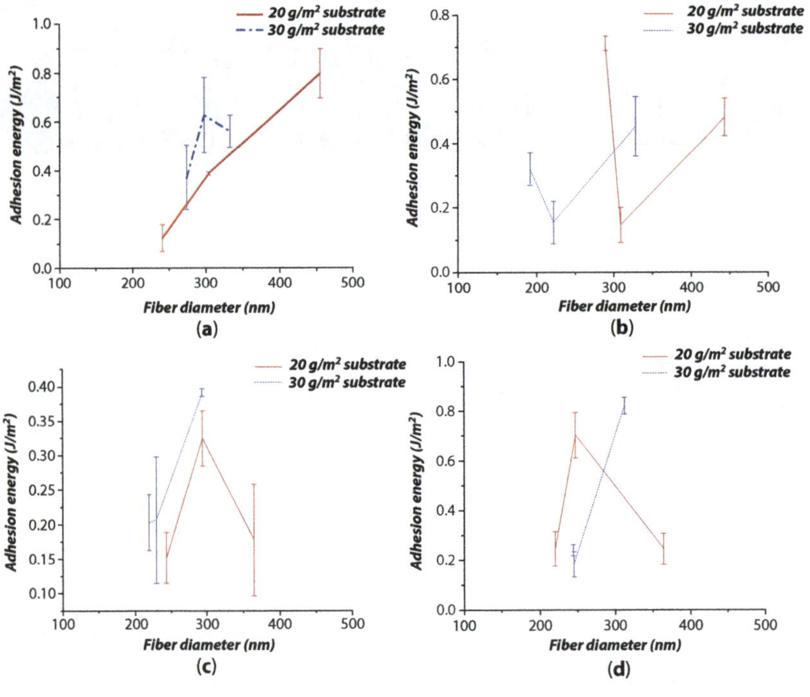

Figure 8.2 Impact of substrate weight (20 and 30 g/m^2) and electrospun fibers diameter on the adhesion energy between electrospun fiber and substrate at various electrospinning times: (a) 15, (b) 30, (c) 60, and (d) 120 minutes.

was established. Figure 8.2 shows that with increasing nanofiber diameter on both substrates, the adhesion values first decrease and then increase consistent with other published work [56]. Gradually, with time, the surface of the nonwoven substrate was covered entirely with nanofibers. With the continuation of the process, the nanofibers produced are deposited on a layer of previously electrospun nanofibers, and the nanofiber layer becomes thicker. The force required to separate the two layers is measured according to the peel test method to measure adhesion. As the thickness of the nanofiber layer increases, the interface and the contact between these two layers becomes weaker, and the nanofibers remain unchanged, the connection between these two layers becomes weaker, causing a decrease in adhesion value. Previous studies also show that the contact area between two electrospun layers depends on the fiber diameter and the number of contact points between the two layers [55, 57].

On the other hand, our investigation into the effect of thickness of electrospun fiber varied by different times of electrospinning, in line with previous studies, has yielded practical implications. Despite the increase in thickness, the effect of increasing the weight of the nanofiber layer becomes apparent. By increasing the time of electrospinning well as the thickness of electrospun layer, the pressure on the underlying nanofibers in locking theory increases [54]. This occurs at a specific time for each layer, depending on the nanofibers' diameter and the substrate's density. As a result a minimum or maximum value is present in Figure 8.2.

8.4 Conclusions

The electrospun nanofiber layer was first fabricated on a nonwoven substrate, and then it was subjected to various evaluations to investigate the adhesion between electrospun fiber and nonwoven substrate. Three evaluated factors, such as the diameter of the nanofibers and the weight per square meter of the substrate and the time of electrospinning affect the adhesion value. It is a very complex field because it does not fit into any single adhesion model or theory. Considering the number of phenomena involved in adhesion, the variety of materials to be bonded, and the variety of bonding conditions, searching for a unique and general theory that can explain all the experimental facts is futile. In practice, several adhesion mechanisms can be used simultaneously. Finally, the measured bond strength of an assembly can be expressed as a function of three terms related to surface molecular interactions, mechanical properties, and interface characteristics. Therefore, the influence coefficient of each factor varies depending on the conditions, and an accurate prediction cannot be made. What is clear is the possibility of analyzing the phenomenon and consequently improving the fabrication conditions. In the overall comparison between the densities of the substrates, at higher densities, the adhesion of polyacrylonitrile nanofibers was about 2 times higher due to a more significant number of fibers involved. On the other hand, with an increase in the diameter of the fibers on the substrate of 20 g/m^2, the adhesion was doubled for the same volume of electrospun fibers. This value increased about 1.5 times for the 30 g/m^2 substrate. These findings help overcome the poor mechanical properties of nanofibers and facilitating their use and application in various industries.

AI disclosure statement: The author acknowledges the use of AI-based tools for minor language editing or grammatical correction. However, all intellectual contributions, analyses, and conclusions presented in this chapter are entirely the author's own.

References

1. R. Roche and F. Yalcinkaya, Incorporation of PVDF nanofibre multilayers into functional structure for filtration applications, *Nanomaterials*, 8, 771, 1-18 (2018).
2. Q. Zheng, Y. Xi, and Y. Weng, Functional electrospun nanofibers: Fabrication, properties, and applications in wound-healing process, *RSC* Advances, 14, 3359–3378 (2024).
3. B. A. Venmathi Maran, S. Jeyachandran, and M. Kimura, A review on the electrospinning of polymer nanofibers and its biomedical applications, *Journal of Composites Science*, 8, 32, 1-19 (2024).
4. A. Barhoum, K. Pal, H. Rahier, H. Uludag, I. S. Kim, and M. Bechelany, Nanofibers as new-generation materials: From spinning and nano-spinning fabrication techniques to emerging applications, *Appl Mater Today*, 17, 1–35 (2019).
5. H. M. Ibrahim and A. Klingner, A review on electrospining polymeric nanofibers: Production parameters and potential applications, *Polym Test*, 90, 100-120 (2020).
6. S. Nemati, S. Kim, Y. M. Shin, and H. Shin, Current progress in application of polymeric nanofibers to tissue engineering, *Nano Convergence*, 6, 1-16 (2019).
7. T. Lu, J. Cui, Q. Qingli, Y. Wang, J. Zhang, R. Xiong, W. Ma, C. Huang, Multistructured electrospun nanofibers for air filtration: a review, *ACS Appl Mater Interfaces*, 13, 23293–23313 (2021).

8. D. Santos, A. E. Antunes, F. V. D. Santos, K. M. Freitas, L. P. S. Pimenta, L. d. O. Andrade, T.A. Marinho, G.F. d. Avelar, A.B. d. Silva, and R. V. Ferreira, Cellulose acetate nanofibers loaded with crude annatto extract: Preparation, characterization, and *in vivo* evaluation for potential wound healing applications, *Materials Science and Engineering: C* 118, 1-13 (2021).
9. W. Han, L. Wang, Q. Li, B. Ma, C. He, X. Guo, J. Nie, and G. Ma, A review: Current status and emerging developments on natural polymer-based electrospun fibers, *Macromol Rapid Commun*, 43, 2200456 (2022).
10. D. D. Camillo, V. Fasano, F. Ruggieri, S. Santucci, L. Lozzi, A. Camposeo, and D. Pisignano Near-field electrospinning of conjugated polymer light-emitting nanofibers. arXiv preprint arXiv:1310.5101 (2013).
11. Y. Huang, J. Song, C. Yang, Y. Long, and H. Wu, Scalable manufacturing and applications of nanofibers, *Materials Today*, 28, 98–113 (2019).
12. V. Guarino, R. Altobelli, V. Cirillo, A. Cummaro, and L. Ambrosio, Additive electrospraying: A route to process electrospun scaffolds for controlled molecular release, *Polym Adv Technol*, 26, 1359–1369 (2015).
13. S. Torres-Giner, Electrospun nanofibers for food packaging applications, in: *Multifunctional and Nanoreinforced Polymers for Food Packaging*, J.M. Lagarón (Ed) pp.108-125 Woodhead Publishing (2011).
14. B. Pant, M. Park, and S.-J. Park, Drug delivery applications of core-sheath nanofibers prepared by coaxial electrospinning: a review, *Pharmaceutics*, 11, 305, 1-21 (2019).
15. M. F. Elahi, W. Lu, G. Guoping, and F. Khan, Core-shell fibers for biomedical applications-a review, *J. Bioeng. Biomed. Sci*, 3, 1–14 (2013).
16. D. George, A. Garcia, Q. Pham, M. R. Perez, J. Deng, M. T. Nguyen, T. Zhou, S. Martinez-Chapa, Y. Won, C. Liu, and R.C. Lo, Fabrication of patterned graphitized carbon wires using low voltage near-field electrospinning, pyrolysis, electrodeposition, and chemical vapor deposition, *Microsystems & Nanoengineering, 6, 7, 1-12* (2020).
17. F. Fadil, N. D. N. Affandi, M. I. Misnon, N. N. Bonnia, A. M. Harun, and M. K. Alam, Review on electrospun nanofiber-applied products, *Polymers*, 13, 2087, 1-29 (2021).
18. J. Liang *et al.*, Recent advances in electrospun nanofibers for supercapacitors, *J Mater Chem A*, 8, 16747–16789 (2020).
19. Z. Zhao, R. Fang, Q. Rong, and M. Liu, Bioinspired nanocomposite hydrogels with highly ordered structures, *Advanced Materials*, 29, 205-227 (2017).
20. G. Chakrapani, S. Ramakrishna, and M. Zare, Functionalization of electrospun nanofiber for biomedical application, *J Appl Polym Sci*, 140, 1-19 (2023).
21. F. T. G. Dias, S. P. Rempel, L. D. Agnol, and O. Bianchi, The main blow spun polymer systems: processing conditions and applications, *Journal of Polymer Research*, 27, 205,1- 8 (2020).
22. G. C. Dadol, A. Kilic, L. D. Tijing, K.J.A. Lim, L.K. Cabatingan, N.P.B. Tan, E. Stojanovska, and Y. Polat, Solution blow spinning (SBS) and SBS-spun

nanofibers: Materials, methods, and applications, *Mater Today Commun*, 25, 1-8 (2020).
23. A. Abdal-Hay, Y. S. Oh, A. Yousef, H. R. Pant, P. Vanegas, and J. K. Lim, *In vitro* deposition of Ca-P nanoparticles on air jet spinning Nylon 6 nanofibers scaffold for bone tissue engineering, *Appl Surf Sci*, 307, 69–76 (2014).
24. J. Venugopal and S. Ramakrishna, Applications of polymer nanofibers in biomedicine and biotechnology, *Appl Biochem Biotechnol*, 125, 147–157 (2005).
25. M. R. Badrossamay, H. A. McIlwee, J. A. Goss, and K. K. Parker, Nanofiber assembly by rotary jet-spinning, *Nano Lett*, 10, 2257–2261 (2010).
26. W. Zou, R. Y. Chen, G. Z. Zhang, H. C. Zhang, and J. P. Qu, Recent advances in centrifugal spinning preparation of nanofibers, *Advanced Materials*, 1015, 170–176 (2014).
27. S. M. H. Marjuban, S.M.H. Marjuban, M. Rahman, S.S. Duza, M.B.Ahmed, D.K.Patel, M.S. Rahman, and K. Lozano, Recent advances in centrifugal spinning and their applications in tissue engineering, *Polymers (Basel)*, 15, 1-15 (2023).
28. F. Mokhtari, M. Salehi, F. Zamani, F. Hajiani, F. Zeighami, and M. Latifi, Advances in electrospinning: The production and application of nanofibres and nanofibrous structures, *Textile Progress*, 48, 119–219 (2016).
29. S. Y. Ahn, C. H. Mun, and S. H. Lee, Microfluidic spinning of fibrous alginate carrier having highly enhanced drug loading capability and delayed release profile, *RSC Adv*, 5, 15172–15181 (2015).
30. Y. Jun, E. Kang, S. Chae, and S.-H. Lee, Microfluidic spinning of micro-and nano-scale fibers for tissue engineering, *Lab Chip*, 14, 2145–2160 (2014).
31. P. Kumari, B. Ghosh, and S. Biswas, Nanocarriers for cancer-targeted drug delivery, *J Drug Targeting*, 24, 179–191 (2016).
32. P. Ye, Q. Guo, Z. Zhang, and Q. Xu, High-speed centrifugal spinning polymer slip mechanism and PEO/PVA composite fiber preparation, *Nanomaterials*, 13, 1277, 1-15 (2023).
33. C. Englert, J.C. Brendel, T.C. Majdanski, T. Yildirim, S. Schubert, M. Gottschaldt, , N. Windhab, and U.S. Schubert, Pharmapolymers in the 21st century: Synthetic polymers in drug delivery applications, *Prog Polym Sci*, 87, 107–164 (2018).
34. J. Wu, Z. Zhang, J.G. Gu, W. Zhou, X. Liang, G. Zhou, C.C. Han, S. Xu, and Y. Liu, Mechanism of a long-term controlled drug release system based on simple blended electrospun fibers, *Journal of Controlled Release*, 320, 337–346 (2020).
35. W. E. Teo and S. Ramakrishna, Electrospun nanofibers as a platform for multifunctional, hierarchically organized nanocomposite, *Composites Sci Technol*, 69, 1804–1817 (2009).
36. Z.-M. Huang, Y.-Z. Zhang, M. Kotaki, and S. Ramakrishna, A review on polymer nanofibers by electrospinning and their applications in nanocomposites, *Composites Sci Technol*, 63, 2223–2253 (2003).

37. C. Burger, B. S. Hsiao, and B. Chu, Nanofibrous materials and their applications, *Annu. Rev. Mater. Res.*, 36, 333–368 (2006).
38. R. Vasita and D. S. Katti, Nanofibers and their applications in tissue engineering, *Int J Nanomedicine*, 1, 15–30 (2006).
39. A. Keirouz, Z. Wang, V.S. Reddy, Z.K. Nagy, P. Vass, M. Buzgo, S. Ramakrishna, and N. Radacsi, The history of electrospinning: past, present, and future developments. *Advanced Materials Technologies*, 8, 11, 1-34 (2023).
40. A. L. Yarin, B. Pourdeyhimi, and S. Ramakrishna, *Fundamentals and Applications of Micro-and Nanofibers*. Cambridge University Press (2014).
41. Z. Li and C. Wang, Introduction of electrospinning. in: *One-dimensional Nanostructures: Electrospinning Technique) and Unique Nanofibers*, pp.1-13, Springer, Heidelberg (2013).
42. W. Zhang, Z. He, Y. Han, Q. Jiang, C. Zhan, K. Zhang, Z. Li and R. Zhang, Structural design and environmental applications of electrospun nanofibers, *Composites Part A* 137, 1-20 (2020).
43. E. P. S. Tan and C. T. Lim, Mechanical characterization of nanofibers–a review, *Composites Sci Technol* 66, 1102–1111 (2006).
44. F. Yalcinkaya, Mechanically enhanced electrospun nanofibers for wastewater treatment, in: Proc. E3S Web of Conferences, EDP Sciences,1-31 (2017).
45. F. Beermann, M. Schwakenberg, A. L Voigt, and A. Ehrmann, Influence of substrate materials on electrospun pan nanofiber mats. in: Proc. *IEEE 8th International Conference Nanomaterials: Application & Properties (NAP)*, 1-5 *(2018)*.
46. V. K. Midha and A. Dakuri, Spun bonding technology and fabric properties: a review, *Journal of Textile Engineering & Fashion Technology*, 1, 4, 1–9 (2017).
47. S. K. Batra and B. Pourdeyhimi, *Introduction to Nonwovens Technology*. DEStech Publications, (2012).
48. K. Chen, A. Ghosal, A. L. Yarin, and B. Pourdeyhimi, Modeling of spunbond formation process of polymer nonwovens, *Polymer*, 187, 1-14 (2020).
49. S. Jayaraman, W. W. Carr, and L. H. Olson, Textile Engineering, Elsevier, 577-589 (2003).
50. F. Rombaldoni, K. Mahmood, A. Varesano, M.B. Songia, A. Aluigi, C. Vineis, and G. Mazzuchetti, Adhesion enhancement of electrospun nanofiber mats to polypropylene nonwoven fabric by low-temperature oxygen plasma treatment, *Surf Coat Technol*, 216, 178–184 (2013).
51. D. K. Owens and R. C. Wendt, Estimation of the surface free energy of polymers, *J Appl Polym Sci*, 13, 1741–1747 (1969).
52. J. Schultz and M. Nardin, Theories and mechanisms of adhesion, in: *Handbook of Adhesive Technology*, A.Pizzi and K.L.Mittal (Eds.) pp.19-34 Marcel Dekker, New York (1994).
53. K.L. Mittal and A. Pizzi *(Eds.) Adhesion Promotion Techniques: Technological Applications*. Marcel Dekker, New York (1999).

54. D. E. Packham, The mechanical theory of adhesion, in: *Handbook of Adhesive Technology*, second edition, A. Pizzi and K. L. Mittal (Eds.), pp. 69-93, CRC Press, Boca Raton, FL (2003).
55. N. Padhye, D. M. Parks, A. H. Slocum, and B. L. Trout, Enhancing the performance of the T-peel test for thin and flexible adhered laminates, *Reviews of Scientific Instruments*, 1-9 (2016).
56. M. Bouchet, A. Ajji, and S. Lerouge, Determination of the adhesion between electrospun mats through peel tests, *Fibers and Polymers*, 22, 1266–1273 (2021).
57. M. D. Bartlett, S. W. Case, A. J. Kinloch, and D. A. Dillard, Peel tests for quantifying adhesion and toughness: A review, Progress in Materials Science, 137, 1-55 (2023).

9

Microplastic Adhesion at Secondary Minimum in Porous Media: The Role of Hydration Forces

Gang Chen

Department of Civil and Environmental Engineering, FAMU-FSU College of Engineering, Florida State University, Tallahassee, FL, USA

Abstract

Various types of microplastics (MPs) are formed by fragmentation of large plastics through mechanical, biological, chemical, and photo-oxidation degradation. The presence of MPs changes the physiochemical and biological characteristics of the ecosystem and affects the organisms, leading to hazardous impacts on the ecosystem. MPs persist in the environment and their fate and transport are commonly controlled by the polymer type, size, and surface characteristics. The initial adhesion of MPs onto solid media surfaces is generally explained by classic Derjaguin-Landau-Verwey-Overbeek (DLVO) theory, which describes the interactions between MPs and the solid media surfaces in terms of van der Waals and electrostatic interactions. However, the classic DLVO theory fails to explain MP attachment in the secondary minimum, and the extended DLVO theory emerges, which reconciles classic DLVO theory by including non-DLVO Lewis acid-base interactions. Lewis acid/base interactions are also called hydration forces, which are initiated by water density difference in the solvation or hydration layer. The transition from secondary minimum to primary minimum aggregation is explained by the extended DLVO theory in terms of van der Waals, electrostatic, and Lewis acid/base interactions. At the primary and secondary minima, interactions are evaluated close to the equilibrium distance, where Lewis acid/base interactions dominate.

***Keywords*:** Microplastic, Derjaguin-Landau-Verwey-Overbeek (DLVO), van der Waals, electrostatic, extended DLVO, Lewis acid/base, primary minimum, secondary minimum

Email: gchen@eng.famu.fsu.edu

Anil N. Netravali and Frank M. Etzler (eds.) The Wide World of Adhesion: A Festschrift in Honor of Dr. Kash Mittal on his 80th Birthday, (313–322) © 2026 Scrivener Publishing LLC

With synthetic polymers continuing to be produced and used, various types of microplastics (MPs) are formed by fragmentation of large plastics through mechanical, biological, chemical, and photo-oxidation degradation [1, 2]. Currently, MPs including polyethylene (PE), polypropylene (PP), polystyrene (PS), polytetrafluoroethylene (PTFE), poly(methyl methacrylate) (PMMA), etc. find their way into the environment *via* various pathways [3]. The presence of MPs changes the physiochemical and biological characteristics of the ecosystem and affects the organisms, leading to hazardous impacts on the ecosystem [4]. Although it is evidenced that MPs exist in the environment, prediction of the fate and transport of MPs is a significant challenge because MP surface characterization varies significantly depending on the shape, size, polymer type, morphology, etc. Specifically, the size of MPs in the environments varies from μm to mm, with fibers and fragments being the most observed shapes and PE as the most prevalent polymer type [5, 6].

MPs persist in the environment and their fate and transport are commonly controlled by the polymer type, size, and surface characteristics [7, 8]. It should be noted that MP spreading in the environment dramatically reveals the possibility of accumulation and transport of other pollutants, such as heavy metals, persistent organic pollutants, pharmaceuticals, etc., to which MPs serve as vectors [9]. The initial adhesion of MPs to solid media surfaces is generally explained by classic Derjaguin-Landau-Verwey-Overbeek (DLVO) theory, which describes the interactions between MPs and the solid media surfaces in terms of van der Waals and electrostatic interactions [10]. Typically, van der Waals interactions are attractive; however, electrostatic interactions can be either attractive or repulsive depending on the MP and solid medium surface charge. Although MPs are commonly negatively charged, which result in repulsive interactions with soil matrices that are also negatively charged, MPs often have low mobility and adsorb strongly to the soil matrices.

van der Waals interactions are long-range forces, which depend on the Hamaker constant determined by the chemical composition and structure of MP and the solid medium surface [11]:

$$\Pi_{vdW}(D) = -\frac{A}{6\pi D^3} \tag{9.1}$$

where $\Pi_{vdW}(D)$ is the Lifshitz-van der Waals force, A is the Hamaker constant, and D is the separation distance between MP and the attachment medium surface.

The electrostatic forces are due to the formation of charges on the MP and solid medium surfaces, which are enhanced when MPs are in close proximity to the solid medium surface, inhibiting the adsorption of MPs [12]:

$$\Pi_{el} = k_B T \sum_i \left(C_i(z) - C_{i,\infty}\right) - \frac{\varepsilon\varepsilon_0}{2}\left(\frac{d\psi}{dz}\right)^2 \tag{9.2}$$

where Π_{el} is the electrostatic force, $C_i(z)$ is the ion concentration of species i at distance z, $C_{i,\infty}$ is the bulk concentration of the corresponding ion, k_B is the Boltzmann constant, ε and ε_0 are the relative dielectric permittivity of water and of vacuum, respectively, and ψ is the surface potential. $C_i(z)$ follows the Boltzmann distribution [13]:

$$C_i(z) = C_{i,\infty} e^{\left(-\frac{Z_i e \psi(z)}{k_B T}\right)} \tag{9.3}$$

where e represents the electron charge, z_i is the valency of corresponding ion, and T is the absolute temperature. ψ can be calculated by [14]:

$$\psi = \xi\left(1 + \frac{z}{R}\right)\exp(\kappa z) \tag{9.4}$$

where ζ is the zeta potential, R is the size of MPs, and $1/\kappa$ is the Debye-Hückel length, which is also an estimation of the effective thickness of the electrical double layer. The potential distribution $\psi(z)$ can be numerically obtained by solving the Poisson–Boltzmann equation using the standard Runge–Kutta algorithm [15]:

$$\frac{d^2}{dz^2}\psi(z) = -\frac{e}{\varepsilon\varepsilon_0}\sum_i Z_i C_{i\infty} \exp\left(-\frac{Z_i e \psi(z)}{k_B T}\right) \tag{9.5}$$

According to the classic DLVO theory, to achieve successful MPs deposition, MPs must overcome the energy barrier and attach in the primary minimum, which has been found to deviate significantly from experimental observations, i.e., MPs are potentially deposited in the secondary minimum (Figure 9.1). In Figure 9.1. The interaction free energy is expressed

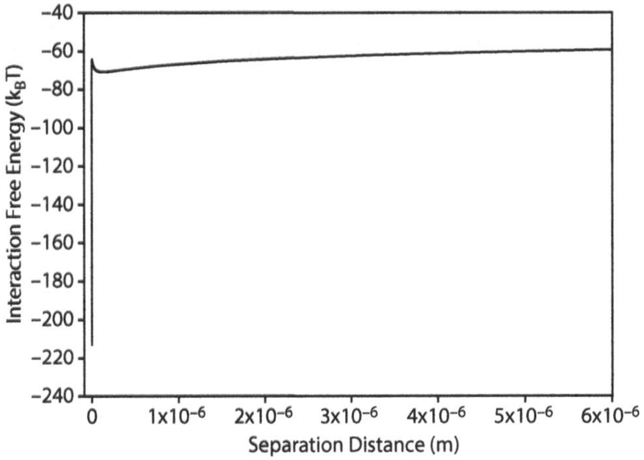

Figure 9.1 MP attachment to a solid porous medium in the secondary minimum.

in terms of k_BT (4.11×10^{-21} J). As defined before, k_B is the Boltzmann constant with a value of 1.38×10^{-23} J/K and T is the absolute temperature.

For MPs typically carrying a negative surface charge under the prevailing environmental pH conditions, the presence of repulsive energy barrier creates a secondary energy minimum located at a finite distance from the medium surface, resulting from the attractive van der Waals interactions [16]. MPs can potentially be retained at the secondary energy minimum, which differs from deposition at the primary energy minimum. Secondary minimum well is not very deep, and theoretically, MP attachment in the secondary minimum is weak and unstable. Subsequently, MP retention at secondary minimum can be dynamically reversible due to Brownian motion or unsteady hydrodynamics [17]. Thus, the retention is chemically reversible when the solution chemical conditions are altered. However, the actual observations indicate that MP deposition in the secondary minimum is strong enough to prevent MPs from being released by Brownian diffusion and hydrodynamic forces. Especially, large-sized MPs have deeper secondary minimum depths than small ones under the same conditions.

The solution chemistry (i.e., pH, ionic strength, types of solutes, and the valence of the ions) controls the deposition of MPs in a porous medium through its impact on electrostatic interactions [18]. Specifically, pH determines the charge density and surface potential of the MP and medium surface. Specifically, ionic strength and valence of the ions determine the shape of the interaction energy curve, and an energy barrier exists resulting from repulsive electrostatic interactions if MP and adsorbent medium surface are negatively charged [19]. Solution chemistry also controls the height of the energy barrier and the presence and depth of the secondary well. Increasing ionic strength decreases the height of the energy barrier and increases the depth of the secondary minimum [17]. Accordingly, deposition of MPs may transition from secondary minima to primary minima when ionic strength is changed from low to high. At some intermediate ionic strength, primary and secondary minimum deposition may occur simultaneously when the energy barrier is relatively small, and the secondary minimum is deep enough. Primary and secondary minimum deposition thus can coexist, leading to fast attachment by secondary minimum deposition and slow attachment by overcoming the energy barrier and deposition in the primary minimum.

The classic DLVO theory fails to explain MP attachment in the secondary minimum, and the extended DLVO theory emerges, which reconciles classic DLVO theory by including non-DLVO Lewis acid-base interactions [10]. Lewis acid/base interactions are also called hydration forces, which are initiated by water density difference in the solvation or hydration layer. Based on the acid-base concept, van Oss, Chaudhury and Good split the surface polar component into an electron-acceptor parameter, γ^+ and an electron-donor parameter, γ^- [20]. For surfaces with both γ^- and γ^+ values greater or smaller than those of water, a thin water layer can potentially balance the difference of γ^- and γ^+ [21]. Thus, the solvation layer has a density less than that of the bulk liquid, leading to a negative surface excess and surfaces being pushed together by the bulk liquid. The interactions between the surfaces are attractive with the strength of the attractive interactions proportional to the absolute value of the negative surface excess. For surfaces with one of γ^- and γ^+ values greater and the other smaller than those of water, more water molecules are needed to balance γ^- and γ^+ potentials and the solvation layer thus has a greater density than that of the bulk liquid, which leads to a positive surface excess. The pressure of the higher water density in the solvation layer separates the surfaces, resulting in repulsive forces between the surfaces with the strength of the repulsive forces proportional to the absolute value of the positive surface excess.

The transition from secondary minimum to primary minimum aggregation is explained by the extended DLVO theory in terms of van der Waals, electrostatic, and Lewis acid/base interactions. At the primary and secondary minima, interactions are evaluated close to the equilibrium distance, where Lewis acid/base interactions dominate. Typically, the interactions between the MP surface and hydrophobic components of the solid medium surface lead to attractive Lewis acid/base, which depends not only on the hydrophilicity of the hydrophilic surface and the hydrophobicity of the hydrophobic surface, but also on the monopolarity of the interacting surfaces [22]:

$$\Delta G_{132}^{AB} = 2[(\sqrt{\gamma_1^+} - \sqrt{\gamma_2^+})(\sqrt{\gamma_1^-} - \sqrt{\gamma_2^-}) - \frac{1}{2}\Delta G_{131}^{AB} - \frac{1}{2}\Delta G_{232}^{AB}] \quad (9.6)$$

where ΔG_{132}^{AB} is the Lewis acid/base interactions between MP surface 1 and solid medium surface 2, immersed in solution 3. ΔG_{132}^{AB} is determined by the hydrophilicity of the hydrophilic surface ΔG_{131}^{AB}, the hydrophobicity of the hydrophobic surface ΔG_{232}^{AB}, and the relative electron-donor and electron-acceptor potentials of the hydrophilic and hydrophobic surfaces.

Interactions of MPs with solid medium surfaces decrease in the presence of surfactants. The decrease in the interactions of MPs with solid medium surfaces with increasing surfactant concentrations is due to variations in solution surface thermodynamic properties [23]. For Lifshitz–van der Waals interactions, the increase in the surfactant concentration results in an increase in solution Lifshitz–van der Waals component of surface tension γ^{LW}, and a consequent increase in Lifshitz–van der Waals interactions between MPs and the media [24]. For Lewis acid/base interactions, the increase in the surfactant concentration leads to increase in γ^- and decrease in γ^+, which contributes to the increase (negative decrease) in Lewis acid/base interactions. The increase (negative decrease) in Lewis acid/base interactions is mainly attributed to the decrease in γ^+, which is actually the primary reason for MP desorption as Lewis acid/base interactions dominate over van der Waals interactions during attachment [25]. The presence of functional groups such as ethers (–CH_2–O–, CH_3–O–, –C–O–C–) and ethenyl groups (–CH=CH–, >C=CH_2, –CH=CH_2), etc. favors the decrease in γ^+. Surfactants are abundant in the above-mentioned functional groups, therefore, the increase in the surfactant concentration favors the decrease in solution γ^+ and decreases interactions between MPs and porous media, leading to MP mobility.

Conclusion

MP spreading in the environment is controlled by the interactions between MP and solid media surfaces. For MPs carrying a negative surface charge to deposit on negatively charged solid surfaces, repulsive energy barrier exists. According to the classic DLVO theory, MPs must overcome the energy barrier to attach in the primary minimum, which has been found to deviate significantly from experimental observations. MPs are believed to potentially deposit in the secondary minimum located at a finite distance from the medium surface. The solution chemistry plays key roles in controlling the deposition of MPs in a porous medium through its impact on electrostatic interactions, i.e., deposition of MPs may transition from secondary minima to primary minima when ionic strength is changed from low to high. The transition from secondary minimum to primary minimum aggregation is explained by the extended DLVO theory in terms of van der Waals, electrostatic, and Lewis acid/base interactions. The presence of surfactants decreases interactions of MPs with solid media surfaces due to variations in solution surface thermodynamic properties, leading to MP mobility.

AI disclosure statement: The author confirms that no artificial intelligence (AI)-generated content or AI-assisted writing tools were used in the preparation of this chapter (including summaries and the abstract). All text, analysis, and conclusions are the result of the author's original work.

References

1. M.A. Dar, P. Palsania, S. Satya, M. Dashora, O.A. Bhat, S. Parveen, S.K. Patidar and G. Kaushik, Microplastic pollution: A global perspective in surface waters, microbial degradation, and corresponding mechanism. Marine Pollution Bull 210, 117344 (2025).
2. O.A. Ahmad, M.T. Jamal, H.S. Almalki, A.H. Alzahrani, A.S. Alatawi and M.F. Haque, Microplastic pollution in the marine environment: Sources, impacts, and degradation. J Adv Veterinary Animal Res 12, 260-279 (2025).
3. M. Majewsky, H. Bitter, E. Eiche and H. Horn, Determination of microplastic polyethylene (PE) and polypropylene (PP) in environmental samples using thermal analysis (TGA-DSC). Sci Total Environ 568, 507-511 (2016).
4. N. Bostan, N. Ilyas, N. Akhtar, S. Mehmood, R.U. Saman, R.Z. Sayyed, A.A. Shatid, M.Y. Alfaifi, S.E.I. Elbehairi and S. Pandiaraj, Toxicity assessment of microplastic (MPs); a threat to the ecosystem. Environ Res 234, 116523 (2023).
5. D. Kim, S.A. Kim, S.H. Nam, J.I. Kwak, L. Kim, T.Y. Lee, H. Kim, S. An and Y.J. An, Microplastic ingestion in aquatic and soil biota: A comprehensive review of laboratory studies on edible size and intake pattern. Marine Pollution Bull 200, 116056 (2024).

6. Z. Taghizadeh Rahmat Abadi, B. Abtahi, M.B. Fathi, N. Mashhadi and H.P. Grossart, Size, shape, and elemental composition as predictors of microplastic surface erosion. J Hazard Mater 476, 134961 (2024).
7. Y. Wang, L. Xu, H. Chen and M. Zhang, Retention and transport behavior of microplastic particles in water-saturated porous media. Sci Total Environ 808, 152154 (2022).
8. J. Xu, R. Zuo, J. Shang, G. Wu, Y. Dong, S. Zheng, Z. Xu, J. Liu, Y. Xu, Z. Wu and C. Huang, Nano- and micro-plastic transport in soil and groundwater environments: Sources, behaviors, theories, and models. Sci Total Environ 904, 166641 (2023).
9. A. Medynska-Juraszek and B. Jadhav, Influence of different microplastic forms on pH and mobility of Cu^{2+} and Pb^{2+} in soil. Molecules 27, 1744 (2022).
10. A. Gomez-Flores, S. Jin, H. Nam, L. Cai, S. Song and H. Kim, Attachment of various-shaped polystyrene microplastics to silica surfaces: Experimental validation of the equivalent Cassini oval extended DLVO model. J Hazard Mater 470, 134146 (2024).
11. L. Perera, U. Essmann and M.L. Berkowitz, The role of water in hydration force - Molecular dynamics simulations, in: *Amphiphiles at Interfaces, Progress in Colloid & Polymer Science*, Vol. 103, J Texter (Ed.), pp. 107-115, Springer, Germany (1997).
12. R.P. Misra, S. Das and S.K. Mitra, Electric double layer force between charged surfaces: Effect of solvent polarization. J Chem Phys 138, 114703 (2013).
13. W. Silvestre-Alcantara, L.B. Bhuiyan, C.W. Outhwaite and D. Henderson, A modified Poisson-Boltzmann study of the singlet ion distribution at contact with the electrode for a planar electric double layer. Collect Czechoslov Chem Commun 75, 425-446 (2010).
14. S.S. Dukhin, Nonequilibrium electric surface phenomena. Adv Colloid Interface Sci 44, 1-134 (1993).
15. J.Y. Yang, C.Y. Yan, J.C. Huang and Z.H. Li, Numerical solutions of the semi-classical Boltzmann ellipsoidal-statistical kinetic model equation. Proc Math Phys Eng Sci 470, 20140061 (2014).
16. J.A. Redman, S.L. Walker and M. Elimelech, Bacterial adhesion and transport in porous media: role of the secondary energy minimum. Environ Sci Technol 38, 1777-1785 (2004).
17. A. Ashok and R.R. Dagastine, Parametric study of colloidal particle confinement near a surface in the presence of DLVO and structural interactions using Brownian dynamic simulations. Langmuir 39, 18380 (2023).
18. W. Zhao, Z. Su, T. Geng, Y. Zhao, Y. Tian and P. Zhao, Effects of ionic strength and particle size on transport of microplastic and humic acid in porous media. Chemosphere 309, 136593 (2022).
19. Z. Adamczyk, Particle adsorption and deposition: Role of electrostatic interactions. Adv Colloid Interface Sci 100-102, 267-347 (2003).

20. C.J. van Oss, L. Ju, M.K. Chaudhury and R.J. Good, Estimation of the polar parameters of the surface tension of liquids by contact angle measurements on gels, J Colloid Interface Sci 128, 313-319 (1989).
21. J. Gading, V. Della Balda, J. Lan, J.J. Konrad, M. Iannuzzi, R.H. Meissner and G. Tocci, The role of the water contact layer on hydration and transport at solid/liquid interfaces. Proc Natl Acad Sci U S A 121, e2407877121 (2024).
22. C.J. van Oss, *Interfacial Forces in Aqueous Media*, Marcel Dekker, New York (1994).
23. J.L. Parker, Surface force measurements in surfactant systems. Prog Surf Sci 47, 205-271 (1994).
24. A. Zdziennicka, K. Szymczyk, J. Krawczyk and B. Jańczuk, Components and parameters of solid/surfactant layer surface tension. Colloids Surfaces A 522, 461-469 (2017).
25. R. Miller, E.V. Aksenenko and V.B. Fainerman, Dynamic interfacial tension of surfactant solutions. Adv Colloid Interface Sci 247, 115-129 (2017).

10

Use of Plasma Technology to Improve Polymer Adhesion Part 1 Plasma Surface Modification to Enhance Polymer Adhesion

Jörg Florian Friedrich

Technical University Berlin, Polymer Materials and Technologies, Berlin, Germany

Abstract

Plasma polymers can be successfully used as adhesion promoters in metal-polymer laminates as discussed in Part 2 of this contribution. Here, in Part 1, the pretreatment of polyolefin surfaces is described. Ideally, the aim of the pretreatment should be to create covalent bonds between the metal and the plasma-generated functional groups on the polymer surface. These covalent bonds between metal and polymer should result in a particularly strong adhesion, which can no longer be separated during the peel strength test.

A special structure of the metal-polyolefin interface is presented, which compensates for the extremely different properties of the metal layer and those of the polymer by means of a flexible transition from one phase to the other.

This flexible transition on the polyolefin side of the interface is created by incorporating polar functional groups in low-pressure or atmospheric-pressure plasma. The most suitable initial configuration of the polyolefin side is the creation of monosort functional groups, which offer a maximum of reactive anchor points on the polymer surface. This enables an optimal structure of the interface design with the help of spacer molecules grafted *via* covalent bonds.

The principal disadvantage of all plasma modifications that cannot be eliminated is the degradation of the polymer molecules by plasma particle bombardment and vacuum UV irradiation, which leads to the formation of a "weak boundary layer" (WBL) on the surface, which reduces the strength of the laminate.

Email: Joerg.florian.friedrich@gmx.de

Anil N. Netravali and Frank M. Etzler (eds.) The Wide World of Adhesion: A Festschrift in Honor of Dr. Kash Mittal on his 80th Birthday, (323–384)

***Keywords*:** Aluminium-polyolefin laminates, plasma pretreatment, covalent bonds, flexible spacer molecules, barrier formation

10.1 Introduction

There are many methods for treating polymer surfaces to promote adhesion between metal overlayers and polymers [1]. The problem is that metal and polymer have very different chemical and mechanical properties.

The metal-polymer bond is also not a conventional covalent bond as we know it from polymers. This bond must bridge and compensate for the different material properties. This is why high forces are exerted on this single bond when the laminate is subjected to mechanical or thermal stress. This metal-polymer bond is also very polar, which makes it susceptible to moisture and therefore to hydrolysis.

Such chemical metal-polymer bonding is only possible by direct contact, by linking *via* functional groups or by thin plasma polymer layers or by plasma polymers that have functional groups capable of chemically reacting with the metal to form a covalent metal-plasma polymer bond. Other

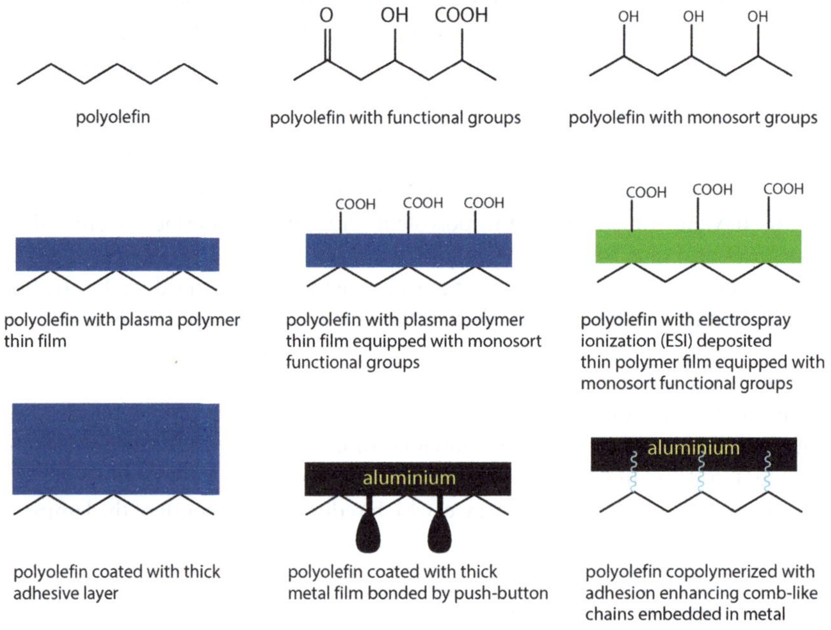

Figure 10.1 Examples of the preparation of polyolefin surfaces for the adhesion of evaporated aluminum layers.

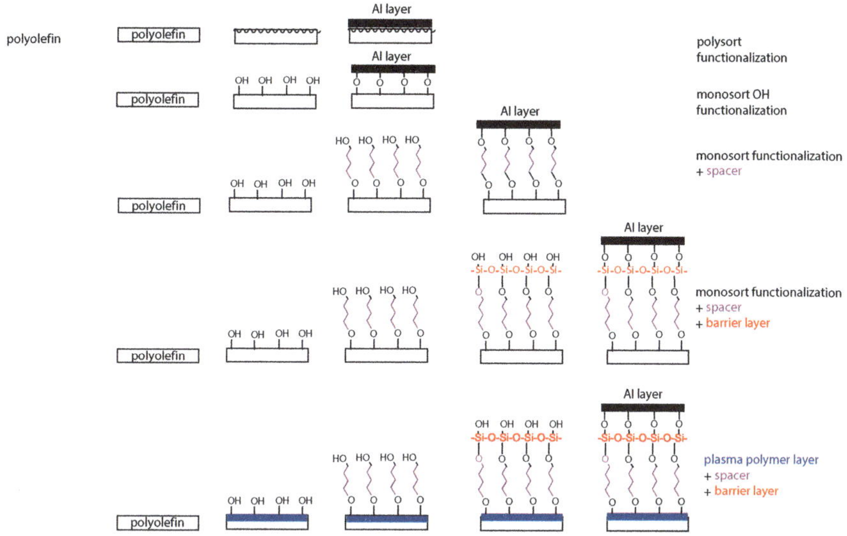

Figure 10.2 Molecular design of an ideal interface based on plasma-modified polyolefin surface or of plasma-deposited adhesion-promoting polymer thin film.

possibilities are the use of adhesion promoters, adhesives or mechanical interlocking after roughening (Figure 10.1).

How can an individual adhesion improvement method be evaluated? What are the underlying adhesion mechanisms? These questions will be answered in the following.

The following principal variants of interface design at the molecular level were tested and compared with each other (Figure 10.2).

10.1.1 Plasma-Based Methods

As is generally known, there are low-pressure and atmospheric-pressure glow discharges for polymer surface treatment. They belong to the non-thermal ("cold") discharges whose constituents are not in energetic equilibrium, i.e. the so-called non-isothermal discharges.

In such non-equilibrium discharges, the electrons have high kinetic energies and therefore high "electron temperatures" (Maxwell equation), while ions, atoms and molecules have low kinetic energies, which is why they are referred to as "cold" plasmas. In this context, it should be mentioned that the fluorescence radiation, recombination of the charge carriers and possibly the bremsstrahlung emit a very shortwave and therefore high-energy vacuum UV radiation (λ<200 nm).

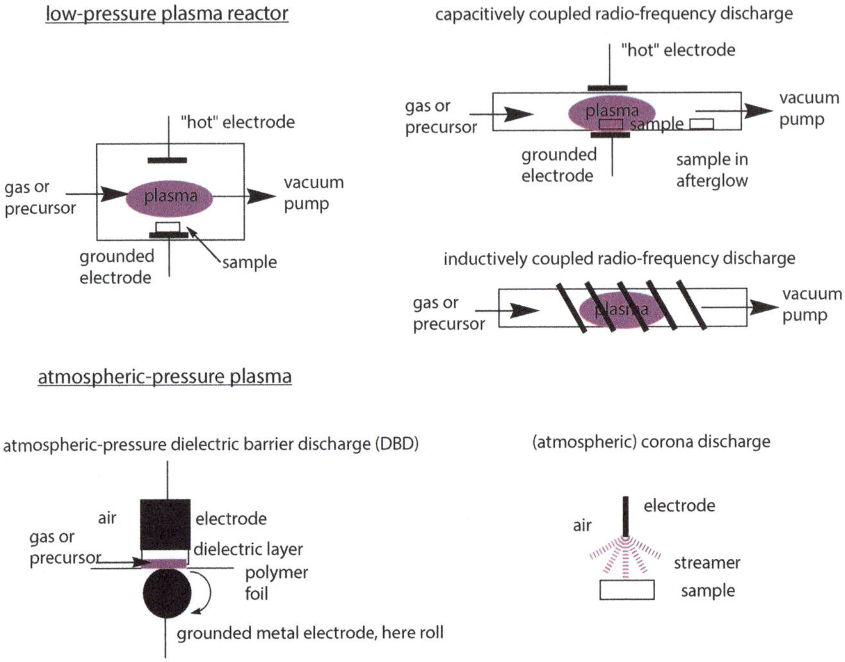

Figure 10.3 Schematics of low-pressure glow discharge equipment for polymer surface modification by direct exposure to the plasma (top left), tube-type reactor driven without internal electrodes by capacitively coupled discharge (top right) or remote plasma (top right, position of sample outside the plasma), inductively coupled discharge (middle right), dielectric barrier discharge (bottom left) and corona discharge (bottom right).

In addition, there are forms of discharge in which the plasma emerges from the actual discharge zone as a plasma jet by convection, referred to as afterglow or remote plasma in the case of low-pressure plasmas. Even thermal ("hot") plasmas are suitable for polymer pretreatment if the dwell time of the polymer in the hot electrical plasma and in the flame plasma can be kept very short.

Two examples of frequently used low-pressure plasma reactors are shown in Figure 10.3 for direct exposure to the plasma and (low-energy) irradiation in the afterglow plasma.

10.1.2 What Happens on the Surface of the Polyolefin When It Comes into Contact with the Plasma?

The polyolefins consist of macromolecules with molar masses greater than 100,000 g/mole (Da). The macromolecules contain the backbone C-C chain and the H atoms, which are bound to the backbone facing outwards, forming a "shield" of H atoms that enclose the zigzag (all-trans) C-C chain. The aliphatic (alkane) structure has 2 different bonds, the C-C and the C-H bonds. These two bonds have slightly different dissociation energies [2]:

$$\Delta_d H_{C\text{-}C} = 348 \text{ kJ/mole (3.4 eV) and}$$

$$\Delta_d H_{C\text{-}H\approx}\ 396 \text{ kJ/mole (3.9 eV)} \quad (10.2).$$

As was shown in the previous Section, the plasma on the polyolefin surface provides a surplus of continuously supplied energy compared to the amount of energy required to cleave a C-H (3.9 eV) or C-C bond (3.4 eV). This continuously supplied electron energy (E= 3-5 eV), which corresponds to the average electron energy of the electron energy distribution function, roughly corresponds to the dissociation energies of the C-C and C-H bonds. Lower energies, below the maximum of the electron energy distribution, which are therefore also below the C-C and C-H bond energies, lead to heating at the polymer surface and can trigger thermochemical reactions. The electron energies beyond the distribution maximum, especially in the higher-energy "tail" of the electron energy distribution (approx. 3-30 eV), can directly cleave covalent bonds in the polymer. This applies to the gas plasmas used for surface modification by means of ion and electron bombardment, interaction with high-energy and/or chemically active plasma species and high-energy plasma vacuum radiation.

The question now is whether, despite these partially higher kinetic energies, the approx. 0.5 eV energy difference between the C-C and C-H bonds does not lead to preferential C-C chain scission.

Ideally, one would expect the plasma effect to lead to H abstraction, leaving behind a radical anchor site for atoms of the plasma gas. After removing an H atom from the C-C chain, a C radical site is formed. This C radical site can then be occupied by an atom or a radical fragment of the plasma gas, leading to plasma gas-specific surface functionalization ($>CH_2$ + *plasma*→ >CH• + H, followed by >CH• + •plasma radical fragment → >CH-plasma fragment).

As it can be seen from the difference in the binding energies of the C-C and C-H bonds, chain degradation should dominate. Chain scission also generates radical sites for the attachment of plasma gas components ($-CH_2-CH_2-$ + *plasma* → 2 $-CH_2$•), which is a positive aspect. Much more unpleasant, however, is the associated molecular degradation, i.e. the reduction in the chain length of the polymer. This leads to dramatic losses in the mechanical strength of the polymer.

Figure 10.4 shows schematically possible reactions, the H-substitution and the C-C scission with decomposition of the polymer chain.

C-radicals have been produced by both reaction pathways. The radicals can recombine immediately after their formation and heat up their environment in the process. Other radicals that cannot recombine immediately because they do not have a corresponding neighboring radical nearby remain hidden in the polymer matrix, but can recombine with oxygen molecules that diffuse into the sample when the sample comes into contact with air.

The direct formation of plasma-specific functional groups can be easily demonstrated, for example, by N incorporation during polymer treatment in ammonia plasma [3]. Direct incorporation of oxygen species of the plasma gas can also be assumed when the sample is oxidized in oxygen or air plasma. However, there is also the reaction of hidden radicals with atmospheric oxygen when the sample comes in contact with the air, for example, if the polyolefin is treated in an ammonia plasma, a significant amount of O-functional groups will always be formed by auto-oxidation after the sample is exposed to air, in addition to the N species introduced

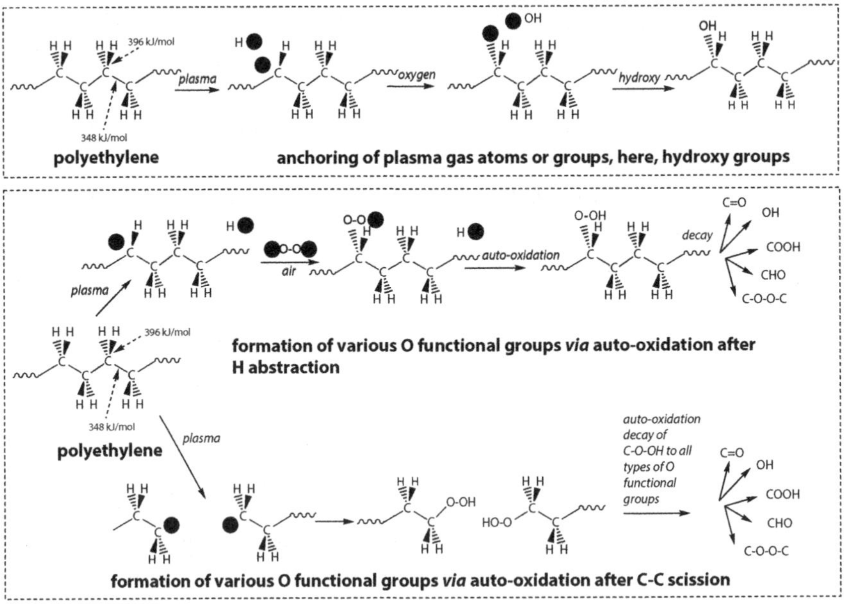

Figure 10.4 Types of oxygen attacks to polyethylene.

into the plasma. To a lesser extent, oxygen-containing groups can also be formed directly in the plasma as a result of leaking reactors or flanges.

If one chooses the ammonia plasma to functionalize the polyolefin surface ideally with NH_2 groups, one will be disappointed. The dissociation of the NH_3 molecule to NH_2 groups is energetically very unfavorable. NH fragments are preferably formed ($\Delta H_{NH3\text{-}NH2}$=440 kJ/mole, $\Delta H_{NH3\text{-}NH}$=209 kJ/mole, $\Delta H_{NH3\text{-}N}$=191 kJ/mole). For this reason, the yield of primary amino groups (NH_2) on the polyolefin surface is low. About 2% NH_2/C groups are usually found, corresponding to about 10% NH_2/N [3].

It is obvious that such a low concentration of primary amino groups after treatment in ammonia plasma is not sufficient to bind enough spacer molecules to the polymer surface [3].

The aforementioned secondary oxygen incorporation, which is actually undesirable, can also be observed during the treatment of polyolefins in noble gas plasmas. Furthermore, many radical sites not only recombine post-plasma with atmospheric oxygen to peroxy groups, but also recombine with each other. This results in a solid cross-linked surface layer, which also has an influence on the adhesion tendency of the polyolefin surface. This effect has been known for about 60 years under the name Crosslinking by Activated Species of Inert Gases (CASING) [4, 5].

In principle, numerous disadvantages of plasma when used to improve the adhesion of metal layers to polyolefin surfaces with polar groups are known, such as polymer degradation and formation of a weak boundary layer, lack of selectivity in formation of only one sort of functional groups, post-plasma auto-oxidation and more. In addition, there are also many fundamental problems in creating a particularly strong and durable bond due to covalent bonds at the metal-polyolefin interface, which will be discussed below.

The following describes how the fundamental incompatibility of metal and polymer can be overcome and how the completely different material properties of the laminate components can be at least partially compensated for [6, 7].

10.2 Ways to High and Durable Adhesion by Polymer Surface Modification

There are various mechanisms by which plasma treatment, chemical treatment or coating with adhesion-promoting layers can improve the adhesion between evaporated metal layers and polyolefin substrates; here are listed some of these and schematically depicted in Figure 10.5 [8]:

1. **Cleaning** of the polyolefin surface from contamination by exposure to noble gas plasmas or oxidation by oxygen or air plasma [9, 10].
2. **Roughening** of the polyolefin surface, up to and including achieving mechanical (push-button) adhesion with acrylonitrile-butadiene-styrene copolymer (ABS) and metal layers [11, 12].
3. **Exposing** the polyolefin surface to a noble gas plasma for **crosslinking** and therefore improvement of adhesion property known as CASING [4, 5].
4. **Irradiation** of polyolefins with vacuum-UV (excimer lamps), laser, electrons, ions, etc. [13].
5. **Sputtering** of polyolefin surfaces, deposition of metal layers and penetration of metal clusters into the polymer, Atomic Layer Deposition (ALD) [14–16]
6. **Plasma spraying**, metal coating due to electrical explosions [17, 18].
7. **Oxidation** of the polyolefin surface by an oxygen or flame plasma or chemically to create various oxygen-containing polar groups, which are capable to spontaneously form dipole-dipole interactions or even covalent bonds with the groups of the metal overlayer [19–21].
8. The introduction of **polar groups** significantly increases the surface energy of the polyolefins and brings it somewhat closer to that of the metal layer, which means improved thermodynamic compatibility and thus improved adhesion [22].
9. Formation of **monosort functional groups** on the polyolefin surface for the targeted formation of chemical metal-polymer bonds [23].
10. **Radical** formation [24–26].
11. **Peroxy groups** as initiators of radical "grafting-from" copolymerization [27, 28].
12. Application of **silanes** and other adhesion-promoting agents [29].
13. **Grafting of polymer** molecules onto functional groups or radicals on the plasma-modified polyolefin surface ("grafting to") [5, 7, 30].

14. **Grafting of monomers** to carbon radicals at the surface and initiation of a chain-growth polymerization ("grafting from") [31–35].
15. Deposition of (monosort) functional groups carrying **plasma polymers** (see Part 2).
16. Plasma or flame deposition of silanes for deposition of thin adhesion-promoting **siloxane (SiO_x) layers** carrying Si-OH groups [36–38].
17. **Plasma-assisted deposition of polymer molecules** onto polyolefin surfaces, Feeding an aerosol of a polymer solution into the plasma space of a dielectric barrier discharge, use of electrospray-ionization of polymer solutions in the atmospheric-pressure chemical ionization (APCI) [39].
18. **Electrospray ionization** (ESI) spray process in a high-voltage field with ionization for the deposition of individual macromolecules, which are usually multiply charged [40–42].
19. Evaporation of adhesion-promoting **adsorption layers** [43].
20. **Dip coating** in polymer solutions **or calendering** of metals with polymer foils [44].
21. **Treatment of polymers with solvents** [45].
22. **Deposition of transition layers** to compensate for different properties of metals and polyolefins by a successive transition with less mechanical stress [6, 7, 46].
23. Utilization of spontaneous or thermally initiated **interfacial redox reactions** between metals and polyolefins to adjust their different surface energies and lead to thermodynamic equilibrium [47–49].
24. Deposition of **barrier layers** to prevent the interfacial diffusion of metal ions into the adjacent polymer and the associated chemical reactions [50].
25. Deposition of **water-repellent interlayers** to prevent hydrolysis of interfacial chemical and physical bonds [51].

A schematic summary of the various methods can be found in Figure 10.5.

Part 1 of the manuscript on the subject of plasma and adhesion focuses on the surface modification of polymers to improve their adhesion to metal overlayers. Another substrate was carbon fibers, which were embedded in a polymer matrix as fiber-resin composite. The surfaces of fibers were also pretreated using plasma.

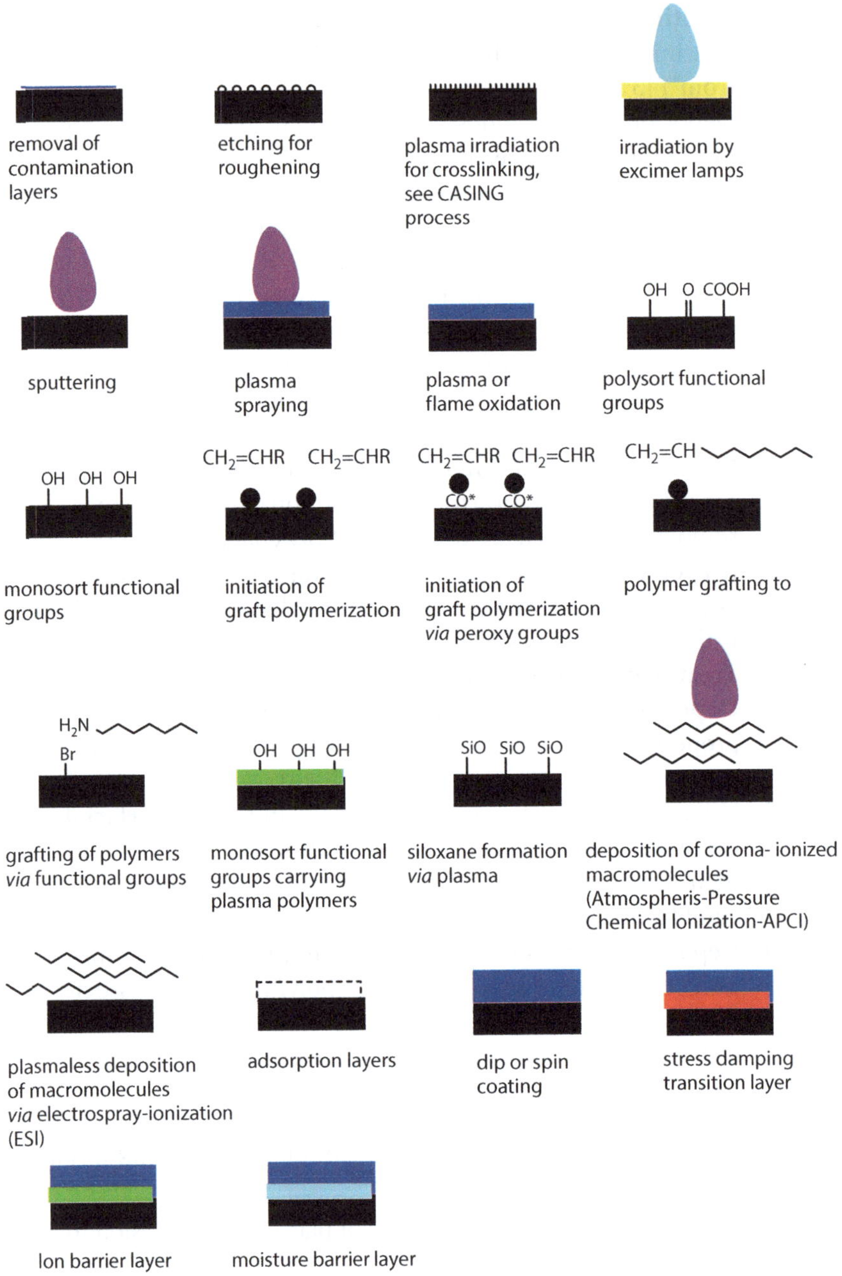

Figure 10.5 Schematics of adhesion improvement procedures.

Part 2 of this article examines how plasma technology can be used to apply thin films (layers) that promote adhesion.

10.3 Principles of Metal-Polymer Adhesion

In this article, the modification of polymer surfaces by means of plasma modification will be discussed with regard to suitable plasma processes, adhesion mechanisms, adhesion strengths achieved, and long-term durability.

10.3.1 General Considerations

The advantages and disadvantages of the various methods are discussed. It is possible to produce an aluminum-polypropylene laminate that is inseparable, not peelable at the metal-polymer interface or in the vicinity of this interface as shown by a complex design of interface and interphase build-up. This maximum adhesion can be maintained even during prolonged storage, exposure to moisture and elevated temperatures. However, the adhesion measurements have limits for technical reasons, so that adhesion failure beyond these measurement limits is possible. It cannot simply be expected that two such different materials, which are connected by strong covalent bonds, will adhere completely without any problems, simply due to individual covalent metal-polymer bonds along the interface, where all forces are concentrated under mechanical stress.

The structure of the interface and adjacent layers should therefore either be achieved with built-in flexible molecules that ensure the most continuous possible transition of material properties between metal and polymer. In polymer-polymer laminates, for example, the necessary successive transition of properties from polymer A to polymer B at the interface is ensured by mutual penetration of both types of macromolecules (interdiffusion) [52]. Metal-polymer laminates as well as carbon fiber-resin composites are characterized by very different surface properties of the components involved. The different coefficients of thermal expansion of metal, carbon fibers and polymers lead to high shear forces at the interface, which can reduce the adhesion between metal or carbon fibers and polymer due to bond scissions.

High concentrations of strong covalent bonds must be created at the metal-polymer interface. In addition, in the event of thermal or mechanical stress on the laminate, it must be possible to compensate for the different thermal expansion coefficients of the metal-polymer laminate or carbon fiber-resin composite. If these forces are not compensated for by intermediate elastic elements, the shear forces are concentrated solely on the covalent metal-polymer bonds at the interface. The shear forces that occur then exceed the metal-polymer bonding forces and the bonds break along the interface. The interface layer must therefore be made more flexible in order to compensate for any shear stresses that occur. If the covalent bonds at the interface are stronger than the intermolecular dispersion and covalent bonding forces in the polymer, the metal-polymer bond at the interface remains intact, but the failure of the laminate occurs near the interface in the polymer (Figure 10.6).

The strong chemical bonds along the metal-polymer interface therefore influence several atomic or even molecular layers of the polymer. In the case of special metal-polymer laminates, for example copper and ABS (acrylonitrile-butadiene-styrene) copolymers, the metal fills small holes in the polymer and thus interlocks mechanically. This mechanical interhooking, so-called push-button adhesion, extends over many molecular layers [53]. However, the strength of the polymer is weakened by the microscopically small holes in the surface layer of the ABS.

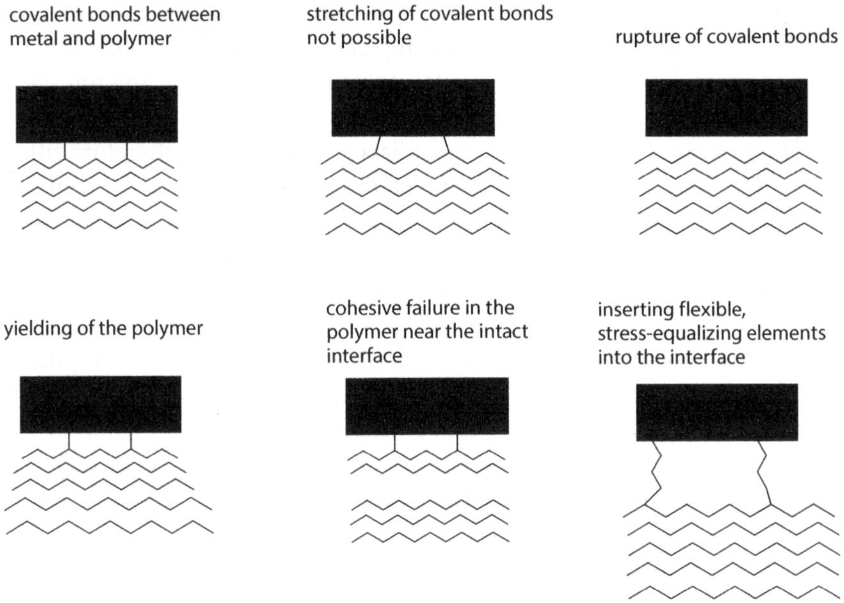

Figure 10.6 Different models of the behavior of the polymer along the interface under thermal stress, assuming different thermal expansion coefficients of metal and polymer.

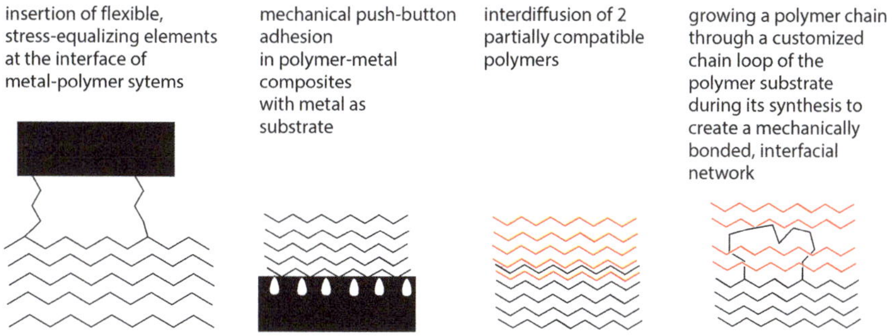

Figure 10.7 Different designs of interfacial layers in which the mechanical stresses are balanced across several macromolecular layers close to the interface.

Another example of the mechanical linking of polymer-polymer compounds is the interpenetrating networks between two polymers across the interface. The loops of the two polymers intertwine mechanically, as is done when knitting sweaters [52]. However, roughening and etching create material weaknesses, while interdiffusion leads to no or only insufficient penetration of the macromolecules at the interface due to the more or less strong incompatibility of almost all polymers with each other.

As already tested on a laboratory scale, a mechanical and penetrating entanglement or "knitting" of two polymers at the interface (interface IPN-interpenetrating networks) can also lead to a firm mechanical interlocking of incompatible polymers. Segments of one of the polymer components can be straightened on the surface as loops using suitable solvents. If further polymerization is then started, the new polymer chain that forms can grow through the loops of the other polymer. This form of mechanical linking at the interface is comparable to the known process of forming interface IPNs on polymer-polymer laminates (Figure 10.7) [30].

Back to the interactions at the metal-polymer interface. First, dipole-dipole based or acid-base and covalent bonds play the most prominent role in adhesion improvement.

10.3.2 Dipole-Based (Physical) Interactions of Polyolefins with Polar Materials

The scientists' first approach to promote the metal-polymer adhesion was to oxidize the hydrophobic polyolefin surface, thereby binding polar, hydrophilic groups to the polyolefin surface, which then gives the polyolefin a higher surface energy, which should improve wettability and thus adhesion [19].

Attractive forces are generated by dipole-dipole interactions, induced dipoles, acid-base interactions and hydrogen bonds [54]. The adhesion in the presence of such interactions can be very high despite very low interaction (bonding) forces, as there is often a high density of such actually weak bonding forces at the metal-polymer interface. Chemical or plasma-chemical polymer surface modification generates high concentrations of necessary polar groups on the hydrophobic polyolefin surface, which is inherently free of polar groups.

The forces emanating from dipoles are called Keesom and Debye forces. They also act on a polyolefin surface without plasma treatment as schematically shown by electron density distribution of this interaction: $^{\delta+}$C▶H$^{\delta-}$ (Figure 10.8 below). Somewhat stronger are dipole-dipole interactions after careful plasma-chemical oxidation of the polyolefin surface ($^{\delta-}$O▶C$^{\delta+}$) (Figure 10.8).

The electrical interactions in the presence of dipoles are very weak ($\Delta H_{dip\text{-}dip}$<20 kJ/mol), decay very quickly with distance ($1/r^3 \approx 3\text{-}5$ nm) and can easily be eliminated by adding water. In addition to the necessary chemical pre-treatment, it must also be taken into account that the physical interactions mentioned here are extremely susceptible to hydrolysis, i.e. to moisture.

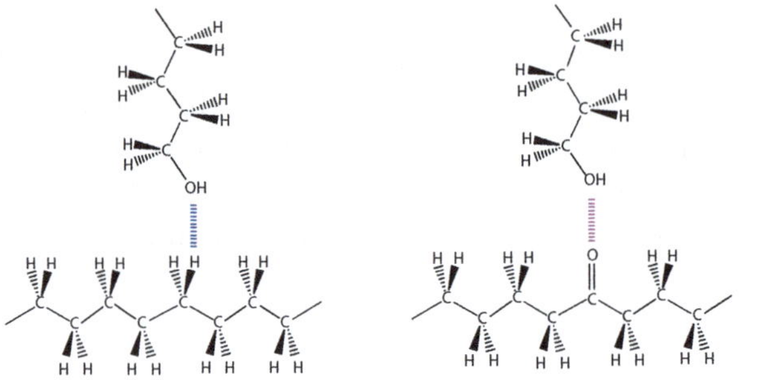

interaction of aliphatic chain with a polar molecule *via* induced dipole-permanent dipole

interaction of a polar group bonded to the aliphatic chain with a polar molecule *via* dipole-dipole or hydrogen bond

Figure 10.8 Dipole-dipole interactions schematically shown and also interactions with induced dipoles of aliphatic polymers with polar organic compounds.

10.3.3 Interactions of Polar Groups at the Polyolefin Surface with Metal Layers

Aluminum as a substrate has an oxide layer on the surface with OH groups on the immediate surface. The surface energy of aluminium with its native oxide/hydroxide covered surface has different surface energy than freshly vacuum evaporated aluminium. While native aluminum has a surface energy of approx. $\gamma_{Al\text{-}O} \approx 70$ mJ/m^2 in the presence of air, metallic aluminum has a surface energy of approx. $\gamma_{Al\text{-}metal} \approx 1200\text{-}1500$ mJ/m^2 in vacuum [55–58]. This difference in surface energy is an expression of the high affinity and reactivity of metallic aluminum to oxygen associated with the rapid oxidation of aluminum.

Polyolefins such as polypropylene and polyethylene have a very low surface energy. Due to the different surface energies of polyolefin and metallic aluminum, optimum wetting of the polyolefin surface is not to be expected when vapor-depositing aluminum, which would be a prerequisite for good aluminum-polyolefin adhesion. However, Mittal as well as Dyckerhoff and Sell have found that maximal adhesion strength was measured if the interface energy (tension) was (near) zero, i.e. the difference between the surface energies of substrate and coating should be zero for maximal adhesion [8, 59, 60]. Oxidative pretreatment of the polyolefin surface can reduce the huge difference in surface energies between aluminum and polyolefin, but only slightly. On the other hand, aluminium can react with the oxygen bound to the surface of the polyolefin after pretreatment and can be oxidized to Al_2O_3. This redox reaction reduces the difference in surface energies from 1200-30=1170 mJ/m^2 to 70-40=30 mJ/m^2, as discussed further in the next section.

The bond dissociation energy in metallic aluminium amounts to $\Delta E_{Al\text{-}Al}$= 167 kJ/mole that of $\Delta E_{Al\text{-}O}$= 485 kJ/mole and of $\Delta E_{Al\text{-}C}$= 268 kJ/mole [61]. Due to its negative redox potential of E^0_{Al}=-1.66 V, in contact with air, aluminium reacts rapidly with oxygen from the air at its surface and forms an oxide/hydroxide (bayerite-aluminium hydroxide [$Al(OH)_3$], boehmite-aluminium oxide hydroxide [AlO(OH)]) layer [62–65]. These AlO-OH hydroxyl groups may serve as anchoring sites for physical interactions or as starting points for the formation of chemical bonds with COOH groups at the surface of polyolefins [66].

An example of such a binding of a polymer to aluminium surfaces is the use of aminosilane to form a covalent bond Al-O-Si. The amino group is then capable to form a covalent bond with polyethylene copolymerized with 1% maleic anhydride (Figure 10.9) [67, 68].

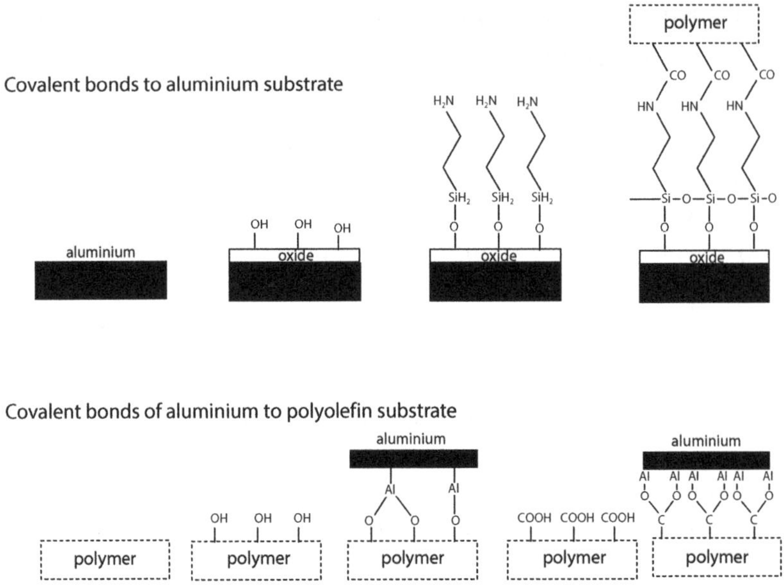

Figure 10.9 Surface chemistry and chemical interactions as a function of substrate type for a combined system of polyethylene and aluminium, either by thermal pressing of the polymer (top) or by thermal evaporation of Al (bottom). The top row shows the grafting of aminosilane onto hydroxyl groups on the Al surface, followed by the grafting of the maleic anhydride (MAA) moiety contained in the PE-MAA copolymer onto the aminosilane on the Al surface. Bottom row, plasma pre-treatment is shown schematically, leading to OH groups on the polyethylene surface, which interact or react with evaporated Al.

This Si-O-Al bond is formed after hydrolysis of the Si alkoxy functions: $R\text{-}Si(OR)_3 + H_2O \rightarrow R\text{-}Si(OH)_3$ with the utilization of a silanol group: $R\text{-}Si(OH)_2\text{-}O\text{-}Al$. The remaining silanol groups can then condense among themselves. Together with neighboring silane units, they form a two-dimensional barrier layer $(Si\text{-}O)_x$ against the passage of water and ions (Figure 10.9) [50].

Al-O-C bonds are stable [69, 70] and surprisingly resistant to hydrolysis in presence of $(Si\text{-}O)_x$ barrier layer, which is beneficial to the ageing resistance of polyolefin-aluminium composites or laminates, where aluminum is the substrate [71].

The next two examples concern the interaction of polymer-bound COOH and alkoxysilane groups with Al atoms of native aluminum surfaces. The findings also help to understand the processes found at the aluminum-polyolefin interface during the vacuum evaporation of aluminum. The adhesion mechanisms of poly(acrylic acid) with pseudoboehmite surfaces were investigated in detail as a model for an aluminum surface in air. To produce chemical bonds (bridged chelate bonds, monodentate, bridging bidentate) to polyolefins, maleic anhydride was often used as comonomer (Figure 10.9) [72].

Styrene monomers were polymerized directly onto a thin aluminum film in the presence of a small amount of a styryl-functional silane coupling agent to achieve high adhesion strength between an aluminum substrate and polystyrene. Some of the silanes adsorb on and react with the aluminum surface *via* its alkoxy group as follows: R-Si-(OCH_3) + HO-Al $\rightarrow$ R-Si-O-Al + CH_3OH. As the silane concentration increases, more silanol groups are available to form Al-O-Si bonds at the interface, and hence increase the probability that they can interact and form a 2D network of -Si-O-Si-O-Si bonds. The 2-dimensional Si-O layer presented in this work as a barrier against water and ion diffusion is also introduced as a barrier layer in the work by Namkanisorn *et al.* cited here [71].

The situation is different if the polyolefin is the substrate. Its surface oxidation by chemical oxidants, flames or oxygen plasma reduces the large difference from the surface energy of the aluminium (1200 mJ/m^2) only slightly, from 30 mJ/m^2 for the untreated polymer to 40 mJ/m^2 for the surface-oxidized polyolefin, as previously discussed. What remains is a weak physical interaction based on dipole-dipole interactions with metal deposits [72]. However, some covalent chemical bonds can also be spontaneously formed between the polymer and the metal overlayer, which can significantly improve adhesion, but a very large difference in surface energies or chemical potentials of the two phases remains. This is why the covalent Al-C and Al-O-C bonds have a very polar character and can be easily hydrolyzed, while Al-O-Si bonds are more stable [70]. The presence of Al-O-C or Al-O-Si bonds is therefore only a metastable non-equilibrium state. Only an interfacial redox reaction leads to the equalization of the different surface energies.

Thermal, electron beam or sputter deposition of metallic aluminium onto the polymer substrate is also able to produce covalent metal-polymer bonds. Such bonds may be carbide (Al-C) or after oxidative pretreatment of the polyolefin alcoholate bonds (Al-O-C). These bonds are hydrolyzed by the attack of moisture as exemplified for aluminium carbide (exothermal reactions): $Al_4C_3 + 12\ H_2O \rightarrow 4\ Al(OH)_3 + 3\ CH_4$ and aluminium alcoholate: Al-O-C + H_2O $\rightarrow$ Al-OH + C-OH.

After metal deposition the polymers show a low binding energy component at 283.4 eV in the C1s signal of XPS spectra, assigned to Al-C bonds [73]. The Al carbide species, formed by electron transfer from aluminium to carbon atoms, have also been observed in other aluminium-doped conjugated polymer systems [75, 76].

In presence of oxygen-containing groups, such as OH groups in case of poly(vinyl alcohol) (PVA), an $AlCO_{1.5}$ complex is formed at the aluminium-PVA interface. This stoichiometry corresponds with that of aluminium oxide (Al_2O_3) [69, 70].

Poly(ethylene terephthalate) as substrate forms also Al-O-C bonds by reaction of the ester or carboxylic group after hydrolysis with metallic aluminium [71–73]. This metal reaction involves one or two Al atoms, mono (C-O-Al) and bidentate ($C<(O\text{-}Al)_2$) bonds [74–76]. On a chemically or plasma-chemically oxidized polyolefin surface, both hydroxyl and carboxyl groups should be present as the most important chemical bond-forming groups.

Al binds preferentially to the carbonyl oxygen in the methylester functional group ($CO\text{-}OCH_3$) until all oxygen is bound to the deposited Al. Al does not react with the methylene backbone [74].

Up to C_{18} chains with silane function (Si-O-R) at the end of the chain increase the adhesion strength, beyond that the accessibility of the silane function for the formation of covalent bonds to aluminum is more difficult [73]. Chain lengths greater than C_{10} also increase the resistance to moisture [73].

Reaction of the COOH groups of self-assembled monolayer continues until slightly more than one atom per reacting group is deposited, on average; thereafter, no further reaction is observed. However, 20–25% of the COOH groups remain unreacted, regardless of the Al coverage [74].

Metallic aluminum reacts with alcohols at elevated temperatures to form aluminum alcoholates: $Al + HO\text{-}R \rightarrow {>}Al\text{-}O\text{-}R + ½\,H_2$. COOH groups react with aluminum to form acetate: $Al + HOOC\text{-}R \rightarrow Al\text{-}O\text{-}CO\text{-}R + ½\,H_2$ (Figure 10.10).

Changes in the line shapes of the C 1s and O 1s peaks and the appearance of a shifted peak in the Al 2p region in the XPS spectra were interpreted in terms of Al interactions with the surface alcoholic group HC-OH, resulting in the formation of an interfacial organometallic complex [69].

The various covalent bonds between aluminium and modified polyolefin surfaces (Al-C, Al-O-C) are susceptible to hydrolysis caused by traces of moisture at the interface. The hydrolysis of Al-O-Si and Al-O-Al is preferred over the hydrolysis of Si-O-Si [77, 78].

On the contrary, the redox reaction of O functional groups and metallic aluminium does not improve the Al-polyolefin adhesion. This reaction must be stopped at the level of the (metastable) Al-O-C or Al-O-Si bonds. The Al-O-Si bonds in particular can be stabilized by condensing

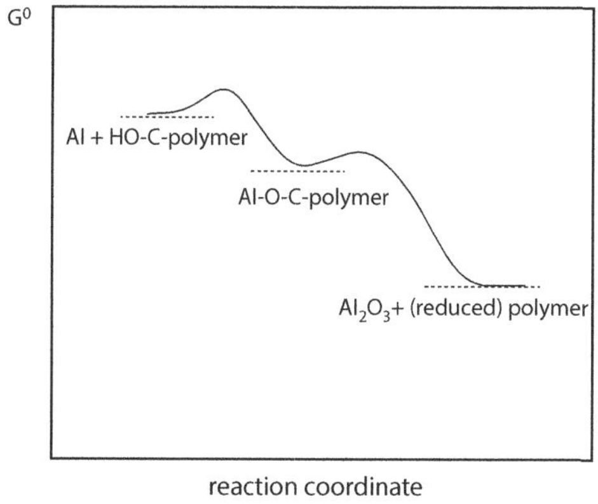

Figure 10.10 Course of Gibbs energy G° of Al interactions and reactions with an oxidized polyolefin surface.

the neighboring silanol groups(Si-OH) to Si-O-Si bonds. This can be illustrated thermodynamically by the reaction coordinate. It represents the progress of Gibbs energy (G^0) along a reaction pathway to stable end products (Figure 10.10).

The formation of covalent Al-O-C or Al-O-Si bonds can only be considered as a metastable transition state in the context of a redox reaction of metallic aluminum with oxygen containing groups at polyolefin surface. The system strives towards the stable final state of Al_2O_3 formation of the strongly electronegative aluminum and the stable final state of an aliphatic polymer chain. In order to stabilize the very strong Al-O-C or Al-O-Si bonds, which are actually thermodynamically unstable intermediate states, their protection against hydrolysis and ion migration must be ensured by a suitable molecular structure at the interface. Otherwise, the redox reaction continues to produce the stable endproducts described.

10.3.4 Redox Reactions across the Metal-Polymer Interface

While the redox potential of the metallic aluminium is strongly negative (E^0_{Al}=-1.66 V), the plasma-oxidized polyolefin surface is assigned a moderately positive redox potential, which is characteristic of organic redox systems, but cannot be determined exactly. These different redox potentials lead to a redox reaction on the partially oxidized surface of the polyolefin with the evaporated aluminium atoms (see also Figure 10.10): polyolefin-O_x + 2 Al $\rightarrow$ polyolefin + Al_2O_3. However, such a redox reaction does not improve the Al-polyolefin adhesion, although the different redox potentials are largely equalized during the reaction. The cohesive strength of the Al_2O_3 formed during the redox reaction is low. In addition, the aluminum oxide is not connected to either the polyolefin or the aluminum layer by strong covalent bonds. Therefore, this redox reaction must be stopped at the level of the (metastable) Al-O-C or Al-O-Si bonds (see Gibbs energy in Figure 10.10). The Al-O-Si bonds can be stabilized by condensing the silane units in a 2-dimensional Si-O-Si monolayer as already explained above (Figure 10.11).

As Al was evaporated onto poly(vinyl alcohol) (PVA), complex and bond formation was found; however, redox reactions between the Al and OH groups from PVA without Al-O-C bond formation were also found. Aluminium was oxidized and formed aluminium oxide (Al_2O_3) while at the same time the OH groups of PVA were reduced, i.e., removed [69, 73, 76, 80, 81].

Such redox reaction between evaporated Al layers and plasma oxidized polyolefin surfaces was found [47] and also with copper [48]. A typical redox reaction is also used for promoting the bondability of poly(tetrafluoroethylene) using Na+NH_3 [82, 83] or Na-naphthalene-tetrahydrofuran (THF) systems [84, 85]. Unfortunately, the polymer chains are partially destroyed in these processes, and the resulting mechanically weakened surface layer limits the achievable adhesion strength. This mechanically weakened surface layer was described by Bikerman as WBL (Weak Boundary Layer) [86].

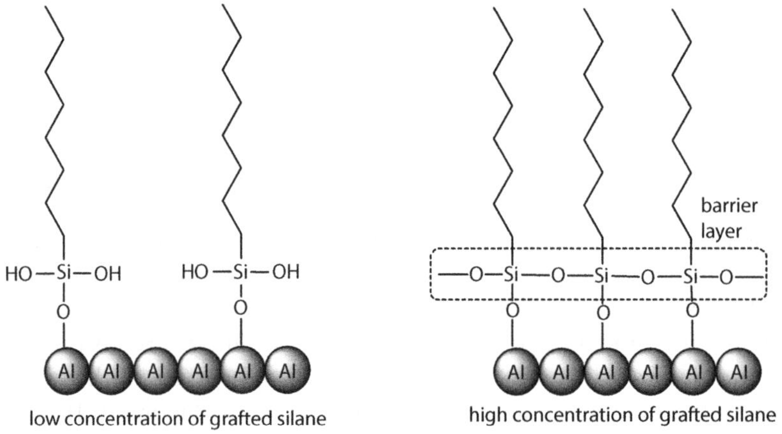

Figure 10.11 2-dimensional Si-O layer with grafted silane for protecting the metastable covalent Si-O-Al bonds from hydrolysis and ion diffusion.

The driving force of a redox reaction is the difference in the electro-chemical potential of the two substances involved in the electron transfer. In one reaction, oxidation takes place with the oxidation potential as the driving force; in the other, reduction takes place with the reduction potential as the driving force. The more negative a redox potential is, the stronger the reducing power of the substance. Aluminum has a strong negative redox potential of E^0_{Al}=-1.66 V and is therefore a strong reducing agent. The oxidized polyolefin surface serves as an oxidizing agent. Its exact oxidation potential is unknown and cannot be easily measured. By comparison with other organic redox systems, the potential can be estimated with $E_{ox} \approx +0.2$ V.

Kinetic hindrance and the need for additional activation energy also determine the initiation of redox reactions [47]. Thus, kinetic inhibition of redox reaction stabilizes the desired but metastable covalent Me-C, Me-O-C and Me-O-Si bonds needed for high adhesion. It can be said that the redox process at least approximates, if not completely equalizes, the different chemical potentials of metal and polymer and thus also their surface energies at the interface, cf. [60].

Non-ionic reactions between metals and polymers are not directly relevant to a redox reaction, but are nevertheless part of the equalization of the chemical potentials of metal and polymer at the interface. Chromium, for example, uses its 3d orbitals to interact with π-electrons of the polymer (dπ-pπ interactions), such as polystyrene vacuum-evaporated with a monoatomic layer of chromium atoms using a Knudsen cell [49]. The chemical (redox) potentials at the interface between the plasma-oxidized polystyrene surface and the vapor-deposited monoatomic chromium layer are balanced by a redox reaction with the formation of chromium(III) oxide (Cr^0 + oxidized polyolefin $\rightarrow Cr_2O_3$ + polyolefin) and simultaneous reduction or abstraction of the O-functional groups on the polystyrene surface. However, evaporated chromium can also interact with the surface of non-pretreated polystyrene and attack its aromatic rings and, under certain circumstances, destroy them with the simultaneous formation of chromium carbide (Cr^0 + -[CH_2-CHØ]- $\approx Cr_3C_4$) [49, 73].

10.3.5 Role of Functional Groups for Physical or Chemical Bonding of Metals

As the stoichiometry of a polyolefin corresponds to that of a hydrocarbon, thus, there are no reactive (polar) functional groups that can form physical or even covalent bonds. To make this possible, plasma-chemical surface modification and flaming are the most common methods.

When exposed to oxygen plasma or plasma in air, many different polar oxygen-containing groups are formed in high density. These different groups, which are produced in high density on the polyolefin surface, allow quite a high adhesion to metal overlayers, which is often sufficient for technical applications, despite low bonding forces. These different O-containing groups are either hardly suitable for spontaneously forming chemical covalent Al-O-C bonds without increasing the temperature. They can often only partially react to form Al-O-C bonds, but they can at least form physical interactions: C-O...Al (Figure 10.12).

With the choice of plasma gas, the type of functional groups formed on the polyolefin surface can be very roughly predetermined. Unfortunately, the oversupply of energy prevents the formation of monotype of functional groups that are particularly favorable. The formation of the thermodynamically particularly stable endproducts of the reaction in the plasma cannot be avoided either. On the other hand, these end products lead precisely to the desired products, as in the case of plasma bromination or the oxidative conversion of silicon compounds, which is an exception in plasma chemistry (Figure 10.12).

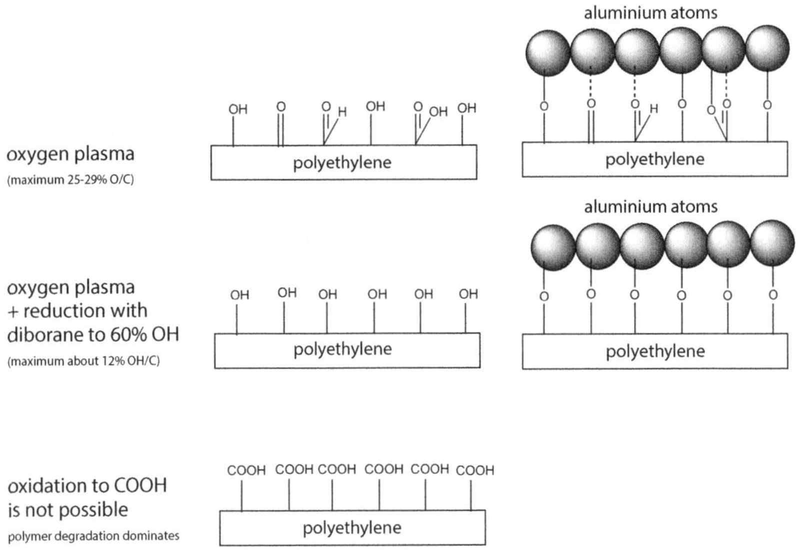

Figure 10.12 Plasma oxidation of polyethylene surface and interaction with evaporated aluminium atoms.

If a chemically reactive plasma gas is selected, the chemical incorporation of atoms of the plasma gas into the polymer surface can be expected. This allows the type of functional groups formed on the polyolefin surface to be influenced, whereby different functional groups can be formed from the atoms of the plasma gas and the atoms abstracted from the polymer substrate. Unfortunately, the oversupply of energy prevents the formation of only monotype of functional groups that are particularly favorable. The treatment of polyolefins in an oxygen plasma leads to the formation of various O-containing functional groups *via* the process of auto-oxidation, such as OH, epoxy, C-O-C, C-O-OH, CO-O-OH, O-CO-O, COOH, COOR, CHO, aryl-OH.... If this oxidation process continues, the thermodynamically stable endproducts CO_2 and H_2O are formed. The formation of the thermodynamically stable endproducts of the reaction in the plasma cannot be avoided. On the other hand, these endproducts lead precisely to the desired products, as in the case of plasma bromination [87, 88]:

$$\text{-CH}_2\text{- + Br}_2 \rightarrow \textit{plasma} \rightarrow \text{-CHBr- + HBr.}$$

The formation of a Br functional group other than C-Br on the polymer surface is not possible. The HBr that is also formed is pumped off. Another advantage is that plasma-generated Br atoms quench all C radical sites generated by the plasma by reaction to C-Br. Since all radical sites are saturated with bromine, the polymer sample can no longer react with oxygen from the air and undergo auto-oxidation.

Other plasma gases form also different types of functional groups, such as ammonia [3]. All functional groups generated by the ammonia plasma contain the key element N in different oxidation states and chemical environments, such as NH^+, N, NH, NH_2, N=N, C≡N, CO-NH, O-CO-NH, CO-NH-CO, NH-CO-NH, NO, NO_2, O-NO_2.... [3].

What is interesting now is which functional group plays which role at the interface. Aluminum can form covalent bonds with OH, COOH and COOR groups. It also reacts with Si-OH or SH groups.

Amino groups hardly interact with aluminum vapor deposition layers. However, if they are covalently bonded to the polyolefin surface, for example, they can enter into chemical reactions with aliphatic dialdehydes and thus covalently bind flexible spacer molecules.

10.3.6 Further Thoughts on Interface Design

For the important flexibilization of the interface, flexible aliphatic spacer molecules are covalently bonded to the polyolefin surface (Figure 10.13).

The chemical bonds between vapor-deposited aluminium metal and functionalized polyolefin surfaces should meet the following requirements (Figure 10.13):

- Plasmachemical formation of covalently bonded chemically reactive functional groups on the polyolefin surface, such as OH, CHO, COOH, NH_2
- Preferred plasma-chemical formation of groups from one species by plasma bromination and formation of C-Br bonds on the polyolefin surface or by treatment in oxygen plasma with subsequent wet-chemical reduction with the help of diborane to predominantly OH groups
- Avoidance of strong plasma-induced degradation phenomenon on the polyolefin surface in the sense of creating a "weak boundary layer" through low power input of the plasma
- Grafting of longer (C_3-C_{10}) spacer molecules, for example diisocyanato hexane to OH groups on the polyolefin surface and hydrolysis of terminal NCO group to NH_2 group
- Chain extension of the grafted aliphatic chain at its reactive endgroup with glutaraldehyde

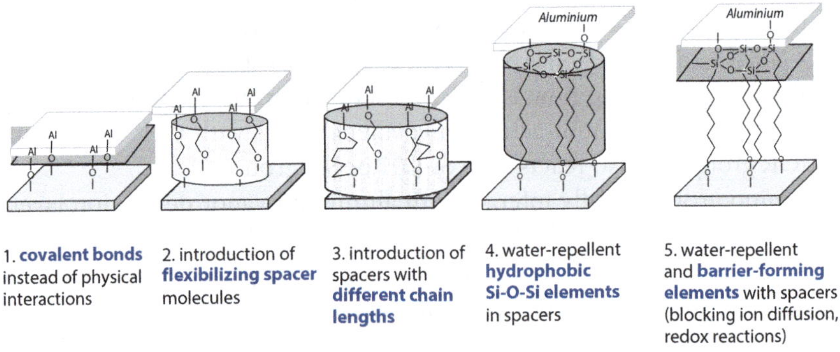

Figure 10.13 Flexibilization of the Al-polyolefin interface by introducing covalently bound, highly mobile aliphatic spacer molecules, which at the same time have the properties of a barrier layer against diffusing water molecules and metal ions (in gray-hydrophobic, water repellent action).

- Grafting of silanes to their reactive endgroups (NH_2, epoxy, SH....) at the endgroups of the chain extender, here exemplified by linking of CHO and NH_2 to Schiff's base
- Hydrolysis of the alkoxy groups of silanes to silanols
- Formation of a 2-dimensional Si-O layer by condensation with silanol groups of neighboring spacers
- Non-reacted silanol groups react with evaporated aluminum atoms to form covalent Al-O-Si bonds.

This creates a flexible, water-repellent and barrier-forming, continuous covalently bonded connection between aluminum and polyolefin. This interface design has made it possible to produce inseparable Al-polyolefin laminates that have not lost their strength even after intensive artificial ageing.

10.3.7 Role of Auto-Oxidation of the Polyolefin Surface During and After Plasma Treatment

The surface oxidation of polyolefins can occur by occupation of a C-radical site by an H atom (recombination) or by chemical substitution of hydrogen in C-H bonds by oxygen atoms. The formation of O-containing groups can also occur through the addition of biradical O_2 molecules by recombination with the C radicals to form peroxy radicals, which initiates auto-oxidation. For example, the unsaturated C radicals generated in the plasma and trapped in the polymer matrix can react with diffusing atmospheric oxygen following plasma treatment.

The ammonia plasma provides evidence of direct incorporation of the plasma gas in various forms as a functional group. In this case, the nitrogen should only be able to bind to the polymer surface with the help of the plasma. However, the nitrogen incorporation by treatment in the NH_3 plasma is significantly lower than the oxygen incorporation in the oxygen plasma [3]. A considerable proportion of the oxygen in the O_2 plasma is converted into H_2O and CO_2 and pumped out, leaving the plasma space in gaseous form. From this it can be concluded that auto-oxidation plays an important role in the formation of the various analytically detected O-functional groups. The high oxygen concentrations of 26-29% O/C often found on polyolefin surfaces after treatment in low-pressure O_2 plasma suggest that auto-oxidation plays an important role in the level of O incorporation and in the formation of the various O-functional groups [8].

Auto-oxidation can take place in the oxygen plasma itself or with a continuous flow of oxygen as well as during (intermediate) storage in air.

If one considers that almost as much (unwanted) oxygen as nitrogen is incorporated on the polymer surface during the treatment in the ammonia plasma, one can certainly speak of a dominance of auto-oxidation.

Based on the purely chemical process of auto-oxidation with atmospheric oxygen and the radicals formed in the plasma, the same types of O-functional groups should occur in approximately the same concentrations for each type of plasma treatment. The auto-oxidation process already starts on exposure to the O_2 plasma, as the radicals can immediately attach to the molecular oxygen and form peroxy radicals. The fact that auto-oxidation is almost complete is shown by the fact that no or hardly any increase in O-functional groups can be observed when the sample is exposed to air. In contrast, the samples treated in ammonia plasma still take up oxygen afterwards. The radicals formed on the polymer surface can react rapidly post-plasma with the biradical O_2 molecules and form peroxy radicals and secondary products. In the case of ammonia plasma, however, the C radicals cannot react with the ammonia molecule. The auto-oxidation of hydrocarbons, e.g. cyclohexane, is a radical chain reaction in which a large number of different partial reactions take place [89–96]. First, an initiator carbon radical reacts with oxygen to form a peroxy radical. This peroxy radical abstracts a hydrogen atom from an alkyl chain, which leads to a hydroperoxide and an alkyl radical. The alkyl radical, in turn, reacts with oxygen again to form a peroxy radical in a chain reaction. After photolytic or thermal decomposition of the hydroperoxides into the intermediate alkoxy and hydroxyl radicals, various O-functional groups are formed, such as alcohols, ethers, ketones, aldehydes, carboxylic acids and peroxy groups. These functional groups are found again and again with every oxidation. [8]. The direct formation of these groups by recombination and rearrangement in the plasma first requires the abstraction of hydrogen from the polymer chain or chain scission, for example in order to be able to react with molecular oxygen or after dissociation of molecular oxygen into atomic oxygen. In a few collisions, may be three-body collisions for dissipating all energy, OH radicals may be formed, which then can recombine with radicals at the polymer surface or can chemically substitute H atoms (S_R mechanism) from polymer chain. The formation of radical ketone (C=O), aldehyde (CHO) or carboxylic acid (COOH) fragments by combining individual oxygen, carbon and hydrogen atoms in the plasma phase through radical recombination and the subsequent binding to the polymer surface through recombination with a surface radical is difficult to imagine (Figure 10.14). This is because the enthalpy released

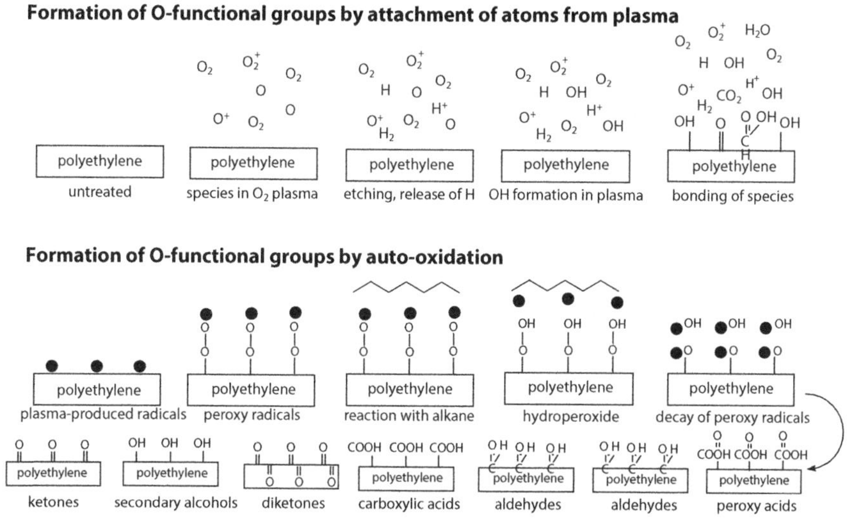

Figure 10.14 Comparison of the plasma-induced formation of functional groups on the surface of polyolefins with those formed by auto-oxidation.

during radical recombination can only be realized by dissipating energy through three-way collisions or by impacting the wall and transferring energy to it. The addition of atoms and fragments to the polyolefin surface as a functional group requires either the addition of further atoms from the plasma or chemical rearrangement reactions on the polymer surface. For example, the addition of CO_2 to a radical on the polyolefin surface does not yet create a COOH group [97]. Only the addition of a H atom from the plasma phase or the abstraction of a H atom from a C-H bond of adjacent polymer chain is necessary:

$$\text{C}\bullet + \text{CO}_2 \rightarrow \text{C-O-C}(\bullet)\text{=O, that means C-O-C}(\bullet)\text{=O} + \text{H}\bullet$$
$$\rightarrow \text{C-O-C(H)=O or C-C(O)-O}\bullet\text{, C-C(O)-O}\bullet + \text{H}\bullet$$
$$\rightarrow \text{C-C(O)OH (i.e. C-COOH).}$$

The addition of atoms and groups can also take place as a chemical substitution of H atoms in the polymer chains at the polyolefin surface. However, H scission from C-H bonds requires energies of approx. 400 kJ/mole. In contrast, unwanted C-C chain scission is energetically more favorable at around 348 kJ/mole [2].

A third variant of the formation of such O-functional groups on the polymer surface using the oxygen plasma consists of the re-deposition of gaseous oxygen-containing etching products (CO_2, H_2O) from the plasma phase onto the polymer surface as ultra-thin O functional groups containing polymer film, however, often observed on surfaces with low plasma exposure. Yasuda described this process of surface degradation of the polymer and the re-deposition of the degradation products from the plasma as a thin layer with CAP (Competitive Ablation Polymerization).mechanism [98].

10.3.8 Roughening

Roughening a surface is a safe and reliable means of increasing the adhesion between two materials. Roughening increases the effective surface area and thus increases the number of possible interactions per unit area. In extreme cases, hooking can add a purely mechanical adhesion component, and, with special roughened shapes, even allow mechanical hooking, sometimes also referred to as push-button adhesion. Mechanical, physical or chemical roughening also leads to the formation of functional groups on the immediate surface through the reaction of radicals from broken polymer chains with atmospheric oxygen and moisture.

Roughening can be caused by preferential etching of weakly bound (amorphous) polymer components, and also by density fluctuations of the plasma, such as caused by the locally different composition of the gaseous polymer etching products in the plasma phase or due to locally different charge distribution on inhomogeneous polymer surfaces.

Plasma-chemical etching of semi-crystalline polyolefins on exposure to oxygen plasma should enable roughening by preferred removal of weakly intermolecularly bonded amorphous regions and the remaining strongly bonded crystalline polymer regions. Selective etching in oxygen plasma should be able to expose such tightly bound crystalline supermolecular structures in polymers, which should lead to an increase in the roughness of the polymer surface. It should be mentioned that with the help of this selective plasma etching, supermolecular structures are exposed in polymer layers close to the surface and can then be examined with the help of electron microscopy [99, 100]. Such attempts were already made at the beginning of the 1950s [99–103]. Indeed a slight increase in crystallinity of the whole polymer sample during oxygen plasma exposure could be observed (1-20 min), which indicates a preferential etching of the amorphous structure, followed at longer etching times by a slight decrease in crystallinity and thus roughness, which is attributed to chain scissions and formation of low-molecular weight oxidized material (LMWOM) (Figure 10.15) [104].

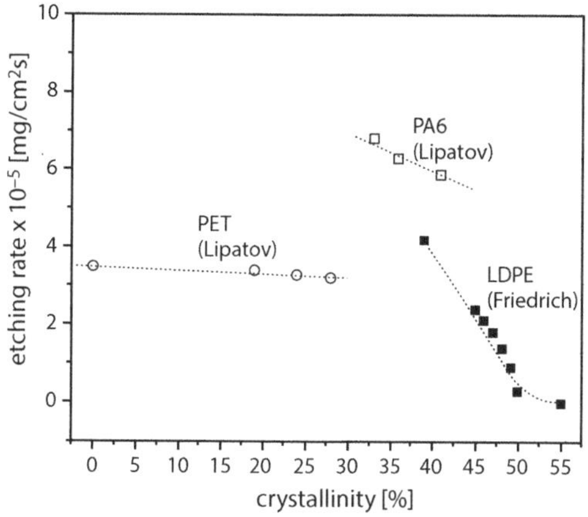

Figure 10.15 Dependence of etching rate on the degree of crystallinity (density measurements) [30, 106].

The usual treatment times of polymers in the plasma are too short to etch away the weakly bonded amorphous polymer regions to such an extent that supermolecular structures with micrometer dimensions remain with such a height that they can contribute to mechanical hooking and thus to improved adhesion. It is also unclear whether the first visible structures actually originate from the supermolecular polymer structure or whether they are secondary structures that were generated by the plasma treatment itself, e.g. by cross-linking [105].

Another explanation for the slight increase in crystallinity on exposure to oxygen plasma is the chemi-crystallization induced by physical interactions between plasma-introduced polar O containing groups anchored at adjacent polymer chains [8].

Confirmation of this hypothesis of roughening by selective etching is difficult. There are only a few results in the literature on the same polymer with identical molar mass, annealed differently to vary and adjust the crystallinity, when exposed to low pressure oxygen plasma [105, 106]. Such selective etching behavior is plotted in Figure 10.15.

However, in most cases the degrees in crystallinity may increase at short exposure to oxygen plasma, and after passing a maximum it decreases (Figure 10.16) [107, 108].

This maximum exposure of crystalline structures of the polyolefin sample by etching in the oxygen plasma after 20 min exposure time is followed by a slow decomposition and thus an amorphization of the crystalline areas by etching, UV irradiation and possibly thermal stress of a surface

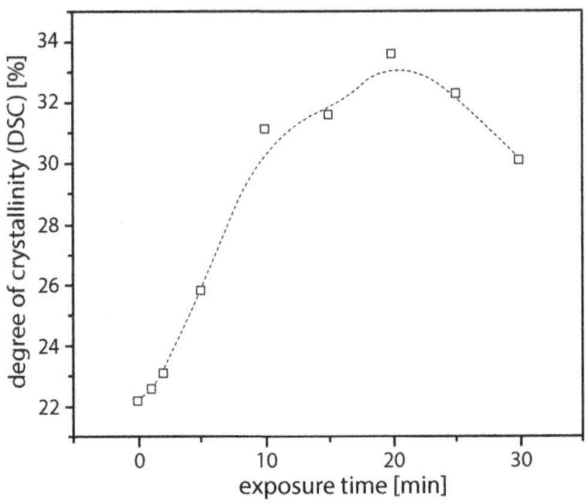

Figure 10.16 Crystallinity of LDPE vs. time of exposure to the oxygen plasma measured by Differential Scanning Calorimetry (DSC) [108].

layer (Figure 10.17). This amorphization of crystalline structures is probably also associated with etching of the loops of the folded polymer chains in the crystal and polymer degradation, which means the formation of a weak boundary layer (WBL) [86].

The etching rates of the amorphous and crystalline phases hardly differ, as the dissociation energies of the C-C and C-H bonds are the same. Only the chain-to-chain or segment-to-segment interactions differ slightly in dependence on the degree of order (crystallinity). Considering the broad energy distribution of electrons in the plasma and the high reactivity of dissociated and ionized oxygen species, these small differences in interactions can certainly not be used for selective plasma etching of exclusively amorphous constituents [109, 110].

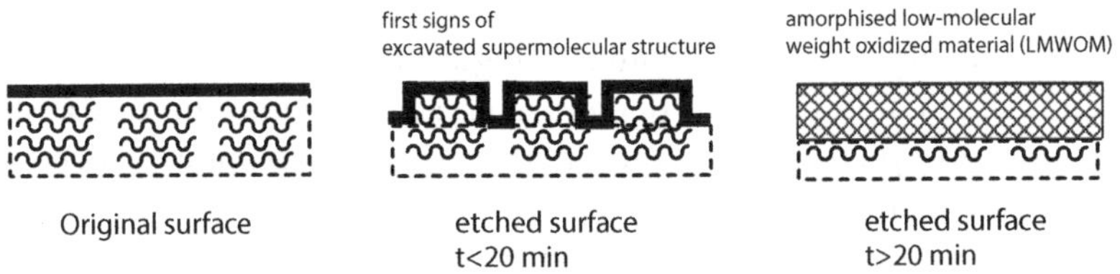

Figure 10.17 Schematic representation of the etching process of semi-crystalline polyolefin in an oxygen plasma.

10.4 Grafting of Molecules and Polymers on Monosort-Equipped Polyolefin Surfaces for Flexibilization of Interface

Functional groups, in this case OH or Br groups, of one type (monosort) with a high density (approx. 10-20% OH/C or Br/C) on polyolefin surfaces are needed for the chemical incorporation of flexible spacer molecules into the metal-polyolefin interface [87, 88, 111, 112]. This flexibilization with aliphatic spacer molecules is a prerequisite for improved adhesion of metal overlayers. The adhesion-promoting monosort functional groups alone are not sufficient to improve the metal-polyolefin adhesion, but serve primarily as anchor points for the covalent attachment of the spacer molecules. As already described, surfaces equipped with one type of functional group can be produced in two ways using plasma chemistry alone or in combination with wet chemistry.

On the one hand, highly selective plasmas can be used. However, there are only two truly selective functionalizing plasmas, i.e., bromination with bromine or bromoform [7, 45, 87, 88, 110, 111] and the use of Si-organic precursors together with the oxygen plasma [29, 35]. The results of these processes are brominated polymer surfaces or these are coated with a SiO_x layer. The second variant is the subsequent (post-plasma) wet-chemical reduction of carbonyl-containing O-functional groups, including olefinic double bonds, uniformly to OH groups [112–118].

The oxidation of O-containing groups to COOH groups, including the direct generation of COOH groups in the CO_2 plasma, is not very successful [2]. The chemical oxidation to COOH groups is not selective enough and cannot be stopped at the COOH group level for all initial carbon oxidation states.

10.4.1 OH Group Formation by Exposure to Oxygen Plasma and Subsequent Wet Chemical Reduction

As mentioned in the Introduction, there is the possibility to produce more or less uniform C-OH groups by means of a post-plasma wet-chemical reduction process according to Brown [112] and Nuzzo and Smolinsky [113]. With the help of preferably diborane, about 11-14% OH/C groups can be generated by reducing all C=O bonds and hydroxylating double bonds [111]. Following Brown´s route, carboxylic, ester, ketone and aldehyde functions can be transformed into OH groups by reduction with diborane as well as C=C double bonds can be hydroborated and hydrolyzed to OH groups [22, 114–117]:

$$>C=O + B_2H_6 \rightarrow >CH\text{-}OH.$$

$NaBH_4$ is a gentler reducing agent and it reduces only ketones, aldehydes and acid chlorides, while stronger reducing agents such as $LiAlH_4$ or Zn/HCl can also cleave ether bonds and convert them to OH groups [2, 111]. However, this is associated with chain scission in case of polyethers, which should be avoided at all cost.

The moderate reducing agent diborane (B_2H_6, dissolved in tetrahydrofuran-THF), reduces all carbonyl features (C=O, CHO, COOH, COOR) but does not attack ether structures (C-O-C); anyway these are not present in polyolefins [111, 117].

As a result of such a reduction with B_2H_6, the oxygen content on the polyolefin surface decreases from a maximum of around 28% O/C to 20% O/C, of which around 60% are OH groups (11-14% OH/C). A significantly higher yield and selectivity is obtained with plasma bromination. The 60% OH/O (≈20% O/C) with up to 97% C-Br/Br bromine groups (≈60% Br/C) is clearly surpassed by all plasma-introduced heteroelements. Since C-Br groups only react spontaneously with amino groups (polyamines, polyimines, melamines), corresponding bifunctional molecules must also first

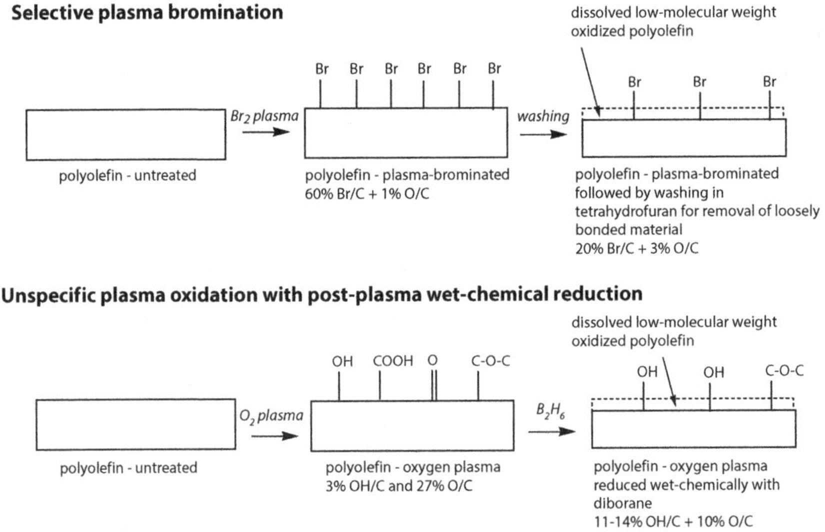

Figure 10.18 Paths to selectively functionalized polyolefin surfaces ("monosort").

be coupled to the C-Br groups in a wet chemical reaction step in the case of adhesion to Al overlayers. The direct connection of (evaporated and thus metallic) aluminium with a brominated polymer surface leads to the formation of $AlBr_3$ in an interfacial redox reaction, similar to that with an oxidized polymer surface (Figure 10.18).

10.4.2 Formation of Metal-Polyolefin Laminates

There are two types of Al-polymer laminates:

a) polyolefins as substrate and
b) aluminum as a substrate.

In both cases there are completely different adhesion issues, which can only be solved incompletely in most cases.

If a polyolefin is the substrate, anchor groups must first be created on its surface, for example by plasma treatment or by coating with thin layers that promote adhesion. The high reactivity of evaporated metallic aluminum layers must be taken into account in the further course of the bond formation. Another problem to be solved is the abrupt transition of the very different material properties in an atomic-molecular monolayer at the interface. A smooth and flexible transition from one set of properties to another must be ensured.

○ Formation of monosort functional groups at polymer surfaces by

- exposure to oxygen plasma and wet-chemical reduction of carbonyl groups with $LiAlH_4$ or B_2H_6 to OH groups [111, 115]
- exposure of polyolefin to underwater plasma forming preferentially OH groups
- low-pressure plasma bromination and formation of C-Br groups

○ Deposition of thin plasma polymer layers (low- and atmospheric-pressure plasma) containing functional groups from

- allylamine (NH_2 groups)
- acrylic acid (COOH groups)
- allyl alcohol (OH groups).

In these ways the desired monosort functional groups can be introduced to polyolefin surface.

There is a significant difference if aluminium with its native oxide/hydroxide covered surface is the substrate under ambient (humid) air conditions or the polyolefin is the substrate and is coated with pure metallic Al under high-vacuum conditions. As mentioned earlier, the surface energies of participating solids (γ_{Al}, $\gamma_{polyolefin}$) are considerably different and, therefore, also the corresponding interface energy ($\gamma_{interface}=\gamma_{Al}-\gamma_{polyolefin}$). This is valid for vacuum vapor deposition of Al onto the polyolefin surface. The other case is using native aluminium as substrate and coating or laminating it with the polyolefin, and then there are no such differences in surface energy (cf. Figure 10.1).

To produce chemical bonds (bridged chelate bonds, monodentate, bridging bidentate) to polyolefins, maleic anhydride is often used as comonomer [79, 81]. The coordinative bonding of a monofunctional carboxylic acid group to the oxide surface is not stable in an aqueous environment [79]. Plasma polymers adhere very well to aluminium and are potential adhesion promoters for polyolefins [77–79]. In case of polyurethanes it could be shown that cleaning, oxide formation and conditioning as well as plasma-assisted primer deposition can improve the adhesion of polymer coatings to metal substrates [63, 80].

10.4.3 Principal Disadvantages

The metal-polymer bond is much more sensitive to chemical attacks than the covalent C-C and C-H bonds in the polyolefin. This bond accomplishes the transition from Al to polyolefin. It must therefore be able to withstand the different properties of both materials. For this reason, we would like to list a few disturbing circumstances here:

- Unavoidable formation of LMWOM (Low-Molecular Weight Oxidized Material) at the polyolefin surface caused by exposure to plasma [104]

The treatment of polymers in the oxygen plasma (slowly) etches away the polymer and leaves a polymer surface with O-functional groups in a permanent steady-state process of etching and permanent formation of new O functional groups. As long as this steady-state persists during the etching of the polyolefin in the oxygen plasma, this surface consists of a thin layer of mostly oxidized and thus degraded, but partly also of newly cross-linked fragments of the polymer or even of a newly deposited plasma polymer layer (LMWOM-Low-Molecular Weight Oxidized Material, Figure 10.19).

At the interface to the metal, the O-functional groups have an adhesion-promoting effect, but they are bound to the thin surface layer restructured by polymer degradation, which, in the sense of Bikerman is a "weak boundary layer (WBL)" [86]. This layer has a negative effect on adhesion at the laminate interface due to a lack of cohesive force. The bond separation during metal peeling then does not take place directly at the interface under corresponding mechanical stress, but in its vicinity in somewhat deeper zones of the polymer, which are weakened by the previous plasma treatment in the sense of the LMWOM (Low-Molecular Weight Oxidized Material) [104].

- Chemical auto-oxidation of polyolefin

The generation of radicals by plasma treatment leads to an immediate recombination of neighboring C-radicals to C-C bonds and thus to the heating of the polymer. Radicals encapsulated in the polymer matrix can survive for a longer period of time if no recombination, disproportionation or radical transfer is possible. Such strongly stabilized radicals can survive stably for a long time due to the corresponding electron-donating substituents and spatial encapsulation with large substituents (trapping), which reduce the accessibility of the radicals for reaction partners. Only reactions

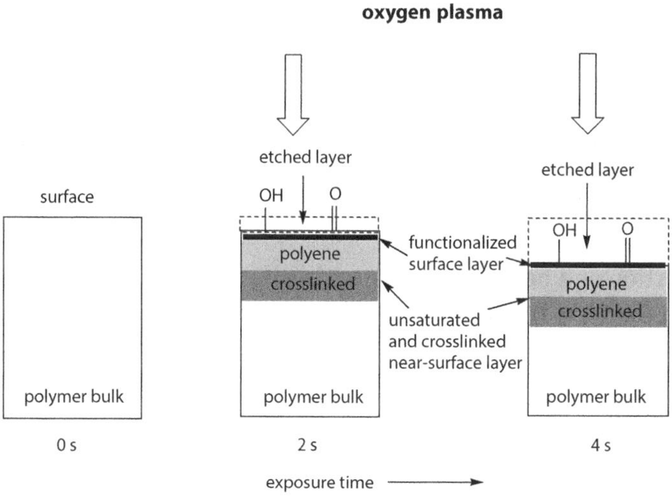

Figure 10.19 Steady-state between etching of polymer and permanent formation of new O functional groups.

in the oxygen plasma or later in the air with the small gaseous oxygen molecules, which diffuse into the polymer *via* pores or dissolved in the polymer solid, saturate these radicals and form peroxy radicals. In the course of this developing auto-oxidation, the same O-functional groups are formed again and again [93]. The most important metastable intermediate products of this process are peroxy radicals and hydroperoxides derived from them. This secondary oxidation can last for days.

- Problems with LMWOM (Low-Molecular Weight Oxidized Material)in wet chemical reactions

The O functional groups, which were introduced into the surface of the polyolefins on exposure to oxygen plasma and those which were secondarily introduced by auto-oxidation, reflect all oxygen species found by XPS and other methods. However, if a wet chemical process is required after plasma modification, the water or solvent used will dissolve the majority of the O-functional groups measured, typically 70-80% of all plasma generated groups. This LMWOM with many O functional groups is detached and is no longer available for the subsequent wet chemical (graft) reaction. About 20-30% of the functional groups remain bonded to the more or less intact bulk molecules, which is sufficient for adhesion. This applies, for example, to the diborane reduction of carbonyl groups or the grafting of molecules containing amino groups onto brominated surfaces.

- Covalent bonds with polar character

The covalent metal-polymer bonds in particular have an additional strong polar component due to the different electronegativities of Al and modified polyolefin surface. This polar part of the covalent bond makes it susceptible to polar substances, especially water. Both metal-carbon bonds and metal-oxygen-carbon bonds hydrolyze very quickly when exposed to moisture, which extremely weakens the bond strength. Bidentate C-O-Al and Si-O-Al bonds are more stable [118].

- Distortion and bending of covalent bonds along the interface

The alignment of the metal crystallites in the evaporated aluminum layer with the metal atoms covalently bonded to the polymer (Al-O-C) does not exactly match the positions of the OH or COOH groups reacting with Al on the polymer backbone, which can lead to a distortion and a bending of the covalent bonds formed. This distortion or bending reduces the bond strength and weakens neighboring layers in the polymer and also in the metal but it can be compensated by flexible spacer molecules.

- Thermal expansion coefficients

The mechanical stress on the metal-polymer bonds is increased by mechanical loading and by changes in temperature. The thermal expansion coefficients of polyolefins and aluminum are very different, as are between carbon fibers and polymer resins. In this case too, the covalent bonds are stressed by strong mechanical stress and possibly split.

- Redox reactions instead of covalent bonds

The formation of covalent metal-polymer bonds can also be understood as an intermediate metastable state. As described above, aluminium has a negative redox potential, while the oxidized polyolefin surface can be formally assigned a weakly positive redox potential [49]. The equalization of the different potentials is the driving force behind the formation of covalent bonds. The reaction actually continues until the formation of aluminium oxide and the reduction of the polymer surface to the non-oxidized polymer. With the transformation of the metal into an oxide, the boundary layer on the metal side is mechanically weakened. The covalent bond between metal and polymer (Al-O-C) then no longer exists. The O-functional group on the polymer side of the interface, which is necessary for the covalent bond, is then reduced and no longer exists. On the other hand, as already described, the redox reaction improves the thermodynamic compatibility of the two phases by minimizing the interfacial energy or by bringing the surface energies of the metal and polymer closer together. Figure 10.10 shows a schematic diagram of the course of the Gibbs enthalpy during the redox reaction of elemental aluminium with an O-functionalized polyolefin surface. The final state represents the equalization of the redox potentials, while the chemical bond Al-O-C is a metastable intermediate.

- Ion diffusion

Under certain circumstances, metal ions, protons or newly formed ionic complexes of metal with degradation products of the polymer such as copper carboxylates can slowly diffuse into the polymer from the boundary layer of the metal to the polymer, which is altered by redox reactions. This phenomenon has been known for a long time and leads to a reduction in the dielectric strength of the polymer insulation of high-voltage cables [74, 119]. A hydrophobic and dense diffusion barrier along the interface that also keeps moisture away from the interface can help here (Figures 10.20) [6].

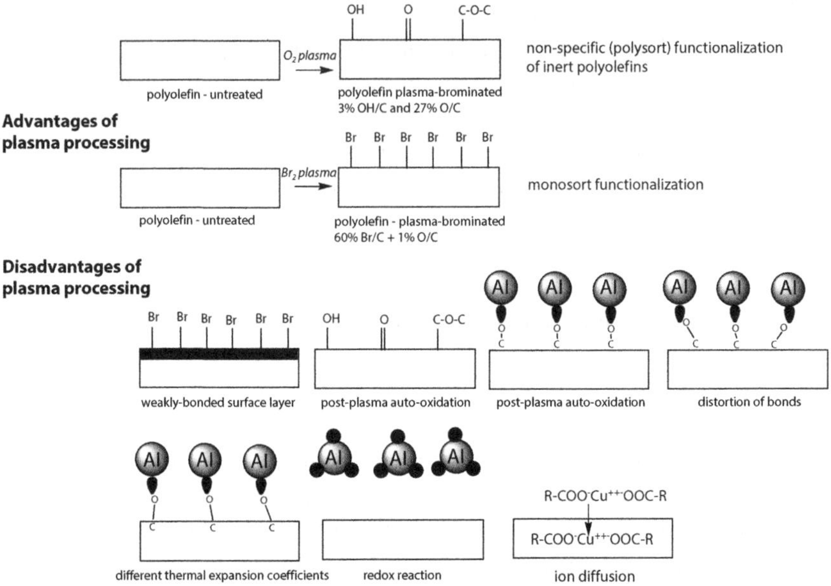

Figure 10.20 Advantages and disadvantages of plasma processed interfaces.

10.4.4 Special Requirements for Maximum Adhesion

As already described at the beginning, the abrupt transition from metal layer to solid polymer or from carbon fiber to embedding resin represents an extreme step from one set of properties to a completely different one. Even if there are strong covalent bonds between metal and polymer, fiber and resin, these bonds at the metal-polymer interface cannot withstand the mechanical forces under high thermal or mechanical stress. The bonds break. Since the abrupt transition from one property to another (metal-polymer) cannot be avoided in principle with covalent bonding, an attempt must be made to compensate for it with flexible elements, ideally to create a continuous transition between the different properties of metals and polyolefins.

a) Covalently bonded flexible molecules between metal layer and polyolefin surface

The task now is to bond flexible elements covalently, i.e. very firmly, between the metal layer and the polyolefin:

- Aliphatic chains offer flexibility due to their ability to rotate around the C-C bond axis.

- The flexible aliphatic chains can be firmly anchored by bifunctional spacer molecules.
- The flexible aliphatic chains are hydrophobic and therefore water-repellent.
- CF_2 chains (Teflon-like) would be even more hydrophobic, but react immediately with Al atoms: Al + 1.5 $CF_2 \rightarrow AlF_3$ + 1.5 C.
- The optimum chain length was tested with n=4-18. Shorter chains do not have sufficient flexibility, longer chains or even polymers entangle and end groups may not find the reaction partner.
- The graft density of the chains on the polyolefin surface also decreases with increasing chain length of the grafted chain due to the ever-increasing space requirement for wet-chemical grafting.
- Alternatively, thin, preferably flexible plasma polymers with monosort functional groups can be used (see Part 2).

b) Requirements for the covalent binding of spacers to polyolefins and of barrier-forming spacers with reactive endgroups

The monosortically functionalized polyolefin surface serves as the basis for the attachment of bifunctional spacer molecules. As previously shown, there are three variants of monosort surface modification of polyolefins:

- Bromination with a density of about 60% Br/C, of which about 20% Br/C remains after washing off, and
- Diborane-reduced oxidized polyolefin surfaces with an OH density of about 10% OH/C,
- use of plasma polymers (see Part 2).

Simple and spontaneous reactions with C-Br groups allow spacers with amino groups. Glutaraldehyde is suitable as a chain extender. Reactions with OH groups require more drastic reactions (Williamson's ether synthesis, using Na metal).

OH groups on polyolefin surfaces can react with carboxyl groups to form esters with the assistance of dehydrating carbodiimides. Isocyanates react spontaneously with alcoholic OH groups to urethanes or with amino groups to urea derivates (Figure 10.21).

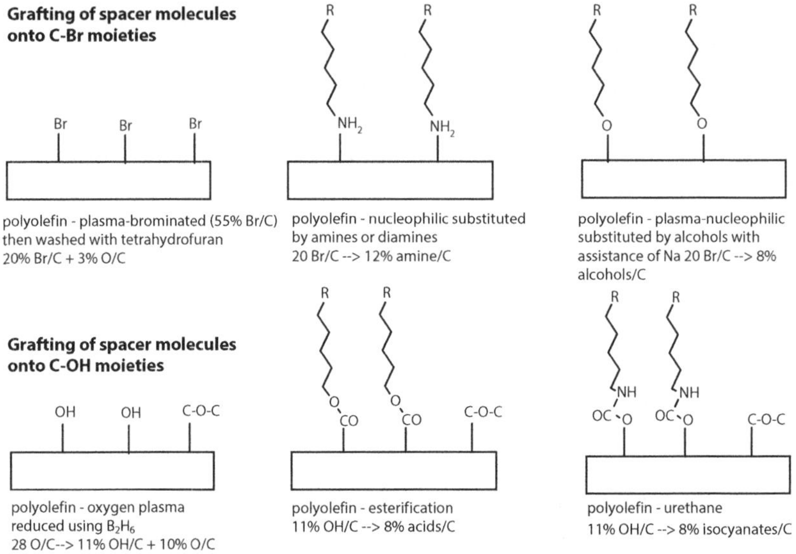

Figure 10.21 Examples of grafting onto Br or OH groups at the surface of polyolefins.

c) Hydrophobic, water-repellent design of the interface

Siloxane structures are well suited for forming a barrier against moisture and diffusion of ions. One Si-O-Si siloxane layer of a methoxy or ethoxy group is sufficient to achieve the desired effect. By chemically coupling the flexible spacer molecules with a silane molecule, the barrier effect can be combined with the flexibilization of the interface. This barrier layer is resistant to strongly reducing metal atoms, such as aluminium, but also to oxidizing metals, such as copper. Silanes and their siloxane reaction products fulfill these conditions very well.

Aminosilane, epoxysilane or mercaptosilane can react with their namesake functional group with the functional group of the spacer molecule (Figure 10.22).

d) Barrier formation and covalent bonding to Al

The 3 alkylated Si-O bonds of the silane hydrolyze easily to Si-OH, of which 2 Si-OH bonds can condense with neighboring Si-OH groups to form a 2D layer (x-y plane), which acts as a barrier against the diffusion of water and metal ions.

The free Si-OH bond in the third dimension (z-axis) is then used to form the Si-O-Al bond when aluminum is evaporated (Figure 10.23).

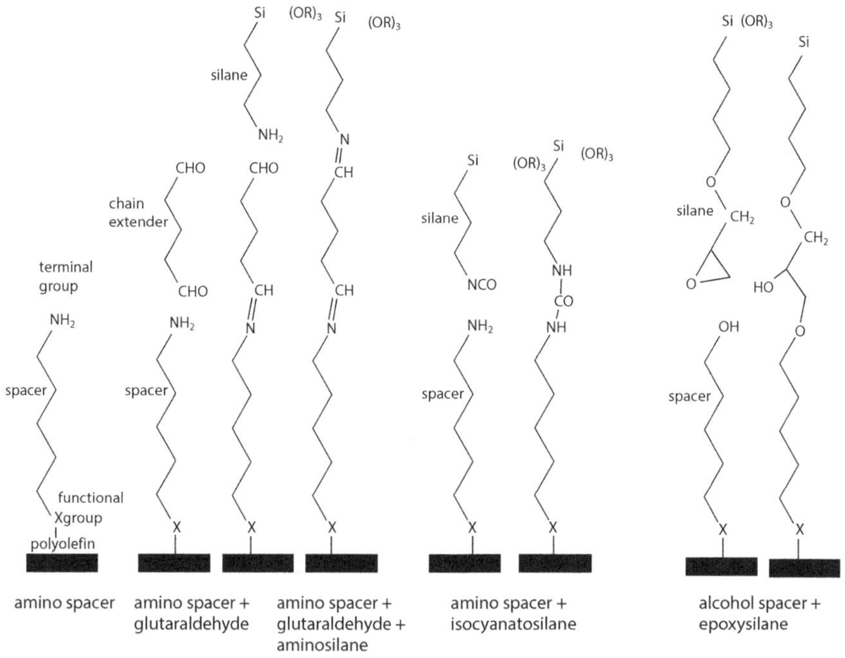

Figure 10.22 Examples of spacer grafting and chain extension.

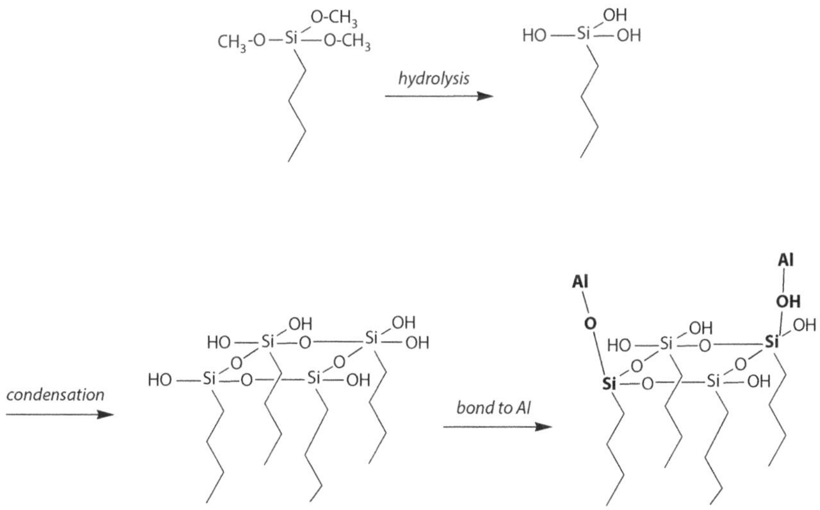

Figure 10.23 Hydrolysis of alkoxysilanes and formation of an Si-O-Si barrier layer as covalent Si-O-Al bonds.

With such an arrangement, the laminate interface could be firmly bonded, protected from hydrolysis and other permanent chemical processes and withstand mechanical and thermal movements without fracture at or near the interface.

The influence of the Si-OH and Si-O-Si structures of silanes on spacers on the adhesion strength of aluminum evaporation layers (150 nm) on polypropylene is shown in Figure 10.24. Various model silanes were synthesized to demonstrate the formation of such a barrier layer by condensation of 2 Si-OH groups of adjacent spacers and the reaction of the third Si-OH group with the evaporated aluminium atoms. The blocking of even just one hydrolyzable alkoxy position on the Si atom.

(Si-O-R + H_2O--> Si-OH + HO-R) by a non-hydrolyzable Si-CH_3 group considerably reduces the adhesion strength compared to a silane with 3 alkoxy groups; the adhesion strength with 2 or even 3 Si-CH_3 groups is correspondingly lower.

After hydrolysis of the Si-O-CH_3 groups, the barrier layer is formed by condensation of neighboring Si-OH groups, which also makes the interface hydrophobic. The remaining Si-OH groups can then form covalent bonds with the vapor-deposited Al atoms, resulting in a strongly adhering Al-polyolefin laminate that cannot be separated either physically or by humidity and temperature.

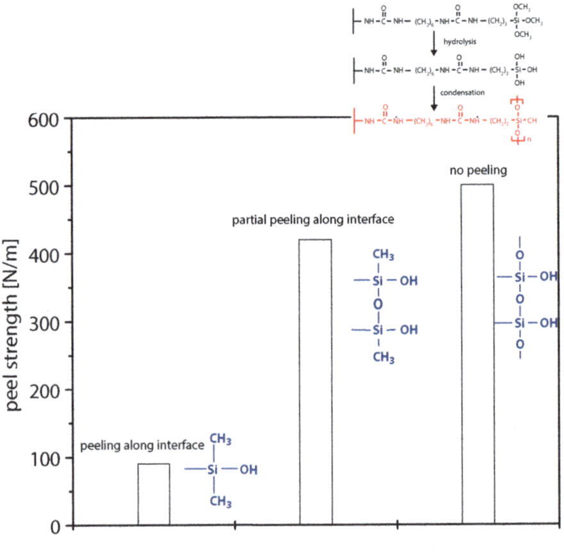

Figure 10.24 Influence of Si-OH and Si-O-Si structures of silanes on spacers on the adhesion strength of aluminum evaporated coatings (150 nm) on polypropylene.

10.5 Which Plasma Treatment Gives Maximum Adhesion?

As explained at the beginning, there are no or only very weak interactions of dipoles with induced dipoles in the polymer between aluminum layers and aliphatic polyolefins. However, propylene and also polypropylene have a very weak dipole moment due to their methyl side group. Therefore, there is a very weak adhesion of aluminum to polypropylene as opposed to no adhesion to polyethylene.

Pretreatment of the polyolefin is needed to achieve a reasonably strong bond between metal and polymer. Particularly strong dipoles on the polyolefin surface are generated by O-containing groups. This is used in technology to generate these polar groups by means of oxidative pre-treatment. This can be achieved using chemical reagents such as chromosulphuric acid, thermal oxidative processes such as flame treatment or plasma oxidative low-temperature treatment. The polar groups generated in this process, which are ideally bound to carbon atoms of the macromolecules, can then form bonds with the metal atoms *via* dipole-dipole interactions or covalent bonds.

10.5.1 Influencing Factors

Concentration of functional groups

If the polyolefin surface is exposed to the oxygen plasma, a preliminary maximum of O-functional groups is already reached after t=2 s treatment time (Figure 10.24). From then on, the O/C concentration increases only slowly. The peel strength of aluminum evaporated layers also initially increases abruptly with increasing Al layer thickness. Then the maximum peel strength no longer changes up to an O/C=20%. At higher O/C ratios, the peel strength then decreases again (Figure 10.25).

The reason for the drop in peel strength values at higher O/C values is the degradation of the polyolefin molecules into a loosely bound layer of low-molecular weight oxidized material (LMWOM), which then forms what Bikerman calls a weak boundary layer [86]. The number of O-functional groups on the LMWOM surface is very high, which is why a large number of interactions and bonds to the aluminium layer are formed. On the other hand, the cohesive strength of the LMWOM layer itself and the unmodified polyolefin matrix drops sharply. The fracture during peeling shifts from the metal-polymer interface to the LMWOM layer.

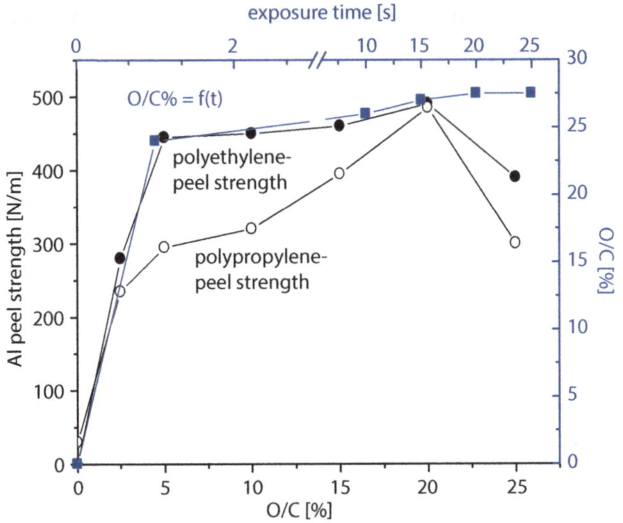

Figure 10.25 Dependence of Al peel strength on O/C of PP and PE and time of exposure of PP to oxygen low-pressure plasma.

<u>Type of functional group</u>

As has been shown, aliphatic structures can form only extremely low interactions (induced dipoles) with Al overlayers, however, Al-C bonds were sometimes detected [120]. If the polyolefin surface is equipped with primary amino groups, there is also only minimal adhesion strength. Methoxy and aldehyde functions result in moderate Al-polymer adhesion, whereas mercapto groups result in increased peel strength. C-OH groups further increase the adhesion strength, and Si-OH groups also show a significantly increased effectiveness in aluminum adhesion strength compared to the functional groups mentioned above. However, they are surpassed by carboxyl groups, which result in high adhesion strengths (Figure 10.26).

<u>Polysort or monosort functionalization</u>

An interesting question is whether the polysort functionalization of the polyolefin surface, produced by flame or oxygen plasma treatment and subsequently reshaped by auto-oxidation, can keep up with the monosort OH functionalization. If one looks at the variety of different O-functional groups after treatment in oxygen plasma, one will find about 10% OH/O with about 28% O/C, thus ≈3% OH/C. If this surface is reduced wet-chemically with diborane, the concentration of OH groups increases only

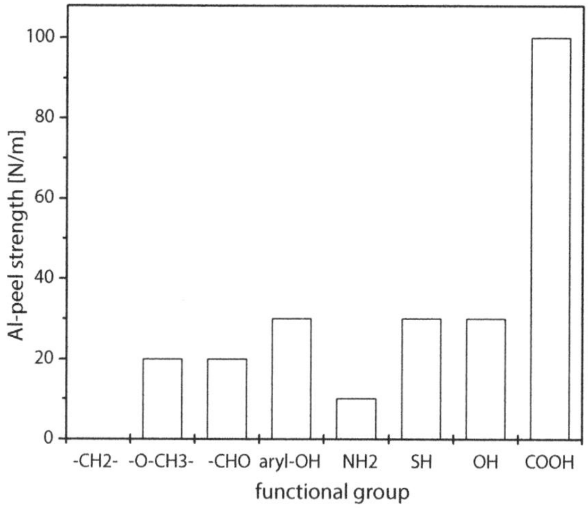

Figure 10.26 Efficiency of functional groups to increase the Al-polyolefin peel strength referenced to 1 functional group.

slightly from 3% OH/C to a maximum of 11 to 14% OH/C, while the O content drops to around 20% O/C. The wet chemical (tetrahydrofuran) diborane treatment has dissolved away about one-third of the weakly bonded layer (Low-Molecular Weight Oxidized Material-LMWOM). It must also be taken into account that the oxygen plasma etches away the polymer very quickly, so that the LMWOM should have been removed immediately. The UV radiation emitted by the plasma into the polymer interior and the barrage of high-energy species from the plasma also produce such an LMWOM layer in the oxygen plasma. This process can be thought of as a steady-state process.

It is, therefore, not surprising that the polysort functionalized polymer surface achieves adhesion strength similar to the monosort, diborane-treated surface (Figure 10.27). Both metal-polymer laminates produced in this way are separated along the interface or near-interface during the peel test. This means that the cohesive strengths of the metal layer and that of polyolefin are greater than the interactions at the metal-polymer interface.

In the next step, flexible spacer molecules are chemically bonded between the aluminum and polyolefin. The influence of the chain length of the spacer molecules and the type of endgroup of the spacer that reacts with the aluminum is investigated.

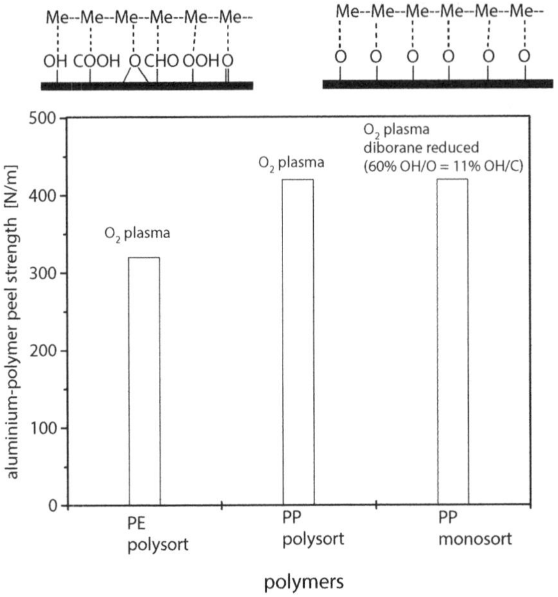

Figure 10.27 Peel strength of 150 nm thick aluminium layer thermally evaporated onto plasma modified polyolefin surfaces with polysort and monosort O-containing groups.

Is the introduction of spacer molecules into the interface absolutely necessary?

First of all, the bond strength between aluminium atoms and OH or COOH groups at the polyolefin surface is independent of whether aluminium is bonded to the functionalized polyolefin surface directly or *via* an intermediate spacer molecule. The bond strengths of the Al-O-C bonds of the two reaction variants (direct or with spacer) do not differ. The C-O-C bond of the spacer on the polyolefin surface should also not be a weak point, as it is a stable and strong covalent bond.

The question now remains whether the flexibility of an aliphatic spacer has an influence on the aluminum-polyolefin adhesion. Figure 10.28 shows clear differences in adhesion if these are referenced to a single covalent bond for comparison. The use of the C_6 spacer resulted in a significantly increased peel strength in relation to a bonding *via* spacer or a direct bonding. The reason for this effect has already been explained several times above, i.e., the flexibilization of the interface and a uniform distribution of shear forces under mechanical or thermal stress.

A branched spacer with 2 carboxylic groups additionally increases the measured peel strength (Figure 10.28).

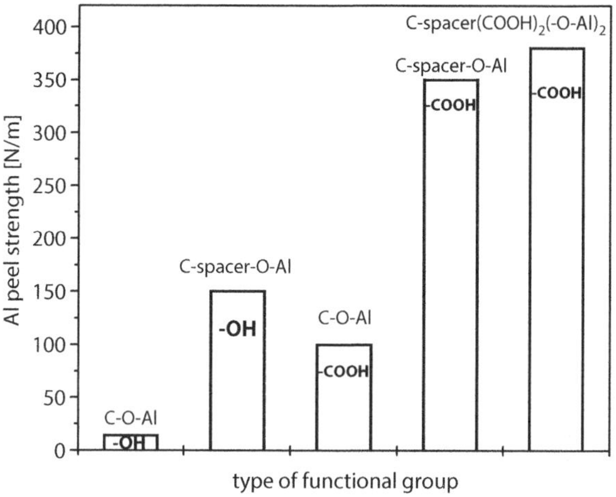

Figure 10.28 Differences in aluminum adhesion to plasma-modified polypropylene when Al is bonded directly to the functional groups or *via* a C6 spacer. In order to compensate for the difference between many OH or COOH groups in the case of direct Al-O-C bond formation with the much fewer spacers present and thus fewer Al-O-C bonds, the peel strengths were related to 1 direct bond or 1 bond *via* spacer in each case.

Spacer length

The working hypothesis was that the different material properties of metal and polyolefin required a flexible rather than a rigid transition at the interface. For this purpose, aliphatic diols of different chain lengths were grafted onto C-Br groups followed by deposition of a vacuum evaporated thin overlayer of Al.

An example of the dependence of the peel strength of Al on polyolefin surfaces grafted with diols of different chain lengths is presented in Figure 10.29. In order to make the results of the measurements comparable, the value was referenced to one spacer molecule in each case.

The longer and larger the spacer molecules are, the more space they take up on the surface and the fewer of them can be chemically anchored on the same surface as with functional groups that are bound to the surfaces without spacers. Oligomers and polymers also tend to become entangled, occupy a large surface area and the reactive end groups are hidden in the polymer coil and are therefore no longer able to react. This can be seen in COOH-terminated spacer molecules with a chain length of 18 CH_2 or -N= links (Figure 10.30).

Figure 10.31 shows the relationship between the chain length and thus also the space required by the spacer and the resulting maximum graft

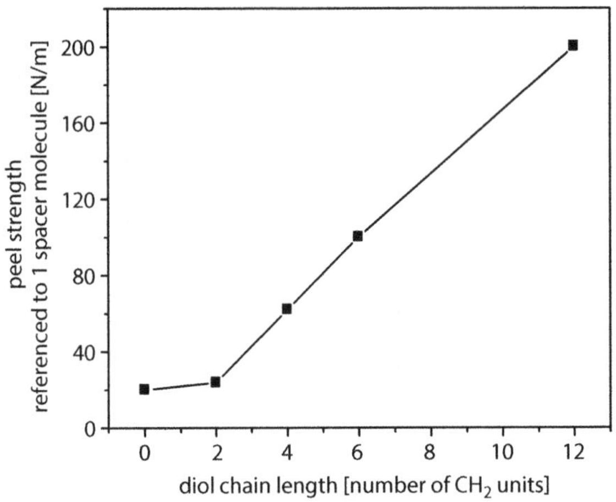

Figure 10.29 Example of the dependence of the peel strength of Al on polyolefin surfaces grafted with diols of different chain lengths. In order to make the results of the measurements comparable, the value was related to one spacer molecule in each case.

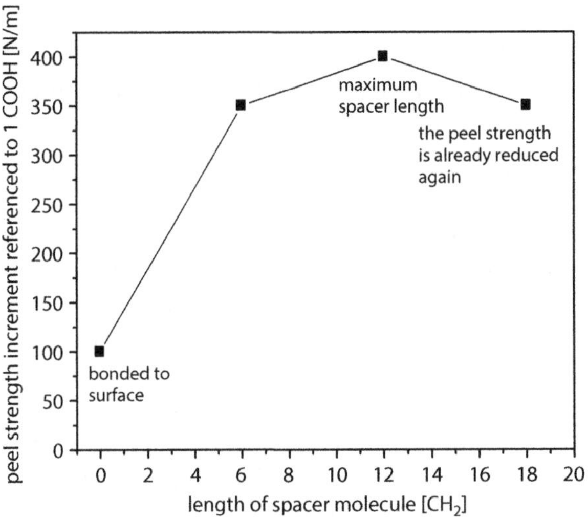

Figure 10.30 Example of the dependence of the peel strength of Al on polyolefin surfaces grafted with COOH terminated spacers of different chain lengths. In order to make the results of the measurements comparable, the value was related to one spacer molecule in each case.

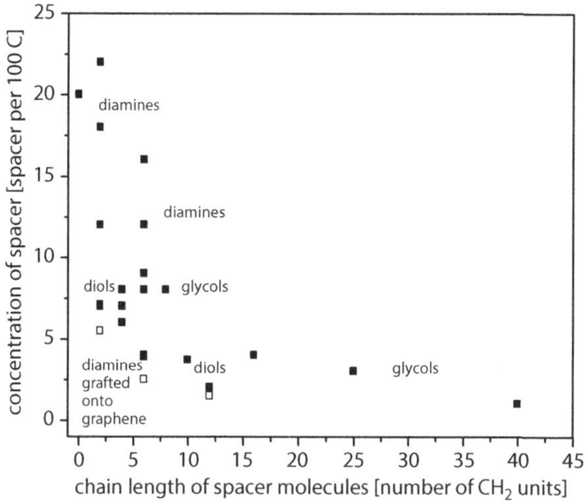

Figure 10.31 Dependence of the achievable density of spacer molecules on the polyolefin surface as a function of the chain length (in CH_2 units) and thus the size of the spacer molecule.

density on the polyolefin surface. The dependence of the graft density on the size of the spacer molecule applies to several different spacer structures.

Changes in the bonding state of metal atoms in the metal lattice after forming covalent Al-polymer bond along the interface and the distortion and bending of macromolecules by bonding to Al

This problem was discussed in the previous section, when the aluminum atoms form covalent bonds with functional groups during the reaction; their integration into the atomic structure of the metal is disrupted. By incorporating flexible spacer molecules, the distortion of the Al-O-C bond can be prevented.

When the aluminum atoms form covalent bonds with functional groups on the polyolefin surface, they are partially detached from the atomic structure of the metal. The spacing of the functional groups on the polyolefin surface means that there is not always an aluminum atom exactly opposite as a reaction partner. In addition, the Al atoms involved in the Al-O-C bond are no longer fully integrated into the crystalline framework of the metal. These disadvantages no longer exist when flexible spacer molecules are inserted. Then there is a transition zone not between covalently bonded Al and metallic Al, but between metal oxide and metallic Al. In the latter case, the bonding forces between Al_2O_3 and Al_{metal} are significantly reduced.

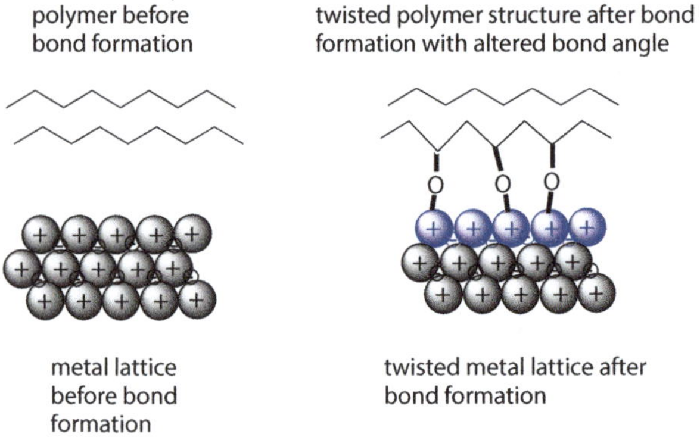

Figure 10.32 Twisting of the metal lattice and distortion of the polymer structure along the interface after bonding with altered Al-O-C bond angles.

Furthermore, despite a certain flexibility of the polymer component, including with a thin metal layer, it must be taken into account that the metal-polymer bonds that form are subject to distortion with regard to the ideal bond angle, especially under thermal and mechanical stresses, which weakens the bond.

The covalent bond to the aluminum also exerts a distorting alignment force on the polymer at the interface. This can lead to the weakening of crystalline structures in particular in the polymer matrix adjacent to the interface. This may explain the frequently observed shift in the location of peeling from the actual interface to areas of the polymer close to the interface (Figure 10.32).

What are the effects of rigid groups in the spacer molecules compared to fully flexible aliphatic spacer molecules?

If the molecules inserted into the interface are intended to make it more flexible and thus reduce stresses, then the "pliability" of the spacer molecules should also have an influence on the achievable adhesion strength of metal-polymer laminates.

All aliphatic chains are flexible due to the 360° rotation possibility of the C-C single bonds. Rigidity is provided by C=C double bonds and triple bonds, C=N bonds (azomethines) and N=N bonds, aromatic rings, condensable groups, ether groups and ring structures.

In fact, the installation of stiffening elements in the spacer chain reduces the resulting Al-polypropylene adhesion strength with the same chain length and referenced to 1 spacer molecule (Figure 10.33).

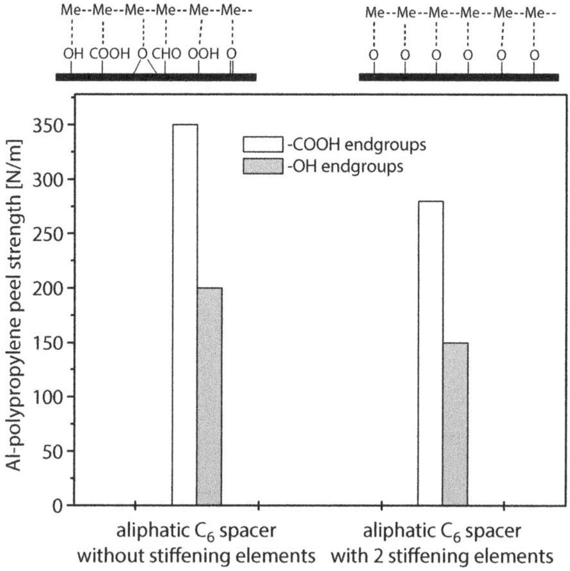

Figure 10.33 Influence of the incorporation of stiffening elements in the C_6 spacer with OH and COOH endgroups on peel strength of Al-polyolefine laminates.

10.5.2 How Must the Interface of a Metal-Polyolefin Laminate be Constructed in Order to Achieve Maximum Adhesion Strength and Aging Resistance?

A number of problems in the design of the metal-polymer interface have been discussed. From these, synthesis strategies were developed for chemically designing the interface (Figure 10.34). Two starting monosort groups on the polymer surface were used as anchor points: the C-Br groups generated by Br_2 plasma as a monosort functional group and the (monosort) OH group generated by plasma chemical oxidation with subsequent wet-chemical reduction using diborane. When using brominated polymer surfaces, the following synthesis strategies were used: nucleophilic substitution by diamine and the formation of Schiff's base with chain extender glutaraldehyde followed by coupling of aminosilane as well as the hydrolysis of its alkoxy groups to silanol groups, the partial condensation of the resulting silanol groups and the bond formation between silanol groups and aluminum.

OH-equipped surfaces were chemically grafted with the spacers after Williamson's ether synthesis, also using diisocyanates, their hydrolysis to amino groups and the above-mentioned reactions. After this complicated

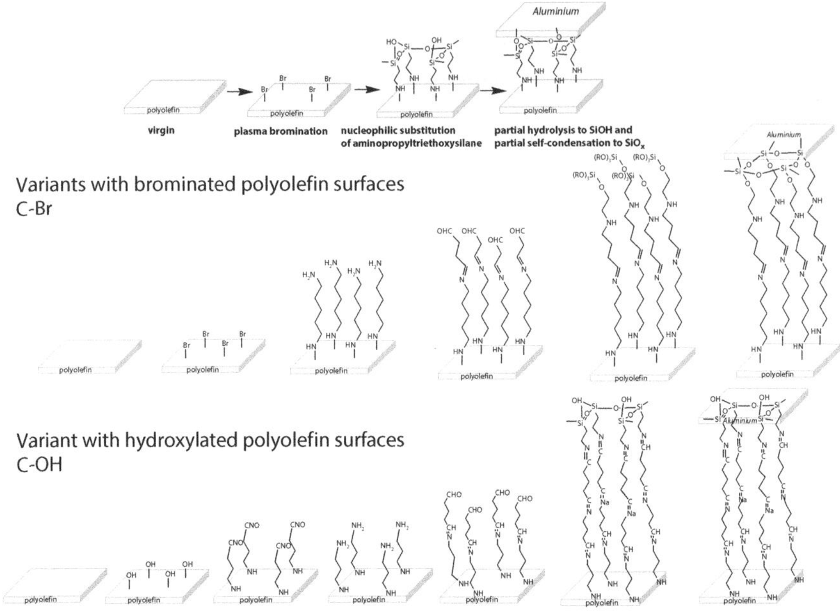

Figure 10.34 Example of an ideally designed aluminum-polypropylene laminate that contains all the structural elements discussed above and could not be mechanically separated even after one year of storage and cyclic thermal stress and irrigation.

surface modification with spacers, the modified polymer surface was thermally evaporated with Al.

Despite the use of various types of adhesive tapes, it was not possible to remove the aluminum layer from the aluminum-polypropylene laminates constructed in this way by peeling the adhesive tape. The aluminum layer remained adhered as a whole to the modified polymer surface. Even the use of different qualities of these adhesive tapes did not change the result. The metal-polymer laminates were not only non-peelable, but also after 6 months of storage at 90% relative humidity and 40°C. Some long-term tests lasting over 1 year also showed that they were not peelable.

10.6 Summary

The polyolefin surfaces were preferably equipped with monosort OH and Br functional groups using plasma chemical processes. Subsequently, the plasma-treated surfaces were further modified wet-chemically with flexible spacer molecules and silanes and then thermally evaporated with thin aluminum layers.

The aluminum-polypropylene laminates constructed in this way was permanently stable (Figure 10.35). Even the use of different qualities of adhesive tapes did not change the result. The metal-polymer laminates could not be peeled immediately after Al evaporation, and also not after 1 year of storage at 90% relative humidity and 40°C.

The main components of the tailor-made structure of a stable and strong interface between aluminum evaporated layer and polyolefin are:

- covalent bonding of a flexible (aliphatic) spacer molecule with the substrate
- covalent bonding of a silane to the spacer
- partial condensation of the hydrolyzed alkoxy groups to form a Si-O barrier layer
- linking of spacer with metal by covalent Si-O-Al bonds.

The density of the spacers (Sp) and thus of the covalent bonds between aluminium and polyolefin was relatively low, approx. 6% Sp/C, in order to

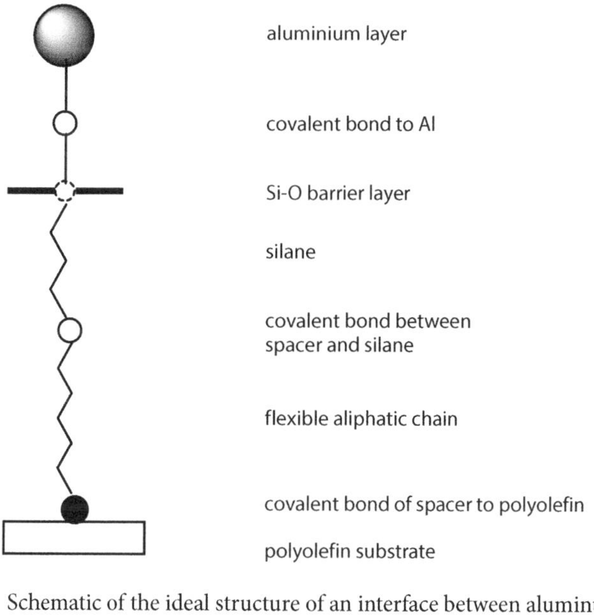

Figure 10.35 Schematic of the ideal structure of an interface between aluminum and polyolefin.

achieve maximum adhesion. The range of optimal length of the spacers was about 6-12 CH_2 groups.

Acknowledgement

A large number of highly qualified coworkers and employees, who were co-authors of various previous publications on the subject of adhesion and plasma, have made this publication possible with their work. I cannot list them all, but you can be rest assured that their work was greatly appreciated and they delivered high-quality results.

AI disclosure statement: The author acknowledges the use of AI-based tools for minor language editing or grammatical correction. However, all intellectual contributions, analyses, and conclusions presented in this chapter are entirely the author's own.

References

1. K. L. Mittal and A. N. Netravali (Eds.), *Polymer Surface Modification to Enhance Adhesion: Techniques and Applications*, Wiley-Scrivener, Beverly, MA (2024).
2. E. Fanghänel (Ed.), *Organikum*, Wiley-VCH, Weinheim, Germany (2004).
3. J. Friedrich, The formation processes of functional groups at polyolefin surfaces on exposure to oxygen or ammonia plasma: A critical review, Rev. Adhesion Adhesives 6, 253-328 (2018).
4. R. H. Hansen and H. Schonhorn, A new technique for preparing low surface energy polymers for adhesive bonding, J. Polym. Sci. C, Polym. Lett. 4, 203-209 (1966).
5. M. Hudis, Surface crosslinking of polyethylene using a hydrogen glow discharge, J. Appl. Polym. Sci. 16, 2397-2414 (1972).
6. J. Friedrich, *Metal-Polymer Systems: Interface Design and Chemical Bonding*, Wiley-VCH, Weinheim, Germany (2017).
7. J. Friedrich, R. Mix and S. Wettmarshausen, A new concept of adhesion promotion in metal-polymer composites by introduction of covalently bonded spacers at interface, J. Adhesion Sci. Technol. 22, 1123-1143 (2008).
8. J. Friedrich and J. Meichsner, *Nonthermal Plasmas for Materials Processing*, Wiley-Scrivener, Beverly, MA (2022).
9. P. Krüger, R. Knes and J. Friedrich, Surface cleaning by plasma-enhanced desorption of contaminants, Surf. Coatings Technol. 112, 240-244 (1999).

10. P. Krüger and J. Friedrich, Feinreinigung von Oberflächen, in: *Vakuum-Plasma-Technologien Beschichtung und Modifizierung von Oberflächen, Teil I und II*, G. Blasek and G. Bräuer (Eds.), pp. 604-621, Eugen-G-Leuze-Verlag, Saulgau, Germany (2010).
11. N. Anagreh and L. Dorn, Haftfestigkeitssteigerungen von Klebeverbindungen aus Aluminiumblechen und ASA/PC-Kunststoffen durch Modifizierung der Oberflächen, Schweiß u. Schneid. 53, 272-279 (2001).
12. S. Wu, *Polymer Interface and Adhesion*, Marcel Dekker, New York (1982).
13. H. Esrom and U. Kogelschatz, Modification of surfaces with new excimer UV sources, Thin Solid Films 218, 231-246 (1992).
14. P. J. Cumpson, J. F. Portoles and N. Sano, Material dependence of argon cluster ion sputter yield in polymers: Method and measurements of relative sputter yields for 19 polymers, J. Vac. Sci. Technol. A 31, 020605 (2013).
15. F. Faupel, R. Willecke and A. Thran, Diffusion of metals in polymers, Mater. Sci. Eng.: R: Reports, 22, 1-55 (1998).
16. H. C. Guo, E. Ye, Z. Li, M. Y. Han and X. J. Loh, Recent progress of atomic layer deposition on polymeric materials, Mater. Sci. Eng. C 70, 1182-1191 (2017).
17. H Koivuluoto, A review of thermally sprayed polymer coatings, J. Thermal Spray Technol. 31, 1750-1767 (2022).
18. R. Melentiev, A. Yudhanto, R. Tao and T. Vuchkov, Metallization of polymers and composites: State-of-the-art approaches, Materials & Design 221, 110958 (2022).
19. K. Rossmann, Improvement of bonding properties of polyethylene, J. Polym. Sci., 19, 141-144 (1956).
20. J. Friedrich, J. Gähde, H. Frommelt and H. Wittrich, Modifizierung von Feststoffoberflächen in einer HF-Entladung. II. Schichtuntersuchungen, Reaktionsmechanismus und Verbundergebnisse, Faserforsch. Textiltechn. / Z. Polymerenforsch. 27, 599-603 (1976).
21. M. Strobel, M. C. Branch, M. Ulsh, R. S. Kapaun, S. Kirk and C. Lyons, Flame surface modification of polypropylene film, J. Adhesion Sci. Technol. 10, 515-539 (1996).
22. K. W. Allen, Polar forces, in: *Handbook of Adhesion*, D. E. Packham (Ed.), pp. 326-328, Wiley, Chichester, UK (2007).
23. J. Friedrich, W. Unger, A. Lippitz, I. Koprinarov, A. Ghode, Sh. Geng and G. Kühn, Plasma-based introduction of monosort functional groups of different type and density onto polymer surfaces, Part I: Behaviour of polymers exposed to oxygen plasma, Composite Interfaces, 10, 139-172 (2003).
24. M. Kuzuya, K. Ito, S.-I. Kondo and Y. Yamauchi, Plasma-induced free radicals of polycrystalline carbohydrates as spin probe for plasma diagnosis of plasma treatment, Thin Solid Films 345, 85-89 (1999).
25. A. M. Plath, Radikalidentifizierung auf Polymeren nach Plasmabehandlung, Vakuum in Forschung und Praxis 16, 118-125 (2004).

26. R. Wilken, A. Holländer and J. Behnisch, Surface radical analysis on plasma-treated polymers, Surf. Coatings Technol. 116–119, 991-995 (1999).
27. M. Suzuki, A. Kishida, H. Iwata, and Y. Ikada, Graft copolymerization of acrylamide onto a polyethylene, Macromolecules 19, 1804-1808 (1986).
28. E. T. Kang, K. G. Neoh, K. L. Tan and D.-J. Liaw, Surface graft copolymerization and grafting of polymers for adhesion improvement, in: *Adhesion Promotion Techniques-Technological Applications*, K. L. Mittal and A. Pizzi (Eds.), pp. 289-321, Marcel Dekker, New York (2002).
29. E. P. Plueddemann, *Silane Coupling Agents*, Springer, Heidelberg, Germany (1991).
30. J. Friedrich, *The Plasma Chemistry of Polymer Surfaces, Advanced Techniques for Surface Design*, Wiley-VCH, Weinheim, Germany (2012).
31. J. R. McCallum and C. T. Rankin, A novel method for modifying surfaces, J. Polym. Sci. B 9, 751-752 (1971).
32. A. Bradley and J. D. Fales, Prospects for industrial applications of electrical discharge, Chem. Technol. 4, 232-245 (1971).
33. C. I. Simionescu, F. Denes, M. M. Macoveanu, M. Totolin and G. Cazakin, Polymerization of some potential "grafting-monomers" for natural and synthetic fibres in cold plasma conditions, Acta Polymerica 30, 26-31 (1979).
34. K. E. Geckeler, R. Gebhardt and H. Grünwald, Surface modification of polyethylene by plasma grafting with styrene for enhanced biocompatibility, Naturwissenschaften 84, 150-151 (1997).
35. I. Gancarz, G. Pozniak, M. Bryjak and A. Frankiewicz, Modification of polysulfone membranes. 2. Plasma grafting and plasma polymerization of acrylic acid, Acta Polymerica 50, 317–326 (1999).
36. J. Prachar, I. Novak, A. Kleinova, P. Gemeiner, M. Ostamova, S. Chmela and E. Borsig, Plasma grafting of polypropylene with organosilanes and its alkylamine treatment, Vacuum 127, 38-44 (2016).
37. Z. R. Farag, S. Krüger, G. Hidde, A. Schimanski, C. Jäger and J. Friedrich, Deposition of thick polymer or inorganic layers with flame-retardant properties by combination of plasma and spray processes, Surf. Coatings Technol. 228, 266-274 (2013).
38. A. Peutzfeld and E. Asmussen, Silicoating: evaluation of a new method of bonding composite resin to metal, Eur. J. Oral Sci. 96, 171-176 (1988).
39. E. C. Horning, M. G. Horning, D. I. Carroll, I. Dzidic and R. N. Stillwell, New picogram detection system based on a mass spectrometer with an external ionization source at atmospheric pressure, Anal. Chem. 45, 936-943 (1973).
40. M. Dole, L. L. Mack, R. L. Hines, R. C. Mobley, L. D. Ferguson and M. B. Alice, Molecular beams of macroions, J. Chem. Phys. 49, 2240-2249 (1968).
41. J. B. Fenn, M. Mann, C. K. Meng, S. F. Wong and C. M. Whitehouse, Electrospray ionization for mass spectrometry of large biomolecules, Science 246, 64-71 (1989).

42. J. B. Fenn, M. Mann, C. K. Meng, S. F. Wong and C. M. Whitehouse, Electrospray ionization-principles and practice, Mass Spectrom. Rev. 9, 37-70 (1990).
43. K. W. Allen, Adsorption theory of adhesion, in: *Handbook of Adhesion*, D. E. Packham (Ed.), pp. 38-40, Wiley, Chichester, UK (2007).
44. G. Habenicht, *Kleben-Grundlagen, Technologien, Anwendungen,* 6th ed., Springer, Heidelberg, Germany (2009).
45. J. M. Martín-Martínez, Improving adhesion of rubber, in: *Current Topics in Elastomers Research*, A. K. Bhowmick (Ed.), Chapter 27, CRC Press, Boca Raton, FL (2008).
46. J. F. Friedrich, R. Mix, R.-D. Schulze, A. Meyer-Plath, R. Joshi and S. Wettmarshausen, New plasma techniques for polymer surface modification with monotype functional groups, Plasma Proc. Polym. 5, 407-423 (2008).
47. J. Friedrich, B. Falk, I. Loeschcke, B. Rutsch, Kh. Richter, H.-D. Reiner, U. Throl and H. Raubach, Untersuchungen zur Haftung zwischen plasmachemisch oxidiertem Kupfer und Polyphenylchinoxalin, Acta Polymerica 36, 310-320 (1985).
48. J. Friedrich, I. Loeschcke and J. Gähde, Zur Adhäsion von Aluminium auf Polypropylen, Acta Polymerica 37, 687-695 (1986).
49. J. F. Friedrich, W. E. S. Unger, A. Lippitz, R. Giebler, I. Koprinarov, St. Weidner and G. Kühn, Significance of interfacial redox reactions and formation of metal-organic complexes for the adhesion of metals on pristine and plasma-treated polymers, in: *Polymer Surface Modification: Relevance to Adhesion, Vol. 2,* K. L. Mittal (Ed.), pp. 137-172, CRC Press, Boca Raton, FL (2000).
50. J. Friedrich, Tailoring of interface/interphase to promote metal-polymer adhesion, in: *Adhesion in Microelectronics*, K. L. Mittal and T. Ahsan (Eds.), pp. 67-135, Scrivener-Wiley, Beverly, MA (2014).
51. J. Friedrich, Competition in adhesion between polysort and monosort functionalized polyolefin surfaces coated with vacuum-evaporated aluminium, in: *Polymer Surface Modification to Enhance Adhesion: Techniques and Applications*, K. L. Mittal and A. N. Netravali (Eds.), pp. 389-458, Wiley-Scrivener, Beverly, MA (2024).
52. S. S. Voyutskii and V. L. Vakula, Effects of self-diffusion and inter-diffusion in polymer systems, Rubber Chem. Technol. 37, 1153-1177 (1964).
53. C. Bischof and W. Possart, *Adhäsion*, Akademie-Verlag, Berlin, Germany (1983).
54. N. A. de Bruyne and A. Houwink (Eds.), *Adhesion and Adhesives*, Elsevier, Houston, TX (1951).
55. D. H. Kaelble, Dispersion-polar surface tension properties of organic solids, J. Adhesion 2, 66–81 (1970).
56. D. K. Owens and R. Wendt, Estimation of the surface free energy of polymers, J. Appl. Polym. Sci. 13, 1741–1747 (1969).

57. W. Rabel, Einige Aspekte der Benetzungstheorie und ihre Anwendung auf die Untersuchung und Veränderung der Oberflächeneigenschaften von Polymeren, Farbe und Lack 77, 997–1005 (1971).
58. Inno tape.de
59. K. L. Mittal, The role of the interface in adhesion phenomena, Polym. Sci. Eng. 17, 467-473 (1977).
60. G. A. Dyckerhoff and P.-J. Sell, Über den Einfluß der Grenzflächenspannung auf die Haftfestigkeit (Influence of surface energy on adhesion), Angew. Makromol. Chem. 21, 169-185 (1972).
61. M. H. Stans, *Bond Dissociation Energies of Simple Molecules*, NIST, Washington D.C. (1970).
62. R. K. Hart, A study of boehmite formation on aluminium by electron diffraction, Trans. Faraday Soc. 59, 269-273 (1953).
63. J. Böhm, Über Aluminium- und Eisenhydroxyde, Z. anorg, allg. Chem. 149, 203-216 (1925).
64. J. Gähde, J. Friedrich, Th. Fischer, W. Unger, A. Lippitz and J. Falkenhagen, Reactions of polymer analogous model substances with metals and oxides, Prog. Colloid Polym. Sci. 101, 194-198 (1996).
65. M. R. Alexander, G. Beamson, C. J. Blomfield, G. Leggett and T. M. Duc, Interaction of carboxylic acids with the oxyhydroxide surface of aluminium: Poly(acrylic acid), acetic acid and propionic acid on pseudoboehmite, J. Electr. Spectr. Rel. Phenom. 121, 19-32 (2001).
66. Q. D. Ling, S. Li, E. T. Kang, K. G. Neoh, B. Liu and W. Huang, Interface formation between the Al electrode and poly[2,7-(9,9-dihexylfluorene)-co-alt-2,5-(decylthiophene). (PFT) investigated *in situ* by XPS, Appl. Surf. Sci. 199, 74–82 (2002).
67. E. T. Kang, K. G. Neoh and K. L. Tan, Conductive polymers: Spectroscopy and physical properties, in: *Handbook of Organic Conductive Molecules and Polymers, Vol. 3,* H. S. Nalwa (Ed.), pp. 121-145, Wiley, Chichester (1997).
68. P. Dannetun, M. Boman, S. Stafström, W. R. Salaneck, R. Lazzaroni, C. Fredriksson, J. L. Bredas, R. Zamboni and C. Taliani, The chemical and electronic structure of the interface between aluminum and polythiophene semiconductors, J. Chem. Phys. 99, 664-672 (1993).
69. P. Stoyanov, S. Akhter and J. M. White, XPS study of metal/polymer interaction: Evaporated aluminum on polyvinyl alcohol. Surf. Interface Anal. 15, 509-515 (1990).
70. J. F. Silvain and J. J. Ehrhardt, An overview on metal/PET adhesion, Thin Solid Films 236, 230-235 (1993).
71. A. Namkanisorn, A. Ghatak, M. K. Chaudhury and D. H. Berry, A kinetic approach to study the hydrolytic stability of polymer–metal adhesion, J. Adhesion Sci. Technol.15, 1725–1745 (2001)].
72. M. R. Alexander and T. M. Duc, A study of the interaction of acrylic acid/1, 7-octadiene plasma deposits with water and other solvents, Polymer 40, 5479-5488 (1999).

73. J. F. Friedrich, I. Koprinarov, R. Giebler, A. Lippitz and W. E. S. Unger, Reactions and intermediates at the metal-polymer interface – Detected by XPS and NEXAFS spectroscopy, J. Adhesion, 71, 297-310 (1999).
74. G. L. Fisher, A. E. Hooper, R. L. Opila, D. L. Allara and N. Winograd, The interaction of vapor-deposited Al atoms with CO_2H groups at the surface of a self-assembled alkanethiolate monolayer on gold, J. Phys. Chem. B104, 3267-3273 (2000).
75. R. Cueff, G. Baud, M. Benmalek, J. P. Besse, J. R. Butruille and M. Jacquet, X-ray photoelectron spectroscopy studies of plasma-modified PET surface and alumina/PET interface, Appl. Surf. Sci. 115, 292-298 (1997).
76. M. R. Alexander, S. Payan and T. M. Duc, Interfacial interactions of plasma-polymerized acrylic acid and an oxidized aluminium surface investigated using XPS, FTIR and poly(acrylic acid) as a model composite, Surf. Interface Anal. 26, 961-973 (1998).
77. G. L. Fisher, A. Hooper, R. L. Opila, D.R. Jung, D. L. Allara and N. Winograd, The interaction between vapor-deposited Al atoms and methylester-terminated self-assembled monolayers studied by time-of-flight secondary ion mass spectrometry, X-ray photoelectron spectroscopy and infrared reflectance spectroscopy, J. Electron Spectr. Rel. Phenom. 98–99, 139–148 (1999).
78. J. Du, J. Kalahe, T. Mahavedan and S. Urata, Reaction Mechanism of Aluminosilicate Glasses in Aqueous Solution from Reactive Molecular Dynamics Simulations, https://ssrn.com/abstract=4883630 or http://dx.doi.org/10.2139/ssrn.4883630.
79. M.-A. Chen, X.-M. Zhang, R. Huang and X.-B. Lu, Mechanism of adhesion promotion between alumium sheet and polypropylene with maleic anhydride-grafted polypropylene by γ-aminopropyltriethoxy silane, Surf. Interf. Anal. 40, 1209-1218 (2008).
80. J. Gähde, J. Friedrich, I. Loeschcke, R. Gehrke and J. Sachse, Adhesion of polyurethane to surface-modified steel, J. Adhesion Sci. Technol. 6, 569-580 (1992).
81. B. Schneider, O.-D. Hennemann and W. Possart, The adhesion of maleic anhydride on native aluminium oxide: An approach by infrared spectroscopy and quantum mechanical modeling, J. Adhesion 78, 779-797 (2002).
82. R. J. Purvis and W. R. Beck, Methods of activating the surface of perfluorocarbon polymers and resultant articles, US patent 2789063 (1957).
83. D. H. Kaelble and E. H. Cirlin, Dispersion and polar contributions to surface tension of poly(methylene oxide) and Na-treated polytetrafluoroethylene, J. Polym. Sci. Part A-2: Polym. Phys. 9, 363-368 (1971).
84. E. R. Nelson, T. J. Kilduff and A. A. Benderly, Bonding of Teflon, Ind. Eng. Chem. 50, 329-334 (1958).
85. M. Morra, E. Occhiello and F. Garbassi, Surface characterization of plasma-treated PTFE, Surf. Interf. Anal. 16, 412-419 (1990).
86. J. J. Bikerman, *The Science of Adhesive Joints*, 2nd edition, Academic Press, New York (1968).

87. S. Wettmarshausen, H.-U. Mittmann, G. Kühn, G. Hidde and J. F. Friedrich, Plasmabromination – the selective way to monotype functionalized polymer surfaces, Plasma Proc. Polym. 4, 832-839 (2007).
88. J. Friedrich, S. Wettmarshausen and M. Hennecke, Haloform plasmas applied to polymers, Surf. Coatings Technol. 203, 3647-3655 (2009).
89. W. Langenbeck and W. Pritzkow, Untersuchungen über den Mechanismus der Paraffinoxydation. Teil 1, Fette u. Seifen 55, 435-440 (1953).
90. G. Wietzel, Herstellung synthetischer Fettsäuren durch Oxydation von paraffinischen Kohlenwasserstoffen mit molekularem Sauerstoff, Chem. Eng. Sci. 3, Supplement 1, 17–30 (1954).
91. F. Asinger, Chemie und Technologie der Paraffinkohlenwasserstoffe, Akademie Verlag, Berlin (1956).
92. A. N. Bashkirov and V. Kamzolkin, Synthesis of higher aliphatic alcohols by direct oxidation of paraffinic hydrocarbons, Proc. World Petroleum Congress 4, pp. 175–183 (1959).
93. C. Engler and W. Wild, Ueber die sogenannte Activirung des Sauerstoffs und über Superoxydbildung, Berichte der Deutschen Chemischen Gesellschaft 30, 1669-1681(1897).
94. A. N. Bach, On the role of peroxides in the processes of slow oxidation, Zh. Russ. Fiz.-Khim. Obschestva 29, 373–395 (1897).
95. A. D. Walsh, Processes in the oxidation of hydrocarbon fuels-1, Trans. Faraday Soc. 42, 269-279 (1946).
96. D. G. Knorre, Some elementary stages in the liquid-phase oxidation of hydrocarbons, in: *The Oxidation of Hydrocarbons in the Liquid Phase*, N. N. Emanuel (Ed.), pp. 45-53, Pergamon Press, Oxford, UK (1965).
97. N. Médard, J.-C. Soutif and F. Poncin-Epaillard, CO_2, H_2O, and CO_2/H_2O plasma chemistry for polyethylene surface modification, Langmuir 18, 2246-2253 (2002).
98. H. Yasuda and T. Yasuda, The competitive ablation and polymerization (CAP) principle and the plasma sensitivity of elements in plasma polymerization and treatment, J. Polym. Sci., Part A, Polym. Chem. 38, 943-953 (2000).
99. B. J. Spit, Gas discharge etching as a new approach in electron microscopy research into high polymers, Polymer 4, 109-117 (1963).
100. J. Friedrich and J. Gähde, Untersuchungen zur Plasmaätzung von Polymeren. II. Einfluß der Plasmaparameter auf die Abbaugeschwindigkeit und die Freilegung übermolekularer Polymerstrukturen, Acta Polymerica 31, 52-58 (1980).
101. W. Goldie, *Metallic Coating of Plastics*, Electrochemical Publications Ltd., Middlesex, UK (1968).
102. R. C. Snogren, *Handbook of Surface Preparation*, Palmerton, New York (1974).
103. K. Kato, Formation of 2,4-dinitrophenylhydrazones in polyethylene film treated with chromic acid mixture, J. Appl. Polym. Sci. 18, 3087-3094 (1974).

104. M. Strobel, C. Dunatov, J. M. Strobel, C. S. Lyons, P. Morgen and M. C. Morgen, Low-molecular-weight materials on corona-treated polypropylene, J. Adhesion Sci. Technol. 3, 321-335 (1989).
105. J. Friedrich, J. Gähde and M. Pohl, Untersuchungen zur Plasmaätzung von Polymeren. IV. Selektives Plasmaätzen von Polyethylenen unterschiedlicher Kristallinität, Acta Polymerica 31, 312-315 (1980).
106. L. I. Bezruk and Yu. S. Lipatov, Improved etching in a glow discharge for electron microscopic identification of structure of polyamides. Vysokomol. Soedinen. A10, 1434-1443 (1968).
107. M. R. Sanchis, V. Blanes, M. Blanes, D. Garcia and R. Balart, Surface modification of low density polyethylene (LDPE) film by low pressure O_2 plasma treatment, Eur. Polym. J. 42, 1558-1568 (2006).
108. K. S. Kim, C. M. Ryu, C. S. Park, G. S. Sur and C. E. Park, Investigation of crystallinity effects on the surface of oxygen plasma treated low density polyethylene using X-ray photoelectron spectroscopy, Polymer 44, 6287–6295 (2003).
109. L. Eggert, J. Friedrich, H. Wittrich, M. von Löwis and W. Abraham, Plasmaätzen dünner Schichten organischer Polymere. Teil III. Zum phänomenologischen Mechanismus des Schichtabtrags, Acta Polymerica 41, 569-574 (1990).
110. R. D. Boyd, A. M. Kenwright, J. P. S. Badyal and D. Briggs, Atmospheric nonequilibrium plasma treatment of biaxially oriented polypropylene, Macromolecules 30, 5429-5436 (1997).
111. G. Kühn, St. Weidner, R. Decker, A. Ghode and J. Friedrich, Selective surface functionalization of polyolefins by plasma treatment followed by chemical reduction, Surf. Coatings Technol. 116-119, 796-801 (1999).
112. H. C. Brown, *From Little Acorns to Tall Oaks - From Boranes Through Organoboranes*, Nobel Lecture (December 8, 1979).
113. R. G. Nuzzo and G. Smolinski, Preparation and characterization of functionalized polyethylene surfaces, Macromolecules 17, 1013-1019 (1984).
114. J. Friedrich, R. Mix and G. Kühn, Polymer surface modification with monotype functional groups of different type and density, in: *Plasma Processing and Polymers*, R. d'Agostino, P. Favia, C. Oehr and M. R. Wertheimer (Eds.), pp. 3-21, Wiley-VCH, Weinheim, Germany (2005).
115. G. Kühn, A. Ghode, St. Weidner, I. Retzko, W. E. S. Unger and J. F. Friedrich, Chemically well-defined polymer layers by grafting or pulsed plasma polymerization, in: *Polymer Surface Modification: Relevance to Adhesion, Vol. 2*, K. L. Mittal (Ed.), pp. 45-64, CRC Press, Boca Raton, FL (2000).
116. G. Kühn, I. Retzko, A. Lippitz, W. Unger and J. Friedrich, Homofunctionalized polymer surfaces formed by selective plasma processes, Surf. Coatings Technol. 142-144, 494-500 (2001).
117. D. S. Everhart and C. N. Reilley, Chemical derivatization in electron spectroscopy for chemical analysis of surface functional groups introduced on low-density polyethylene film, Anal. Chem. 53, 665-676 (1981).

118. J. van den Brand, O. Blajiev, P. C. J. Beentjes, H. Terryn and J. H. W. de Wit, Interaction of anhydride and carboxylic acid compounds with aluminum oxide surfaces studied using infrared reflection absorption spectroscopy, Langmuir 20, 6308-6317 (2004).
119. D. L. Allara, C. W. White, R. L. Meek and T. H. Briggs, Mechanism of oxidation at a copper–polyethylene interface. II. Penetration of copper ions in the polyethylene matrix, J. Polym. Sci.: Polym. Chem. Ed. 14, 93-104 (1976).
120. B. M. DeKoven and P. L. Hagans, XPS studies of metal/polymer interfaces — Thin films of Al on polyacrylic acid and polyethylene, Appl. Surf. Sci. 27, 199-213 (1986).

11

Use of Plasma Technology to Improve Polymer Adhesion Part 2 Plasma Polymers as Adhesion Promoters

Jörg Florian Friedrich

Technical University Berlin, Polymer Materials and Technologies, Berlin, Germany

Abstract

There are two ways in which the completely unreactive surfaces of polyolefins can be prepared in such a way that the reaction of incorporated functional groups can generate covalent and thus very strong bonds to polymeric as well as metallic overlayers with high adhesion strength. The direct functionalization of polyolefins in suitable plasmas such as oxygen or bromine vapor plasma is possible (see Part 1) as well as the alternative deposition of ultra-thin plasma polymer layers, which carry correspondingly reactive functional groups. These monosort groups are well suited for forming covalent bonds to overlayers.

The irregular structure of the plasma polymers, which is not comparable with conventional polymers, could be partially or even completely eliminated by appropriate processes. Above all, the ageing resistance of plasma polymers to auto-oxidation has been enhanced.

Keywords: Plasma polymerization, plasma polymers as adhesion promoters, atmospheric plasma, electrospray ionization

11.1 Introduction

There are two variants of adhesion promotion through covalent chemical bonds between thermally evaporated aluminium and polyolefins by using plasma technology in combination with wet chemical processes:

Email: Joerg.florian.friedrich@gmx.de

Anil N. Netravali and Frank M. Etzler (eds.) The Wide World of Adhesion: A Festschrift in Honor of Dr. Kash Mittal on his 80th Birthday, (385–428) © 2026 Scrivener Publishing LLC

a) H-substitution on polyolefin chains by atoms or groups of the plasma gas with the formation of functional groups.
b) Application of adhesion-promoting inorganic layers or plasma polymers that have functional groups.

The plasma-chemically generated groups directly on the polymer surface and their effectiveness as anchor groups for a preferably covalent bond between the metal layer and polyolefin were discussed in Part 1 of this article. Ways have been found to create non-separable laminates. However, it turned out that these synthesis routes were very complicated and involved a great deal of effort.

It is shown that the coating with plasma polymers, which have a high density of monosort functional groups, nearly corresponding to the monomer composition, can also replace the covalent metal-polymer linkage *via* the monosort functional groups located directly on the polymer surface.

The example of the schematic comparison of the oxygen plasma treatment with subsequent diborane reduction of the carbonyl component and grafting of an isocyanate and its hydrolysis to NH_2 with the deposition of a thin plasma polymer layer of allyl alcohol is shown in Figure 11.1.

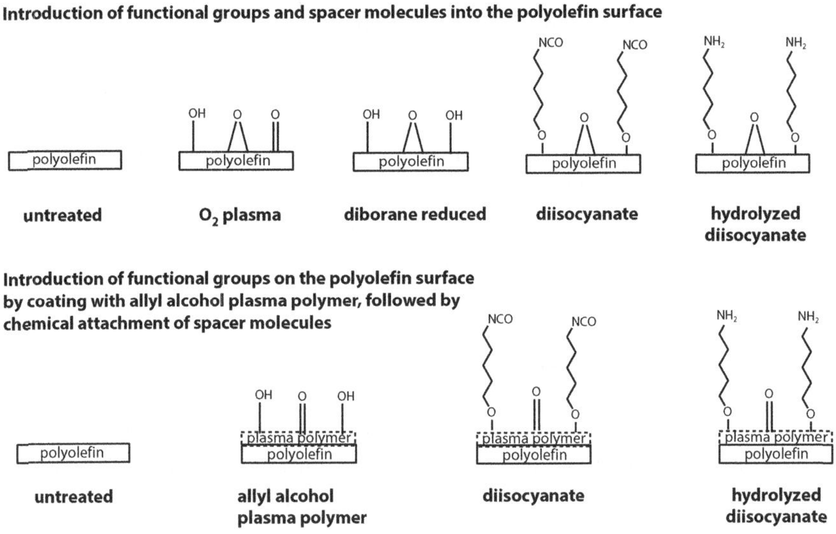

Figure 11.1 Schematic comparison of surface modification through O_2 plasma followed by spacer grafting with deposition of allyl alcohol plasma polymer and spacer grafting of 1,6 hexanediisocyanate.

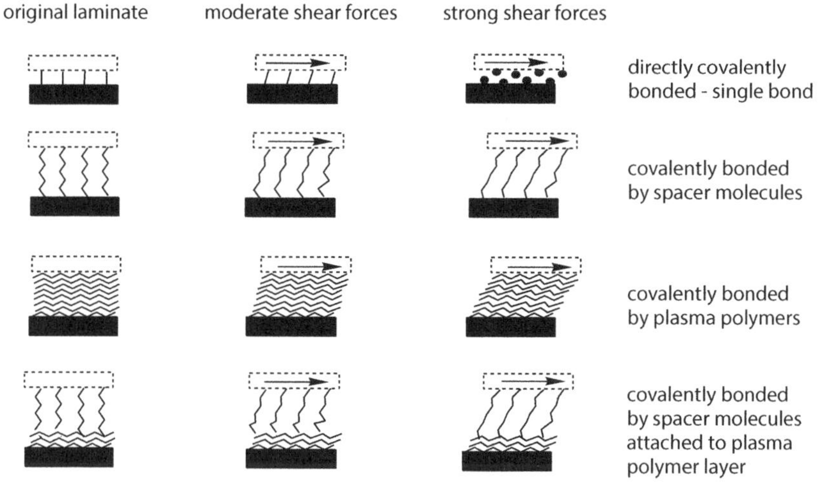

Figure 11.2 Possibilities for chemical bonding of polymers to other polymers or to metals.

The amino endgroup of the spacer can be used to extend the chain or to connect the reactive group to the aluminum layer, e.g. using a silane.

One could initially assume that the plasma polymer layer is flexible and can absorb all material stresses occurring at the metal-polymer interface. However, as will be shown, the plasma polymer layer is flexible only to a limited extent (Figure 11.2). However, it is more or less monosort functional groups which are suitable for attaching flexible spacer molecules, whose tasks were described in detail in Part 1 of this article.

Figure 11.2 shows schematically the behavior of the molecules at the interface when shear forces occur. In the case of direct bonding to the polymer surface, the metal-polymer bond is exposed to the total shear force; in the case of the use of flexible spacers, the shear forces are distributed, even somewhat when using a plasma polymer layer that yields only slightly, and optimally in the case of a plasma polymer layer with covalently bonded flexible spacer molecules.

11.1.1 Comparison of Classical Polymers and Plasma Polymers

14 years ago, the author had described in detail in an article how extremely different so-called plasma polymers were from conventional classic polymers [1]. The current article is a continuation of these earlier considerations [2–5], but now with a new focus on how the properties of plasma

polymers can be improved so that they are comparable to those of conventional polymers and can be used without limitations as adhesion promoters in metal-polymer laminates. Plasma polymers should actually be characterized as a-CH layers in contrast to classic polymers (Figure 11.3). Possible alternatives to plasma polymerization are also discussed below [6]. The existence of classic polymers in the form of entangled macromolecules composed of many repeat subunits was first postulated by Herrmann Staudinger in 1920 [7].

Subsequently, many researchers developed different ideas about the structure and organization of macromolecules in solids as well as in solution. For example, Kanig *et al.* demonstrated some supermolecular structures in semi-crystalline polymers [8].

With the introduction of stereo-regulated polymerization by Kaminsky and coworkers [9] using zirconocene in 1980, they described the catalytic polymerization reactions of mixtures of metallocene dihalides (type 1) with methylaluminoxane (MAO). These Kaminsky catalysts allow the polymerization of ethylene, propylene or other olefins and their mixtures with very high productivity and selectivity [10].

From the work of Staudinger to the groundbreaking discoveries of Kaminsky (who died on November 23, 2024), there were a number of other scientists who made major contributions to research into polymer chemistry. The Nobel Prizes awarded underline the scientific and economic importance of polymer science.

In contrast to such classical polymerization, plasma-chemical polymerization was astonishing in that classical monomers, evaporable, non-polymerizable organic molecules and even inorganic gases could form polymers. The first experiments on plasma polymerization are already listed in [1]. One could demonstrate as early as 1953, the outstanding and

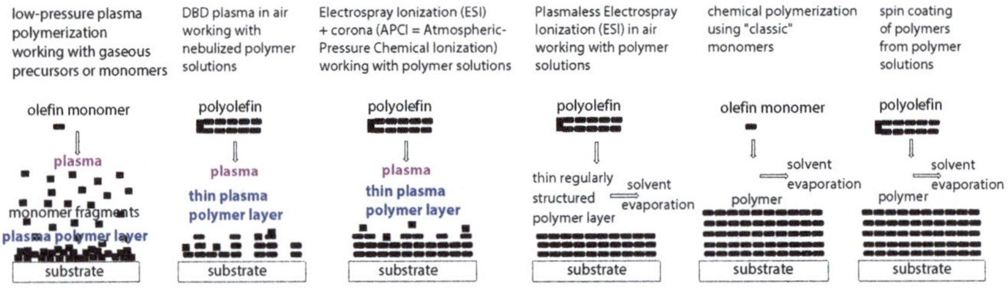

Figure 11.3 Schematic comparison of classical polymerization to long-chain macromolecules composed of identical repeat units with plasma polymerization, which consists of a conglomerate of randomly recombined fragments and atoms that form from precursors or monomers by dissociation and fragmentation when exposed to plasma.

unconventional character of plasma polymerization; the pioneering Miller-Urey experiment showed that inorganic gases in the earth's primal atmosphere were formed under the influence of electrical discharges to form biologically important molecules, which could be continued until the formation of tar-like bio-polymer layers [11, 12]. The Miller-Urey experiment showed very succinctly the advantages, and also the benefits of chemical synthesis with the assistance of electric plasma.

Many, albeit small and simple biochemical building blocks of life, such as amino acids, were synthesized in this experiment. This experiment may be a clue as to how life originated on Earth with the help of plasma. At the same time, however, not only a single building block or sequence of polymer is selectively synthesized, as happens in an organic or biochemical reaction, but dozens or even hundreds of different substances are synthesized. This shows the lack of selectivity of plasma-chemical processes in the organic and biochemical world in contrast to reactions with and on inorganic materials, as is known, for example, from microelectronics. And there are also the undefined tar-like (bio) polymers, which do not fit at all into the world of well-defined classical polymers (Figure 11.4).

Classical polymers, such as polyolefins, ideally have long linear macromolecule chains with only a few short chains (Figure 11.4). In addition, there are various molecular structures, such as linear, branched, comb-shaped and star-shaped polymers and copolymers, crystalline and amorphous regions as well as supermolecular structures such as fringed micelles, spherulites or rods.

Plasma polymers are not actually polymers, but rather a tar or resin-like mixture of a cross-linked polymer backbone of infinite molecular weight with oligomeric or monomeric reaction products of different molecular

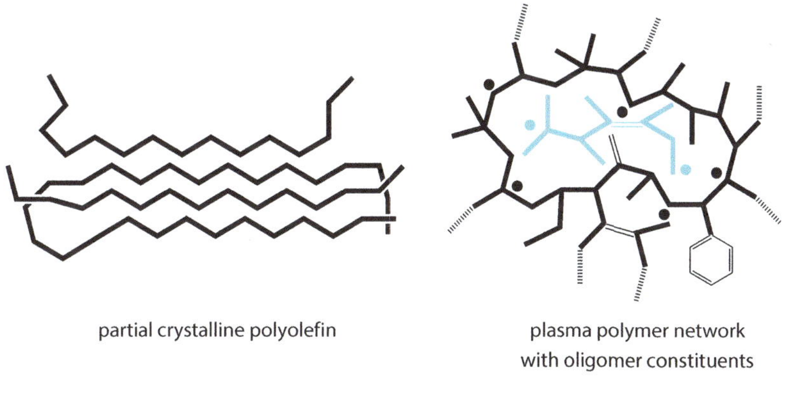

Figure 11.4 Simplified schemes of classical polyolefin and plasma polymer.

weights (Figure 11.4) [13]. Linear polymer chains are not formed. In addition, they have numerous trapped metastable radicals [14], which are not saturated during the polyrecombination process of the plasma-dissociated fragments, but which can also be formed by irradiating the growing polymer with high-energy plasma radiation 20 nm< λ < 200 nm during the deposition process [15].

Yasuda tried to bring order to the many hypotheses of plasma polymerization. He tried in vain to confirm mechanisms such as chain growth polymerization, instead he demonstrated a complete fragmentation of the monomers into atoms and their assembly into a plasma polymer layer [2]. He also recognized the partial degradation of the newly deposited plasma polymer layer and the immediate reintegration of the split-off fragments into a layer by the action of the plasma and its radiation [16]. He established the first approximate formal kinetic laws of polymer deposition [17, 18].

The monomer or organic precursor molecule is immediately dissociated in the plasma, producing fragments [19] and hydrogen [6]. The fragmentation goes so far that Yasuda speaks of "atomic polymerization" [19]. Friedrich describes the layer formation as polyrecombination of atoms and fragments, which leads to a hydrogen-poor a-CH layer [1]. This layer consists of a porous network in which oligomeric molecules are also enclosed (Figure 11.4) [6, 13, 20].

Figure 11.5 gives a schematic overview of the plasma polymerization with extensive dissociation of the ethylene monomer shown here and the random recombination of these radical fragments and atoms to form an undefined conglomerate with many unsaturated and metastable trapped radicals. This resulting plasma polymer with entrapped monomers and

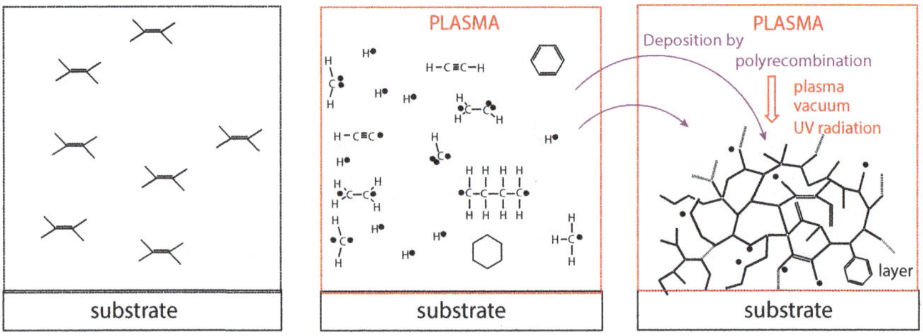

Figure 11.5 Schematics of plasma polymerization of ethylene on a molecular level characterized by fragmentation and polyrecombination of atoms and fragments to an irregular network with trapped radicals.

oligomers as a skeleton, which is formed from an infinite carbon network, bears no resemblance to conventional polymers and can be regarded as amorphous hydrogen-poor carbon (a-C:H).

11.1.2 Are Irregularly Structured Plasma Polymers Suitable as Adhesion Promoters in Metal-Polymer Laminates?

Chemically irregularly structured plasma polymers, which, as already explained, consist of a randomly structured carbon network with embedded low-molecular undefined components and oils (see Figure 11.4) [13]. Such amorphous carbon with a low hydrogen content has good adhesion-promoting property as well as some disadvantages, such as low thermal, mechanical and oxidative resistance.

11.1.2.1 Limitations of Plasma Polymers for Using as Adhesion Promoters

Some important properties of plasma polymers influence their use as adhesion promoter layers in metal-polyolefin laminates:

1. The structure consists of a carbon network with enclosed monomers and low-molecular oil-like components (cf. Figure 11.4) [13, 21], i.e. the structure of no plasma polymer corresponds to that of a classical polymer.
2. Due to the extremely broad molecular weight distribution of the individual components that make up the plasma polymer, especially the low molecular weight and network fractions, thermal and mechanical instability occurs (see Figure 11.5) [22, 23]. This is a major disadvantage for many applications. Mechanical and thermal stresses at the interface between the metal-plasma polymer layer and the polyolefin substrate cannot be relieved as required.
3. High radical content, which is associated with long-term oxidation when stored in air (auto-oxidation *via* peroxide formation) [24–28], this means the layer can strongly and rapidly age; however, radicals can react with the polyolefin surface, form a chemical bond and thus make a small contribution to adhesion.
4. The plasma polymers have different types of unsaturations in high concentration, partly conjugated polyene structures,

which is why the polymer layers are yellow to dark brown in color. The unsaturations can also react with atmospheric oxygen in the long term and thus can impair adhesion [6, 29].

5. Unsaturation and a high degree of crosslinking lead to a high hydrogen deficit in the plasma polymer [2], which can therefore be described as amorphous carbon with residual hydrogen content (a-CH-). In particular, a-CH- formation occurs at high coupled power into the discharge and at low pressure.
6. The hydrogen released during plasma polymerization dominates the plasma as a "quasi-hydrogen plasma" and makes the plasma more energetic [6].
7. Coatings produced at low energy input and high pressure have an increased proportion of low-molecular oil-like components, as a kind of "plasticizers" for the plasma polymer [13], however, the mechanical strength of such soft coatings is not sufficient to produce a high-strength bond with the polyolefin substrate. Furthermore, such low-molecular components can be leached out by the use of solvents during subsequent chemical syntheses, for example when flexible spacer molecules are to be covalently bonded.
8. The structure of the plasma polymers has little or no similarity to the structure of the monomeric building blocks, this leads to poor mechanical strength and loss of functional groups [2, 5].
9. Sequences in plasma polymers whose structures indicate a chain-growth polymerization, i.e. a regular radical polymerization of olefinic, vinyl, acrylic or diene-containing or ring-opening monomers, are limited to a maximum of 2 monomer units. Longer linear chain segments have not yet been found [24], which confirms the irregular structure of the plasma polymers.
10. During deposition, the plasma polymer layer is additionally modified by the short-wavelength UV radiation from the plasma, in particular hydrogen is split off [30], which results in additional ageing and deterioration of mechanical properties.
11. Functional, e.g. adhesion-promoting, groups are only partially retained, which reduces the number of possible adhesion-promoting interactions [31].

12. The ageing resistance of the layers is limited, which reduces the service life and reliability of a laminate.
13. The layer structure is characterized by agglomerates (cauliflower structure), a rough surface structure enables stronger mechanical anchoring of all possible surface layers and increases the number of interaction possibilities.
14. The layer structure changes greatly when the plasma parameters are varied, that is why the properties of the plasma polymer layer can be controlled to a certain extent by the choice of plasma parameters [2, 32].
15. The exact layer structure and composition is difficult to reproduce experimentally. The layer compositions vary somewhat from experiment to experiment despite setting the same plasma parameters.

All the disadvantages of plasma polymers compared to conventionally produced polymers are reflected particularly characteristically in the comparison of their thermograms. Widespread technical application of plasma polymers is hampered by their poor thermal stability, as shown by thermo-gravimetry analysis (TGA) (Figure 11.6) [1].

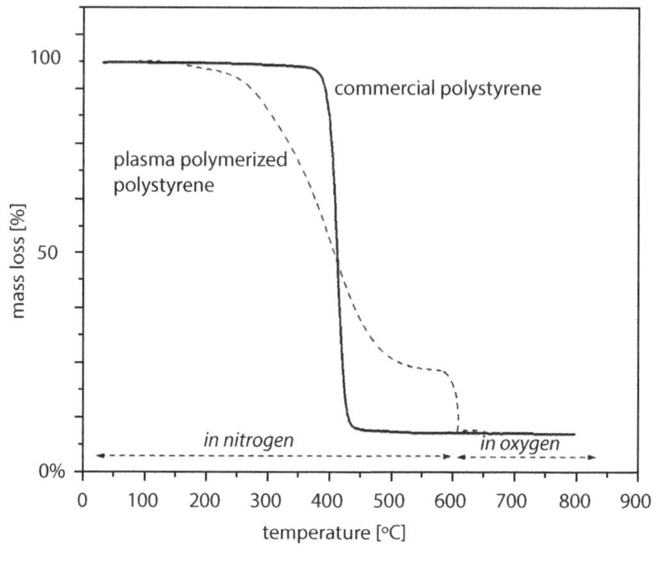

Figure 11.6 Thermal gravimetric analysis (TGA) of commercial polystyrene and low-pressure plasma polymerized (300 W, 10 Pa) polystyrene.

Even at low temperatures (approx. 100°C), the plasma polymer shows mass loss due to trapped monomer and oligomer molecules. At higher temperatures, degradation takes place more slowly. A carbon network is present, which can only be degraded (burnt) in an oxygen atmosphere.

If there are so many disadvantageous properties of plasma polymers, why are they still used frequently and, above all, successfully? What are the key advantages of plasma polymers as adhesion promoters?

11.1.2.2 Advantages of Plasma Polymers

After such a long list of drawbacks, one might have the impression that plasma polymers are worthless and have no significance whatsoever. Nevertheless, plasma polymer layers do have a few important advantages for adhesion promotion:

1. Since a predominant proportion of functional groups of the monomer is often retained in the plasma polymer layer, therefore chemical reactions can be carried out with this dominant group ("predominantly monosort") [22, 33, 34] after plasma treatment with the aim of creating covalent bonds between the substrate and the grafted material.
2. They can be applied very thinly (≤10 nm) [21].
3. The layers are hole-free even for a thickness of less than 10 nm.
4. The coating composition can be varied over a wide range by selecting the appropriate monomers, precursors and gases and can therefore be tailored to requirements [35].
5. During the deposition of plasma polymers, the substrate surface is also briefly exposed to the plasma and activated at the beginning of the deposition. After beginning of deposition the coating then prevents interaction between the plasma and substrate surface. As a result, plasma polymers often adhere quite well to various substrate surfaces.
6. The inherent reactivity of plasma polymers, the tendency for auto-oxidation caused by the reaction of trapped radicals with atmospheric oxygen, can be prevented easily by heating (radical recombination) or by quenching with permanent radicals (NO fumigation) or by exposure to bromine vapors (Br_2) [21].
7. The reaction of the trapped radicals with atmospheric oxygen to form peroxy groups and hydroperoxides can also

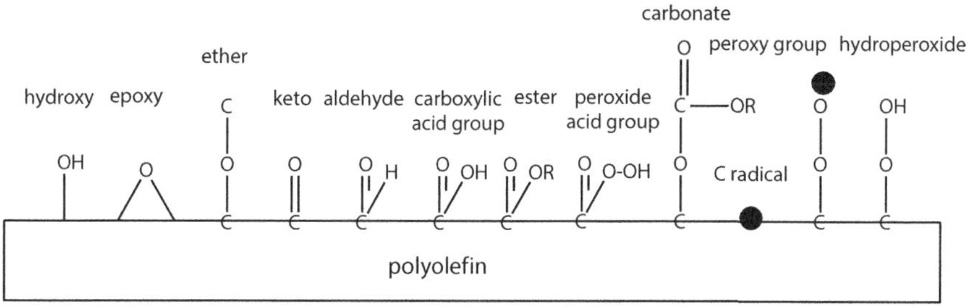

Figure 11.7 Functional groups that can be formed by autooxidation on polyolefin surfaces.

be used to create a polymer layer covalently bonded to the substrate. A polymer layer can be grafted onto the resulting alkyl radicals by thermally or radiochemically induced decomposition of the hydroperoxy group ("grafting from"). The decay of the intermediately formed hydroperoxides (R-O-OH→R-O• + •OH) produces a variety of O functional groups (Figure 11.7) [36].

11.1.2.3 *What is the Reason for the Totally Different Structure of Plasma Polymers in Comparison to Classical Polymers?*

Selectivity of chemical reactions and, therefore, also of polymerization in the plasma is impossible because of the permanent energy supply and the energy excess in comparison to needed energies for initiation of polymer formation. The following factors are responsible for the emergence of disorganized, random structures:

1. Continuous (inexhaustible) supply of electrical energy (Figure 11.8).
2. High-energy tail of electron energy distribution (E>4 eV).
3. Bond dissociation energies (3-4 eV) are much smaller than the electron energy of the high-energy tail [5].
4. Ionization energies are higher than bond dissociation energies [37].
5. Permanent ion bombardment from plasma [38].
6. Permanent short-wave UV-photon irradiation [15].
7. Hydrogen enrichment in plasma phase increases electron temperature and therefore the aggressivity of plasma; the

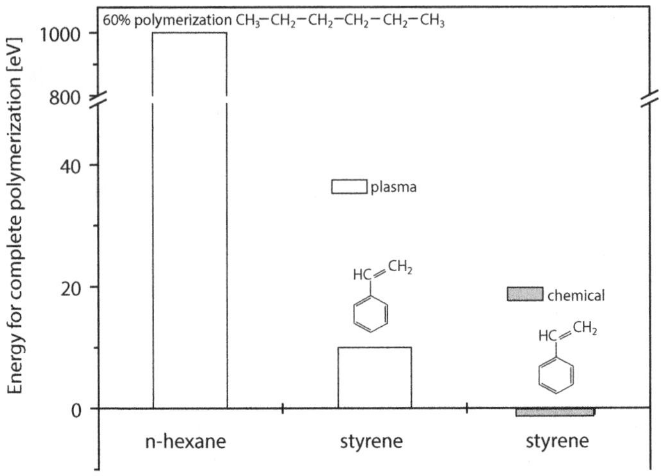

Figure 11.8 Illustration of the exorbitant energy input required for 1 styrene or 1 molecule of n-hexane for plasma polymerization compared to the energy released during chemical polymerization.

released hydrogen accumulates in the plasma phase, where it even forms "quasi-hydrogen plasma" [6].

8. Partial degradation of the deposited plasma polymer layer to gaseous degradation products and their reincorporation into the growing plasma polymer layer takes place continuously, which was first observed by Friedrich [4] and described by Yasuda [16] as the CAP mechanism (competing ablation and polymerization). This means that polymer degradation products, such as acetylene, ethylene, methane, etc., are knocked out by plasma action on the deposited plasma polymer layer, which are redeposited as plasma polymers by the same plasma at the same or a different location.

11.1.2.4 Options for Eliminating the Disadvantages of Plasma Polymers

The obvious thing to do is to eliminate the predominantly negative features of plasma polymers through various measures and to further improve the existing advantages, such as largely monosort functionality. The following changes were made to the deposition process:

1) Lowering the wattage

In the 1970s, Yasuda discovered that the wattage of the plasma had a major influence on the structure of the deposited polymer [39]. The higher the wattage, the more cross-linked the plasma polymer; the lower the wattage, the more tar-like or oil-like low-molecular products are present in the deposited plasma polymers. Based on these findings, Yasuda developed a simple formula (the so-called "Yasuda factor"-YF=W/MF, with W=Wattage in J/s, F=monomer flow rate in mole/s, M= molecular weight of monomer in kg/mole) [2]. The hope was that low input power (W/MF) would result in chain-growth polymerization to form at least partially classical polymer sequences [2, 13, 40, 41]. In fact, the polymerization of allyl alcohol (Figure 11.9) shows that a large proportion of the OH groups can actually be found in the plasma polymer layer at extremely low wattages. However, it must be taken into account that the auto-oxidation of the plasma polymer in air results in the secondary formation of new OH groups, which are added to the OH groups that originate from the monomer allyl alcohol and have survived the plasma polymerization process: $OH_{measured}=OH_{survived} + OH_{auto\text{-}oxidation}$. In addition, the polymer layers produced at very low plasma powers are really greasy in nature and do not have a solid consistency. Figure 11.9

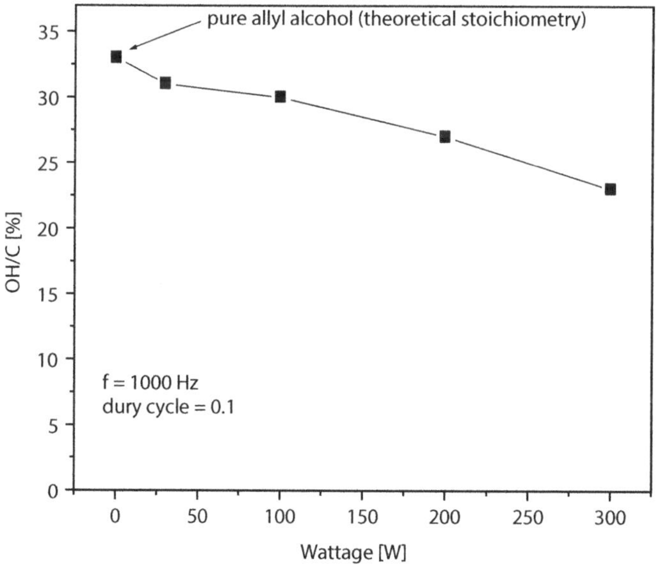

Figure 11.9 Preservation of the OH groups during polymerization in the pulsed plasma of allyl alcohol as a function of the plasma power fed in under otherwise identical plasma conditions.

shows the dependence of the retention of OH groups in the plasma polymerization of allyl alcohol. At the lowest power input into the plasma, the retention of hydroxyl groups, which is important for adhesion, is highest. However, at this low wattage, the cohesive strength of the (greasy) plasma polymer layer is very low. A compromise must therefore be found between maximum preservation of OH functionality and sufficient strength of the plasma deposited layer.

2) Advantage of pulsing the plasma

The idea was to initiate a classic polymerization with short plasma pulses and to allow it to continue as a classical chain reaction in the subsequent long plasma-off periods. This should produce predominantly regular classical structures in the plasma polymer layer.

However, the repeated destruction of the structure of the newly deposited polymer layer with each new plasma pulse by particle bombardment and plasma radiation hardly disturbed the regularity of the structure of the deposited plasma polymer layers, especially since each new plasma pulse required a greatly increased voltage to ignite the discharge, which produced a particularly high-energy plasma for a short time [42–44]. Another reason for the irregular structures of the plasma polymers is the immediate termination of the growing polymer chain, because in the low-pressure plasma not enough new monomer molecules can be bound to the reaction center (radical) in time to continue the chain growth. With each plasma pulse only one monomer molecule can be attached here and there, so that layer growth under these conditions of extremely short kinetic chain lengths requires frequent re-starting of the polymerization by new plasma pulses.

3) Pressure-pulsed plasma

An electrically and pressure-pulsed plasma was developed. In order to ignite a new plasma pulse, the pressure must be low. At the end of the plasma pulse to the beginning of the plasmaless period, the pressure in the reactor must be immediately increased to such an extent that purely chemical chain propagation to linear products is possible.

The reactor was then evacuated again and the next plasma pulse was used to initiate new chain growth (Figure 11.10) [45].

4) Gassing of plasma polymers with NO and Br_2

Numerous radicals are contained in the plasma-polymerized styrene [14], as shown by the second heating of the plasma polymer during the dielectric relaxation recording. These radicals react with atmospheric oxygen up

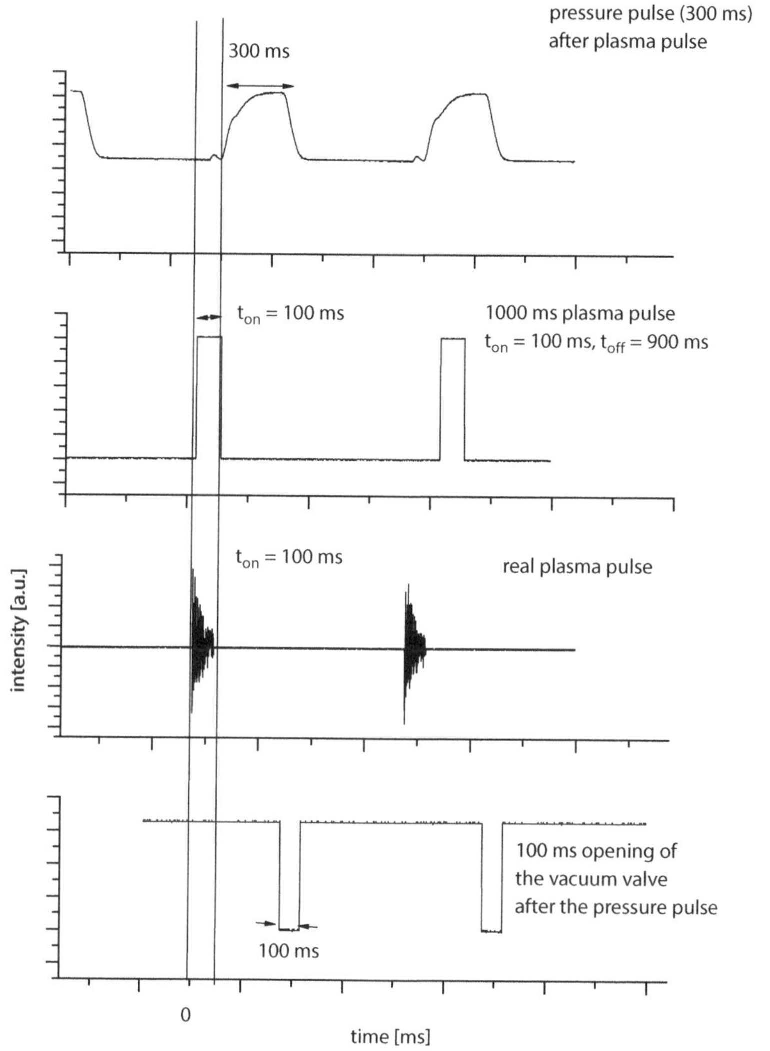

Figure 11.10 Schematics of pressure-pulsed plasma (real pulse=actually measured current).

to 20% O/C, as shown in the example of the plasma polymerization of allylamine [41, 46, 47].

An appreciable amount of radicals cannot be quenched for steric reasons and survive the deposition process. The lifetime of the trapped radicals can be days, weeks and even months. However, if the layer comes into contact with air and thus with oxygen, which can slowly diffuse into the

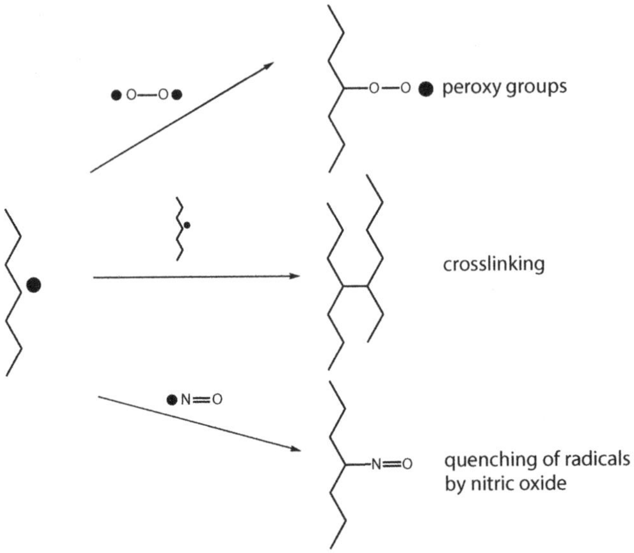

Figure 11.11 Reactions of radicals with atmospheric oxygen, with neighboring radicals and with nitric oxide when it is added for quenching.

plasma polymer due to its small dimensions of molecular size, then these trapped radicals react with oxygen to form peroxy radicals and hydroperoxides (auto-oxidation): C• + •O-O• → C-O-O• (Figure 11.11).

The auto-oxidation leads to a rapid ageing of the plasma polymer layer. The addition of gases with a radical character (NO) or easily radical-forming vapors (Br_2), which can easily diffuse into the plasma polymer, where they recombine with C-radicals to form stable compounds, can stop auto-oxidation of plasma polymers (Figure 11.12) [48, 49]:

$$R\bullet + \bullet NO \rightarrow R\text{-}NO \text{ and } R\bullet + Br_2 \rightarrow R\text{-}Br + \bullet Br$$

5) Use of polymers as precursors instead of monomers for deposition of ultra-thin adhesion-promoting plasma polymer layers

The idea was to feed polymers instead of monomers into the plasma. This was to ensure that the plasma polymer layer did not become an irregularly structured mixture due to random recombination of small monomer fragments and atoms. It was hoped that larger, regularly structured segments of the injected polymer would survive the plasma deposition undisturbed at very low energy input and thus give the layer a more regular structure.

The problem here was and is to transfer the polymer molecules undecomposed into the gas or plasma phase. The simplest solution to this

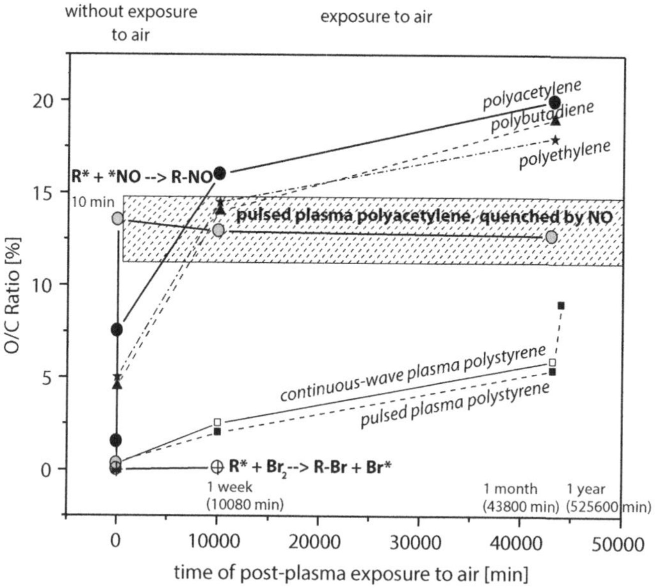

Figure 11.12 Total elimination of oxygen uptake and thus the auto-oxidation of plasma polymers by gassing with NO or bromine vapor.

problem was to spray a polymer solution into the discharge using a nebulizer. However, this creates large droplets of solution, which are the equivalent of painting a car. Initially, the plasma encounters an excess of solvent molecules as fine droplets produced by the nebulizer. The deposited polymer layers are then also modified by plasma until the structure of the macromolecules is destroyed (Figure 11.13).

The Atmospheric-Pressure Chemical-Ionization (APCI) is a process that generates extra-small droplets, which are then activated and modified in the plasma. A corona discharge is then applied to the droplet mist, the solvent evaporates and individual macromolecule ions are left behind by charge transfer from the solvent to the macromolecule (Figure 11.14) [50, 51].

This ESI principle (see next Section) is also the basis for the APCI process, in which the corona plasma is used to support the ionization of the macromolecules. As has been shown, the additional application of a corona plasma to the ESI spray to enable the ionization of macromolecules that are difficult to ionize also leads to massive degradation of the polymers (Figure 11.14).

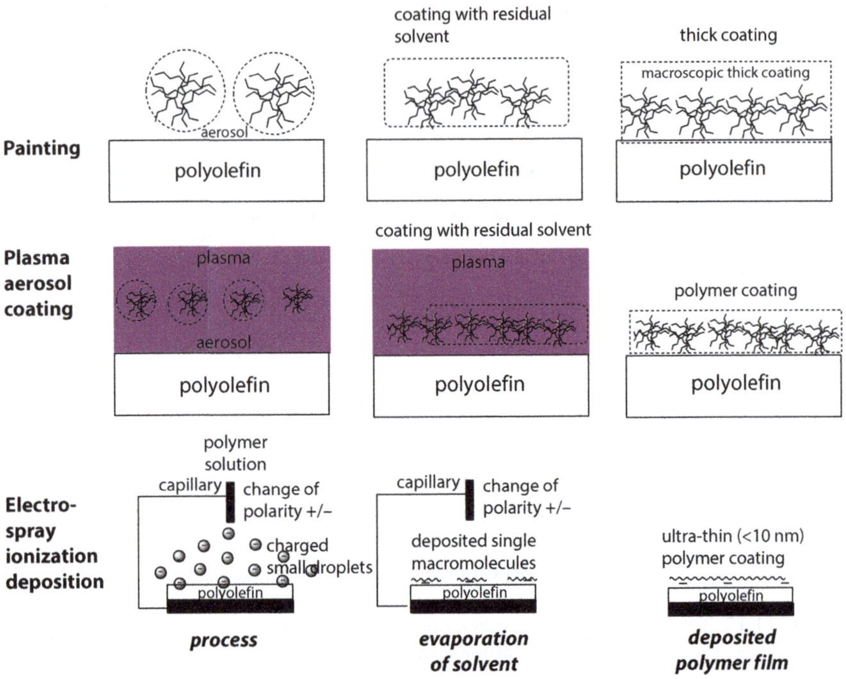

Figure 11.13 Schematic comparison of the droplet dimensions for ordinary paint processes, plasma-assisted deposition of polymers by means of aerosol and electrospray ionization (ESI) deposition of individual macromolecules.

6) Complete preservation of the classical structure in the thin adhesion-promoting polymer layer by plasmaless deposition using electrospray ionization

Here, the droplet mist is generated by the Electrospray Ionization (ESI) process. In a high-voltage field, a mist of the polymer solution is sucked from a capillary, which is also an electrode. The small droplets become charged in the high voltage field, solvent evaporates on the way to the counter electrode, the charges at the surface of droplets come so close together that repulsions occur (Coulomb fissions or Coulomb explosions) until the droplets have lost their solvent due to the ever-increasing surface area during atomization and individual macromolecular ions can be deposited as an ultra-thin polymer layer after their discharge. This basic principle works completely without plasma and, therefore, without any polymer degradation and fragmentation (Figure 11.15) [52–55].

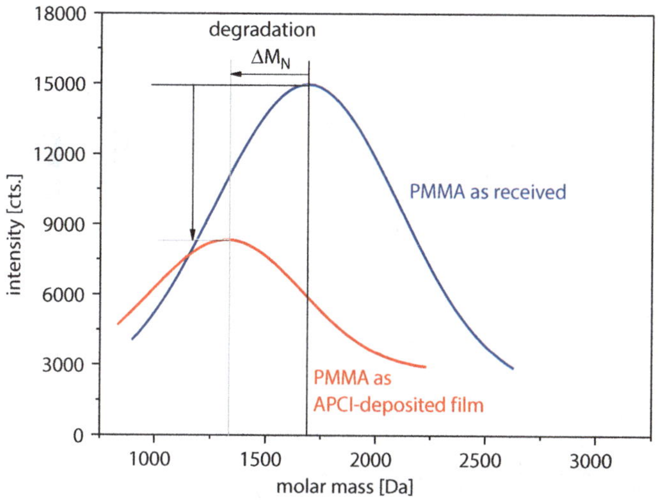

Figure 11.14 Shift in the molecular weight distribution of poly(methyl methacrylate) (PMMA) with an average molecular weight of 1700 g/mol to significantly lower values and reduction in the concentration of PMMA molecules measurable with MALDI-ToF-MS.

7) Conclusions on optimal adhesion promotion through thin layers

Summarizing the results of the various methods for adjusting the structure of plasma polymers in terms of a more regular, more classical structure, there is no plasma-associated method that ensures all the adhesion-promoting functional groups of the monomer and the classical regular structure during polymer formation. Even the injection of complete polymers instead of a large number of monomer molecules in the hope of "survival" of larger defined polymer segments in the plasma polymer layer did not lead to the detection of longer regularly structured sequences in the layers (Figure 11.16).

Only a complete omission of any plasma with the introduction of the ESI process made it possible to realize a regular layer structure and preservation of all functional groups. The C1s peaks of casted commercial poly(acrylic acid) and that of ESI-deposited are completely identical (Figure 11.17) [55–58].

These ultra-thin ESI-deposited layers showed surprisingly good adhesion to the polyolefins. Oxygen plasma treatment of the polyolefin substrate prior to ESI deposition further improved the adhesion strength.

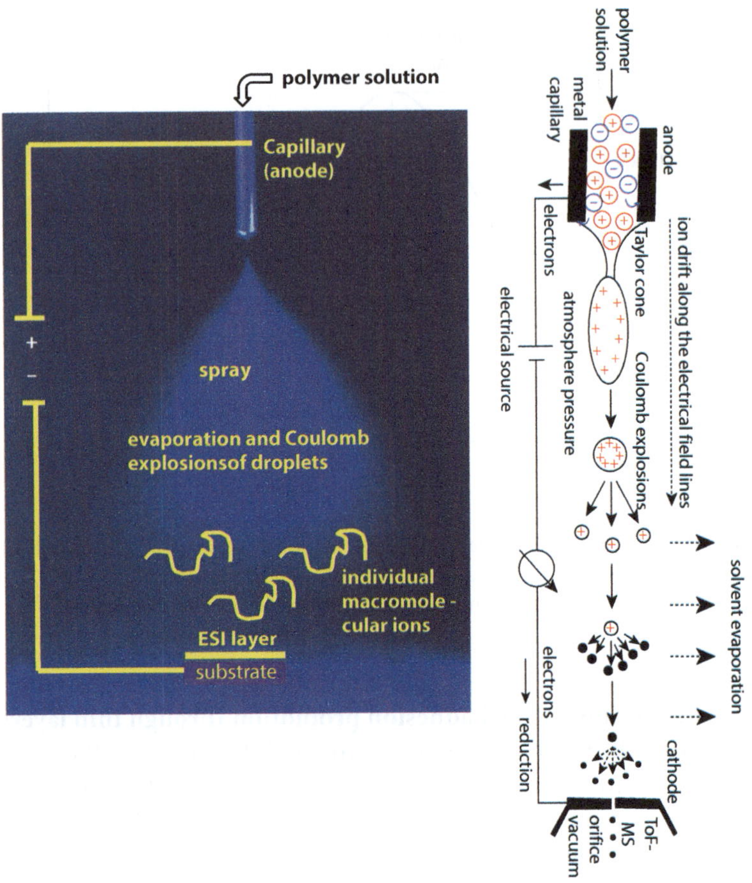

Figure 11.15 Left: Photograph of electrospray with capillary, electrode, substrate and schematically depicted macromolecular ions deposited as an ESI layer Right: Schematic mechanism of spray with droplet ionization, solvent evaporation, Coulomb fissions or explosions and transport to the orifice of a time-of-flight mass spectrometer.

The concentration of the (monosort) functional groups in the plasma polymer, which is determined by the functional group of the monomer (or the polymer precursor), is decisive for the use of the plasma polymer layers as adhesion promoters. Many of these functional groups of the monomer are converted into other functional groups during plasma polymerization.

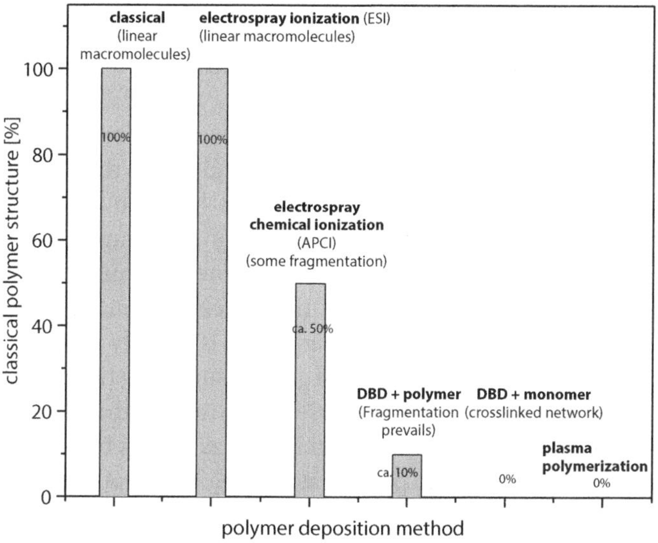

Figure 11.16 Percentage of classic linear polymer segments in the layer estimated after various analyses of the respective plasma polymer layers, which is subject to a high degree of uncertainty.

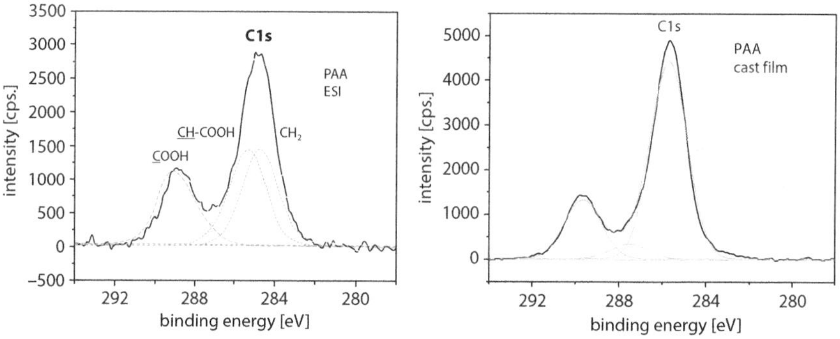

Figure 11.17 C1s peaks in the X-ray photoelectron spectra (XPS) of commercial poly(acrylic acid) (PAA) and ESI-deposited poly(acrylic acid) (PAA) showing no differences.

11.2 Possible Uses of Plasma Polymers Despite Their Chemically Irregular Structure

When using plasma polymers, especially for adhesion promotion, the different initial forms of plasma polymers and other thin films must be compared. A distinction must be made here between irregularly structured plasma polymers with adhesion-promoting groups and chemically regular layers with a maximum, i.e. stoichiometric number of adhesion-promoting groups. The difference in the concentration of adhesion-promoting groups between the two layer types is usually not much (Table 11.1). The lower number of functional groups surviving the plasma process is possibly compensated by an increased reactivity of the layer of oxygen-containing groups formed by auto-oxidation, as well as by an activation of the polyolefin surface by the plasma at the beginning of the plasma polymer deposition.

11.2.1 Broad Variety of Applications

After their discovery in 1960 [35], plasma polymers were initially tested as membrane coatings for technical applications [2]. Another possible

Table 11.1 Concentrations of functional groups in ultra-thin polymer layers deposited by plasma polymerization or by electrospray-ionization (ESI) deposition [22].

Functional group of monomer forming plasma polymer	Functional group of injected polymer	Concentration of functional groups in plasma polymers	Concentration of functional groups in ESI layers
Allylamine NH_2	Poly(allylamine) NH_2	40-55%	100%
Allyl alcohol OH	Poly(allyl alcohol) OH	80-90%	100%
Acrylic acid COOH	Poly(acrylic acid) COOH	60-70%	100%
Allylbromide Br	Plasma-polymerized allylbromide Br	70%	not yet sprayed as classical polymer

application should be mentioned as anti-corrosive coatings; however, they did not prove successful on steel and other metals [59]. A few further examples should be mentioned. Siloxane-containing coatings are required as barrier layers and are used industrially [60–62].

Layers of hexamethyldisiloxane (HMDSO) deposited in the presence of a stoichiometric excess of oxygen resulted in SiO_2-like layers with flame-retardant property [63].

11.2.2 Which Functional Groups in Plasma Polymers or Other Ultra-Thin Polymer Layers are Important for Adhesion Promotion?

For the selection of adhesion-promoting plasma polymers, the reactivity to aluminum or the reactivity to spacer molecules to be grafted, which then form bonds to aluminum with corresponding end groups or silane modification, plays a role.

A distinction must be made between 2 types of functional groups:

a) groups that can form covalent and thus strong bonds with aluminum, such as OH, COOH and Si-OH, and
b) groups to which covalent spacer molecules and adhesion-promoting groups can be grafted, such as Br, NH_2, NCO, epoxy or CHO (Figure 11.18).

Silanes such as vinylsilane or functional silanes such as aminosilane, mercaptosilane or epoxysilane, which can also have an adhesion-enhancing effect as a plasma polymer layer or as an endgroup on spacers, are also used.

11.2.3 Reactions of Functional Groups

OH and COOH groups react to form Al-O-C mono- or bidentate compounds or with silanes to form Si-O-C bonds. Amino groups do not bond to aluminum, but bond to copper as a complex [64, 65].

Primary amino groups are excellently suited to react spontaneously with bromine, aldehyde, acid chloride, isocyanate and carboxyl groups of spacer molecules [66]. Diols, diamines, dialdehydes or glycols are particularly suitable as spacer molecules, up to poly(ethylene glycol)s of higher chain length [67, 68].

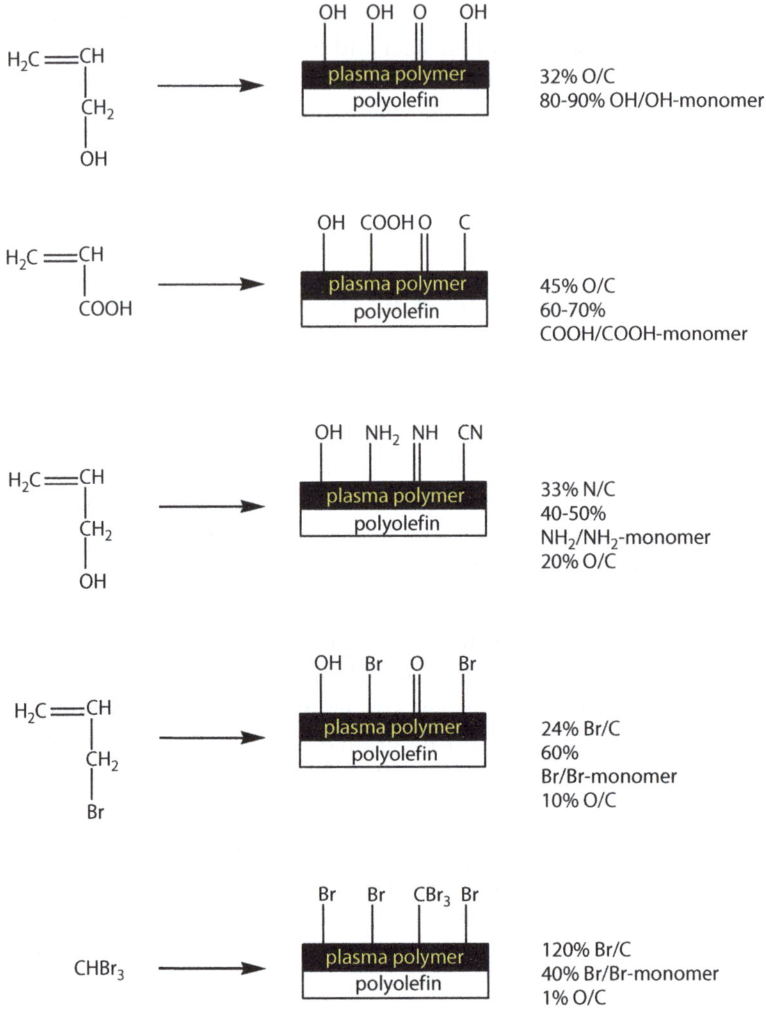

Figure 11.18 Schematics of selected functional groups on surfaces of plasma polymers of different precursors and their concentrations, yields and auto-oxidation, which were often used as adhesion promoters.

Plasma polymer coatings with monosort or dominance of monosort adhesion-promoting functional groups (-OH, -COOH, -NH_2, -Br) have been tested many times [44]. They proved successful in solving adhesion problems in laminate materials in some cases. Plasma polymer layers with a high content of functional groups (■-X) are well suited for chemical grafting reactions with functional groups of molecules or polymers (~) at high density such as [31, 66]:

■-OH + ONC~ → ■-O-CO-NH~

■-COOH + HO~ → ■-CO-O~

■-NH_2 + OHC~ → ■-N=CH~

■-Br + NH_2~ → ■-NH~

The incorporation of flexible longer-chain aliphatic graft molecules and the simultaneous formation of diffusion barrier layers at the interface of metal-polyolefin laminates led to highly adhesive laminate materials, such as referred in [69, 70]:

■-OH + ONC~~~CNO → ■-O-CO-NH~~~CNO → ■-O-CO-NH~~~NH_2

■-COOH + HO~~~OH → ■-CO-O~~~OH

■-OH + HOOC~~~COOH → ■-CO-O~~~OH

■-NH_2 + OHC~~~CHO → ■-N=CH~~~CHO → ■-N=CH~~~C=N-~~~NH_2

■-NH_2 + ONC~~~CNO → ■-NH-CO-NH~~~CNO → ■-NH-CO-NH~~~NH_2

■-Br + NH_2~~~NH_2 → ■-NH~~~NH_2

■-Br + HO~~~OH → ■-O~~~OH

Additional insertion of moisture and ion barrier function at the interface produces highly adhering, non-separable metal-polymer laminates [65, 71]:

■-NH_2 + OHC-(CH_2)-CHO → ■-N=CH-$(CH_2)_3$-CHO

■-N=CH-$(CH_2)_3$-CHO + H_2N-$(CH_2)_3$-Si-$(OCH_3)_3$→

■-N=CH-$(CH_2)_3$-CH=N-$(CH_2)_3$-Si-$(OCH_3)_3$

■-N=CH-$(CH_2)_3$-CH=N-$(CH_2)_3$-Si-$(OCH_3)_3$ + HO-Al< →

■-N=CH-$(CH_2)_3$-CH=N-$(CH_2)_3$-Si-$(OCH_3)_2$-O-Al< or

■-Br + NH_2-$(CH_2)_3$-Si-$(OCH_3)_3$ → ■-NH-$(CH_2)_3$-Si-$(OCH_3)_2$-O-Al<

11.2.4 Variation of Group Density through Copolymerization

For adhesion promotion, one usually tries to create as many functional (monosort) groups as possible on the polyolefin surface. However, it has been shown that a much lower number of adhesion-promoting COOH groups than the maximum concentration that can be generated is sufficient to achieve maximum metal-polyolefin adhesion. Maximum concentrations of OH and COOH groups lead to internal hydrogen bonding and thus to the unwanted weakening of the metal-polymer chemical bonds.

The variation in group density during plasma polymerization is achieved by plasma-initiated, more or less chemically dominated copolymerization. A monomer that carries the desired functional group is copolymerized with a "neutral" co-monomer that does not carry a functional group. This is a copolymerization governed by chemical copolymerization laws. The chemically determined copolymerization of two co-monomers is superimposed by the fragmenting-recombining plasma polymerization. While plasma polymerization produces randomly composed products, chemical copolymerization produces classical, chemically precisely defined copolymer structures. These components can still be seen in some structures of such plasma-initiated copolymers produced. The copolymerization capacity of the individual comonomers and the deposition rate in plasma-initiated copolymerization also show some similarities to chemical copolymerization laws [72]. When the ratio of the initial concentrations of the copolymers is varied, there is no linear relationship between the initial concentration ratio and the copolymer composition (Figure 11.19). Depending on the type of comonomers, homopolymerization of a

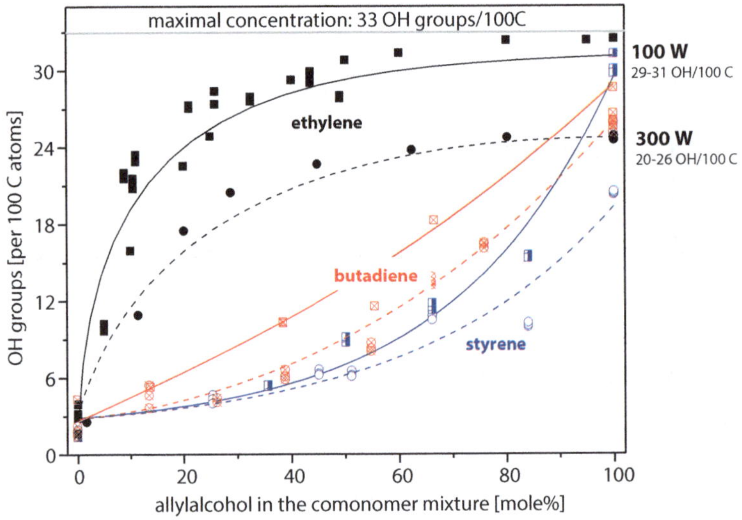

Figure 11.19 Plasma-initiated copolymerization of allyl alcohol with styrene, 1,4-butadiene and ethylene using two different wattages and the resulting OH concentration in dependence on different ratios in comonomer mixture.

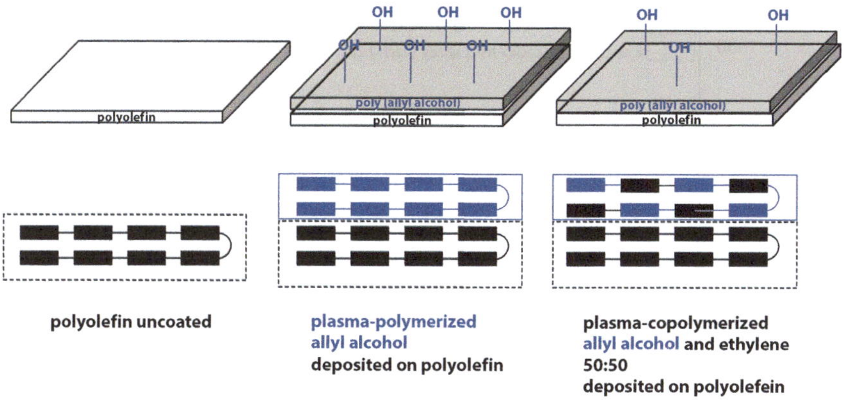

Figure 11.20 Schematic of plasma-initiated 50:50 copolymer formed from allyl alcohol and ethylene.

comonomer dominates at certain concentration ratios or homopolymers are generally formed. Only if the so-called "copolymerization factors" are considered, true copolymers are formed in a pure chemical way [73].

To illustrate the plasma-initiated copolymerization, a schematic and thus highly idealized image of such a 50:50 copolymer is shown (Figure 11.20).

11.3 Efficiency of Plasma Polymers as Adhesion Promoters

11.3.1 Aluminium Bonding to Functional Groups of the Plasma Polymer Layer

While in Part 1 "Use of Plasma Technology to Improve Polymer Adhesion" the polyolefin surface was modified in plasma with polysort and monosort functional groups, the Part 2 also compares the results of adhesion strength measurements of aluminum-polyolefin laminates by adhesion promotion due to plasma polymer layers (Figure 11.21).

All precursors were injected into the atmospheric plasma as low molecular weight substances or as polymer solution sprayed as aerosol, then deposited as plasma polymer and coated by thermally evaporated 150 nm thick aluminum layers. Low-pressure plasma polymerization used only evaporated monomers.

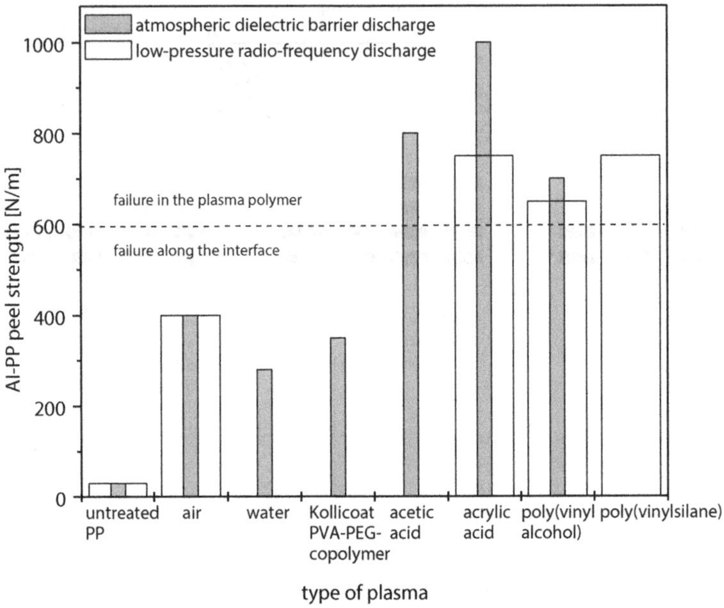

Figure 11.21 Peel strength of aluminium from polypropylene covered with plasma polymers as adhesion promoter layers. The plasma polymers were deposited with a dielectric barrier (DBD) under atmospheric pressure or in vacuum using RF discharge.

The few comparable measured values for atmospheric and low-pressure plasma treatment show that the peel strength is approximately the same for low-pressure plasma treatment as for polymer aerosols and gas or vapor, depending on the monomer or gas. It therefore appears that the type of plasma, low or atmospheric-pressure, does not play a role.

At higher bond strengths, mechanical separation no longer occurs at the metal-polymer interface, as is the case at lower peel strengths, but in the plasma polymer, which then represents the locus of failure.

Figure 11.2 shows different variants of bonding. It was originally expected that plasma polymers would be able to adequately compensate for mechanical stresses occurring at the metal-polymer interface. However, a more detailed characterization of the structure of the plasma polymers showed that they were highly cross-linked inflexible networks (see Figure 11.4) if appropriate mechanical and, above all, thermal stability is required.

11.3.2 Aluminum Layer Adhesion after Additional Binding of Spacer Molecules to the Plasma Polymer Layer

As the required flexibility of the plasma polymer layer was not available, flexible spacer molecules were additionally grafted to the monosort functional groups of the plasma polymer layer, as they had also been bonded directly to the polyolefin surface such as explained in Part 1. The results of the peel strength measurements confirmed that this additional flexibility of aliphatic spacer elements was necessary to achieve maximum adhesion even at low concentrations of spacers with adhesion-promoting functional groups. As will be shown later, with optimal plasma polymerization, bonding of suitable spacers with a terminal strong bonding group that forms a covalent bond to Al, Al-polymer laminates can be produced that are practically inseparable (Figure 11.22).

The attachment of spacer molecules to the monosort functional groups of the plasma polymer layer brings advantages. If the maximum peel strength of the metal-polyolefin laminate is measured, then the laminate has failed

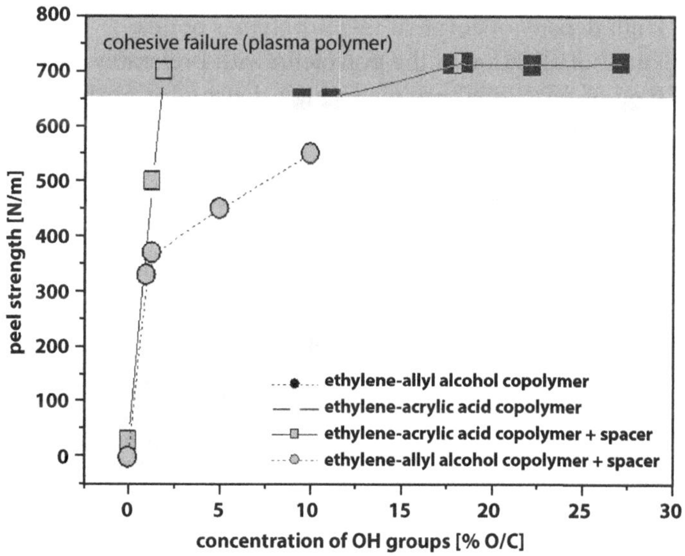

Figure 11.22 Peel strength of Al coatings in dependence on the composition of the adhesion-promoting plasma polymer layer composed of ethylene and allyl alcohol or ethylene and acrylic acid as well as additionally with spacers having OH and COOH terminal groups.

mechanically in the plasma polymer layer as mentioned before. This is the case with a higher concentration of OH or COOH groups. If only a few spacers with the same endgroup are then introduced, the maximum peel strength can be achieved with significantly fewer bonds between aluminum and polyolefin (Figure 11.22). In the case of maximum peel strength by means of COOH-terminated spacer molecules chemically bound to the plasma polymer layer, 2% COOH/C is sufficient to cause cohesive failure in the underlying plasma polymer layer. In contrast, about 10% COOH/C groups of the acrylic acid plasma polymer are required to achieve the maximum peel strength.

11.3.3 Protection of the Al Bond to the Spacer Terminal Group from Hydrolysis

As mentioned earlier, good metal-polyolefin adhesion depends on various factors:

1) Preference for covalent metal-polymer bonds over physical interactions.
2) High density of covalent metal-polymer bonds.
3) Functionalization of the polyolefin with preferably only one type of functional groups in high density for anchoring of spacer molecules (Part 1).
4) Deposition of adhesion-promoting plasma polymer layers with more or less only one type of functional groups dominating.
5) The functional group must react spontaneously with the evaporated aluminum layer; the COOH group is very reactive here, followed by OH groups.
6) Silanol groups (Si-OH) of hydrolyzed silanes are also very suitable for firmly bonding aluminum layers.
7) Silanol groups can also condense with each other to form a 2d Si-O-S layer and thus form a molecular and hydrophobic barrier against the diffusion of water molecules and metal ions.
8) Additionally flexible spacer molecules can be covalently anchored onto the functional groups of the adhesion-promoting plasma polymer layer. Thus, the flexibility of interface is much increased.

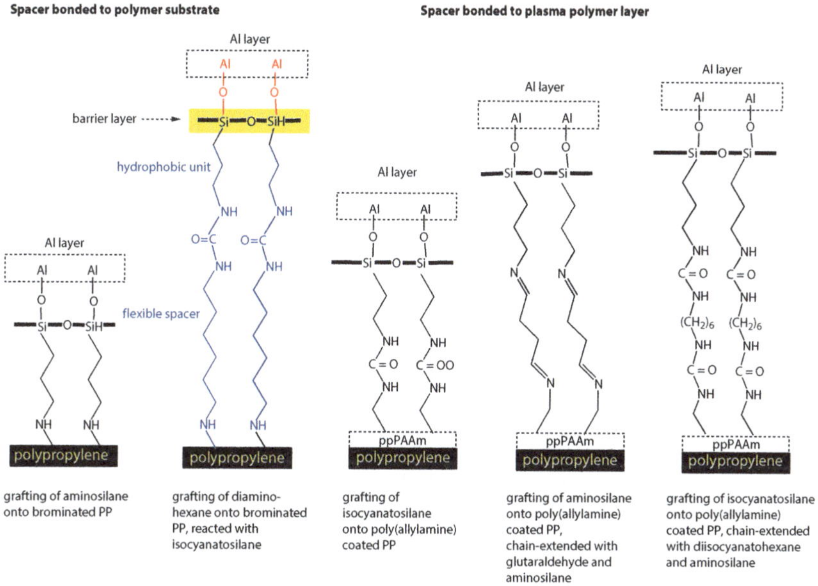

Figure 11.23 Schematic representation of idealized variants of an aluminium-polypropylene interface bridged with spacer molecules, each synthesized starting from NH_2 groups on the polymer substrate surface (ppPAAm=plasma polymerized poly(allylamine).

9) The spacer molecules must have flexible chains. Aliphatic chains are well suited because of the possibility of unlimited rotation around the C-C bond axis. They are also hydrophobic and therefore moisture-repellent.

Differently structured interfaces with and without a plasma polymer layer are shown schematically in Figure 11.23.

11.3.4 Results of Peel Strength Measurements on Aluminum-Polypropylene Laminates with Specially Designed Interface

The schematic structures shown in Figure 11.23 were synthesized. First the modified polyolefin surfaces were coated with 150 nm aluminum and then the adhesive tape was peeled off together with the aluminum layer. The peel strengths determined in the process are shown in Figure 11.24.

When the adhesion of the plasma polymer-polyolefin and plasma polymer-spacer with silane endgroup and this endgroup reacts with the

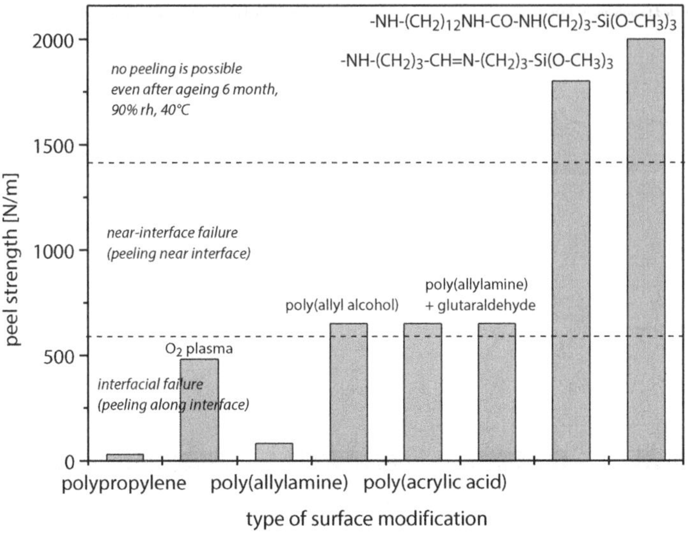

Figure 11.24 Survey on maximum peel strength results for differently modified aluminium-polypropylene laminates.

aluminium layer to form covalent bond, the metal-polyolefin laminate could no longer be mechanically separated. The adhesive tape at the aluminium layer was pulled off at around 1400 N/m without peeling the laminate.

This particularly high, no longer measurable adhesion could be reproduced many times. This extremely high adhesion strength was maintained even after 0.5 and 1.0 years of artificial ageing according to specifications (alternating temperatures, humidity) [21, 34, 65, 71].

The results confirm that the plasma polymer layer alone is not sufficient as an adhesion promoter. Peeling along the interface is no longer observed, but failure within the plasma polymer layer is observed. Only the additional incorporation of stress-peak relieving spacer molecules, the incorporation of the Si-O-Si barrier layer and the covalent Si-O-Al bond makes it possible to produce durable and inseparable Al-polymer laminates.

11.3.5 Alternative Methods

The flame-assisted application of ultra-thin SiOx layers of silane/air mixtures as adhesion promoters also leads to good adhesion of the Al-polyolefin laminate. Such SiO_x layers can also be produced using plasma, which are used as barrier coatings on the inner walls of plastic bottles [74, 75].

In principle, sputtering of aluminum targets is also possible and the deposition of correspondingly thin aluminum layers on polymers is in industrial use [76]. However, this would correspond to a direct bond between aluminum and polyolefin without the involvement of an intermediate adhesion-promoting layer [77–79]. However, the aluminum atoms can penetrate deeper into the polymer and thus anchor themselves better.

A particularly modern non-plasma process for the deposition of thin adhesion-promoting polymer layers is electrospray ionization (ESI) technology [52, 53]. It functions as ionization of mini-droplets and even of individual macromolecules, which are generated in a particularly strong direct current or in a constantly polarity-changing field in the case of application to insulating polymers. This process can work entirely without plasma. Corona plasma can support the ionization of difficult to ionize apolar polymer molecules as already mentioned in the description of the APCI (Atmospheric Plasma Chemical Ionization) process [50]. ESI and APCI work under action of a strong high voltage field. The high charge concentration on the surface of the droplets and the fact that the droplet diameters on the way to the cathode are reduced by the evaporation of the solvent, and the charges therefore come closer and closer together, lead to smaller droplets due to Coulomb fissions or also known as Coulomb explosions. These fissions continue in a cascade until individual macromolecule ions are formed. The polymer solution is sucked through a metal capillary, which acts also as electrode, usually as anode.

It is interesting to note that no polymer deposition occurs in the straight direction when the ESI spray is deflected by 90° (Figure 11.25). This means that there are no significant quantity of neutral droplets in the spray [80].

The advantages of polymer deposition using the ESI method compared to the deposition of a plasma polymer layer are listed below:

- 100% preservation of all functional groups.
- No polymer degradation.
- No cross-linking.
- No radical formation.
- No auto-oxidation.
- Hole-free up to approx. <5 nm layer thickness.
- The "electrophoretic effect" coats all remaining holes in the deposited ESI layer. It is also responsible for the internal coating of pores or for coating deposition on all shadowed back surfaces that are not directly exposed to the spray. An increased current flows to these uncovered areas or holes, which means that more macromolecular ions reach the

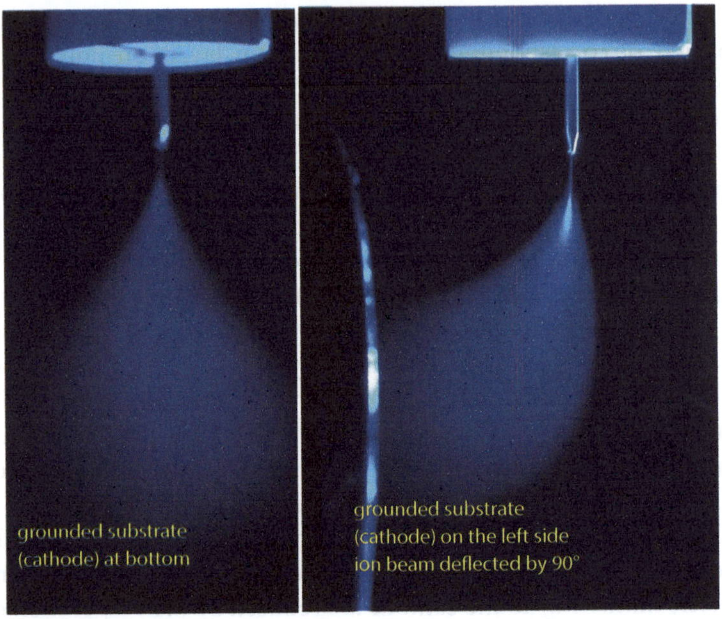

Figure 11.25 ESI spray to grounded substrate and 90° deflected spray showing the charged character of droplets.

uncovered areas and holes and close them. In contrast, areas with a thicker polymer layer have an electrically insulating effect. They therefore become charged, so that newly arriving macromolecule ions are repelled. Thus, ESI layers can be deposited onto conducting substrates up to a maximum thickness of 30 nm. Beyond this, the electrostatic repulsion of the charged deposited ESI polymer layer dominates.

- The deposition of thicker layers therefore requires a permanent polarity change of the applied high-voltage electric field, such as deposition on polyolefin surfaces because of the surface charge just mentioned.
- Backside deposition with ESI sprayed thin films, i.e. shadowed areas through "electrophoretic effect" is possible.
- Macromolecular ions penetrate up to 100 μm deep into carbon fiber bundles (rovings) using ESI spraying. Up to this depth, the fibers are covered with a hole-free polymer film.
- ESI works at atmospheric pressure.
- The technical complexity of ESI compared to low-pressure and atmospheric pressure plasma is significantly lower. Costs are lower.

- Very low current flow makes the applied high voltage safe for humans.
- Polymers suitable for ESI, however, must be soluble in water or in solvents.
- To be used as adhesion-promoting thin polymer layers, the ESI-sprayed polymers must have reactive functional groups that can react as spontaneously as possible with spacer molecules, silanes or aluminum, such as poly(allylamine) or poly(acrylic acid).
- A possible disadvantage of ESI can be the lack of plasma activation of the polymer substrate at the beginning of the deposition of the plasma polymer layer. However, it has been shown that the short moment of plasma exposure of the substrate until the formation of a closed polymer layer that cannot be penetrated by the plasma is sufficient to achieve good plasma polymer layer-substrate adhesion.
- A separate ultrashort (<2 s) plasma pretreatment can enhance adhesion to ESI layer.
- A single ESI spray spot is comparable in size to a plasma treatment spot or a gas burner spot. Since industrial polymer films 2 m wide can be easily oxidized with gas burners, this is also possible with ESI capillaries, which are arranged in a row next to each other and overlap their deposition areas. A similar arrangement is used, for example, when flaming wide polymer films.

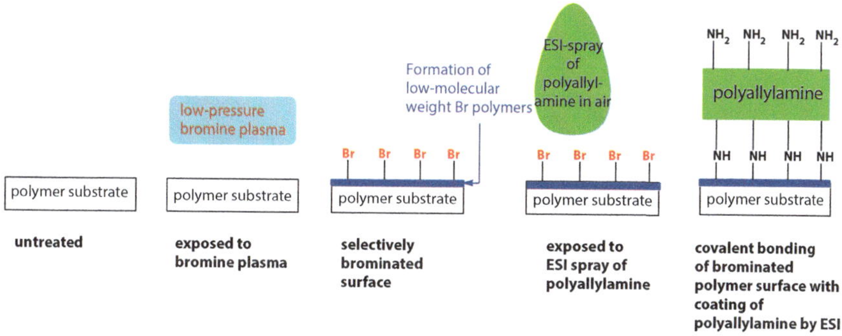

Figure 11.26 Covalently bonded laminate of polyolefin after plasma bromination or deposition of bromine-containing plasma polymer layer and spontaneous reaction of ESI sprayed polyallylamine between C-Br and NH_2 groups.

An example of the combination of plasma treatment with ESI polymer deposition, which is suitable for chemically binding spacer molecules to the polymer surface, is the bromine or layer-forming bromoform vapor plasma combined with the ESI deposition of polyallylamine (Figure 11.26).

Corresponding spacer molecules can then be attached to the amino groups of the ESI polymer layer.

11.4 Summary

Depending on the monomer, plasma polymers offer the possibility of possessing various more or less monosort groups. For adhesion of evaporated thin aluminum layers to polyolefin surfaces the deposition of adhesion-promoting plasma-polymerized poly(acrylic acid) and poly(allyl alcohol) produce the highest peel strength. Silanes are also very suitable for adhesion to aluminium and as bridging agents to flexibilizing spacer molecules, which can reduce the mechanical and thermal stresses along the interface.

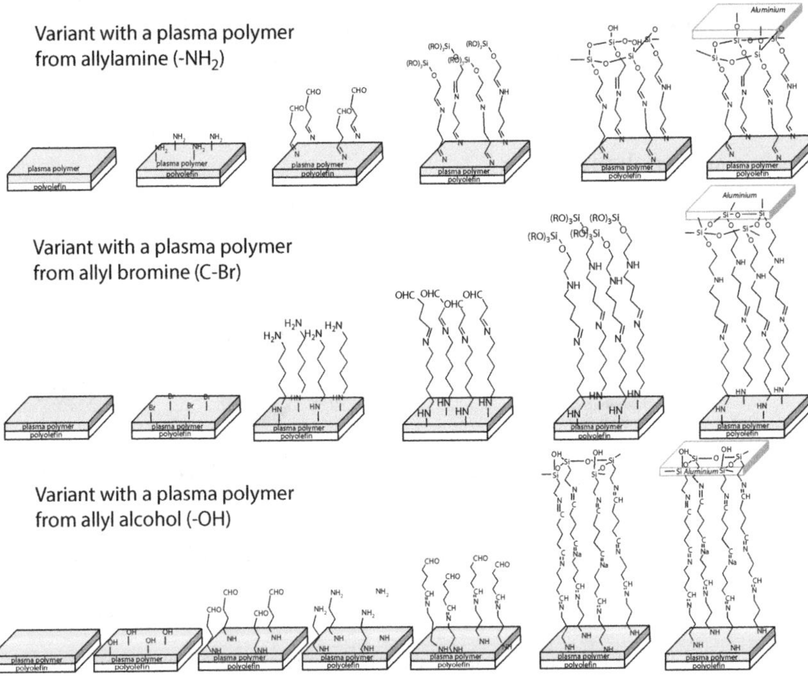

Figure 11.27 Variants to realize the "ideal" interface structure.

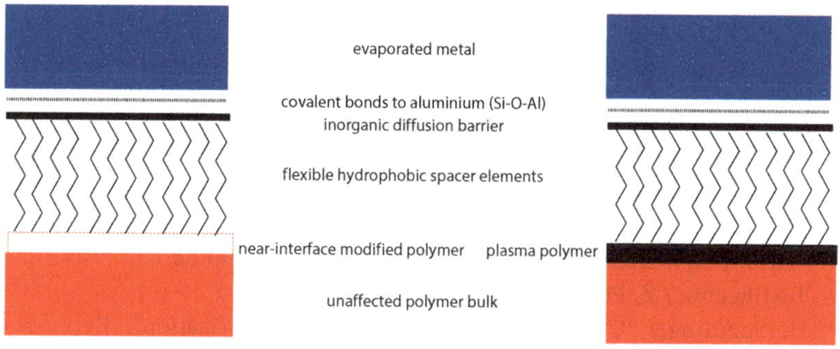

Figure 11.28 Schemes of the ideally designed interface for a metal-polymer laminate.

The combination of plasma (or ESI polymers) layer-spacer-silane-covalent bonding to aluminum overlayer has proven to be the best. It enables aluminium-polyolefin laminates which can no longer be separated (peeled), even under long-term thermal and moisture stress.

The use of plasma polymers only as adhesion promoters leads to lower adhesion, only marginally better than the adhesion between aluminium overlayer and surface-modified polyolefin.

Figure 11.27 summarizes at a glance which variants give the best bonded Al-polyolefin laminates, more precisely, laminates that could not be separated with conventional tapes even after 1 year of artificial aging.

This results in the following layer sequences in the cross section through the interface, which must be present in an "ideally" designed interface in order to achieve maximum adhesion, as shown schematically in Figure 11.28. The "ideal" interface design for the case of plasma modification of the polyolefin surface is shown on the left, while the "ideal" interface design for the case of using an adhesion-promoting plasma polymer layer is shown on the right.

AI disclosure statement: The author acknowledges the use of AI-based tools for minor language editing or grammatical correction. However, all intellectual contributions, analyses, and conclusions presented in this chapter are entirely the author's own.

References

1. J. Friedrich, Mechanisms of plasma polymerization –a chemist's point of view, Plasma Proc. Polymers 8, 783-802 (2011).
2. H. Yasuda, *Plasma Polymerization*, Academic Press, Orlando, FL (1985).

3. N. Inagaki, *Plasma Surface Modification and Plasma Polymerization*, CRC Press, Boca Raton, FL (1996).
4. J. Friedrich, *The Plasma Chemistry of Polymer Surfaces - Advanced Techniques for Surface Design*, Wiley-VCH, Weinheim, Germany (2012).
5. J. F. Friedrich and J. Meichsner, *Nonthermal Plasmas for Materials Processing*, Wiley-Scrivener, Beverly, MA (2022).
6. J. Friedrich, H. Wittrich and J. Gähde, Untersuchungen zur Bildung von Glimmpolymerschichten aus deuterierten Monomeren, Faserforsch. Textiltechn. / Z. Polymerenforsch. 29, 481-489 (1978).
7. H. Staudinger, Über Polymerisation [On polymerization], Berichte der Deutschen Chemischen Gesellschaft, 53, 1073–1085 (1920).
8. G. Kanig, H. Bauer, H. G. Kilian, H. G. Zachmann and R. Hosemann, Kristallisier- und Schmelzvorgänge bei Polymeren, Colloid Polym. Sci. 260, 356-377 (1982).
9. H. Sinn, W. Kaminsky, H.-J. Vollmer and R. Woldt, "Living polymers" with Ziegler catalysts of extreme productivity, Angew. Chem. 92, 396-402 (1980).
10. W. Kaminsky, K. Külper, H. H. Brintzinger and F. R. W. P. Wild, Polymerisation von Propen und Buten mit einem chiralen Zirconocen und Methylaluminoxan als Cokatalysator, Angew. Chem. 97, 507-506 (1985).
11. S. L. Miller, A production of amino acids under possible primitive earth conditions, Science 117, 528–529 (1953).
12. S. L. Miller and H. C. Urey, Organic compound synthesis on the primitive earth, Science 130, 245–251 (1959).
13. J. M. Tibbitt, M. Shen and A. T. Bell, Structural characterization of plasma-polymerized hydrocarbons, J. Macromol. Sci., Chem. A10, 1623-1648 (1976).
14. A. R. Denaro, P. A. Owens and A. Crawshaw, Glow discharge polymerization—styrene, Europ. Polym. J. 4, 93-106 (1968).
15. D. T. Clark and A. Dilks, An investigation of the vacuum UV spectra of inductivity coupled RF plasmas excited in inert gases as a function of some of the operating parameters, J. Polym. Sci.: Polym. Chem. Ed. 18, 1233-1246 (1980).
16. H. Yasuda and T. Yasuda, The competitive ablation and polymerization (CAP) principle and the plasma sensitivity of elements in plasma polymerization and treatment, J. Polym. Sci., Part A, Polym. Chem. 38, 943-953 (2000).
17. H. Yasuda and C. E. Lamaze, Polymerization of styrene in an electrodeless glow discharge. J. Appl. Polym. Sci. 15, 2277-2292 (1971).
18. H. Yasuda and C. E. Lamaze, Polymerization of styrene in an electrodeless glow discharge. II. Olefinic monomers. J. Appl. Polym. Sci. 17, 1519-1531 (1973).
19. H. Yasuda, Glow discharge polymerization, J. Polym. Sci. Macromol. Rev., 16, 199-293 (1981).

20. A. Fahmy, T. A. Mohamed and J. F. Friedrich, XPS and IR studies of plasma polymerization of allylamine with addition of NH_3 for deposition of N-rich layers, Appl. Surf. Sci. 458, 1006-1017 (2018).
21. J. Friedrich, Plasma polymers as adhesion promoters, in: *Progress in Adhesion and Adhesives*, K. L. Mittal (Ed.), Vol. 8, pp.389-457, Wiley-Scrivener, Beverly, MA (2024).
22. J. Friedrich, R. Mix, G. Kühn, I. Retzko, A. Schönhals and W. Unger, Plasma-based introduction of monosort functional groups of different type and density onto polymer surfaces, Part II: Pulsed plasma polymerization, Composite Interfaces 10, 173-223 (2003).
23. L. F. Thompson and K. G. Mayhan, The plasma polymerization of vinyl monomers. II. A detailed study of the plasma polymerization of styrene, J. Appl. Polym. Sci. 16, 2317-2340 (1972).
24. K. Jesch, J. E. Bloor and P. L. Kronick, Structure and physical properties of glow discharge polymers. I. Polymers from hydrocarbons, J. Polym. Sci. A1, 1487-1497 (1966).
25. P. L. Kronick, K. Jesch and J. E. Bloor, Spectra of glow-discharge polymers from aromatic compounds, J. Polym. Sci. A1, 767-772 (1969).
26. J. Friedrich, G. Kühn and J. Gähde, Untersuchungen zur Plasmaätzung von Polymeren. I. Strukturveränderungen von Polymeren nach Plasmaätzung, Acta Polymerica 30, 470-477 (1979).
27. A. R. Denaro, P. A. Owens and A. Crawshaw, Glow discharge polymerization—styrene, Eur. Polym. J. 4, 93-106 (1968).
28. A. R. Westwood, Glow discharge polymerization—I. Rates and mechanisms of polymer formation, Eur. Polym. J. 7, 363-375 (1971).
29. J. F. Friedrich, W. E. S. Unger, A. Lippitz, I. Koprinarov, St. Weidner, G. Kühn and L. Vogel, Plasma pretreatment of polymers - relevance to adhesion of metals, in: *Metallized Plastics 5&6: Fundamental and Applied Aspects*, K. L. Mittal (Ed.), pp. 271-293, CRC Press, Boca Raton, FL (1998).
30. R. H. Hansen and H. Schonhorn, Surface treatment of polymers for adhesive bonding, J. Appl.Polym. Sci. 11, 1461-1474 (1967).
31. J. Friedrich, R. Mix and G. Kühn, Functional groups bearing plasma homo and copolymer layers as adhesion promoters in metal-polymer composites, Surface Coatings Technol. 174-175, 811-815 (2003).
32. H. Kobayashi, A. T. Bell and M. Shen, Plasma polymerization of saturated and unsaturated hydrocarbons, Macromolecules 7, 277-283 (1974).
33. J. Friedrich, R. Mix and G. Hidde, Ways to generate monosort functionalized polyolefin surfaces, Rev. Adhesion Adhesives 2, 311-370 (2014).
34. J. Friedrich, Competition in adhesion between polysort and monosort functionalized polyolefin surfaces coated with vacuum-evaporated aluminium, in: *Polymer Surface Modification to Enhance Adhesion: Techniques and Applications*, K. L. Mittal and A. N. Netravali (Eds.), pp. 389-458, Wiley-Scrivener, Beverly, MA (2024).

35. J. Goodman, The formation of polymer thin films in the gas discharge, J. Polym. Sci. 44, 51-52 (1960).
36. M. Suzuki, A. Kishida, H. Iwata, and Y. Ikada, Graft copolymerization of acrylamide onto a polyethylene, Macromolecules 19, 1804-1808 (1986).
37. W. A. Wedenejew, L. W. Gurwitsch, W. H. Kondratjew, W. A. Medwedew and E. L. Frankewitsch, Energien chemischer Bindungen, Ionisationspotentiale und Elektronenaffinitäten (Energies of chemical bonds, ionisation potentials and electron affinities), VEB Deutscher Verlag für Grundstoffindustrie, Leipzig, Germany (1971).
38. F. K. McTaggart, *Plasma Chemistry in Electrical Discharges,* Elsevier, New York (1967).
39. H. Yasuda and T. Hirotsu, Critical evaluation of conditions of plasma polymerization. J. Polym. Sci., Polym. Chem. Ed. 16, 743-759 (1978).
40. J. Friedrich, J. Gähde, H. Frommelt and H. Wittrich, Modifizierung von Feststoffoberflächen in einer HF-Entladung. II. Schichtuntersuchungen, Reaktionsmechanismus und Verbundergebnisse. Faserforsch. Textiltechn. / Z. Polymerenforsch. 27, 599-603 (1976).
41. J. Friedrich, The formation processes of functional groups at polyolefin surfaces on exposure to oxygen or ammonia plasma: A critical review, Rev. Adhesion Adhesives 6, 253-328 (2018).
42. H. Yasuda and T. Hsu, Some aspects of plasma polymerization investigated by pulsed R.F. discharge, J. Polym. Sci.: Polym. Chem. Ed. 15, 81-97 (1977).
43. R. B. Timmons and A. J. Griggs, Pulsed plasma polymerizations, in: *Plasma Polymer Films,* H. Biederman (Ed.), Imperial College Press, London, UK (2004).
44. J. Friedrich, G. Kühn and R. Mix, Tailoring of polymer surfaces with different monotype functional groups of variable density using chemical and plasma chemical processes, Prog. Colloid Polym. Sci. 132, 62-71 (2006).
45. J. F. Friedrich, R. Mix, R.-D. Schulze, A. Meyer-Plath, R. Joshi and S. Wettmarshausen, New plasma techniques for polymer surface modification with monotype functional groups, Plasma Proc. Polym. 5, 407-423 (2008).
46. S. Krüger, R.-D. Schulze, K. Brademann-Jock and J. Friedrich, Characterization of plasmapolymers by thermoluminescence, Surf. Coatings Technol. 201, 543-552 (2006).
47. S. Swaraj, U. Oran, A. Lippitz, J. F. Friedrich and W. E. S. Unger, Study of influence of external plasma parameters on plasma polymerised films prepared from organic molecules (acrylic acid, allyl alcohol, allylamine) using XPS and NEXAFS, Surf. Coatings Technol. 200, 494-497 (2005).
48. M. Ramanujam, R. Mix, M. Wagner and J. F. Friedrich, Investigation of C-radical production at polyolefins after Ar plasma treatment followed by bromine incorporation, J. Adhesion Sci. Technol. 27, 1828-1839 (2013).
49. M. Ramanujam, V. Wachtendorf, P. J. Purohit, R. Mix, A. Schönhals and J. F. Friedrich, Monitoring the effects of dry artificial UV weathering on low

density polyethylene by dielectric relaxation spectroscopy, Thermochimica Acta 530, 73-78 (2012).
50. E. C. Horning, M. G. Horning, D. I. Carroll, I. Dzidic and R. N. Stillwell, New picogram detection system based on a mass spectrometer with an external ionization source at atmospheric pressure, Anal. Chem. 45, 936-943 (1973).
51. E.C. Horning, D.I. Carroll, I. Dzidic, K.D. Haegele and M.G. Horning, Atmospheric pressure ionization mass spectrometry, solvent-mediated ionization of samples introduced in solution and in a liquid chromatograph effluent system, J. Chromatogr. Sci, 12, 725-729 (1974).
52. M. Dole, L. L. Mack, R. L. Hines, R. C. Mobley, L. D. Ferguson and M. B. Alice, Molecular beams of macroions, J. Chem. Phys. 49, 2240-2249 (1968).
53. J. B. Fenn, M. Mann, C. K. Meng, S. F. Wong and C. M. Whitehouse, Electrospray ionization for mass spectrometry of large biomolecules, Science 246, 64-71 (1989).
54. J. B. Fenn, M. Mann, C. K. Meng, S. F. Wong and C. M. Whitehouse, Electrospray ionization - principles and practice, Mass Spectrom. Rev. 9, 37-70 (1990).
55. K. Altmann, R.-D. Schulze, G. Hidde and J. Friedrich, Electrospray-ionization deposition of polymers – Principle, electrophoretic effect and applications, J. Adhesion Sci. Technol. 27, 988-1005 (2013).
56. K. Altmann, R.-D. Schulze and J. F. Friedrich, Polymer deposition morphology by electrospray deposition - Modifications through distance variation, Thin Solid Films 564, 269-276 (2014).
57. J. F. Friedrich, S. Wettmarshausen, K. Altmann and G, Hidde, Adhesion promotion by ESI deposition of ultra-thin polymer layers onto fibers, Surface Innovations 3, 215-227 (2015).
58. J. Friedrich, K. Altmann, S. Wettmarshausen and G. Hidde, Coating of carbon fibers with adhesion-promoting thin polymer layers using plasma polymerization or electrospray ionization technique - a comparison, Plasma Process. Polym. 16(3), 1600074 (2017).
59. G. Grundmeier and M. Stratmann, Plasma polymerization – a new and promising way for the corrosion protection of steel, Materials and Corrosion 49, 150-160 (1998).
60. W. Möller and M. Schmidt, Untersuchungen zur chemischen Struktur von Glimmpolymerschichten aus Hexamethyldisiloxan (HMDS), Beiträge aus der Plasmaphysik 17, 121-134 (1977)
61. A. S. da Silva Sobrinho, M. Latrèche, G. Czeremuszkin, J. E. Klemberg-Sapieha and M. R. Wertheimer, Transparent barrier coatings on polyethylene terephthalate by single- and dual-frequency plasma-enhanced chemical vapor deposition, J. Vac. Sci. Technol. A16, 3190-3198 (1998).
62. H. Grünwald, W. Dicken, K. Neuenburg, R. Adam, J. Bartella, T. Gebele, S. Mitzlaff, U. Patz and J. Snyder, Integration of plasma-pretreatment, metallization, and plasma polymerization for improved aluminium mirror quality

in large-scale lighting production, in: *Metallized Plastics 7: Fundamental and Applied Aspects*, K. L. Mittal (Ed.), CRC Press, Boca Raton, FL (2001).
63. Z. R. Farag, S. Krüger, G. Hidde, A. Schimanski, C. Jäger and J. Friedrich, Deposition of thick polymer or inorganic layers with flame-retardant properties by combination of plasma and spray processes, Surf. Coatings Technol. 228, 266-274 (2013).
64. J. Friedrich, B. Falk, I. Loeschcke, B. Rutsch, Kh. Richter, H.-D. Reiner, U. Throl and H. Raubach, Untersuchungen zur Haftung zwischen plasmachemisch oxidiertem Kupfer und Polyphenylchinoxalin, Acta Polymerica 36, 310-320 (1085).
65. J. Friedrich, Tailoring of interface/interphase to promote metal-polymer adhesion, in: *Adhesion in Microelectronics*, K. L. Mittal and T. Ahsan (Eds.), pp. 67-135, Wiley-Scrivener, Beverly, MA (2014).
66. S. Wettmarshausen, R. Mix, A. Meyer-Plath, H.-U. Mittmann and J. Friedrich, Plasmabromination – the selective way to monotype functionalized polymer surfaces, in: *Polymer Surface Modification, Vol. 5., K. L. Mittal (Ed.),* pp.3-18, CRC Press, Boca Raton, FL (2009).
67. J. Friedrich, R. Mix and S. Wettmarshausen, A new concept of adhesion promotion in metal-polymer composites by introduction of covalently bonded spacers at interface, J. Adhesion Sci. Technol. 22, 1123-1143 (2008).
68. J. Friedrich, S. Wettmarshausen and M. Hennecke, Haloform plasmas applied to polymers, Surf. Coatings Technol. 203, 3647-3655 (2009).
69. J. F. Friedrich, R. Mix, R.-D. Schulze, A. Meyer-Plath, R. Joshi and S. Wettmarshausen, New plasma techniques for polymer surface modification with monotype functional groups, Plasma Proc. Polym. 5, 407-423 (2008).
70. J. Friedrich and G. Hidde, Ultra-thin polymer films as novel class of adhesion promoters, Rev. Adhesion Adhesives 3, 1-52 (2015).
71. J. Friedrich, Study of metal-polymer interfaces: Relevance to adhesion, in: *Recent Advances in Adhesion Science and Technology in Honor of Dr. Kash Mittal,* V. Gutowski and H. Dodiuk (Eds.), pp. 3-33, CRC Press, Boca Raton, FL (2014).
72. R. Mix, J. Falkenhagen, R.-D. Schulze, V. Gerstung and J. F. Friedrich, Plasma homo- and copolymerization of allyl alcohol and styrene, in: *Polymer Surface Modification, Relevance to Adhesion*, Vol. 5., K. L. Mittal (Ed.), pp. 317-340, CRC Press, Boca Raton, FL (2009).
73. H.-G. Elias, *Makromoleküle*, 5th ed., Hüthig&Wepf, Basel, CH (1990).
74. G. Czeremuszkin, M. Latreche, M. R. Wertheimer and A. S. da Silva Sobrinho, Ultrathin silicon-compound barrier coatings for polymeric packaging materials: An industrial perspective, Plasmas Polym. 6, 107-120 (2001).
75. H. Grünwald, M. Jung, R. Kukla, R. Adam and S. Kunkel, New generation sputtering and plasma polymerization plant for reflector production, Surf. Coatings Technol. 93, 99-104 (1997).

76. J. A. Thornton, Influence of apparatus geometry and deposition conditions on the structure and topography of thick sputtered coatings, J. Vac. Sci. Technol. 11, 666–670 (1974).
77. P. S. Ho, Chemistry and adhesion of metal-polymer interfaces, Appl. Surf. Sci. 41-42, 559-566 (1990).
78. P. Marcus, C. Hinnen, D. Imbert and J. M. Siffre, An *in situ* XPS study of the formation of aluminium–polymer interfaces, Surf. Interf. Anal. 19, 127-132 (1992).
79. M. Bou, J. M. Martin, Th. Le Mogne and L. Vovelle, Chemistry of the interface between aluminium and polyethyleneterephthalate by XPS, Appl. Surf. Sci. 47, 149-161 (1991).
80. S. Wettmarshausen, J. F. Friedrich, A. Meyer-Plath, G. Kalinka, G. Hidde and S. Weidner, Coating of carbon fibers with adhesion-promoting thin poly(acrylic acid) and poly(hydroxyethylmethacrylate) layers using electrospray ionization, J. Adhesion Sci. Technol. 29, 1628-1650 (2015).

12

Surface Engineering for Enhanced Adhesion in Polymer Materials

Ankit Rawal[1], Dipankar Pal[1] and Sudarsan Neogi[2*]

[1]*Department of Chemical Engineering, Indian Institute of Petroleum and Energy, Visakhapatnam, Andhra Pradesh, India*
[2]*Department of Chemical Engineering, Indian Institute of Technology Indore, Indore, Madhya Pradesh, India*

Abstract

Polymer materials are widely used across various industries due to their superior mechanical properties, chemical resistance, and lightweight nature. However, despite these advantages, they inherently exhibit low adhesion owing to their low surface free energy, chemical inertness, smooth surface morphology, and lack of polar functional groups. These characteristics hinder effective wetting and chemical bonding with adhesives or other materials. Polymeric materials require surface engineering to enhance interfacial adhesion for various applications for improved performance. This work provides a detailed account of adhesion theories to explain the interaction mechanisms between polymer substrates and adhesives. Surface engineering methods are discussed, which are used to modify surface chemistry and topography, thereby promoting improved wettability, chemical compatibility, and interfacial strength. Analytical characterization techniques are employed, which are used to observe changes in surface functionalization and morphology directly related to adhesion behavior. The study highlights detailed case studies wherein these combined strategies have led to quantifiable improvements in the adhesion performance of polymeric systems, thereby validating theoretical principles and advancing material design for demanding structural and functional applications.

Keywords: Polymer materials, interfacial adhesion, surface engineering, wettability

**Corresponding author*: sneogiiit@gmail.com; sneogi@iiti.ac.in

Anil N. Netravali and Frank M. Etzler (eds.) The Wide World of Adhesion: A Festschrift in Honor of Dr. Kash Mittal on his 80th Birthday, (429–468) © 2026 Scrivener Publishing LLC

12.1 Introduction

Polymeric materials are widely employed in diverse applications, from everyday utilization as plastic bags and bottles to advanced industrial operations, including filtration systems and the aerospace sector, where they serve as thermal and protective coatings for aircraft and satellite components. The widespread adoption of polymeric materials is attributed to their excellent mechanical strength, chemical and thermal resistance, ease of fabrication, and surface modifications for the desired application [1]. The historical background of polymer chemistry is significantly marked by the contributions of researchers in the 19th century. Leo Baekeland synthesized Bakelite in 1907 [2], establishing it as the first fully synthetic polymer. In 1920, Hermann Staudinger further advanced the field by introducing the concept of macromolecules [3]. The development of nylon by Wallace Carothers in 1935 also had a transformative impact on polymer materials, as the unique characteristics of nylon garnered considerable attention and facilitated a broader adoption and development of synthetic polymers among researchers [4]. Subsequently, a diverse range of synthetic polymers have emerged, characterized by enhanced durability, customized properties, and efficient mass production techniques. Polyethylene (PE), Polypropylene (PP), Polytetrafluoroethylene (PTFE), Polysulfone (PSf), Polyacrylonitrile (PAN), and poly(vinylidene fluoride) (PVDF) are some prominent polymers recognized for their extensive applications across diverse industries [5]. These materials are commonly employed in packaging, membrane technology, coatings, medical devices, electronics, and filtration systems. Polymer materials are characterized by excellent chemical resistance, mechanical durability, and the ability to maintain performance at elevated temperatures [6]. Despite the advantages, one inherent limitation of polymer materials is the intrinsically low surface adhesion [7]. This phenomenon can be attributed to several factors, including low surface free energy, high chemical inertness, the deficiency of polar functional groups, and a smooth or non-porous surface morphology [8]. The poor surface adhesion of polymers can potentially affect the performance of polymer materials in various applications, particularly those that require strong interfacial bonding or adhesion strength. It is to be noted that the polymer adhesion is a highly application-dependent phenomenon. For instance, in polymeric membrane-based filtration of oil-water emulsions, the low adhesion or, more specifically, the poor affinity of hydrophobic polymer surfaces with water leads to inadequate wettability, making it more susceptible to oil fouling and reducing filtration efficiency. Similarly, in biomedical implants, polymers like PTFE exhibit poor cell adhesion, which can hinder tissue integration. In another example, automotive coatings on polyolefin-based plastic parts often suffer from peeling or poor bonding due to low surface adhesion, affecting both aesthetics and durability. Thus, these examples illustrate that the influence of adhesion depends on the specific application, whether it is filtration, biomedical compatibility, or protective coatings. However, in general, low adhesion of polymer materials becomes a critical issue in areas where strong interfacial bonding is required, such as in coating, printing, lamination processes, etc. To overcome this limitation, surface engineering of polymer materials is required to enhance the surface free energy, imparting desired functional groups, or altering the surface morphology to promote better adhesion for different applications [9]. Surface engineering of polymer materials can potentially modify the polymer surface to enhance adhesion without altering the bulk characteristics [10]. Various chemical or physical treatments can effectively alter the surface properties of polymers, enhancing the adhesion, such as plasma treatment, ultraviolet/ozone treatment, flame treatment, grafting, etc. [11, 12]. By tuning the surface characteristics, either through a morphological approach or by imparting surface functionalization, the adhesion behavior of the polymer can be significantly improved. This chapter provides a comprehensive discussion on the fundamentals of polymer adhesion, challenges associated with low adhesion in polymers, various techniques used to modify polymer surfaces, and case studies performed by various researchers for adhesion enhancement and future outlooks. It also covers the theoretical basis of adhesion, including mechanical interlocking, molecular interactions, and interfacial energy considerations, as well as the latest advancements in surface engineering for polymer adhesion enhancement.

12.2 Fundamentals of Adhesion in Polymers

Adhesion in polymers plays a critical role in determining the performance and durability of polymer-based composites, coatings, and interfaces. A deep understanding of adhesion mechanisms is essential for tailoring surface properties and enhancing interfacial strength. In this section, the theories of adhesion, techniques for adhesion measurement in polymers, and various surface engineering techniques to improve adhesion are highlighted.

12.2.1 Theories of Adhesion

The fundamental understanding of adhesion mechanisms between two different materials is crucial for optimizing the performance in diverse applications, such as protective coatings, structural adhesives, polymer composites, biomedical implants, and electronic devices. In a polymer material, adhesion is governed by an interplay of interfacial, mechanical, and chemical phenomena. Due to the complex nature of adhesive and polymer interactions, no single theory can fully explain all aspects of how adhesion works. As a result, several theories have been established, each emphasizing different factors influencing adhesion, ranging from surface energies and molecular interactions to mechanical interlocking and thermodynamic compatibility [13–17]. Figure 12.1 presents the seven predominant theories of adhesion, which are broadly categorized according to the scale of interaction involved: macroscopic, molecular, and atomic.

The two macroscopic theories, mechanical interlocking and electrostatic, explain adhesion at the macroscopic level, and the schematic significance of these two theories is shown in Figure 12.2.

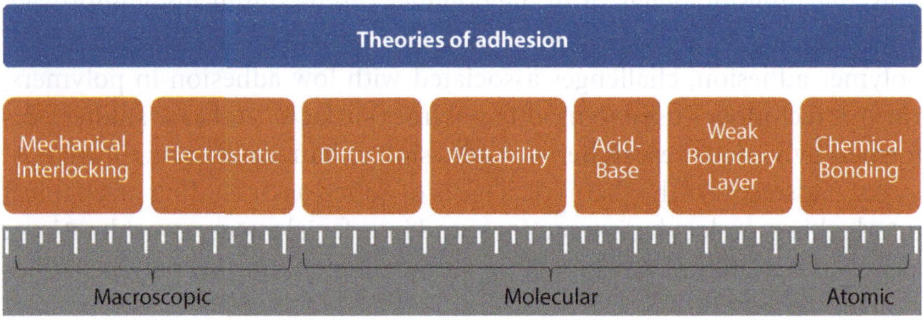

Figure 12.1 Theories of adhesion, as suggested by Ghanem, A. and Lang, Y. Introduction to Polymer Adhesion, 2017 [48].

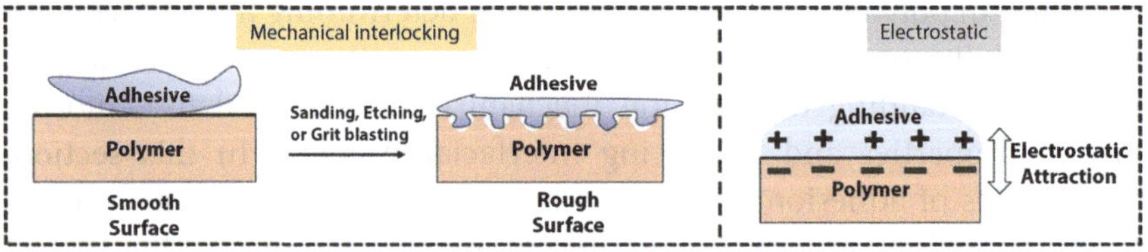

Figure 12.2 Schemes of macroscopic scale theories of adhesion.

12.2.1.1 Mechanical Interlocking Theory

The Mechanical Interlocking Theory, classified within macroscopic-scale adhesion mechanisms, attributes interfacial bonding to the physical entrapment of an adhesive within the morphological features of a substrate surface. First introduced by McBain in 1925 [18], this theory postulates that adhesion results from the adhesive's ability to penetrate and conform to surface irregularities such as pores, fissures, and asperities followed by solidification, thereby creating a mechanical interlock at the interface [19]. For effective adhesion *via* this mechanism, it is imperative that the adhesive exhibits adequate wetting behavior, eliminates interfacial air, and penetrates the surface irregularities. Mechanical interlocking is mainly influenced by the surface structure (including roughness and porosity), the actual contact area between materials, and how well the adhesive chemically and physically interacts with the substrate, especially in terms of wettability. Empirically, substrates exhibiting increased surface roughness tend to facilitate superior mechanical anchorage, thereby enhancing bond strength. Consequently, pre-treatment techniques such as abrasion, grit blasting, or chemical etching are routinely employed to optimize surface morphology before adhesive application.

As illustrated in Figure 12.2, the enhancement of surface roughness directly correlates with improved adhesive performance, particularly in materials such as wood, textiles, and abraded metals. The practical relevance of this theory is evident in a broad range of industrial sectors, including aerospace, automotive, and construction, where surface engineering strategies are deliberately employed to promote mechanical adhesion. However, despite its practical relevance, the Mechanical Interlocking Theory presents certain limitations. It does not consider the contribution of molecular or chemical interactions at the adhesive–substrate interface, which are often critical to the overall adhesion process. Experimental evidence has shown that an effective adhesion can still be achieved on smooth, non-porous surfaces where mechanical interlocking is minimal, suggesting that additional mechanisms must also be involved.

12.2.1.2 Electrostatic Theory

The Electrostatic Theory of adhesion suggests that when two dissimilar materials, particularly having different electronic properties, come into contact, an interfacial charge imbalance may arise due to the redistribution or transfer of electrons. This phenomenon leads to the formation of an electrical double layer at the interface, wherein one surface acquires a net positive charge and the other a net negative charge. The resulting Coulombic attraction between these oppositely charged regions generates an electrostatic force that contributes to the adhesion between the materials [20]. The theory is especially related to systems involving electronically incompatible materials, such as polymer–metal interfaces, where disparities in electronic band structures facilitate interfacial charge separation. The electrostatic attraction developed in such cases enhances adhesion strength even in the absence of mechanical interlocking or chemical bonding.

As illustrated in Figure 12.2, this mechanism can play a significant role in promoting interfacial adhesion, particularly in applications where conventional bonding mechanisms are limited or ineffective. This theory is widely recognized in adhesion science for its ability to explain strong interfacial bonding between certain dissimilar materials that neither undergo chemical reactions nor exhibit significant mechanical interlocking. It offers valuable insights, particularly for metal/polymer interfaces, which are prevalent in fields such as packaging, electronics, and automotive engineering. However, the applicability of this theory is limited to cases involving materials with contrasting electrical conductivities or electron affinities. It does not adequately describe adhesion phenomena in systems where such charge separation is unlikely or where other mechanisms, such as covalent bonding or mechanical anchoring, dominate. In essence, the Electrostatic Theory underscores the significance of interfacial charge separation and

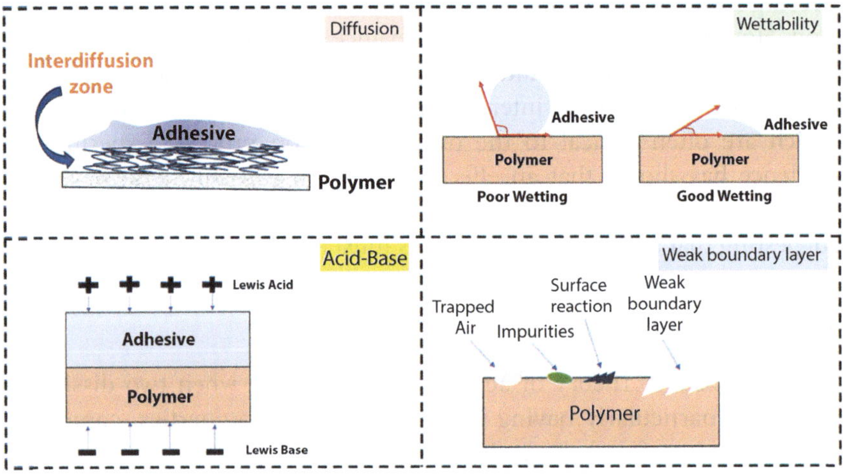

Figure 12.3 Schemes of molecular scale theories of adhesion, as suggested by Ghanem, A. and Lang, Y. Introduction to Polymer Adhesion, 2017 [48].

Coulombic attraction in adhesion and is particularly relevant in hybrid systems comprising organic and inorganic components.

After the macroscopic scale theories, the four molecular theories: diffusion, wettability, acid-base and weak boundary layer, explain adhesion at the molecular level. The schematic significance of molecular scale theories of adhesion is shown in Figure 12.3.

12.2.1.3 Diffusion Theory

The diffusion theory of adhesion was introduced by Voyutskii in 1963 and is based on the principle that macromolecules from both the polymer adhesive and the polymer adherend can diffuse into each other at the interface [21]. According to this theory, adhesion occurs when long-chain molecules from the adhesive and the surface of the substrate interpenetrate and entangle with each other, forming a stable and strong bond. This theory is applicable primarily when both the adhesive and the adherend are mutually miscible and compatible. In other words, they should be able to blend at the molecular level and allow their polymer chains to move freely. This movement is essential for diffusion to take place. As a result of this interdiffusion, a transitional or interphase zone is formed at the interface, where the molecules of the adhesive and the substrate mix, leading to interlocking at a molecular level rather than just physical entrapment, as shown in Figure 12.3.

The strength and effectiveness of this adhesion depend on several important factors. These include the length of the polymer chains, their concentration, and the temperature during the bonding process. Higher temperatures, for example, can increase chain mobility and enhance diffusion. A longer chain length allows for more entanglement, which typically results in stronger adhesion. The significance of the diffusion theory lies in its ability to explain the adhesion behavior of rubbery polymers and materials bonded through solvent welding: a technique used to join thermoplastics by softening them with a solvent.

This theory helps researchers understand how similar polymer materials can form a durable bond through molecular-level interaction, which is especially important in industries like packaging, medical devices, and soft polymer-based composites. However, it is important to note that the diffusion theory has limitations. It cannot explain adhesion when the materials are not compatible or when they lack sufficient molecular mobility. Additionally, the theory mainly applies to polymeric systems, and, therefore, it is not suitable for explaining adhesion at metal or ceramic interfaces. Even within polymer science, this theory remains under ongoing study and debate as researchers continue to explore the microscopic behavior of macromolecules at interfaces.

12.2.1.4 Wettability Theory

The wettability theory of adhesion explains that the adhesive must be able to spread and make close contact with the surface of the adherend to form a strong bond. In simple terms, wetting is the ability of a liquid adhesive to cover the surface of a solid substrate. The better the adhesive spreads on (or wets) the surface, the more likely it is to form strong molecular interactions and, therefore, a better adhesive bond. According to this theory, molecular-level attractions, such as van der Waals forces or hydrogen bonding, occur when the adhesive comes into close contact with the substrate. For such interactions to take place effectively, the adhesive must penetrate all the microscopic valleys and crevices on the surface, thereby maximizing the contact area. Good wetting results in a continuous adhesive film without gaps or air pockets, creating a strong and uniform bond. In contrast, poor wetting leads to incomplete surface coverage, leaving voids or interfacial defects that weaken the bond and make it more prone to failure. The angle at which a droplet of adhesive contacts the surface, called the contact angle, is an indicator of wetting [22, 23].

A low contact angle indicates good wetting, while a high contact angle indicates poor wetting, as shown in Figure 12.3. One of the most important factors in achieving good wetting is the relationship between the surface tension of the adhesive and the surface free energy of the adherend. For effective wetting, the adhesive must have a lower surface tension than the surface free energy of the substrate [24]. This ensures that the adhesive spreads easily and completely across the surface, allowing intimate contact at the molecular level. The wettability theory is significant because it highlights the role of surface free energy, surface preparation, and adhesive properties in adhesion. This theory is especially relevant in applications involving paints, coatings, and adhesives where flow and spreadability are crucial. It also explains why certain surface treatments like plasma treatment or corona discharge are used to increase surface free energy and improve the adhesion of low-energy surfaces such as polymers.

12.2.1.5 Acid–Base Theory

The acid–base theory of adhesion is a relatively modern approach compared to traditional adhesion models. It offers a deeper understanding of the chemical interactions that occur at the interface between an adhesive and a substrate, especially in systems involving polymers. This theory is grounded in the classical chemical principles proposed by J.N. Brønsted and G.N. Lewis in the early 20th century [25]. According to this theory, adhesion occurs due to the interaction between acidic and basic functional groups present on the surfaces of the adhesive and the adherend. In Lewis terms, an acid is an electron pair acceptor, while a base is an electron pair donor. When these two types of chemical groups come into contact at the interface, they form acid–base interactions or adducts, which result in a strong interfacial attraction and contribute significantly to the bond strength, as shown in Figure 12.3.

Unlike the mechanical interlocking or diffusion models that focus on physical bonding, the acid–base theory emphasizes chemical affinity and surface free energy. It explains that polar interactions, such as hydrogen bonding and dipole–dipole forces, can substantially enhance adhesion when the surfaces involved have complementary acidic and basic properties. These interactions are often much stronger than simple van der Waals forces and can lead to more durable and chemically stable adhesive bonds. The extent of acid–base interaction can be evaluated based on the enthalpy of adduct formation, which measures the energy released when an acid and a base combine. A higher enthalpy value indicates stronger interactions and, thus, better adhesion. This theory is especially useful in understanding polymer–polymer and metal–polymer interfaces, where functional groups like -OH, -COOH, $-NH_2$, and others play a key role in bonding.

It also forms the basis for many surface modification techniques aimed at introducing or enhancing surface polarity, such as plasma treatment, chemical oxidation, and surface grafting. The acid–base theory is significant because it highlights the importance of surface chemistry and functional group interactions in adhesion science. It helps researchers design better adhesives and surface treatments by targeting specific chemical affinities at the interface. Van Oss, Chaudhury, and Good explained how surfaces stick together by breaking surface free energy into two parts: one from general attraction and one from acid-base interactions [26].

12.2.1.6 Weak Boundary Layer Theory

The Weak Boundary Layer Theory was proposed by Bikerman [27] and offers an important perspective on why adhesive bonds sometimes fail, even when everything else appears to be suitable [28]. Unlike other adhesion theories that explain how bonding happens, this theory focuses on the causes of adhesion failure, especially at the interface between the adhesive and the adherend. According to this theory, the presence of a weak layer either on the surface of the adhesive or the substrate can drastically reduce the strength of adhesion. This weak boundary layer may be caused by surface contamination, impurities, air entrapment, incompatible materials, or improper processing conditions. It can act as a fragile barrier that prevents the adhesive from establishing a strong bond with the underlying substrate, thereby compromising the adhesion strength.

A weak boundary layer often results in what is known as a cohesive failure, where the break occurs within the boundary layer itself rather than at the true interface. This is problematic because even if the adhesive wets the surface well or if chemical or mechanical bonding occurs, the presence of this unstable or poorly bonded intermediate layer weakens the entire joint. Multiple sources can lead to the formation of weak boundary layers: (a) Surface impurities such as oils, dust, or oxides; (b) Moisture or air pockets trapped during the bonding process; (c) Degraded surfaces due to plasma treatment or environmental exposure; and (d) Incompatible surface treatments or residues from previous processes (Figure 12.3).

Even something as subtle as a slight thermal or chemical degradation of the substrate during surface preparation can create a weak boundary layer. For instance, overexposure to plasma treatment can break down polymer chains at the surface, making them brittle or poorly connected to the bulk material. The significance of the weak boundary layer theory lies in its ability to explain why bond strength is poor despite the proper application of adhesives. It emphasizes the critical need for thorough surface cleaning, controlled bonding environments, and appropriate surface treatments. The theory serves as a strong reminder that successful adhesion is not just about strong bonding mechanisms it is also about ensuring that nothing weak or unstable exists at the bonding interface.

12.2.1.7 Chemical or Molecular Bonding Theory

The Chemical or Molecular Bonding Theory is one of the earliest and most widely accepted models for explaining how adhesion occurs between two surfaces at atomic scale. Unlike theories that rely purely on physical contact or interlocking, this theory is based on chemical interactions at the interface of the adhesive and the substrate. According to this theory, adhesion takes place due to molecular forces and chemical bonds that form when two surfaces are brought into intimate contact. These forces can include dipole–dipole interactions, hydrogen bonding, van der Waals forces, and even strong chemical bonds like ionic, covalent, or metallic bonds, as shown in Figure 12.4.

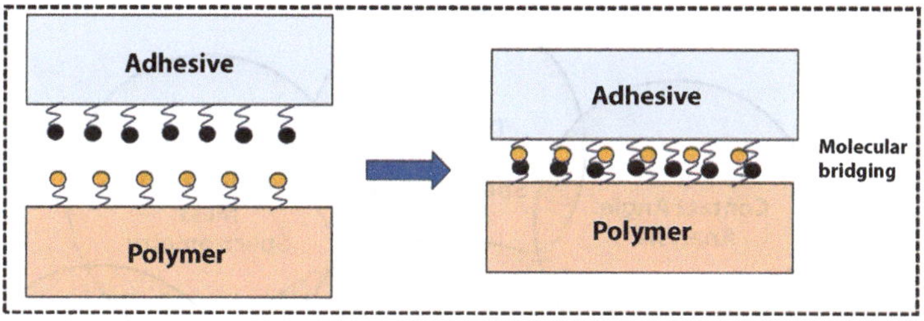

Figure 12.4 Scheme of atomic scale theory of adhesion.

The strength and type of interaction depend heavily on the chemical composition of the adhesive and the substrate. Researchers like Hutchinson and Iglauer [28] conducted peel and tack tests on construction sealants and found that chemical bonding played a more dominant role in adhesion than other mechanisms like diffusion or mechanical interlocking. This highlights the importance of surface chemistry in forming durable and high-strength bonds. Real surfaces often have defects, cracks, and trapped air bubbles, which can hinder proper bonding. To overcome these challenges, adhesion promoters, also known as coupling agents, are used. These are specialized molecules that can chemically react with both the adhesive and the substrate, forming a chemical bridge that strengthens the bond at the interface. One of the most common and effective types of coupling agents are silane-based molecules. These are widely used in adhesives and coatings to improve compatibility between dissimilar materials (like polymers and metals or glass). The silane groups react with hydroxyl groups on surfaces and also link to the adhesive resin, resulting in a chemically stable and strong bond. The significance of the chemical bonding theory lies in its ability to explain strong, durable, and stable bonds, particularly in structural applications where physical or mechanical bonding alone would not be sufficient.

12.2.2 Characterization Techniques for Adhesion in Polymers

A critical aspect of the adhesion of two surfaces lies in characterizing surface properties that directly influence interfacial bonding. Parameters such as surface roughness, surface free energy, polarity, and chemical composition suggest how well a polymer can adhere to another material. A variety of analytical techniques are available to evaluate these characteristics, each offering unique insights into the surface behavior of polymeric materials, as shown in Figure 12.5.

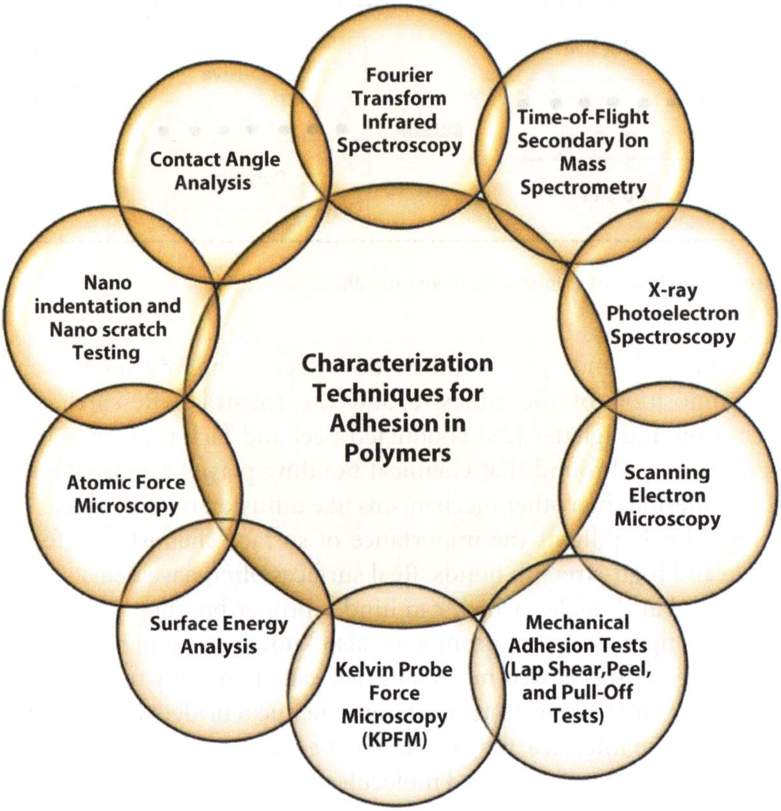

Figure 12.5 Characterization techniques for evaluating adhesion in polymeric materials.

12.2.2.1 Contact Angle Analysis

Contact angle analysis is a widely used technique to evaluate the wettability of a polymer surface, which is directly related to surface free energy and adhesion potential. In this method, a small droplet of a liquid (commonly water) is deposited on the surface, and the angle formed between the liquid and the solid surface is measured. A low contact angle indicates good wettability and high surface free energy, which are favorable for adhesion. By using multiple test liquids and applying models such as Owens-Wendt or Fowkes, both polar and dispersion components of the surface free energy can be calculated. This technique is quick and provides valuable insight into how surface treatments or modifications alter adhesion property.

12.2.2.2 Atomic Force Microscopy (AFM)

Atomic Force Microscopy (AFM) is an essential tool for the nanoscale analysis of surface morphology and roughness. A cantilever with a sharp tip scans across the surface of the polymer, detecting atomic-scale interactions between the tip and the surface. The deflections of the cantilever are recorded to generate a topographical map. AFM can quantitatively measure surface roughness parameters such as average roughness (Ra) and root mean square roughness (Rq), which influence mechanical interlocking during adhesion. Additionally, this technique allows local adhesion forces between the cantilever tip and the surface to be measured, providing detailed insights into surface interactions at the nanoscale.

12.2.2.3 Scanning Electron Microscopy (SEM)

Scanning Electron Microscopy (SEM) provides high-resolution imaging of the polymer surface to reveal topographical features and morphology. An electron beam is directed onto the surface, causing the emission of secondary or backscattered electrons, which are then detected to create detailed images. SEM is particularly useful for analyzing surface features before and after adhesion tests, helping identify interfacial defects, crack propagation, or failure modes. Though it requires conductive coating for non-conductive polymers, it remains a powerful tool for visualizing the effectiveness of surface modifications.

12.2.2.4 X-Ray Photoelectron Spectroscopy (XPS)

XPS is a surface-sensitive analytical technique used to determine the elemental composition and chemical state of elements in the top few nanometers of a polymer surface. The method involves irradiating the surface with X-rays, causing the emission of photoelectrons. By measuring their kinetic energies, the binding energies of specific elements are identified. XPS can detect functional groups introduced during surface modification, such as hydroxyl, carboxyl, or amine groups, which are crucial for improving chemical adhesion. This technique is essential for confirming the presence of reactive sites or assessing contamination that could hinder adhesion.

12.2.2.5 *Time-of-Flight Secondary Ion Mass Spectrometry (ToF-SIMS)*

ToF-SIMS offers molecular-level insight into the surface composition of polymers. A focused ion beam bombards the surface, dislodging secondary ions that are analyzed based on their mass-to-charge ratio using a time-of-flight detector. This method provides both qualitative and semi-quantitative information about the surface's molecular structure, including carbon-based compounds, contaminants, and functional layers. ToF-SIMS is particularly effective in depth profiling and detecting very thin surface modifications, making it ideal for studying adhesion promoters or functional coatings on polymers [29].

12.2.2.6 *Fourier Transform Infrared Spectroscopy – Attenuated Total Reflectance (FTIR-ATR)*

FTIR-ATR is employed to identify chemical bonds and functional groups present on the surface of polymer materials. The technique works by passing infrared radiation through a crystal in contact with the polymer surface; the evanescent wave penetrates a few micrometers into the sample, where it is absorbed depending on the molecular vibrations of the surface compounds. FTIR-ATR is highly effective in verifying chemical modifications introduced during surface treatments and confirming the presence of polar groups that enhance adhesion through hydrogen bonding or covalent interactions.

12.2.2.7 *Surface Free Energy Analysis*

There are various ways to determine surface free energy of solids using various approaches [30, 31]. This involves measuring the contact angles of various test liquids with known surface tensions on the polymer surface and using the data to calculate the surface free energy and its components. High surface free energy is typically associated with better adhesive performance due to improved wetting and interaction at the interface. This analysis is particularly valuable when comparing treated and untreated polymer surfaces or evaluating the impact of plasma and chemical treatments.

12.2.2.8 Mechanical Adhesion Tests (Peel, Lap Shear, and Pull-Off Tests)

Mechanical test methods are essential for directly quantifying adhesion strength. In a peel test, the force required to peel an adhesive from the substrate is measured, whereas lap shear tests evaluate the resistance to sliding forces along the adhesive joint. Pull-off tests assess the tensile strength needed to detach a bonded surface perpendicularly [32]. These tests are standardized (e.g., ASTM or ISO protocols) and provide real-world performance metrics, locus of failure, which are crucial for validating the success of surface treatments.

12.2.2.9 Kelvin Probe Force Microscopy (KPFM)

KPFM is a variation of AFM used to map the surface potential and work function differences across a polymer surface [33]. This technique measures the electrostatic forces between a conductive AFM tip and the surface, allowing for visualization of charge distribution and local electrical properties. Variations in surface potential can indicate the presence of functional groups or charge-trapping sites that affect adhesion, especially in applications involving electronics or dielectric polymers.

12.2.2.10 Nanoindentation and Nanoscratch Testing

Nanoindentation and nanoscratch tests are valuable for evaluating the mechanical integrity and adhesion performance of thin films or coatings on polymers [34]. In nanoindentation, a sharp indenter applies a controlled force to the surface, and the resulting depth of penetration is used to determine the hardness and elastic modulus. Nanoscratch tests involve dragging the tip across the surface under increasing load to observe failure onset. These methods are particularly useful in assessing the cohesive strength of modified layers and the adhesion between multi-layered polymer systems.

12.2.3 Surface Engineering Techniques for Enhanced Adhesion of Polymers

To improve the adhesion characteristics of polymers, a variety of surface engineering strategies have been developed. These techniques, grouped by their underlying mechanisms, modify surface cleanliness, texture, chemistry, energy, or morphology to enhance interfacial bonding. Each method

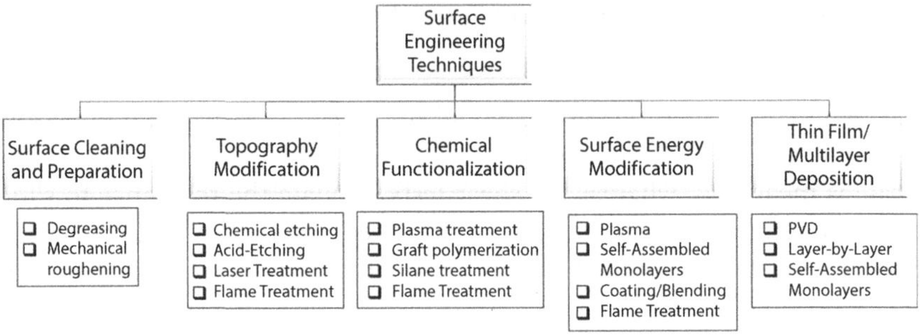

Figure 12.6 Surface engineering strategies for improving the adhesion of polymers.

offers unique benefits tailored to specific materials and applications. As shown in Figure 12.6, these approaches can be broadly categorized for better understanding and implementation.

12.2.3.1 Surface Cleaning and Preparation

These techniques remove contaminants and prepare the polymer for further modification, ensuring that adhesion is not hindered by surface impurities or low surface free energy.

(a) Degreasing & Solvent Cleaning: This method involves washing the polymer surface using organic solvents like acetone, ethanol, or isopropanol to remove oils, dust, and processing residues. A clean surface is critical for effective adhesion, as any residual contaminants can inhibit adhesive bonding.
(b) Mechanical Roughening (Grit Blasting/Sanding): Mechanical abrasion increases the surface roughness and, hence, the effective surface area. This creates micro-anchoring sites, allowing adhesives to grip the substrate better through mechanical interlocking.

12.2.3.2 Surface Topography Modification

These methods enhance mechanical interlocking by modifying the surface morphology.

(a) Chemical Etching/Acid Etching: Chemical etching uses strong acids (e.g., chromic acid, nitric acid) or oxidants to dissolve a thin surface layer of the polymer. This process increases surface roughness and introduces functional groups, enhancing both mechanical and chemical bonding [35].

(b) Laser Treatment: High-energy laser beams ablate polymer surfaces in a controlled manner. This modifies surface texture and can initiate surface grafting reactions by breaking polymer bonds, resulting in micro-roughness and increased reactivity.

(c) Flame Treatment: It involves exposing the polymer to an oxidizing flame. The heat and reactive radicals (like O and OH) generated in the flame alter surface morphology and introduce polar groups. It enhances both topographical and chemical properties conducive to adhesion.

12.2.3.3 *Surface Chemical Functionalization*

These techniques introduce functional chemical groups (e.g., -OH, -COOH, -NH_2) that promote covalent or hydrogen bonding with adhesives.

(a) Plasma Treatment: Utilizes ionized gases at low pressures to activate the surface by incorporating oxygen, nitrogen, or other functional groups. It modifies only the top few nanometers of the surface without affecting the bulk. Enhanced wettability and adhesion are achieved due to increased surface polarity.

(b) Graft Polymerization: This method involves the covalent attachment of monomers or polymer chains to the existing polymer surface using chemical initiators or radiation. The newly grafted chains can introduce functionalities tailored for specific interactions, improving compatibility and interfacial bonding.

(c) Silane Coupling Agent Treatment: Silanes contain bifunctional molecules that react with both the polymer and the adhesive. One end of the molecule forms covalent bonds with hydroxyl or carboxyl groups on the polymer surface, while the other end binds to adhesives, forming a molecular bridge [36].

12.2.3.4 *Surface Free Energy and Wettability Enhancement*

These methods modify surface polarity and energy, improving adhesive wetting and spreading.

(a) Plasma Treatment: Not only does it introduce polar groups, but plasma treatment also significantly increases surface free energy, promoting better wetting of adhesives and coatings.
(b) Flame Treatment: Flame-induced oxidation elevates surface free energy, aiding in improved wettability and adhesion.
(c) Self-Assembled Monolayers (SAMs): SAMs are thin molecular films formed spontaneously *via* adsorption of functional molecules (e.g., alkanethiols or silanes). These monolayers can precisely control surface chemistry, polarity, and reactivity. They are used to impart hydrophobic, hydrophilic, or bioactive properties [37].
(d) Polymer Blending and Coating: Involves blending the base polymer with functional additives or applying a surface coating (e.g., hydrophilic layers) to improve interfacial interaction. For instance, blending with polar polymers can elevate surface free energy and chemical affinity.

12.2.3.5 Thin Film and Multilayer Deposition

These approaches engineer surface layers with specific properties (mechanical, chemical, electrical) that facilitate stronger bonding.

(a) Physical Vapor Deposition (PVD): This high-precision technique deposits thin films (metals, oxides) *via* evaporation or sputtering. These films improve surface hardness, adhesion, and chemical resistance without altering the polymer bulk.
(b) Layer-by-Layer (LbL) Assembly: Alternating deposition of oppositely charged polyelectrolytes builds up nanostructured films with controlled thickness and composition. This enables functionalities such as selective permeability, antifouling, and bioactivity [38].
(c) Self-Assembled Monolayers (SAMs): As a form of molecular film deposition, SAMs also serve as nanoscale interfacial layers that modify surface chemistry in a precise and reproducible way.

12.3 Case Studies and Applications

12.3.1 Polymer Adhesion with Ceramics

As discussed earlier, improving the adhesion of polymers is application-dependent because different fields demand different performance characteristics. One area where this becomes critically important is when polymers need to be bonded to ceramics. Strong adhesion between these two materials is essential in several practical and industrial applications. For instance, in dentistry, ceramic materials are used for crowns, bridges, and implants, where they need to be securely bonded with polymer-based adhesives or composites. In electronics, ceramics are often used as substrates or insulators, while polymers serve as coatings or protective layers. Here, poor adhesion can lead to delamination or device failure. In the aerospace and automotive industries, ceramic coatings are applied over polymer composites for thermal protection, corrosion resistance, and mechanical reinforcement, where even minor adhesion failure can lead to significant damage or safety risks. However, achieving strong adhesion between polymers and ceramics is technically challenging due to their different surface properties. Ceramics typically possess high surface free energy and are easily wetted (contact angles < 90°), whereas most polymers have low surface free energy and are hydrophobic in nature (contact angles > 90°). This results in poor interfacial contact and weak bonding. Additionally, ceramics are brittle and chemically inert, while polymers are flexible and chemically active, making it even more difficult to establish a stable chemical or mechanical bond. To overcome these issues, surface engineering techniques are essential. These methods aim to modify the polymer material to make it more compatible for adhesion to the ceramic.

12.3.1.1 Surface Engineering of PEEK and PEKK for Improved Adhesion to Ceramic Materials

In a study conducted by Tosun and Yanıkoğlu [39], poly(ether ether ketone) (PEEK) poly(ether ketone ketone) (PEKK) specimens were surface-engineered to improve the adhesion to ceramic materials used in prefabricated computer-aided design and computer-aided manufacturing materials (CAD/CAM) used in dental restorations. The polymer specimens were initially prepared by polishing with 600-grit silicon carbide abrasive paper in water to standardize surface texture. Following this, the samples were ultrasonically cleaned in distilled water to ensure the removal of surface contaminants and to achieve consistent surface smoothness.

Subsequently, two distinct surface modification protocols were applied:

(a) Air Abrasion (AA): Utilizing aluminum oxide (Al_2O_3) particles.
(b) Silica Coating (SC): Utilizing 30 µm silica-coated aluminum oxide particles (Cojet Sand Treatment).

The cojet sand treatment refers to a tribochemical silica coating method where silica-coated alumina particles are blasted onto the surface of the material. This process creates micromechanical roughness (like sandblasting) and deposits a silica layer that enables chemical bonding with silane-based adhesives. It is a chairside-friendly method (simple, time-efficient, and suitable for direct application in clinical or practical settings) used to enhance both mechanical and chemical adhesion. Post-treatment, AA-modified specimens were subjected to ultrasonic cleaning, while SC-treated samples were air-dried only to preserve the integrity of the surface before adhesive application. Each surface-treated group was further subdivided into two subgroups: one with the application of an adhesive and one without. The adhesive used was Tokuyama Bond Force II, a dental adhesive, applied using a disposable brush in a circular motion, followed by air-drying and light curing using a Valo GRAND dental LED curing unit. This categorization resulted in four surface conditions for each polymer:

- AA-treated with adhesive,
- AA-treated without adhesive,
- SC-treated with adhesive, and
- SC-treated without adhesive.

The surface roughness analysis demonstrated substantial variations across both polymer types and surface treatments, as illustrated in Figure 12.7(a). For the PEEK specimens, AA treatment resulted in a mean surface roughness of approximately 3 µm, with negligible difference observed between adhesive and non-adhesive subgroups. Conversely, SC-treated PE samples exhibited higher roughness values of approximately 3.5 µm without adhesive and an increased value of approximately 4.2 µm following adhesive application. The PEKK specimens exhibited a more pronounced response to surface treatment. AA treatment in the absence of adhesive yielded a significantly higher roughness value of approximately 8 µm. However, adhesive application substantially reduced this roughness by approximately 50% (around 4 μm). In the SC-treated PEKK group, surface roughness remained relatively stable, measuring around 4.8 μm without adhesive and decreasing slightly to approximately 3.8 μm with adhesive application. Contact angle measurements, depicted in Figure 12.7(b), revealed the influence of surface treatment and adhesive on surface wettability. For both PEEK and PEKK polymers, AA treatment without adhesive resulted in high contact angles of approximately 110° and 115°, respectively. Adhesive application showed significantly reduced contact angles to around 65° for both polymers, indicative of enhanced surface wettability and improved adhesive spreading. Interestingly, SC-treated samples demonstrated an inverse trend, with higher contact angles post-adhesive application compared to their untreated counterparts. This observation suggests that the adhesive may not have sufficiently infiltrated or wetted the roughened SC-modified surfaces, which could potentially impair interfacial adhesion. Shear bond strength (SBS), a critical mechanical indicator of adhesive efficacy under applied stress, varied distinctly across treatment groups, as illustrated in Figure 12.7(c). The AA-treated PE polymer without adhesive exhibited the lowest SBS value of approximately 3 MPa, however, the application of adhesive significantly improved SBS to approximately 6 MPa. Notably, the highest SBS values around 10 MPa were consistently observed for both PEEK and PEKK polymers when SC treatment was combined with adhesive application. This underscores the synergistic effect of surface treatment and adhesive application in optimizing polymer-ceramic adhesion. Environmental Scanning Electron Microscopy (ESEM) analysis, shown in Figures 12.7(d) and (e), further substantiated the influence of treatment on surface morphology. SC treatment followed by adhesive application produced surfaces with more homogeneously distributed micro-level pores and smoother topography. This suggests enhanced surface uniformity and, by extension, improved adhesion performance, particularly for the PEKK polymer. The combination of SC treatment with adhesive application appears to render the polymer surface more conducive to durable and reliable bonding with ceramic materials. After the application of adhesive, the surface appeared to be smooth for all the cases except for AA-treated PE polymer. Thus, the study affirms that tailored surface treatments, particularly when coupled with adhesive application, significantly enhance the interfacial adhesion of PEEK and PEKK polymers to ceramics. Among the evaluated conditions, the SC treatment followed by adhesive application provided superior surface uniformity and shear bond strength, especially in the case of PEKK, highlighting its potential as a preferred strategy for polymer-ceramic integration in CAD/CAM applications.

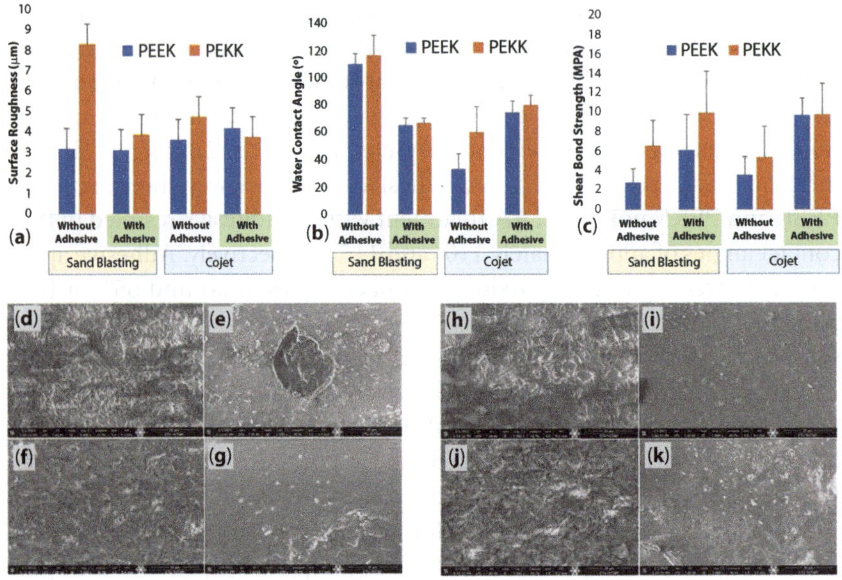

Figure 12.7 (a) Roughness values (mean ± SD); (b) water contact angle values (mean ± SD); (c) shear bond strength values (mean ± SD); environmental scanning electron microscope (ESEM) images of PEEK at 5000X: (d) air Abrasion treated, (e) air abrasion treated and adhesive, (f) SC treated, (g) SC treated and adhesive; ESEM images of PEKK at 5000X: (h) air abrasion treated, (i) air abrasion treated and adhesive, (j) SC treated, (k) SC treated and adhesive, as suggested by Tosun and Yanıkoğlu [39], with copyright permission from Springer Nature.

12.3.1.2 Influence of Ceramic Surface Roughness and PDMS Viscosity on Interfacial Adhesion in PDMS/Ceramic Composite Membranes

This case study by Wei *et al.* [40] investigates the role of surface engineering strategies in enhancing adhesion between polydimethylsiloxane (PDMS) and ceramic supports for the development of high-performance composite membranes. The study focuses on two key parameters: the surface roughness of the ceramic substrate and the viscosity of the PDMS solution, both of which critically influence interfacial bond strength and the overall polymeric membrane quality. To evaluate the adhesion and mechanical properties of the resulting membranes in real time, a nanoindentation-based methodology was employed. The membranes were fabricated using tubular asymmetric ceramic supports composed of zirconium dioxide/aluminum oxide (ZrO_2/Al_2O_3) with an initial pore size of 0.2 μm. To modulate surface roughness, the ceramic tubes were polished using 600 mesh sandpaper, followed by rigorous cleaning and drying procedures. PDMS, a silicone-based elastomer with a molecular weight of 50,000 and a polydispersity index of 1.13, was dissolved in n-heptane. A crosslinking agent, tetraethylorthosilicate (TEOS), along with 0.2 wt% of the catalyst dibutyltin dilaurate, was incorporated into the solution. After thorough degassing, the polymer solution was applied to the ceramic supports *via* dip-coating, followed by sequential drying at ambient temperature and under vacuum to remove residual solvents. The morphological characterization of the composite membranes was conducted *via* SEM. Morphological images of membranes fabricated with 7.5 wt% PDMS and dipped for 1-minute reveal significant differences based on the surface treatment of the ceramic support. As shown in Figure 12.8a, membranes prepared using unpolished supports displayed incomplete surface coverage, attributed to excessive surface roughness. In contrast, Figure 12.8b illustrates that polishing the ceramic support resulted in a smooth, continuous, and defect-free PDMS layer. These results underscore the importance of surface roughness reduction in promoting effective and uniform polymer coating. Further SEM analysis, as presented in Figure 12.8c1 and Figure 12.8c2, compared the surface and cross-sectional morphology of unpolished versus polished

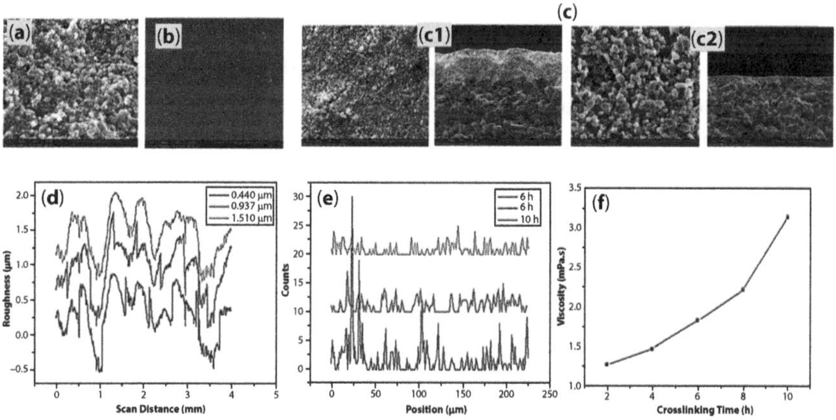

Figure 12.8 SEM images of PDMS/ceramic membrane on (a) unpolished support, (b) of polished support (roughness: 0.937 μm); (c) surface and cross section of (c1) unpolished support, (c2) polished support; (d) roughness results of three polished supports; (e) EDX Si line scan across membrane cross sections; (f) viscosity of PDMS vs. crosslinking time (h), as suggested by Wei *et al.* [40], with copyright permission from John Wiley and Sons.

ceramic supports. The unpolished sample exhibited a coarse and flawed surface texture, while polishing with 600 mesh sandpaper yielded a substantially smoother and more homogeneous surface. This improved topography enabled superior PDMS adhesion, contributing to the formation of defect-free and reproducible composite membranes. Surface roughness measurements, conducted using a profilometer, confirmed that increasing polishing time progressively reduced surface roughness. The arithmetic mean roughness values for each surface treated with different degrees of polishing are presented. As shown in Figure 12.8d, the roughness test results are provided for two selected ceramic supports. Adhesion was enhanced due to the formation of hydrogen bonds between surface hydroxyl (–OH) groups on the ceramic support and the PDMS during heat treatment. Since intermolecular interactions at the interface are highly dependent on surface topography, the degree of roughness was found to directly impact the adhesion strength. In this study, the roughness of the ceramic supports was modulated in the range of 0.44 μm to 1.51 μm, depending on the polishing duration. In addition to surface roughness, the viscosity of the PDMS solution was identified as a crucial parameter influencing membrane performance. Energy-dispersive X-ray (EDX) line scans across the membrane cross section (Figure 12.8e) demonstrated that as the viscosity of the PDMS increased, its penetration into the ceramic support decreased. While slight infiltration can enhance interfacial bonding by improving mechanical interlocking, excessive penetration is detrimental as it increases mass transfer resistance and reduces flux. Therefore, a careful adjustment of PDMS viscosity is necessary to balance adhesion and membrane permeability. Scratch testing was used to quantify the improvement in adhesion strength as a function of surface roughness. Results showed that the critical load required to initiate film delamination increased from 27.6 mN for unpolished surfaces to 50.7 mN for polished ones, indicating a significant enhancement in interfacial adhesion due to surface engineering. Moreover, the study evaluated the mechanical properties of the PDMS separation layers across a range of viscosities (1.28, 1.48, and 2.22 mPa.s). The elastic modulus values recorded were 86 ± 10 MPa, 91 ± 8 MPa, and 98 ± 11 MPa, respectively, while the corresponding hardness values were 15 ± 2.9 MPa, 13 ± 2.1 MPa, and 13 ± 1.9 MPa. These minimal variations and low standard deviations indicate that changes in PDMS viscosity do not significantly affect the mechanical integrity of the separation layer, ensuring reliable membrane performance under varying fabrication conditions. Thus, this study demonstrates the critical role of surface engineering in improving the adhesion of PDMS to ceramic supports. By simultaneously optimizing ceramic surface roughness and PDMS viscosity, it is possible to fabricate composite membranes that are structurally robust, defect-free, and capable of consistent performance. Figure 12.8f illustrates that increasing the crosslinking time leads to higher PDMS viscosity, while the mechanical properties of the PDMS separation layers remain largely unchanged. These findings are particularly relevant for the development of polymer/ceramic hybrid membranes in applications where interfacial durability and reliability are paramount, such as in advanced separation technologies across environmental, energy, and biomedical sectors.

12.3.2 Polymer–Metal Adhesion

Metals are known to possess high surface free energy, which naturally makes them more favorable for strong adhesion. However, polymers and metals differ significantly in surface properties, leading to incompatibility in many bonding situations. While metals may offer good wettability, polymers are often chemically inert or have low surface free energy, limiting effective bonding. To overcome this mismatch, various surface treatment techniques have been developed to modify either the metal or the polymer surface or both to enhance adhesion. These treatments improve surface chemistry, surface free energy, and surface roughness and promote chemical compatibility.

12.3.2.1 Surface Engineering of Polypropylene to Enhance Adhesion to Copper

This case study by Dayss *et al.* [41] investigates the effectiveness of polymer surface engineering strategies in enhancing the interfacial adhesion between polypropylene (PP) and copper (Cu), a critical requirement for fabricating reliable polymer–metal laminates. Given the inherently low surface free energy and chemical inertness of PP, achieving strong and durable adhesion to metallic substrates remains a significant challenge. To address this, a combination of surface engineering techniques was employed, including mechanical roughening, low-pressure plasma treatment using different gases, and deposition of functional interlayers such as poly(3,4-ethylene dioxythiophene) (PEDOT) and plasma-polymerized acrylic acid (PPAA).

Initially, mechanical pre-treatment was carried out by abrading the PP substrates with silicon carbide (SiC) abrasive papers of 320 and 500 grit. This process significantly enhanced the surface roughness from minimal values (untreated) to 0.76 μm and 1.55 μm, respectively, thereby improving the surface topography for better mechanical interlocking. Following this, the roughened surfaces were subjected to low-pressure microwave plasma treatment using a system operating at 2.46 GHz and 300 W, supported by an electron cyclotron resonance (ECR) magnet. The plasma process employed argon (Ar), nitrogen (N_2), and oxygen (O_2) gases at a flow rate of 50 mL/min for durations ranging from 30 seconds to 5 minutes. Among these, argon plasma treatment proved most effective, elevating the surface free energy of PP to 57 mJ/m^2 (Figure 12.9a) and significantly lowering the water contact angle, indicating improved wettability. It was also observed that after 1 minute of plasma exposure, there was not much change in the surface free energy. In contrast, O_2 plasma caused undesirable over-etching even only after 30 seconds, and at higher plasma powers, the reduction of

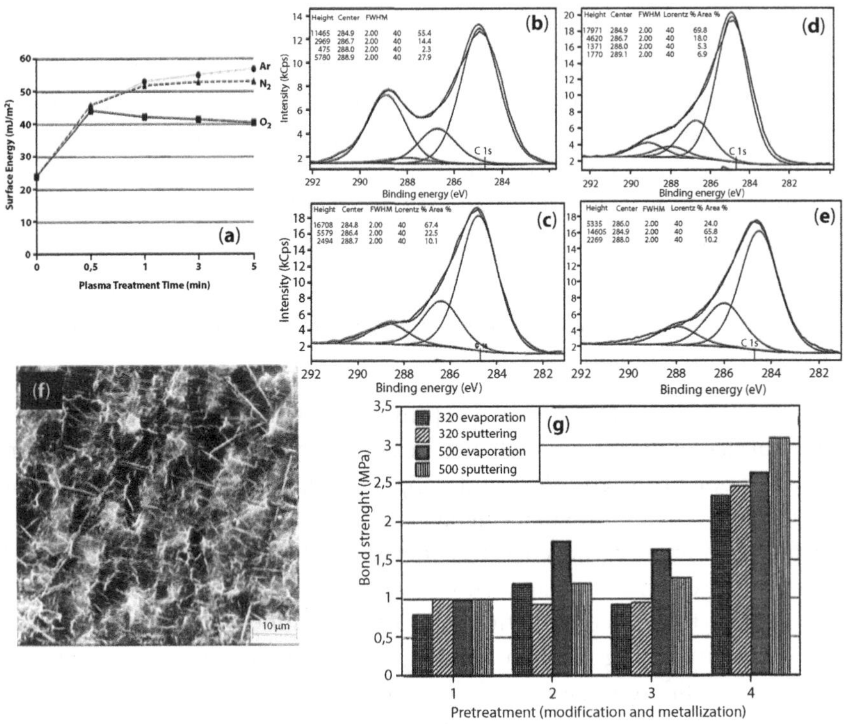

Figure 12.9 (a) Surface energy variation of PP after plasma treatment, (b–e) XPS spectra of plasma-polymerized acrylic acid films at different RF powers: (b) 30 W, (c) 60W, (d) 100 W, (e) 200W, (f) SEM image of plasma polymerized acrylic acid film (treated at RF power of 200 W) showing cross-linked structure, (g) the final metallization using vacuum based method (evaporation or sputtering), after sanding the polymer substrate using abrasive papers (grit 320 and 500) made of silicon carbide (SiC), as suggested by Dayss *et al.* [41], with copyright permission from Elsevier.

surface oxygen functional groups compromised long-term adhesion property. To further modify the surface chemistry, a conductive PEDOT layer was synthesized *via* a three-step chemical process: (1) oxidative activation of the PP surface using 60 g/L potassium permanganate ($KMnO_4$) solution at 90°C, (2) immersion in a monomer solution of 3,4-ethylene dioxythiophene, and (3) *in-situ* polymerization using different types and concentrations of sulfuric acid. The resulting PEDOT-coated surfaces displayed surface resistance values ranging from 7 kΩ to 2 MΩ, and high surface free energy values between 60 and 72 mJ/m^2, indicating favorable hydrophilicity and compatibility for metal deposition. In parallel, plasma-polymerized acrylic acid (PPAA) films were deposited on PP using radiofrequency (RF) plasma under powers of 30 W, 60 W, 100 W, and 200 W, with a fixed monomer flow rate of 2.5 mL/min. XPS analysis (Figures 12.9b and 12.9c) revealed that higher RF powers increased the incorporation of carboxylic and other polar functional groups into the films. At 200 W, SEM image (Figure 12.9f) confirmed the formation of a dense, highly cross-linked, and insoluble film structure, ideal for robust interfacial bonding with metallic layers. Subsequent Cu metallization was performed using both vacuum evaporation and sputtering methods, and adhesion strength was measured using the DIN ISO 4624 pull-off test. Mechanical roughening alone yielded an improvement in bonding force, with 500-grit abrasion achieving a maximum of 1 MPa. When combined with Ar plasma treatment, adhesion strength increased further to 1.75 MPa. PEDOT-modified surfaces alone achieved 1.64 MPa, although failure predominantly occurred at the interface, indicating weak interfacial adhesion. The highest adhesion strength, 3 MPa, was recorded when PEDOT coating was combined with mechanical abrasion, demonstrating a synergistic enhancement from both topographical and chemical modifications. Interestingly, vacuum-evaporated Cu exhibited superior adhesion compared to sputtered Cu, except in the presence of PPAA layers, where enhanced chemical compatibility between Cu and functionalized PP improved the bonding. Thus, the study clearly demonstrated that the combination of mechanical abrasion, plasma activation (especially with argon), and functional interlayer deposition (PEDOT and PPAA) significantly enhances the adhesion of Cu to PP. This was achieved through increased surface roughness, improved wettability, and the incorporation of polar functional groups, all contributing to stronger interfacial interactions. The results underscore the importance of multi-step polymer surface engineering in developing high-performance, durable metal-polymer hybrid systems.

12.3.2.2 UV–Ozone Surface Modification of PMMA for Enhanced Metal Adhesion and Durable Biosensor Fabrication

In this study, Liu *et al.* [42] aimed to develop a simple, cost-effective, and efficient method to enhance the adhesion between metal layers and the polymer substrate PMMA (poly (methyl methacrylate)). This improvement was crucial for fabricating stable and reliable electrochemical biosensors, where metal adhesion failure often limits performance. To address this issue, the researchers used a surface engineering technique involving ultraviolet–ozone (UV–ozone) treatment to activate the PMMA surface prior to metal deposition. Unlike traditional adhesion enhancement techniques such as oxygen plasma or chemical solvent cleaning, UV–ozone treatment offered a clean, non-invasive, and chemical-free method for surface modification. The PMMA surface was exposed to UV light at 185 nm and 254 nm wavelengths for 5 minutes, during which ozone (O_3) was generated from oxygen by 185 nm UV and then split into highly reactive atomic oxygen species by 254 nm UV. These reactive species introduced oxygen-containing polar functional groups and increased surface roughness, significantly enhancing surface hydrophilicity and chemical reactivity. Following this treatment, thin metal films: either copper (Cu) or a titanium/gold (Ti/Au) stack were sputter-deposited, and photolithography was used to pattern the metal electrodes for biosensor fabrication. The schematic illustration of the whole treatment is shown in Figure 12.10a. The UV–ozone treatment resulted in dramatic improvements in metal adhesion, as confirmed by standard tape tests: Cu film adhesion to PMMA increased from only 15.6% on untreated (virgin) PMMA to 94.1% after UV–ozone surface treatment, showing a 6-fold improvement. This is even better than Cu film adhesion to glass substrates (86.9%), Ti/Au film adhesion to PMMA improved from 9.5% on untreated PMMA to 96.0% on UV-ozone treated PMMA (a 10-fold increase), which is much higher than the adhesion seen with oxygen plasma-treated PMMA (26.3%). Au without Ti layer on UV-ozone treated PMMA showed 89.9% adhesion. Compared to conventional methods like oxygen plasma or acetone soaking, UV–ozone yielded superior and more stable adhesion. Stability tests further verified the robustness of the treatment. A chip made from PMMA with Cu and Au dots was immersed in deionized (DI) water to simulate long-term use. While Cu electrodes showed oxidation and eventually cracked after 7 days due to water-induced swelling and internal stress, Au electrodes remained intact and free from delamination even after 30 days, demonstrating the long-term stability of the UV–ozone-modified PMMA surface. Characterization of the modified surfaces provided further

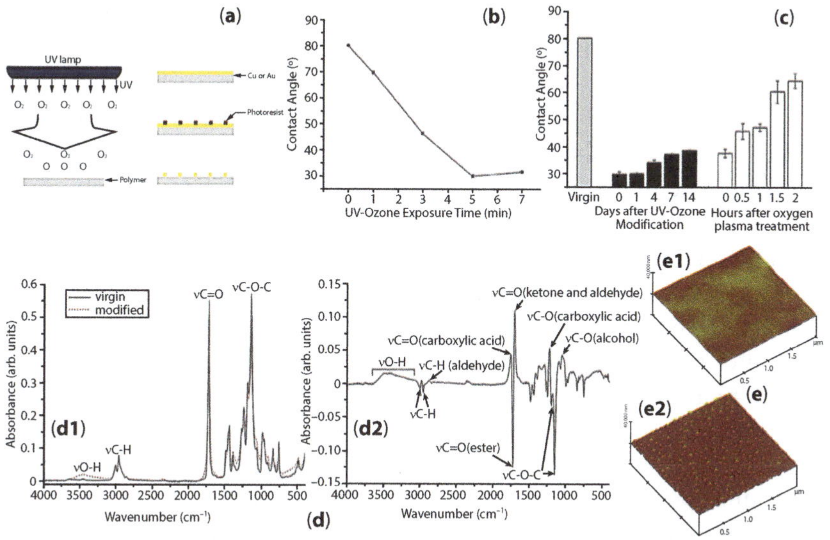

Figure 12.10 (a) Schematic of the UV–ozone treatment process, (b) water contact angle reduction of polymer (PMMA) surface with exposure time, (c) ageing study after UV ozone and oxygen plasma treatments, (d) ATR–FTIR spectra indicating oxygen-containing groups on PMMA surface: (d1) before modification and after UV-ozone treatment, (d2) corresponding difference spectra highlighting changes after the treatment, and (e) AFM images of PMMA showing increased surface roughness virgin (e1) and UV/ozone treated (e2), as suggested by Liu *et al.* [42], with copyright permission from American Chemical Society.

insight into the adhesion improvement mechanism. Water contact angle measurements (Figure 12.10b) showed a decrease from 80.0° (untreated, hydrophobic) to 29.9° after 5 minutes of UV–ozone exposure, indicating increased surface wettability and energy; beyond 5 minutes, no significant changes were observed. Treatment beyond 7 minutes caused heat-induced distortion, establishing 5 minutes as the optimal exposure time. Notably, this contact angle was even lower than that achieved by oxygen plasma (37.7°), confirming better surface activation. Ageing studies (Figure 12.10c) revealed excellent temporal stability, with contact angles at 30.1° (1 day), 37.3° (7 days), and 38.6° (14 days) post-treatment. In comparison, oxygen plasma-treated PMMA rapidly lost its hydrophilicity, with contact angles rising to 45.8° after 30 minutes and 64.2° after 2 hours, highlighting UV–ozone's advantage in preserving surface activation during typical device preparation timelines.

Further surface chemical analysis using attenuated total reflectance (ATR–FTIR) (Figure 12.10d) confirmed the formation of hydroxyl (OH), carbonyl (C=O), and carboxylic acid (COOH) groups, which promote strong chemical bonding with metals. Additionally, atomic force microscopy (AFM) (Figure 12.10e) showed an increase in surface roughness from 1.02 nm to 5.15 nm, contributing to mechanical interlocking that further improved adhesion. This dual effect chemical functionalization and nanoscale topographic change was the main mechanism behind the enhanced metal-polymer bonding. Finally, electrochemical testing validated the practical relevance of the modified PMMA surface. The biosensors demonstrated sensitive and reproducible dopamine detection with low relative standard deviations (RSDs) in both intra- and inter-device tests. Even after 6 days of immersion in phosphate-buffered saline (PBS), the sensors retained their performance, confirming both adhesion durability and functional stability. Thus, the study demonstrated that UV–ozone surface modification is an effective and practical method for enhancing metal films adhesion to polymer substrate, especially for PMMA. By introducing polar functional groups and increasing surface roughness, it achieved significant improvements in both immediate adhesion and long-term stability: outperforming traditional methods and offering a promising route for the fabrication of robust polymer-based microelectronic and biosensing devices.

12.3.3 Polymer-Polymer Adhesion

Polymer-to-polymer adhesion is a significant challenge due to the inherent low surface free energy, chemical inertness, and smooth morphology of most polymers, which prevent effective interfacial bonding. For example, bonding polyethylene (PE) to polyamide (PA) or polytetrafluoroethylene (PTFE) to other polymers is difficult without surface modification. In conventional juice pouches, various polymers are laminated to create a multi-layer structure—polyethylene (PE) is typically employed for sealing, poly(ethylene terephthalate) (PET) provides structural integrity, and polyamide (PA) serves as a barrier layer. However, due to their low surface free energy and inherent chemical incompatibility, these polymers exhibit poor interfacial adhesion. To enhance adhesion between layers, manufacturers often employ corona treatment on the surfaces of PE or PET films before bonding. This surface modification introduces polar functional groups, such as hydroxyl (-OH) and carboxyl (-COOH), and induces a slight surface roughness, thereby enhancing the interfacial bonding between the substrate and adhesives or co-extruded layers. Consequently, this treatment facilitates robust bonding among the layers, which effectively withstands the mechanical stresses associated with bending, filling, or sealing of the pouch, thereby mitigating issues related to delamination and leakage. Thus, surface engineering plays a crucial role in achieving durable, high-strength polymer–polymer laminates in real-world applications.

12.3.3.1 Enhancing Polymer-to-Polymer Adhesion of LDPE Films Using Plasma Surface Treatment for Automotive Applications

In this case study, Sanchis *et al.* [43] aimed to improve polymer-polymer adhesion between a low-density polyethylene (LDPE) film and polyolefin foam using a reactive polyurethane adhesive for automotive applications. The adhesive used was a polyurethane (PUR) grade (XPU 18288), activated at a temperature of 25°C and relative humidity of 60–90%. The lamination was performed using a laboratory-scale hot-melt laminator, with a cylinder temperature of 120°C, cylinder rate of 6 mm/min, and cylinder pressure of 2 bar. To enhance the adhesion between the LDPE film and the polyolefin foam, a surface engineering technique involving low-pressure RF plasma treatment with oxygen (O_2) and nitrogen (N_2) gases was employed. Plasma treatment, particularly O_2 plasma, is known for its ability to introduce highly reactive polar groups, such as hydroxyl (-OH), carbonyl (C=O), and carboxyl (-COOH) groups, onto the polymer surface, thus improving wettability and adhesion by increasing surface free energy. The treatment was conducted using an RF glow discharge system operating at 13.56 MHz with a power of 300W, exposing the samples for 1–20 minutes for both O_2 and N_2. Surface modification was characterized by measuring contact angle variations and surface elemental composition using XPS. Plasma treatment led to a significant increase in the wettability of the LDPE film, as reflected in the contact angle. The contact angle for water (89.3°) and glycerol (77.9°) decreased significantly after plasma treatment, reaching values of 36.5° and 39.0°, respectively, with O_2 plasma for 20 minutes, and 31.5° and 38.0° with N_2 plasma. XPS analysis revealed an increase in surface oxygen content (atomic %) from 3% in untreated films to 21% and 22% after O_2 and N_2 plasma treatments, respectively. This functionalization by oxygen-containing species was crucial for improving the adhesion property of the polymer materials. Surface roughness measurements showed only slight changes in roughness after plasma treatment, with RMS values increasing from 21.3 nm (untreated) to 24.6 nm for O_2 and 22.6 nm for N_2 plasma-treated films (Figure 12.11a). The minimal increase in roughness indicated that the major mechanism for improving adhesion was surface functionalization rather than topographical modification. The bond performance was evaluated through T-peel strength tests, which revealed an increase in bond strength after plasma treatment. Initially, the T-peel strength for untreated films was around 0.01 N/cm, but after only 2 minutes of plasma exposure, the strength improved to 1.37–1.40 N/cm for both O_2 and N_2 plasmas (Figure 12.11b). Prolonged exposure times (10–20 minutes) led to a slight increase, reaching 1.46 N/cm for both gases, demonstrating that the most significant improvement occurred with short plasma exposure. Furthermore, the durability of the adhering systems was assessed under ageing conditions, with two different temperature and humidity profiles: c1 (T = 38°C, RH = 95%) and c2 (T = 70°C, RH = 55%). In the c1 condition, the T-peel strength decreased by 6.5% for O_2-treated films after 15 days, from 1.39 N/cm to 1.31 N/cm (Figure 12.11c). Similarly, the T-peel strength for N_2-treated films showed a reduction of 5.1% in c1 and 11.7% in c2 after 15 days of aging (Figure 12.11d). These results confirm that both plasma treatments significantly enhanced the adhesion strength, with O_2 and N_2 plasma treatments showing comparable performance. The slight decrease in T-peel strength under aggressive ageing conditions demonstrated the durability of the adhering

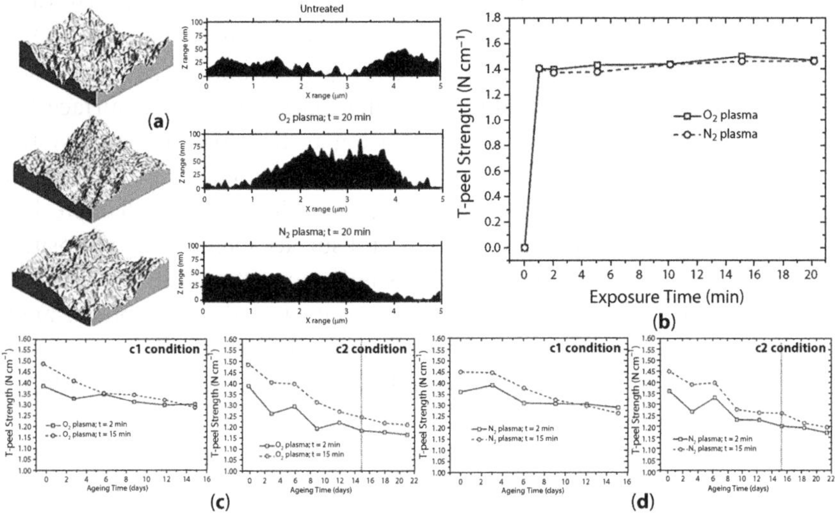

Figure 12.11 (a) Surface topography profiles of different plasma-treated LDPE films with various gases, obtained by AFM analysis. (b) Evolution of the T-peel strength of LDPE film–polyolefin foam adhesive joints as a function of exposure time to low-pressure plasma with different gases. (c) Evolution of the T-peel strength of oxygen plasma-treated LDPE films with aging time under storage conditions c1: T = 38°C, RH = 95%, and c2: T = 70°C, RH = 55%. (d) Evolution of the T-peel strength of nitrogen plasma-treated LDPE films with aging time under storage conditions c1: T = 38°C, RH = 95%, and c2: T = 70°C, RH = 55%, as suggested by Sanchis *et al.* [43], with copyright permission from Elsevier.

systems formed on plasma-treated surfaces, with a loss of less than 12–13%, which is within acceptable limits for automotive applications, particularly for use in steering wheels. This study concludes that plasma treatment with both O_2 and N_2 gases significantly enhances the adhesion between LDPE films and polyolefin foam adhesive, offering a reliable surface engineering approach for improving polymer-polymer adhesion in automotive applications.

12.3.3.2 Enhanced Adhesion of Engineering Polymers via Diffuse Coplanar Surface Barrier Discharge Plasma Treatment: A Comparative Study on PA6 and POM-C

This case study explores the enhancement of adhesion between Polyamide 6 (PA6) and Polyoxymethylene-Copolymer (POM-C) through surface engineering using diffuse coplanar surface barrier discharge (DCSBD) plasma treatment performed by Károly *et al.* [44]. These polymers are widely used in structural and automotive applications where reliable bonding is essential, yet they often suffer from poor adhesion due to their low surface free energy and chemical inertness. In this study, PA6 and POM-C samples were subjected to DCSBD plasma treatment under ambient air conditions. Prior to treatment, the samples were mechanically polished and cleaned to ensure consistent surface preparation. The plasma treatment was carried out for a short duration, using a DCSBD system designed to provide uniform discharge over a wide surface area. The treatment setup is shown Figure 12.12a. After treatment, the surface wettability of both polymers improved considerably, as evident from reduced water contact angles and increased surface free energy. These changes were more pronounced for PA6, suggesting a greater introduction of polar functional groups, which are known to promote adhesion. Surface chemical analysis using XPS confirmed the formation of oxygen-containing functionalities such as carboxyl and urethane-like groups on PA6, while POM-C mainly exhibited the removal of surface contaminants with minimal chemical changes (see Figure 12.12b). In terms of surface morphology, profilometric analysis showed a reduction in surface roughness for both polymers after plasma treatment, likely due to localized melting or smoothing effect caused by the plasma exposure. PA6, having initially higher surface roughness, showed more significant topographical changes compared to POM-C (Figure 12.12d). Mechanical lap-shear testing revealed a substantial improvement in the adhesion performance of PA6 when bonded with different types of adhesives, especially acrylic-based systems. Plasma-treated PA6 joints not

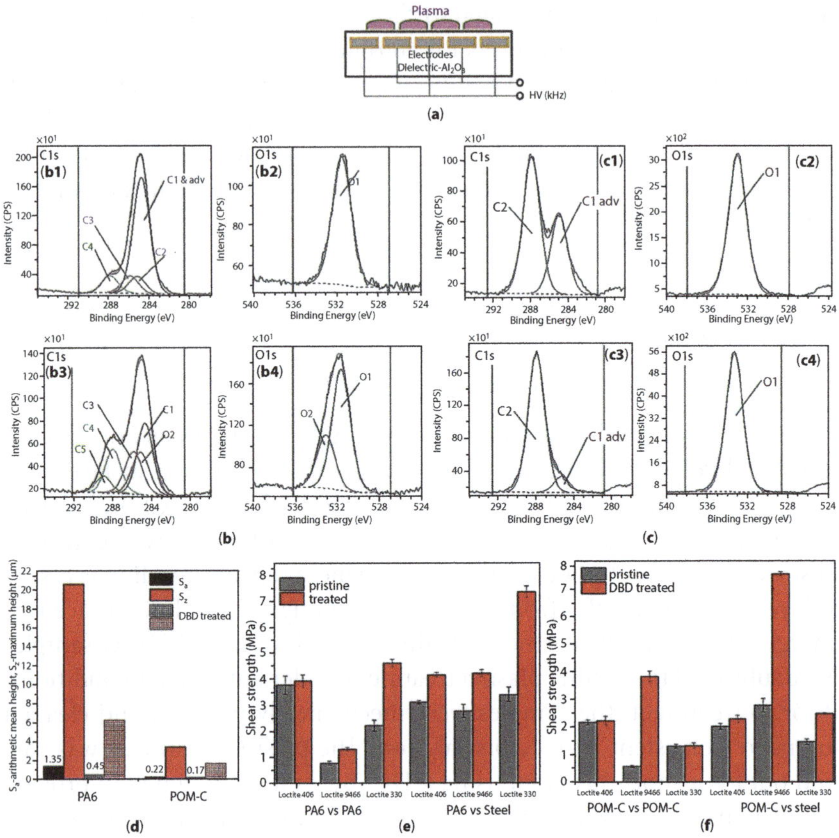

Figure 12.12 (a) Experimental diffuse coplanar surface barrier discharge (DCSBD) plasma setup; (b) C1s and O1s XPS spectra of (b1, b2) untreated and (b3, b4) plasma-treated PA6 showing reduced adventitious carbon; (c) XPS spectra of POM-C samples: (c1, c2) before and (c3, c4) after treatment; (d) surface roughness changes in PA6 and POM-C; Adhesive shear strength of pristine and plasma-treated: (e) PA6 and (f) POM-C with various adhesives, as suggested by Károly *et al.* [44], with copyright permission from MDPI.

only demonstrated higher bond strength but also showed more consistent results, indicating improved reliability (Figure 12.12e). In contrast, the improvement in POM-C adhesion was relatively limited and was mostly observed when bonded with epoxy adhesives (Figure 12.12f). The differences in performance between PA6 and POM-C highlight the importance of surface chemistry in adhesion, as the extent of plasma-induced modification directly influenced the bond strength. Overall, the study demonstrates that DCSBD plasma treatment is an effective method for enhancing the adhesion of PA6 by modifying its surface chemistry and wettability.

However, for polymers like POM-C with highly inert surfaces, additional or alternative surface modification techniques may be required to achieve comparable adhesion performance.

12.4 Prospects

Polymer adhesion to substrates such as fibers, glass, and wood continues to pose persistent challenges due to interfacial incompatibilities. Polymer-fiber systems often suffer from poor wettability and weak interfacial bonding, compromising load transfer and mechanical integrity. Iqbal *et al.* [45] aimed to improve the adhesion of PEEK and carbon fiber (CF) and glass fiber (GF) reinforced poly(phenylene sulfide) using atmospheric pressure plasma treatment, which significantly increased the polar component of surface free energy and surface roughness. Compared to low-pressure plasma, atmospheric plasma showed superior performance, with 3-4 times improvement in the lap shear strength of adhering systems. In polymer-glass systems, chemical dissimilarity and the absence of reactive sites hinder the formation of stable bonds. Benard *et al.* [46] improved the adhesion of glass/epoxy and carbon/epoxy composites using excimer laser treatment, which selectively ablates the polymer matrix without damaging the fiber reinforcements. This surface engineering approach enhances surface roughness, wettability, and the oxygen-to-carbon (O/C) ratio, leading to improved single lap shear strength, especially when laser parameters are optimized. Meanwhile, the inherent porosity, surface variability, and moisture content of wood complicate uniform adhesion and long-term performance. Oksman [47] improved adhesion in wood-polymer composites made of linear low-density polyethylene (LLDPE) and wood flour (WF) by incorporating compatibilizers such as maleic anhydride-modified Styrene-Ethylene-Butylene-Styrene (SEBS) triblock copolymer and Surlyn ionomer (a partially neutralized ethylene copolymer with ionic groups). These compatibilizers enhanced mechanical properties and interfacial bonding, as confirmed by tensile/impact tests and SEM analysis. These challenges necessitate tailored surface engineering strategies to meet specific application requirements. Further research on enhancing the adhesion of polymers through advanced surface engineering techniques remains essential, particularly for improving bonding performance with substrates such as wood, glass, and fiber in critical applications, including aerospace, automotive, biomedical devices, and structural composites. All surface modification techniques exhibit critical limitations that affect their effectiveness and durability. Plasma treatment, though effective in enhancing surface free energy, is often limited by hydrophobic recovery and ageing degradation, which reduce long-term adhesion performance. Surface coatings may introduce compatibility at the interface but are prone to delamination if not well-anchored or if subjected to environmental stress. Chemical grafting offers durable bonding through covalent modification but is hindered by complex protocols, toxicity of reagents, and scalability concerns. Flame treatment lacks precision and may induce thermal degradation, particularly on heat-sensitive polymers. Similarly, UV/ozone treatment is restricted by shallow penetration, short-lived effects, and ineffectiveness on non-transparent polymers. Future advancements in polymer surface engineering are expected to focus on hybrid treatments that integrate physical activation with chemical functionalization to enhance interfacial stability. The development of environmentally-benign and scalable techniques, such as low-temperature plasma or aqueous grafting methods, is also critical. Additionally, the incorporation of *in-situ* characterization tools can enable precise control over surface modification processes. Emerging strategies involving responsive surfaces and nanoscale patterning offer further potential to achieve durable, adaptive, and high-performance adhesion across diverse applications.

12.5 Summary

The surface engineering of polymers is a critical strategy for enhancing adhesion property to diverse substrates such as ceramics, metals, and composite materials. Various surface modification techniques have been established to elicit favorable alterations in the morphological and chemical attributes of polymer surfaces, significantly improving interfacial interactions and bond strength. These approaches range from conventional methods to more advanced techniques, including chemical etching, plasma treatment, laser irradiation, and the integration of functional nanoparticles. Despite significant advancements, the current landscape of surface engineering presents several noteworthy challenges. Many existing modification methods suffer from limitations concerning long-term stability or raise environmental concerns due to the employment of hazardous chemicals and energy-intensive processes. This underscores the urgent need for sustainable, environmentally-friendly, and durable surface modification technologies. Moreover, the investigation of novel approaches, particularly those leveraging nanotechnology and plasma-based treatments, opens promising research avenues. To propel the field forward, interdisciplinary collaborations that combine insights from materials science, surface chemistry, and applied engineering are crucial. A deeper understanding of the molecular mechanisms governing adhesion, optimization of surface modification parameters, and assessment of long-term performance under practical conditions are imperative. Such initiatives will significantly contribute to developing reliable, application-specific adhesion enhancement strategies, ultimately expanding the utilization of polymers in high-performance and multifunctional material systems.

12.6 Acknowledgements

The authors gratefully acknowledge the Indian Institute of Petroleum and Energy (IIPE) and the Indian Institute of Technology Indore for providing access to institutional research databases, which greatly supported the completion of this work.

AI disclosure statement: The author confirms that no artificial intelligence (AI)-generated content or AI-assisted writing tools were used in the preparation of this chapter (including summaries and the abstract). All text, analysis, and conclusions are the result of the author's original work.

References

1. J. P. Candlin, Polymeric materials: composition, uses and applications, *Comprehensive Analytical Chemistry* 53, 65–119 (2008).
2. J. Mercelis, Leo Baekeland's transatlantic struggle for Bakelite: patenting inside and outside of America, *Technology and Culture* 53, 366–400 (2012).
3. R. Mülhaupt, Hermann Staudinger and the origin of macromolecular chemistry, *Angewandte Chemie International Edition* 43, 1054–1063 (2004).
4. E.T. Strom, Wallace Carothers and polymer chemistry: a partnership ended too soon, in: The Posthumous Nobel Prize in Chemistry. Volume 1. *Correcting the Errors and Oversights of the Nobel Prize Committee*, pp. 121–163, American Chemical Society (2017).
5. X. Tan and D. Rodrigue, A review on porous polymeric membrane preparation. Part II: Production techniques with polyethylene, polydimethylsiloxane, polypropylene, polyimide, and polytetrafluoroethylene, *Polymers* 11, 1310 (2019).
6. R. Hsissou, R. Seghiri, Z. Benzekri, M. Hilali, M. Rafik and A. Elharfi, Polymer composite materials: A comprehensive review, *Composite Structures* 262, 113640 (2021).

7. F. Awaja, M. Gilbert, G. Kelly, B. Fox and P.J. Pigram, Adhesion of polymers, *Progress in Polymer Science* 34, 948–968 (2009).
8. G. Fourche, An overview of the basic aspects of polymer adhesion. Part I: Fundamentals, *Polymer Engineering and Science* 35, 957–967 (1995).
9. E.T. Kang, K.G. Neoh, J.L. Shi, K.L. Tan and D.J. Liaw, Surface modification of polymers for adhesion enhancement, *Polymer International* 10, 20–29 (1999).
10. S.K. Nemani, R.K. Annavarapu, B. Mohammadian, A. Raiyan, J. Heil, M.A. Haque, A. Abdelaal and H. Sojoudi, Surface modification of polymers: methods and applications, *Advanced Materials Interfaces* 5, 1801247 (2018).
11. P. Fabbri and M. Messori, Surface modification of polymers: chemical, physical, and biological routes, in: *Modification of Polymer Properties*, C. F. Jasso-Gastinel and J. M. Kenny (Eds), pp. 109–130, William Andrew Publishing (2017).
12. K.L. Mittal and A.N. Netravali (Eds.), *Polymer Surface Modification to Enhance Adhesion: Techniques and Applications*, Wiley-Scrivener, Beverly, MA (2024).
13. J. Schulz and M. Nardin, Theories and mechanisms of adhesion, in: *Handbook of Adhesive Technology*, A. Pizzi and K.L. Mittal (Eds.), pp. 19–34, Marcel Dekker, New York (1994).
14. D.E. Packham, The mechanical theory of adhesion, in: *Handbook of Adhesive Technology*, Second Edition, A. Pizzi and K.L. Mittal (Eds.), pp. 69–93, CRC Press, Boca Raton, FL (2003).
15. M.M. Chehimi, A. Azioune, and E. Cabet-Deliry, Acid-base interactions: relevance to adhesion and adhesive bonding, in: *Handbook of Adhesive Technology*, Second Edition, A. Pizzi and K.L. Mittal (Eds.), pp. 95–144, CRC Press, Boca Raton, FL (2003).
16. F.M. Fowkes, Acid-base interactions in polymer adhesion, in: *Physicochemical Aspects of Polymer Surfaces*, Vol. 2, K.L. Mittal (Ed.), pp. 583–603, Plenum Press, New York (1983).
17. K.L. Mittal (Ed.), Acid-Base Interactions: *Relevance to Adhesion Science and Technology*, Vol. 2, CRC Press, Boca Raton, FL (2000).
18. D.E. Packham and C. Johnston, Mechanical adhesion: were McBain and Hopkins right? An empirical study, *International Journal of Adhesion and Adhesives* 14, 131–135 (1994).
19. W.C. Wake, Theories of adhesion and uses of adhesives: a review, *Polymer* 19, 291–308 (1978).
20. D.A. Hays, Role of electrostatics in adhesion, in: *Fundamentals of Adhesion*, L. H. Lee (Ed.) pp. 249–278, Springer US, Boston, MA (1991).).
21. S.S. Voyutskii and V.L. Vakula, The role of diffusion phenomena in polymer-to-polymer adhesion, *Journal of Applied Polymer Science* 7, 475–491 (1963).
22. M. Geoghegan and G. Krausch, Wetting at polymer surfaces and interfaces, *Progress in Polymer Science* 28, 261–302 (2003).

23. R.J. Good, Contact angle, wetting, and adhesion: a critical review, *Journal of Adhesion Science and Technology* 6, 1269–1302 (1992).
24. K.L. Mittal, The role of the interface in adhesion phenomena, Polym. Eng. Sci., 17, 467–473, (1977)
25. J. Comyn, Theories of adhesion, in: *Handbook of Adhesives and Sealants*, P. Cognard (Ed.), pp. 1–50, Elsevier Science Ltd. (2006).
26. R.J. Good, M.K. Chaudhury, and C.J. van Oss, Theory of adhesive forces across interfaces: 2. Interfacial hydrogen bonds as acid–base phenomena and as factors enhancing adhesion, in *Fundamentals of Adhesion*, L. H. Lee (Ed.), pp. 153–172, Springer US, Boston, MA (1991).
27. J.J. Bikerman, Causes of poor adhesion: weak boundary layers, *Industrial & Engineering Chemistry*, 59, 40–44 (1967).
28. A.R. Hutchinson and S. Iglauer, Adhesion of construction sealants to polymer foam backer rod used in building construction, *International Journal of Adhesion and Adhesives*, 26, 555–566 (2006).
29. R. Galas, M. Omasta, L.B. Shi, H. Ding, W.J. Wang, I. Krupka and M. Hartl, The low adhesion problem: The effect of environmental conditions on adhesion in rolling-sliding contact, *Tribology International* 151, 106521 (2020).
30. F.M. Etzler, Characterization of surface free energies and surface chemistry of solids, in: *Contact Angle, Wettability and Adhesion*, Vol. 3, K.L. Mittal (Ed.), pp. 219–264, CRC Press, Boca Raton, FL (2003).
31. F.M. Etzler, Determination of the surface free energy of solids: A critical review, Rev. Adhesion Adhesives, 1, 3–45 (2013).J. Domińczuk and A. Krawczuk, Comparison of surface free energy calculation methods, *Applied Mechanics and Materials* 791, 259–265 (2015).
32. M. Lee, C.H. Wang and E. Yeo, Effects of adherend thickness and taper on adhesive bond strength measured by portable pull-off tests, *International Journal of Adhesion and Adhesives* 44, 259–268 (2013).
33. H. Zhang, D. Hussain, X. Meng, J. Song and H. Xie, Measurement of surface potential and adhesion with Kelvin Probe Force Microscopy, in: Proc. *2016 International Conference on Manipulation, Automation and Robotics at Small Scales (MARSS)*, pp. 1–6, IEEE (2016).
34. M. Toparli and S. Sasaki, Evaluation of the adhesion of TiN films using nanoindentation and scratch testing, *Philosophical Magazine A* 82, 2191–2197 (2002).
35. R.C. Dehaghani, M.M. Shokrieh and F. Taheri-Behrooz, An investigation on effects of acid etching duration on adhesive bonding of polyethylene to E-glass/epoxy composites, *International Journal of Adhesion and Adhesives* 85, 177–183 (2018).
36. E.P. Plueddemann, Adhesion through silane coupling agents, *Journal of Adhesion* 2, 184–201 (1970).
37. A.V. Zhuk, A.G. Evans, J.W. Hutchinson and G.M. Whitesides, The adhesion energy between polymer thin films and self-assembled monolayers, *Journal of Materials Research* 13, 3555–3564 (1998).

38. X. Zhang, Y. Xu, X. Zhang, H. Wu, J. Shen, R. Chen, Y. Xiong, J. Li and S. Guo, Progress on the layer-by-layer assembly of multilayered polymer composites: Strategy, structural control and applications, *Progress in Polymer Science* 89, 76–107 (2019).
39. B. Tosun and N. Yanıkoğlu, Evaluation of the effects of different surface modification methods on the bond strength of high-performance polymers and resin matrix ceramics, *Clinical Oral Investigations* 26, 3781–3790 (2022).
40. W. Wei, S. Xia, G. Liu, X. Gu, W. Jin and N. Xu, Interfacial adhesion between polymer separation layer and ceramic support for composite membrane, *AIChE Journal* 56, 1584–1592 (2010).
41. E. Dayss, G. Leps and J. Meinhardt, Surface modification for improved adhesion of a polymer–metal compound, *Surface and Coatings Technology* 116, 986–990 (1999).
42. J. Liu, L. He, L. Wang, Y. Man, L. Huang, Z. Xu, D. Ge, J. Li, C. Liu and L. Wang, Significant enhancement of the adhesion between metal films and polymer substrates by UV–ozone surface modification in nanoscale, *ACS Applied Materials & Interfaces* 8, 30576–30582 (2016).
43. R. Sanchis, O. Fenollar, D. García, L. Sánchez and R. Balart, Improved adhesion of LDPE films to polyolefin foams for automotive industry using low-pressure plasma, *International Journal of Adhesion and Adhesives* 28, 445–451 (2008).
44. K. Z. Károly, G. Kalácska, L. Zsidai, M. Mohai and S. Klébert, Improvement of adhesion properties of polyamide 6 and polyoxymethylene-copolymer by atmospheric cold plasma treatment, *Polymers* 10, 1380 (2018).
45. H.M.S. Iqbal, S. Bhowmik and R. Benedictus, Surface modification of high performance polymers by atmospheric pressure plasma and failure mechanism of adhesive bonded joints, *International Journal of Adhesion and Adhesives* 30, 418–424 (2010).
46. Q. Bénard, M. Fois, M. Grisel and P. Laurens, Surface treatment of carbon/epoxy and glass/epoxy composites with an excimer laser beam, *International Journal of Adhesion and Adhesives* 26, 543–549 (2006).
47. K. Oksman, Improved interaction between wood and synthetic polymers in wood/polymer composites, *Wood Science and Technology* 30, 197–205 (1996).
48. Ghanem, A., and Lang, Y., Introduction to Polymer Adhesion. 2017. Available online: https://www.researchgate.net/publication/319098070_Introduction_to_polymer_adhesion.

13

Interfacial Studies on Cellulose/ Biopolymer Composites

Merin Sara Thomas[1], Surya S. Nair[1], Sathaiah Gunaseelan[2], Prasanth K. S. Pillai[2], Gayathry Sreekumar[1], Sabu Thomas[3,4] and Laly A. Pothen[3,4*]

[1]Department of Chemistry, Mar Thoma College, Tiruvalla, Kerala, India
[2]Department of Food Science and Nutrition, University of Minnesota, St. Paul, MN, USA
[3]School of Energy Materials, Mahatma Gandhi University, Kottayam, Kerala, India
[4]School of Nanoscience and Nanotechnology, Mahatma Gandhi University, Kottayam, Kerala, India

Abstract
Interfacial studies play a critical role in determining the mechanical and functional properties of biopolymer composites. Due to the inherent hydrophilicity of cellulose, attaining strong interfacial bonding to relatively hydrophobic biopolymers is often challenging. Various approaches, including surface modifications, chemical treatments, and use of compatibilizers, have been employed to boost interfacial adhesion in composite systems. This chapter explores the fundamental interactions at the interface, common adhesion mechanisms, and applications of cellulose biocomposites.

***Keywords*:** Cellulose, interfacial adhesion, biopolymers, cellulose/biopolymer composites

13.1 Introduction

Cellulose is a naturally occurring biomaterial, first discovered by French chemist Anselme Payen in 1838. Cellulose, primarily made up of glucose

**Corresponding author*: lapothan@gmail.com

Anil N. Netravali and Frank M. Etzler (eds.) The Wide World of Adhesion: A Festschrift in Honor of Dr. Kash Mittal on his 80th Birthday, (469–488) © 2026 Scrivener Publishing LLC

units linked by beta-1,4-glycosidic bonds, imparts strength and rigidity to plants, making it vital for their structure and function [1]. Its structure is given in Figure 13.1. Cellulose is abundantly obtained in higher plants but also is produced by some algae, oomycetes, bacteria, the amoeboid protozoan *Dictyostelium discoideum*, and a group of marine animals such as tunicates [2]. Nearly 7.5 x 10^{10} tons of cellulose are synthesized each year in Nature, making it the most abundant revivable material on Earth. Cellulose is a unique and versatile biomaterial because of its several important properties. Some of these properties include high strength, biodegradability, insolubility in water, crystallinity, chemical reactivity, flexibility and hydrophilicity [3]. Due to its abundance, cellulose finds extensive use in a wide range of applications including food and medicine packaging, selective membranes, drug delivery systems, antimicrobial and transparent films, biomedical implants, reinforcing fillers for polymer composites, fibers and textiles, batteries and more [4]. By altering the properties, cellulose can also be used in sensors and environmental remediation [4].

Natural fiber reinforced biopolymers can occur as environmentally degradable biopolymer composites which are disintegrated by the influence of air, light, heat or microorganisms [5]. In addition, the production of various forms of cellulose-based materials has increasingly attracted researchers worldwide because of their renewability, biodegradability, biocompatibility and plentiful availability of hydroxyl and carboxyl groups for functionalization [6].

Cellulose-based biocomposite systems are made from cellulose which can act as both filler and matrix. Biodegradable polymer composites can be reinforced by natural cellulose fibers and have been widely used as a substitute to talcum and glass fibers for plastic reinforcement [7]. The strength of cellulose fiber composites is lower as compared to glass fibers due to the incompatibility between the hydrophobic host polymer matrix and hydrophilic natural fibers as well as lower thermal resistance of cellulose material. So, the structural modification of the cellulose fiber surfaces by chemical grafting and physicochemical treatments is essential to improve the mechanical properties and interfacial adhesion between the two phases [7, 8].

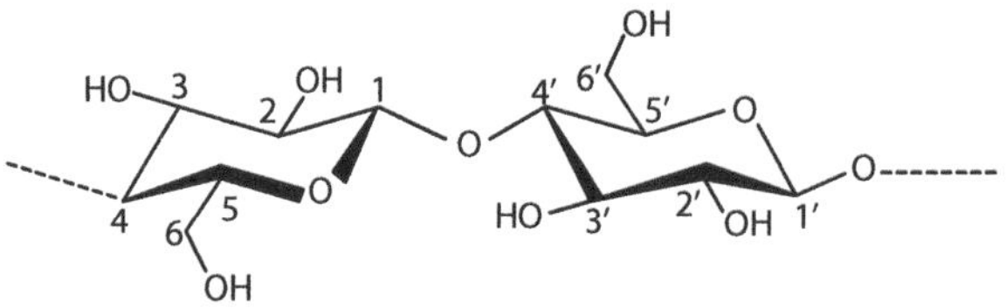

Figure 13.1 Schematic representation of cellulose [1].

To improve interfacial adhesion, it is essential that the materials in combination must have similar properties. For instance, hydrophilic fillers should be paired with hydrophilic matrices, or hydrophobic materials should be combined with other hydrophobic materials, to create a strong bond between them. The structural characteristic of a plant cellulose fiber composite is optimized by interfacial bonding between the polymer and reinforcing fiber. The interface plays a crucial role in controlling the mechanical properties of a composite [9].

13.2 Biopolymer/Cellulose Composites

The exploration of biopolymer/cellulose composites has gained significant interest in recent years due to their potential applications in various fields, including packaging, biomedical engineering, and sustainable materials development. This section delves into the characteristics and interfacial properties of several biopolymers when combined with cellulose, focusing on starch, chitin, proteins (collagen and gelatin), alginate, poly(lactic acid) (PLA), and polyhydroxyalkanoates (PHAs).

13.2.1 Starch

Starch is a widely used biopolymer that when combined with cellulose can enhance the mechanical and barrier properties of composites. The incorporation of microcrystalline cellulose (MCC) into starch films has been shown to improve their tensile strength and thermal stability, making them suitable for food packaging applications [10]. The interaction between starch and cellulose can lead to the formation of a network structure that enhances the mechanical properties of the composite. Moreover, the hydrophilic nature of both starch and cellulose facilitates good interfacial adhesion, which is crucial for effective load transfer within the composite. However, the processing conditions must be carefully controlled to avoid degradation of the biopolymers, which otherwise can adversely affect the mechanical performance [11].

13.2.2 Chitin

Chitin, a biopolymer derived from crustacean shells, exhibits excellent mechanical properties and biodegradability. When combined with cellulose, chitin can enhance the composite's strength and stiffness. Studies have demonstrated that chitin/cellulose composites possess improved tensile strength compared to pure cellulose or chitin alone [12]. The interfacial adhesion between chitin and cellulose can be optimized through chemical modifications, such as grafting or the use of compatibilizers, which improve the dispersion of chitin within the cellulose matrix [7]. Moreover, the distinctive properties of chitin, including its antibacterial property, can provide functional benefits to the composite, enhancing its suitability for use in biomedical applications [13].

13.2.3 Proteins

Proteins such as collagen and gelatin are biopolymers that can be effectively combined with cellulose to create composites with enhanced mechanical properties and biocompatibility. Collagen, being a natural polymer found in connective tissues, can improve the flexibility and toughness of cellulose-based composites [14]. Gelatin, on the other hand, can enhance the moisture retention and biodegradability of the composites. The use of crosslinking agents can improve the interfacial adhesion, leading to better load transfer and enhanced mechanical properties [12]. Furthermore, the incorporation of these proteins can also provide bioactive property, making the composites suitable for drug delivery systems and tissue engineering applications [13].

13.2.4 Alginate

Alginate, a polysaccharide derived from brown seaweed, is known for its gel-forming ability and biocompatibility. When combined with cellulose, alginate can enhance the mechanical and barrier properties of the resulting composites. The interaction between alginate and cellulose leads to the formation of a stable network structure, which improves the overall mechanical performance [15]. Alginate's ability to form hydrogels can also be beneficial for applications in controlled drug release and wound healing [13]. The interfacial adhesion between alginate and cellulose can be enhanced using crosslinking agents or by modifying the cellulose surface to improve compatibility [7]. This synergy between alginate and cellulose opens new avenues for developing functional materials with tailored properties.

13.2.5 Poly(lactic Acid)

PLA is a biodegradable thermoplastic that has gained popularity as a sustainable alternative to petroleum-based plastics. When reinforced with cellulose, PLA composites exhibit improved mechanical properties and thermal stability [14]. The interfacial adhesion between PLA and cellulose is crucial for achieving optimal performance, and it can be enhanced with compatibilizers or by modifying the cellulose surface [16]. Studies have shown that the incorporation of cellulose into PLA matrix can significantly improve the tensile strength and modulus of the composites, making them suitable for various applications, including packaging and automotive components [17]. Furthermore, the biodegradability of both PLA and cellulose makes these composites environmentally - friendly options for sustainable materials [13].

13.2.6 Polyhydroxyalkanoates (PHAs)

PHAs are a class of biodegradable polyesters produced by microbial fermentation of renewable resources. When combined with cellulose, PHA composites exhibit enhanced mechanical properties and biodegradability [18]. A high degree of interfacial adhesion between PHA and cellulose is essential for achieving high-performance composites, and it can be improved through various modification techniques, such as grafting or the use of coupling agents [7]. The unique properties of PHA, such as its good moisture resistance and flexibility, make it a highly effective material suitable for a wide range of applications, including packaging and biomedical devices [13]. Additionally, the sustainable nature of both PHA and cellulose makes these composites attractive for environmentally - conscious consumers and industries.

13.3 Interfacial Properties

13.3.1 Interfacial Adhesion

Interfacial adhesion is a critical factor that influences the mechanical properties of cellulose/biopolymer composites. The use of coupling agents such as maleic anhydride grafted polypropylene (MAPP) has been shown to significantly improve interfacial bonding between cellulose fibers and hydrophobic polymer matrices like polypropylene (PP) [19]. This enhancement in adhesion leads to better load transfer and improved mechanical performance [16]. Furthermore, the modification of cellulose surface to reduce its hydrophilicity can also enhance interfacial adhesion, as it minimizes aggregation and promotes better dispersion within the polymer matrix [20, 21].

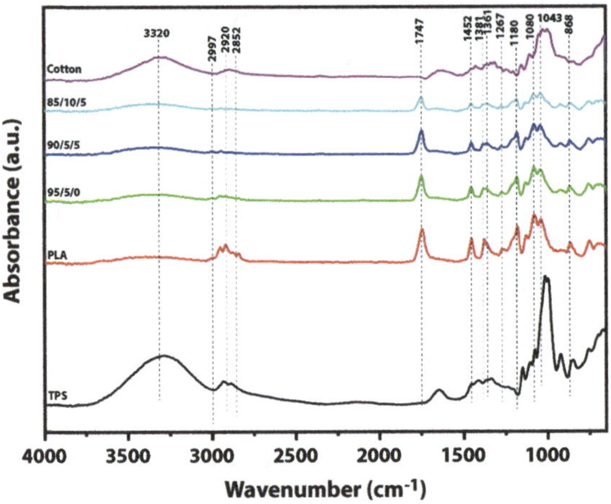

Figure 13.2 FTIR-ATIR spectra of PLA, TPS, cotton and PLA/TPS/cotton composites of different ratios [23].

Ruz-Cruz *et al.* [22] studied the effect of the incorporation of cellulose nanocrystals (CNCs) on the PLA biocomposites reinforced with cellulose microfibers (CMFs). They reported that the higher the CNCs content, the lower the crystallization of the multiscale biocomposites. In addition, CNCs improve the adhesion between cellulosic fibers and the PLA matrix by acting as adhesion promoters.

In a study by Silva *et al.* [23], the FT-IR analysis of the PLA/thermoplastic starch (TPS)/cotton composite showed distinct signals for PLA, TPS, and cotton, which confirmed the presence of physical interactions between the components. Figure 13.2 indicates that the extrusion process facilitates the integration of composite elements resulting in the formation of distinct bands related to TPS and cotton within the polymer matrix.

13.3.2 Interfacial Tension

The interfacial tension between cellulose and biopolymer matrices is influenced by the surface characteristics of the cellulose fibers. For instance, the dispersion component of surface energy is similar across various biopolymers, indicating that the interactions at the interface can be optimized through surface modifications [24]. This optimization is essential for achieving a balance between hydrophilic and hydrophobic properties, which directly affects the interfacial tension and, consequently, the overall performance of the composite [21].

13.3.3 Interfacial Morphology

The morphology at the interface can be significantly affected by the processing method and the nature of the cellulose used. Studies have shown that a homogeneous dispersion of cellulose within the polymer matrix leads to enhanced interfacial properties [25]. The presence of interfacial defects, often due to poor distribution of cellulose fibers, can compromise the mechanical and thermal properties of the composites [20]. Techniques such as dynamic mechanical analysis (DMA) and scanning electron microscopy (SEM) are commonly employed to characterize the interfacial morphology and assess the effectiveness of various modifications [11].

Khakalo *et al.* [26] provide an excellent example (Figure 13.3) of how SEM analysis effectively assesses the interface quality in cellulose nanofibrils (CNFs)/protein/PLA composites. The SEM analysis suggests that the hydrogen bonds play an insignificant role in the interfacial adhesion between CNFs and PLA, since the cracks appear in the protein-free nanocomposite indicative of poor PLA film formation. Due to these cracks, there occurs an inadequate interfacial adhesion and induced stress during composite formation, storage and handling [26].

Al *et al.* [27] in a study revealed that the increased mechanical properties of biopolymers such as polyhydroxybutyrate (PHB) and PLA with nanoclay

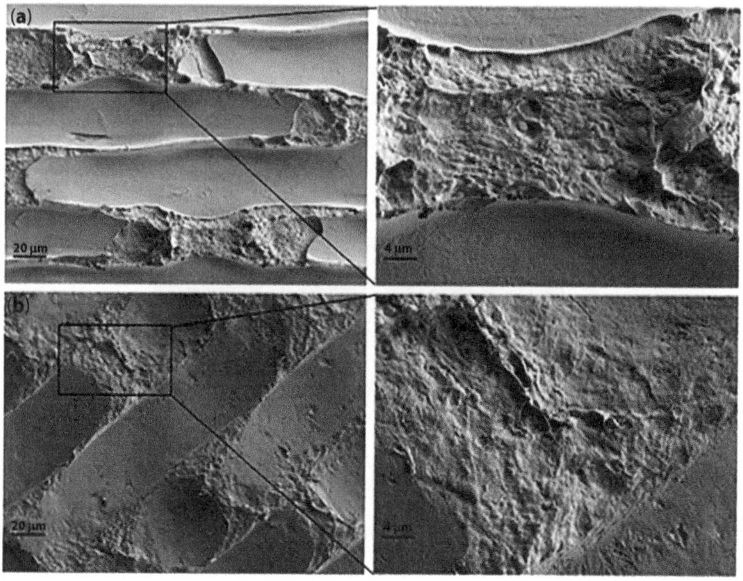

Figure 13.3 SEM images of CNF/PLA composite films prepared in the absence of protein (a) and with casein used as compatibilizer between CNF and PLA (b) [26].

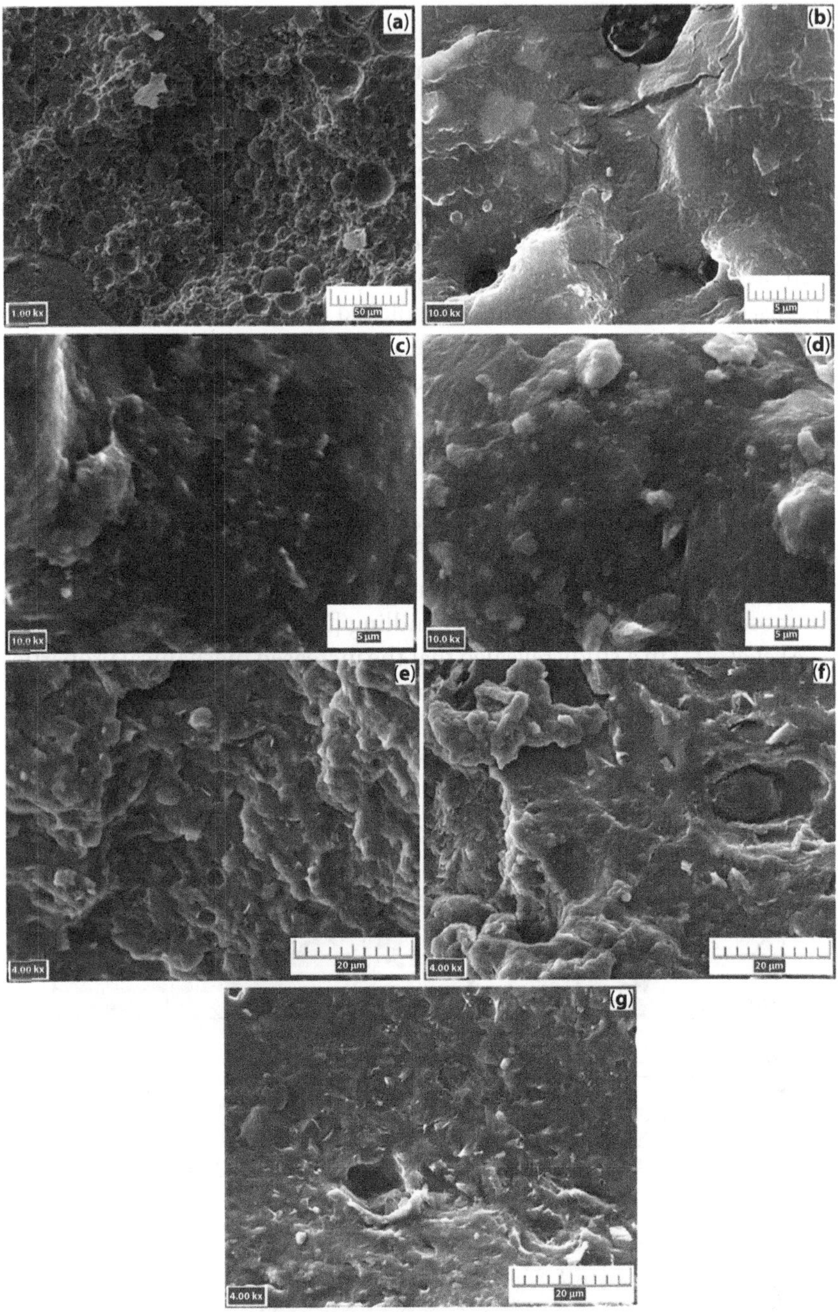

Figure 13.4 NC and CNF filled PHB biopolymer nanocomposites: (a) neat PHB, (b) 0.5% NC, (c) 2% NC, (d) 4% NC, (e) 0.5% CNF, (f) 2% CNF, and (g) 4% CNF [27].

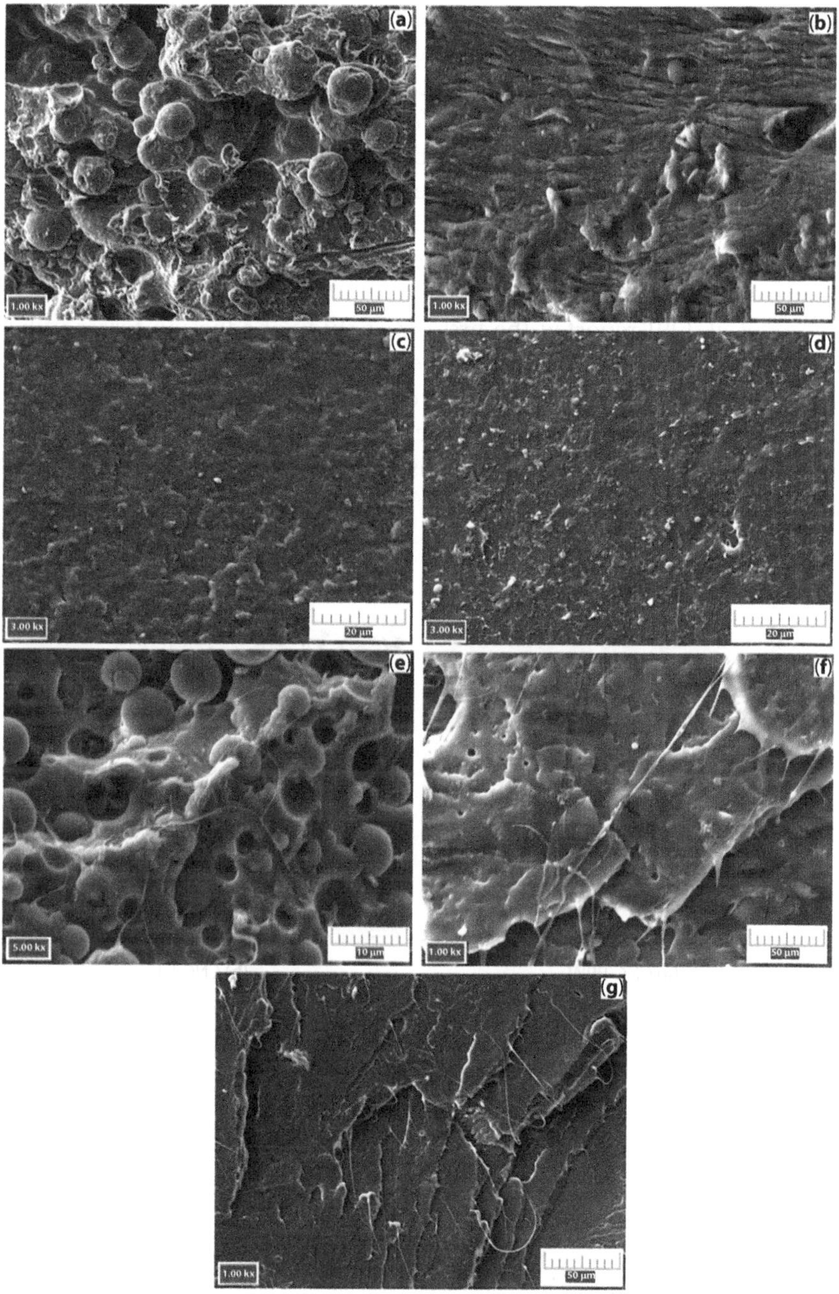

Figure 13.5 NC and CNF filled PLA biopolymer nanocomposites: (a) neat PLA, (b) 0.5% NC, (c) 2% NC, (d) 4% NC, (e) 0.5% CNF, (f) 2% CNF, and (g) 4% CNF [27].

(NC) and CNFs are due to good dispersion and resulting increased interfacial area between polymer matrix and nanofibers. This was explained with the help of SEM micrographs depicted in Figure 13.4 and Figure 13.5.

Agarwal *et al.* [28] studied the effect of CNCs from the pineapple peel waste (PPW), sisal fiber (SiF) and commercially available CNCs (CNC-C) on polypropylene nanocomposites. The TEM micrographs are shown in Figure 13.6. The aspect ratios of CNC-C and CNC-SiF were around 28 and 30 respectively, and facilitate effective stress transfer at the matrix-fiber interface.

Raj *et al.* [29] reported the first direct study of adhesion forces utilizing colloidal force microscopy between cellulose microbeads and soft PLA films. The results of the experiments conducted under dry air revealed the contribution of van der Waals forces which show that the forces in the PLA/cellulose/air system are lower as compared to those in the cellulose/cellulose/air system.

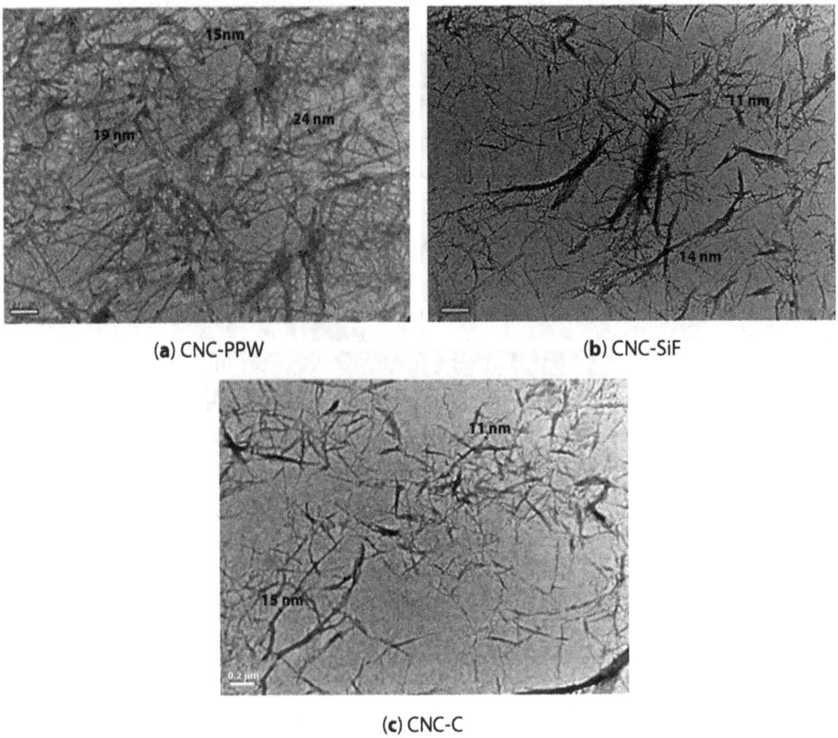

Figure 13.6 TEM micrographs of (a) CNC-PPW (b) CNC-SiF and (c) CNC-C [28].

13.3.4 Mechanical Properties

The mechanical properties, including tensile strength and modulus, are directly influenced by the interfacial adhesion and morphology. The interfacial adhesion between the filler and the matrix significantly impacts the mechanical properties of composites, and a strong interaction is essential to provide good stress transfer from the matrix to the fibres [30]. Composites reinforced with cellulose have demonstrated significant improvements in Young's modulus, although tensile strength may not always show a corresponding increase due to poor interfacial interactions [31, 32]. The mechanical performance can be further enhanced by optimizing the cellulose content and ensuring its proper dispersion within the matrix [33].

13.3.5 Thermal Properties

The thermal stability of cellulose/biopolymer composites is another important aspect influenced by interfacial properties. The degradation temperature and glass transition temperature (Tg) can be affected by the interaction between the cellulose and the polymer matrix. For instance, the incorporation of cellulose can lead to an increase in thermal stability, but excessive processing temperatures may degrade cellulose, adversely affecting its mechanical property [11]. Moreover, the thermal properties can be tailored *via* the selection of appropriate biopolymers and processing conditions [34].

13.3.6 Barrier Properties

The barrier properties of cellulose/biopolymer composites, such as water vapor permeability and oxygen transmission rate, are also influenced by interfacial characteristics. The concentration of cellulose fibers plays an important role in the formation of tortuous pathways within the matrix, enhancing barrier performance [10]. Additionally, the use of compatibilizers can improve the dispersion of cellulose, leading to better barrier property [35]. The interplay between the hydrophilic nature of cellulose and the hydrophobic characteristics of the polymer matrix is critical in determining the overall barrier performance [13].

Recent research reports show that nanocellulose can serve as an oxygen barrier material due to its ability to form a dense microstructure. TEMPO (2,2,6,6-tetramethylpiperidine-1-oxyl)-oxidised nanocellulose coated PLA films exhibit transparency and extraordinary oxygen barrier property [36].

13.4 Strategies for Interfacial Enhancement

13.4.1 Chemical Modification

Chemical modification can either be achieved by grafting or cross-linking. Grafting involves the attachment of functional groups or polymer chains onto the cellulose surface, enhancing compatibility and adhesion with the polymer matrix. Kelly *et al.* [37] fabricated polymer-grafted cellulose nanofibrils with enhanced interfacial compatibility. In the study, CNFs are surface-grafted with polymer chains that are chemically compatible with PLA by aqueous surface grafting. Thus, grafting creates a continuous chemical interface, by chemically attaching polymer chains to the cellulose surface, which can enhance the compatibility and properties of the composite material. This approach preserves the fibrillar architecture of the CNFs during spray drying and alter their interfacial interactions with the composite matrix. The resulting spray-dried CNFs exhibit significantly larger specific surface areas compared to unmodified CNFs and are comparable to values typically achieved through conventional freeze-drying method. This method retains the fibrillar structure of the CNFs and improves their dispersion within the PLA matrix. This results in improved strength and stiffness of the composites.

Cross-linking creates a network of covalent bonds between cellulose fibers and the polymer matrix, improving mechanical strength and thermal stability. Crosslinking can be achieved using crosslinking agents such as epoxy compounds, diisocyanates, maleic anhydride, polycarboxylic acids, etc. Zha *et al.* [38] studied the influence of CNF surface characteristics on CNF/alginate composite films. To understand the influence of alginate structure on the properties of the composites, they extracted two types of alginates with different guluronic acid to mannuronic acid (G/M) ratios and molecular weights from cultivated *Alaria esculenta* (AE) and harvested *Laminaria hyperborea* (LH). The study focuses on how modifying the surface charge of CNFs plays a crucial role in forming strong interactions with alginates. By creating an interpenetrating double network reinforced with interfacial cross-links (achieved through calcium ion cross-linking between alginate and the carboxylate groups on the CNF surface) they accomplished a synergistic improvement in the mechanical properties of the CNF/alginate composite. They also found that the G/M ratio and the molecular weight of alginate are critical factors. Alginates with a higher molecular weight and a lower G/M ratio exhibited stronger interactions with the modified CNFs, leading to composites with enhanced mechanical strength and water stability.

13.4.2 Physical Treatment

Physical treatment of CNFs refers to methods that modify their structure, surface properties, or dispersion behaviour without the use of chemical reactions. These techniques are often employed to improve the mechanical performance, compatibility, or processability of CNF-based composites. High pressure homogenization and ultrasonication are some of the physical treatments for improving interfacial interactions. During high pressure homogenization, CNF dispersions are subjected to intense shear, impact, and cavitation forces. As a result, the size of fibrils reduces, dispersion improves creating enhanced surface area and better interfacial interactions with the polymer matrix. Ultrasonication involves the application of high-frequency ultrasonic waves to disintegrate CNF aggregates, enhancing the dispersion and individualization of nanofibrils.

Jin *et al.* [39] studied the mechanics of CNC-reinforced bio-nanocomposites through micromechanical modelling and identified the critical roles of CNC aggregation and interphase properties in influencing composite performance. They achieved improvement in CNC dispersion and reduced defects by advanced processing techniques, such as high-pressure homogenization.

Ni *et al.* [40] studied the effects of ultrasonic conditions on the interfacial property and emulsifying property of cellulose nanoparticles from ginkgo seed shells and revealed that extending the duration of ultrasonic treatment is more effective than increasing ultrasonic power for producing nano-sized cellulose particles. Ultrasound treatment not only reduced the particle size but also decreased the crystallinity index of the cellulose, indicating structural modifications. Furthermore, it significantly enhanced the hydrophobicity of the cellulose nanoparticles, allowing them to adsorb more efficiently at the oil/water interface. The results showed that the emulsions stabilized by cellulose nanoparticles with small size and high hydrophobicity exhibited a better storage stability due to the formation of a viscoelastic interfacial film.

13.4.3 Use of Interfacial Compatibilizers

Interfacial compatibilizers are used to improve the compatibility between the two phases in a composite material, particularly between a hydrophilic filler (like cellulose fibers or nanoparticles) and a hydrophobic polymer matrix [41]. Interfacial compatibilizers work by either forming chemical bonds or physically modifying the interface between the two phases. These compatibilizers modify the interfacial interactions between the filler and the matrix, leading to better dispersion, stronger bonding, and enhanced mechanical properties of the final composite. Coupling agents and surfactants can be used as compatibilizers in cellulose composite systems [42, 43].

Silane coupling agents are usually used to improve the bonding between hydrophilic cellulose and polymer matrices. This mechanism may be chemical, intermolecular, diffusion, or physical [44]. Silanes, such as aminopropyltriethoxysilane (APTES), chemically bond to both the filler surface and the matrix, facilitating better adhesion [45]. Compounds like toluene diisocyanate (TDI), methylene diphenyl diisocyanate (MDI) or hexamethylene diisocyanate (HDI) react with hydroxyl groups on cellulose to form urethane linkages [46]. This enhances the interfacial bonding between the matrix and filler.

Non-ionic surfactants, such as poly(ethylene glycol) (PEG), or anionic surfactants, such as sodium dodecyl sulfate (SDS), can be used to reduce the surface tension between the matrix and filler, thus enhancing dispersion and adhesion [47, 48]. Olszewska [47] suggested an irreversible adsorption mechanism of a carboxymethylated cellulose (CMC) derivative onto nanofibrillar cellulose (NFC) fibrils surface in order to obtain strong hydrogen bonding between the two phases. This is achieved by the functionalization of native cellulose fibril surface by adsorption of PEG *via* non-ionic interactions.

13.5 Applications

13.5.1 Biodegradable Packaging

Biodegradable packaging often uses cellulose in combination with other biopolymers or additives, such as PLA, starch, or lignin. Strong interfacial adhesion between cellulose fibers and the matrix ensures efficient load transfer, enhancing the mechanical properties such as tensile strength and flexibility of the packaging material. Nanocellulose (e.g., cellulose nanofibers or cellulose nanocrystals) has a high surface area and superior mechanical properties, making it an excellent reinforcement in biodegradable packaging.

The interfacial adhesion between nanocellulose and other biodegradable materials significantly impacts the overall strength and toughness of the composite. Proper dispersion and interaction at the interface are vital to achieving these improvements. Han *et al.* [49] prepared an antimicrobial and degradable cellulose composite for functional and sustainable food packaging. They fabricated the composite films by solution casting method with a mixture of amino cellulose and carboxymethyl cellulose. The composite films were found to effectively extend the shelf-life of shrimp by inhibiting microbial growth and oxidative deterioration of the shrimp meat.

Di Liberto and Dintcheva [50] developed biobased films based on chitosan and microcrystalline cellulose for packaging applications by solvent casting method.

13.5.2 Biomedical Applications

Cellulose composites are revolutionizing the biomedical field by providing sustainable, biocompatible, and versatile materials for applications ranging from implants to drug delivery systems. They hold immense potential for advancing personalized medicine and in 'green' healthcare.

Cellulose composites hold great promise for tissue engineering due to their versatility and adaptability. PLA/CNC nanofiber electrospun membranes [51] have great potential in the fields of biological engineering, drug delivery, and tissue engineering scaffolds. Cellulose nanocomposites mimic extracellular matrix (ECM) and create an ideal environment for the adhesion and proliferation of cells [52]. Cellulose-ceramic composites, such as cellulose with hydroxyapatite, mimic the natural bone ECM [53]. Phatchayawat *et al.* fabricated three-dimensional (3D) hydrogel scaffolds for tissue engineering applications by combining bacterial cellulose, chitosan, alginate and gelatin [54]. These scaffolds promote wound healing by maintaining a moist environment and allowing oxygen permeability [54].

Nanocellulose can serve as an effective carrier for drugs, enabling targeted and controlled delivery. Its porous structure enables efficient loading and release of drugs, facilitating precise treatment.

13.5.3 Water Purification

Cellulose-based materials are widely used for water purification application because of their renewability, biodegradability, high surface area, and tunable functionalization [55]. Cellulose-based materials effectively reduce organic pollutants and heavy metal ions from water by electrostatic attraction, complexation or ion exchange. The high degree of biodegradability, adsorption capacity, and functional adaptability makes chitosan-based cellulose composites a good material for the removal of heavy metals from wastewater.

13.6 Summary

Biocomposites exhibit a lower carbon footprint compared to traditional composites. Cellulose-based composites are emerging as a sustainable alternative to conventional materials due to their eco-friendliness, renewable nature, and excellent mechanical properties. The integration of cellulose with various biopolymers such as starch, chitin, proteins, alginate, PLA, etc. presents significant opportunities for developing high-performance, sustainable materials. The interfacial properties between cellulose and these biopolymers play a crucial role in determining the mechanical, thermal, and barrier characteristics of the resulting composites. By optimizing these interfacial interactions through chemical modifications and processing techniques, it is possible to create composites that meet the demands of various applications while promoting sustainability.

AI disclosure statement: The author confirms that no artificial intelligence (AI)-generated content or AI-assisted writing tools were used in the preparation of this chapter (including summaries and the abstract). All text, analysis, and conclusions are the result of the author's original work.

References

1. O.S. Manoukian, N. Sardashti, T. Stedman, K. Gailiunas, A. Ojha, A. Penalosa, C. Mancuso, M. Hobert, and S.G. Kumbar, Biomaterials for tissue engineering and regenerative medicine. in: *Encyclopedia of Biomedical Engineering*,Vol. 1, R. Narayan (Ed), pp.462-482, Elsevier, Netherlands (2019).
2. A.D. French, S. Pérez, V. Bulone, T. Rosenau and D. Gray, Cellulose.in: *Encyclopedia of Polymer Science and Technology*, H. F. Mark (Ed.), pp.1–69, John Wiley & Sons, (2002).
3. T.F. Cooke, Biodegradability of polymers and fibers-A review of the literature. *J. Polym. Eng.* 9, 171–212 (1990).
4. P. Grzybek, G. Dudek and B. van der Bruggen, Cellulose-based films and membranes: A comprehensive review on preparation and applications. *Chem. Eng. J.* 495,153500 (2024).
5. B. Aaliya, K.V. Sunooj and M. Lackner, Biopolymer composites: a review. *Int. J. Biobased Plastics* 3, 40–84 (2021).
6. A.M. Mhatre, A.S.M. Raja, S. Saxena and P.G. Patil, Environmentally benign and sustainable green composites: current developments and challenges. in: *Green Composites: Sustainable Raw Materials*, S. Muthu (Ed.), pp. 53–90, Springer, Singapore (2019).

7. L. Wei and A.G. McDonald, A review on grafting of biofibers for biocomposites. *Materials (Basel).* 9, 303 (2016).
8. E.B. Ly, M.J. Lette, A.K. Diallo, A. Gassama, A. Takasaki and D. Ndiaye, Effect of reinforcing fillers and fibres treatment on morphological and mechanical properties of typha-phenolic resin composites. *Fibers Polymers* 20, 1046–1053 (2019).
9. H. Mohit and V. Arul Mozhi Selvan, A comprehensive review on surface modification, structure interface and bonding mechanism of plant cellulose fiber reinforced polymer based composites. *Composite Interfaces* 25, 629–667 (2018).
10. S.H. Othman, N.A. Majid, I.S.M.A Tawakkal,R. K. Basha, N. Nordin and R. A. Shapi'i, Tapioca starch films reinforced with microcrystalline cellulose for potential food packaging application. *Food Sci. Technol.* 39, 605–612 (2019).
11. A. Kiziltas, D.J. Gardner, Y. Han and H.-S. Yang, Dynamic mechanical behavior and thermal properties of microcrystalline cellulose (MCC)-filled nylon 6 composites. *Thermochim. Acta* 519, 38–43 (2011).
12. A.M. Adel, A. Dupont, H. Abou-Yousef, A. El-Gendy, S. Paris and N. El-Shinnawy, A study of wet and dry strength properties of unaged and hygrothermally aged paper sheets reinforced with biopolymer composites. *J. Appl. Polym. Sci.* 131, 40761- 40774 (2014).
13. R.M. Cigala, G. De Luca, I. Ielo and F. Crea, Biopolymeric nanocomposites for CO_2 capture. *Polymers (Basel).* 16, 1063 (2024).
14. T. Jamnongkan, N. Boonjuban, J. Sangkhachat, A. Wattanakornsiri and R. Mongkholrattanasit, Mechanical properties of biocomposite films based on poly (lactic acid) reinforced with cellulose fibers. *Solid State Phenom.* 280, 410–414 (2018).
15. R.J.B. Pinto, M.C. Neves, C.P. Neto and T. Trindade, Composites of cellulose and metal nanoparticles. *Nanocomposites–New Trends Dev.* 2, 1–25 (2012).
16. H. Wang, X. Liu, J. Liu, M. Wu and Y. Huang, Tailoring interfacial adhesion between PBAT matrix and PTFE-modified microcrystalline cellulose additive for advanced composites. *Polymers (Basel).* 14, 1973 (2022).
17. A. Gumowska, E. Robles, A. Bikoro, A. Wronka and G. Kowaluk, Selected properties of bio-based layered hybrid composites with biopolymer blends for structural applications. *Polymers (Basel).* 14, 4393 (2022).
18. S. Torres-Giner, M.J. Ocio and J.M. Lagaron, Development of active antimicrobial fiber-based chitosan polysaccharide nanostructures using electrospinning. *Engineering in Life Sciences* 8, 303–314 (2008).
19. N.I. Zulkifli and N. Samat, Mechanical properties of green recycled polypropylene composites: effect of maleic anhydride grafted polypropylene (MAPP) coupling agent. *Adv. Mater. Res.* 812, 187–191 (2013).
20. L. Zha, M. Yan, L.A. Berglund and Q. Zhou, Tailoring the holocellulose fiber/acrylic resin composite interface with hydrophobic carboxymethyl cellulose to enhance optical and mechanical properties. *Biomacromolecules* 25, 3731–3740 (2024).

21. Z. Ren, R. Guo, H. Bi, X. Jia, M. Xu and L. Cai, Interfacial adhesion of polylactic acid on cellulose surface: a molecular dynamics study. *ACS Appl. Mater. Interfaces* 12, 3236–3244 (2019).
22. M.A. Ruz-Cruz, P.J. Herrera-Franco, E.A. Flores-Johnson, M. V Moreno-Chulim, L.M. Galera-Manzano and A. Valadez-González, Thermal and mechanical properties of PLA-based multiscale cellulosic biocomposites. *J. Mater. Res. Technol.* 18, 485–495 (2022).
23. L.F. Silva, P.H.P.M. da Silveira, A.C.B. Rodrigues, S.N. Monteiro, S.F. Santos, J.P.S. Morais and D.C.Bastos, Cotton incorporated poly (lactic acid)/thermoplastic starch based composites used as flexible packing for short shelf life products. *Mater. Res.* 27, e20230366 (2024).
24. E. Gustafsson, E. Johansson, L. Wagberg and T. Pettersson, Direct adhesive measurements between wood biopolymer model surfaces. *Biomacromolecules* 13, 3046–3053 (2012).
25. B. Kaynak, M. Spoerk, A. Shirole, W. Ziegler and J. Sapkota, Polypropylene/cellulose composites for material extrusion additive manufacturing. *Macromol. Mater. Eng.* 303, 1800037 (2018).
26. A. Khakalo, I. Filpponen and O.J. Rojas, Protein-mediated interfacial adhesion in composites of cellulose nanofibrils and polylactide: Enhanced toughness towards material development. *Composites Sci. Technol.* 160, 145–151 (2018).
27. G. Al, D. Aydemir, B. Kaygin, N. Ayrilmis and G. Gunduz, Preparation and characterization of biopolymer nanocomposites from cellulose nanofibrils and nanoclays. *J. Composite Mater.* 52, 689–700 (2018).
28. J. Agarwal, S. Mohanty and S.K. Nayak, Valorization of pineapple peel waste and sisal fiber: Study of cellulose nanocrystals on polypropylene nanocomposites. *J. Appl. Polym. Sci.* 137, 49291 (2020).
29. G. Raj, E. Balnois, C. Baley and Y. Grohens, Probing cellulose/polylactic acid interactions in model biocomposite by colloidal force microscopy. *Colloids Surfaces A* 352, 47–55 (2009).
30. M.R.M. Asyraf, M.R. Ishak, A. Syamsir, N.M. Nurazzi, F.A. Sabaruddin, S.S. Shazleen, M.N.F. Norrrahim, M. Rafidah, R.A. Ilyas, M.Z. Abd Rashid and M.R. Razman, Mechanical properties of oil palm fibre-reinforced polymer composites: A review. *J. Mater. Res. Technol.* 17, 33–65 (2022).
31. S. Wang, Y. Lin, X. Zhang and C. Lu, Towards mechanically robust cellulose fiber-reinforced polypropylene composites with strong interfacial interaction through dual modification. *RSC Adv.* 5, 50660–50667 (2015).
32. S. Boran, A. Kiziltas, E.E. Kiziltas and D.J. Gardner, Characterization of ultrafine cellulose-filled high-density polyethylene composites prepared using different compounding methods. *BioRes.*11, 8178-8199 (2016).
33. K. Sakakibara, Y. Moriki, H. Yano and Y. Tsujii, Strategy for the improvement of the mechanical properties of cellulose nanofiber-reinforced high-density polyethylene nanocomposites using diblock copolymer dispersants. *ACS Appl. Mater. Interfaces* 9, 44079–44087 (2017).

34. S. Torres-Giner, N. Montanes, V. Fombuena, T. Boronat and L. Sanchez-Nacher, Preparation and characterization of compression-molded green composite sheets made of poly (3-hydroxybutyrate) reinforced with long pita fibers. *Adv. Polym. Technol.* 37, 1305–1315 (2018).
35. P.G. Ponnusamy, S. Sharma and S. Mani, Cotton noil based cellulose microfibers reinforced polylactic acid composite films for improved water vapor and ultraviolet light barrier properties. *J. Appl. Polym. Sci.* 139, 52329 (2022).
36. H. Fukuzumi, T. Saito, T. Iwata, Y. Kumamoto and A. Isogai, Transparent and high gas barrier films of cellulose nanofibers prepared by TEMPO-mediated oxidation. *Biomacromolecules* 10, 162–165 (2009).
37. P. V Kelly, S. Shams Es-haghi, M.E. Lamm, K. Copenhaver, S. Ozcan, D.J. Gardner and W.M.Gramlich, Polymer-grafted cellulose nanofibrils with enhanced interfacial compatibility for stronger poly (lactic acid) composites. *ACS Appl. Polym. Mater.* 5, 3661–3676 (2023).
38. L. Zha, F.L. Aachmann, H. Sletta, Ø. Arlov and Q. Zhou, Cellulose nanofibrils/alginates double-network composites: effects of interfibrillar interaction and g/m ratio of alginates on mechanical performance. *Biomacromolecules* 25, 4797–4808 (2024).
39. H. Jin, W. Goldberg, Z. Wang, H. Li, Y. Huang, M. Foston and G.M.Genin, Domain structure and interface control of mechanical stiffness in sustainable cellulose bio-nanocomposites. *arXiv Prepr. arXiv* 2412.13210 (2024).
40. Y. Ni, J. Li and L. Fan, Effects of ultrasonic conditions on the interfacial property and emulsifying property of cellulose nanoparticles from ginkgo seed shells. *Ultrasonics Sonochem.* 70, 105335 (2021).
41. M. Mohammed, R. Rahman, A.M. Mohammed, T. Adam, B.O. Betar, A.F. Osman and O.S. Dahham, Surface treatment to improve water repellence and compatibility of natural fiber with polymer matrix: Recent advancement. *Polym. Test.* 115, 107707 (2022).
42. C. Clemons and R. Sabo, A review of wet compounding of cellulose nanocomposites. *Polymers (Basel).* 13, 911 (2021).
43. A. Ferrer Carrera, Coupling agent usage in the preparation of cellulose nanofibril (CNF)-and cellulose nanocrystal (CNC)-based nanocomposites.in *Handbook of Nanocellulose and Cellulose Nanocomposites,* Vol.1, H. Kargarzadeh, I. Ahmad, S. Thomas and A. Dufresne (Eds.), pp. 335–364, John Wiley & Sons (2017).
44. K. Olonisakin, M. Fan, Z. Xin-Xiang, L. Ran, W. Lin, W. Zhang and Y. Wenbin, Key improvements in interfacial adhesion and dispersion of fibers/fillers in polymer matrix composites; focus on PLA matrix composites. *Composite Interfaces* 29, 1071–1120 (2022).
45. S.I. Voicu and V.K. Thakur, Aminopropyltriethoxysilane as a linker for cellulose-based functional materials: New horizons and future challenges. *Curr. Opin. Green Sustain. Chem.* 30, 100480 (2021).

46. A. Hejna, M. Marć and J. Korol, Modification of cellulosic filler with diisocyanates–volatile organic compounds emission assessment and stability of chemical structure over time. *Nord. Pulp Paper Res. J.* 36, 353–372 (2021).
47. A.M. Olszewska, Interfacial forces in nanocellulose based composite materials.Ph.D thesis, Aalto University, Espoo, Finland (2013).
48. A.R. Rahman, P.L. Teh, K.Y. Cheow and K. Suppiah, Modified carboxymethyl cellulose/halloysite nanotube (CMC/HNT) using sodium dodecyl sulfate (SDS), in:*Mineral-Filled Polymer Composites*, H. Ismail, S. M. Sapuan and R. A. Ilyas (Eds.), pp. 211–229, CRC Press, Boca Raton, Fl (2022).
49. J. Han, M. Chen, H. Liu, D. Zhang, Q. Shi, X. Xie and Y. Guo, Antimicrobial and degradable all-cellulose composite for functional and sustainable food packaging. *Industrial Crops Prod.* 222, 119966 (2024).
50. E.A. Di Liberto and N.T. Dintcheva, Biobased films based on chitosan and microcrystalline cellulose for sustainable packaging applications. *Polymers (Basel).* 16, 568 (2024).
51. M. Madani, C.D. Cruz, Z. Gounani, H. Baniasadi, P. Tammela, T. Laaksonen, J. Niskanen, and J. Seppälä, Functionalized cellulose nanocrystals reinforced PLA-gelatin electrospun fibers for potential antibacterial wound dressing and coating applications. *Int. J. Biol. Macromol.* 287, 138389 (2025).
52. F. V Ferreira, A.G. Souza, R. Ajdary, L.P. de Souza, J.H. Lopes, D.S. Correa , G. Siqueira, H.S. Barud, D.D.S. Rosa, L.H. Mattoso, and O.J. Rojas, Nanocellulose-based porous materials: Regulation and pathway to commercialization in regenerative medicine. *Bioact. Mater.* 29, 151–176 (2023).
53. J. Deng, Q. Song, S. Liu, W. Pei, P. Wang, L. Zheng, C.Huang, M. Ma, Q. Jiang and K. Zhang, Advanced applications of cellulose-based composites in fighting bone diseases. *Composites Part B.* 245, 110221 (2022).
54. P.P. Phatchayawat, A. Khamkeaw, S. Yodmuang and M. Phisalaphong, 3D bacterial cellulose-chitosan-alginate-gelatin hydrogel scaffold for cartilage tissue engineering. *Biochem. Eng. J.* 184, 108476 (2022).
55. A. Abdel-Hakim and R.M. Mouradb, Nanocellulose and its polymer composites: preparation, characterization, and applications. *Russ. Chem. Rev* 92, 5076 (2023).

14

Fiber/Resin Interfacial Shear Strength (IFSS) in Composites: A Brief Review of Micromechanical Methods of IFSS Measurement

Anil N. Netravali

Human Centered Design, Cornell University, Ithaca, NY, USA

Abstract

Advanced composites that use high strength fibers such as graphite (carbon), Kevlar®, glass, etc., to reinforce resins such as epoxies, esters, and others have been used in a wide range of fields such as military, aerospace, sports, automobiles, and many others for several decades. Their high strength and stiffness coupled with their low density make them suitable to replace metal parts and structures where lightweighting is critical. While fibers and resins are the critical components of such composite materials, the fiber/resin interfacial bonding plays a critical role. Strong fiber/resin bond makes these composites strong and stiff while weaker bond enhances their toughness. Depending on the application and the properties needed, the fiber/resin bond characteristics may be altered. As a result, measuring bond strength is critical in obtaining desired composite properties. This chapter presents a brief review of the most common micromechanical tests that are used to quantitatively assess the fiber/resin bond. These tests include single-fiber-composite test, single fiber pull-out test, and single fiber push-out and push-in tests. Other tests such as microbond and single fiber pull-out from a micro-composite, which have evolved from single fiber pull-out test are also described. Many examples of these tests carried out on different fiber/resin combinations and under a variety of test conditions have also been discussed.

Keywords: Fiber/resin interface measurement, fiber/resin bond, single fiber pull-out test, single fiber push-out test, single fiber push-in test, single-fiber-composite test

Email: ann2@cornell.edu

Anil N. Netravali and Frank M. Etzler (eds.) The Wide World of Adhesion: A Festschrift in Honor of Dr. Kash Mittal on his 80th Birthday, (489–520) © 2026 Scrivener Publishing LLC

14.1 Introduction

Advanced composites that combine high strength fibers such as glass, graphite, aramids, and others with resins such as epoxies, esters, etc., have been used in military and commercial airplanes as well as space systems (e.g. satellites) for the past 4-5 decades. Their high strength and stiffness (modulus) coupled with lower density compared to metals results in their extraordinarily high specific properties, making these composites a suitable material for lightweighting them. Fiber reinforced composite technology has now become common and trickled down to several diverse fields such as sports gear, electronics, wind turbine blades, and many others. They are now increasingly being incorporated into automobiles to reduce their weight, make them safer, as well as increase their fuel efficiency.

While the superior mechanical properties of the advanced composites are a result of high strength fibers (discontinuous phase) and resin (continuous phase) used to make them, the fiber/resin interfacial bonding plays a significant role [1]. Resin maintains the individual fiber positions and allows the stress to be transferred from broken fibers to the intact fibers, thus, allowing the composite to sustain the applied load. This stress transfer, from broken to intact fibers, can occur through the fiber/resin interface only when the fibers are bonded to the resin [1].

While polymeric resins do not rust, they do undergo physical aging and/or environmental degradation. Stresses created in the resin due to processing, environmental degradation, physical aging and fatigue generated through stress and/or thermal cycling and sustained mechanical loading during use can create microcracks in the resin. These microcracks, as they propagate under continuous load, can initiate fiber/resin interfacial debonding as well as fiber fracture, reducing the mechanical properties and the useful life of the composite.

When a composite is loaded, i.e., under tensile stress, fibers at their weakest points, i.e., at their largest flaws, start to break. The broken fibers do not carry any load at the fracture ends and that load is transferred to the surviving fibers in the neighborhood, by shear in the matrix, through the fiber/resin interface [1, 2]. It can, therefore, be safely assumed that fiber/resin bonding is key to this load transfer. An increase in the fiber/resin interfacial bonding, that prevents fiber from debonding, generally results in an increase in both composite strength and stiffness but reduces its toughness as well as impact strength [3, 4]. On the flip side, weaker fiber/resin bonding results in energy absorbing modes of fiber debonding and crack propagation along the fiber/resin interface resulting in increased energy absorption (toughness) of the composite [2–6]. However, weak bonding also reduces its tensile strength, transverse strength and modulus of unidirectional fiber composites. In addition, the interlaminar shear strength (ILSS) and transverse flexural strength are also directly related to the fiber/resin interfacial bonding [6, 7].

Fiber/resin interface has been traditionally defined as the 2-dimensional boundary between the resin and the fiber surface [3]. Chamis [3] calculated the amount of fiber/resin interface in 16.38 cm^3 (1 in^3) volume of a fiber composite consisting of 50% fiber volume with fiber diameter of 7.6 μm to be approximately 42,000 cm^2. Fiber diameters of present-day carbon fibers are even smaller, in the range of 5 μm. Also, in today's composites, the fibers are much densely packed, to take advantage of their strength, creating fiber volume of about 70%. Smaller fiber diameters and higher fiber volume means that the interfacial area would be significantly larger than 42,000 cm^2 calculated by Chamis [3]. Having such large fiber/resin interfacial area within a small composite volume indicates that a small change in the fiber/resin interface bonding characteristic, in either direction, can make a significant difference in the mechanical properties of the composites. From this point alone it should be clear that the fiber/resin interfacial characteristic plays a significant role in controlling the mechanical properties of the composites.

Three fundamental factors have been shown to determine or control the fiber/resin interfacial bonding: 1) micromechanical interlocking, also known as "lock and key" type mechanical coupling (bonding), between fibers and resin, 2) surface energetics that includes secondary bonds such as hydrogen bonds, van der Waals or electrostatic interactions, and 3) covalent bonding between fibers and resin [1, 7–12]. These are shown schematically in Figures 14.1 (a) to (c) [6, 10, 11]. Mechanical lock and key type of interlocking is created by fiber surface roughness that allows the resin to occupy the fiber surface valleys and holds the fibers mechanically, i.e., provides mechanical bonding. Once the fiber debonds from resin, fiber surface roughness enhances the fiber/resin friction and increases the energy absorbed as the fiber is pulled out during composite fracture. A couple of cautionary points to keep in mind are; 1) for the resin to flow into the fiber surface valleys effectively, the viscosity of the resin should be low enough, and 2) for fibers such as carbon, glass, boron, etc., which are brittle and sensitive to surface roughness, i.e., defects, introducing roughness may not be a good idea. It can decrease fiber strength significantly which, in turn, decreases mechanical properties of the composite. Matching surface energies of fibers and resin allows quick and even spreading of the resin on the surface of the fiber, further enhancing the mechanical bonding.

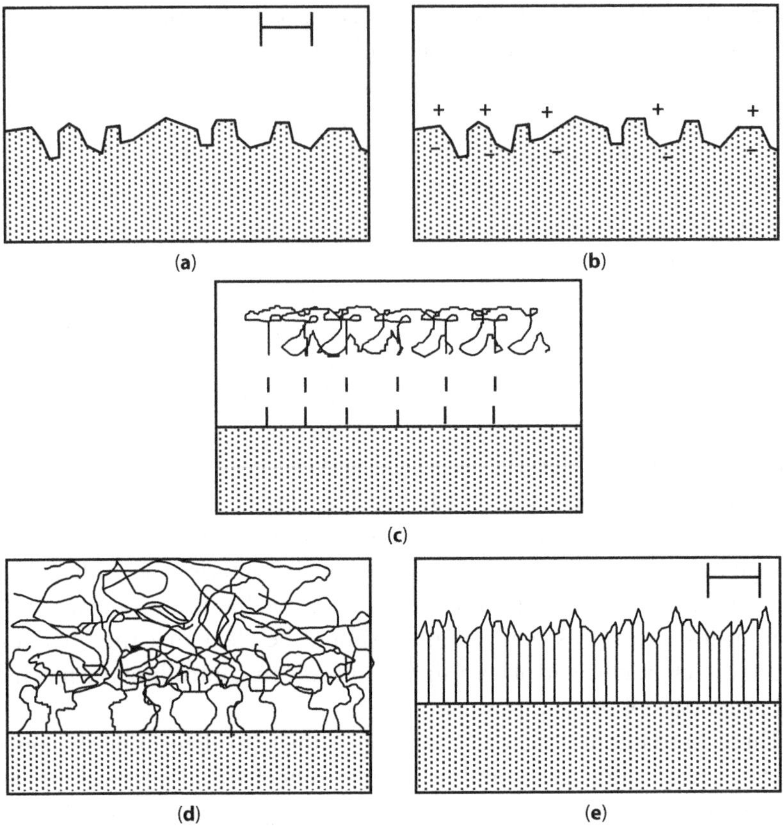

Figure 14.1 Schematic depiction of various interactions at the fiber/resin interface. (a) micromechanical interlocking, (b) permanent or induced dipole interactions, (c) chemical bonding, (d) chain entanglement, and (e) transcrystallinity on fiber surface [6, 11].

Close fiber/resin contact also increases secondary bonding between them. Covalent bonding between fibers and resin obviously provides the strongest bonding that requires significant energy to break the bonds. Matching fiber and resin surface energetics and providing possibilities for covalent bonding can be achieved by surface coatings or treating fibers by plasma, corona, oxidation, etc., to introduce desired chemical groups to match their surface chemistry with the resin [10, 11, 13, 14].

In the case of thermoset resins, e.g., epoxies, where heat is generated during exothermic curing (crosslinking) reaction, fibers act as heat sink and can create a thin 3-dimensional (3-D) layer of resin surrounding the fibers with a different, generally lower, crosslink density and, hence, different properties than the bulk resin. This 3-D region of resin, having a measurable thickness, is often termed as the 'interphase'. In thermoset resins, interphase can also be formed due to preferential adsorption of either the curing agent or the resin at the fiber surface, leading to a region of higher or lower crosslink density, respectively [13, 14]. In composites where fibers are very densely packed such that the distance between individual fibers is small, most of the resin may be considered as interphase.

In case of thermoplastic resins, depending on the fiber and resin chemistries, the fiber surface may act as a nucleating agent and help create transcrystallinity of measurable thickness on its surface when the composite (or resin) is cooled. This crystalline resin formed on the fiber surface obviously has different properties than bulk resin and acts as an interphase. Figures 14.1 (d) and (e) show the schematics of the entangled polymer molecules prior to crystallization and transcrystallinity formed on the fiber surface, respectively [6, 11].

To restate, the 'interface' is defined as the 2-D surface boundary, with no thickness, between the fiber and the resin, whereas the 'interphase' is defined as the 3-D region, of a certain thickness, that extends from the fiber surface inside the resin. Interphase has been shown to have different properties than the bulk resin.

14.2 Fiber/Resin Interface Studies

The importance of fiber/resin interface in advanced composites has been well recognized by researchers since 1960s. As a result, fiber/resin interface has been studied extensively. These studies can be classified into two distinct groups. The first group of studies consists of theoretical modeling of stress profile along the fiber, when the composite is loaded or is under stress and is characterized using Raman technique (spectroscopy) as well as optical microscopy. Micro-composite specimens in which a single fiber or single layer of multiple parallel fibers are embedded in resin have been particularly useful in such studies [15, 16]. These micro-composites are then strained to characterize the stress profile along the embedded fibers and compared to values predicted by the theoretical models. Fiber strength at each fracture location, during the tests, can also be calculated. The micro-composite techniques have been used for hybrid composites having different kinds of fibers as well [15–17]. The second category consists of a variety of micromechanical experimental methods that include single-fiber-composite (SFC) or fiber fragmentation tests [18–24], single fiber push-through (push-out) and push-in tests [25–33] and single fiber pull-out tests [34–40] which can provide a direct and quantitative measure of the fiber/resin interfacial shear strength (IFSS). While majority of these tests have been conducted on thermoset resins, primarily epoxy, researchers have modified some of these tests to make them suitable for different fiber/resin combinations and physical size and shape of the specimen. While all these tests are called direct tests, there are also indirect tests, such as short beam shear test, transverse shear test, vibration damping test, etc., that are conducted using real composite specimens that give some information about the fiber/resin bond characteristics and failure modes [10, 11]. This chapter deals with the most common direct micromechanical test methods that include SFC, single fiber pull-out, and single fiber push-out and push-in tests and their modified versions that have been used to measure the fiber/resin interfacial bond strength quantitatively.

14.3 Single Fiber Composite (SFC) or Fragmentation Test

The SFC test is a straightforward experimental procedure that involves embedding a single fiber, usually with a diameter in the range of 5-10 μm along the centerline of a relatively large dogbone-shaped specimen made of matrix materials (resins) such as epoxy or others [1]. The dogbone, 3-6 mm wide and about 1 mm thick in the central part (gauge section), containing the fiber is strained incrementally in very small steps or continually at a low strain rate. A key condition for the SFC test is that the matrix (resin) being used should have 3 to 5 times the fracture strain of the fiber. This is simply to assure that the specimen (resin) can be strained without breaking until the fiber fragmentation process stops. As the specimen is strained, the fiber breaks repeatedly, first at its weakest point (i.e. largest flaw). The break occurs because the resin shear transfer along the fiber length builds enough tensile stress to cause it to break at its flaw. While the fractured ends of the fiber do not carry any load (stress), the tensile stress starts to build along the length as the specimen is strained further. The stress continues to build until it reaches the 2nd weakest point and the fiber fractures at that location. This process continues as the specimen continues to get strained. At some point, when the fiber has fractured many times, the fragments become so short that the shear transfer along the fragment length can no longer build enough tensile stress to cause any further failure with increasing strain [1]. At this point, the test is complete, and the resulting fiber fragment length distribution makes up the raw data from the SFC test [1].

Kelly and Tyson [41] came up with the micromechanical model for the SFC test and argued that the maximum length, *lc*, of the fiber fragment can be given by

$$lc = S_f\, d/2t_y \tag{14.1}$$

where S_f is the fiber stress (taken as a constant), t_y is the yield stress in shear of the matrix, and d is the fiber diameter. The *lc* was also called the 'critical fragment length'. They modeled the matrix as a rigid, perfectly plastic material so that t_y is constant. This yielded a linear tensile stress buildup in the fiber fragment starting from zero at the broken end up to the applied stress. The S_f was also assumed as constant. However, this assumption is particularly untrue as the fibers have different flaws along the length, and each flaw is associated with a different strength or fracture stress. Since most high strength fibers follow Weibull distribution for their tensile strength, Netravali *et al.* [1] used this relation to restate the Eq. (14.1) as

$$t_y = d\, S_{peak}/2l \tag{14.2}$$

where l is the average fragment length and S_{peak} is the expected value of peak tensile stress achieved in that same fragment during the test. Further, they used Monte-Carlo simulation that employed commonly used Poisson/Weibull model to find the relation between S_{peak} and l [1].

Netravali *et al.* [1] also carried out extensive experiments using IM6-G carbon fibers and epoxy resins that were plasticized to six different extents. The main epoxy resin DER 331, bisphenol-A based (DGEBA) liquid epoxy, was blended with DER 732 (polyglycol diepoxide) that was used as a plasticizer (both from Dow Chemical Company). The stress vs strain plots of the 6 epoxy resins are presented in Figure 14.2.

In Figure 14.3 graphic representation of the progressive failures along a single fiber within the test gauge section during the SFC test on the left side and corresponding stress profiles in the fiber fragments are shown on the right side. As mentioned earlier, the ends of the fiber fragment do not carry any stress (load). However, the stress increases from both ends and levels out forming a trapezoidal shape. How fast the stress increases from the fragment end is a function of the resin modulus as well as the fiber/resin bond strength. For example, higher resin modulus along with good fiber/resin bonding will result in higher slope of the stress. As the specimen is strained further, the stress on the fiber keeps increasing until it reaches another weak point where it breaks. If the fragment is short, the

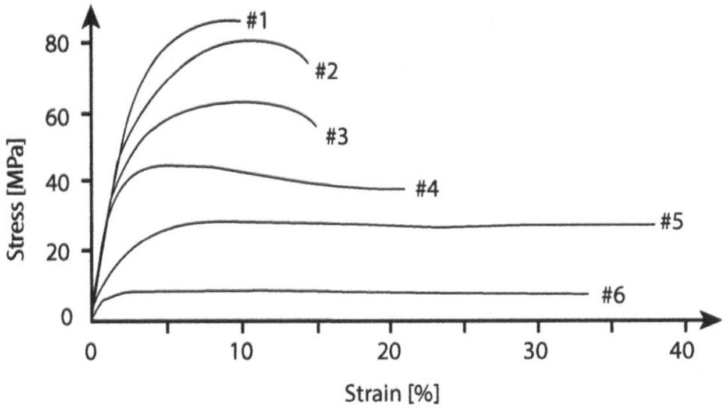

Figure 14.2 Stress vs strain plots of 6 epoxy blends, resin #1 (stiff) containing no DER 732 and resin #6 (fully plasticized) containing 50% DER 732 [1].

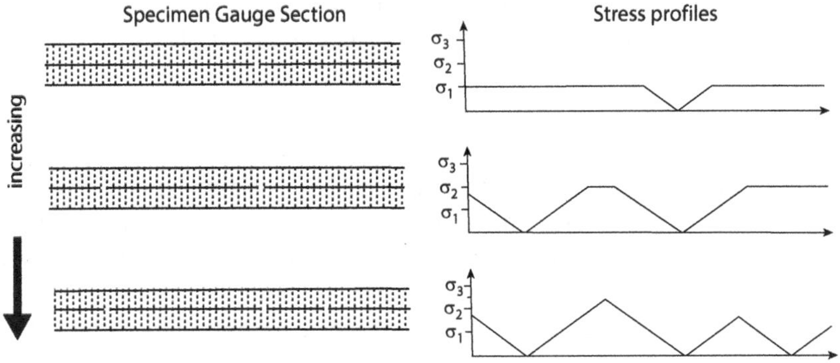

Figure 14.3 Graphical representation of the specimen gauge section (left) and the corresponding stress profile on the fiber (right) [1].

stress profile becomes a triangle. At this point, maximum possible stress on that fragment has been achieved and further straining the specimen has no effect on the stress profile. However, stress on longer fragments can keep on increasing until they break. Once all fragments along the entire gauge length show such triangular stress profile, no further breaks can be possible. At that point the SFC test is complete. The fragment lengths are then measured and the IFSS can be calculated by using Equation (14.2).

For matrix materials that are transparent such as epoxy resins, the SFC test can be carried out on an optical microscope using a small strain frame, and the entire fragment length distribution can be obtained for further processing. Many of these epoxy resins tend to be birefringent materials. In such cases, stress (or shear) profile in and around the fiber fragments, in the form of birefringence patterns, can be visually observed and recorded. Netravali *et al.* [1] who used plasticized epoxy resins were clearly able to show the different behaviors of the stiff and soft epoxies. In soft, heavily plasticized, epoxies every time the fiber broke, the fracture energy was enough to initiate a microcrack in the weak epoxy at the fracture location. However, resin being more flexible, the fiber/resin bond was intact. As the SFC test progressed and the specimen was strained further, not only did the fiber continue to break at different locations as expected, but the cracks formed at prior fiber breaks kept growing. At some point, one of those cracks grew big enough to break the specimen and, thus, the test was complete [1, 13]. Figure 14.4 (a) shows the birefringence patterns on two sides of the fragment end along with a small crack in the middle [13]. Figure 14.4 (b) shows the cracks grown significantly larger at each fiber break, indicated by the birefringence patterns moving away, towards the end of the SFC test [13].

In the case of stiff epoxy resin, the situation is very different. In this case, fiber break energy was insufficient to crack the stiff and strong resin. Instead, the fiber/resin bond that is weaker than the resin failed, i.e. fiber debonding occurred. While the initial debond length was small and thus not quite visible, as the specimen was strained further, the debond length kept growing. Since only friction existed at the debonded length, the shear stress in the resin was quite low compared to the stress that existed at the bonded part. Figure 14.5 (a) shows the birefringence pattern at the initial break and Figure 14.5 (b) shows the birefringence pattern at a later stage of the SFC test with much longer debond length. The low (weak) birefringence, (i.e., low shear stress in the matrix) seen around the fiber fragment break is due to the fiber/resin friction, whereas high (strong) birefringence

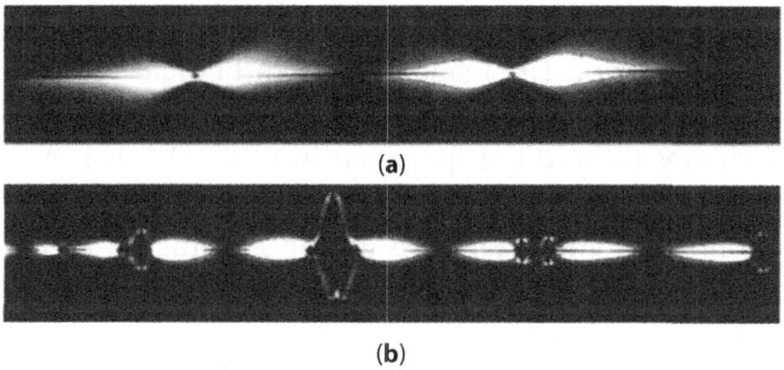

Figure 14.4 (a) Single fiber break creating a crack in soft (plasticized) resin and (b) Cracks grown at the end of the SFC test [13].

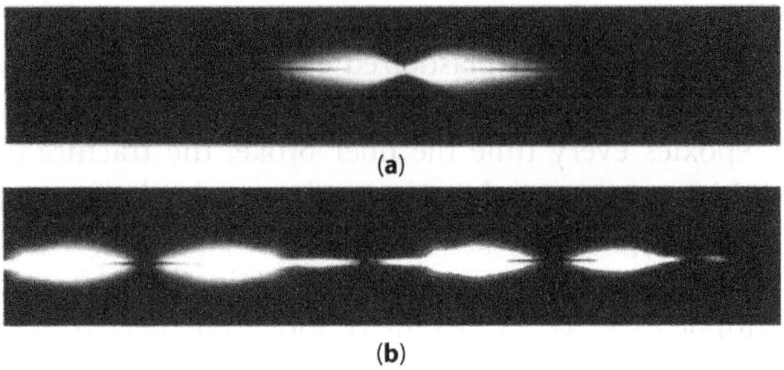

Figure 14.5 (a) Single fiber break in a stiff epoxy resin and (b) debonding seen around the fiber close to the fiber fragment ends and no debonding away from the fragment end [13].

pattern (high shear stress in the matrix), seen away from the fragment end, is where the fiber is still bonded to the resin.

It is important to note that when the resin is transparent and birefringent and the fiber fractures are clean, i.e. perfectly transverse, measuring the fragment lengths and observing the shear (birefringence) in the resin is easy and the obtained data are reliable. Conducting the SFC test and get fragment length data becomes difficult when 1) the resin is not transparent such as particle-filled or dark or completely opaque (e.g., metal) and fiber breaks cannot be seen and located, and 2) when the fiber fracture is not clean i.e., not transverse such as that obtained with glass and carbon fibers, so that the exact fracture location cannot be easily pointed out. For example, when fibrillar fibers such as Kevlar® or natural fibers such as jute fracture, the fracture cannot be easily detected. For a single break, instead of all fibrils fracturing at a single location, different fibrils or fibril bundles fracture at different locations depending on the flaws within the fibrils. For such fibers the birefringence pattern is not clear or spread over a large distance and hence, exact fiber break location cannot be pinpointed. However, part of the energy released during fracture is converted to ultrasound acoustic emission (AE) signal that can be detected using suitable sensors. Netravali and coworkers developed an AE-instrumented straining frame with clamps to hold the specimen in place [16, 17]. A motor with speed controller was used to strain the specimen at any desired strain rate. The elongation/strain and load/stress on the specimen were continuously monitored by a digital extensometer and a load cell, respectively. Both extensometer and load cell were interfaced to a computer (PC) system. A schematic diagram of the AE-instrumented SFC test system is shown in Figure 14.6 [16]. To run the SFC test, the specimen was clamped to the straining frame and AE sensors (miniature piezoelectric transducers), labeled A and B in Figure 14.6, were kept in contact with the specimen at two ends of the specimen gauge length. The specimen was then strained at a predetermined strain rate. When the fiber fractured at any point along the gauge length, both AE sensors picked up the AE signal and from the time lag (difference) between the two AE sensors, the exact fiber fracture location could be determined. As mentioned earlier, in the case of fibrillar fibers such as Kevlar®, many fibrils break during a single break. However, majority of fibrils break at a single location. While that fracture point is almost impossible to pinpoint by optical birefringence, the AE sensors are able to locate that point. The AE-instrumented SFC test system proved to work equally well for both transparent and opaque resins. At the end of the test when all the fiber fracture points are located, fragment lengths and distribution could be easily calculated. Further, the IFSS could be calculated using Eq. (14.2) as explained earlier. Netravali and coworkers were successful in characterizing glass [18], carbon and Nicalon fiber [19] IFSS with epoxy resin using the AE technique.

Holmes *et al.* [18] developed a fully automated testing machine for characterizing the evolution of fiber breaks during the SFC test and, thus, to evaluate the fiber/resin IFSS. Instead of using the AE from the fiber breaks, in this case, they recorded the fiber fractures using a camera. The archived fiber fracture data could then be analyzed using the algorithm

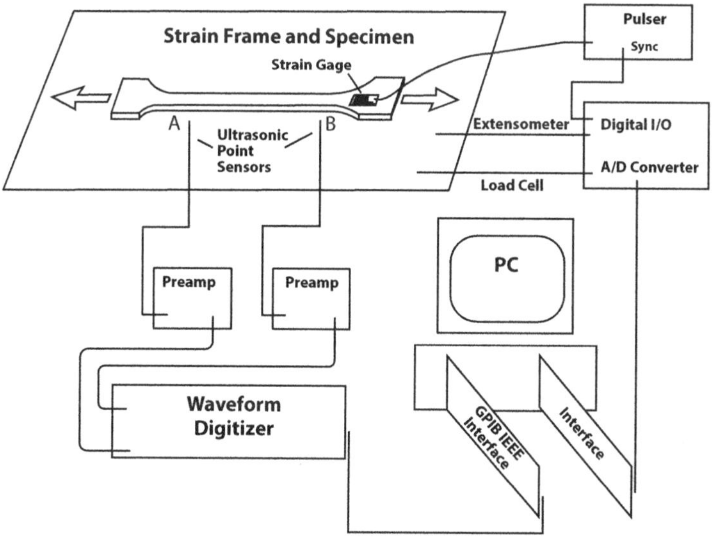

Figure 14.6 A schematic diagram of the AE-instrumented SFC test system [16].

they developed to obtain fiber/resin IFSS. Since the camera could record the stress field (birefringence patterns), fiber/resin debonding could be easily seen and characterized as well. They found their IFSS results to be comparable to those obtained by other researchers for similar fiber/resin systems. Holmes *et al.* [18] also used parallel glass fiber arrays in epoxy resin to characterize stress fields between the fibers and to observe how it varied with the distance between the fibers. They also studied stress fields near fiber breaks and how the intact fiber next to a fiber break was being loaded in that region, as speculated by theory. Their testing machine was also equipped with a load cell which allowed them to characterize the drop in load at each fiber break in the multi-fiber array during the entire fragmentation process. A major advantage of the automated system was to significantly reduce the time of the SFC tests while obtaining accurate data.

Ahure Powell *et al.* [19] recently developed a new and improved automated testing machine called 'Snappy', to conduct single fiber fragmentation tests (SFFT). This was achieved by increasing the image acquisition rate during the step-strain SFC experiment. This enhancement allowed easy observation of the fiber break evolution even during matrix relaxation periods, after each strain step. It also created critical record of the position, time, and local strain for each fiber break. Being an automated system, it minimized time and manpower needed for the tests. In this system, they developed computer software that processed large image data sets acquired during the test and automatically located fiber breaks and stored the information in a database. The automated fiber detection algorithm developed for the Snappy allowed rapid acquisition of a record of fiber break location as well as the stress at which that break occurred. This allowed them to calculate the strength of the fiber at any length and more importantly, at the critical length, that is needed to calculate the IFSS. Since the fiber fracture stress at every fragment length is obtained, Weibull distribution parameters for the fiber strength were also obtained.

Motta de Castro *et al.* [20] recently demonstrated a novel data processing approach for the SFC test on glass/epoxy system. Their primary objective was to maintain practicality by using the fiber fragmentation data as-is while deriving interfacial characteristics. Their two-step process involved using Monte Carlo simulations to establish accurate boundary conditions which were then used for subsequent numerical analysis that used cohesive zone model (CZM). They used this method to derive average IFSS (τ_{ave}) and critical fiber length which provided an estimation of the link between fiber/matrix properties and the fragmentation process. The CZM incorporated sample fabrication and thermal cure stresses, friction at the interface, and plasticity effects. This also allowed estimating useful composite properties such as Mode II critical energy release rate, and maximum shear traction (τ_{max}). Their results were comparable to those obtained by other researchers.

Several researchers have used the SFC technique to calculate the IFSS of various fiber/resin combinations and fiber surface treatments. Zhou and Netravali [21] characterized AS4 graphite fiber/PEEK (polyetheretherketone) resin after ethylene and ethylene/ammonia gas mixture plasma treatment using SFC technique. Both these plasmas deposit a thin polymeric coating on the fiber surface. However, ethylene plasma polymer is mostly hydrocarbon (low surface energy) in nature whereas ethylene/ammonia gas mixture deposits polymer that has many nitrogen containing groups which significantly increase the surface polarity and energy and provides possibilities for covalent bonding. Their results showed that graphite fiber/PEEK IFSS did not change after 100% ethylene plasma deposition. However, ethylene/ammonia gas mixture plasma increased the IFSS from about 50 MPa to about 91 MPa, an increase of 82%. Both these results were in line with the expectations.

Recently Umemura *et al.* [22] studied carbon fiber/polyetheretherketone (PEEK) resin interfaces using the AE technique. PEEK is a high temperature and chemically resistant thermoplastic resin used in many applications. They found the AE technique to be useful in determining the critical fragment length. They also obtained the Weibull parameters of fiber strength from fiber fragmentation (SFC) tests and compared them with the Weibull parameters obtained by testing dry tow (fibers) which turned out to be slightly different. However, they concluded that the Weibull parameter estimation from AE data analysis from the SFC test was more relevant than dry tow tensile tests as it incorporated the influence of the direct contact between the carbon fiber and the PEEK matrix. Umemura *et al.* also used fibers with the chemical surface treatment and found that the IFSS was much greater than for fibers that were not surface treated [22]. In addition, their study involving observation of fiber fragmentation under synchrotron X-ray microtomography allowed the visualization of the fragmentation mechanisms *in situ* and to estimate the effect of fiber surface roughness on the IFSS. They concluded that higher fiber surface roughness implied a faster IFSS decrease, probably because of higher wear of the polymer resin because of the fiber movement relative to the resin.

Weng *et al.* [23] characterized the carbon fiber/biobased epoxy resin IFSS using both SFC and fiber push-out (discussed later) techniques. They found IFSS values of 40.5 MPa and 42.2 MPa from SFC and push-out tests, respectively, for unsized carbon fibers. They also characterized the effect of atmospheric air plasma treatment of carbon fibers on their

IFSS with a biobased epoxy resin. Their result for plasma-treated carbon fiber/resin IFSS of about 75 MPa was about 87% improvement over control (untreated) carbon fiber. They indicated three reasons for this significant increase in IFSS after plasma treatment: (1) higher van der Waals bonding from enhanced fiber surface area; (2) better mechanical interlocking from the rough fiber surface; and (3) enhanced fiber/resin chemical bonding due to the oxygen-containing groups on the treated carbon fiber surface [23].

Girard *et al.* [24] used model single fiber composites made using larger diameter glass fiber/epoxy resin specimens and investigated fiber/resin interface. Displacement and strain fields, developed during the test, around the interface were measured experimentally using high magnification optical digital image-stress correlation. Debonding initiation and propagation stages were also characterized by using these images. Experimental measurements were verified and supplemented by finite element analysis (FEA). They concluded that the experimental results were relatively consistent than the FEA analysis.

Nakamura *et al.* [25] characterized the effect of halloysite nanotube (HNT) loading in epoxy resin on its IFSS with carbon fibers. Some of the carbon fibers, in their study, were prepared with a thin coating of ammonia/ethylene plasma polymer. They found that the modulus of the resin increased with HNT loading. Carbon fiber/epoxy IFSS also increased slightly with HNT loading. The ethylene/ammonia plasma polymer coated fibers, however, exhibited significantly higher IFSS by over 160%, independent of the HNT loading. The highest IFSS obtained in their study was about 79 MPa for plasma coated fibers while the IFSS for control fibers was around 30 MPa. They concluded that the ethylene/ammonia plasma polymer coating containing amine groups could react with epoxy resin to form covalent bonds and thus improve the interfacial bonding.

Fiber/resin interfacial debonding has also been characterized by loading the SFC specimen in transverse direction rather than in the direction of the fiber. This would debond the fiber from resin in tensile mode, rather than in shear mode as in the normal SFC tests. Girard *et al.* [26] prepared glass fiber/epoxy resin SFC specimens and observed fiber debonding under transverse loading. They observed abrupt debonding initiation followed by stable debonding propagation. They found the debonding to be either driven by 1) energy only or 2) both stress and energy conditions. Their main conclusion was that loading required for debonding to initiate depended on the opening (mode I) critical energy release rate (ERR) and tensile and shear strengths of the resin. These results would be useful when unidirectional composite is loaded in transverse direction or when the fibers are placed 0/90 degrees so that half of the fibers are in transverse direction to the applied stress.

14.4 Single Fiber Pull-Out Test

The single fiber pull-out test was initially used for larger diameter glass fiber/reinforced concrete and then for carbon filament/aluminum systems before being used for fiber reinforced polymeric (FRP) composites [27]. As has been demonstrated with many systems, fiber pull-out tests are relatively easy to carry out with rod-like or larger diameter fibers where handling such fibers is not too difficult [27]. However, when fiber diameter gets small, in the range of 4-6 μm, or when fiber/resin interfacial adhesion is strong, the embedded fiber length must be very small, in the range of 100-200 μm. In such cases, specimen preparation can get tricky. Also, if the fiber is not perpendicular to the resin surface, i.e., aligned in the pull direction, or the specimen thickness is larger than needed, the pull-out force will be larger than the fiber fracture stress and hence it will result in fiber breakage.

In a single fiber pull-out test, the fiber/resin interfacial shear strength (or bond strength) is determined by measuring the force needed to pull the fiber axially out of the solid/bulk resin [11, 13, 14]. The specimen preparation involves embedding part of the fiber (generally several 100 μm to less than a mm) in liquid resin with the fiber axis perpendicular to the resin surface, and then hardening the resin, either by cooling in the case of thermoplastic resin or by crosslinking reaction in the case of thermoset resin. The fiber is pulled out axially using tensile testing machine such as Instron Universal Tester, MTS or similar equipment. The fiber diameter (*D*) (measured prior to embedding) and embedded length (*L*) are measured using an optical microscope. The interfacial shear strength (IFSS), τ, is calculated using the following equation:

$$\text{Interfacial shear strength } (\tau) = F/\pi DL \tag{14.3}$$

where F is the force required to debond the fiber and πDL is the fiber/resin interfacial area [11, 13, 14]. In the fiber pull-out test while the fiber alignment and axial pull are critical, it does not matter if the resin is transparent or opaque as in the SFC test. Fiber pull-out test and its modified version, the microbond (or microbead) test that is discussed later, have been found to be more versatile and researchers have used them for a variety of fiber/resin combinations.

Li and Netravali [28] characterized the effect of allylamine plasma treatment on ultrahigh strength polyethylene (UHSPE) fibers, Spectra® 900, on their adhesion to bisphenol A-based epoxy resin. They used silicone mold to prepare the specimens with controlled embedded length. The silicone mold, specimen and the schematic of the fiber pull-out test are shown in Figure 14.7. Their results suggested that while there was no change in the fiber surface topography after plasma treatment, interfacial shear strength increased significantly from 0.56 MPa (control) to 1.82 MPa, a 225% increase. They explained that allylamine plasma forms a polymer coating with nitrogen containing groups which not only increase the surface energy that allows better spreading of epoxy resin but these groups also allow crosslinking reactions with epoxide groups and form covalent bonds [10]. They further tested the bond strength after immersing the specimens in water and found that for control fibers the IFSS reduced to 0.4 MPa (~30% reduction) and plasma treated fibers showed 1.4 MPa (~23% reduction). Li and Netravali concluded that this smaller reduction in IFSS was a result of the covalent bonds between the plasma polymer coating and epoxy resin that remained unbroken, at least partially, after water immersion [28].

Song and Netravali [29] used single fiber pull-out technique to measure the IFSS of XeCl (308 nm) UV pulsed excimer laser treated UHSPE fiber/epoxy resin. The excimer laser treatment was carried out in both air and ammonia gas. They found that the laser treatment was able to

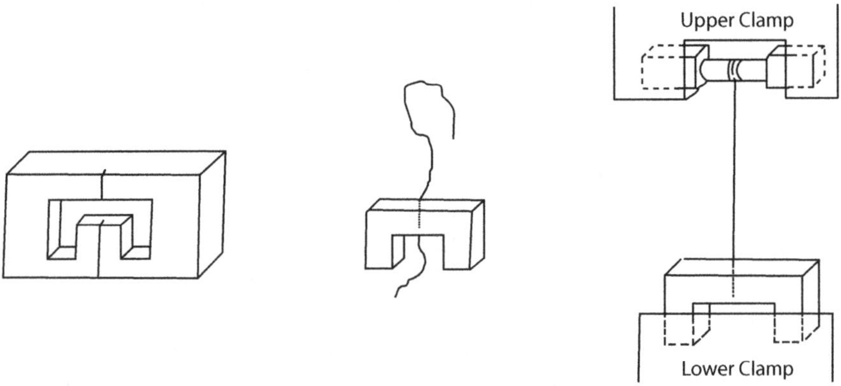

Figure 14.7 Silicone mold (left), epoxy resin specimen with UHSPE fiber embedded (middle) and fiber being pulled out (right) [28].

incorporate both oxygen and nitrogen on the fiber surface, and that the oxygen and nitrogen contents depended on the type of gas environment used. Obviously, more nitrogen was incorporated when the fiber was treated in ammonia making the fiber surface more polar and with higher surface energy. They also observed that the fiber surface became significantly rougher after the treatment. They explained that the rough surface was due to the XeCl laser melting the top layer of the fiber and reorganizing with different surface topography during cooling. These three factors, higher surface energy, covalent bonding and rougher surface, together increased the IFSS by close to 300%. While the control fiber showed IFSS of 0.96 MPa, IFSS values for fibers that were laser treated in air showed 3.13 MPa and those treated in ammonia showed 3.76 MPa. They also observed the IFSS values increase as a function of both laser energy density and the number of excimer laser pulses.

Yang and Thomason [30] used both single fiber pull-out and microbond tests (discussed later) to characterize glass fiber/polypropylene (GF/PP) IFSS. Both pure PP and commonly used maleic anhydride grafted PP (PPm) blend were used in their study. The single fiber pull-out test for GF/PP showed IFSS of 8.2 MPa while for GF/PPm, the IFSS was more than twice at 17.3 MPa as was expected. In the load vs extension plots, depending on the IFSS values, they were able to observe different types of failure modes from sudden drop in loads resulting from fiber fracture to fiber/resin friction.

The change in IFSS as a function of resin properties was shown by Netravali *et al.* using SFC tests on plasticized epoxy resins and discussed earlier [1]. It is well known that as the polymer or resin temperature is raised from well below their glass transition temperature (Tg) to their Tg, they soften and their modulus decreases. Above their Tg they become soft and rubbery. The effect of temperature is somewhat similar to resin being plasticized i.e., softened. As a result, it can be expected that the temperature would also affect the fiber/resin bonding. Wang *et al.* [31] studied the polyheteroarylene-*co*-p-phenyleneterephthalamide (PBO) fiber/epoxy resin IFSS as a function of temperature. The fiber diameter was 12 μm and the Tg of the epoxy resin was 71°C. They characterized the IFSS using single fiber pull-out. The PBO/epoxy IFSS at room temperature (25°C) was about 25 MPa. At 40°C the IFSS decreased slightly to about 22 MPa but at 80°C, above the Tg of epoxy resin, the IFSS decreased to 6 MPa, a decrease of about 72%. They also simulated the single fiber pull-out test using the COMSOL, finite element method. The structural mechanics module of COMSOL software was used to construct the single fiber pull-out model at different temperatures using the cohesion model and applying the linear traction separation law. The results of the simulation matched well with the experimental results.

Christ *et al.* [32] investigated the effect of both temperature and humidity on carbon fiber/polyamide 6 (PA-6) resin IFSS. They characterized both debonding fracture toughness and surface-specific work of friction. Single fiber pull-out test was used to obtain the IFSS values and other characteristics. Main conclusion of their study was that both temperature and humidity swell the PA-6 resin which results in a significant relaxation of radial stress at the interface. As a result, IFSS decreased as a function of both temperature and humidity. Furthermore, debonding fracture toughness also followed a declining trend with increase in both temperature and humidity. However, it displayed non-linear characteristics. This implied that the effect of temperature and humidity was coupled or synergetic. The IFSS decreased from 23 MPa at 25°C to about 15 MPa at 45°C, a decrease of about 35%. In addition, about 15% decline in IFSS was seen as the RH was increased from 10% RH to 75% RH at both temperatures, 25°C and 45°C.

Nuhoglu *et al.* [33] used single fiber pull-out test to study carbon fiber/epoxy resin system. They used AS4/BR102 carbon fibers (Hexcel Corp.) and EPON 826 epoxy resin (Dow Corp.) modified with nanoclay and chopped carbon fiber in their study. They used novel 3-D printing to fabricate the composite specimens from which the fibers could be pulled out. While the 3-D specimen preparation was faster, authors stated that specimens showed physical imperfections and defects, such as porosity and fiber misalignment. They also found that controlling embedded fiber length was difficult. The average IFSS was only 2.4 MPa with a maximum of 5 MPa, showing large variation for individual fibers. While 3-D method of specimen preparation was faster, it could be improved further to avoid the defects.

Qiu and Schwartz [34] were the first to introduce a novel modified version of the single fiber pull-out test. They argued that pulling a single fiber from bulk resin provides no information about the fiber volume fraction effect on interfacial properties. In fact, bulk properties of the resin may have significantly different properties than in a composite. This is particularly true when the fiber volume content is high, and the resin behaves much stiffer. They developed a facile technique to prepare micro-composite specimens. They prepared glass and Kevlar® fiber bundles, separately, that contained 7 fibers, one in the middle and 6 others forming a hexagonal packing that is generally observed in typical composites. These fibers were tied in the middle with a tie fiber. Resin (epoxy in this case) was then applied at the tie fiber to form the micro-composite that covered all 7 fibers. Once the resin was cured, six fibers surrounding the central fiber were cut from top side. The fiber assembly was held from the bottom and the central fiber was pulled out. The geometry of the specimen, including fiber volume within the micro-composite and spacing between the fibers, could be easily altered. They found slight increase in the IFSS values as the fiber volume content increased from 30 to 50%.

14.5 Microbond Test (Microbead Test)

From the previous section, it is clear that the single fiber pull-out test works well for obtaining IFSS values for various fiber/resin systems and in different environmental conditions such as temperature and humidity. However, the length of the embedded fiber must be small enough, about 100-200 μm, so that the interface area is small and the pull-out force is not large to break the fiber before being pulled free from the resin. Miller *et al.* [35] developed a novel single fiber pull-out test where a small drop of resin (called microbead) is deposited on the fiber. The resin microbead is then cured, if thermoset, held under the plates of a microvice and the fiber is pulled free from the resin. The schematic of the microvice arrangement for holding the resin microbead is shown in Figure 14.8 [35]. They did the initial tests using E-glass fibers (~13 μm diameter), Kevlar® fibers and Celion® 3000 carbon fibers. The epoxy resin and hardener used were Epon 828 and methylene dianiline, respectively. They obtained IFSS of about 33 MPa for control (untreated) E-glass and about 49 MPa for silane treated (0.1%) E-glass. In the case of Kevlar® and Celion® fibers, the IFSS values were 38 MPa and 57 MPa, respectively. They found high variation within

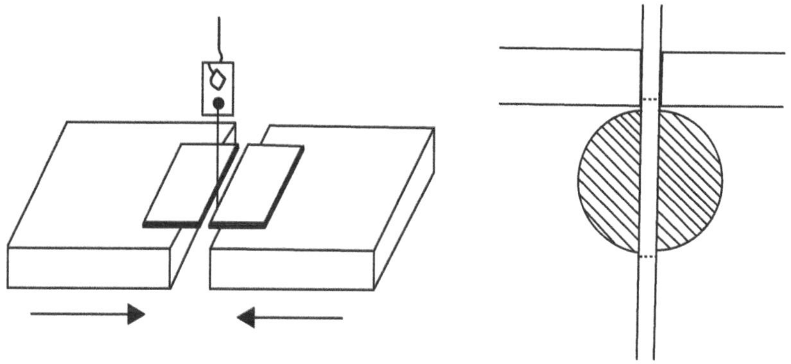

Figure 14.8 The schematic view of the microvice arrangement for microbond test (left) and enlarged view of the resin droplet held under the microvice plates (right) [35].

each group and attributed the variation to experimental errors such as fiber diameter and embedment length measurements.

Luo and Netravali [36] used the microbond test to measure the IFSS of pineapple fibers with poly(hydroxybutyrate-co-valerate) (PHBV). While pineapple fibers are cellulosic with high surface energy, PHBV is a naturally occurring biodegradable aliphatic polyester produced by microorganisms and has low surface energy. It has mechanical properties similar to polyethylene (PE) and polypropylene (PP). Specimens in their study were prepared by placing small width (~½ mm) film strips (~4 mm long) on top of the pineapple fibers and heating to melt and form PHBV microbeads. Pineapple fiber/PHBV resin IFSS was 8.23 MPa. The SEM images of the failure surface showed interfacial failure (failure at the interface). They argued that the low IFSS was due to the surface energy difference between them.

Zhi *et al.* [37] studied the effects of various parameters (fiber diameter, microbead size, blade position, etc.) through experimental study as well as finite element simulation. They found that the fiber diameter has the largest effect on IFSS. Microbead size. i.e. embedment length and blade position also have significant influences on the IFSS results from microbead tests. A smaller diameter fiber, smaller microbead or larger distance between the fiber and the microvice blade edge generally result in larger IFSS values. It is, therefore, important that testing parameters be kept constant or as close to constants as possible when conducting comparative microbond tests.

Tung *et al.* [38] studied the bamboo fiber/PP IFSS using microbead tests. They used small fiber bundles obtained using steam explosion method and also treated some fiber bundles with silane coupling agent. Their results showed bamboo fiber/PP IFSS of around 30 MPa. Surprisingly, they found no effect of the silane coupling agent. Further, they indicated that smaller bamboo fibril bundles resulted in higher IFSS than larger bundles, as a result of higher surface roughness that provided mechanical bonding.

Choi and Park [39] used a quasi-disk shaped microbead, instead of a conventional spherical microbead, to measure carbon fiber/epoxy IFSS. A new method was developed to obtain quasi-disk shaped microbead. This involved inserting a Teflon® film with a pinhole just under the liquid epoxy resin droplet mounted on a single vertical carbon fiber. The resin microbead rested on the Teflon® film which allowed the formation of quasi-disk shaped microbead with flat bottom. While the IFSS values were 34 MPa for conventional microbeads and much lower, 20 MPa, for quasi-disk specimens, significantly lower variability in IFSS was obtained in the case of quasi-disk type specimens.

Zu *et al.* [40] were the first to study the interfacial properties of continuous carbon nanotube (CNT) fibers (yarns) spun from CNT carpet consisting of double- and triple-wall tubes in an epoxy matrix using microbead technique. The IFSS value obtained was rather high, 14.4 MPa, for the hydrophobic CNT yarns. The SEM analysis of the interfacial failure surface indicated that a few CNT bundles were torn from the bulk indicating partial fiber damage and resin infiltrated the CNT bundle during testing, pushing the IFSS value higher than expected.

Many researchers have studied interfacial bonding of natural fibers with a variety of resins using the microbond technique in an effort to improve the properties of green composites. For example, Souzandeh and Netravali [42] studied sisal fiber/zein protein resin IFSS as a function of plasticizer (sorbitol) content and microfibrillated cellulose (MFC). While plasticizer reduced modulus of the zein resin and made it flexible, MFC increased the modulus. As expected, IFSS decreased with sorbitol and increased with MFC content. At 20% sorbitol content, there was a 53% decrease in the IFSS whereas with 15% MFC the IFSS increased by about 81%. Lower IFSS for plasticized resin and higher IFSS for higher modulus resin are consistent with the results obtained by SFC technique for carbon/epoxy system [1]. Sarpanahi and Ahmadi [43] studied bonding of date palm fibers to unsaturated polyester, epoxy, and phenolic resol resins. Their results showed that the epoxy resin provided the highest IFSS, in the range of 6-9 MPa and the phenolic resol resin, with lower surface energy, resulted in the lowest IFSS of about 3 MPa. They also found a clear relation between the contact angle and the IFSS values, as the contact angle decreased the IFSS increased. Yao *et al.* [44] studied the enzymatic treatments on the interfacial behavior of sisal fiber (SF) reinforced bio-epoxy resin. They treated the sisal fibers with xylanase and pectinase for 6 h. The treatment mainly caused the hydrolysis and cleavage of hemicellulose and pectin of SFs. SEM images showed increased surface grooves and surface roughness of sisal fibers after the enzyme treatments which allowed mechanical interlocking. The effect of the mechanical interlocking was not only seen experimentally in increased IFSS from 8 MPa to over 17 MPa but was also proven by the finite element model the authors used.

Many natural fibers are being considered for reinforcing cement in recent years. In such cases, their bonding with cement is critical in obtaining better cement properties. Fu *et al.* [45] characterized the jute fiber/cement interfacial failure mechanism at elevated temperatures. Their results suggest that at high temperatures, the mismatch between the anisotropic radial shrinkage of jute fibers and increased thermal strain in cement make it easy for the jute fiber to debond and progressively cause larger voids at the interface. The authors also noticed that the covalent bonds between the jute fibers and cement gradually decreased; this weakened the jute fiber/cement interfacial strength. However, at the same time, elevated temperature caused surface roughness in jute fibers which increased the fiber/cement friction during pull-out procedure.

Minty *et al.* [46] focused on characterizing the effect of temperature and resin chemistry on glass fiber/epoxy resin. As expected, they found an inverse dependence on test temperature for the same reason, radial thermal expansion of the epoxy resin, compared to glass fibers. They also found the IFSS to be influenced by the hardener-to-resin ratio, which changed not only the chemistry and crosslink density of the resin but mechanical properties as well. They hypothesized that residual radial compressive stresses at the interface are a function of the chemistry of the resin, higher crosslink density tends to increase the compressive stresses and relax away when the test temperature is raised due to thermal expansion.

Since its introduction, the microbond test has become popular and has been widely used by researchers to characterize the fiber/matrix IFSS for a variety of fiber/polymer systems. Since the fiber diameters are in the range of 4-6 μm, polymer microdroplets (microbeads) with a mass as little as 20 nanograms could be used in testing. This, however, can bring the possibility of scaling issues that can cause unanticipated but significant differences to the properties and performance of these microbeads in comparison to macroscale samples. Thomason [47] looked at data from multiple studies and cautioned the researchers about data obtained in microbond tests. According to him, the results of these tests show a clear dependence on the properties of the microdroplet such as modulus and Tg. The changes in the droplet properties can influence the results than the actual changes at the fiber/polymer interface. Another point is the small mass of the microbead. The structure and properties of the tiny microdroplets may not represent the same polymer as in its bulk form even though same thermal and environmental exposure history was used to prepare the microbead. As a result, the reliability of the microbead test results in many cases may not be as good as other test methods. However, the data may be used for comparing data of similar fiber/resin systems.

14.6 Fiber Push-Out and Push-In Tests

Single fiber push-out and push-in tests are *in situ* micromechanical tests that have been also known as push-through, push-down, indentation, micro-indentation or nano-indentation tests. However, push-out tests are slightly different than push-in tests. In the push-out test, the specimen thickness is small and the fiber is axially pushed down using an indenter (flat, spheroconical or other shape) until it debonds through the entire embedded length and pushed out [48]. In the push-in test, the specimen thickness is much larger. The fiber is axially pushed until it debonds. However, the fiber may not debond through the entire embedded length which can be large [48]. Both tests can be carried out on specimens taken from an actual composite.

It should be noted upfront that for both push-out and push-in tests, the fiber must be hard (e.g. glass, carbon, silicon carbide, etc.) to handle the indenter pressure. Soft fibers such as nylon or UHSPE fibers do not work as they can easily get axially deformed or squeezed under indenter pressure. Also, fibrillar fibers such as Kevlar® or natural fibers such as jute, sisal, etc., do not work as well as they also get axially deformed or undergo fibril separation.

Netravali *et al.* [49] used continuous microindenter push-out technique to measure glass fiber/epoxy resin IFSS. Glass fibers used in this study were E-glass with diameter of 23 μm. A few unidirectional fibers were embedded in a cube of DGEBA-based epoxy resin and a thin transversely cut section of the cube was used as the specimen. The specimen was polished using standard metallography technique to obtain thicknesses of about 500 μm. The spheroconical microindenter was driven by a piezoelectric translator (PZT) at the rate of 50 nm/s. From the measured displacements and load, indentation depths were calculated. Experiments were done on untreated (control) as well as silane coupling agent treated glass fibers. The typical load-indentation depth plots for loading-unloading sequence for control and silane treated glass fibers are shown in Figure 14.9. Initially the load increased linearly with the indentation depth corresponding to the elastic deflection of the specimen, but later it exhibits a plateau as the fiber debonds and is slowly pushed out. Upon removal of the indenter, a large amount of non-recoverable deformation was observed. Using SEM, the authors were able to confirm slight deformation of the fiber and the fiber being pushed out of the resin.

Debonding load was taken where slope of the plot changes as indicated in Figure 14.9 (right). At this point, the fiber debonds and starts to slide down. On the basis of the debonding load value, F, the thickness of the specimen, T, and the diameter of the fiber, D, the average shear stress τ, required to debond the fiber is given by

$$(\tau) = F/\pi DT \tag{14.4}$$

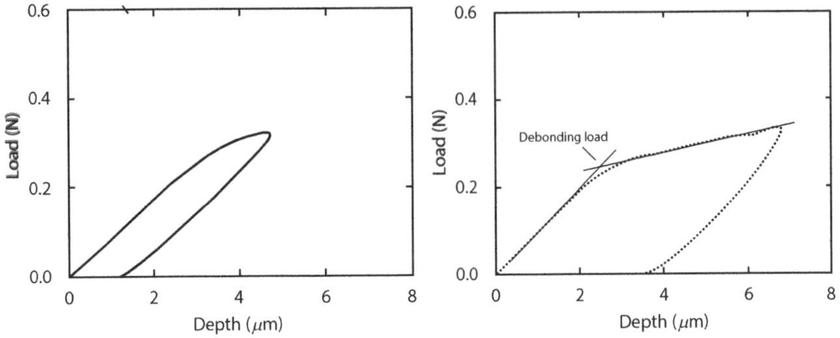

Figure 14.9 Typical load vs indentation depth plot (left) of control fiber (no coupling agent) showing a small plateau in the load indicating debonding (right) of a fiber with coupling agent showing a larger plateau [49].

where πDT is the interfacial area. The IFSS values obtained were around 15-20 MPa, consistent with those obtained by other researchers for similar fiber/resin system.

Zhou *et al.* [50] conducted both SFC and fiber push-out tests using E-glass/epoxy resin system. They used both silane-treated and untreated (control) glass fibers. They found that the average IFSS values were different for the two tests. However, in both cases, silane coupling agent treated fibers resulted in about 1.7 times the IFSS for control fibers. It should be noted that many researchers have found that the silane treatment works well for glass fibers as has been discussed in earlier section. According to the authors, the differences observed between SFC and push-out tests were because of the fundamental differences in the test mechanics as well as the limitations of the analytical schemes for interpreting the calculated values.

Sha *et al.* [51] used single fiber push-out tests to measure carbon fiber/phenolic resin IFSS. They used both unidirectional and two-directional (0/90) composites. Load-displacement plots were used to characterize debonding and fiber sliding at the interface as discussed above. They found that the IFSS in unidirectional composites was slightly higher (~90 MPa) than the two-directional composites (~85 MPa). They attributed this difference to thermally induced residual stresses caused by the difference in the coefficient of thermal expansion of fibers and the resin. Also, better fiber packing may be obtained in unidirectional composites, reducing the fiber-to-fiber distance which can increase the resin modulus.

Fan and Zhou [52] characterized the debonding behavior of carbon coated silicon carbide (SiC) fibers in Ti-17 aluminum alloy composites. The IFSS in their case was just over 108 MPa. The SEM analysis of the fracture surface indicated that the debonding occurred along the interface between the carbon coating on the fiber surface and the reaction zone (RZ). They defined RZ as the space or volume around the fiber where the fiber coating reacts with the aluminum alloy matrix forming a zone of a few mm thickness. RZ may also be termed as the interphase. In their case they found that the thickness of RZ was about 4 μm. They further found that the RZ was well bonded to the matrix (aluminum alloy) and that the SiC fiber was well bonded to the carbon coating.

Rohrmuller *et al.* [53] worked with short glass fiber reinforced sheet molding compound (SMC). The resin was a thermoset matrix material consisting of an unsaturated polyester polyurethane hybrid resin (UPPH). Fiber push-out tests were conducted using a diamond conical flat tip Triboindenter (TI-950 indenter) and at a displacement rate of 50 nm/s until the fiber debonded. In addition, cyclic tests were performed with loading and unloading cycles with 200 nm per cycle increments. The cyclic push-out tests were used to obtain the fracture toughness. The obtained glass fiber/UPPH IFSS values varied between 50 and 65 MPa and the fracture toughness was around 107 J/m^2. They also found that the fiber debonds before the maximum load is reached. The numerical simulations of the push-out tests fit well with the experimental plots and revealed strong interaction with the neighboring fibers. Further, the authors noticed that during the push-out test the neighboring fibers could get damaged if they were too close together.

Karakoc *et al.* [54] studied 3rd generation SiC composites reinforced with SiC fibers (SiC_f). These SiC_f/SiC composites are used in next generation nuclear reactors because of their high strength and chemical inertness at high temperatures. Their investigation was focused on chemical vapor infiltrated (CVI) SiC_f/SiC composites. While most push-out tests are carried out by axially indenting, and compressing, the SiC fibers (micropillar) in this case were slanted at a predetermined angle while being indented (micropillar tests). They argued that micro-compression testing at an angle successfully quantifies both bond shear strength (IFSS) as well as fiber/resin friction coefficient. The micropillar specimens containing an inclined pyrolytic carbon (PyC) interphase were prepared using focused ion beam. They found that the interfacial properties and debonding shear stress were significantly affected by the PyC layer thickness. The authors found micropillar test results to be closely correlated with the macroscopic mechanical properties of the SiC_f/SiC composites.

Weng *et al.* [23] prepared carbon fiber/bio-based epoxy composites using cold vacuum-assisted resin infusion method. Sections of the composites were used for push-out tests using a nanoindenter. In addition to push-out tests, they also carried out SFC tests and compared the IFSS values obtained by the two methods. The IFSS obtained by SFC test was 40.5 MPa whereas the push-out tests resulted in a comparable value of 42.2 MPa. They also treated carbon fibers by atmospheric air plasma to alter their surface chemistry as well as topography to increase the IFSS. The push-out tests of plasma treated carbon fiber specimens resulted in IFSS of just over 76 MPa, an increase of 88%, as a result of significantly higher oxygen (O-C=O peak) and significantly lower carbon content (C-C peak) on the fiber surface. While nitrogen on the surface also increased, the increase was not so significant as oxygen. Adding both oxygen and nitrogen on the surface not only increased the surface energy for better spreading of the resin on the fiber but also presented opportunity for covalent bonding.

Medina *et al.* [48] compared the IFSS of glass fiber/epoxy composites using both push-out and push-in tests. The epoxy resin was loaded with carbon nanotubes (CNTs). By varying the CNT loading from 0 to 0.8%, they were able to assess the effect of CNT loading on the glass fiber/epoxy IFSS. The CNTs were multiwalled and non-functionalized (untreated). Their results indicated that both push-out and push-in tests give equivalent results in terms of IFSS. Further, they found that the IFSS increased by about 19% with 0.3% CNT loading; this is because of the stiffening of epoxy resin. However, increasing the CNT loading to 0.8% did not increase the IFSS any further, perhaps due to aggregation of CNTs in the resin.

14.7 Summary and Concluding Remarks

Fiber reinforced composites are being used in innumerable applications from airplanes to satellites, automobiles to sporting goods and from windmill blades to furniture. All these applications require application-dependent specific properties for proper operation. While most composite properties are a function of the main constituents, i.e., fibers and resin, fiber/resin interfacial bonding plays an important role. For example, properties such as toughness and transverse strength are directly related to the fiber/resin interfacial characteristics. As a result, it is important to be able to quantitatively measure the fiber/resin bond strength. As discussed in this review chapter, several micromechanical tests such as single fiber composite, single fiber pull-out, single fiber push-in, single fiber push-out, microbond, etc., are available that can provide a good estimate of the fiber/resin bond strength. Many of these tests also yield other relevant and important information such as nature of debonding, stress-field around the fiber or fiber break, fiber stress at different locations, etc. Which test to perform depends on the fiber/resin combination of interest and how easy it is to prepare the test specimens as well as what information needs to be obtained. Since the stress field is different in different tests, one must also select the test that closely resembles the kind of stress the composite will experience in real life. However, as done by many researchers, the tests can be modified to suit the fiber/resin combination and the stress situation the composite will experience. Researchers have also expressed some issues that need to be considered when using some of the tests. For example, it has been shown that there are issues related to microbond test when scaling it up. The microbead can be tiny and may not have the properties of the bulk resin, even when it has experienced the same thermal history. Hence, the results obtained may not be reliable. Nevertheless, the test may be used for comparative purposes.

AI disclosure statement: The author confirms that no artificial intelligence (AI)-generated content or AI-assisted writing tools were used in the preparation of this chapter (including summaries and the abstract). All text, analysis, and conclusions are the result of the author's original work.

References

1. A. N. Netravali, R. Henstenburg, S. L. Phoenix. and P. Schwartz, Interfacial shear strength studies using single-filament-composite test. I: Experiments on graphite fibers in epoxy, Polymer Composites, 10, 226-241 (1989).
2. M.R. Piggott, Micromechanics of fibre-polymer interfaces, in: *Interfacial Phenomena in Composite Materials*, I. Verpoest and F. Jones (Eds.), Butterworth-Heinemann, Oxford (1991).
3. C. C. Chamis, Mechanics of load transfer at the interface, in: *Interfaces in Polymer Matrix Composites*, E. P. Pluddemann (Ed.), pp. 31-77, Academic Press, New York (1974).
4. W. D. Bascom and R. M. Jensen, Stress transfer in single fiber/resin tensile tests, J. Adhesion, 19, 219-239 (1986).
5. M. R. Piggott, Progress in Science and Engineering of Composites, Proceedings of 4th International Conference on Composite Materials, pp. 193-202 (1982).
6. Committee on High-Performance Structural Fibers for Advanced Polymer Matrix Composites, National Research Council, The National Academy Press, Washington DC, ISBN: 978-0-309-09614-0 (2005).
7. K. L. Mittal, The role of the interface in adhesion phenomena, Polym. Eng. Sci., 17, 467-473 (1977).

8. D.N. Saheb and J.P. Jog, Natural fiber polymer composites: A review. Adv. Polym. Technol. 18, 351–363 (1999).
9. M.R. Piggott, Micromechanics of fibre-polymer interfaces, in: *Interfacial Phenomena in Composite Materials*, I. Verpoest and F. Jones (Eds.), Butterworth-Heinemann, Oxford (1991).
10. A. N. Netravali and K. L. Mittal (Eds), *Interface/Interphase in Polymer Nanocomposites*, Wiley-Scrivener, Beverly, MA (2017).
11. D. Kalita and A. N. Netravali, Interfaces in green composites: A critical review, Rev. Adhesion Adhesives, 3, 386-443 (2015). DOI: 10.7569/RAA.2015.097311
12. Z. Lei, X. Li, F. Qin, and Q. Wei, Interfacial micromechanics in fibrous composites: Design, evaluation, and models. Scientific World J., 36, 1–9 (2014). http://dx.doi.org/10.1155/2014/282436
13. A.N. Netravali and K.L. Mittal, Fiber surface treatment: Relevance to interfacial characteristics, in *Wiley Encyclopedia of Composites*, 2nd Edition, L. Nicolais, A. Borzacchiello and S. M. Lee (Eds.), pp. 980-998 (2012).
14. S. Ahmadvashaghbash, I. Verpoest, Y. Swolfs and M. Mehdikhani, Methods and models for fibre–matrix interface characterisation in fibre-reinforced polymers: a review, International Materials Reviews, 68, 1245–1319 (2023). https://doi.org/10.1080/09506608.2023.2265701
15. R. Gulino and S. L. Phoenix, Weibull strength statistics for graphite fibers measured from the break progression in a model graphite/glass/epoxy composite, J. Mater. Sci., 26, 3107-3118 (1991).
16. A. N. Netravali, Z. F. Li, W. Sachse and H. F. Wu, Determination of fiber/matrix interfacial shear strength by an acoustic emission technique, J. Mater. Sci., 24, 6631-6638 (1991).
17. W. Sachse, A. N. Netravali and A. R. Baker, An enhanced, acoustic emission-based single-fiber-composite test, J. Nondestructive Evaluation, 11, 251-261 (1992).
18. G. A. Holmes, S. Wesson, W. G. McDonough, J. H. Kim, A. Netravali, J. N. Walker and R. A. Johnson, An automated testing machine for monitoring the evolution of fiber breaks, J. Mater. Sci., 44, 2007-2015 (2009). https://doi.org/10.1007/s10853-009-3294-1
19. L. A. Ahure Powell, R. J. Sheridan, S. Yucel, B. Yucel, E. Rust, S. R. Kalidindi, J. Woodcock, J. W. Gilman and G. A. Holmes, Snappy: A new automated testing machine for monitoring the break evolution process during single fiber fragmentation test, Experimental Techniques, 47, 1073-1084 (2023). https://doi.org/10.1007/s40799-022-00611-3
20. E. Motta de Castro, A. Tabei, D. B. H. Cline, E. Haque, L. B. Chambers, K. Song, L. Perez, K. Kalaitzidou and A. Asadi, New insights in understanding the fiber-matrix interface and its reinforcement behavior using single fiber fragmentation data, Advanced Composites and Hybrid Materials, 8 (5), 1-24 (2025) https://doi.org/10.1007/s42114-024-01054-7

21. Y. L. Zhou and A. N. Netravali, Ethylene/ammonia plasma deposition for controlled adhesion of graphite fibers to PEEK part II: Effect on fiber and fiber/matrix interface, J. Adhesion Sci. Technol., 9, 1505-1520 (1995).
22. M. T. Umemura, F. Dalmas, P. Reynaud, D. Long, G. Fantozzi, N. Godin, L. Trouillet-Fonti and J. Bikard, Fiber-matrix interface characterization in CF/PEEK composites by analysis of single fiber fragmentation, Composites Sci. Technol., Vol 270, 2025, 111301. https://doi.org/10.1016/j.compscitech.2025.111301
23. P. Weng, V. Prasad, P. S. Rao, N. O'Dowd, N. Murphy, A. Ivankovic and G. Scarselli, Interfacial properties of carbon fiber-reinforced biobased resin composites by single fiber fragmentation, fiber pull-out, and interlaminar shear strength, Polymer Composites, 45, 3355-3369 (2024). DOI: 10.1002/pc.27995
24. H. Girard, B. Koohbor, A. Doitrand and R. Livingston, Experimental characterization of in-plane debonding at fiber-matrix interface using single glass macro fiber samples, Composites Part A, 171, 107573 (2023). https://doi.org/10.1016/j.compositesa.2023.107573
25. R. Nakamura, A. N. Netravali and M. V. Hosur, Effect of halloysite nanotube incorporation in epoxy resin and carbon fiber ethylene/ammonia plasma treatment on their interfacial property, J. Adhesion Sci. Technol. 26, 1295-1312 (2012).
26. H. Girard, A. Doitrand, B. Koohbor, R. G. Rinaldi, N. Goldin, D. Long, J. Bikard and L. Trouillet-Fonti, Numerical simulation of fiber-matrix debonding: Inverse identification of interface properties, Engineering Fracture Mechanics, 286, 1092454 (2023). https://doi.org/10.1016/j.engfracmech.2023.109254
27. P. Favre and J. Perrin, Carbon fibre adhesion to organic matrices. J. Mater. Sci., 7, 1113–1118 (1972).
28. Z. F. Li and A. N. Netravali, Surface modification of UHSPE fibers through allylamine plasma deposition. II. Effect on fiber and fiber/epoxy interface, J. Appl. Polym. Sci., 44, 333-346 (1992).
29. Q. Song and A. N. Netravali, Excimer laser surface modification of ultra-high-strength polyethylene fibers for enhanced adhesion with epoxy resins. Part 2. Effect of treatment environment, J. Adhesion Sci. Technol. 12, 983-998 (1998).
30. L. Yang and J. L. Thomason, Development and application of micromechanical techniques characterizing interfacial shear strength in fibre-thermoplastic composites, Polymer Testing, 31, 895-903 (2012). http://dx.doi.org/10.1016/j.polymertesting.2012.07.001
31. J. Wang, S. Xu, X. Xiong, C. Lu and R. Ren, Characterization of the effect of testing temperature on the interfacial shear strength of fiber/resin single-fiber-composite, Materials Letters, 330, 133367 (2023). https://doi.org/10.1016/j.matlet.2022.133367

32. N. Christ, P. Gumbsch and J. Hohe, Single fiber pull-out investigation at different temperature and humidity conditions: Experimental characterization of the fiber-matrix interface in carbon fiber reinforced polyamide 6, J. Thermoplastic Composite Mater., 0, 1-28 (2025). DOI: 10.1177/08927057251314436
33. K. Nuhoglu, N. M. Baltodano, and E. Celik, Investigation of fiber-matrix interface strength *via* single-fiber pull-out test in 3D-printed thermoset composites: A simplified methodology, Materials, 17, 2433 (2024). https://doi.org/10.3390/ma17102433
34. Y. Qiu and P. Schwartz, Single-fiber pull-out from a microcomposite test, Composites Sci. Technol., 48, 5-10 (1993).
35. B. Miller, P. Muri and L Rebenfeld, A microbond method for determination of the shear strength of a fiber/resin interface, Composites Science and Technology, 28, 17-32 (1987).
36. S. Luo and A. N. Netravali, Interfacial and mechanical properties of environment-friendly "green" composites made from pineapple fibers and poly(hydroxybutyrate-co-valerate) resin, J. Mater. Sci., 34, 3709-3719 (1999).
37. C. Zhi, H. Long and M. Miao, Influence of microbond test parameters on interfacial shear strength of fiber reinforced polymer-matrix composites, Composites: Part A, 100, 55-63 (2017). http://dx.doi.org/10.1016/j.compositesa.2017.05.004
38. N. H. Tung, H. Yamamoto, T. Matsuoka and T. Fujii, Effect of surface treatment on interfacial strength between bamboo fiber and PP resin, JSME International J.Series A Solid Mechanics and Material Engineering, 47, 561-565 (2004).
39. N-S. Choi and J-E Park, Fiber/matrix interfacial shear strength measured by a quasi-disk microbond specimen, Composites Sci. Technol., 69, 1615-1622 (2009). doi:10.1016/j.compscitech.2009.03.012
40. M. Zu, Q. Li, Y. Zhu, M. Dey, G. Wang, W. Lu, J. M. Deitzel, J. W. Gilespie, J. H. Byun and T-W. Chou, The effective interfacial shear strength of carbon nanotube fibers in an epoxy matrix characterized by a microdroplet test, Carbon, 50, 1271-1279 (2012). doi:10.1016/j.carbon.2011.10.047
41. A. Kelly and W. R. Tyson, Tensile properties of fibre-reinforced metals: copper/tungsten and copper/molybdenum, J. Mech. Phys. Solids, 13, 329-338 (1965).
42. H. Souzandeh and A. N. Netravali, Study and characterization of sisal fiber/zein resin interface, Rev. Adhesion Adhesives pp 1-19. (2018) DOI: 10.7569/RAA.2018.097307
43. R. Sarpanahi and M. S. Ahmadi, Experimental study on interfacial shear strength between date palm fibers and thermoset resins, International J. Adhesion Adhesives, 127, 103524 (2023). https://doi.org/10.1016/j.ijadhadh.2023.103524
44. Y. Yao, H. Dou, T. Liu, S. Wang, Y. Gao, J. Kang, X. Gao, C. Xia, Y. Lu and W. Fan, Micro- and nano-scale mechanisms of enzymatic treatment on

the interfacial behaviors of sisal fiber reinforced bio-based epoxy resin, Industrial Crops Products, 194, 116319 (2023). https://doi.org/10.1016/j.indcrop.2023.116319
45. Z. Fu, Y Yao, Y. Duan, B. Wang and X. Li, Failure mechanism of bonding between natural fiber and cement matrix at high temperature, Construction and Building Mater., 412, 134724 (2024). https://doi.org/10.1016/j.conbuildmat.2023.134724
46. R. F. Minty, L. Yang and J. L. Thomason The dependence of interfacial shear strength on temperature and matrix chemistry in glass fibre epoxy composites, Composites Part A, 164, 107303 (2023). https://doi.org/10.1016/j.compositesa.2022.107303
47. J. Thomason, An overview of some scaling issues in the sample preparation and data interpretation of the microbond test for fibre-matrix interface characterization, Polym. Testing, 111, 107591 (2022). https://doi.org/10.1016/j.polymertesting.2022.107591
48. C. M. Medina, J. M. Molina-Aldareguia, C. Gonzalez, M. F. Melendrez, P. Flores and J. Llorca, Comparison of push-in and push-out tests for measuring interfacial shear strength in nano-reinforced composite materials, J. Composite Mater. 50, 1651-1659 (2016). DOI: 10.1177/0021998315595115
49. A. N. Netravali, D. Stone, S. Ruoff and L. T. T. Topoleski, Continuous micro-indenter push-through technique for measuring interfacial shear strength of fiber composites, Composite Sci. Technol. 34, 289-303 (1989).
50. X.-F. Zhou H. D. Wagner and S. R. Nutt, Interfacial properties of polymer composites measured by push-out and fragmentation tests, Composites Part A, 32, 1543-1551 (2001).
51. J. J. Sha, J. X. Dai, J. Li, Z. Q. Wei, J-M. Hausherr and W. Krenkel, Measurement and analysis of fiber-matrix interface strength of carbon fiber-reinforced phenolic resin matrix composites, J. Composite Mater., 48, 1303-1311 (2014). DOI: 10.1177/0021998313485264
52. Y. Fan and X. Zhou, Investigation of interface debonding behaviour in a unidirectional SiC fibre reinforced Ti-17 composite by a single fibre push-out technique, J. Composite Mater., 55, 3945-3953 (2021). DOI: 10.1177/00219983211031653
53. B. Rohrmuller, P. Gumbusch and Jorg Hohe, Calibrating a fibre-matrix interface failure model to single fiber push-out tests and numerical simulation, Composites Part A, 150, 106607 (2021). https://doi.org/10.1016/j.compositesa.2021.106607
54. O. Karakoc, T. Koyanagi, T. Nozawa and Y. Katoh, Fiber/matrix debonding evaluation of SiC_f/SiC composites using micropillar compression technique, Composites Part B, 224, 109189 (2021). https://doi.org/10.1016/j.compositesb.2021.109189

Part 3

ADHESIVES AND ADHESIVE JOINTS

15

Switchable Thermosets in the World of Adhesives

Natanel Jarach[1,2]*, Rachel Avshalomov[1] and Hanna Dodiuk[1]

[1]Department of Polymer Materials Engineering, Pernick Faculty of Engineering, Shenkar - Engineering. Design. Art, Ramat-Gan, Tel Aviv District, Israel
[2]Institute of Chemistry, The Hebrew University of Jerusalem, Jerusalem District, Israel

Abstract

Adhesives are widely used by industries and everyday users as an approach for joining different parts. However, concerns regarding environmental pollution have led scientists and engineers to seek sustainable ways of recycling and reusing attached components. Delamination and debonding of polymeric adhesives present significant challenges, particularly with cross-linked thermoset and elastomeric adhesive systems, which once cured, unlike thermoplastics, are infusible and insoluble. As most adhesives are either thermosets or elastomers, finding a solution to this challenge is necessary. One promising approach is covalent adaptable networks (CANs), which can produce switchable thermosets or elastomeric adhesives that do not require mechanical disruption for detaching. This chapter analyses various debonding stimuli used with thermosets and elastomeric CAN adhesives. It covers critical aspects such as the bonding and debonding processes and how these stimuli affect the adhesive strength. This chapter aims to provide valuable insights into the practical application of CANs and how they can produce desirable switchable attributes in thermoset adhesives.

Keywords: Debonding-on-demand, switchable adhesives, reversible covalent bonds containing polymers, covalent adaptable networks, vitrimers

**Corresponding author*: natanel.jarach@mail.huji.ac.il

Anil N. Netravali and Frank M. Etzler (eds.) The Wide World of Adhesion: A Festschrift in Honor of Dr. Kash Mittal on his 80th Birthday, (523–564) © 2026 Scrivener Publishing LLC

15.1 Introduction

Adhesive bonding is a critical joining technique across multiple industrial applications. Two primary mechanisms emerge when examining potential bond failures (Figure 15.1): cohesive and adhesion failure modes [1]. Cohesive failure occurs when external forces surpass the intrinsic strength of the adhesive material [2]. Conversely, in adhesion failure, bonds - whether covalent or intermolecular - break at the interface between the adhesive and the substrate materials, resulting in the adhesive completely separating from one of the substrates [2].

Debondable adhesives are unique adhesives that maintain a specific bond strength during their operational life but can be deliberately disassembled by applying a specific stimulus. Key factors to consider in their design include the adhesive type (low-strength or structural), material compatibility (e.g., metal, plastic), and the size and geometry of the bonded area. Accessibility to the bonded area is critical, as some techniques, such as light-induced debonding, may not be suitable for inaccessible regions [3].

15.2 Covalent Adaptable Networks – A Quick Overview

To tackle "debonding on demand" safely and without causing harm to the surface or generating polymeric waste, one approach is to utilize reversible

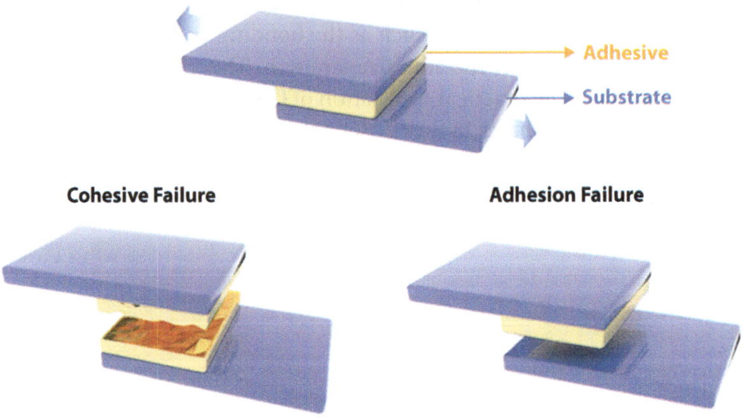

Figure 15.1 Adhesion failure mechanisms: interfacial adhesion and cohesion. Reproduced from Z. Liu, F. Yan, Switchable Adhesion: On-Demand Bonding and Debonding (2022), with permission from Wiley [1].

or dynamic bonds. These bonds can undergo a series of chemical processes, allowing the molecular structure of Reversible Covalent Bonds-Containing Polymers (RCBPs), also known as CANs or reorganizable networks [4] to change over time or with specific stimuli. This feature enables recycling while preserving most of the polymer's properties. Despite the various dynamic and reversible bonds, only a few have been studied and incorporated in adhesives. The other bonds, like transamination of vinylogous urethane [5], olefin metathesis [6, 7], and siloxane and silyl ether exchanges [8, 9], would not be further discussed.

15.2.1 [2 + 2] and [4 + 4] Cycloaddition Reactions

Cycloaddition reactions are reactions involving double bonds that create new sigma bonds, leading to a product known as an adduct. Characterized by Woodward–Hoffmann principles, these reactions vary depending on initiation method and conditions, including thermal (like Diels-Alder [4+2]) or photochemical pathways. [2 + 2] and [4 + 4] cycloadditions (Figure 15.2) typically require electronic excitation, involving orbital interactions between excited HOMO and unexcited LUMO states to form new molecular bonds – both σ and π. These molecular transformations depend on wavelength, and occur only under specific wavelengths [10–13].

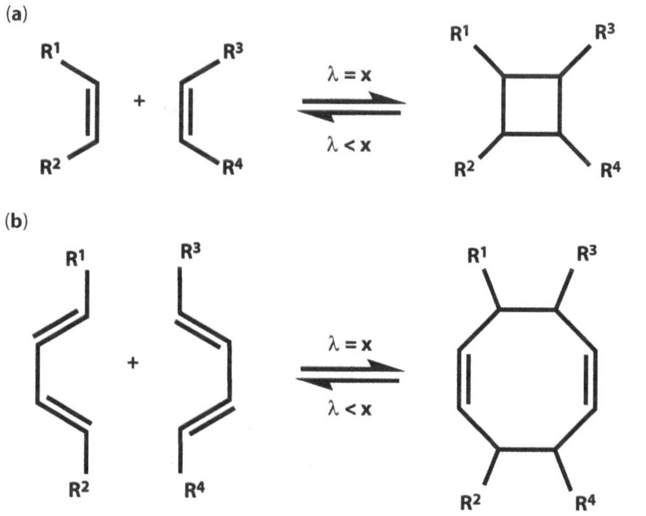

Figure 15.2 A schematic illustration of photocycloaddition reactions: [2 + 2] (a) and [4 + 4] (b). X represents the required wavelength for cycloaddition. At this wavelength, a reverse-cycloaddition occurs.

15.2.2 Diels-Alder (DA) Reaction

The [4 + 2] cycloaddition known as the Diels-Alder reaction involves molecular interactions between a diene and dienophile, generating a cycloadduct without requiring external catalysts. Temperature significantly modulates this reaction's equilibrium, with forward reactions typically occurring between 60-90°C and reverse (retro-) reactions initiating above 120°C. Researchers have leveraged these thermal characteristics to develop cross-linkable network materials with inherent reprocessing capabilities [14–21], frequently employing furan-maleimide moieties as representative structural elements [4, 22, 23] (Figure 15.3).

15.2.3 Transalkylation of Triazolium Salts

Another dynamic reaction that has been studied in the field of CANs is transalkylation, a reaction involving transferring an alkyl group from one compound to another [24] (Figure 15.4). Drockenmuller and coworkers have proposed a plausible two-step mechanism governing trans-N-alkylation exchanges [25], which can facilitate the rearrangement of covalent cross-linking. Their studies have led to the first known example of ion-conductive vitrimers.

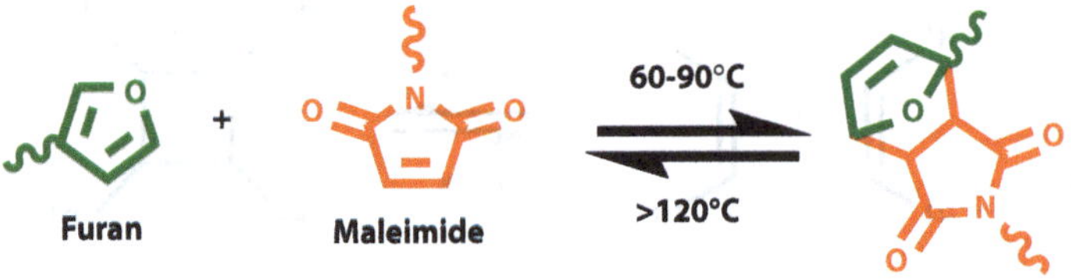

Figure 15.3 Diels-Alder (DA) thermo-reversible reaction between furan and maleimide.

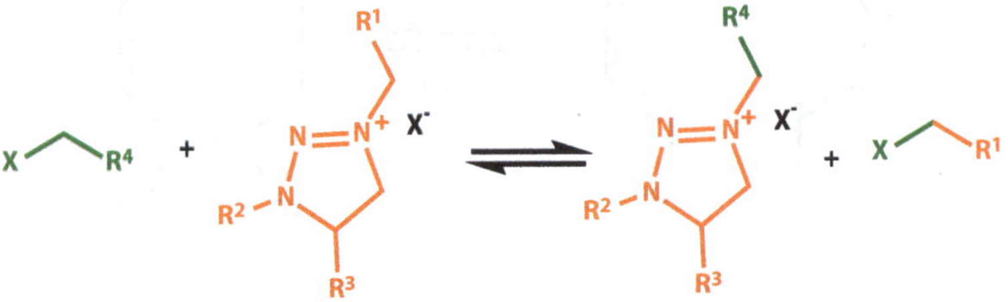

Figure 15.4 Transalkylation exchange reaction of triazolium salts.

15.2.4 Disulfide Exchanges

Disulfide chemistry encompasses intricate transformational mechanisms profoundly influenced by environmental and operational conditions. These molecular entities can undergo exchange reactions with other disulfides or thiols, or alternatively undergo reduction to thiol forms, which subsequently may re-oxidize through a dissociative pathway [4]. The research conducted by Klumperman and coworkers provides examples of both mechanisms [26], which are illustrated in Figure 15.5.

15.2.5 Transcarbamoylation and Transthiocarbamoylation

Polyurethanes represent a prominent polymer class with extensive applications, particularly in adhesive technologies. Motivated by environmental sustainability concerns, researchers have explored the dynamic characteristics of carbamate bonds to mitigate their ecological footprint [8, 27, 28]. Urethane bonds demonstrate two distinct reaction mechanisms (Figure 15.6): associative and dissociative [29]. Catalyst concentration, specifically dibutyltin dilaurate (DBTDL), significantly influences polymerization and transcarbamoylation processes, with higher catalyst levels correlating to reduced polyurethane relaxation times [30].

Like transcarbamoylation, the thiol derivatives have also been studied. For example, Torkelson and coworkers [31] employed small-molecule model systems to investigate the dynamic behaviour of thiourethane linkages, revealing the presence of two exchange pathways, as depicted in Figure 15.7.

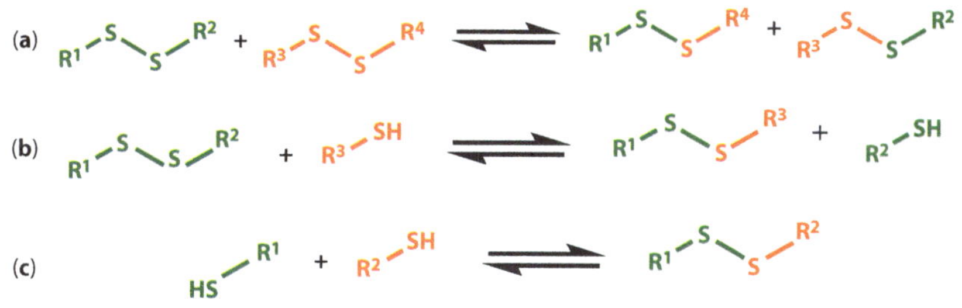

Figure 15.5 The disulfides' two associative mechanisms: an exchange reaction between two disulfides (a) and between thiol and disulfide (b). Oxidation of thiols to disulfide (c).

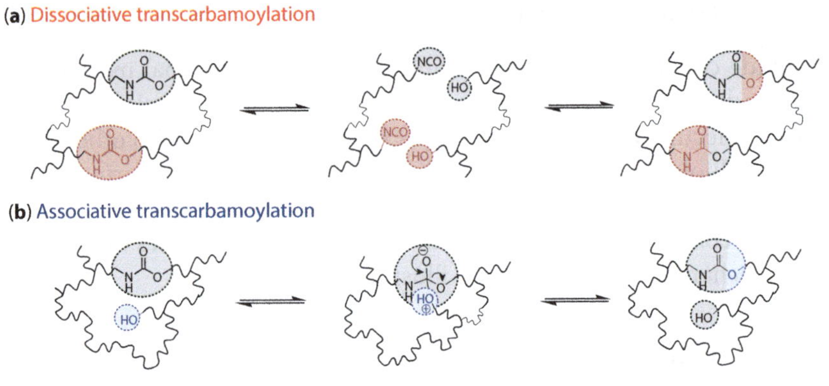

Figure 15.6 (a) Dissociative (no free alcohol groups) and (b) Associative (free alcohol groups) transcarbamoylation exchange reactions. Reproduced from [29] with permission from the Royal Society of Chemistry.

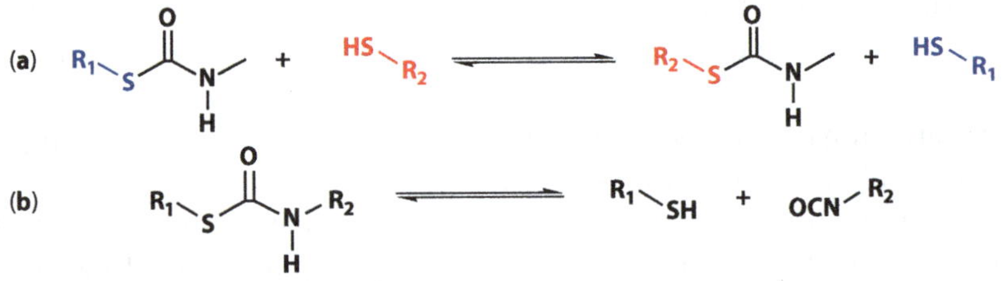

Figure 15.7 Thiourethane dynamic chemistry *via* (a) associative and (b) dissociative mechanism. Reproduced with permission from L. Li, X. Chen, J.M. Torkelson, Reprocessable Polymer Networks *via* Thiourethane Dynamic Chemistry: Recovery of Cross-link Density after Recycling and Proof-of-Principle Solvolysis Leading to Monomer Recovery, Macromolecules. Copyright 2019 American Chemical Society [31].

15.2.6 Transimination

Imine-containing polymers can undergo two types of associative mechanisms: imine metathesis and imine exchange (Figure 15.8), known together as transimination. The first one involves the reaction of imine and amine, whereas the latter involves the reaction of two imines. Furthermore, imine can undergo hydrolysis, leading to the re-formation of the original aldehyde and amine moieties. Consequently, it may also be described as a dissociative process [32, 33].

Imine-based polymers exhibit a distinctive behavior diverging from conventional polymer systems, characterized by the ability to flow at temperatures below the topology freezing transition temperature (Tv), indicating a

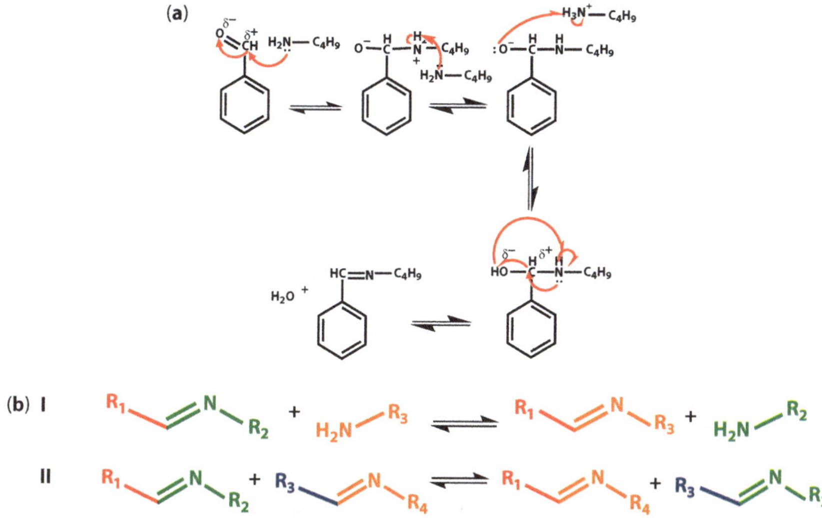

Figure 15.8 Imine formation (a) and the two transimination mechanisms (b): imine exchange (I) and imine metathesis (II). Reproduced from N. Jarach, H. Dodiuk, S. Kenig, N. Naveh, Rheology—Composition relationship of vitrimers based on polyethyleneimine, J. Appl. Polym. Sci. 139 (2022) e52353 with permision from Wiley [32].

transformative shift between viscoelastic solid and liquid states. Research demonstrates that elevated cross-linking density correlates with reduced viscosity, attributable to the predominance of imine metathesis mechanisms over imine exchange processes [32, 33]. Whereas a catalyst is not necessary in these processes, some studies have shown that polar solvents can accelerate the imine exchange reaction [34].

15.2.7 Boronic Ester (Boronate Ester) and Dioxaborolane Metathesis

Boronic esters are formed through condensation reactions between boronic acids and alcohols, resulting in cyclic ester structures (Figure 15.9a) [35]. Like the mechanisms of transesterification involving carbonyl esters, which are widely discussed in the literature [4, 36], boronic ester bonds can undergo multiple exchange reactions. These pathways are referred to as dioxaborolane metathesis or trans-boronate esterification (Figure 15.9b). Importantly, these reactions exhibit greater efficiency than carbonyl-based alternatives due to their lower activation energy, allowing them to occur at reduced temperatures without needing free diols [37].

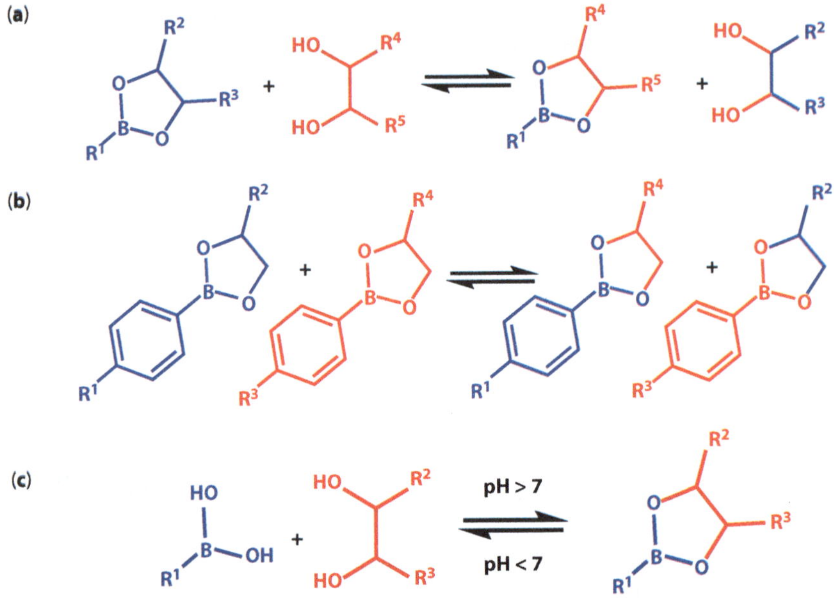

Figure 15.9 Boronate esters dynamic mechanisms: exchange reaction between boronic esters and diols (a), Metathesis reaction between two dioxaborolanes (b), and association and dissociation under different pHs (c).

Boronate esters can also dissociate or associate depending on the pH level (Figure 15.9c) [4]. This sets them apart from carbonyl-based esters. The boronate esters break down into their original diols and boronic acid components in acidic environments. Conversely, when exposed to basic conditions, boronic acids and diols combine to form stable boronate esters. Although this property shows potential, no research has been conducted on its application as adhesive. Therefore, this chapter will not delve into it any further.

15.3 Switchable Adhesives

Responsive stimuli heavily influence the selection of adhesives in industrial, biomedical, and other applications. In the following section, we have organized the CANs studied as adhesives based on their response to light, thermal, and pH stimuli (Figure 15.10) to aid in this selection process (Table 15.1). While there are dynamic bonds that can respond to other

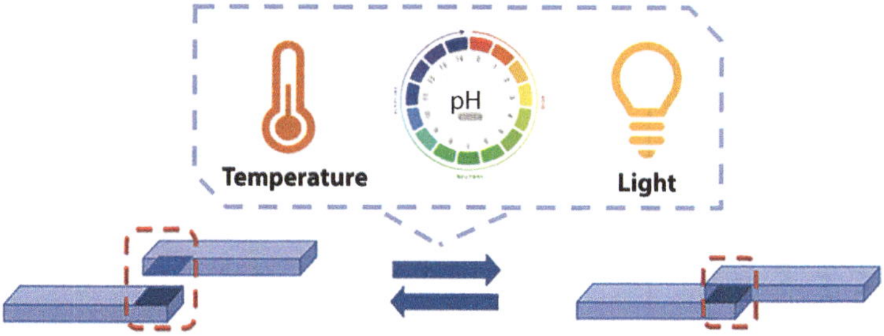

Figure 15.10 Bonding stimuli discussed in this chapter.

stimuli, like electrical charges, we have omitted them from this section due to a lack of relevant studies within the field of adhesives.

15.3.1 Light-Responsive Adhesives

Recently, researchers have been investigating the use of light-sensitive components in traditional adhesives to allow for light-triggered debonding. This method involves two primary mechanisms: photoinduced overcuring and photodegradation. However, this discussion will focus on alternative strategies, such as isomerization and dimerization, which have been extensively studied and can provide bonding and debonding capabilities at specific wavelengths [2, 38]. Although these mechanisms can result in fast, on-demand debonding, they do have limitations. Firstly, they necessitate transparent substrates that transmit light at the appropriate wavelengths. Additionally, they may not be suitable for applications where the debonding wavelength will be present throughout the product's lifespan.

15.3.1.1 Photoinduced Isomerization

As previously discussed, adhesives can fail in two ways: cohesive failure or adhesion failure. In the case of a decrease in the adhesive modulus, cohesive failure will occur. By inducing an on-demand decrease in the modulus, debonding-on-demand can be achieved. Photoinduced isomerization allows for controlled changes in modulus within adhesive systems, enabling precise debonding through light-activated molecular transformations. Several light-induced isomerization bonds have been suggested, with spiropyran as an example. This functional group can change between ring-open and closed isomers, altering its electrostatic charge and reducing its adhesive strength, as demonstrated by Branda and coworkers in their study [39].

Table 15.1 Summary of stimuli-responsive reversible adhesives.

Stimulus	Dynamic bond	Bonding and debonding [a] conditions	Components	Tested substrates	Lap-Shear (or peel) strength during various bonding and debonding cycles [b]	Reference
Light	Isomerization (open and closed rings)	Bonding: UV light 365 nm (60s) Debonding: Visible light >530 nm	Spiropyran-based materials (microstructured PDMS film doped with spiropyran)	Glass sheets	Cycle 1: 365 nm- ~ 105 mN >530 nm- ~102 mN Cycle 5: 365 nm- ~ 106 mN >530 nm- ~ 103 mN	[39]
	Isomerization (Cis-Trans)	Bonding: Visible light 530 nm (15 min) Debonding: UV light 365 nm (15 min)	Azobenzene groups incorporated into acrylate polymers	Quartz sheets	Cycle 1: 365 nm- ~ 0.08 MPa 530 nm- ~ 1.02 MPa Cycle 3: 365 nm- ~ 0.2 MPa 530 nm- ~ 0.9 MPa	[41]

(Continued)

Table 15.1 Summary of stimuli-responsive reversible adhesives. (*Continued*)

Stimulus	Dynamic bond	Bonding and debonding [a] conditions	Components	Tested substrates	Lap-Shear (or peel) strength during various bonding and debonding cycles [b]	Reference
		Bonding: Visible light 510 nm Debonding: UV light 365 nm	Compounds based on sugar derivatives with multi-azobenzene arms.	Quartz sheets	Cycle 1: 365 nm- 0.003 MPa 510 nm- 0.5 MPa Cycle 3: 510 nm- ~ 0.5 MPa	[42]
		Bonding: UV light 350 nm (30s) Debonding: Visible light 440 nm (1 min)	Acrylic polymers containing azobenzene groups	Silicon wafers	180° Peel strength: (4% Azo-M): 350 nm- ~ 8.1 N/25mm 440 nm- ~ 4.3 N/25mm	[43]
	Dimerization [2 + 2] cycloaddition	Bonding: UVA light >300 nm Debonding: UVC light <290 nm	Acrylate polymers containing coumarin moieties	Glass sheets	Peel strength: Initial- 1.62 N/mm UVA- 0.05 N/mm UVC- 0.1 N/mm	[44]

(*Continued*)

Table 15.1 Summary of stimuli-responsive reversible adhesives. (*Continued*)

Stimulus	Dynamic bond	Bonding and debonding [a] conditions	Components	Tested substrates	Lap-Shear (or peel) strength during various bonding and debonding cycles [b]	Reference
		Bonding: UV light 365 nm (15 min) Debonding: 254 nm light (5 min)	Siloxane backbone with coumarin groups	Quartz sheets	Cycle 1: 365 nm- 1.1 MPa 254 nm- 0.2 MPa Cycle 2: 365 nm- 0.8 MPa 254 nm- 0.1 MPa	[45]
		Bonding: UV light 365 nm Debonding: 254 nm light	Bio-based soybean oil with coumarin groups	Quartz sheets	Cycle 1: 365 nm- 3.1 MPa 254 nm- 0.1 MPa Cycle 3: 365 nm- 2.4 MPa 254 nm- 0.1 MPa	[46]
		Bonding: UV light 365 nm + 120°C Debonding: 254 nm light	Caffeic acid containing catechol and cinnamoyl units	Aluminium sheets	Before UV- 3.2 MPa After UV- 7.2 MPa	[47]

(*Continued*)

Table 15.1 Summary of stimuli-responsive reversible adhesives. (*Continued*)

Stimulus	**Dynamic bond**	**Bonding and debonding [a] conditions**	**Components**	**Tested substrates**	**Lap-Shear (or peel) strength during various bonding and debonding cycles [b]**	**Reference**
	Dimerization [4 + 4] cycloaddition	Bonding: UV light 365 nm (20 min) Debonding: 254 nm light (20 h)	Anthracene-based epoxy resin	Quartz sheets	365 nm- 2.4 MPa 254 nm- caused lap shear debonding	[48]
		Bonding: UV light 365 nm (5 min) Debonding: Heat	Anthracene-based acrylate polymer	Glass sheets	Cycle 1: 365 nm- 1.67 MPa Heat- 0.15 MPa Cycle 2: 365 nm- 2.37 MPa Heat- 0.43 MPa	[49]

(*Continued*)

Table 15.1 Summary of stimuli-responsive reversible adhesives. (*Continued*)

Stimulus	Dynamic bond	Bonding and debonding [a] conditions	Components	Tested substrates	Lap-Shear (or peel) strength during various bonding and debonding cycles [b]	Reference
Thermal	Diels-Alder (DA) reaction	Bonding: 80°C (2 h) Debonding: 150°C (5 min)	Materials containing furan and maleimide derivatives	Aluminium sheets	Initial: 7.9 MPa Cycle 1: 7.0 MPa Cycle 2: 7.5 MPa	[55]
		Bonding: 25°C (2-10 min) Debonding: 80-100°C (< 5 min)	Materials with incorporation of furan and pyridinyldithioformate	-	-	[56]
	Transalkylation	Bonding: 180°C (1-2 hr) Debonding: by damage [a] (shear force)	1,2,3-triazolium-containing polymers	Aluminium sheets	Initial: 23.7 MPa Cycle 1: 21.7 MPa Cycle 20: 12.5 MPa	[57]

(*Continued*)

Table 15.1 Summary of stimuli-responsive reversible adhesives. (*Continued*)

Stimulus	**Dynamic bond**	**Bonding and debonding [a] conditions**	**Components**	**Tested substrates**	**Lap-Shear (or peel) strength during various bonding and debonding cycles [b]**	**Reference**
		Bonding: 140°C Debonding: by damage (shear force)	1,2,3-triazolium-containing polymers	Glass sheets	Initial: ~ 1.08 MPa Cycle 1: ~ 1.0 MPa Cycle 3: ~1.06 MPa	[58]
		Bonding: hot pressing Debonding: by damage (tensile/shear force)	Natural rubber (NR)	NR samples	Failure in the substrate rather than in the bonded area	[59]

(*Continued*)

Table 15.1 Summary of stimuli-responsive reversible adhesives. (*Continued*)

Stimulus	Dynamic bond	Bonding and debonding [a] conditions	Components	Tested substrates	Lap-Shear (or peel) strength during various bonding and debonding cycles [b]	Reference
	Disulfides	Bonding: hot pressing (20 kN; 200°C ; 20 min) Debonding: by damage (shear force)	Epoxy resin containing disulfide bonds	Aluminium sheets	Initial: ~ 13.5 MPa Rebonded: ~ 12.5 MPa	[60]
		Bonding: 80°C and then heating (150°C) or UV light (320-390 nm) for strong adhesion. Debonding: by damage (shear force)	Polymer containing disulfide bonds (by oxidation of thiol groups)	Glass sheets	80°C- ~ 2.9 MPa 150°C- ~ 5.4 MPa UV light- ~6.5 MPa	[61]

(*Continued*)

Table 15.1 Summary of stimuli-responsive reversible adhesives. (*Continued*)

Stimulus	Dynamic bond	Bonding and debonding [a] conditions	Components	Tested substrates	Lap-Shear (or peel) strength during various bonding and debonding cycles [b]	Reference
		Bonding: 130°C (1 h) Debonding: by damage (tensile/shear force)	Epoxy resin containing disulfide bonds and polycaprolactone (PCL)	Two films of epoxy + disulfide + polycaprol-actone (PCL)	Failure at the substrate rather than in the bonding area	[62]
	Thiocarbamates	Bonding: hot pressing (3 MPa; 100°C; 30 min) Debonding: by damage (shear force)	Polythiourethane containing thiocarbamate	Glass/Glass Steel/steel Aluminium/Aluminium Oak/oak	~ 2.9 MPa Initial: 4.8 MPa Rebonded: 4.2 MPa ~ 5.3 MPa ~ 6.2 MPa	[63]

(*Continued*)

Table 15.1 Summary of stimuli-responsive reversible adhesives. (*Continued*)

Stimulus	Dynamic bond	Bonding and debonding [a] conditions	Components	Tested substrates	Lap-Shear (or peel) strength during various bonding and debonding cycles [b]	Reference
	Imine Metathesis	Bonding: 120°C (30 min) Debonding: by damage (shear force)	Biobased polyimine vitrimer	Iron sheets	Initial: 6.79 N/mm^2 Rebonded: 5.76 N/mm^2	[64]
Thermal	Boronic Transesterification	Bonding: 120°C (20 min) Debonding: by damage (shear force)	Polyurethane containing boronic esters bonds	Glass sheets Steel sheets Wood sheets	1st- 4.2 MPa; 2nd- 3.9 MPa; 3rd- 3.6 MPa 1st- 7.1 MPa; 2nd- 3.9 MPa; 3rd- 3.6 MPa 1st- 8.9 MPa; 2nd- 6.8 MPa; 3rd- 6.5 MPa ** 1st – initial[c] ** 2nd – cycle[b] 1 ** 3rd – cycle[b] 2	[65]

(*Continued*)

Table 15.1 Summary of stimuli-responsive reversible adhesives. (*Continued*)

Stimulus	Dynamic bond	Bonding and debonding [a] conditions	Components	Tested substrates	Lap-Shear (or peel) strength during various bonding and debonding cycles [b]	Reference
		Bonding: 60°C (2 hr) Debonding: by damage (shear force)	Polyurethane containing boronic ester bonds derived from catechol	Glass sheets	0.58-0.96 MPa	[66]
		Bonding: 215°C Debonding: by damage (shear force)	Polystyrene-b-poly(ethylene-co-butylene)-b-polystyrene (SEBS) containing boronic esters bonds	Aluminium/Aluminium Steel/Steel Glass/Glass	25.1 MPa 28.54 MPa 39.6 MPa	[67]

(*Continued*)

Table 15.1 Summary of stimuli-responsive reversible adhesives. (*Continued*)

Stimulus	Dynamic bond	Bonding and debonding [a] conditions	Components	Tested substrates	Lap-Shear (or peel) strength during various bonding and debonding cycles [b]	Reference
pH	Dissociation of boronate esters	Bonding: basic pH 10 Debonding: acidic pH >7.4	Alginate-boronic acid hydrogel	Glass sheets	Alginate with boronic acid- 32 kPa Alginate without boronic acid- 3 kPa	[69]
		Bonding: acidic pH 3 Debonding: basic pH 9	Catechol-boronic acid	Glass sheets	-	[70, 71]

[a] Debonding caused by damage occurs when the sample is subjected to shear or tensile force.
[b] The term "cycle" refers to the process of bonding, debonding, and rebonding of the sample.
[c] The term "initial" refers to the initially measured bond strength.

They looked at spiropyran-containing polymer films as electrostatic dry adhesives, which show increased adhesion after ultraviolet (UV) light exposure (365 nm for 60s), but visible light exposure (>530 nm) decreases their adhesive property (Figure 15.11). Exposure of the micro-structured films to UV light (365 nm) significantly increased adhesion values [2, 39].

Despite its advantages, side-reactions and sensitivity to temperature, pH, and mechanical forces limit spiropyran's isomerization applicability

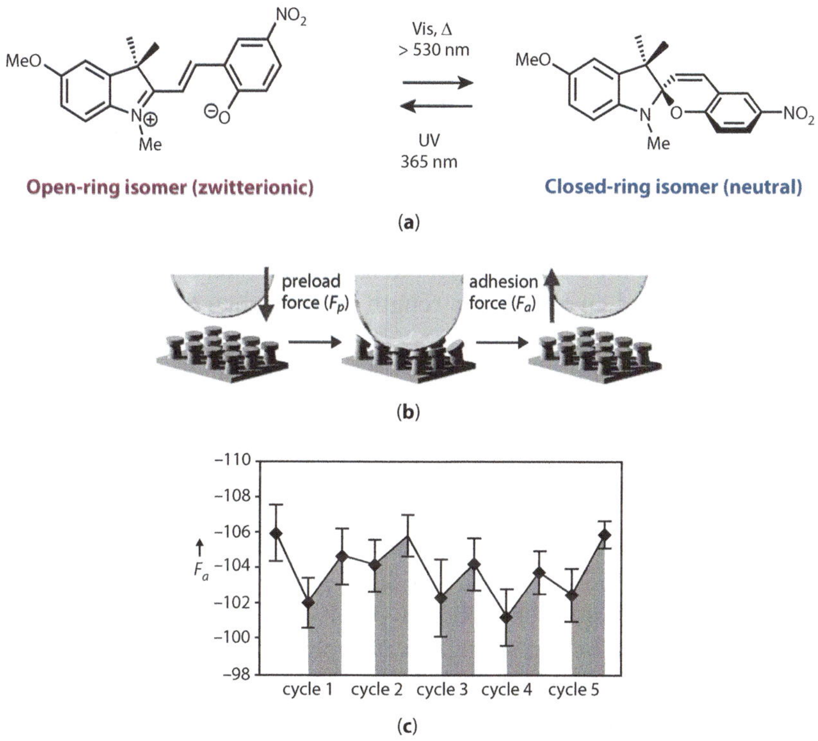

Figure 15.11 (a) – Photoinduced isomerization of spiropyran. (b) – Schematic illustration of the adhesion strength measurement between the films and a glass sphere. After a preload force (Fp) is applied, the force (Fa, in mN) required to retract the sphere from the film reflects the adhesion between the two surfaces. (c) – Adhesion strength measurements for micro-textured polydimethylsiloxane (PDMS) films incorporating spiropyran, characterized across varying wavelength exposures including ultraviolet (365 nm) and visible light (>530 nm) spectral ranges. Reproduced from K.R. Mulcahy, A.F.R. Kilpatrick, G.D.J. Harper, A. Walton, A.P. Abbott, Debondable Adhesives and Their Use in Recycling (2022) with permission from Royal Society of Chemistry and reproduced with permission from P. Tannouri, K.M. Arafeh, J.M. Krahn, S.L. Beaupré,Beaupré, C. Menon, N.R. Branda, A Photoresponsive Biomimetic Dry Adhesive Based on Doped PDMS Microstructures .Copyright 2014 American Chemical Society [2, 39].

to switchable adhesives [40]. A more stable alternative is the trans/cis isomerization of azobenzene, which can be incorporated into polymers to produce solid trans-azopolymers. Trans-azobenzene is a solid material that is capable of bonding two materials together with a similar level of stiffness as conventional adhesives. Under UV, isomerization of the trans-azobenzene occurs, changing conformation from *trans* to *cis*, causing reduction in melting point and liquefaction, thus softening the adhesive. This transformation allows for the separation of the bonded materials (Figure 15.12) [2, 41]. Zhang and coworkers showed in their study [41] that a poly(6-(4-(p-tolyldiazenyl) phenoxy) hexyl acrylate), containing azobenzene groups, can bond two quartz substrates using the *trans*-form of the polymer. In this form, the resulting adhesion strength was approximately 1.02 MPa after exposure to 530 nm for 15 minutes. However, when the *trans*-polymer was exposed to UV light at 365 nm for 15 minutes, it underwent isomerization to the *cis* form. This isomerization led to a significant decrease in adhesion strength to approximately 0.08 MPa, which was only 7.8% of the strength observed in the *trans* state. The reused adhesive demonstrated an adhesion strength of more than 68% of its original strength.

Akiyama and Yoshida [42] have also demonstrated an azobenzene-based adhesive, using sugar alcohol derivatives-based compounds with multi-azobenzene arms. These compounds could undergo photochemical and isothermal liquefaction from a powdered solid state when exposed to UV light (365 nm) at room temperature, followed by solidification upon irradiation with visible light (510 nm). Measuring adhesion strength on quartz plates showed 0.5 MPa in the solid state (510 nm irradiation) and 0.003 MPa in the liquid state (365 nm irradiation). After the adhesive was re-solidified, the adhesion strength was approximately 0.5 MPa. Kim and coworkers [43] also studied these bonds, though in a pressure-sensitive adhesive (PSA) in combination with acrylic polymer. To evaluate the adhesive property of the switchable PSA, a lap shear test was conducted using poly(methyl methacrylate) (PMMA) as the substrate, resulting in a shear modulus of 2 GPa (Figure 15.13-1). A 180° peel test evaluated adhesion switchability on silicon wafer substrates, revealing that Azo-M-4 demonstrated maximum peel strength in the activated state, with peel strength inversely correlating to increasing Azo-M-4 concentration, and universal adhesion reduction upon visible light exposure (Figure 15.13-2). The switchable PSAs demonstrated rapid adhesion and exceptional on-demand adhesion control, allowing them to alternate between adhesive and non-adhesive states without undergoing a phase transition between the solid and liquid states.

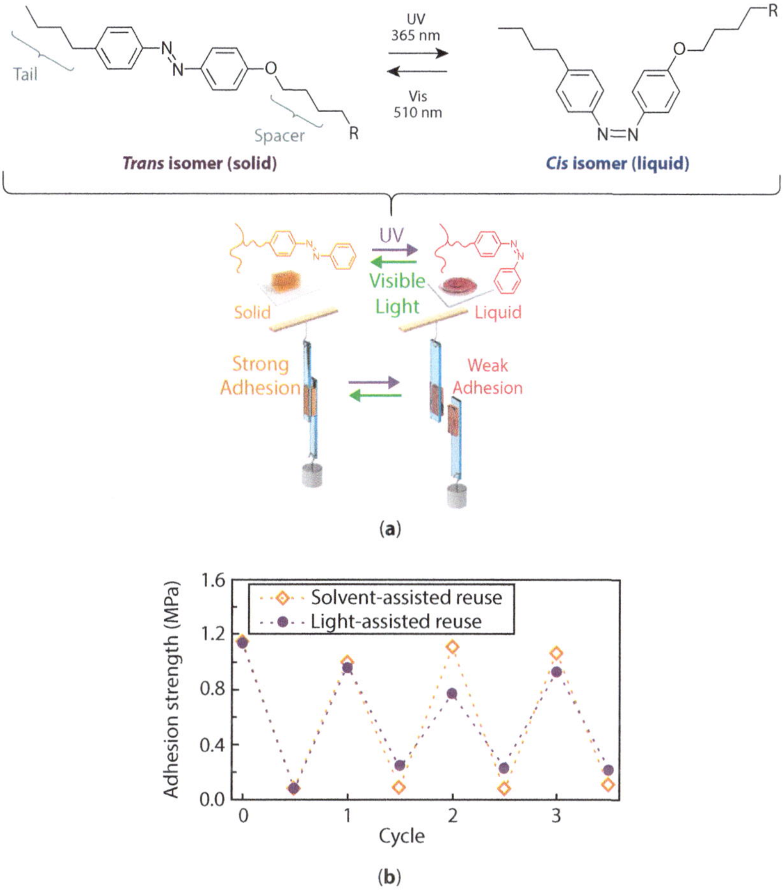

Figure 15.12 (a)- Photoinduced isomerization of azobenzene derivatives. (b)- light-assisted and solvent-assisted reuse of the adhesive. Reproduced from K.R. Mulcahy, A.F.R. Kilpatrick, G.D.J. Harper, A. Walton, A.P. Abbott, Debondable Adhesives and Their Use in Recycling (2022) and Y. Zhou, M. Chen, Q. Ban, Z. Zhang, S. Shuang, K. Koynov, H.J. Butt, J. Kong, S. Wu, Light-Switchable Polymer Adhesive Based on Photoinduced Reversible Solid-to-Liquid Transitions (2019) with permission from Royal Society of Chemistry and ACS Publications [2, 41].

15.3.1.2 Reversible Photodimerization

Several functional groups were studied in the field of photodimerization. One example of functional group that can undergo dimerization and has been widely studied is coumarin and its derivatives, which undergo [2 + 2] cycloaddition photo-dimerization upon irradiation at ~300 nm and dissociate upon exposure to ~250 nm. Trenor and coworkers, for example,

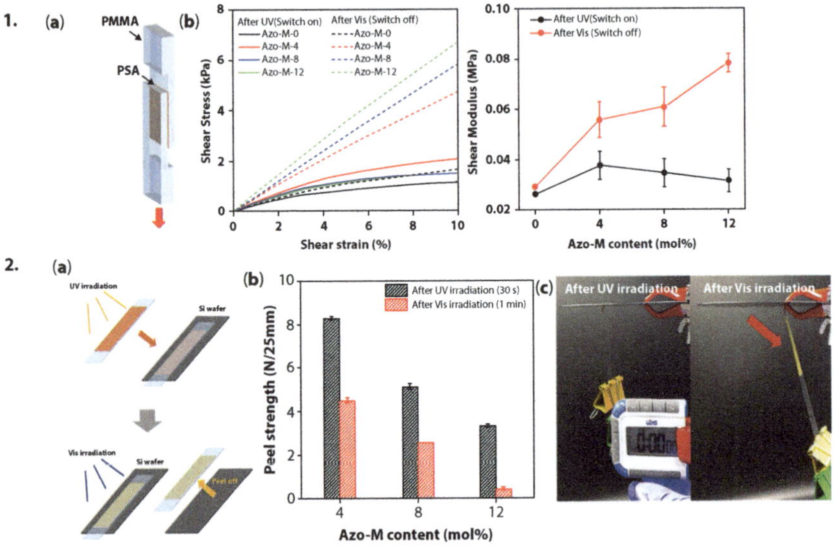

Figure 15.13 1. (a) Schematic representation of the lap shear test specimen, (b) changes in the lap shear stress and shear modulus of the switchable PSAs with UV and Vis irradiation. 2. (a) Schematic representation of the peel strength switching evaluation method, (b) changes in the 180° peel strength of the switchable PSAs with UV and Vis irradiation, (c) images capturing the effect of the irradiation during the 90° peel test conducted with Azo-M-12. Reproduced with permission from T.H. Lee, G.Y. Han, M.B. Yi, H.J. Kim, J.H. Lee, S. Kim, Rapid Photoresponsive Switchable Pressure-Sensitive Adhesive Containing Azobenzene for the Mini-Light Emitting Diode Transfer Process. Copyright 2021 American Chemical Society [43].

developed pressure-sensitive adhesives (PSAs) based on a polymer with 7-hydroxycoumarin as pendant groups [44], which undergo [2 + 2] cycloaddition reaction under >300 nm. Unlike other cross-linking methods, this process does not require the presence of a photoinitiator (PI), and it can be reversed by irradiation with UVC light (<290 nm). However, the peel strength of the PSA decreased from 1.62 N/mm to 0.05 N/mm. Although a full recovery of the initial peel strength was not achieved, the cleavage of the dimers by UVC irradiation increased the value to 0.1 N/mm.

Saito and colleagues also conducted a study exploring the use of coumarin derivatives in adhesives [45] (Figure 15.14). Siloxane-based adhesive systems were evaluated by bonding quartz glass substrates using UV light (365 nm) and applied pressure to assess adhesive strength characteristics. The debonding process was achieved by irradiation with 254 nm light and applying a load. Comprehensive evaluations assessed the influence of irradiation duration and cyclic bonding-debonding processes on shear

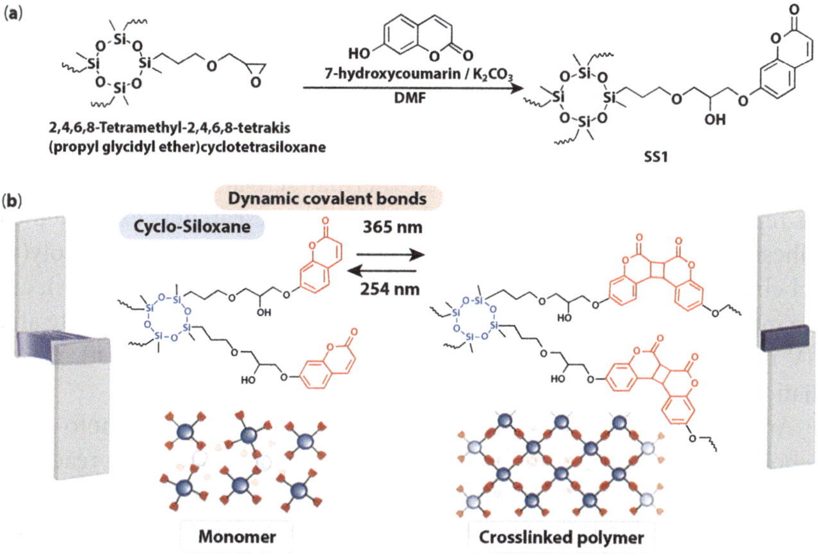

Figure 15.14 The adhesive system that was demonstrated by Saito and coworkers: synthesis scheme for adhesive (a), and a schematic diagram of a photoinduced debonding-on-demand system (b). Reproduced from T. Asahi, K. Saito, M. Inada, T. Horii, T. Fujie, T. Nakanishi, Debonding-on-Demand Adhesives Based on Photo-Reversible Cycloaddition Reactions (2023), with permission from Royal Society of Chemistry [45].

adhesion strength, demonstrating progressive strength enhancement with increased 365 nm light exposure, culminating in 1.0-1.7 MPa strength after 15 min irradiation. However, after 30 minutes of irradiation, the adhesion force increased to such an extent that it caused the substrate to fail, rather than resulting in cohesive or adhesion failure. Re-bonding investigations revealed dynamic adhesion strength modulation: the material initially exhibited a high adhesion strength of 1.1 MPa in its crosslinked state, which diminished to 0.2 MPa upon de-cross-linking. Subsequent exposure to 365 nm irradiation triggered re-cross-linking and partially restored the adhesion strength to 0.8 MPa, validating the material's reversible and reworkable adhesive characteristics. However, each debonding and re-bonding cycle led to a gradual decrease in bond strength. In a separate study, they also successfully demonstrated the incorporation of coumarin groups into bio-based adhesives derived from soybean oil [46].

Naito and co-workers [47] investigated a photoreactive, mussel-inspired adhesive derived from caffeic acid. They utilized its two functional moieties: catechol, which facilitates hydrogen and polar bonding, and cinnamoyl, a photoreactive unit capable of undergoing a [2 + 2] cycloaddition reaction. UV irradiation at specific wavelengths (365 nm and 254 nm) triggered [2 + 2] photo-cycloaddition and retro-cycloaddition mechanisms (Figure 15.15). The adhesion strategy involved two-stage processing: initial polymer application with caffeic acid functionalization and UV (365 nm) irradiation to generate a robust pre-applied adhesive layer *via* inter-molecular cross-linking. Later, thermal activation (120°C) allowed for repeated adhesion to various substrates. The researchers examined the adhesive strength to various substrates (Al, Fe, Cu, wood, carbon, poly(vinyl chloride) (PVC), PMMA, polypropylene (PP), polyethylene(PE), and the maximum adhesive strength achieved for an aluminum substrate after UV irradiation was 7.2 MPa, while the adhesive strength before UV irradiation was 3.2 MPa.

Anthracenes were also studied as functional groups for photo-based debonding-on-demand due to their [4 + 4] photocycloaddition reactions. Zhang and coworkers, for example, synthesized an epoxy resin containing anthracene. First, they reacted the epoxy groups with polyetheramine curing agent to obtain prepolymers (pre-AER/D230, pre-AER/D400, pre-AER/D2000), where "AER" stands for 9-Anthracene Epoxy Monomer)

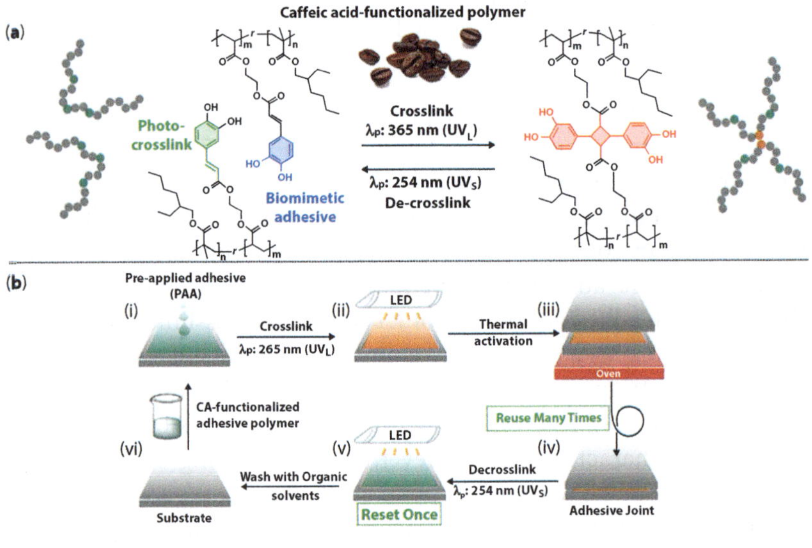

Figure 15.15 (a) The adhesive contains a catechol moiety that mimics mussel l-DOPA and a light-sensitive moiety that is present in cinnamates extracted from plant materials such as coffee beans. (b) Preparation of a light-induced pre-applied adhesive layer and its closed-loop sustainability. Reproduced from S. Wang, W.H. Hu, Y. Nakamura, N. Fujisawa, A.E. Herlyng, M. Ebara, M. Naito, Bio-Inspired Adhesive with Reset-On Demand, Adv. Funct. Mater. (2023) 2215064, with permission from Wiley [47].

that had four anthracene end groups [48]. Then, while exposed to 365 nm light, the anthracene groups underwent the cycloaddition reaction (Figure 15.16). The lap shear strength of the samples was measured using quartz glass substrates. Samples were prepared at different temperatures (pre-AER/D230 and pre-AER/D400 at 120 and 80°C, respectively, and pre-AER/D2000 at room temperature) by exposing them to 365 nm UV light for 20 minutes, followed by 254 nm UV light. Whereas the adhesive exhibited lap shear strength of 2.4 MPa, but after exposure to 254 nm UV light for 20 hours, the shear strength significantly decreased, and the adhesive reversed into its liquid form. Ultimately, the corresponding shear strength decreased until the bonded joint was separated. An almost similar method

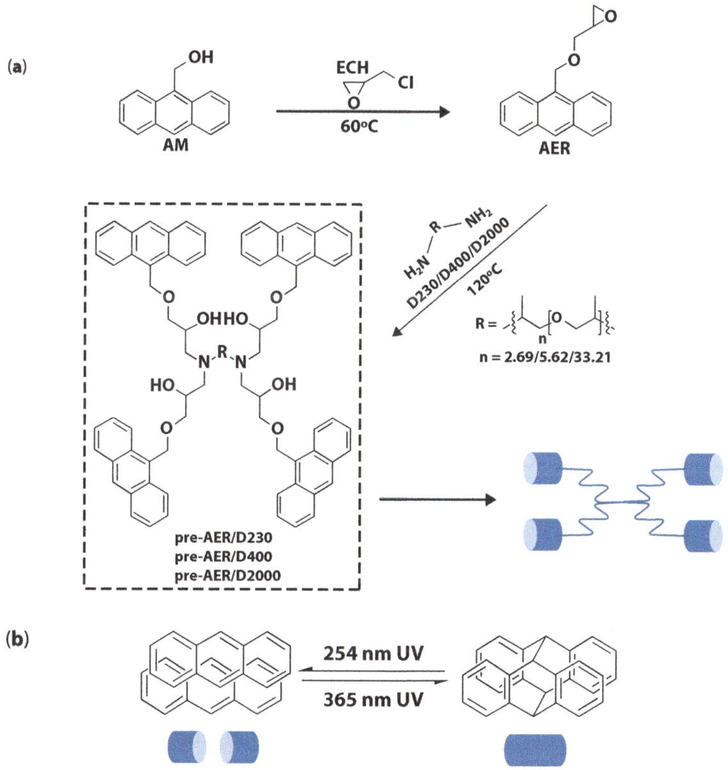

Figure 15.16 Synthesis of epoxy-based anthracene pre-polymers (a) and light-induced dimerization of anthracene aromatics (b). Reproduced from Z. Liu, J. Cheng, J. Zhang, An Efficiently Reworkable Thermosetting Adhesive Based on Photoreversible [4+4] Cycloaddition Reaction of Epoxy-Based Prepolymer with Four Anthracene End Groups, Macromol. Chem. Phys. 222 (2021) 2000298, with permission from Wiley [48].

was also reported by Liu and coworkers, though with different backbone, resulting in slightly different mechanical properties [49].

15.3.2 Thermally Responsive Adhesives

15.3.2.1 Diels-Alder (DA) Reaction

In the past years, DA reactions have been among the most studied reversible reactions [50–54]. Das and coworkers, for example, developed thermally responsive adhesives using furyl-modified copolymers (FMPs) and commercially available 1,1'-(methylenedi-4,1-phenylene)bismaleimide (BMI). DA reaction, and therefore the adhesion process, occurred at 80°C (for 2 hr) and debonding occurred at 150°C (for 5 min) [55]. At the optimal components ratio, they observed a maximum lap shear strength to an aluminum substrate of 7.9 MPa. The lap shear strengths after the first and second re-attachments obtained were 7.0 and 7.5 MPa, respectively, which corresponded to 88% and 95% of the initial value. Whereas most examples include the use of furan and maleimide derivatives, some, like Barner-Kowollik and coworkers [56], used different functional groups. In their study, they demonstrated the incorporation of cyclopentadiene (Cp) into a PMMA chain and the use of a trifunctional linker, pyridinyldithioformate, which facilitated rapid and reversible cross-linking. The bonding occurred at room temperature (25°C) and debonding at 80–100°C [56].

15.3.2.2 Transalkylation

Poly(1,2,3-triazolium) vitrimers (PTAVs) have been studied by several groups as potential debondable adhesives. Huang and coworkers, for example, investigated the adhesion property of the PTAV adhesive to an aluminum surface [57]. After being thermally-treated at 180°C, above Tv (Tv ≈ 150°C), for 2 hours, the bonding process was completed. The bond between the two adherends exhibited a lap-shear strength as high as 23.7 MPa. Additionally, to demonstrate the self-healing property of the PTAV adhesive, they used a lap-shear test apparatus to break an adhesive joint into two parts. After a simple process of re-assembling the parts and heating them at 180°C for 1 hour, the joint was repaired through self-healing, with a new lap-shear strength of 21.7 MPa. Repeated damage and repair cycles were executed 20 times, demonstrating a gradual strength reduction while maintaining sufficiently high mechanical performance for structural adhesive applications. Jeong and coworkers have also studied 1,2,3-triazolium-containing polymers, and showed that (re)adhesion can also occur at lower temperatures (~ 140°C), demonstrating a lap shear strength of 1.06 MPa to glass slides [58]. In an additional study conducted by Luo and coworkers, they also studied these bonds with other matrices, such as natural rubber, thus also incorporating different types of dynamic bonds, including disulfides, which will be discussed below [59].

15.3.2.3 Disulfides

Disulfides have been extensively researched in the field of adhesives. However, unlike other types of dynamic and reversible bonds, the cross-linking process occurs before the adhesion. After the polymer is cured, heat is applied to soften it and allow for exchange reactions, which help in wetting the surface and causing adhesion. Naito and coworkers reported one example of a disulfide-containing adhesive that incorporated disulfide bonds into an epoxy-based adhesive. After a failure occurred, heating the two parts allowed re-bonding due to exchanges between disulfides of the polymer on both sides, resulting in a recovery rate of 83-95% of the initial properties [60]. Meanwhile, Rowan and coworkers employed a different approach by oxidizing thiol groups to incorporate the disulfides. Heating the disulfide-containing polymer to 150°C triggered the exchange reactions, resulting in adhesion to glass slides [61]. However, due to poor

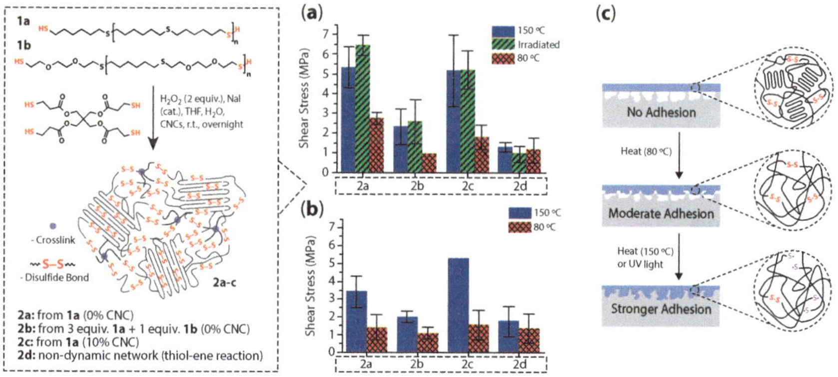

Figure 15.17 Shear stress results from different stimuli by applying different contact pressures between the adhesive and substrate during bonding, achieved through either (a) using two 1" binder clips (contact pressure = 1.2 MPa) or (b) applying a 500 g weight (contact pressure = 0.136 MPa). (c) Schematic illustration of the two-level adhesion. Reproduced with permission from B.T. Michal, E.J. Spencer, S.J. Rowan, Stimuli-Responsive Reversible Two-Level Adhesion from a Structurally Dynamic Shape-Memory Polymer. Copyright 2016 American Chemical Society [61].

surface wetting, adhesion strength decreased at lower temperatures (Figure 15.17). One suggestion to reduce softening temperature was studied by Yang and coworkers, who added soft long segments such as polycaprolactone (PCL), lowering T_g and reducing softening temperature [62].

15.3.2.4 *Thiocarbamates*

Although promising, only a few studies have been conducted on thiocarbamate bonds. One such study was conducted by Zhang and coworkers, who developed a re-processable thiocarbamate-containing polythiourethane [63]. They conducted a study to confirm the dynamic behavior of the adhesive and created cross-linked polymers by combining trimethylolpropane-tris(3-mercaptopropionate) (TMMP) with various diisocyanates (Figure 15.18). Evaluation of the adhesive property was conducted on different substrates including glass, metal, and wood. Compared to commercial thermoplastic polyurethanes (TPUs) and reactive polyurethane (PUR) hot-melt adhesives (HMAs), the new polymer containing m-tetramethyl xylenediisocyanate (TMXDI) as a cross-linker displayed superior bonding property, particularly on red oak, owing to its rough surface texture and the formation of urethane bonds facilitated by the reactive hydroxyl groups and hydrogen bonds on the red oak's surface. The adhesive also successfully bonded two stainless steel substrates, withstanding a force of ~325 N. This team also assessed the re-workability of TMMP-TMXDI through a lap shear test by rejoining and pressing two broken pieces at 100°C with a force

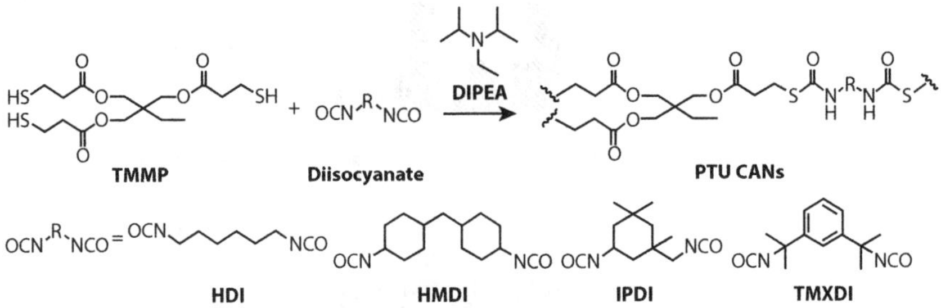

Figure 15.18 Synthesis of Zhang and coworkers' polyurethanes: trimethylolpropane tris(3-mercaptopropionate) (**TMMP**), N,N-Diisopropylethylamine (**DIPEA**), hexamethylene diisocyanate (**HDI**), 4,4′-methylene bis(cyclohexyl-isocyanate) (**HMDI**), isophorone diisocyanate (**IPDI**), and m-tetramethyl xylenediisocyanate (**TMXDI**). Reproduced with permission from C. Cui, X. Chen, L. Ma, Q. Zhong, Z. Li, A. Mariappan, Q. Zhang, Y. Cheng, G. He, X. Chen, Z. Dong, L. An, Y. Zhang, Polythiourethane Covalent Adaptable Networks for Strong and Reworkable Adhesives and Fully Recyclable Carbon Fiber-Reinforced Composites. Copyright 2020 American Chemical Society [63].

of 3 MPa for 0.5 hours, which showed that healed stainless steel samples exhibited a slightly lower adhesive strength of 4.2 MPa, comparable to the original bond strength of 4.8 MPa.

15.3.2.5 Imine Metathesis

Though imine exchanges can practically undergo three types of reversible\dynamic reactions, only imine metathesis has been investigated in the adhesives field. Liu and coworkers, for example, synthesized biobased polyimine vitrimers from vanillin-derived aldehydes with commercial biobased aliphatic amines [64]. These polymers were also tested as adhesives by applying them as thin layers between iron sheets. Depending on the cross-linking density, they achieved excellent adhesion (6.79 MPa). They also tested these adhesives on plastic substrates, although the obtained shear strength was lower than that observed with iron plates. After separation of the iron plates, they re-bonded them by heating them to 120°C for 30 minutes, achieving a strength of 5.76 MPa.

15.3.2.6 Boronic Transesterification

Boronic esters undergo two types of dynamic reactions: transesterification and pH-dissociation. Whereas the first is a thermally responsive reaction, the latter is pH responsive. Therefore, only transesterification will be discussed in this section. Testing the adhesive property of boronic esters-containing polyurethanes, Zeng and coworkers conducted several lap-shear tests on wood, stainless steel, and glass substrates [65]. Results (Figure 15.19) revealed substrate-dependent adhesive performance, with wood specimens achieving peak strength at 8.9 MPa, while glass substrates demonstrated the lowest adhesion strength of 4.2 MPa, attributed to surface topography variations. After healing through the transesterification reaction, the adhesive strength across specimen types exhibited minimal reduction, substantiating the material's robust recoverable adhesion characteristics.

Li and coworkers have also studied boronic esters-containing polyurethanes as adhesives, finding that elongation and adhesion increased when having residual hydroxyl groups. Moreover, by incorporating silver flakes, they demonstrated the polymers' application as strain sensors [66]. Savara and coworkers studied the incorporation of these bonds into a different elastomeric backbone: polystyrene-b-poly(ethylene-co-butylene)-b-polystyrene (was labeled in the original publication as SEBS) [67]. They have established a robust chemical connection between silica nanoparticles and

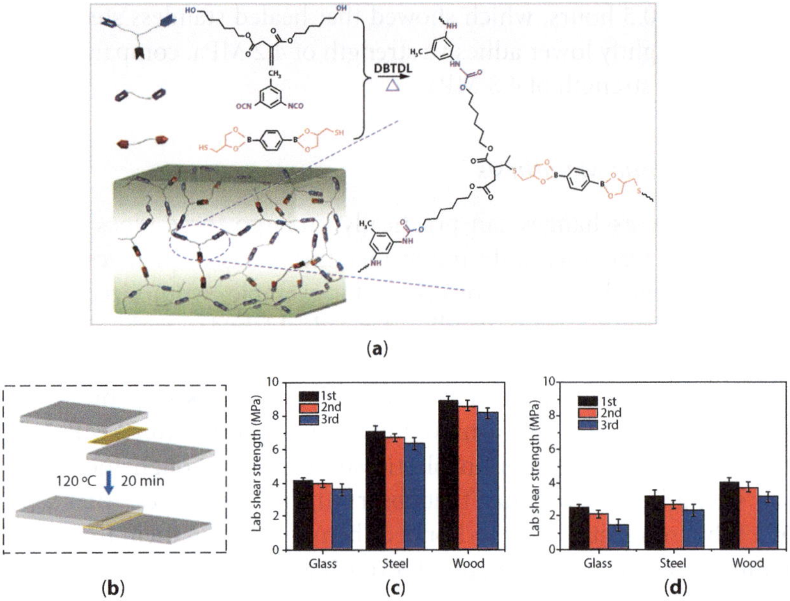

Figure 15.19 (a) A schematic illustration of the boronate ester-containing polyurethane reported by Zeng and coworkers. (b) A schematic illustration of the bonding process. Lap shear results of two polyurethanes tested across three bonding and debonding cycles: one containing dynamic boronic ester bonds (c) and the other without (d). Reproduced from J. Li, C. Hu, B. Yang, Z. Ning, Y. Zeng, Recyclable, Self-Healing Itaconic Acid-Based Polyurethane Networks With Dynamic Boronic Ester Bonds for Recoverable Adhesion Application, Polymer (Guildf). 256 (2022) 125227 with permission from Elsevier [65].

a triblock thermoplastic elastomer by incorporating boronic ester groups. These boronic ester functionalities enabled strategic bonding with the hydroxyl-terminated surface of the nanoparticles, ultimately generating a crosslinked nanocomposite material. The research also explored the adhesion strength of the borylated triblock copolymer to hydroxyl-rich substrates. It was found that covalent bonds formed between the boron ester moieties of the polymer and the hydroxyl-containing surfaces of aluminum, steel, and glass. While comparing the samples' performances, S-Bpin exhibited higher lap shear strength (4 MPa) on aluminum (Al) surfaces than SEBS (2.5 MPa). Introducing silica nanoparticles (SiNPs) increased the lap shear strength from 4 to 25 MPa with 20 wt% particles. On the glass, the adhesion was exceptionally strong, possibly due to the higher density of hydroxyl groups per given area, leading to stronger adhesion and potentially the formation of additional hydrogen bonding between hydroxyl groups of the polymer nanocomposite and the glass surface. The re-bonding capability was evaluated through repetitive bonding and re-bonding to the Al surface at 215°C, showing high re-bondability.

Whereas most studies on boronate-ester-containing adhesives have focused on elastomeric-like polymers, some, like Zeng and coworkers, have also demonstrated their potential in stiffer and stronger adhesives, like epoxy-based ones [68]. Despite being in a more rigid matrix, the re-bonding did not occur even at higher temperatures (~ 150°C).

15.3.3 pH-Responsive Adhesives

Although thermal and light stimuli are the primary focus in developing re-bondable adhesives, pH-responsive adhesives have also been investigated. These adhesive systems leverage the breakdown of boronate ester bonds into boronic acids and hydroxyl groups when exposed to acidic environments. Ryu and coworkers developed an alginate-boronic acid (alginate-BA) hydrogel [69]. Frequency sweep analysis revealed the hydrogel's complex viscoelastic behavior, characterized by an incomplete network structure at lower frequencies (below 3 Hz) and a well-defined gel-like mechanical response at frequencies exceeding 3 Hz, specifically observed under physiological pH conditions (7.4). Increasing the pH to 10, which forms boronic esters rather than just intermolecular bonds, led to a fivefold increase in both storage and loss moduli, expanding the gel-like region. The hydrogel exhibited reversible pH-dependent changes in mechanical properties, including sol-gel transitions. The researchers also assessed the adhesion property of the hydrogel using glass substrates. The detachment force of the hydrogel was approximately 32 kPa, much higher than alginate without boronic acid (3 kPa) [69].

A different method to benefit the transformation between boronic ester and acid was suggested by Lee and coworkers [70]. In their study, they used a copolymer of dopamine methacrylamide (DMA) and 3-acrylamido phenylboronic acid (AAPBA). While under acidic conditions, boronic acid was favored, and its attachment to the glass surface occurred through intermolecular interactions with a work of adhesion (W_{adh}) of 2000 mJ/m^2. Under basic conditions, however, the polymer was cross-linked and thus detached from the surface (Figure 15.20). In complementary research, Narkar and Lee demonstrated that the addition of acrylic acid as a co-monomer effectively modulated the pH conditions necessary for catechol-boronate complex formation, expanding the range towards more alkaline environments [71].

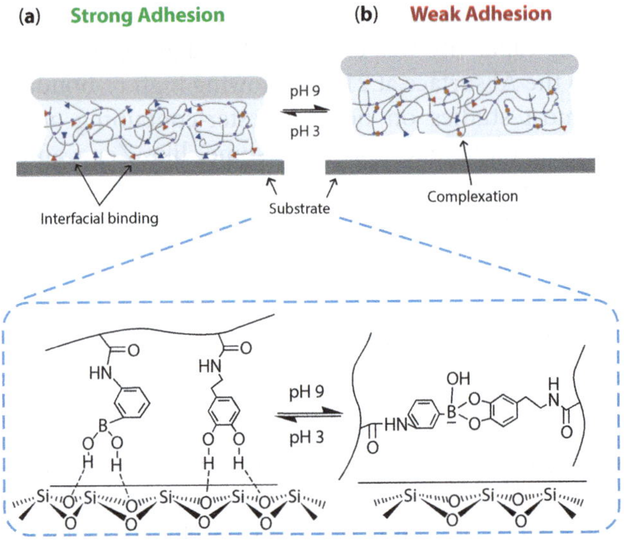

Figure 15.20 A schematic representation of the smart adhesive containing catechol and phenylboronic acid functional groups as suggested by Lee and coworkers. Reproduced from A.R. Narkar, B. Barker, M. Clisch, J. Jiang, B.P. Lee, pH Responsive and Oxidation Resistant Wet Adhesive based on Reversible Catechol-Boronate Complexation (2016) [70] and B. M. Dufort and M. W. Tibbitt, Design of moldable hydrogels for biomedical applications using dynamic covalent boronic esters (2019) with permission from ACS Publications and Elsevier [72].

15.4 Summary

This review explores debonding stimuli in thermoset adhesives and highlights the breakthrough achieved by using CANs / vitrimers in their recycling process. The review covers significant stimuli such as light, heat, and pH, each playing a crucial role. It discusses switchable adhesives that utilize light-induced mechanisms like cis-trans isomerization and [2 + 2] or [4 + 4] cycloaddition. Additionally, it examines mechanisms triggered by thermal stimulus, like Diels-Alder reaction, transalkylation, disulfides, and boronic esters. The review also emphasizes the pH sensitivity of hydrogel-based adhesives with boronate ester bonds.

These materials possess unique characteristics, such as varying mechanical properties and the ability to be de-bonded and re-bonded on-demand multiple times while maintaining adhesion strength. These properties distinguish them from traditional permanent adhesives. Although thermal and light-induced debonding has received much research attention, comprehensive case studies on other debonding stimuli are still needed.

Each debonding stimulus has specific mechanisms, applications, and limitations that require further exploration.

Based on the findings of this review, it is apparent that there is potential for further research on different stimuli and improvements in the current bonding and debonding mechanisms of adhesively bonded joints. For instance, in the case of light stimulus, reducing irradiation times should be explored. In thermal stimulus, efforts should be on optimizing mechanisms that rely on high temperatures. Furthermore, research should continue regarding pH stimulation, particularly in exploring adhesives that are not based on hydrogels.

List of Abbreviations

1,1'-(Methylenedi-4,1-phenylene)bismaleimide	BMI
1-Phenyl-2-(4-(tetradecyloxy)phenyl)diazene	Azo-M
3-Acrylamido phenylboronic acid	AAPBA
4,4′-Methylene bis(cyclohexyl-isocyanate)	HMDI
Aluminum	Al
Boronic acid	BA
Borylated SEBS triblock copolymer	S-Bpin
Covalent Adaptable Networks	CANs
Cyclopentadiene	Cp
Dibutyltin dilaurate	DBTDL
Diels-Alder	DA
Dopamine methacrylamide	DMA
Furyl-modified copolymers	FMPs
Hexamethylene diisocyanate	HDI
Hot-melt adhesives	HMAs
Isophorone diisocyanate	IPDI
m-Tetramethyl xylenediisocyanate	TMXDI
N,N-Diisopropylethylamine	DIPEA
Natural Rubber	NR
Photoinitiator	PI
Poly(1,2,3-triazolium) vitrimers	PTAVs

Poly(dimethyl siloxane)	PDMS
Poly(methyl methacrylate)	PMMA
Polycaprolactone	PCL
Polyethylene	PE
Polypropylene	PP
Polystyrene-b-poly(ethylene-co-butylene)-b-polystyrene	SEBS
Polyurethanes	PUs
Poly(vinyl chloride)	PVC
Pre-applied adhesive	PAA
Pressure-sensitive adhesive	PSA
Reactive polyurethane	PUR
Reversible covalent bonds containing polymers	RCBPs
Silica nanoparticles	SiNPs
Thermoplastic polyurethanes	TPUs
Topology freezing transition temperature	Tv
Trimethylolpropane-tris(3-mercaptopropionate)	TMMP
Ultraviolet	UV
Work of adhesion	W_{adh}

AI disclosure statement: The author acknowledges the use of AI-based tools for minor language editing or grammatical correction. However, all intellectual contributions, analyses, and conclusions presented in this chapter are entirely the author's own.

References

1. Z. Liu and F. Yan, Switchable adhesion: On-demand bonding and debonding. *Advanced Science*, 9, 2200264 (2022).
2. K.R. Mulcahy, A.F.R. Kilpatrick, G.D.J. Harper, A. Walton, and A.P. Abbott, Debondable adhesives and their use in recycling. *Green Chemistry*, 24, 36–61 (2022).
3. M.D. Banea, Debonding on demand of adhesively bonded joints, in: *Progress in Adhesion and Adhesives*, K.L. Mittal (Ed.), Vol. 5, pp. 33–50, Wiley-Scrivener, Beverly, MA (2020).

4. N. Jarach, D. Golani, H. Dodiuk, N. Naveh, and S. Kenig, Thermosets based on reversible covalent bonds (Vitrimers), in: *Handbook of Thermoset Plastics*, Fourth edition, H. Dodiuk (Ed.), pp. 656–800, Elsevier (2022).
5. W. Denissen, G. Rivero, R. Nicolaÿ, L. Leibler, J.M. Winne, and F.E. Du Prez, Vinylogous urethane vitrimers. *Adv Funct Mater*, 25, 2451–2457 (2015).
6. Y.X. Lu, and Z. Guan, Olefin metathesis for effective polymer healing *via* dynamic exchange of strong carbon-carbon double bonds. *J Am Chem Soc*, 134, 14226–14231 (2012).
7. G.C. Vougioukalakis and R.H. Grubbs, Ruthenium-based heterocyclic carbene-coordinated olefin metathesis catalysts. *Chem Rev*, 110, 1746–1787 (2010).
8. T. Bartosz, M. Xavier, W. Karolina, and J. Renata, (Eds.) *Polymer Engineering*, De Gruyter (2022).
9. C.A. Tretbar, J.A. Neal, and Z. Guan, Direct silyl ether metathesis for vitrimers with exceptional thermal stability. *J Am Chem Soc*, 141, 16595–16599 (2019).
10. J. Zhao, J.L. Brosmer, Q. Tang, Z. Yang, K.N. Houk, P.L. Diaconescu, and O. Kwon, Intramolecular crossed [2+2] photocycloaddition through visible light-induced energy transfer. *J Am Chem Soc*, 139, 9807–9810 (2017).
11. S. Ha, Y. Lee, Y. Kwak, A. Mishra, E. Yu, B. Ryou, and C.M. Park, Alkyne-Alkene [2 + 2] cycloaddition based on visible light photocatalysis. *Nature Communications*, 11, 1–12 (2020).
12. P. Vollhardt, and N. Schore *Organic Chemistry- Structure and Function*, seventh edition, W.H. Freeman and Company, New York (2014).
13. M.B. Smith, *March's Advanced Organic Chemistry: Reactions, Mechanisms, and Structure*. 8th edition, John Wiley & Sons, Hoboken, New Jersey (2020).
14. H. Kwart and K. King, The reverse Diels-Alder or retrodiene reaction. *Chem Rev*, 68, 415–447 (1968).
15. Q. Li, S. Ma, P. Li, B. Wang, H. Feng, N. Lu, S. Wang, Y. Liu, X. Xu, and J. Zhu, Biosourced acetal and Diels–Alder adduct concurrent polyurethane covalent adaptable network. *Macromolecules*, 54, 1742–1753 (2021).
16. J.W. Kim, D.H. Lee, H.J. Jeon, S.I. Jang, H.M. Cho, and Y. Kim, Recyclable thermosetting thermal pad using silicone-based polyurethane crosslinked by Diels-Alder adduct. *Appl Surf Sci*, 429, 128–133 (2018).
17. P. Wu, L. Liu, and Z. Wu, Synthesis of Diels-Alder reaction-based remendable epoxy matrix and corresponding self-healing efficiency to fibrous composites. *Macromol Mater Eng*, 305, 2000359 (2020).
18. F. Orozco, J. Li, U. Ezekiel, Z. Niyazov, L. Floyd, G.M.R. Lima, J.G.M. Winkelman, I. Moreno-Villoslada, F. Picchioni, and R.K. Bose, Diels-Alder-based thermo-reversibly crosslinked polymers: Interplay of crosslinking density, network mobility, kinetics and stereoisomerism. *Eur Polym J*, 135, 109882 (2020).
19. F. Orozco, Z. Niyazov, T. Garnier, N. Migliore, A.T. Zdvizhkov, P. Raffa, I. Moreno-Villoslada, F. Picchioni, and R.K. Bose, Maleimide self-reaction in

furan/maleimide-based reversibly crosslinked polyketones: Processing limitation or potential advantage? *Molecules*,26, 2230 (2021).
20. T.T. Truong, H.T. Nguyen, M.N. Phan, and L.T.T. Nguyen, Study of Diels–Alder reactions between furan and maleimide model compounds and the preparation of a healable thermo-reversible polyurethane. *J Polym Sci A Polym Chem*, 56, 1806–1814 (2018).
21. L. Li, X. Qin, H. Mei, L. Liu, and S. Zheng, Reprocessed and shape memory networks involving poly(hydroxyl ether ester) and polydimethylsiloxane through Diels-Alder reaction. *Eur Polym J*, 160, 110811 (2021).
22. N. Jarach, and H. Dodiuk, Debondable, recyclable and/or biodegradable naturally-based adhesives, in: *Biobased Adhesives: Sources, Characteristics and Applications*, M. Dunky and K.L. Mittal (Eds.), pp. 427–462, Wiley-Scrivener, Beverly, MA (2023).
23. N. Jarach, R. Zuckerman, N. Naveh, H. Dodiuk, and S. Kenig, Bio- and water-based reversible covalent bonds containing polymers (Vitrimers) and their relevance to adhesives – A critical review, in: *Progress in Adhesion and Adhesives, Volume 6*, K.L. Mittal (Ed.), pp. 587–619, Wiley-Scrivener, Beverly, MA (2021).
24. H.H. Moon, I. Cha, H.A. Hegazy, K. Kim, and C. Song, Covalent adaptive polymer networks for switchable adhesives. *Bull Korean Chem Soc*, 44, 750–767 (2023).
25. M.M. Obadia, A. Jourdain, P. Cassagnau, D. Montarnal, and E. Drockenmuller, Tuning the viscosity profile of ionic Vitrimers incorporating 1,2,3-triazolium cross-links. *Adv Funct Mater*, 27, 1703258 (2017).
26. M. Pepels, I. Filot, B. Klumperman, and H. Goossens, Self-healing systems based on disulfide–thiol exchange reactions. *Polym Chem*, 4, 4955–4965 (2013).
27. L. Zhao, and V. Semetey, Recycling polyurethanes through transcarbamoylation. *ACS Omega*, 6, 4175–4183 (2021).
28. C. Bakkali-Hassani, D. Berne, V. Ladmiral, and S. Caillol, Transcarbamoylation in polyurethanes: Underestimated exchange reactions? *Macromolecules*, 55, 7974–7991 (2022).
29. F. Elizalde, R.H. Aguirresarobe, A. Gonzalez, and H. Sardon, Dynamic polyurethane thermosets: Tuning associative/dissociative behavior by catalyst selection. *Polym Chem*, 11, 5386–5396 (2020).
30. P. Yan, W. Zhao, X. Fu, Z. Liu, W. Kong, C. Zhou, and J. Lei, Multifunctional polyurethane-Vitrimers completely based on transcarbamoylation of carbamates: Thermally-induced dual-shape memory effect and self-welding. *RSC Adv*, 7, 26858–26866 (2017).
31. L. Li, X. Chen, and J.M. Torkelson, Reprocessable polymer networks *via* thiourethane dynamic chemistry: Recovery of cross-link density after recycling and proof-of-principle solvolysis leading to monomer recovery. *Macromolecules*, 52, 8207–8216 (2019).

32. N. Jarach, H. Dodiuk, S. Kenig, and N. Naveh, Rheology—composition relationship of vitrimers based on polyethyleneimine. *J Appl Polym Sci*, 139, e52353 (2022).
33. N. Jarach, D. Golani, O. Asaf, H. Dodiuk, Y. Shamir, A. Goldbourt, S. Kenig, and N. Naveh, Composition processing properties relationship of Vitrimers based on polyethyleneimine. *Polym Chem*, 12, 3307–3320 (2021).
34. A. Chao, I. Negulescu, and D. Zhang, Dynamic covalent polymer networks based on degenerative imine bond exchange: Tuning the malleability and self-healing properties by solvent. *Macromolecules*, 49, 6277–6284 (2016).
35. M. Gosecki and M. Gosecka, Boronic acid esters and anhydrates as dynamic cross-links in Vitrimers. *Polymers* , 14, 842 (2022).
36. F. Cuminet, D. Berne, S. Lemouzy, É. Dantras, C. Joly-Duhamel, S. Caillol, É. Leclerc, and V. Ladmiral, Catalyst-free transesterification Vitrimers: Activation *via* α-difluoroesters. *Polym Chem*, 13, 2651–2658 (2022).
37. M. Röttger, T. Domenech, R. Van Der Weegen, A. Breuillac, R. Nicolaÿ, and L. Leibler, High-performance Vitrimers from commodity thermoplastics through dioxaborolane metathesis. *Science*, 356, 62–65 (2017).
38. C. Bandl, W. Kern, and S. Schlögl, Adhesives for "Debonding-on-Demand": Triggered release mechanisms and typical applications. *Int J Adhesion Adhesives*, 99, 102585 (2020).
39. P. Tannouri, K.M. Arafeh, J.M. Krahn , S.L. Beaupré, C. Menon, and N.R. Branda, A photoresponsive biomimetic dry adhesive based on doped PDMS microstructures. *Chem Mater*, 26, 4330–4333 (2014).
40. R. Klajn, Spiropyran-based dynamic materials. *Chem Soc Rev*, 43, 148–184 (2013).
41. Y. Zhou, M. Chen, Q. Ban, Z. Zhang, S. Shuang, K. Koynov, H.J. Butt, J. Kong, and S. Wu, Light-switchable polymer adhesive based on photoinduced reversible solid-to-liquid transitions. *ACS Macro Lett*, 8, 968–972 (2019).
42. H. Akiyama and M. Yoshida, Photochemically reversible liquefaction and solidification of single compounds based on a sugar alcohol scaffold with multi azo-arms. *Advanced Materials*, 24, 2353–2356 (2012).
43. T.H. Lee, G.Y. Han, M.B. Yi, H.J. Kim, J.H. Lee, and S. Kim, Rapid photoresponsive switchable pressure-sensitive adhesive containing azobenzene for the mini-light emitting diode transfer process. *ACS Appl Mater Interfaces*, 13, 43364–43373 (2021).
44. S.R. Trenor, T.E. Long, and B.J. Love, Development of a light-deactivatable PSA *via* photodimerization. *J. Adhesion*, 81, 213–229 (2005).
45. T. Asahi, K. Saito, M. Inada, T. Horii, T. Fujie, and T. Nakanishi, Debonding-on-Demand adhesives based on photo-reversible cycloaddition reactions. *Materials Advances*, 4, 1289–1296 (2023).
46. K. Boga, A.F. Patti, J.C. Warner, G.P. Simon, and K. Saito, Eco-Friendly photoreversible adhesives derived from coumarin-functionalized epoxy soybean oil. *ACS Appl Polym Mater*, 5, 4644–4653 (2023).

47. S. Wang, W.H. Hu, Y. Nakamura, N. Fujisawa, A.E. Herlyng, M. Ebara, and M. Naito, Bio-Inspired adhesive with reset-on demand, reuse-many (RORM) modes. *Adv Funct Mater*, 33, 2215064 (2023).
48. Z. Liu, J. Cheng, and J. Zhang, An efficiently reworkable thermosetting adhesive based on photoreversible [4+4] cycloaddition reaction of epoxy-based prepolymer with four anthracene end groups. *Macromol Chem Phys*, 222, 2000298 (2021).
49. Y. Feng, X. Han, Y. Ji, S. Li, K. Dong, S. Liang, Y. Ma, Y. Yang, and F. Liu, The reversible modification of acrylate adhesive system by introducing anthracene groups. J *Adhesion*, 100, 200–218 (2023).
50. D.H. Turkenburg, H. van Bracht, B. Funke, M. Schmider, D. Janke, and H.R. Fischer, Polyurethane adhesives containing Diels–Alder-based thermoreversible bonds. *J Appl Polym Sci*, 134, 44972 (2017).
51. K.K. Oehlenschlaeger, N.K. Guimard, J. Brandt, J.O. Mueller, C.Y. Lin, S. Hilf, A. Lederer, M.L. Coote, F.G. Schmidt, and C. Barner-Kowollik, Fast and catalyst-free hetero-Diels–Alder chemistry for on demand cyclable bonding/debonding materials. *Polym Chem*, 4, 4348–4355 (2013).
52. K.M.A. Kaiser, Recycling of multilayer packaging using a reversible cross-linking adhesive. *J Appl Polym Sci*, 137, 49230 (2020).
53. J.H. Aubert, Thermally removable epoxy adhesives incorporating thermally reversible Diels-Alder adducts. *J. Adhesion*, 79, 609–616 (2003).
54. A. Gandini, The furan/maleimide Diels–Alder reaction: A versatile click–unclick tool in macromolecular synthesis. *Prog Polym Sci*, 38, 1–29 (2013).
55. S. Das, S. Samitsu, Y. Nakamura, Y. Yamauchi, D. Payra, K. Kato, and M. Naito, Thermo-Resettable cross-linked polymers for reusable/removable adhesives. *Polym Chem*, 9, 5559–5565 (2018).
56. A.J. Inglis, L. Nebhani, O. Altintas, F.G. Schmidt, and C. Barner-Kowollik, Rapid bonding/debonding on demand: Reversibly cross-linked functional polymers *via* Diels-Alder chemistry. *Macromolecules*, 43, 5515–5520 (2010).
57. J. Tang, L. Wan, Y. Zhou, H. Pan, and F. Huang, Strong and efficient self-healing adhesives based on dynamic quaternization cross-links. *J Mater Chem A*, 5, 21169–21177 (2017).
58. Y. Oh, J. Park, J.J. Park, S. Jeong, and H. Kim, Dual cross-linked polymer thermosets: Modular design, reversible transformation, and triggered debonding. *Chem Mater*, 32, 6384–6391 (2020).
59. W.Z. Xu, W.W. Yu, X. Chen, S. Liao, and M.C. Luo, Based on transalkylation reaction the rearrangeable conventional sulfur network facile design for vulcanized diolefin elastomers. *J Appl Polym Sci*, 138, 51182 (2021).
60. H.Y. Tsai, Y. Nakamura, T. Fujita, and M. Naito, Strengthening epoxy adhesives at elevated temperatures based on dynamic disulfide bonds. *Materials Advances*, 1, 3182–3188 (2020).
61. B.T. Michal, E.J. Spencer, and S.J. Rowan, Stimuli-responsive reversible two-level adhesion from a structurally dynamic shape-memory polymer. *ACS Appl Polym Mater Interfaces*, 8, 11041–11049 (2016).

62. F. Ji, X. Liu, D. Sheng, and Y. Yang, Epoxy-Vitrimer composites based on exchangeable aromatic disulfide bonds: Reprocessibility, adhesive, multi-shape memory effect. *Polymer (Guildf)*, 197, 122514 (2020).
63. C. Cui, X. Chen, L. Ma, Q. Zhong, Z. Li, A. Mariappan, Q. Zhang, Y. Cheng, G. He, X. Chen, Z. Dong, L. An, and Y. Zhang, Polythiourethane covalent adaptable networks for strong and reworkable adhesives and fully recyclable carbon fiber-reinforced composites. *ACS Appl Polym Mater Interfaces*, 12, 47975–47983 (2020).
64. Z. Zhou, X. Su, J. Liu, and R. Liu, Synthesis of vanillin-based polyimine Vitrimers with excellent reprocessability, fast chemical degradability, and adhesion. *ACS Appl Polym Mater*, 2, 5716–5725 (2020).
65. J. Li, C. Hu, B. Yang, Z. Ning, and Y. Zeng, Recyclable, self-healing itaconic acid-based polyurethane networks with dynamic boronic ester bonds for recoverable adhesion application. *Polymer (Guildf)*, 256, 125227 (2022).
66. Y. Yang, F.S. Du, and Z.C. Li, Highly stretchable, self-healable, and adhesive polyurethane elastomers based on boronic ester bonds. *ACS Appl Polym Mater*, 2, 5630–5640 (2020).
67. M.A. Rahman, C. Bowland, S. Ge, S.R. Acharya, S. Kim, V.R. Cooper, X.C. Chen, S. Irle, A.P. Sokolov, A. Savara, and T. Saito, Design of tough adhesive from commodity thermoplastics through dynamic crosslinking. *Science Advances*, 7 (2021).
68. W. Yang, Q. Yi, F. Liu, X. Pan, and Y. Zeng, Fabrication of malleable, repairable, weldable, recyclable and robust epoxy vitrimers from itaconic acid for recycled adhesion. *Eur Polym J*,196, 112278 (2023).
69. S.H. Hong, S. Kim, J.P. Park, M. Shin, K. Kim, J.H. Ryu, and H. Lee, Dynamic bonds between boronic acid and alginate: Hydrogels with stretchable, self-healing, stimuli-responsive, remoldable, and adhesive properties. *Biomacromolecules*, 19, 2053–2061 (2018).
70. A.R. Narkar, B. Barker, M. Clisch, J. Jiang, and B.P. Lee, pH responsive and oxidation resistant wet adhesive based on reversible catechol-boronate complexation. *Chem Mater*, 28, 5432–5439 (2016).
71. A.R. Narkar and B.P. Lee, Incorporation of anionic monomer to tune the reversible catechol-boronate complex for pH-responsive, reversible adhesion. *Langmuir*, 34, 9410–9417 (2018).
72. B.M. Dufort and M.W. Tibbitt, Design of moldable hydrogels for biomedical applications using dynamic covalent boronic esters. *Mater Today Chem*, 12, 16–33 (2019).

16

Investigation of Pressure-Sensitive Adhesives Solely from Epoxidized Soybean Oil

Yunfei Liang and Kaichang Li*

School of Chemical, Biological, and Environmental Engineering, Oregon State University, Corvallis, OR, US

Abstract

Epoxidized soybean oil (ESO) was selectively hydrolyzed by NaOH to give raw epoxidized fatty acid (REFA) that was further purified by crystallization in hexane to form crystallized epoxidized fatty acid (CEFA). Neat CEFA was polymerized in the presence of a catalyst, 1-methylimidazole, to give a pre-polymer that was further crosslinked with toluene diisocyanate (TDI) or epoxidized soybean oil (ESO) to give pressure-sensitive adhesives (PSAs) with superior adhesion property. Cohesive failure, peel strength, shear adhesion, and tack were evaluated for the resulting PSA. The surface of PSA with TDI as the crosslinking agent was rough. In the preparation of PSAs with ESO as the crosslinking agent, the usage of ESO, reaction temperature and time in the pre-polymerization and the curing reactions were investigated in detail. PSA with the highest peel strength of 6.4 N/cm and tack of 13.8 N was obtained when CEFA was pre-polymerized at 100°C for 2 h, followed by addition of 3% ESO, and the curing reaction at 125°C for 2 h. After one week of aging at 60°C, the peel strength and tack of PSA prepared from the optimal conditions decreased from 6.4 N/cm to 5.3 N/cm and from 13.8 N to 6.9 N, respectively. PSAs with ESO as the crosslinking agent were prepared solely from ESO without use of any organic solvent or any toxic chemical.

***Keywords*:** Epoxidized soybean oil, pressure-sensitive adhesives, polyester, epoxidized fatty acids

**Corresponding author*: Kaichang.li@oregonstate.edu

Anil N. Netravali and Frank M. Etzler (eds.) The Wide World of Adhesion: A Festschrift in Honor of Dr. Kash Mittal on his 80th Birthday, (565–584) © 2026 Scrivener Publishing LLC

16.1 Introduction

Pressure-sensitive adhesives (PSAs) adhere to various substrates just upon application of light pressure. PSAs are widely used in various industries, ranging from automotive and electronics to packaging and medical applications [1]. PSAs typically contain an elastomeric material and small amounts of tackifiers, processing oils, and waxes [1]. At present, the most commonly available elastomeric materials are petrochemical-based polymers such as polyacrylates and styrenic block copolymers. Tackifiers are typically based on five-carbon (C5) or nine-carbon (C9) aliphatic monomers which are derived from the cracking of petroleum fractions. Petroleum is not renewable and thus not sustainable. As industries increasingly seek sustainable and bio-based materials, fatty acids and vegetable oils have emerged as promising renewable materials for the production of PSAs [2–4].

Preparation of PSAs from vegetable oils or fatty acid esters generally follows the following approaches. The first approach involves the introduction of acryloyl groups onto the fatty acid chains, followed by free-radical-initiated polymerization of the acrylates [5–8]. In this approach, fatty acid esters or vegetable oils are first epoxidized, followed by the ring-opening reactions of the epoxy groups with acrylic acid or methacrylic acid to generate acrylated or methacrylated fatty acid esters or vegetable oils [5–8]. Fatty acid esters or vegetable oils basically replace the butyl group and 2-ethylhexyl group in butyl acrylate and 2-ethylhexyl acrylate that are commonly used in production of petrochemical-based PSAs. Petrochemical-based acrylic acid or methacrylic acid is still required and the resulting PSAs are not readily biodegradable. The free-radical-initiated polymerization rates for acrylated fatty acid esters or vegetable oils are typically lower than that for butyl acrylate and 2-ethylhexyl acrylate because of the bulkiness of fatty acid esters or vegetable oils [5–8]. The second approach involves the direct polymerization of epoxidized plant oils or fatty acid esters *via* cationically catalyzed ring-opening polymerization of the epoxy groups [9]. Some petrochemical-based epoxy compounds or vinyl ethers were required as comonomers in these polymerizations [9]. The third approach involves the direct polymerization of the carbon-carbon double bonds (C=C) in vegetable oils or fatty acid derivatives and acrylic monomers [9]. In this approach, however, the plant oils had to be modified to form conjugated double bonds prior to the free radical polymerization [9]. The fourth approach involves the development of PSAs from polymerization of epoxidized fatty acid esters or vegetable oils with dibasic acids [10, 11]. It has also been reported that epoxidized soybean oil (ESO) was selectively hydrolyzed to form epoxidized fatty acids (EFAs) [12]. Polymerization of EFAs with various dibasic acids such as adipic acid could generate PSAs. Most recently, a PSA was prepared from castor oil-based polyricinoleate using ESO as crosslinker [13].

In this study, a novel approach has been investigated for preparation of PSAs from ESO. ESO was first selectively hydrolyzed in an alkaline solution to obtain a mixture of epoxidized fatty acids that were further purified and then polymerized to form a pre-polymer. The pre-polymer was further crosslinked with ESO to form superior PSAs. The adhesive property of the resulting PSAs was evaluated in detail.

16.2 Experimental

16.2.1 Materials

Sodium hydroxide and hydrochloric acid were purchased from Aventor Performance Materials, LLC. (Center Valley, PA). Dibutyltin dilaurate (DBTDL) and 1-methylimidazole were purchased from Sigma-Aldrich Corp. (St. Louis, MO). Toluene diisocyanate (TDI) was purchased from Alfa Aesar (Ward Hill, MA). Epoxidized soybean oil (ESO) was manufactured by Spectrum Chemical Mfg. Corp. (Gardena, CA) and purchased from VWR International (Radnor, PA). The BOPP (biaxially oriented polypropylene)-based backing film (thickness: ~50 μm), PET poly(ethylene terephthalate)-based release film (thickness: ~50 μm), and a release liner were obtained from CleanMark Labels (Portland, OR). All common chemicals and solvents were purchased from commercial sources and used as received.

16.2.2 Preparation of Crystallized Epoxidized Fatty Acids (CEFAs)

ESO (60 g) was added to a solution consisting of sodium hydroxide (8.9 g), water (50 mL) and ethanol (150 mL). The resulting mixture was stirred with a magnetic stirrer at 60°C for 1 h. The ethanol and water were removed at 40°C using a Rotovap under diminished pressure to form a white solid. The resulting solid was dissolved in water (150 mL), followed by addition of a 2 N HCl solution (120 mL) to neutralize the sodium hydroxide. The aqueous phase was removed and the organic phase was washed with hot brine three times. The resulting wax-like solid was designated as raw epoxidized fatty acid (REFA). The yield of REFA was 90% based on ESO.

The REFA (100 g) was dissolved in hexane (150 mL) at 60°C. The resulting solution was then cooled in a refrigerator (around 4°C) overnight. The crystals were filtered and washed with cooled hexane (50 mL) three times. The collected solids were then placed under high vacuum overnight to form white powder that is designated as crystallized epoxidized fatty acid (CEFA). The yield of CEFA was 33% based on ESO.

16.2.3 Preparation of PSAs from CEFA and Use of TDI as a Crosslinker

CEFA (2.0 g) and 1-methylimidazole (0.030 g) were added to a 20-mL glass vial. The mixture was purged with nitrogen gas and agitated at 100°C for a pre-determined reaction time, followed by addition of TDI (0.45 g) and DBTDL (0.015 g). The resulting mixture was then coated on a sheet of BOPP film with the HLCL-1000 coater. The coated film was heated at 125°C for 2 h in an oven. The resulting products were designated as PSA-TDI-A, PSA-TDI-B, and PSA-TDI-C when the pre-determined reaction times were 2.5 h, 2.0 h, and 1.0 h, respectively.

16.2.4 Preparation of PSAs with Different Pre-Polymerization Temperatures

CEFA (5.0 g) and 1-methylimidazole (0.075 g) were added to a 20-mL glass vial. The mixture was purged with nitrogen gas. After the reaction at a pre-determined pre-polymerization temperature for 2 h, ESO (0.469 g) was added to the mixture. The resulting mixture was mixed at the pre-determined pre-polymerization temperature for another 10 min before being coated on a sheet of BOPP film with the HLCL-1000 coater. The coated film was heated at 125°C for 2 h in an oven. The resulting products were designated as PSA-ESO-P90C, PSA-ESO-P100C, PSA-ESO-P110C, and PSA-ESO-P120C when the pre-determined pre-polymerization temperatures were 90°C, 100°C, 110°C, and 120°C, respectively.

16.2.5 Preparation of PSAs with Different Pre-Polymerization Times

CEFA (5.0 g) and 1-methylimidazole (0.075 g) were added to a 20-mL glass vial. The mixture was purged with nitrogen gas. After the reaction at 100°C for a pre-determined pre-polymerization time, ESO (0.469 g) was added to the mixture. The resulting mixture was mixed at 100°C for another 10 min before being coated on a sheet of BOPP film with the HLCL-1000 coater. The coated film was heated at 125°C for 2 h in an oven. The resulting products were designated as PSA-ESO-P10H, PSA-ESO-P15H, PSA-ESO-P20H, and PSA-ESO-P25H when the pre-determined pre-polymerization times were 1.0 h, 1.5 h, 2.0 h, and 2.5 h, respectively.

16.2.6 Preparation of PSAs with Different Curing Temperatures

CEFA (5.0 g) and 1-methylimidazole (0.075 g) were added to a 20-mL glass vial. The mixture was purged with nitrogen gas and mixed at 100°C for 2 h, followed by addition of ESO (0.469 g). The resulting mixture was mixed at 100°C for another 10 min before being coated on a sheet of BOPP film with the HLCL-1000 coater. The coated film was heated at a pre-determined curing temperature for 2 h in an oven. The resulting products were designated as PSA-ESO-C110C, PSA-ESO-C120C, PSA-ESO-C125C, PSA-ESO-C130C and PSA-ESO-C135C when the pre-determined curing temperatures were 110°C, 120°C, 125°C, 130°C and 135°C, respectively.

16.2.7 Preparation of PSAs with Different Curing Times

CEFA (5.0 g) and 1-methylimidazole (0.075 g) were added to a 20-mL glass vial. The mixture was purged with nitrogen gas and mixed at 100°C for 2 h, followed by addition of ESO (0.469 g). The resulting mixture was mixed at 100°C for another 10 min before being coated on a sheet of BOPP film with the HLCL-1000 coater. The coated film was heated at 125°C for a pre-determined curing time in an oven. The resulting products were designated as PSA-ESO-C2H, PSA-ESO-C3H, PSA-ESO-C4H, and PSA-ESO-C8H when the pre-determined curing times were 2.0 h, 3.0 h, 4.0 h, and 8.0 h, respectively.

16.2.8 Preparation of PSAs with Different Amounts of ESO as the Crosslinker

CEFA (5.0 g) and 1-methylimidazole (0.075 g) were added to a 20-mL glass vial. The mixture was purged with nitrogen gas and mixed at 100°C for 2 h, followed by addition of a pre-determined amount of ESO. The resulting mixture was mixed at 100°C for another 10 min before being coated on a sheet of BOPP film with the HLCL-1000 coater. The coated film was heated at 125°C for 2 h in an oven. The resulting products were designated as PSA-ESO-A, PSA-ESO-B, PSA-ESO-C, PSA-ESO-D, PSA-ESO-E, and PSA-ESO-F when the pre-determined amounts of ESO were 0.094 g, 0.156 g, 0.312 g, 0.469 g, 0.781 g, and 1.09 g, respectively.

16.3 Testing

16.3.1 Measurement of Peel Strength

The peel strength was measured in accordance with Test Method A of ASTM D3330/D3330M-04 standard (Reapproved 2010). The measurement was performed with a Tinius Olsen Testing Machine (model H10KT) at room temperature (23 ± 1°C). One end of a PSA specimen tape (24 mm by 200 mm) was applied onto the stainless-steel panel and then pressed twice with a roller with a weight of 2040 g in the lengthwise direction. After the applied tape was left at room temperature for one day, the other free end of the tape was attached to the grip of the load cell of the testing machine. The tape was then peeled upwards at 180° angle at a speed of 5 mm/s; the force required to peel off the tape from the test panel was recorded as peel strength in the unit of N/cm.

16.3.2 Measurement of Tack

The tack measurement was performed in accordance with Test Method A of ASTM D6195-03 standard (Reapproved 2011). The measurement was performed with a Tinius Olsen Testing Machine (model H10KT) at room temperature (23 ± 1°C). A PSA specimen was looped and attached to the grip of the testing machine. The grip was allowed to move downward until the loop was brought into contact to the test panel surface with a contact area of 25 mm × 25 mm. Immediately after the contact, the grip and the loop were allowed to move upward at a speed of 5 mm/s until the loop was completely separated from the test panel. The maximum force in newton (N) required to break the adhesive bond was recorded.

16.3.3 Measurement of Cohesive Failure

Cohesive failure was recorded during the measurement of peel strength. The degree of cohesive failure (DCF) was used for quantifying cohesive failure and was defined as the ratio of the surface area covered by adhesive residues to the total contact area of the adhesive. The surface area covered by adhesive residues was estimated by visual inspection. The reported DCF was the average of three replicates. Cohesive failure describes the ability of a PSA to remove cleanly from its applied surface. Commercial products typically require clean removal, i.e., a DCF of 0%.

16.3.4 Measurement of Shear Adhesion

The shear adhesion was measured in accordance with the Procedure A of ASTM D3654/D3654M-06 standard. The measurements were performed at 23 ± 1°C and 40 ± 5 % RH. A strip of PSA tape with 24 mm in width and 130 mm in length was applied with its one end onto the test panel with a contact area of 24 by 24 mm; the tape on the panel was then pressed with a roller with a weight of 2040 g twice in lengthwise direction. The other end of the PSA tape was fixed to a constant mass of 1000 g. The distance from the mass to the test panel was about 50 mm. After 20 min, the test panel along with the PSA tape and mass was held by a metallic stand at an angle of 2° relative to the vertical direction, allowing the mass to hang in the air. A timing system was used to record the time when the tape separated from the panel. The time from the point that the tape was held onto the metallic stand to the point that the tape completely separated from the test panel was used as the indication of shear adhesion.

16.3.5 Aging Test

Specimens were aged at 60°C in an Isotemp 625D Incubator (Thermo Fisher Scientific Inc., Waltham, MA). After being aged for one week, the specimens were conditioned at 23 ± 1°C and 40 ± 5 % RH for 10 h and then evaluated for peel strength and tack according to the methods previously described.

16.3.6 Measurement of Gel Fraction

The gel fraction of the PSA was determined by soaking the sample in chloroform for one day at room temperature. The insoluble polymer was recovered by filtration and dried at 70°C to constant weight. The fraction of the insoluble polymer over the total weight of the PSA was used as the gel fraction of the PSA.

16.3.7 Characterization with NMR and FTIR

^{1}H NMR spectra of ESO, REFA, and CEFA were obtained at room temperature with a Bruker 700 MHz spectrometer with $CDCl_3$ as solvent and showed: ESO (δ, in ppm): 4.2-5.3 (associated with protons of esterified glycerol [14]), 3.4 (CH of diols [10]), 3.0–3.2 (CH of epoxy groups separated by a methylene group [14]), 2.95 (CH of standalone epoxy group [14]), 2.28–2.43 (α–CH_2 adjacent to -COOH), 1.72-1.90 (CH_2 protons between two epoxy groups), 1.60–1.72 (CH_2 β to–COOH), 1.45–1.60 (CH_2 α to the epoxy group), 1.20–1.45 (protons of other methylene groups), and 0.85–1.12 (protons of methyl groups); REFA (δ, in ppm): 3.0–3.2 (CH of epoxy groups separated by a methylene group [14]), 2.95 (CH of standalone epoxy group [14]), 2.28–2.43 (α–CH_2 adjacent to -COOH), 1.72-1.90 (CH_2 protons between the epoxy groups), 1.60–1.72 (CH_2 β to–COOH), 1.45–1.60 (CH_2 α to the epoxy groups), 1.20–1.45 (protons of other methylene groups), 0.85–1.12 (protons of methyl groups); CEFA: 3.0–3.2 (CH of epoxy groups separated by a methylene group [14]), 2.95 (CH of standalone epoxy group [14]), 2.28–2.43 (α–CH_2 adjacent to -COOH), 1.72-1.90 (CH_2 protons between the epoxy groups), 1.60–1.72 (CH_2 β to–COOH), 1.45–1.60 (CH_2 α to the epoxy groups), 1.20–1.45 (protons of other methylene groups), 0.85–1.12 (protons of methyl groups).

FTIR spectra were recorded on a Nexus 470 FTIR spectrometer (Thermo Fisher Scientific Inc., Waltham, MA) equipped with a Golden Gate diamond ATR (attenuated total reflectance) accessory in a scanning range of 650-4000 cm^{-1} for 32 scans at a spectral resolution of 2 cm^{-1}.

16.3.8 Statistical Analysis of Data

All data for adhesive property of PSA were statistically analyzed by a two-sample *t* test using RStudio software (RStudio, Inc., Boston, MA). All comparisons were based on a 95 % confidence interval.

16.4 Results and Discussion

ESO was hydrolyzed with sodium hydroxide in a mixture of water and ethanol to give REFA with a yield of 90% [12]. The REFA was an off-white and wax-like solid and further purified by crystallization in hexane to give CEFA. CEFA is an AB monomer, i.e., it contains at least one epoxy group and a carboxylic acid group on the same fatty acid molecule. ^{1}H NMR spectra of ESO, REFA and CEFA are shown in Figure 16.1. REC (the molar ratio of epoxy groups to carboxylic acid groups in REFA and CEFA or the molar ratio of epoxy groups to ester groups in ESO) was determined from the ratio of the peak area of the epoxy protons at 2.9–3.2 ppm to the peak area of the α–CH_2 protons adjacent to –COOH or -COOR at 2.3–2.4 ppm. The REC of ESO, REFA and CEFA were 1.35, 1.11, and 1.56, respectively. ESO had a higher REC than REFA, which could mean two scenarios: 1) some epoxy groups were hydrolyzed to form di-hydroxyl groups in the alkaline hydrolysis of ESO; 2) a high portion of fatty acids containing more than one epoxy group did not remain in the organic phase in the neutralization of the reaction mixture with HCl and the washing of REFA with brine (fatty acids with multiple epoxy groups may be more soluble in water than fatty acids with zero or one epoxy group). Analysis of ^1H NMR spectra of REFA reveals that the epoxy group in 2.2% of the fatty acids in REFA was hydrolyzed by NaOH to form diols (di-hydroxyl groups) (the percentage of the fatty acids was determined by the ratio of the peak areas of the two CH protons of the diols at 3.4 ppm (the peak is not visible in Figure 16.1) [10] to the peak area of the α-CH_2 adjacent to –COOH at 2.34 ppm). This minor hydrolysis of epoxy group cannot account for the significant change in REC. Soybean oil contains about 14% saturated fatty acids, 18% fatty acid with one C=C bond, 55% fatty acids with two C=C bonds, and 13% fatty acids with three C=C bonds [15]. REFA was thus expected to contain

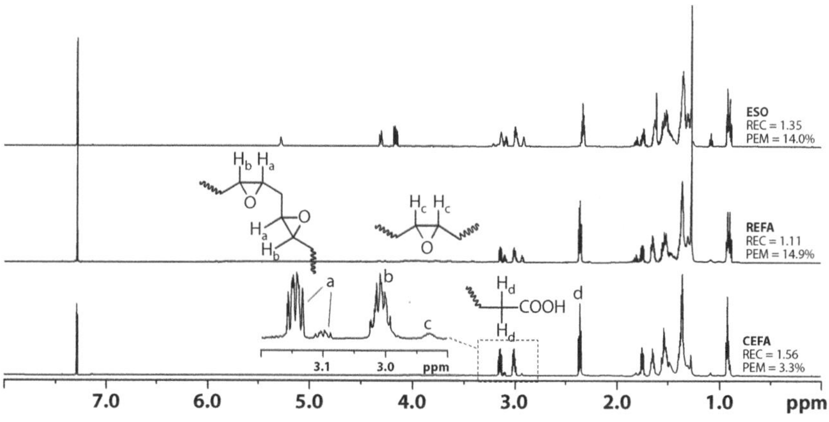

Figure 16.1 ^1H NMR spectra of ESO (top), REFA (middle), and CEFA (bottom). with close-up views of the characteristic absorption signals related to the epoxy groups (REC, the molar ratio of epoxy groups to carboxylic groups or ester groups; PEM, the molar ratio of epoxy groups derived from monounsaturated fatty acid to the total amount of epoxy groups, shown as percentage).

a higher percentage of saturated fatty acids than ESO. The second scenario is more likely to be the main cause of the lower REC of REFA.

In ESO, epoxy groups in fatty acids with more than one epoxy group are all separated by one methylene group, which exhibit different chemical shifts in ^{1}H NMR from the epoxy groups in fatty acids containing only one epoxy group. PEM (the percentage of epoxy groups in mono-epoxidized fatty acids over the total amount of epoxy groups) was determined by the ratio of the peak areas of protons on the standalone epoxy rings at 2.95 ppm to the peak area of protons on all epoxy group rings (Figure 16.1). PEM of REFA and CEFA was 3.3% and 14.9%, respectively, which indicates that REFA has a lower percentage of fatty acids containing only one epoxy group than CEFA. This explains why CEFA has a higher REC value than REFA.

Fatty acids containing no epoxy group cannot polymerize by themselves to form polyesters, but can consume epoxy groups in epoxidized fatty acids, thus terminating the polymerization process and shortening the polymer chains. Polymerization of REFA in the presence of the catalyst, 1-methylimidazole, resulted in polymers that were not tacky but very weak. It was speculated that the molecular weight of the resulting polymers from REFA was too low to have good PSA properties because REFA had a high amount of fatty acids containing no epoxy group. For removal of the fatty acids containing no epoxy group, REFA was purified to form CEFA through the crystallization of REFA in hexane. CEFA was further investigated for preparation of PSAs.

CEFA was first polymerized to form flowable pre-polymers that were crosslinked with a crosslinking agent. TDI was investigated as the first crosslinking agent. In the polymerization of CEFA, the carboxylic acid groups reacted with epoxy groups to generate polymers containing hydroxyl groups that can react with TDI, thus crosslinking the polymers. Properties of the pre-polymers and TDI-cured PSAs are shown in Table 16.1. Polymerization of CEFA at 100°C for 2.5 h generated pre-polymers that were highly viscous and thus were hard to spread onto the BOPP film. When the pre-polymerization time was longer than 2.5 h, the resulting pre-polymers gelled and could not spread onto the BOPP film. PSA-TDI-A had a very rough surface that was obviously from big bubbles. We initially thought the bubbles were from reaction of moisture in air and isocyanate groups of TDI because the reaction releases carbon dioxide. For excluding moisture in air from reaction with TDI, the TDI-containing pre-polymer on the PET film was covered with another sheet of a PET film before the curing in an oven. However, this approach did not significantly reduce the roughness of the PSA-TDI-A surface. Reaction of a carboxylic acid group

Table 16.1 Properties of PSAs with TDI as the crosslinking agent.

Cured PSAs	Pre-polymerization time (h)	Property of pre-polymers	Appearance of cured PSAs	Peel strength (N/cm)	DCF (%)
PSA-TDI-A	2.5	very viscous	rough	4.0 ± 0.1	100
PSA-TDI-B	2	viscous	rough	2.8 ± 0.1	80
PSA-TDI-C	1	not viscous	rough	2.8 ± 0.3	10

Note: DCF, degree of cohesive failure. The ± values are based on 95% confidence interval.

with an isocyanate would also release carbon dioxide. It is highly likely that the bubbles in PSA-TDI-A were from the reaction of isocyanate groups of TDI and residual carboxylic acid groups in the pre-polymers of CEFA. PSA-TDI-A had a higher peel strength than PSA-TDI-B and PSA-TDI-C. PSA-TDI-A had 100% cohesive failure. Reduction of the pre-polymerization time from 2.5 h to 2.0 h generated pre-polymers that were viscous, yet not hard to spread onto a PET film. PSA-TDI-B also had a rough surface that was obviously from small bubbles. The PSA-TDI-B also had 80% cohesive failure. Further reduction of the pre-polymerization time from 2.0 h to 1.0 h generated pre-polymers that were not viscous and were easy to be spread onto a PET film. PSA-TDI-C also had a rough surface that was obviously from tiny bubbles. It appeared that the size of bubbles was closely correlated with the viscosity of the pre-polymers. The higher the viscosity the bigger the bubbles. The PSA-TDI-C has about 10% cohesive failure. PSA-TDI-B and PSA-TDI-C have comparable peel strength.

As previously described, REC of CEFA was 1.56, i.e., the molar ratio of epoxy groups vs carboxylic acid groups in CEFA was larger than one. Polymerization of CEFA would generate non-flowable gel if the polymerization time is long enough. In other words, the polymerization time has to be relatively short if the resulting pre-polymer has to be flowable. Because stoichiometric imbalance of epoxy groups and carboxylic acid groups, the flowable pre-polymers will always have some carboxylic acid groups remaining as indicated by FTIR analysis (Figure 16.2). The fully cured PSAs will always have a rough surface if isocyanates such as TDI are used as the crosslinking agent. All these prompted us to investigate ESO as a crosslinking agent. The crosslinking reactions are proposed to occur

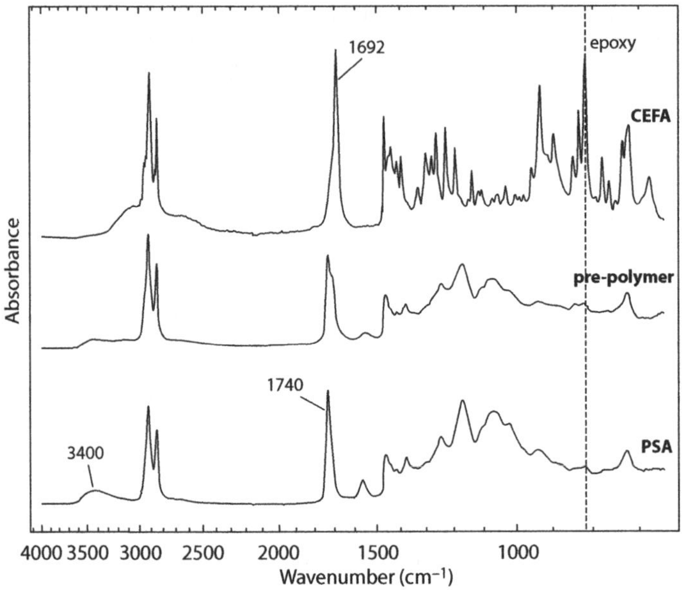

Figure 16.2 FTIR spectra of CEFA (top), EFA pre-polymer after reaction at 100°C for 2 h (middle), and PSA-ESO-D (bottom).

between epoxy groups in ESO and the remaining carboxylic acid groups in the pre-polymers.

Properties of PSAs from different pre-polymerization temperatures are shown in Table 16.2. PSA-ESO-P90C at 90°C had 100% cohesive failure. The peel strength increased when the pre-polymerization temperature was increased from 90°C to 100°C. PSA-ESO-100C had a good tack and had no cohesive failure. The pre-polymers of CEFA became not flowable after CEFA was polymerized at 110°C for 2 h. The pre-polymers of CEFA became a gel after CEFA was polymerized at 120°C for 1 h. The pre-polymerization temperature of 100°C appeared to be the best and was used for further investigation.

Properties of PSAs from different pre-polymerization times are shown in Table 16.3. Pre-polymers from pre-polymerization times of 1.0 h, 1.5 h, and 2.0 h were all flowable and could be coated onto the BOPP film for making PSAs. However, the pre-polymer from the pre-polymerization time of 2.5 h was not flowable and could not be coated onto the BOPP film for making PSAs. The peel strength of PSAs remained the same, but the tack increased when the pre-polymerization time was increased from 1.0 h to 1.5 h. In addition, PSA-ESO-P15H has a decreased cohesive failure from 80% to 30% when compared to PSA-ESO-P10H. Both peel strength

Table 16.2 Properties of PSAs from different pre-polymerization temperatures.

Cured PSAs*	Pre-polymerization temperature	Property of pre-polymers	DCF (%)	Peel strength (N/cm)	Tack (N)
PSA-ESO-P90C	90°C	Flowable	80	5.6 ± 0.1	N/A
PSA-ESO-P100C	100°C	Flowable	0	6.4 ± 0.3	13.8 ± 0.5
PSA-ESO-P110C	110°C	Not flowable	N/A	N/A	N/A
PSA-ESO-P120C	120°C	Gelled	N/A	N/A	N/A

*Conditions for preparation of PSAs: ESO usage, 3%; pre-polymerization time, 2h; curing temperature, 125°C; and curing time, 2h. DCF, degree of cohesive failure. N/A, not determined. The ± values are based on 95% confidence interval.

Table 16.3 Properties of PSAs from different pre-polymerization times.

Cured PSAs*	Pre-polymerization time (h)	Property of pre-polymers	DCF (%)	Peel strength (N/cm)	Tack (N)
PSA-ESO-P10H	1.0	Flowable	80	4.1 ± 0.5	1.5 ± 1.1
PSA-ESO-P15H	1.5	Flowable	30	4.2 ± 0.5	5.4 ± 0.9
PSA-ESO-P20H	2.0	Flowable	0	6.4 ± 0.3	13.8 ± 0.5
PSA-ESO-P25H	2.5	Not flowable	N/A	N/A	N/A

*Conditions for preparation of PSAs: ESO usage, 3 mol%; pre-polymerization temperature, 100°C; curing temperature, 125°C; and curing time, 2h. DCF, degree of cohesive failure. N/A, not determined. The ± values are based on 95% confidence interval.

and tack of PSAs significantly increased when the pre-polymerization time was increased from 1.5 h to 2.0 h. PSA-ESO-P20H had no cohesive failure. The pre-polymerization time of 2 h appeared to be the best and was used for further investigation.

Properties of PSAs from different curing temperatures are shown in Table 16.4. PSA-ESO-C110C had 80% cohesive failure, which means that the curing temperature of 110°C is not suitable for making PSAs. The peel strength slightly decreased, and tack significantly increased when the curing temperature was increased from 110°C to 120°C. Both peel strength and tack significantly increased when the curing temperature was increased from 120°C to 125°C. Both peel strength and tack decreased significantly when the curing temperature was increased from 125°C to 130°C. Both peel strength and tack remained the same when the curing temperature was further increased from 130°C to 135°C. The curing temperature of 125°C appeared to be the best and was used for further investigation.

All PSAs from the curing times of 2 h, 3 h, 4 h, and 8 h had no cohesive failure (Table 16.5). Both peel strength and tack significantly decreased when the curing time was increased from 2 h to 3 h. Both peel strength and tack remained the same when the curing time was increased from 3 h to 4 h. The peel strength decreased slightly, and tack remained the same when the curing time was further increased from 4 h to 8 h. The curing time of 2 h appeared to be the best and was used for further investigation.

Effects of ESO usage on properties of the resulting PSAs are shown in Table 16.6. PSA-ESO-A and PSA-ESO-B had comparable peel strength and tack, i.e., increase in the molar percentage of ESO over CEFA from 0.6% to 1.0% did not significantly change the peel strength and tack. The peel strength significantly increased, but the tack remained the same when the molar percentage of ESO was increased from 1.0% to 2.0%. Both peel

Table 16.4 Properties of PSAs from different curing temperatures.

Cured PSAs*	Curing temperature	DCF (%)	Peel strength (N/cm)	Tack (N)
PSA-ESO-C110C	110°C	80	5.3 ± 1.1	1.3 ± 1.8
PSA-ESO-C120C	120°C	0	4.6 ± 0.5	8.1 ± 2.6
PSA-ESO-C125C	125°C	0	6.4 ± 0.3	13.8 ± 0.5
PSA-ESO-C130C	130°C	0	5.2 ± 0.6	8.9 ± 1.0
PSA-ESO-C135C	135°C	0	4.9 ± 0.7	7.3 ± 1.4

*Conditions for preparation of PSAs: ESO usage, 3 mol%; pre-polymerization temperature, 100°C; pre-polymerization time, 2 h; and curing time, 2h. DCF, degree of cohesive failure. N/A, not determined. The ± values are based on 95% confidence interval.

Table 16.5 Properties of PSAs from different curing times.

Cured PSAs*	Curing time (h)	DCF (%)	Peel strength (N/cm)	Tack (N)
PSA-ESO-C2H	2	0	6.4 ± 0.3	13.8 ± 0.5
PSA-ESO-C3H	3	0	4.3 ± 1.1	8.6 ± 1.4
PSA-ESO-C4H	4	0	4.3 ± 0.9	9.3 ± 0.7
PSA-ESO-C8H	8	0	3.1 ± 0.4	8.6 ± 1.2

*Conditions for preparation of PSAs: ESO usage, 3 mol%; pre-polymerization temperature, 100°C; pre-polymerization time, 2 h; and curing temperature, 125°C. DCF, degree of cohesive failure. The ± values are based on 95% confidence interval.

Table 16.6 Effects of ESO usage on the properties of the resulting PSAs.

Cured PSAs*	% of ESO**	DCF (%)	Peel strength (N/cm)	Tack (N)
PSA-ESO-A	0.6 %	0	0.2 ± 0.0	6.1 ± 4.4
PSA-ESO-B	1.0 %	0	0.2 ± 0.1	6.8 ± 4.8
PSA-ESO-C	2.0 %	0	3.8 ± 0.5	4.6 ± 0.5
PSA-ESO-D	3.0 %	0	6.4 ± 0.3	13.8 ± 0.5
PSA-ESO-E	5.0 %	0	3.7 ± 0.9	3.1 ± 0.6
PSA-ESO-F	7.0 %	80	3.8 ± 0.5	N/A

*Conditions for preparation of PSAs: pre-polymerization temperature, 100°C; pre-polymerization time, 2h; curing temperature, 125°C; and curing time, 2 h. DCF, degree of cohesive failure. N/A, not determined. Values are given as 95% confidence intervals.
** the % is defined as the molar percentage of ESO to CEFA.

strength and tack significantly increased when the molar percentage of ESO was increased from 2.0% to 3.0%. Both peel strength and tack significantly decreased when the molar percentage of ESO was increased from 3.0% to 5.0%. The peel strength remained the same when the molar percentage of ESO was increased from 5.0% to 7.0%. All PSAs had no cohesive failure when the molar percentage of ESO ranged from 0.6% to 5.0%. PSA-ESO-F had 80% cohesive failure. There is only a certain amount of free carboxylic acid groups in the CEFA pre-polymers. The amount of ESO at a lower than 3% appeared not to be sufficient for reactions with the free carboxylic acid groups to provide sufficient crosslinking. Some of ESO may remain unreacted, thus reducing the peel strength and tack of the resulting PSAs when the amount of ESO was higher than 3%. It appeared that 3% of ESO was optimum for generating the highest peel strength and tack.

Results from Tables 16.2 to 6 reveal that PSA-ESO-D was the best PSA and the optimal conditions for making PSA-ESO-D were as follows: ESO usage, 3%; pre-polymerization temperature, 100°C; pre-polymerization time, 2h; curing temperature, 125°C; and curing time, 2 h. PSA-ESO-D was further evaluated for its shear adhesion, gel fraction, and aging properties (Table 16.7). The peel strength of PSA-ESO-D slightly decreased and tack significantly decreased after PSA-ESO-D was aged at 60°C for one week. PSA-ESO-D had the shear adhesion longer than 28 days and the gel fraction of 74%.

FTIR spectra of CEFA, pre-polymers of CEFA, and PSA-ESO-D are shown in Figure 16.2. The FTIR spectrum of CEFA had a strong peak at 1693 cm^{-1} that corresponds to the carbonyl group (C=O) of -COOH groups. However, the peak at 1740 cm^{-1} that corresponds to the carbonyl group (C=O) of ester groups is nearly invisible, which indicates that the ester bonds have been hydrolyzed to form -COOH groups. The peaks at 842 to 818 cm^{-1} represent epoxy groups [16], suggesting that most epoxy groups remained intact during the hydrolysis. The results from this study are consistent with the previous report [12].

The FTIR spectrum of CEFA pre-polymers shows both peaks at 1692 cm^{-1} and 1740 cm^{-1}, suggesting that a part of -COOH groups indeed reacted with epoxy groups to form new ester bonds. The epoxy peaks at 842 cm^{-1} to 818 cm^{-1} are weakened, further confirming the participation of epoxy groups in the reaction. A new broad peak centered at 3400 cm^{-1} represents

Table 16.7 Effect of aging on properties of PSA-ESO-D.

Aging time (week)	DCF (%)	Peel strength (N/cm)	Tack (N)	Shear adhesion	Gel fraction (%)
no aging	0	6.4 ± 0.3	13.8 ± 0.5	> 28 days	74 ± 3
1	0	5.3 ± 0.2	6.9 ± 3.3	N/A	N/A

Notes: DCF, degree of cohesive failure. N/A, not determined. The ± values are based on 95% confidence interval.

the newly generated -OH groups from the ring-opening reaction of epoxy groups with carboxylic acid groups.

In the FTIR spectrum of PSA-ESO-D, the -COOH peak at 1692 cm^{-1} is not observed and the ester peak at 1740 cm^{-1} is very strong, which indicates that the crosslinking reactions between -COOH groups of the CEFA pre-polymers and epoxy groups of ESO indeed occurred. The epoxy peaks ranging from 842 cm^{-1} to 818 cm^{-1} are weakened, yet still observable, confirming the crosslinking reactions, and indicating that a considerable amount of epoxy groups remained in the PSA product.

16.5 Conclusions

CEFA was prepared by selective hydrolysis of ESO with NaOH, followed by purification through crystallization in hexane. Heating of CEFA in the presence of a base catalyst generated flowable pre-polymers that were crosslinked with TDI and ESO to form PSAs. The surface of PSAs with TDI as the crosslinking agent was always rough, which reduced its viability to make PSAs. PSA-ESO-D had the highest peel strength of 6.4 ± 0.3 N/cm, and the highest tack of 13.8 ± 0.5 N. PSA-ESO-D had the peel strength of 5.3 ± 0.2 N/cm after one week of aging at 60°C, and the shear adhesion of greater than 28 days. PSA-ESO-D was prepared under the following optimum conditions: ESO usage, 3%; pre-polymerization temperature, 100°C; pre-polymerization time, 2 h; curing temperature, 125°C; and curing time, 2 h.

16.6 Acknowledgement

This project was supported by the return royalty fee income from patented adhesives of Li's group.

AI disclosure statement: The author confirms that no artificial intelligence (AI)-generated content or AI-assisted writing tools were used in the preparation of this chapter (including summaries and the abstract). All text, analysis, and conclusions are the result of the author's original work.

References

1. D. Satas, Pressure sensitive adhesives and adhesive products in the United States, in: *Handbook of Pressure-Sensitive Adhesive Technology, 2nd ed.* D. Satas (Ed.), pp. 1–21, Van Nostrand Reinhold, New York (1989).
2. M. A. R. Meier, J. O. Metzger, and U. S. Schubert. Plant oil renewable resources as green alternatives in polymer science, *Chem. Soc. Rev.* 36, 1788–1802(2007).
3. U. Biermann, W. Friedt, S. Lang, W. Lühs, G. Machmüller, J. O. Metzger, M. Rüsch Gen Klaas, H. J. Schäfer and M. P. Schneider, New syntheses with oils and fats as renewable raw materials for the chemical industry. *Angew. Chem. Int. Engl. Ed.* 39, 2206–2224(2000).
4. H. Baumann, M. Bühler, H. Fochem, F. Hirsinger, H. Zoebelein and J. Falbe, Natural fats and oils—renewable raw materials for the chemical industry. *Angew. Chem. Int. Engl. Ed.* 27, 41–62(1988).
5. S. P. Bunker and R. P. Wool, Synthesis and characterization of monomers and polymers for adhesives from methyl oleate. *J. Polym. Sci. Part A. Polym. Chem.* 40, 451–458(2002).
6. S. Bunker, C. Staller, N. Willenbacher and R. Wool, Miniemulsion polymerization of acrylated methyl oleate for pressure sensitive adhesives. *Int. J. Adhesion Adhesives.* 23, 29–38(2003).
7. C. M. Klapperich, C. L. Noack, J. D. Kaufman, L. Zhu, L. Bonnaillie and R. P. Wool, A novel biocompatible adhesive incorporating plant-derived monomers. *J. Biomed. Mater. Res. A.* 91, 378–384(2009).
8. S. B. David, K. Sathiyalekshmi and G. A. Gnana Raj, Studies on acrylated epoxydised triglyceride resin-co-butyl methacrylate towards the development of biodegradable pressure sensitive adhesives. *J. Mater. Sci. Mater. Med.* 20, S61-S70 (2009).
9. C. Koch, Pressure sensitive adhesives made from renewable resources and related methods, World Intellectual Property Organization WO2008144703, assigned to Avery Dennison Corporation (2008).
10. Y. Wu, A. Li, and K. Li, Development and evaluation of pressure sensitive adhesives from a fatty ester. *J. Appl. Polym. Sci.* 131, 41143(2014).
11. A. Li and K. Li, Pressure-sensitive adhesives based on epoxidized soybean oil and dicarboxylic acids. *ACS Sustain. Chem. Eng.* 2, 2090–2096(2014).
12. A. Li and K. Li, Pressure-sensitive adhesives based on soybean fatty acids. *RSC Advances.* 4, 21521–21530(2014).
13. Y.-F. Lei, X.-L. Wang, B.-W. Liu, L. Chen and Y.-Z. Wang, Bio-based removable pressure-sensitive adhesives derived from carboxyl-terminated polyricinoleate and epoxidized soybean oil. *Chin. Chem. Lett.* 32, 875–879(2021).

14. W. Xia, S. M. Budge and M. D. Lumsden, 1H-NMR characterization of epoxides derived from polyunsaturated fatty acids. *J. Am. Oil Chem. Soc.* 93, 467–478(2016).
15. T. E. Clemente and E. B. Cahoon, Soybean oil: genetic approaches for modification of functionality and total content. *Plant Physiol.* 151, 1030–1040(2009).
16. O. D. Shreve, M. R. Heether, H. B. Knight and D. Swern, Infrared absorption spectra of some epoxy compounds. *Anal. Chem.* 23, 277–282(1951).

17

Wood Adhesion and Wood Adhesives: What Remains to Be Investigated?

Manfred Dunky

Chemical and Wood-based Panels Industry, St.Poelten, Austria

Abstract

This article summarizes topics which require further scientific investigation and development. Work on wood bonding has shown huge progress in the last two decades, using many new methods and techniques for investigating wood adhesives, the bonding process, the prepared joints, and the wood-based panels (wood composites).

The questions and recommendations at the end of each section in this article summarize ideas and requirements for future work by researchers, scientists, and engineers. They also highlight the need for secured financial support to perform this work. This especially is important and justified, because wood-based panels are the best examples for a sustainable use of natural raw materials, including cascade use of wooden biomass and recycling of used and old wood.

Keywords: Wood adhesion, adhesive, wood-based panels, bondline, microstructure, weak boundary layer, wetting, penetration

17.1 Introduction

This paper is a contribution to the "Festschrift" in honor of Dr. Kash Mittal's 80th birthday and shall serve two aspects. The first aspect points to the great effort of Dr. Mittal in editing books about adhesion and adhesives ("Progress in Adhesion and Adhesives") and to have been the founder and Editor-in-Chief for many years of the journal "Reviews of Adhesion and Adhesives". In both "Progress" and "Reviews" he offered the author of this article the chance to publish articles about wood adhesives based on natural

Email: manfred.dunky@gmx.at

Anil N. Netravali and Frank M. Etzler (eds.) The Wide World of Adhesion: A Festschrift in Honor of Dr. Kash Mittal on his 80th Birthday, (585–622) © 2026 Scrivener Publishing LLC

resources, as well as on various topics in wood bonding. The second aspect is related to an idea I had already long time ago, meaning to summarize various aspects in wood bonding which still need more investigation. More than 20 years ago, the author was leader of a working group about "Wood adhesives" in the COST Action E13 of the European Commission [1]. In this European-wide cooperation, a State-of-the-Art Report [2] on wood adhesives was published, and many topics as the basis for further investigations and scientific work were listed. This procedure now shall also be implemented in this article of the "Festschrift" again. At the end of each section, questions are posed, which point to possible research topics and further demand on research work performed. The questions are based on the long-term experience of the author in wood bonding and production of wood adhesives and wood-based panels. The author claims to know the scientific and technical literature on wood adhesives and wood-based panels quite well [3–10]; nevertheless, some questions cannot be answered yet. Hence, they shall serve as a guide for further work.

Though extensive information is available on adhesives and adhesion, further basic research on the adhesion behavior of wood including the evaluation of the microstructure of the bondline formed as the key element within the produced bonded products needs to be promoted. This includes a fundamental approach as well as an applied approach to industrial bonding processes in their large variety, from single laminated wood preparation in carpenter's shop to industrial production of wood-based panels of several thousand m^3 per day.

The article highlights only a few topics, which are important in the eyes of the author, based on his experience. A comprehensive description of wood bonding and wood-based panels in their huge variation and complexity is left to comprehensive textbooks, e.g., [3, 4], where additional references also can be found.

17.2 Basics of Wood Bonding

Wood bonding is based on forming adhesive bonds between individual structural components, including the cohesive strength development of the adhesive itself and the adhesion between wood as adherend and the adhesive. When two pieces of wood are joined using an adhesive, the basic objective is to hold the two pieces in a fixed position; when applying stresses exceeding the strength of the joint, the failure zone preferably should occur in the wood material far from the bondline but not in the bulk of the adhesive. To achieve this goal, both the molecular attraction between the adhesive and the wood surface (adhesion forces) as well as the molecular attraction within the adhesive in the solid state (cohesion forces) must exceed the strength of the wood. It needs to show not only a certain cohesive bond strength as such, but also some elastic behavior in order to avoid excessive stresses within the bondline due to movement of the wood material, e.g., under external forces or at changing climate conditions inducing wood swelling or shrinkage due to water uptake or release. Depending on the application, either high strength or higher elasticity might be required. In addition, the cohesive strength of the cured or solidified adhesive must not suffer from chemical degradation.

The production of wood-based panels is based on the bondlines formed (i) by the distribution of the adhesive on the surfaces of the individual wood components and (ii) by its physical and/or chemical solidification (curing). The size of the bond areas can be very different, from several m^2 in plywood down to few cm^2 for oriented strand board (OSB) or even only a few or single mm^2 in case of particleboard. Nevertheless, these are still bondlines between two wood surfaces, irrespective of their size. The entirety of all these individual and small bondlines especially in particleboard, then determines, among other parameters, the macroscopic properties of the panels.

Questions and recommendations:

a) Investigate the influence of the microstructure of the bondlines formed as important influential parameter for the macroscopic properties of wood-based panels.
b) Investigate and describe the microstructure of the wood-based composites and the individual bondlines between particles, strands, or fibers.
c) The two terms “interphase” (as 3D region of wood and adhesive) and “interface” (as the 2D region of direct and individual contact between reactive sites of the wood surface and reactive sites on the adhesive molecules) have been often mixed up in literature. A clear and commonly accepted definition and understanding is important.
d) How can the border between the interphase and the cohesive bondline (adhesive layer) be determined? Compare the thicknesses of the interphase and of this cohesive adhesive layer for various examples of bonding two wood pieces.

17.2.1 Weak Boundary Layers

The concept of weak boundary layers (WBL) in adhesion science was first mentioned in 1961 [11]. Three different classes of WBLs were specified at that time, i.e., air bubbles, impurities at the interface, and products from reactions between the components and the atmosphere. The WBL on the surface of adherends was further seen as responsible for lower mechanical strength in bonded systems [12].

The general term of weak boundary layer (WBL) can be distinguished into:

- chemical weak boundary layer (CWBL): lubricant contamination, aged or inactivated surface, weathered surface, extractives migrating to the surface:
  - hydrophobic: resin acids as components of wood rosin, fats/fatty acids, sterols, waxes
  - sugars, phenols, tannins, proteins

- mechanical weak boundary layer (MWBL): damaged surface structure due to machining, trapped air bubbles in the interphase, deformation and destruction of wood cells close to the bondline during pressing.

17.2.1.1 Mechanical Weak Boundary Layer (MWBL)

The quality of the bonded wood joints and furniture products is influenced by the surface characteristics of wood and the physicochemical characteristics of adhesive. MWBLs are formed by different kinds of machining of the surface due to mechanical destruction of the uppermost wood layer causing deformed or damaged wood cells, e.g., by using coarse or blunt tools [11, 13, 14], and lower bond strengths. Fracture often shows very low wood failure or even only a fine fiber layer similar to a sealskin. MWBLs can partly be supported or repaired by the adhesive. Removing the weakened zones (damaged wood cells) before bonding can improve significantly the bond performance.

It is remarkable that there is no easy and widely used technique to assess the presence and the severity of a MWBL in wood [15]. One option of quantifying a MWBL is the so-called surface integrity test, consisting of attaching a metal tab to a veneer surface with a double-sided tape and tearing the tab off. The tape is then imaged to determine the number and size of wood particles removed from the wood surface [16].

Questions and recommendations:

a) Which causal connections exist between MWBL and the failure strength of wood joints?
b) How can MWBLs be detected and investigated on the surface of strands and particles?
c) Can the term of MWBL also be used for fibers after defibrillation, e.g., for cut fibers, deformed fibers, or fibers with partly affected or destroyed surfaces?
d) How can adhesives repair a MWBL? Which chemical or physical reactions contribute to this effect?

17.2.1.2 In Situ - MWBL

A second type of MWBL can occur directly during the hot-press process due to the rheological properties of wood, i.e., deformation and relaxation occur under the influence of temperature, high moisture content (MC), and pressure. Water as a component of the applied adhesive penetrates into the top layer of the wood surface. Hence, the MC of the wood substance close to the bondlines can increase significantly after application of an aqueous adhesive and during the assembly time before the press cycle. Usually the MC close to the bondline is about 15 - 20%, but can even reach the fiber saturation point for the closest layers (i.e., a few tenths of a millimeter). Wood strength during the hot-press cycle is affected by temperature, MC, and the densification of the wood structure close to the bondline. Often, the thin earlywood cell walls near the bondline are deformed (micro-buckling) or densified due to the combined influence of moisture, heat and mechanical pressure. This can even result in a destruction of the upper cell rows directly beneath the bondline, weakening the joint and causing such a MWBL. These effects not only happen during solid wood bonding, but also when particles and strands are bonded in the production of wood-based panels, since their surfaces still reflect the behavior of solid wood.

Questions and recommendations:

a) How can *in situ*-MWBLs be detected?
b) Can *in situ*-MWBLs occur also at "usual" wood bonding conditions, or only at extreme conditions?
c) How can *in situ*-MWBLs be avoided?

17.2.1.3 Chemical Weak Boundary Layer (CWBL)

In wood adhesion, "aged" (old) surfaces are often chemically altered because of environmental influences such as exposure to moisture, ultraviolet light, oxygen, or heat. In addition, migration of hydrophobic wood-inherent chemicals (extractives) significantly worsens the wettability of wood surfaces, causing stronger hydrophobic behavior; this especially is the case with wood species of high content of extractives, e.g., birch. Aged surfaces tend to have lower surface free energies and are therefore more difficult to be wetted by polar liquids, such as adhesives; the higher the age of the surface, the stronger the changes in the surface chemistry are [17].

Typical processes causing wood surface inactivation are wood aging, wood weathering, and wood heating or drying. Immediately after mechanical preparation, the wood surfaces are exposed to atmospheric influences, including oxidation, temperature, humidity, light irradiation, or airborne contaminants, causing the inactivation of the surface [18].

Especially exposure to high temperatures, as it is the case in several steps of the wood-based panel production, can cause wood surface inactivation. The drying process, as well as a certain intermediate storage time of the dried particles and strands before the application of the adhesive (blending step) affect the surface of the particles and strands in the production of particleboards and OSB. This means that fresh surfaces may not be available. The time regimes in the particleboard and the OSB production can vary widely, depending on the individual production conditions at different sites. The delay time for OSB strands and particles from the preparation of the strands to the dryer and finally to the blender is usually in the range of few hours, also including some short intermediate storage between the various process steps. Such time ranges are still on the short side and no special effects of aging should occur. The situation with sawdust and fine recycled wood, however, is different. These components are mixed with the particles without the need for flaking or grinding, hence without the generation of new surfaces. The age of such material and its surface is undefined, and aged surfaces will be present.

Generating fresh surfaces just before bonding helps to overcome the problem with the inactivation of the wood surface, however this process step is not always easily achievable or even not possible at all, e.g., in the particleboard production. Removal of surface layers impeding or inhibiting bonding before the application of adhesives can be performed in solid wood bonding, though even here this is not a common technology; rather the time regime will be adjusted accordingly, i.e., using only fresh surfaces maximum one day after preparation. For the standard particleboard or OSB production, a treatment of the surfaces of the particles or strands after the flaker or especially after the dryer is not performed.

Questions and recommendations:

a) How big is the risk of wetting problem in the production of particleboards or OSB?
b) Can the age of surfaces of particles and strands be estimated, despite the very different raw materials and their history?
c) How can contact angles and their changes with time measured at the surfaces of particles or strands? For coarse particles and strands, this should not be an unsolvable task, as long as a flat surface is present in the same size as the droplets of water or other liquids applied. For fine particles, other test methods than the sessile drop method must be used.
d) Which technologies or pretreatments might improve wetting or avoid wetting problems in the production of particleboards or OSB?

17.2.2 Wetting and Penetration

Two important aspects in the production of wood-based panels (WBPs) are wetting of the surfaces and the penetration of the adhesive into the wood furnish, forming the interphase.

17.2.2.1 Wetting Behavior

Wetting by a liquid or a liquefied adhesive is a precondition for the adhesive bonding of two wood surfaces. It enables the creation of high adhesion forces between the solid surface and the cured or solidified adhesive. Wetting includes formation of a contact angle and the spreading of the liquid on the surface of the wood as porous adherend; in most cases, this is linked with partial penetration of the liquid into the porous wood adherend and gradual absorption by the wood [19–21]. The wetting phenomenon can be characterized by contact angles, surface free energy, or work of adhesion (work required to separate a certain area of the solid-liquid interface).

Determination of contact angles on wood surfaces is an important measurement. A preferred method is the advancing contact angle measurement (dynamic contact angle method), because it compensates for the penetration of the adhesive in the wood furnish [22, 23]. Good wetting is a precondition for strong adhesive bonding. Low contact angles (< 45°) indicate such good wetting of the surface. However, direct correlations between the contact angles and bond strengths are rare [24] or seem not to be universal.

For any bonding to take place, the adhesive must spread across and wet the wood surface to develop molecular interactions between the adhesive and wood. Maximal surface interactions occur when the surface energies of the adhesive and the wood are nearly equal [25].

Questions and recommendations:

a) Does a correlation exist between the wetting behavior of a surface (e.g., expressed by the contact angle of the adhesive on this surface to be bonded) and (i) the possible adhesion and (ii) the resulting bond strength?
b) Which effects do occur at the wood surface during aging, which influence wetting?

17.2.2.2 Penetration into Wood Tissue

Movement of adhesives into the wood tissue is distinguished (i) into penetration as bulk flow and (ii) into cell wall infiltration (see section 17.2.2.3.) as the ability of an adhesive to enter into cell walls involving movement through the micro-voids within the wood cell walls [26].

Bulk penetration (void penetration) is the ability of an adhesive to enter into the large voids in the porous wood structure such as vessels and rays and into the cell lumen as a process of fluid movement [27]. Penetration is governed by pore size (wood anatomy, permeability and porosity), size of the adhesive molecules and adhesive viscosity, external compression forces from mechanical pressure applied to the wood, and the wetting behavior [28, 29]. The resulting large area of close contact between the adhesive and the internal surface of the substrate plays an important role in developing good adhesion, including mechanical interlocking as well as any type of surface adhesion through dispersion, polar, hydrogen, or ionic bonds [30, 31]. Penetration creates the interphase consisting of adhesive and wood, with the interface as microscopic boundary between wood and adhesive.

The lower molar masses of condensation resins with their higher proportion of hydrophilic moieties can easily penetrate into the wood furnish and are responsible for the good wettability. In addition, penetration is enhanced by the decrease in viscosity due to the increased temperature in the bondline during hot-pressing, before curing of the thermosetting adhesive or the physical solidification starts. However, a too high portion of low molar masses can cause overpenetration and so-called starved bondlines with low or no bond strength obtained. The higher molar masses preferably remain at the wood surface and form the adhesive layer. Too high viscosity causes no or very low penetration into the word surface, with a narrow and uneven interphase and less mechanical interlocking of the adhesive into the substrate, again causing poor bond strengths.

Within this interphase, the interface as a two-dimensional contact zone between the internal surface of wood and the adhesive is formed, determining strongly the bonding effect. For a strong bond the adhesive must penetrate deep enough into the wood substrate in order to be able to reinforce weakened cells and to obtain large contact surfaces. Penetration is needed to repair processing damages of the wood surface and to allow better stress transfer between laminates (plies) [28].

Adhesive bond performance between wood elements is presumed to be significantly influenced by the degree of penetration of the adhesive into the porous network of interconnected cells in the wood material. However, the complex relationship between adhesive penetration and joint performance is difficult to describe quantitatively. Shallow penetration means that the adhesive does not to move into the wood substance enough in order to create a large active bonding surface (interface) within the interphase and hence strong interaction between wood and the adhesive; this may result in poor stress transfer and reduced dimensional stability [32].

When inspecting cross sections, often adhesive penetration into a lumen three or more cells deep from the surface is observed, but without adhesive in the first or second cell row. The logical explanation is that such deep penetration is due to capillary action of the adhesive within a lumen of a cell cut above or below the plane of the particular cross section investigated [32]. X-ray CT allows visualization of wood, adhesive, and voids in 3D using labeled adhesives; in this way, the 3D-wood anatomy, bondline, and the adhesive distribution on a micrometer scale can be investigated, and models for adhesive-wood interactions can be established.

The degree of condensation (DOC) of three urea-formaldehyde (UF) resins UF I, UF II, and UF III showed strong influence on the tangential and the radial penetration into various wood species; the higher the DOC, the less is the penetration (Figure 17.1) [33–37].

Overpenetration can occur when using low-density hardwood in particleboard manufacturing; the open vessels of the hardwood structure easily induce penetration of part of the adhesive, causing higher resin demand compared to resins with higher molar masses. Higher viscosity of the adhesive (higher molar mass), a certain waiting time (if possible from the technological sequence of the process), and an earlier solidification of the applied adhesive (either by higher cure rate or by an earlier start of the hardening process) can overcome these problems.

Most investigations of adhesive penetration in the past have focused on the two-dimensional analysis of microtome sections prepared from bonded wood samples. The preparation of such microtome sections, however, might affect the bondline. Furthermore, these studies can only speculate about the pathways of liquid adhesive flow in the wood substance during bond formation. New analysis methods like high-resolution computer tomography (such as sub-µm-X ray CT) with resolutions of a few micrometers (and aims to achieve values < 1 µm) have significantly improved the knowledge about the adhesive penetration.

Questions and recommendations:

a) How can penetration be described in the best way?
b) What connections or even correlations do exist between the penetration behavior and bond strength or bond performance?
c) What is the role of the microstructure of the bondline, including (i) the adhesive penetration into the wood surface, (ii) the effect of ageing of a bondline, and (iii) the description of the cohesive strength of the bondline in terms of an optimization of its brittle and elastic behaviors?

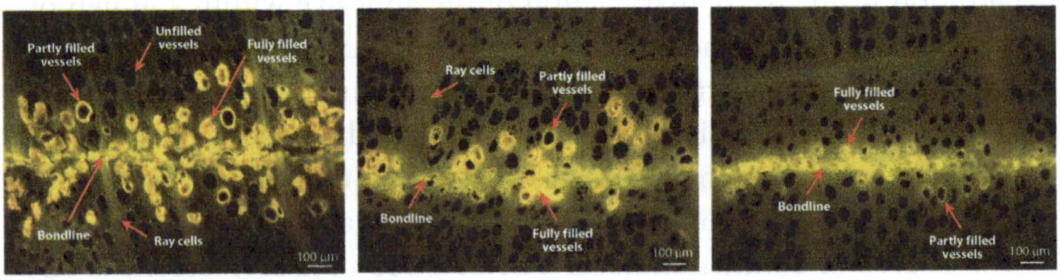

Figure 17.1 Photomicrographs of the three UF resins for radial penetration into beech, using epi-fluorescence. Left side: UF I (lowest DOC); middle: UF II; right side: UF III (highest DOC) [33].

d) The penetration of an adhesive into the porous wood structure has strong influence on bond strength, durability, and performance. However, there is limited direct experimental evidence demonstrating the link between adhesive penetration depth and bond performance in wood joints, as the optimum depth of penetration is not known [38, 39].
e) Which key numbers describe the penetration in the best way, also in view of formation of bond strength and performance of the joints under various conditions, such as changing relative humidity, causing variation in the moisture content of the joint (adherend + bondline)? Possible key numbers can be the percentage of the lumen cross section area filled with adhesive, the average penetration depth, the maximum penetration depth, the total width of the interphase, or the percentage of the total interphase region filled with adhesive.
f) What are the reasons for the good bonding behavior of polymeric methylene-diisocyanate (PMDI): good wetting behavior, good penetration partly into the cell walls, therefore forming an interphase of impregnated wood cells?
g) How to clarify the pathways of liquid adhesive flow in the wood substance during bond formation? What contribution can be expected from high-resolution computer tomography (such as sub-μm-X ray CT)?
h) Low bond strengths can result from either under- or over-penetration. How can these two effects be explained and avoided?
i) To what extent can penetration take place through the various forms of pits? What is the ratio of penetration between flow through pits and flow into cut tracheids or vessels?

17.2.2.3 Cell Wall Infiltration

Cell wall penetration (infiltration) is possible for substances with low molar masses, like aminoplastic impregnating resins or low-molar-mass portions of the adhesive; it is highly controlled by the size of the molecules of the adhesive; the effect can be enhanced by the decrease in viscosity due to the increased temperature in the bondline during hot-pressing. Infiltration of adhesive generally fortifies the cell wall, increasing strength and lowering cell wall swelling. Penetration of resin compounds into the wood cell wall can enhance the durability of adhesive bonds. The diffusion of adhesive into undamaged cell walls clearly increased the hardness of phenol-formaldehyde (PF) resin-impregnated cell walls [40].

In-situ polymerized adhesives like PF resins with low molar masses alter the cell wall properties and stress transition in the interphase by infiltrating the cell wall; hence, the wood surface could better tolerate induced strains [28]. The dimensional stability caused by the infiltration treatment correlated with the amount of resin infiltrated, which as such depends on the molar mass of the resins [41].

Questions and recommendations:

a) What effects from cell wall infiltration (stiffening the cell wall, chemical modification of the cell wall, implementation of an interpenetrating network within the cell wall, or others) yield improvement in bond strength and bonding performance?
b) How large is the contribution of cell wall infiltration to the bonding effect?
c) Can cell wall infiltration repair MWBL? How will it work on a microscopic level?
d) Although it is generally known that a certain penetration helps in creating strong bonds [27, 42], it is not clear whether penetration into the lumens or infiltration into the cell walls is more beneficial for wood bonding. How can this question be answered?

17.2.3 Bonding Theories or "How Does Wood Bonding Really Work?"

Adhesion is the sum of attractive forces operating at the interface between adhesive and substrate. This effect is based on a number of different possible interactions at the adhesive-adherend interface. Several theories are used (and are also necessary) to explain the adhesion aspect of bonding two adherends, such as secondary forces/physical bonding and mechanical anchoring. None of the theories has been found to describe the full pattern of bonding, considering that all different theories might contribute to the final bond strength. Due to the porous nature of a wood surface, the number of theories is even higher compared to other materials with a closed surface. These different adhesion theories attempt to find an explanation for the mechanisms involved in wood bonding, especially how bonding works and how it may fail [43]. According to Frihart [44], the mechanism of bond failure is not as well understood as are most other adhesive applications. This is not surprising given the complexity of wood; both its structural and chemical complexities hinder the development and evaluation of models for bond formation and fracture [44].

Questions and recommendations:

a) What are the contributions of the individual bonding theories?
b) What type of bonding would be the optimum in the sense of efficiency but also in sense of easy implementation during joining two pieces of wood?
c) What conclusions can be drawn from the failure behavior of a bond?

17.2.4 Bondlines under Strain, Stress and Fracture

Understanding why wood bonds fail is an important aspect and an excellent route for understanding how to make them better [15]. Understanding failure often requires going beyond what meets the naked eye, conducting additional tests, probing the wood surface, the fracture surface, adhesive properties, and the interaction of wood and adhesive during bond formation and service [15].

a) Fracture behavior of a bondline and of wood-based panels

The quality of the wood-adhesive bond is usually determined by its shear strength and the percentage of wood failure; the latter usually is determined by visual inspection and (to some extent) is prone to the examiner's subjective evaluation. For objective evaluation of the percentage of wood failure, device supported methods, such as 3D scanning or image analysis techniques, can be used.

The best performing adhesive for enhanced fracture toughness (i) does not develop shrinkage stresses during cure, (ii) has a modulus close to that of wood perpendicular to the grain, (iii) has a modulus that changes in parallel with the wood modulus as the moisture content varies, (iv) penetrates into small-lumen thick cell walls but does not overpenetrate into large-lumen thin-walled cells, and (v) can infiltrate the cell wall to reinforce the weak interphase between cell wall layers [45].

Particleboards consist of a network of crossing and overlapping particles, which are bonded together by adhesive bridges. Between the particles, macroscopic holes exist, whose proportion can be up to 30% of the volume of the board. Based on this structural model, the deformation and macroscopic failure as the endpoint of such deformation can be seen as the sum of elastic and plastic deformations of the particles and of bondlines, micro-cracks, and interparticle displacement [46].

b) Wood failure
The quality of a wood adhesive joint is usually characterized by two values: shear strength and percentage of wood failure. A high percentage of wood failure represents a high quality of bonding, showing that the solid adhesive bondline is stronger than the wood adherend. Different requirements concerning the shear strength of bondlines (in most cases in combination with requirements for the average percentage of wood failure, in case expressed as various classes) are stated in the various quality standards for wood products when shear-tests are required for quality assessment.

The percentage of wood failure as such is a numerical result, which is based on the various tests and quality standards. In addition, however, the thickness and the appearance of the wood layer in case of wood failure is an important aspect, which is neglected if just the proportion of wood failure area is evaluated and reported, and with fine fibrous layers being calculated as wood failure. The thickness of the wood layer of a wood failure gives important additional information about the quality of bonding. The same wood failure (as percentage), but with different average thickness of the wood failure (e.g., just a very thin fiber layer on the one hand, and a really clear and "thick" wood failure on the other hand), should be seen as different results, yielding some extra information. However, standards (e.g., EN 314-2 [47]) ask just for the percentage of wood failure, even sometimes as classes in certain range steps of, e.g., 10 or 25%.

An attempt to determine the cohesive strength of the adhesive layer is possible using the so-called fracture cleavage test [48, 49]. In this test, the fracture area predominantly occurs either in the adhesive layer or in the interphase. Using a small angle between bond area and the fiber direction, the cleavage fracture is redirected into the adhesive layer during the test.

c) Strain and stress distributions in adhesive joints and wood-based panels
Changing climate conditions induce deformation and stress concentrations in bonded wood joints. Reducing the extent and equalizing the distribution of internal stresses in wood-based panels should improve stability at alternating moisture content due to climate changes. However, the more rigorous the press conditions, the higher the residual stresses in the boards. Chances to achieve boards with low internal stresses can be realized by proper pressure programs during the press process or certain post-treatments.

Depending on the difference in the shear modulus of the adhesive and the adherend, various distributions of shear strain and shear stress occur. A higher shear modulus of the adhesive enables more evenly distributed shear strain, whereas a lower shear modulus rather gives a more even distribution of the shear stress. In general, bondlines showing higher shear deformations show lower shear stress concentrations [50].

Questions and recommendations:

a) The extent of the wood failure is an important quality criterion in several test methods for wood bonding. How to get maximum information from determination of type and appearance of the wood failure?
b) Most tests for assessing the quality of a wood bond are destructive aiming at determining strength values. Usually an indication for good quality is a fracture of undamaged wood far away from the bondline. In this case, however, no information about the cohesive bond strength of the adhesive layer is possible. How can information about the cohesive strength of an adhesive be obtained?
c) How would it be possible to use the fracture cleavage test as a quick and routine test?
d) How to achieve wood-based panels with low internal stresses in an optimized way, without losing press time and capacity of the production line?
e) More durable adhesive joints can be achieved with adhesives less prone to stress concentrations than with stiff adhesives. How can this be handled in the praxis of wood bonding, especially when producing particleboard, OSB, or plywood?
f) How can "real" wood failure (in the solid wood) and a failure in the interphase be distinguished?
g) What failure mode is given when testing particleboards and OSB?
h) How can wood failure be determined when testing the internal bond strength of particleboards or OSB?
i) What percentage of wood failure can be expected in particleboards and OSB?
j) Which failure mode is given when testing medium density fiberboards (MDFs)? How does "fiber failure" look like?

17.3 Bio-Based Wood Adhesives

Some years ago I was asked by a good friend of mine (who is no chemical or technical expert at all): "Of what substances do wood adhesives usually are made of, if not of natural raw materials? And why do all the experts promulgate only nowadays to use such natural substances?" [51]. Though adhesives based on natural resources [3, 5–10] have been in the research focus again since approx. three decades, either alone or in combination with synthetic adhesives, they only show very small industrial presence without real breakthrough. The main motivation for this renewed interest in natural adhesives is based on the sensitivity of the public opinion towards environment and its protection, the expectation of cost increases in petrochemicals, threatening more stringent legislative regulations, concerns related to human health risk, and interest in resource recycling and sustainability.

The main drawbacks in most cases are low curing and crosslinking reactivity based on low concentration of reactive sites, causing partly significantly elongated press times (hence reducing the capacity of existing production plants), low cohesive strength, and low moisture resistance due to missing or insufficient crosslinking (with remaining thermoplastic behavior if not crosslinked). Such use of synthetic or bio-based crosslinkers significantly increases the cohesive strength of the adhesive in the bondline and the moisture resistance, e.g., for proteins. Only cured tannins show excellent hydrolysis resistance due to their phenolic nature. An exciting aspect is the potential for so-called reversible adhesives, which in future might play a role when recycling will be more and more important.

Options for bio-based adhesives are the partial replacement of synthetic raw materials or adhesives as well as the use as classical naturally based adhesives, in most cases after pre-treatment (e.g., denaturation of proteins) and in many cases with additional crosslinking. Highest chances for industrial implementation as adhesives are given for modified lignin, for soy protein, and for tannin, together with suitable crosslinking. Carbohydrates rather will act as raw material for monomers, e.g., 5-hydroxymethylfurfural (5-HMF) [52, 53] and less as adhesives as such. So far, only a low number of applications has been transferred into industrial practice, such as the partial replacement of phenol in PF resins using (pre-methylolated) lignin (reportedly up to 80%) or the use of soy protein with additional crosslinking for plywood.

Looking to a possible reduced use of fossil resources, the wood-based panels industry would be a perfect opportunity, as this industry reutilizes already large amounts of natural secondary products from other industries (wood, agriculture) as well as waste wood and recycling wood in its operations. Approximately 80 - 90% of a wood-based panel are already such natural organic materials (wood, wood fibers, etc.); in few cases, such as hardboards, this percentage is even higher.

The high demand for adhesives with millions of tons per year is most challenging for bio-based adhesives, including consistency in availability. The list of natural raw materials available in such huge volumes and at reasonable price is short: lignin from pulping, carbohydrates from major crops, and protein from the high-volume oilseed crops, e.g., soy (353 million tons/year), rapeseed, sunflower seed, peanut, and cottonseed (40 to 70 million tons/year) [1, 54]. However, highly abundant and relatively inexpensive raw materials such as lignin or soybeans have so far required significant effort of manipulation or additional and higher-cost ingredients to deliver acceptable performance [55].

Questions and recommendations:

a) How can the limitations with bio-based adhesives, such as the low reactivity and the need for further crosslinking be overcome?
b) How can the reactivity of lignin enhanced when used as sole adhesive?
c) What activities and progress will be necessary to come to a real breakthrough using bio-based adhesives?

17.4 "Ecologically-Correct" (Synthetic) Adhesives, such as Aminoplastic Wood Adhesives Based on "Green Hydrogen"

Well and already long established synthetic adhesive resins might be an alternative in longer-term future, if they are produced in an "ecologically correct way" from the well-known monomers [56], meaning without any fossil raw materials, such as crude oil or natural gas. Such ecological monomers can be produced through natural synthesis paths, or they are based on, e.g., "green hydrogen" instead of so-called "synthesis gas" based on methane (natural gas).

One example could be a UF resin with extremely low formaldehyde emission, based on (i) methanol from fermentation (bio-methanol) or *via* "green hydrogen" and (ii) ammonia (based on "green hydrogen", for the urea production). Another alternative raw material in the future might be bio-gas instead of fossil-based methane (natural gas). Biogas is formed by decomposition of organic material in the absence of oxygen (air); however, due to the composition of biogas (50 – 80% CH_4 + 15 – 50% CO_2) still purification and up-concentration are necessary.

Questions and recommendations:

a) What will be the costs and prices for such adhesives? How can they be competitive to the established synthetic adhesives and bio-based adhesives?
b) What are and will be the capacities for the production of "green hydrogen", bio-methanol, and bio-gas?
c) Will the purity of bio-methanol meet the requirements for further process steps, e.g., the catalytic conversion to formaldehyde?

17.5 Activation of the Surface, Binderless Boards, Activation of Wood-Inherent Chemicals as Adhesives (Lignin, Carbohydrates/Hemicellulose)

Wood modification describes the application of chemical, physical, thermal, or biological methods (or combinations of such methods) to alter the properties especially at the wood surface. The aim is to increase the performance of wood, resulting in improvements in dimensional stability, decay resistance, or weathering resistance [57, 58].

17.5.1 Modification and Activation of the Wood Surface by Chemical and Thermochemical Treatments

Chemical treatments are performed not only to enhance dimensional stability of the panel, but also to improve physical and mechanical properties, as well as to yield a higher resistance against physical, chemical, and biological attacks and the following degradation of the various wood components. Chemical wood modification can be achieved by (i) bulk modification of the wood substance, including the lumen or (ii) modification only of the cell wall. An important aspect is bonding of modified wood. This can be solid wood, particles, fibers prepared from modified wood, or particles and fibers modified after their preparation.

Several technologies, such as acetylation, furfurylation, impregnation, thermal modification (e.g., thermo-hydro and thermo-hydro-mechanical processing), and surface densification have been successfully implemented in industry and demonstrate their potential [59, 60].

Surface modifications are restricted to treatment only of the first few cell layers of the wood surface, including changes in the chemical composition of wood building polymers and other compounds in wood. The target property to be improved is mostly the chemical activity of the surface in order to achieve good adhesion bonding [58]. Improving the dimensional stability of composite materials has long been a focus for the panel products industry. Conventional methods involve the chemical modification of starting materials, such as by acetylation.

Direct reaction with chemicals can activate the surface, forming free radicals on the surface of fibers, and yielding additional reactive sites for self-bonding. This improves the performance of binderless boards, such as hardboards, which can be produced without use of external adhesives. Such chemical surface activation of wood, i.e., of lignin and hemicellulose on the surface, uses oxidizing agents, strong acids, and caustic substances.

Activation of the wood surface means increasing bonding efficiency between the wood polymer chains and the adhesive [61]. This increase may be achieved by the addition of small amounts of reactive chemicals or by the application of radiation energy before spreading the adhesive. The oxidative removal of the low-energy surface film on wood, formed during storage by low-molecular oleophilic materials (e.g., waxes and fatty acids) is considered to be the most effective pre-treatment (removal of the CWBL). Part of the process of activation relates to the creation of reactive groups, primarily in the lignin of wood and, to a less extent, in the (hemi-) cellulose polymers. The ability for bonding is promoted by improved wettability and penetration of the adhesive and by enlarging the adhesive contact area [61].

Questions and recommendations:

a) Graft polymerization techniques were applied in order to activate wood surfaces chemically for the following cross-linking. Can this procedure be used in a cost competitive way for wood-based panels?
b) Chemical wood modification, such as acetylation, is seldom used in the production of medium density fiberboard (MDF). Can this procedure be used more often and for additional applications? What are the costs compared to, e.g.,

higher adhesive consumption when producing the same board quality?

17.5.2 Thermal and Physical Pre-Treatments of Wood

Thermal modification is a well-established commercial technology for improving the dimensional stability and durability of wood. With the increase of the heat treatment intensity, the mass loss is attributed to loss of moisture and degradation of volatile compounds as well as thermally less stable components. The thermal modification process makes wood more hydrophobic and dimensionally stable, providing protection against decomposition and thermo-degradation. This increases its chances for utilization and for further applications as solid wood with enhanced properties, or as wood particles and fibers in wood-based panels [62].

Questions and recommendations:

a) How can this procedure be used in a cost-competitive way for wood-based panels, since drying of particles, strands or fibers is performed anyway?
b) What problems can occur when bonding heat-treated wood?

17.5.3 Hydrothermal Treatment (Steam Pre-Treatment)

Hydrothermal treatment [63] refers to the application of high-temperature pressurized steam. During the steaming process, hemicelluloses are degraded into their constituent parts; in addition, lignin may be activated during this process. The use of hydrothermal treatment as a pre-treatment in the production of fiberboards might improve dimensional stability, and minimize irreversible swelling by permanent fixation of compressive deformation of wood fibers.

Steam treatment improves the dimensional stability of densified wood-based panels:

- pre-treatment of the wood causes a partial degradation of the hemicelluloses and reduces internal stresses,
- steam injection process, which means treatment of the wood (relaxing of internal stresses) and acceleration of the warming up process, but also influencing hardening of the resins,
- post-treatment of the boards.

Physical phenomena involving thermal softening of lignin have been suggested to be partially responsible for fiberboards formation. Mechanical properties of binderless fiberboards correlate to the hydrothermal history of the fibers during the board production process.

Hydrothermal treatment of wood leads to the formation of acetic acid liberated from the hemicelluloses, which further catalyzes carbohydrates cleavage, with final formation of formaldehyde, furfural and other aldehydes as well as of some aldehydes from lignin cleavage; this reaction is followed by lignin auto-condensation with formation of methylene bridge-connected aromatic rings. These reactions lead to an increase in crosslinking with consequent improvement in dimensional stability and decreased hygroscopicity of wood [64].

Questions and recommendations:

a) Can this procedure be used in a cost-competitive way for wood-based panels?
b) In fiberboard production, steam treatment takes place before the defibrillation. What effects are given? Can these effects be increased or improved?

17.5.4 Enzymatic Pre-Treatment of the Wood Surface

The objectives of biological wood modification are (i) the reduction or even avoidance of external adhesives by enhanced use of the inherent bonding effect of wood chemicals and (ii) the reduction of the energy consumption for the refining process by mycological disintegration of the wood structure. The activation of various components of the wood fibers by oxidizing enzymes, like laccases and peroxidases (phenoloxidases) can improve autoadhesion and enable binderless bonding (i.e., bonding without use of external adhesives) [65, 66]. Such treatment induces an increase in the reactivity of functional groups at the wood surface (especially formation of phenoxy-radicals) as well as the accessibility to these reactive sites for further bonding. The possible bond strength depends on the number of reactive groups present during bonding.

Enzyme-catalyzed adhesive bonding can be achieved (i) as two-component-system, where, e.g., lignin activated *via* oxidases is used as an adhesive and applied to the particles or the fibers, or (ii) as one-component system, where the oxidases react directly on the surface of the fibers before the hot-press process takes place, enhancing the level of autoadhesion between wood fibers.

Laccase (with oxygen as oxidant) or peroxidase enzymes (using peroxide as oxidant) can oxidize lignin. In both reactions phenoxy radicals are formed, hence creating fiber - fiber bonding (self-bonding) by polymerization and crosslinking of lignin, by creating covalent bonds and higher molar masses without disruption of the aromatic lignin backbone [67]. However, industrial-scale production has not yet been realized [68].

Questions and recommendations:

a) Will there be new and more efficient enzymes available?
b) What is the necessary time regime in the enzymatic pre-treatment of the wood surface?
c) How can the enzymatic pre-treatment be implemented in the usual production process of boards?

17.6 Binderless Boards, Activation of Wood-Inherent Adhesives (Lignin, Carbohydrates = Mainly Hemicellulose)

Binderless bonding means bonding without use of an external adhesive (adhesive-free bonding). Self-adhesion and binderless boards use inherent wood bonding forces based on *in situ* generation of adhesives and adhesive bondlines directly during the wood bonding process. The necessary adhesive is generated *in situ* from the wood material and its components during the bonding process, e.g., by activation of the various wood components on the surface, especially lignin and hemicellulose. This includes chemical activation as such (e.g., by forming radicals) as well as degradation of wood components to small and reactive compounds, such as furanic moieties; these are then the starting points for polymerization and polycondensation, creating polymers and resin-like substances.

The main bottleneck in all binderless bonding, however, is the very low reactivity, causing significant and sometimes even extreme prolongation of the necessary solidification times (curing and hardening times). Consequently, the needed press times often are extremely long, compared to the usual industrial production of wood-based panels using established synthetic wood adhesives. So far, the binderless technology is in use only for the production of so-called hardboards using the wet fiber process, with their high density and much longer press times compared to thin MDFs.

High temperatures can cause various reactions of the wood components and of inherent wood chemicals (extractives), such as degradation of hemicelluloses to furanic moieties or cleavage of bonds within lignin. Some of these reactions are the basis for binderless boards, whereby the activation of the wood components enables bonding. Pressing two wood surfaces against each other for long enough time and at sufficiently high temperatures (> 100°C) will always give some bonding effect, irrespective of the applied adhesive system or without external adhesives. This might be the chemical activation of the wood surface creating electric charges or radicals, or also melting of wood components enabling bonding effects by interpenetration (autoadhesion). Achieving industrially acceptable press times, however, requires strong acceleration of these processes.

Questions and recommendations:

a) How can binderless boards be produced in a cost-competitive way? This especially addresses the extremely long press times, as they are reported in literature.
b) How can the formation of the *in situ*-adhesive be accelerated?

17.7 Covalent Chemical Bonding between Wood Surface and Adhesive

The formation of covalent bonds would be the optimum for wood bonding, with the highest bond strengths on molecular level between the adhesive layer and the wood adherend. Covalent bonds might be formed at the interfaces between the adhesives and the wood surfaces or by directly linking the two wood surfaces together, preferably in case of longer-chain adhesive molecules. Chemical bonds might be formed, e.g., (i) between hydroxyl groups of the adhesive and of the cellulose or the hemicellulose, or (ii) between NCO-groups of isocyanate-based adhesives and various chemical moieties of the wood surface, mainly such as OH-groups. However, the bonds between the adhesive and the surface of the adherend are predominantly based on physical effects, such as secondary forces or hydrogen bonds. There is some evidence that adhesive-to-wood covalent bonds can form under specific experimental conditions; however, the conditions favoring such covalent bonds generally differ from the conditions commonly used in the industrial production of wood-based panels. This refers to the needed temperatures to enable chemical bonds as well as long press times, which in many cases are far outside of any industrial range. Only in the face layers (outer layers) of a panel with their direct contact with the hot-press platens, the temperatures in the mat/board reach up to 200°C, at least for a short time. Due to evaporation and the steam flow to the core, the moisture content is very low in these outer layers during this step. Under such conditions, covalent bonds might be formed in much higher proportion, e.g., based on reactions of NCO-groups with hydroxyl groups at the wood surface.

The existence as such and the proportion of chemical bonds in relation to the overall bonds in wood joining is still unclear. Modification of the wood surface might enhance formation of chemical bonds between an adhesive and the wood surface, e.g., by creating a higher number of polar sites at the surface of the adherend. It can be assumed that chemical bonds, if at all, will rather be reached for fiberboards due to the greater surface than for particleboards.

Questions and recommendations:

a) It is assumed that PMDI can bond to the primary hydroxyl groups on the wood structure, hence leading to covalent interactions [69]. How can the existence of such covalent links be verified?
b) Which processing conditions can help to increase the proportion of covalent bonds when joining wood surfaces by adhesives?
c) Can covalent links be achieved with a thermosetting adhesive resin? What pre-treatments of the wood surface would be beneficial?

17.8 Citric Acid

Citric acid has three carboxyl groups that can be ester-linked with the hydroxyl groups in wood. It activates and mobilizes polymeric wood constituents in dependence of temperature, forming ester bonds with the hydroxyl groups in the wood carbohydrates or with hydroxyl groups generated from carbohydrates by heat-induced decomposition, and thus forming crosslinks between wood constituents and at the interface. In addition, reactions between both the phenolic and aliphatic hydroxyl groups of lignin and citric acid are possible.

Since starch also contains considerable amount of hydroxyl groups and is more easily accessible to chemicals than cellulose, acids with several carboxyl groups, such as citric acid, can be expected to crosslink starch, improving mechanical strength and thermal stability and increasing the resistance against dissolution of starch films in water and formic acid [70].

The main problem is the low rates of the various reactions during board pressing in order to create a solid bondline by (i) establishing chemical bonds between the adhesive and the wood surface and (ii) forming an adhesive layer with sufficient cohesive strength. This causes extremely long press times.

Citric acid can also be directly applied onto wood surfaces for subsequent pressing to form boards, showing similar reactions as with sugar or starch. The basic chemistry of the activation mainly of the carbohydrates on the wood surface is their decomposition to furanic moieties, especially 5-HMF. The main disadvantages when using this treatment are (i) the high acidity (low pH) remaining in the bondline and the adjacent wood tissue, causing risk of acidic deterioration of the wood substance, and (ii) the extremely long press times.

Questions and recommendations:

a) How can the reaction of citric acid be accelerated, in order to enable a cost-efficient production of wood-based panels?
b) The remaining amount of citric acid, which has not reacted with the hydroxyl groups of the wood surface, causes a low pH in the bondline. Can wood degradation occur? How could such a low pH be neutralized?

17.9 Simulation of the Hot-Press Process

The hot-press operation is the most important and cost-intensive step in the manufacture of wood-based panels, where a mat of fibers/particles is compressed and heated to promote the cure of the adhesive. The properties of the boards strongly depend on the press conditions. Three independent mechanisms take place during hot-pressing, which are considered in computer-based simulation models [6, 39]:

a) coupled heat and water/moisture/steam transport through the porous mat of wood particles or fibers, causing quick changes of temperature, moisture content (MC), and steam pressure
b) densification of the material and development of internal stresses, and also relaxation processes of the structural elements (particles, fibers) due to the high temperatures attained during heating up

c) adhesion between the particles or fibers, followed by hardening of the adhesive (cohesion) and forming of the bond strength in the bondlines within the mat; even chemical reactions between the adhesive and the wood surface might occur.

· Due to the multiplicity and interdependency of various phenomena, such as heat and mass transfer, adhesive curing, and rheological behavior, the development of mathematical models that enable the simulation of the press process will enhance the understanding of the interactions and the effects of operating variables and parameters on the product properties and consumption of energy and time. This helps optimization, control, and scheduling of the press process [71].

Critical for the use of hot-pressing models is, among others, the availability of appropriate input data. Several physical-mathematical models describing effects in a panel during hot-pressing and simulating the hot-press process have been published [72–78]. Such simulation models have been established for batch processes (daylight presses) as well as continuous presses [76, 79, 80]. Simulation models can display all variables included in the model for any location and for the entire time span of the simulation.

Internal bond strengths were simulated using regression equations describing bond strength development with press time as measured by the Integrated Pressing and Testing System (IPATES) for a variety of different parameters, such as temperature, mat density, and adhesive content. The aim was to predict the bond strength development for a hot-press cycle [81].

Questions and recommendations:

a) Can simulations really describe the processes ongoing in reality? Or do simulation models rather help by qualitative support, but not delivering real absolute values?
b) Can simulation be used for online control in the production process?

17.10 Use of Wood-Based Panels (WBPs) in Construction, Resistance against Swelling and Shrinking, Elasticity of the Bondline

Changing climate conditions induce deformation and stress concentrations in bonded wood joints and WBPs. The wood is subjected to stress and strain because of moisture induced swelling and shrinking. The adhesive layer is affected by these stresses, responding with elastic and visco-elastic deformations and showing MC-dependent mechanical behavior (40). Destruction of boards under the influence of moisture and water can occur due to chemical degradation of the adhesive (hydrolysis) or due to movement of particles or fibers, causing stress peaks and micro-cracks.

The micrometer-scale structure of the interphase plays a critical role in stress distribution and load transfer from one part to another in a joint. As a general rule, the failure in bulk wood is preferred over other modes, which happens when adequate loads can be transferred between two bonded parts by the bondline from and into the adjacent wood [82]. Adhesion is determined as a force required for the separation of adhesive joints at their weakest point. Weak points can be the interface or the interphase. Besides bond strength, holding wood surfaces together by adhesion, the second role of adhesive bonds is to provide bond stiffness that could be influenced by the interphase formed due to the penetration of the adhesive. This important property could affect other properties like modulus or coefficient of moisture expansion [83]. In addition, adhesives can mitigate the negative influence of weak boundary layers.

Questions and recommendations:

a) What is the optimal compromise between stiffness and flexibility of a bondline?
b) How does the flexibility of the bondline influence the elasticity of wood-based panels?

17.11 Denaturation and Thermal Treatment of Proteins

Various treatments and modes of denaturation of proteins have been mentioned in literature, including various physical, chemical, and biological aspects, such as heat, acid/alkali treatment, urea (as efficient hydrogen bond breaker) or enzymes (causing denaturing by cleavage of chemical bonds). Extensive scientific dispute still exists on many aspects of denaturation, and fundamental clarification is needed [84–86].

The amino acids as protein components have many types of functionalized side chains. However, many of these reactive groups are not exposed to the surface, but are hidden inside the globules, making reactions difficult. Furthermore, the polar groups inside the protein form interactions with other polar groups to stabilize the protein structure, making reactions even more difficult with these polar side chains; hence, the native state cannot be expected to provide the best adhesive property. Denaturation is still a vague term and needs more clarification of the ongoing processes, before the impact on adhesive performance can be estimated.

Questions and recommendations:

a) What is the importance of the thermal history when considering the use of soy protein in adhesive systems?
b) What is the extent and efficiency of Maillard reactions between proteins and carbohydrates in terms of the need for higher bond strength?
c) What are the influences of the degree of denaturation and the proportion of soy protein (protein concentrate after removal of the soluble carbohydrates, protein isolate after removal of insoluble and soluble carbohydrates) on the bond strength?
d) Since denaturation is a vague and broad term, clear procedures and effects need to be clarified. This especially means monitoring of the structural changes in proteins and, hence, of the changes in the surface chemistry of the proteins.

17.12 Tailor-Made Wooden Products (Functional Boards)

Already 30 years ago, P.E. Humphrey thought about mimicking nature for the construction of wood-based panels and wood composites [87]. Most plant fibers with rigid walls consist of highly evolved arrangements of lignocellulose sub-elements. However, the impressive physical properties of these fibers are rather poorly utilized in present-day composites [87]. Nature's solutions to its engineering problems have tailored the internal microstructure and external shape of mammalian bones, e.g., to the mechanical and biological demands placed upon them, manipulating overall shape and the spatial distribution of properties therein. New types of molded engineering components incorporating spatially oriented and modified lignocellulose fibers could be the result. Products so derived may partially replace environmentally less attractive materials in a diverse range of engineering applications [87].

The structures of conventional wood-based composites (mainly panel products) are strongly influenced by physical mechanisms that occur during their consolidation (hot-pressing). Most components used in engineering applications are designed using shape as the principal variable. This usually follows some type of materials selection in light of the anticipated demands in service versus the attributes of suitable materials. Desired combinations of attributes can be very diverse, usually including load-bearing property, density and cost, but also often including thermal property, factors such as machinability, finishing, wear, and means of connection, as well as environmental aspects including life-cycle analysis. Mass-to-strength and mass-to-stiffness ratios under specific demand conditions are common performance indices when selecting materials [87].

Materials are in most cases considered homogeneous in nature, or at least orthotropic. This is in contrast to many materials and microstructures which occur naturally in plants and animals and which perform highly specialized functions, with even the most primitive plant genera showing nature's sophisticated solutions to the task of efficiency in design with some multi-functional and tailored solutions [88]. Examples to control both shape and properties in human-made objects are, e.g., fiber reinforced polymer products with mandrel windings or fabrics of carbon, glass, or boron. The lay-up configurations within such objects are usually selected in the light of stresses anticipated in service [87].

It is the most important aspect of structural design to use material wherever it is needed to fulfil the relevant task of the product or the construction (in large variety of sizes, from huge halls or roof constructions with laminated beams down to microscopic optimization of the structure of special wood joints). Wherever only limited contribution to mechanical strength is required, the material use (e.g., the density in a board) can be reduced. Well-known examples are the high densities in the face layers of wood-based panels, increasing modulus of elasticity (MOE) and modulus of rupture (MOR), or the holes in tubular particleboards. Another example is the density variation across the width of panels, with zones of higher density (e.g., at the cutting zones, when preparing smaller panels) and zones of lower density in between. Preferably, this can be implemented in continuous press systems with specialized forming stations [89].

Questions and recommendations:

a) Which new and unconventional ideas can be thought of when talking about future types of wood-based panels, looking at increased efficiency of the material input?

b) Can mimicking nature help in the optimization of adhesives and wood-based panels?
c) Which additional raw materials could be used in wood-based panels?
d) How to achieve sufficient performance of the boards (high homogeneity, targeted anisotropy) at the lowest overall weight (density) of boards?

17.13 Recycling and Debonding

The wood-based panel industry is in the best position for recycling wood and wood-based panels (post-consumer wood) as part of the circular economy, where CO_2 is kept within products as long as possible before emission due to burning. Waste furniture and other wood products have been successfully transformed into particleboards for many decades.

Difficulties in the recycling process can be caused by (small) portions of adhesive in the recycled material, and due to the fact that using polymers for adhesive bonds produces complex material combinations that are difficult to separate and to recycle. De-bonding or disassembly means the breaking of the polymer chains into shorter sub-units and monomers to aid maintenance or recycling operations. Disassembly at an appropriate time may be triggered either by chemically or physically-assisted approaches [90, 91].

De-bonding on-demand is based on the use of adhesive systems that allow de-bonding. "On-demand" means the adhesive joint stays stable under working conditions, but in the event of maintenance, repair, upgrading, or recycling, an external trigger can weaken the bonding resulting in de-bonding without damaging the substrates. Ideally, there is a clean separation of the adhesive from the substrate without remaining residues on the adherend surface, so lower amounts of residual adhesive need be removed [90]. Efforts to disintegrate waste panels into recycled particles or fibers can be classified into two approaches, i.e., mechanical [92, 93] and hydrothermal [94, 95]. The hydrothermal method separates the particles by hydrolyzing the adhesive between the particles [96]; hence, the recycled particles are intact with less damage [97].

Questions and recommendations:

a) Degradation of the adhesives used in wood-based panels is a precondition for proper recycling. What degradation procedures are suitable for wood-based panels?

b) Recycling of fibers from waste MDF is still in development and implementation stage on industrial scale, whereby various technologies for the disintegration of MDF are investigated; the aim is to gain fibers from this process, which are able to be used again in the panel production. How can MDF recycling be implemented on industrial scale?

17.14 Summary

This contribution to the "Festschrift" in honor of Dr. Kash Mittal's 80th birthday summarizes various topics, which are important in wood bonding. The selection of these topics was made based on the more than 40 years experience of the author in the adhesive and the wood-based panels industry, from his activities in many research projects, as lecturer at universities, and as author/editor of scientific papers and textbooks.

Despite the fact that wood bonding has been investigated for so many decades and has reached an incredible efficiency, thus enabling huge volumes of wood-based panels produced globally each year (393 million m^3 in 2024 [98]), still many questions are not yet answered; they should be motivation for researchers and scientists to work intensively on adhesives and wood joining. May this paper generate some additional ideas and projects in the future, in order to further optimize wood bonding and the production of wood-based panels.

AI disclosure statement: The author confirms that no artificial intelligence (AI)-generated content or AI-assisted writing tools were used in the preparation of this chapter (including summaries and the abstract). All text, analysis, and conclusions are the result of the author's original work.

References

1. European Concerted Research Action E13 (COST Action E13): *Wood Adhesion and Glued Products*. COST Association, Brussels (1998). https://www.cost.eu/actions/E13/
2. M. Dunky, A. Pizzi and M. van Leemput (Eds.), *Wood Adhesion and Glued Products*, Vol.1, Wood Adhesives, State-of-the-Art Report, COST Action E13, Office for Official Publications of the European Communities, Luxembourg (2003).
3. M. Dunky and P. Niemz, *Wood-based Panels and Resins: Technology and Influence Parameters*, Springer, Heidelberg, Germany (2002).

4. M. Dunky, *Technology of Adhesives and Wood-based Panels*, Wiley-Scrivener, Beverly, MA (2025).
5. M. Dunky and K.L. Mittal (Eds.), *Biobased Adhesives: Sources, Characteristics and Applications*, Wiley-Scrivener, Beverly, MA (2023).
6. M. Dunky, Wood adhesives based on natural resources: A critical review, Part I. Protein-based adhesives. *Rev. Adhesion Adhesives* 8, 199-332 (2020) and also in: *Progress in Adhesion and Adhesives*, Vol. 6, K.L. Mittal (Ed.), pp. 203-336, Wiley-Scrivener, Beverly, MA (2021).
7. M. Dunky, Wood adhesives based on natural resources: A critical review, Part II. Carbohydrate-based adhesives. *Rev. Adhesion Adhesives* 8, 333-378 (2020) and also in: *Progress in Adhesion and Adhesives*, Vol. 6, K.L. Mittal (Ed.), pp. 337-382, Wiley-Scrivener, Beverly, MA (2021).
8. M. Dunky, Wood adhesives based on natural resources: A critical review, Part III. Tannin-and lignin-based adhesives. *Rev. Adhesion Adhesives* 8, 379-525 (2020) and also in: *Progress in Adhesion and Adhesives*, Vol. 6, K.L. Mittal (Ed.), pp. 383-529, Wiley-Scrivener, Beverly, MA (2021).
9. M. Dunky, Wood adhesives based on natural resources: A critical review, Part IV. Special topics. *Rev. Adhesion Adhesives* 9, 189-268 (2021) and also in: *Progress in Adhesion and Adhesives*, Vol. 6, K.L. Mittal (Ed.), pp. 761-840, Wiley-Scrivener, Beverly, MA (2021).
10. M. Dunky, Wood adhesives based on natural resources: A critical review, Part V. Wood welding and binderless boards. *Rev. Adhesion Adhesives* 9, 308-367 (2021).
11. J.J. Bikerman, *The Science of Adhesive Joints*, 2nd edition, Academic Press, New York (1968).
12. R.J. Good, Theory of "cohesive" vs. "adhesive" separation in an adhering system. *J. Adhesion* 4, 133-154 (1972).
13. M. Stehr and I. Johansson, Weak boundary layers on wood surfaces. *J. Adhesion Sci. Technol.* 14, 1211-1224 (2000).
14. M. Stehr, J. Seltman and I. Johansson, Laser ablation of machined wood surfaces. 1. Effect on end-grain gluing of pine (*Pinus silvestris L.*) and spruce (*Picea abies Karst.*). *Holzforschung* 53, 93-103 (1999).
15. C.G. Hunt, C.R. Frihart, M. Dunky and A. Rohumaa, Understanding wood bonds – going beyond what meets the eye: A critical review. *Rev. Adhesion Adhesives* 6, 369-440 (2018), and also in: *Progress in Adhesion and Adhesives*, Vol. 4, K.L. Mittal (Ed.), pp. 353-419, Wiley-Scrivener, Beverly, MA (2019).
16. A. Rohumaa, T. Antikainen, C.G. Hunt, C.R. Frihart and M. Hughes, The influence of log soaking temperature on surface quality and integrity performance of birch (*Betula pendula* Roth) veneer. *Wood Sci. Technol.* 50, 463-474 (2016).
17. M. Gindl, A. Reiterer, G. Sinn and S.E. Stanzl-Tschegg, Effects of surface ageing on wettability, surface chemistry, and adhesion of wood. *Holz Roh Werkst.* 62, 273–280 (2004).

18. R.M. Nussbaum, Natural surface inactivation of Scots pine and Norway spruce evaluated by contact angle measurements. *Holz Roh Werkst.* 57, 419-424 (1999).
19. S.Q. Shi and D.J. Gardner, Dynamic adhesive wettability of wood. *Wood Fiber Sci.* 33, 58-68 (2001).
20. M. Petric and P. Oven, Determination of wettability of wood and its significance in wood science and technology: A critical review. *Rev. Adhesion Adhesives* 3, 121-187 (2015).
21. C. Piao, J.E. Winandy and T.F. Shupe, From hydrophilicity to hydrophobicity: A critical review: Part I. Wettability and surface behavior. *Wood Fiber Sci.* 42, 490-510 (2010).
22. M. Scheikl and M. Dunky, Measurement of dynamic and static contact angles on wood for the determination of its surface tension and the penetration of liquids into the wood surface. *Holzforschung* 52, 89-94 (1998).
23. M. Ohyama, B. Tomita and C.Y. Hse, Curing property and plywood adhesive performance of resol-type phenol-urea-formaldehyde cocondensed resins. *Holzforschung* 49, 87-91 (1995).
24. C.M. Chen, Effect of extractive removal on adhesion and wettability of some tropical woods. *Forest Prod. J.* 20(1), 36-41 (1970).
25. K.L. Mittal, The role of the interface in adhesion phenomena. *Polym. Eng. Sci.* 17, 467-473 (1977).
26. J. Konnerth, D. Harper, S.H. Lee, T.G. Rials and W. Gindl, Adhesive penetration of wood cell walls investigated by scanning thermal microscopy (SThM). *Holzforschung* 62, 91-98 (2008).
27. A. Marra, *Technology of Wood Bonding Principles in Practice*, Van Nostrand Reinhold, New York (1992).
28. F. Kamke and J.N. Lee, Adhesive penetration in wood – A review. *Wood Fiber Sci.* 39, 205-220 (2007).
29. C.R. Frihart, Wood structure and adhesive bond strength, in: *Characterization of the Cellulosic Cell Wall*, D. Stokke and L. Groom (Eds.), pp. 241-253, Blackwell Publishing, Ames, IA (2006).
30. D.J. Gardner, Adhesion mechanisms of durable wood adhesive bonds, in: *Characterization of the Cellulosic Cell Wall*, D. Stokke and L. Groom (Eds.), pp. 254-265, Blackwell Publishing, Ames, IA (2006).
31. D.J. Gardner, M. Blumentritt, L. Wang and N. Yildirim, Adhesion theories in wood adhesive bonding. *Rev. Adhesion Adhesives* 2, 127-172 (2014) and also in: *Progress in Adhesion and Adhesives*, Vol. 1, K.L. Mittal (Ed.), pp. 125-168, Wiley-Scrivener, Beverly, MA (2015).
32. J.B. Wilson, G.L. Jay and R.L. Krahmer, Using resin properties to predict bond strength of oak particleboard. *Adhesives Age* 22 (6), 26-30 (1979).
33. I. Gavrilovic-Grmusa, M. Dunky, J. Miljkovic and M. Djiporovic-Momcilovic, Radial penetration of urea-formaldehyde adhesive resins into beech (*Fagus Moesiaca*). *J. Adhesion Sci. Technol.* 24, 1753-1768 (2010).

34. I. Gavrilovic-Grmusa, J. Miljkovic and M. Djiporovic-Momcilovic, Influence of the degree of condensation on the radial penetration of urea-formaldehyde adhesives into silver fir (*Abies alba*, Mill.) wood tissue. *J. Adhesion Sci. Technol.* 24, 1437-1453 (2010).
35. I. Gavrilovic-Grmusa, M. Dunky, J. Miljkovic and M. Djiporovic-Momcilovic, Influence of the degree of condensation of urea-formaldehyde adhesive on radial and tangential penetration into poplar wood and on the shear strength of adhesive joints. *Holzforschung* 66, 849-856 (2012).
36. I. Gavrilovic-Grmusa, M. Dunky, J. Miljkovic and M. Djiporovic-Momcilovic, Influence of the degree of condensation of urea-formaldehyde adhesives on the tangential penetration into beech and fir and on the shear strength of the adhesive joints. *Eur. J. Wood Prod.* 70, 655-665 (2012).
37. I. Gavrilovic-Grmusa, M. Dunky, M. Djiporovic-Momcilovic, M. Popovic and J. Popovic, Influence of pressure on the radial and tangential penetration of adhesive resin into poplar wood and on the shear strength of adhesive joints. *BioResources* 11, 2238–2255 (2016).
38. M. Sernek, J. Resnik and F.A. Kamke, Penetration of liquid urea-formaldehyde adhesive into beech wood. *Wood Fiber Sci.* 31, 41-48 (1999).
39. A. Bastani, S. Adamopoulos, T. Koddenberg and H. Militz, Study of adhesive bondlines in modified wood with fluorescence microscopy and X-ray micro-computed tomography. *Int. J. Adhesion Adhesives* 68, 351-358 (2016).
40. W. Gindl, T. Schöberl and G. Jeronimidis, The interphase in phenol–formaldehyde and polymeric methylene di-phenyl-di-isocyanate glue lines in wood. *Int. J. Adhesion Adhesives* 24, 279-286 (2004).
41. D. Lukowsky, Influence of the formaldehyde content of water-based melamine formaldehyde resins on physical properties of Scots pine impregnated therewith. *Holz Roh Werkst.* 60, 349-355 (2002).
42. W. Wang and N. Yan, Characterizing liquid resin penetration in wood using a mercury intrusion porosimeter. *Wood Fiber Sci.* 37, 505-513 (2005).
43. C.R. Frihart, How wood adhesives work and where are the areas for improvement, in Proceedings of the International Conference on Wood Adhesives, pp. 50-63 (2013).
44. C.R. Frihart, Utility of Horioka's and Marra's models for adhesive failure, in Proceedings of the International Conference on Wood Adhesives, pp. 233-238 (2005).
45. B.H. River, Fracture of adhesive-bonded wood joints, in: *Handbook of Adhesive Technology*, A. Pizzi and K.L. Mittal (Eds.), pp. 151-177, Marcel Dekker, New York (1994) and *Handbook of Adhesive Technology*, second edition, A. Pizzi and K.L. Mittal (Eds.), pp. 325-350, CRC Press, Boca Raton, FL (2003).
46. P. Niemz and A. Hänsel, Deformation and rupture behavior of particleboards. *Holztechnologie* 3, 139-142 (1987).
47. Plywood – Bonding quality – Part 2: Requirements, EN 314-2 (1993).

48. J.M. Gagliano and C.E. Frazier, Improvements in the fracture cleavage testing of adhesively-bonded wood, *Wood Fiber Sci.* 33, 377-385 (2001).
49. P. Niemz and M. Diener, Comparative investigations for the determination of the fracture toughness of wood-based panels. *Holz Roh Werkst.* 57, 222-224 (1999).
50. U. Müller, A. Sretenovic, A. Vincenti and W. Gindl, Direct measurement of strain distribution along a wood bond line. Part 1: Shear strain concentration in a lap joint specimen by means of electronic speckle pattern interferometry. *Holzforschung* 59, 300-306 (2005).
51. D. Oberlercher, personal communication (2018).
52. C. Thoma, J. Konnerth, W. Sailer-Kronlachner, T. Rosenau, A. Potthast, P. Solt and H.W. van Herwijnen, Hydroxymethylfurfural and its derivatives: Potential as key reactant in adhesives. *ChemSusChem* 13, 5408-5422 (2020).
53. W. Sailer-Kronlachner, C. Rosenfeld, J. Konnerth and H.W. van Herwijnen, 5-Hydroxymethylfurfural-based adhesives: Challenges and opportunities, in: *Biobased Adhesives: Sources, Characteristics and Applications*, M. Dunky and K.L. Mittal (Eds.), pp. 477-498, Wiley-Scrivener, Beverly, MA (2023).
54. Statista, Worldwide oilseed production in 2023/2024. https://www.statista.com/statistics/267271/worldwide-oilseed-production-since-2008 (2022).
55. C.G. Hunt and M. Dunky, Analysis of future prospects and opportunities for wood adhesives: A review. *Forest Prod. J.* 72 (s2), 14-22 (2022).
56. M. Dunky, Wood adhesives based on natural resources: Challenges, in Proceedings 20th Munich Wood Colloquium about "Wood Adhesion: Fundamental Scientific Concepts" (2021).
57. D. Sandberg and A. Kutnar, *Wood Modification Technologies: Principles, Sustainability, and the Need for Innovation*, CRC Press, Boca Raton, FL (2021).
58. M. Petric, Surface modification of wood: A critical review. *Rev. Adhesion Adhesives* 1, 216-247 (2013).
59. D. Sandberg, A. Kutnar and G.I. Mantanis, Wood modification technologies - A review. *iForest* 10, 895-908 (2017).
60. G.I. Mantanis, Chemical modification of wood by acetylation or furfurylation: A review of the present scaled-up technologies. *BioResources* 12, 4478-4489 (2017).
61. D. Fan, J. Chang, J. Li, A. Mao and L. Zhang, ^{13}C-NMR study on the structure of phenol-urea-formaldehyde resins prepared by methylolureas and phenol. *J. Appl. Polym. Sci.* 112, 2195-2202 (2009).
62. V. Kamperidou, Chemical and structural characterization of poplar and black pine wood exposed to short thermal modification. *Drvna Ind.* 72, 155-167 (2021).
63. Z. Gao and R. Huang, Effects of pressurized superheated steam treatment on dimensional stability and its mechanisms in surface-compressed wood. *Forests* 13, 1230 (2022).

64. B.F. Tjeerdsma and H. Militz, Chemical changes in hydrothermal treated wood: FTIR analysis of combined hydrothermal and dry heat-treated wood. *Holz Roh Werkst.* 63, 102-111 (2005).
65. H.H. Nimz, Lignin-based adhesives, in: *Wood Adhesives, Chemistry and Technology*, Vol.1, A. Pizzi (Ed.), pp. 247-288, CRC Press, Boca Raton, FL (1983).
66. M. Dunky, Activation of wood surfaces and "binderless" wood composites, in: *Biobased Adhesives: Sources, Characteristics and Applications*, M. Dunky and K.L. Mittal (Eds.), pp.579-620, Wiley-Scrivener, Beverly, MA (2023).
67. C. Felby, M. Hasslingboe and M. Lund, Pilot-scale production of fiberboards made by laccase oxidized wood fibers: Board properties and evidence for crosslinking of lignin. *Enzyme Microbial Technol.* 31, 736-741 (2002).
68. A. Hüttermann, K. Nonninger and A. Kharazipour, Intermediate for the production of lignin polymerizates and their use in the production of derived timber products. WO Patent application 1998/31729, assigned to Pfleiderer Unternehmensverwaltung, Neumarkt, Germany (1998).
69. M.F. Sonnenschein, B.L. Wendt and G.F. Sonnenschein, Interfacial factors affecting polymeric diphenylmethane diisocyanate/wood bond strength. *J. Appl. Polym. Sci.* 98, 449-455 (2005).
70. N. Reddy and Y. Yang, Citric acid cross-linking of starch films. *Food Chem.* 118, 702-711 (2010).
71. L.M.H. Carvalho, Modelling of the hot pressing process, in: *Bonding of Timber, Working Group 3: Timber and Wood Bonding Process*, M. Properzi, D. Jones and C.R. Frihart (Eds.), Core Document COST Action E34, pp. 197-201, Lignovisionen, University of Natural Resources and Applied Life Sciences, Vienna, Austria (2008).
72. P.E. Humphrey and S. Ren, Bonding kinetics of thermosetting adhesive systems used in wood-based composites: The combined effect of temperature and moisture content. *J. Adhesion Sci. Technol.* 3, 397-413 (1989).
73. P.E. Humphrey and D. Zavala, A technique to evaluate the bonding reactivity of thermosetting adhesives. *J. Testing Evaluation* 17, 323-328 (1989).
74. L.M.H. Carvalho, M.R.N. Costa and C.A.V. Costa, A global model for the hot-pressing of MDF. *Wood Sci. Technol.* 37, 241-258 (2003).
75. B.G. Zombori, F.A. Kamke and L.T. Watson, Simulation of the internal conditions during the hot-pressing process. *Wood Fiber Sci.* 35, 2-23 (2003).
76. H. Thoemen and P.E. Humphrey, Modeling the continuous pressing process for wood-based composites. *Wood Fiber Sci.* 35, 456-468 (2003).
77. L.M.H. Carvalho, J. Martins and C. Costa, Transport phenomena, in: *Wood-based Panels: An Introduction for Specialists*, H. Thoemen, M. Irle and M. Sernek (Eds.), pp.123-176, Brunel University Press, London, UK (2010).
78. C. Goncalves, J. Pereira, N.T. Paiva, J.M. Ferra, J. Martins, F.D. Magalhaes, A. Barros-Timmons and L.M.H. Carvalho, A study of the influence of press parameters on particleboards' performance. *Eur. J. Wood Prod.* 78, 333-341 (2020).

79. H. Thoemen and P.E. Humphrey, Hot pressing of wood-based composites: Selected aspects of the physics investigated by means of simulation, in Proceedings of the 5th European Panel Products Symposium, pp. 38-49 (2001).
80. T.E. Fenton, H.M. Budman, M.D. Pritzker, E. Bernard and G. Broderick, Modeling and simulation of oriented strand board pressing. *Ind. Eng Chem. Res.* 42, 5229-5238 (2003).
81. C. Heinemann, Characterization of the curing of amino resins in a wood particle matrix by evaluation of strengths and reaction kinetic, Ph.D. Thesis, University of Hamburg, Hamburg, Germany (2004).
82. J.E. Jakes, C.R. Frihart, C.G. Hunt, D.J. Yelle, N.Z. Plaza, L. Lorenz, W. Grigsby, D.J. Ching, F. Kamke, S.C. Gleber, S. Vogt and X. Xiao, X-ray methods to observe and quantify adhesive penetration into wood. *J. Mater. Sci.* 54, 705-718 (2019).
83. E.A. Le and J.A. Nairn, Measuring interfacial stiffness of adhesively bonded wood. *Wood Sci. Technol.* 48, 1109-1121 (2014).
84. C.R. Frihart and L.F. Lorenz, Protein modifiers generally provide limited improvement in wood bond strength of soy flour adhesives. *Forest Prod. J.* 63, 138-142 (2013).
85. C.R. Frihart and L.F. Lorenz, Protein adhesives, in: *Handbook of Adhesive Technology*, A. Pizzi and K.L. Mittal (Eds.), 3rd edition, pp. 145-175, CRC Press, Boca Raton, FL (2018).
86. C.R. Frihart, Protein adhesives – Composition, structure and performance, in: *Biobased Adhesives: Sources, Characteristics and Applications*, M. Dunky and K.A. Mittal (Eds.), pp. 307-324, Wiley-Scrivener, Beverly, MA (2023).
87. P.E. Humphrey, Engineering composites from oriented natural fibers: A strategy, in: *The Chemistry and Processing of Wood and Plant Fibrous Material*, Cellucon '94 Proceedings, J.F. Kennedy, G.O. Phillips and P.A. Williams (Eds.), pp. 213-220, Woodhead Publishing, Elsevier, Amsterdam, The Netherlands (1996).
88. C. Mattheck, Why they grow, how they grow: The mechanics of trees. *Arboricultural J.* 14, 1-17 (1990).
89. A. Hemmilä, Method and apparatus for making particleboards having a non-uniform density distribution. EP 2653279, assigned to IKEA Supply AG, Pratteln, Switzerland (2016).
90. A.C. Marques, A. Mocanu, N.Z. Tomic, S. Balos, E. Stammen, A. Lundevall, S.T. Abrahami, R. Günther, J.M.M. de Kok and S. Teixeira de Freitas, Review on adhesives and surface treatments for structural applications: Recent developments on sustainability and implementation for metal and composite substrates. *Materials* 13, 5590 (2020).
91. M.D. Banea, Debonding on demand of adhesively bonded joints: A critical review. *Rev. Adhesion Adhesives* 7, 33-50 (2019).
92. J. Izdinsky, Z. Vidholdova and L. Reinprecht, Particleboards from recycled wood. *Forests* 11, 1166 (2020).

93. A. Laskowska and M. Maminski, The properties of particles produced from waste plywood by shredding in a single-shaft shredder. *Maderas, Ciencia Tecnol.* 22, 197-204 (2020).
94. F.Y. Bütün Buschalsky and C. Mai, Repeated thermo-hydrolytic disintegration of medium density fiberboards (MDF) for the production of new MDF. *Eur. J. Wood Prod.* 79, 1451-1459 (2021).
95. A. Nuryawan, K.D. Tambun, I. Risnasari and N. Masruchin, Hydrolysis of particleboard bonded with urea-formaldehyde resin for recycling. *Heliyon* 6, e03936 (2020).
96. B.D. Park and H.W. Jeong, Hydrolytic stability and crystallinity of cured urea-formaldehyde resin adhesives with different formaldehyde/urea mole ratios. *Int. J. Adhesion Adhesives* 31, 524-529 (2011).
97. C. Lykidis and A. Grigoriou, Quality characteristics of hydrothermally recycled particleboards using various wood recovery parameters. *Int. Wood Prod. J.* 2, 38-43 (2011).
98. Food and Agriculture Organization of the United Nations (UN FAO), Production volumes, https://www.fao.org/faostat/en/ (2025).

18

Soy Protein Structures and Wood Adhesive Performance Properties

Charles R. Frihart

USDA Forest Products Laboratory, Madison, WI, USA

Abstract

Formaldehyde emissions from urea-formaldehyde bonded wood products have led to consumer health concerns and to companies looking for no-added formaldehyde adhesives. Most of soy flour/meal is not consumed by humans and has a reasonable cost leading to research on soy wood adhesives. In the literature there is considerable valuable information about soy modification, but there are also important weaknesses in the understanding of fundamental protein structure. Nature developed non-structural proteins to be compact dispersions (rice-like) for dealing with very crowded cell environments making the common speculation that the hydrophobic proteins can become completely uncoiled (spaghetti-like) in water unreasonable. Changes in the internal protein structure (denaturation) can influence the soy colloidal properties and improve the wood adhesive performance by further hydrophobic/hydrophilic aggregation of proteins, but these changes do not eliminate the ability of the protein to absorb water and lower wet strength for the bonded wood products. Crosslinking of the proteins in soy products by adding other chemicals has been shown to play an important role in improving the wet strength of soy-bonded wood products at commercially viable levels. Thus, the model of tying together the dispersed particles, like sticky rice, has been accepted by food researchers and fits with the results in the adhesive literature.

***Keywords*:** Soy, protein, wood, adhesive, agglomeration, water-resistance, folding, dispersion

Email: charles.frihart@usda.gov; cfrihart4@gmail.com

Anil N. Netravali and Frank M. Etzler (eds.) The Wide World of Adhesion: A Festschrift in Honor of Dr. Kash Mittal on his 80th Birthday, (623–642) © 2026 Scrivener Publishing LLC

18.1 Background

The very low-cost urea-formaldehyde adhesives have worked well for decades in producing interior wood panels (plywood, particleboard, and fiberboard) that meet the needs of producers and consumers, but these adhesives release formaldehyde by depolymerization under elevated heat and humidity conditions, providing an opportunity for protein adhesives to displace the urea-formaldehyde adhesives [1]. Of the available proteins, soy leads the list because the flour is available in large volumes at reasonable cost for a high protein content agricultural product [2, 3]. With proteins being approximately 45 % of the flour by weight, the remainder contains a variety of carbohydrates and other compounds that contribute to dry bond strength but lower wet bond strength. The protein content can be enhanced by aqueous ethanol extraction to yield soy concentrate (approximately 75% protein) or a two-step purification that terminates with precipitation of the protein to yield soy protein isolate (SPI) (>90% protein) [4, 5]. Although the flour is inexpensive enough for use as a wood adhesive, the high cost of the purified proteins due to the processing steps and the low yield make them too expensive for use in wood bonding. While examining the properties of purified components often helps to understand the chemistry and properties of mixtures [6], the commercial soy protein isolates (CSPI) have been so highly modified for performance in food applications that they do not provide a model for the native proteins in the flour [5]. The research on native soy protein isolate has mainly been limited to two groups: Sun's group at Kansas State University and the wood adhesive group at the Forest Products Laboratory. For understanding soy protein adhesives, an appreciation of the fundamentals of protein behavior using prior research in food science and general biochemistry is valuable [6] as are other general protein publications [7–9]. The food research is important since commercial soy products were developed to meet the desired properties for the food applications, and most of the general literature on protein structure is also related to this application.

18.2 Protein Structure

Each native protein has a well-defined folded structure due to its sequence of amino acids [6–9]. There can be some natural variation of the amino acid sequences if it does not alter the overall protein folding [10–14]. This sequence of amino acids is known as the primary structure and it controls the secondary structure of helices and sheets, and the overall folding of the tertiary structure [7, 12–16]. The α-helix and β-sheets are controlled by certain sequences of amino acids [9, 16, 17]; the helices are stabilized by hydrogen bonds within a segment, and the sheets use hydrogen bonds between different segments of the same protein chain. The helices and sheets provide stabilization to the overall tertiary structure and are not disturbed until enough energy is applied to unfold these elements, which is referred to as the denaturation temperature [18]. In addition to the sheets and helices, after the hydrophobic collapse the tertiary protein structure is stabilized by hydrogen, polar, hydrophobic, disulfide, and acid-base bonding between individual pairs of atoms within the chain and can be altered by the pH, ionic strength, and chaotropic agents in the external aqueous environment. Because the proteins are mainly hydrophobic due to many of amino acids' side chains being aliphatic or aromatic groups and the backbone amide groups being shielded [7], they minimize their water interactions by folding into a globular shape [9–18]. The entropy of developing this tertiary structure is overcome by minimizing the interactions of hydrophobic residues with water environment and developing internal hydrogen bonding in addition to acid-base and disulfide bonds. The development of this native structure is described as an energy well [10, 12, 13, 19]. However, the sequence of amino acids does not let all the hydrophobic groups fold to the interior, leaving some of them on the surface [14, 17].

The individual chains can lower their energy further by associating with other protein globules to form a stable association classified as the quaternary structure. In the case of soy proteins, these are mainly the glycinin and conglycinin, which are, respectively, the 11S and 7S by size separation [20–22] (S is a Svedberg unit measured as sedimentation coefficient). These aggregates stay together under the basic and acidic conditions used to purify the SPI. X-rays can be used to generate the 3-D image of the 11S protein crystals with their well-ordered structure of 2-stacked hexagons, with each having six proteins (3 acidic and 3-basic) [20–24]. The 7S is not simple since it is made of 3 different protein aggregates, consisting of individual proteins which are often glycosylated [20, 22–24].

Although the helices and sheets are important structural elements of the protein configuration, these elements are physically connected to the rest of the protein chain by polar bonds and lipophilic attractions. About 50% of the amino acids in a globular protein are in a regular secondary structure (helices and sheets) with the remainder in defined structures such as turns and loops, and other connectors [25, 26]; the turns are often aided by proline amino acids due to their cyclic structure. In the wood adhesive literature, the tendency is to consider that the proteins are completely unfolded to become a free-floating spaghetti-like polymer without explaining how the hydrophobic proteins can be stabilized in this extended form surrounded by water molecules [27–32]. When the temperature is very high or other extreme conditions, they can become uncoiled, but upon cooling return to the colloidal state [33, 34].

If the proteins find it difficult to completely unfold, how do the small protein globules (rice-like particles) bridge the millimeter distances between wood surfaces. First, as pointed out previously, the individual proteins aggregate with other proteins to form quaternary structures [17, 25, 34–38]. The driving force for these structures, in some cases, is interactions of polar groups, but in general polar surface interactions can be easily disrupted by water. On the other hand, the folding of the proteins results in some hydrophobic patches on the globular surface as illustrated in Figure 18.1 for more water-resistant hydrophobic attractions that result in the main protein aggregates of glycinin and conglycinin. Even these protein aggregates are not large enough to bridge the mm gap between the wood surfaces, since the individual crystals are about 10 nm in diameter [24]. At room temperature, the soy protein colloids are like sticky rice due to the hydrophobic surface interactions with high affinity for each other until enough shear is provided. Additionally, they are likely to form larger aggregates in more concentrated dispersions. The associations of proteins have been attributed to both hydrophilic and hydrophobic interactions between the surfaces of different proteins [25, 39–41]. It seems natural that the hydrophobic associations are likely to be less water sensitive because these interaction sites are kept away from water by the large bulk of the proteins. As illustrated in Figure 18.1, the protein surfaces are not smooth so that

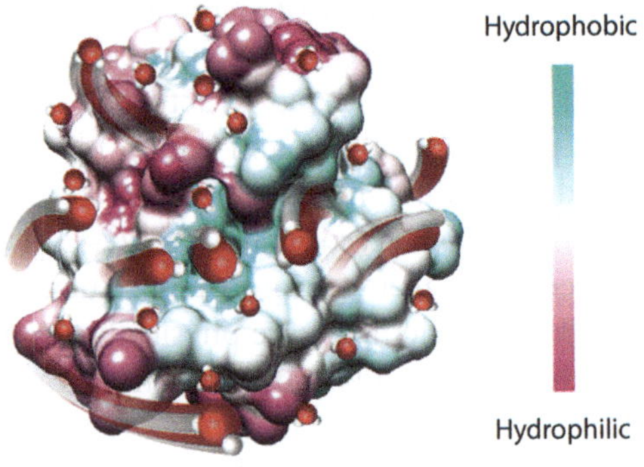

Figure 18.1 Protein surface showing polar and non-polar domains along with irregular surface and water molecules (red balls) with blue representing hydrophobic areas and purple representing hydrophilic domains (https://han.chem.ucsb.edu/research/role-hydration-water-molecular-recognition).

surfaces with the right topologies can form a tighter fit in between protein globules. An important property is that the protein globules develop higher order associations to form chains of beads [21]. This issue is explored more fully in another paper on soy protein adhesives [37].

18.3 Wood Adhesive Bond Strength

The largest adhesives market is in bonding wood for making panel products, such as particleboard and plywood, and structural products, like wood I-joists and glulam, and interior products, including cabinets and furniture. In some applications, the products not only need to meet dry strength, but in many applications, there are also wet strength requirements to account for the fact that wood expands and shrinks with variation in water exposure, causing internal strain on the bondline and adding to any externally applied force [78].

Proteins are natural adhesives, with some having good wet strength such as in the case of attaching mussels to different surfaces [39, 40]. These adhesives use a specific amino acid dihydroxyphenylalanine (L-dopa) that would be hard to obtain and expensive to use commercially for bonding wood. In contrast, plant proteins are widely available, especially soy proteins [2, 37, 42]. In the literature, several different tests have been used to measure the strength of wood bonds under dry conditions and sometimes after water exposure. The Forest Products Laboratory has been using the ABES (Automated Bond Evaluation System) test as described in ASTM D-7998 to evaluate protein-wood bond strength [43, 44] due to its use of thin, smooth veneers, rapid attainment of the set bondline temperature, low sensitivity to dispersion viscosity, and less time consuming to run, additionally giving a similar failure mode as other standard shear wood bond tests for interior plywood products [45, 46], but it is not a replacement for established product performance tests. The protein and other chemicals are dispersed in water at predetermined concentrations, spread on the wood and then bonded in a hot press before testing dry and wet after water soaking. However, because the wet tests are more difficult for adhesives to pass than the dry tests, with the wet tests often leading to bondline failure rather than the wood failure with the dry tests, the reported data has concentrated on the wet strengths [5, 45–47].

Soy flours come in several degrees of denaturation with those from Cargill being the native 90 PDI (protein dispersibility index), slightly denatured 70 PDI, and highly denatured 20 PDI, based on dry heat exposure conditions. PDI measures the percent of the protein that is replaced by non-settling soluble in water after vigorous dispersion in water, followed by centrifugation as described in the American Oil Chemists Society's Protein Dispersibility Index (PDI) Ba 10b-09 test. As illustrated in Figure 18.2, there is not much difference in bond strength with varying degree of denaturation although the denatured proteins showed slightly improved wet strengths [48]. These flours have good dry strengths but do not come close to lower limit of 3 MPa for suitable wet strength; however, improved performance is observed at higher bonding temperatures [46], but bonding at 120°C is the upper limit for 2-5 minutes used for commercial applications. Additionally, the bondline temperatures are lower than 120°C due to greater insulation effect for thicker wood veneers, use of greater number of plies, heat required for evaporating the water in the adhesive

Thus, the soy flour is not suitable by itself under typical commercial bonding conditions, but some improvement was observed if the flour was processed to raise the protein content from about 45% to greater than 90% by removing the soluble and insoluble carbohydrates [5, 47]. However, an improvement was observed, especially in wet strength, again if the bonding temperature was increased (Figure 18.2) [46]. Two soy protein isolates SPIs were tested; the laboratory one isolated under milder conditions and a commercial one that had been jet cooked [5]. The higher temperature in improving the wet bond strength is a protein effect as shown in Figure 18.3 for both the native and jet-cooked soy protein isolates compared to Figure 18.2 for the flours; the effect is present also with added PAE (polyamidoamine-epichlorohydrin) crosslinker [46]. The explanations for this effect

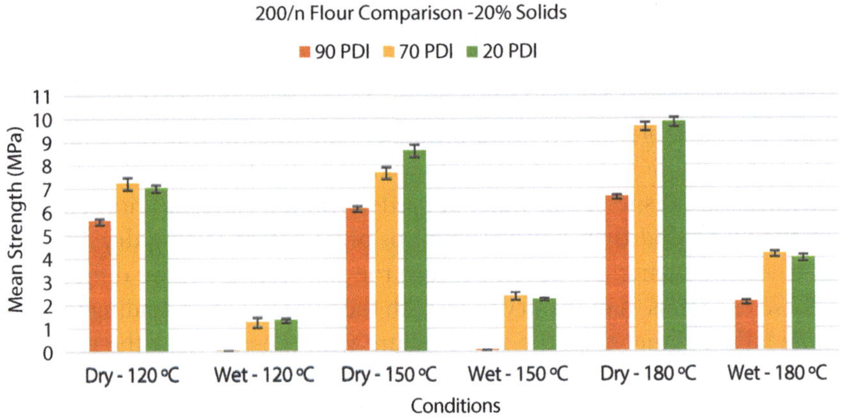

Figure 18.2 ABES bond strength before and after water soaking for soy flour of n=20, 70 and 90 PDI of 200 mesh particle size and at 20 wt% flour aqueous dispersion, using bonding temperatures of 120, 150 and 180°C [46].

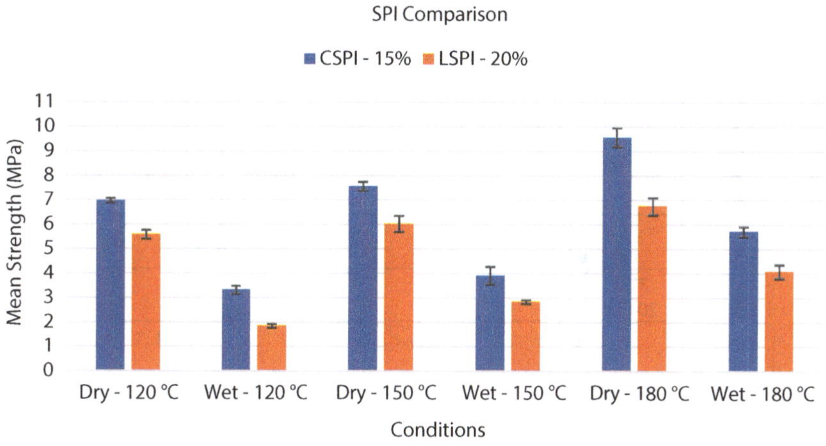

Figure 18.3 The effect of bonding temperature on dry (blue bars) and wet (red bars) strength, tested with maple veneers using ABES with both native (laboratory prepared soy protein isolate, LSPI) and commercial soy protein isolates (CSPI) [46].

can be that the proteins are rearranging either to form stronger bonds between individual protein globules, to entangle the protein chains near the surface, or to provide some protein crosslinking reaction. Since the purified proteins have a similar result with higher temperature bonding as the flour, the result is unlikely to involve carbohydrate-protein reactions: thus, Maillard crosslinking is unlikely.

Since increased heating in the bonding step was shown to increase the bond strength, as well as dry heating of the flour before bonding had an effect (denatured 20 PDI versus native 90 PDI in Figure 18.2), we wondered why did heating affect the bonding of the soy products and did this explain the difference between commercial and native SPI (soy protein isolate) [5, 47]. Figure 18.4 summarizes our findings about the difference between flours and isolates before and after two types of heating (wet autoclaving and jet cooking) [5]. The data show that while wet autoclaving improved the strength, this strength is considerably below that of the jet cooking results. Jet cooking also involves wet heating, but the heating time is shorter, the shearing of the sample is much higher, and the cooling of the sample is very rapid compared to the autoclave heating and cooling. Thus, jet cooking provides a very different soy protein structure that results in the strongest soy bond with no additives. However, this strength is unlikely to be due to entanglement of the protein chains since autoclave heating of the commercial SPI causes a significant reduction of the wet bond strength of the jet cooked SPI. Figure 18.4 also shows that either adding or

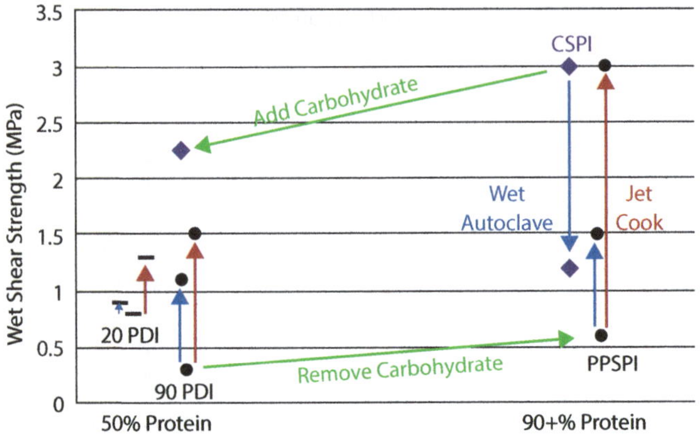

Figure 18.4 The wet strengths of bonds made with different soy formulations with emphasis on protein content from 45% for flour to 95% for SPI, and exposure to different heating conditions (autoclave and jet cooking prior to adhesive formulation). Starting materials: Black circle, 90 PDI flour; black bar, 20 PDI flour; purple diamond, CSPI (PROFAM® 974) and native isolate PPSPI made in the pilot plant. The arrows show the effect of different heating conditions and either adding carbohydrates to SPI or removal from soy flower. Source: [5].

removing the carbohydrates causes only a small change in the wet strength. Both of these heating processes increased the size of the gel particles of the uncured soy as exemplified by the higher viscosities for the dispersions [5].

18.4 Soy Dispersion Rheology

Protein rheology leads to a better understanding of the higher-order structures of the proteins [48–53]. The standard Brookfield cup-and-bob viscosity can indicate differences in the soy flours, but the shear history needs to be kept constant to be able to compare the measured values due to the agitation altering the viscosity value [48]. A better understanding of the protein structure is obtained using dynamic rheology that measures the properties starting from low shear that is ramped over time and separates the solid-like elastic data from the liquid-like flow data [51, 52]. This method shows that soy dispersions are initially thinned by both time-dependent and shear-dependent factors until a steady state condition when measured at low shear rates [53]. These changes are assumed to be caused by increased dispersion of the soy flour into smaller particles. These amplitude-sweep data were obtained using an Anton Paar MCR302

(Graz, Austria) with a CC27 cup and bob measuring system. The 90 PDI native, partially denatured 70 PDI, and the highly denatured 20 PDI flour show a similar steady state storage and loss modulus before dramatically decreasing (Figure 18.5) [53]. The globular structure of native state 90 PDI and a hypothetical extended linear (uncoiled) structure for the denatured 20 PDI soy would not be expected to have very similar curves contrary to what was observed; the uncoiled protein would not have a large initial plateau but show a steady decrease in viscosity with increasing shear. This data is consistent with the native and denatured soy containing larger associations of globular proteins held together by polar and non-polar surface interactions until enough force is applied to break them apart into component globules. The 20 and 70 PDI flours showed a more elastic response (difference between storage and loss moduli) initially compared to the 90 PDI flour as would be expected by a less compact globule after loss of the minimum volume for the native state. Additionally, both loss and storage moduli of the dispersion increased with the denaturation. One explanation is that the less compact denatured protein is taking up a larger volume, leaving less solvent available for flowing, making the entire mixture more rigid. Another interesting point is that the denaturing of the flour increases the amount of shearing needed to break up the initial rigid gel structure. The unusual second shear-thinning step indicates that while the first step breaks apart the gel structure to create a flowing liquid, the second

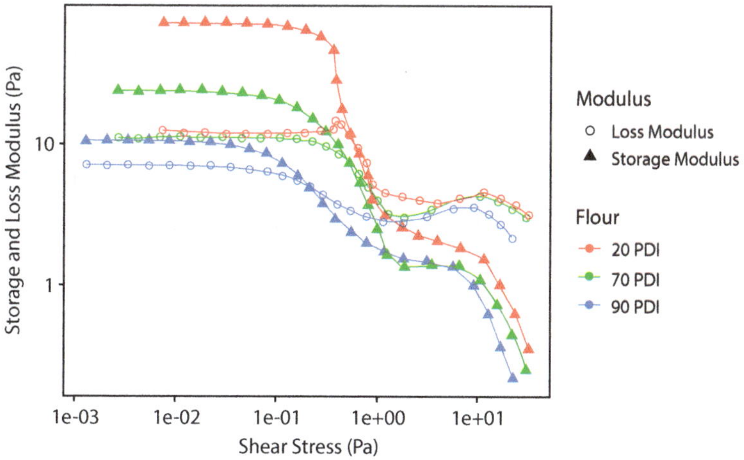

Figure 18.5 Amplitude sweep data for 20 (red), 70 (green), and 90 (blue) PDI flours at 20% solids in dynamic rheology experiments with the closed triangles showing the storage moduli and the open circles showing loss moduli. Only single replicates are shown for clarity, but the curves are quite reproducible [53].

transition mainly affects the storage aspect, but not the loss (flow) character of the dispersion, indicating a further fracture of the protein globules to give smaller particles.

One question is: whether the curve for the flours represents only the effect of the protein, or whether the other components of the flour have an effect on the rheology. Besides the protein, the other main components of the flour are the soluble and insoluble carbohydrates [5, 47, 48]. The soluble fraction consisting of low molecular-weight sugars and proteins is not likely to significantly influence the overall rheology. On the other hand, the insolubles, consisting of higher molecular-weight carbohydrates and proteins, are likely contributors to overall rheological properties of the flour because these insolubles showed significant storage and loss moduli when the insolubles were tested at the same concentration as in the flour (Figure 18.6) [53]. The native SPI made in the pilot plant (PPSPI) tested at the same concentration as in the flour looks a lot like the 90 PDI flour in the shear energy required to break the solid-like structure although the moduli are lower probably due to the overall lower solids content. This supports the assertion that the protein contributes most to the flour rheological properties.

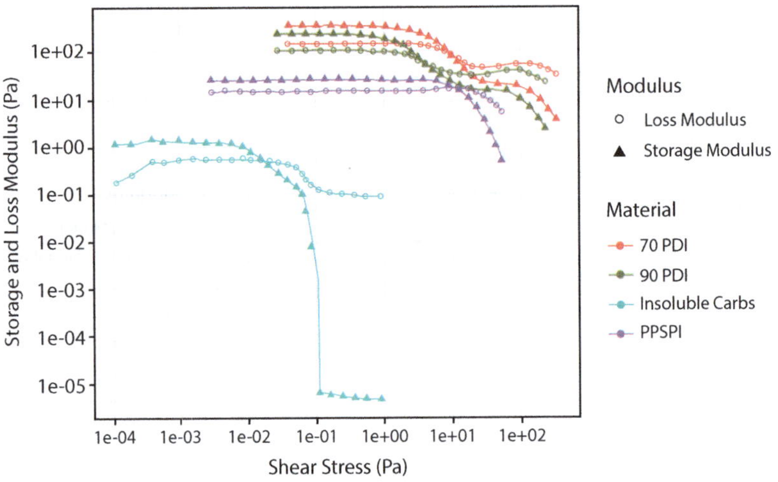

Figure 18.6 Storage and loss moduli versus applied stress for two soy flours (red for 70PDI and green for 90 PDI)) , at 30 wt. % native soy protein isolate in purple, and soy insoluble carbohydrate fraction in blue (at 15 and 7.5 wt. %, respectively, representing their proportional contribution to the soy flour). The closed triangles show the storage moduli and the open circles show loss moduli [53].

18.5 Protein Structure Revisited

Despite some wood adhesive literature claiming that the soy protein is completely unfolded in the denatured state [21, 27–32], most protein research has considered the unfolding to be a transient state between the original state and a new denatured state [7, 19, 54–63]. The arguments against soy protein staying unfolded are listed below:

- The proteins are naturally predominately hydrophobic, and many of the treatments, especially the most common thermal conditions, do not alter the portion of hydrophobic groups, but these may move more hydrophobic groups to the surface leading to greater aggregation and decreased solubility.
- For an unfolded protein to be stable, all the hydrophobic sections of the protein would need to be bound to water, resulting in the loss of the naturally strong water-water hydrogen bonds as the protein penetrates into the water domain. This would be thermodynamically very unfavorable.
- If the protein were extended, the entangled chains with the many interchain hydrogen bonds would provide a very strong bond as is exhibited by silk which does not need added crosslinkers to provide strength.
- Because of the molecular weight of 20 kDa or greater for the protein subunits [20, 42], unfolding would lead to very high viscosities for highly unfolded proteins, which are only observed at very high pHs.
- The dynamic rheology of the 20 and 70 PDI denatured flours have similar profiles to the native 90 PDI flour, which would not be expected if the denatured structures were unfolded [50, 51].
- The thermal denaturation of soy has shown that individual proteins form new combinations of aggregated subunits [62].
- The general protein literature supports the idea of the unfolded protein as just a transient state until conditions are appropriate for refolding [12–14, 19, 63–70].

Denaturation often leads to the formation of new and larger gel aggregates [25, 26, 55, 67, 69–71]. Thermal denaturation has been well studied since it is used in food preparation and leads to gelation [36, 56, 65–69, 71]. The common model is that the intermediate cluster is a string of protein beads that becomes an interconnected gel network, as illustrated in Figure 18.7. The hydrated gel network for the cured soybean was observed by carrying out nanoindentation on a water soaked bondline made with a soybean adhesive. Despite the soy being cured under normal conditions, the indent in the bondline disappeared after being made; this is what would be expected for a gel structure where the indenter pushed water into adjacent cells and flowed back after removal of the indentation pressure. This makes more sense than the proposal of just increasing aggregate sizes [25].

Although bond strength can be formed from the linkage of proteins either by polar or hydrophobic bonds, the polar bonds are more likely to be broken by water exposure. The hydrophobic gelation hypothesis for the soy hydrocolloid is supported by the studies that show that the curing of ovalbumin from egg whites occurs by flipping the structure's surface from a polar to a non-polar configuration [68, 69, 72, 73]. In addition, from Figure 18.7, it is easy to visualize how the addition of a cross-linker can improve the strength of the gel structure by making additional or reinforcing existing connections between gel particles. This allows the data in the literature to be correct even though the papers incorrectly portray the soy protein structure.

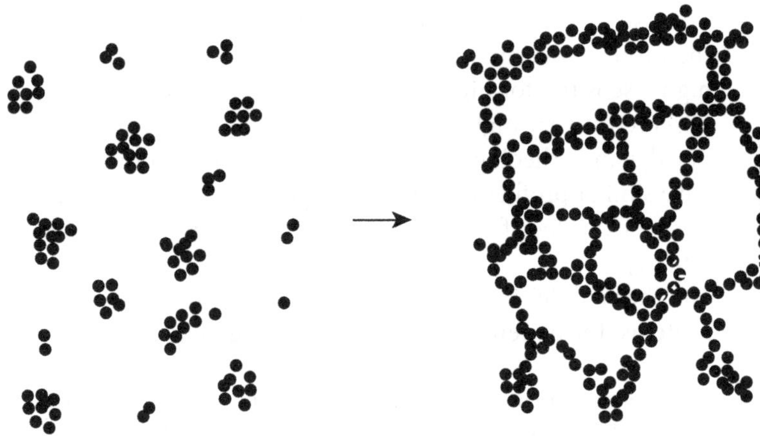

Figure 18.7 A hypothesis of soy protein curing by forming cluster of protein globules that becomes a porous gel with water filling the voids, taken from Oakenfull *et al.* [68].

18.6 Conclusion

Much of the wood adhesive literature contains incorrect information about the three-dimensional structure of denatured soy proteins, but in certain regards this does not negate the experimental data on performance properties using various crosslinking methods [37]. The unaltered fact is that the soy proteins do not form strong water-resistant wood bonds within the adhesive without some cross-linking between the proteins unless jet cooked. Some have claimed that the use of denaturing chemicals is sufficient to form strong bonds [21, 42, 74], but this has been contradicted by studies at the Forest Products Laboratory [75].

The reversible folding reaction differs from an ordinary chemical reaction in that generally no covalent bonds are made or broken when a protein unfolds and re-folds except for that some proteins are stabilized by covalently formed disulfide bonds that can be broken; the protein just re-equilibrates to different three-dimensional states in response to changed chemical and/or physical conditions that either disfavor or favor certain folded states [6, 12, 13, 18, 19, 36, 62–66]. Some unfolded state is a necessary step in going from the native structure to one of many folded denatured structures. Because proteins are predominately hydrophobic, the structures try to minimize water exposure of the hydrophobic domains by adopting globular shapes and by forming quaternary and higher-order structures to shield the surface hydrophobic domains from the water. Contrary to some of the literature, a fully extended state is unlikely to be feasible, and if it did exist, the proteins would form a highly physically crosslinked structure eliminating the need for cross-linking chemicals for water resistance.

Commercially available dry thermally denatured soy, such as 20 PDI flour, give only some improvement in wet bond strength. In contrast, the wet denatured soy isolate from heating in an autoclave or jet cooking gave improved wet strength, especially the latter [5]. It is not surprising that the presence of water makes it easier to increase gel size. Dynamic rheology showed that the native 90 PDI flour gave shear-thinning curves similar to the partially denatured 70 PDI and highly denatured 20 PDI, except that the latter ones had higher moduli, probably by occupying a greater volume [50].

The literature describes the denatured proteins as existing as a gel made using strings of beads of protein globules [68], with gelation existing as a lattice structure that can swell and weaken in the presence of water.

Crosslinking chemicals can improve wet strength by reinforcing the lattice and continue to be an active area of research [76, 77].

18.7 Summary

The importance of using soy proteins as wood adhesives has been hindered by an incorrect vision of denatured soy protein structures. Based on food and general protein literature, denaturation does not produce an uncoiled, linear structure but produces a highly aggregated structure of more loosely coiled globules. Understanding the food protein literature is important since the commercial soy products were developed for that market. In contrast, much of the soy adhesive literature incorrectly assumes complete opening of the protein structure. As an analogy, the soy adhesives in the normal pH range are rice-like rather than spaghetti-like. Thus, studies on soy adhesives should focus on dealing with the hydrocolloidal and gel properties of soy.

AI disclosure statement: The author confirms that no artificial intelligence (AI)-generated content or AI-assisted writing tools were used in the preparation of this chapter (including summaries and the abstract). All text, analysis, and conclusions are the result of the author's original work.

References

1. C.R. Frihart, J.M. Wescott, T.L. Chaffee and K.M. Gonner, Formaldehyde emissions from urea-formaldehyde- and no-added-formaldehyde- bonded particleboard as influenced by temperature and relative humidity, *Forest Prod. J.* 62, 551-558. (2012).
2. B. Adhikari, P. Appadu, M. Chae and D. Bressler, Protein-based wood adhesives: Current trends of preparation and application, in: *Bio-based Wood Adhesives: Preparation, Characterization, and Testing*, Z. He (Ed.), pp. 1-58, CRC Press, Boca Raton, FL (2017).
3. X. Sun, Isolation and processing of plant materials, in: *Bio-Based Polymers and Composites*, R.P. Wool and X. Sun (Eds.) pp. 33-55 Elsevier, New York (2005).
4. G. Qi, N. Li, X.S. Sun and D. Wang, Adhesion properties of soy protein subunits and protein adhesive modification, in: *Bio-based Wood Adhesives: Preparation, Characterization, and Testing*, Z. He (Ed.) pp. 59-85, CRC Press, Boca Raton, FL (2017).

5. C.G. Hunt, L.F. Lorenz, C.J.Houtman, E. Valle, T. Coolidge, C. Mock and C.R. Frihart, Jet cooking dramatically improves the wet strength of soy adhesives. *J. Amer. Oil Chem. Soc. 100,69–79* (2023).
6. S. Damodaran, Food proteins: An overview, in: *Food Proteins and Their Applications*, S.Damodaran and A.Paraf (Eds.), pp. 1-24, Marcel Dekker, New York (1997).
7. H. Jakubowski and P. Flatt, *Fundamentals of Biochemistry Vol. I - Structure and Catalysis, Libre Texts*, UC-Davis, CA (2022).
8. B. Alberts, A. Johnson, J. Lewis, D. Morgan, M. Raff, K. Roberts and P. Walter, Proteins. *Molecular Biology of the Cell*, 120-172 (2015).
9. C. Branden and J. Tooze, *Introduction to Protein Structure,* 2nd ed. p. 410, Garland Science, New York (1999).
10. K.A. Dill and J.L. MacCallum, The protein-folding problem, 50 years on. *Science, 338*, 1042-1046 (2012).
11. S.W. Englander and L. Mayne, The nature of protein folding pathways. *Proc. Natl. Acad. Sci. U.S.A., 111*, 15873-15880 (2014).
12. R. Nassar, G.L. Dignon, R.M. Razban and K.A. Dill, The protein folding problem: The role of theory. *Journal of Molecular Biology 433*, 167126 (2021).
13. R.H. Pain, *Mechanisms of Protein Folding*; Oxford University Press USA, New York (2000).
14. H. Jakubowski and P. Flatt, The three-dimensional structure of proteins. in: *Fundamentals of Biochemistry, Vol. I - Structure and Catalysis*, H. Jakubowski and P. Flatt (Eds.) Libre Texts, UC-Davis, CA (2022).
15. H.H.J. de Jongh, Globular proteins, in: *Progress in Biotechnology*, W.Y. Aalbersberg, R.J. Hamer, P. Jasperse, H.H.J. de Jongh, C.G. de Kruif, P. Walstra and F.A. de Wolf (Eds.), Vol. 23, pp. 31-86. Elsevier, Amsterdam, Netherlands (2003).
16. H. Jakubowski and P. Flatt, Secondary structure and loops, in: *Fundamentals of Biochemistry,* Vol. *I - Structure and Catalysis*, H. Jakubowski and P. Flatt (Eds.) Libre Texts, UC-Davis, CA (2022).
17. H. Jakubowski and P. Flatt, Tertiary and quaternary structures, in: *Fundamentals of Biochemistry Vol. I - Structure and Catalysis*, H. Jakubowski and P. Flatt (Eds.) Libre Texts, UC-Davis, CA (2022).
18. H. Jakubowski and P. Flatt, Protein folding and unfolding (denaturation) – dynamics. in: *Fundamentals of Biochemistry Vol. I - Structure and Catalysis*,. H. Jakubowski and P. Flatt (Eds.) Libre Texts, UC-Davis, CA (2022).
19. R.D. Schaeffer and V. Daggett, Protein folds and protein folding. *Protein Engineering, Design & Selection, 24*, 11-19 (2011).
20. X.S. Sun, Thermal and mechanical properties of soy proteins, in: *Bio-Based Polymers and Composites*, R.P. Wool and X.S. Sun (Eds.) pp. 292-326, Elsevier-Academic Press, Burlington, MA (2005).
21. X.S. Sun, Soy protein polymers and adhesion properties. *Journal of Biobased Materials and Bioenergy 5*, 1–24 (2011).

22. S. Utsumi, Y. Matsumura and T. Mori, Structure-function relationships of soy proteins, in: *Food Proteins and Their Applications*, S. Damodaran and A .Paraf (Eds.), pp. 257–291, Marcel Dekker, New York (1997).
23. X. Sui, T. Zhang and L. Jiang, Soy protein: molecular structure: Revisited and recent advances in processing technologies. *Annual Review of Food Science and Technology, 12*, 119-147 (2021).
24. S. Petruccelli and M.C. Anon, Thermal aggregation of soy protein isolates. *J. of Agricultural and Food Chem., 43*, 3035-3041 (1995).
25. X.S. Sun, D.H. Wang, L. Zhang, X.Q. Mo, L. Zhu and D. Bolye, Morphology and phase separation of hydrophobic clusters of soy globular protein polymers. *Macromolecular Bioscience*, 8, 295-303 (2008).
26. H. Jakubowski and P. Flatt, Main chain conformations, in: *Fundamentals of Biochemistry Vol. I - Structure and Catalysis*, H. Jakubowski and P. Flatt (Eds.) Libre Texts, UC-Davis, CA (2022).
27. Z. Cui, Y. Xu, G. Sun, L. Peng, J. Li, J. Luo and Q. Gao, Improving bond performance and reducing cross-linker dosage of soy protein adhesive *via* hyperbranched and organic and inorganic hybrid structures. *Nanomaterials, 13*, 203 (2023).
28. X. Huang, Y. Chen, X. Lin, J. Li and Q. Gao, A strong soy protein-based adhesive with excellent water retention. *Chemical Engineering Journal, 472*, 145037 (2023).
29. K .Jiang, Y. Chen, J. Qi, J. Xie, X. Huang, Y. Jiang, S. Zhang, S. Jia, Q. Chen and H. Xiao, A high wet strength and toughness soy-based adhesive prepared by single cross-linking. *Int. J. Adhesion Adhesives, 125*, 103406 (2023).
30. Z. Liu, T. Liu, H.X. Jiang, Zhang, J. Li, S.Q. Shi and Q. Gao, Biomimetic lignin-protein adhesive with dynamic covalent/hydrogen hybrid networks enables high bonding performance and wood-based panel recycling, *Int. J. Biol. Macromol., 214*, 230-240 (2022).
31. Y. Xiong, X. Xi, Z. Wu, H. Lei and G.Du, Application of a self-crosslinking strategy based on oxidization modification for the preparation of soy-based adhesive. *Int. J. Adhesion Adhesives.* 126, 103433 (2023).
32. Y. Xu, Y. Han, Y. Li, J. Luo, J. Li, J. Li and Q. Gao, Nacre-inspired construction of soft–hard double network structure to prepare strong, tough, and water-resistant soy protein adhesive, *J. Appl. Polymer Sci., 139*, 52202 (2022).
33. L. Amagliani and C. Schmitt, Globular plant protein aggregates for stabilization of food foams and emulsions. *Trends in Food Sci. & Technol. 67*, 248-259 (2017).
34. C. Akkermans, A.J. Van Der Goot, P. Venema, H. Gruppen, J.M. Vereijken, E. Van Der Linden and R.M. Boom, Micrometer-sized fibrillar protein aggregates from soy glycinin and soy protein isolate. *J. Agricultural Food Chem., 55*, 9877-9882 (2007).
35. X. Wang, Z. He, M. Zeng, F. Qin, B. Adhikari and J. Chen, Effects of the size and content of protein aggregates on the rheological and structural properties

of soy protein isolate emulsion gels induced by CaSO4. *Food Chem.*, *221*, 130-138 (2017).
36. K. Nishinari, Y. Fang, S. Guo and G.O. Phillips, Soy proteins: A review on composition, aggregation and emulsification, *Food Hydrocolloids*, *39*, 301-318 (2014).
37. C.R. Frihart and C. Hunt, Protein structure and wood adhesives. *Industrial Crops and Products* (2025).
38. S.-I. Han and T. Webber, Role of hydration water on molecular recognition. Available online: https://han.chem.ucsb.edu/research/role-hydration-water-molecular-recognition (accessed on Jan. 20, 2022).
39. N. Bandara, H. Zeng and J. Wu, Marine mussel adhesion: biochemistry, mechanisms, and biomimetics, *J. Adhesion Sci. Technol.*, *27*, 2139-2162 (2013).
40. H.G. Silverman and F.F. Roberto, Understanding marine mussel adhesion. *Marine Biotechnology* 9, 661-681 (2007).
41. D. Fukushima, Structures of plant storage proteins and their functions. *Food Rev. Int.*, *7*, 353-381 (1991).
42. X.S. Sun, Soy protein adhesives, in: *Bio-Based Polymers and Composites*, R.P. Wool and X.S. Sun (Eds.), pp. 327-368, Elsevier-Academic Press, Burlington, MA (2005)
43. ASTM International. D 7998-19 Standard Test Method for Measuring the Effect of Temperature on the Cohesive Strength Development of Adhesives using Lap Shear Bonds under Tensile Loading. *Vol. 15.06* (2020).
44. C.R. Frihart and L. Lorenz, Standard Test Method ASTM D 7998-19 for the Cohesive Strength Development of Wood Adhesives. *J. Visualized Experiments, 159* (2020).
45. C.R. Frihart, B.N.Dally, J.M. Wescott and M.J. Birkeland, Bio-based adhesives and reliable rapid small scale bond strength testing, in: Proceedings of the International Symposium on Advanced Biomass Science and Technology for Bio-based Products, pp.364-370 (2009).
46. C.R. Frihart, T. Coolidge, C. Mock and E. Valle, High bonding temperatures greatly improve soy adhesive wet strength. *Polymers* 8, 394 (2016).
47. L. Lorenz, M. Birkeland, C. Daurio and C.R. Frihart, Soy flour adhesive strength compared with that of purified soy proteins, *Forest Prod. J.*, *65*, 26-30 (2015).
48. C.R. Frihart and H. Satori, Soy flour dispersibility and performance as wood adhesive, *J. Adhesion Sci. Technol.*, 27, 2043-2052 (2013).
49. S. Amin, G.V. Barnett, J.A. Pathak, C.J. Roberts and P.S. Sarangapani, Protein aggregation, particle formation, characterization & rheology. *Current Opinion in Colloid & Interface Sci.*, *19*, 438-449 (2014).
50. V.L. Dharmaraj, P.D. Godfrin, Y. Liu and S.D. Hudson, Rheology of clustering protein solutions. *Biomicrofluidics*, *10*, 043509 (2016).
51. T.G. Mezger, *Applied Rheology*, Vincentz Network, Hannover, Germany (2014).

52. T.G. Mezger, *The Rheology Handbook*, Vincentz Network, Hannover, Germany (2014).
53. C.R. Frihart and M. Gargulak, Use of dynamic shear rheology to understand soy protein dispersion properties. *Polymers, 14*, 5490 (2022).
54. J.I. Boye, C. Ma and V.R. Harwalkar, Thermal denaturaton and coagulation of proteins, in: *Food Proteins and Their Applications*, S. Damodaran and A. Paraf (Eds.), pp. 25–56, Marcel Dekker, New York (1997).
55. J.N. Pedersen, J. Lyngsø, T. Zinn, D.E. Otzen and J.S. Pedersen, A complete picture of protein unfolding and refolding in surfactants. *Chemical Science, 11*, 699-712 (2020)
56. S. Damodaran, Refolding of thermally unfolded soy proteins during the cooling regime of the gelation process: Effect on gelation. *J. Agricultural and Food Chem., 36*, 262-269 (1988).
57. A.V. Glyakina and O.V. Galzitskaya, How quickly do proteins fold and unfold, and what structural parameters correlate with these values? *Biomolecules, 10*, 197 (2020).
58. S.-J. Chen, M. Hassan, Jernigan, K. Jia, D. Kihara, A. Kloczkowski, S. Kotelnikov, D. Kozakov, J. Liang, A. Liwo, S. Matysiak, J. Meller, C. Micheletti, J.C. Mitchell, S. Mondal, R. Nussinov, K. Okazaki, D. Padhorny J. Skolnick, T.R. Sosnick, G. Stan, Il. Vakser, X. Zou and G.D. Rose, Protein folds vs. protein folding: Differing questions, different challenges. *Proc. Natl. Acad. Sci. U.S.A.*, 120, 1-4 (2023).
59. L.J. Smith, K.M. Fiebig, H. Schwalbe and C.M. Dobson, The concept of a random coil: Residual structure in peptides and denatured proteins. *Folding and Design 1*, R95-R106 (1996).
60. H. Jakubowski and P. Flatt, Aggregates - amyloids, prions and intracellular granules in: *Fundamentals of Biochemistry, Vol. I - Structure and Catalysis*, H. Jakubowski and P. Flatt (Eds.) Libre Texts, UC-Davis, CA (2022).
61. D. March, B. Valentino and F. Giancarlo, Protein unfolding and aggregation near a hydrophobic interface. *Polymers, 13*, 156 (2021).
62. B. German, S. Damodaran, and J. Kinsella, Thermal dissociation and association behavior of soy proteins, *J. Agricultural Food Chem.* 30, 807-811 (1982).
63. S. Damodaran, Refolding of thermally unfolded soy proteins during the cooling regime of the gelation process: Effect on gelation, *J. Agricultural Food Chem.* 36, 262-269 (1988).
64. Q. Ruan, Y. Chen, X. Kong and Y. Hua, Heat-induced aggregation and sulphydryl/disulphide reaction products of soy protein with different sulphydryl contents. *Food Chem., 156*, 14-22 (2014).
65. I.S. Chronakis, Network formation and viscoelastic properties of commercial soy protein dispersions: effect of heat treatment, pH and calcium ions. *Food Res. Int. 29*, 123-134 (1996).
66. J.E. Kinsella, Functional properties of soy proteins. *J. Am. Oil Chem. Soc. 56*, 242-258 (1979).

67. T. Nakamura, S. Utsumi and T. Mori, Network structure formation in thermally-induced gelation of glycinin. J. Agric. Food Chem., *32*, 349-352 (1984).
68. D. Oakenfull, J. Pearce and R.W. Burley, Protein gelation, in: *Food Proteins and Their Applications*, S. Damodaran and A. Paraf (Eds.), pp. 111-142, Marcel Dekker: New York (1997).
69. J.M.S. Renkema, Relations between rheological properties and network structure of soy protein gels, *Food Hydrocolloids*, *18*, 39-47 (2004).
70. S. Utsumi and J.E. Kinsella, Forces involved in soy protein gelation - effects of various reagents on the formation, hardness and solubility of heat-induced gels made from 7s, 11s, and soy isolate, *J. Food Sci.*, *50*, 1278-1282 (1985).
71. L. Zheng, J.M. Regenstein, L. Zhou and Z. Wang, Soy protein isolates: A review of their composition, aggregation, and gelation. *Comprehensive Rev. Food Sci. Food Safety*, *21*, 1940-1957(2022).
72. L.F. Lorenz and C.R. Frihart, Ovalbumin has unusually good wood adhesive strength and water resistance, *J. Appl. Polym. Sci.*, *140*, e53332 (2023).
73. H.Y. Hu and H.N. Du, α-to-β Structural transformation of ovalbumin: heat and pH effects, *J. Protein Chem.*, *19*, 177-183 (2000).
74. Z. Zhong, X. Sun, X. Fang and J. Ratto, Adhesion strength of sodium dodecyl sulfate-modified soy protein to fiberboard. *J. Adhesion Sci. Technol.*, 15, 1417-1427 (2001).
75. C.R. Frihart and L. Lorenz, Protein modifiers generally provide limited improvement in wood bond strength of soy flour adhesives, *Forest Prod. J.*, *63*, 138-142 (2013).
76. M. Dunky, Natural crosslinkers for naturally-based adhesives, in: *Biobased Adhesives: Sources, Characteristics, and Applications,* M. Dunky and K.L. Mittal (Eds.), pp. 207-254, Wiley-Scrivener , Beverly, MA (2023).
77. M. Dunky, Wood adhesives based on natural resources: A critical review, Part I. Protein-based adhesives. *Rev. Adhesion Adhesives* 8, 199-332 (2020).
78. C.R.Frihart, Adhesive groups and how they relate to the durability of bonded wood, *J. Adhes. Sci. Tech.*, 23, 601–617, 2009.

19

Advances in Dental Adhesives: Current Innovations, Challenges, and Future Directions

Isadora Martini Garcia[1], Lamia Sami Mokeem[2], Abdulrahman A. Balhaddad[3], Fabrício Mezzomo Collares[4] and Mary Anne S. Melo[1*]

[1]Department of Comprehensive Dentistry, Division of Cariology and Operative Dentistry, University of Maryland School of Dentistry, Baltimore, MD, USA
[2]Department of Restorative Dentistry, Dental University Hospital, King Saud University Medical City, Riyadh, Riyadh Province, Saudi Arabia
[3]Department of Restorative Dental Sciences, College of Dentistry, Imam Abdulrahman Bin Faisal University, Dammam, Eastern Province Saudi Arabia
[4]Dental Materials Department, School of Dentistry, Federal University of Rio Grande do Sul, Porto Alegre, RS, Brazil

Abstract
Failure of tooth-colored restorations due to secondary caries and material fractures remains a major challenge in restorative dentistry. Hydrolytic degradation and bacterial acid attacks at the tooth-restoration interface compromise the durability of bonded restorations. This chapter explores the evolution and current advancements in dental adhesives, emphasizing key developments to improve their performance and longevity. Special attention is given to bioactive dental adhesives incorporating antibacterial compounds, highlighting their clinical relevance in preventing recurrent caries and enhancing restoration durability. Furthermore, the chapter highlights the need for interdisciplinary collaboration to develop next-generation adhesive materials capable of addressing the complex biological and mechanical challenges of the oral cavity.

***Keywords*:** Dental adhesives, nanotechnology, dental materials, composites, dental caries

**Corresponding author*: MMelo@umaryland.edu

Anil N. Netravali and Frank M. Etzler (eds.) The Wide World of Adhesion: A Festschrift in Honor of Dr. Kash Mittal on his 80th Birthday, (643–668) © 2026 Scrivener Publishing LLC

19.1 Introduction

Dental caries remains one of the most widespread oral health conditions globally. In 2021, it was among the leading oral diseases, with untreated caries of permanent teeth affecting an estimated 3.69 billion people worldwide. The global age-standardized prevalence reached approximately 27,500 cases per 100,000 population, highlighting the substantial burden this condition poses to public health and overall well-being [1]. In addition, it causes a heavy financial burden globally [2]. For example, in 2016, around $124 billion was spent on dental treatments in the United States alone [3, 4].

Dental caries is a complex disease caused by multiple factors, with a key role played by the buildup of biofilm and an imbalance in the natural community of microorganisms in the mouth—known as the oral microbiome [2]. When this balance is disrupted, certain bacteria in dental plaque produce acids that lead to the gradual loss of minerals from the tooth surface, a process called demineralization [5–7]. This acidic environment favors the growth of acid-tolerant bacteria, known as cariogenic bacteria, which further intensifies the damage. These harmful bacteria thrive in low-pH (acidic) conditions and contribute to dysbiosis, an unhealthy shift in the microbial ecosystem [8]. Oral biofilms, where this process unfolds, are made up of over 700 different bacterial species, along with a mix of salivary proteins, peptides, and food particles that form a protective matrix around the bacteria [9].

Data from the National Health and Nutrition Examination Survey (2005–2008) indicate that more than one in five individuals had untreated dental caries, while approximately 75% had existing dental restorations. The prevalence of untreated dental caries varied significantly by poverty level across all age groups; however, the prevalence of dental restorations showed slight variation with poverty status among children and adolescents aged 5–19 years. Additionally, 27% of children and adolescents in this age group received at least one dental sealant. Among adults, nearly 38% of non-Hispanic Black individuals had not lost a permanent tooth, compared with 51% of non-Hispanic White individuals and 52% of Mexican-American individuals. Furthermore, almost 23% of adults aged 65 and over were edentulous [10]. However, despite advancements in adhesive dentistry, many restorations still fail over time due to secondary caries, polymerization shrinkage, and degradation at the adhesive interface.

In dentistry, the "adhesive interface" or "bonding interface" refers to the junction between the tooth structure and the resin composite (Figure 19.1). In most restorations, this interface is formed by etched dentin that has been infiltrated with resin monomers from an adhesive system. These monomers polymerize *in situ*, creating a micromechanical interlock within the collagen fiber scaffold of the dentin. Unfortunately, this interface is prone to degradation over time, and frequent replacement of failed restorations contributes to substantial healthcare costs, with billions of dollars spent annually on restorative and replacement procedures [11, 12].

Resin composites are widely used to restore missing tooth structures due to their excellent aesthetic qualities, mechanical strength, and biocompatibility. However, their tendency to accumulate biofilms can compromise their durability, often leading to secondary caries. Also known as recurrent caries, this type of decay develops at the interface between the restoration and the natural tooth, where bacterial infiltration can occur if the seal is compromised. Studies indicate that a significant portion of these restorations fail within ten years [14, 15], contributing to the growing challenge of replacement dentistry, which costs an estimated $5 billion annually in the USA alone [16, 17]. The primary cause of failure is this type of caries, typically forming at the restoration margins where bacteria can thrive and demineralize the surrounding tooth structure [12, 18].

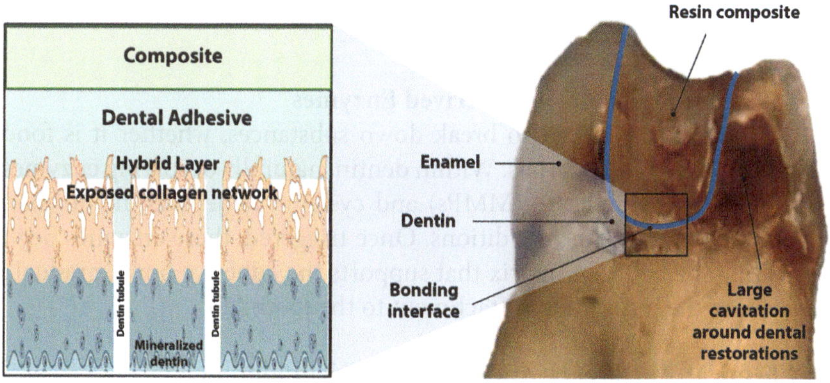

Figure 19.1 A schematic representation illustrating a failed bonded restoration (right). The image highlights the key components of a restored tooth and shows the occurrence of secondary caries surrounding the restoration. The blue line marks the bonding interface between the composite and the tooth structure. On the left, a magnified view details the typical elements of the dentin bonding interface [13].

19.2 Understanding the Challenges of Dental Adhesives inside the Mouth

Imagine a construction project in one of Earth's most hostile environments, an area constantly exposed to moisture, extreme temperature changes, and relentless bacterial attacks. Now, shrink that environment down to the size of a tooth; this is the harsh reality dental adhesives face daily inside the mouth.

Dental restorations use adhesive materials to bond them securely to the tooth structure. However, the oral cavity is far from ideal for maintaining these bonds. It is a warm environment, with temperatures ranging from 0°C to 70°C, and is constantly exposed to saliva and to frequent pH shifts caused by acidic foods and drinks. On top of this, it endures significant mechanical stress from chewing, grinding, and even speaking [11, 12, 18, 19].

A critical factor affecting the longevity of dental adhesives is their interaction with the teeth. Teeth are not smooth, dry surfaces like glass or metal; they are dynamic, porous, and hydrated structures. Dentin, the inner layer of the tooth, is like a sponge filled with microscopic tubules that transport fluids from the pulp, making it a challenging surface for adhesives to adhere to. This starkly contrasts with enamel, the outermost protective layer, which provides a more predictable bonding surface. The actual test of dental adhesives is their ability to withstand the harsh realities of the oral environment over time. Unfortunately, several factors conspire against them (Figure 19.2):

a) The Enemy Within: Host-Derived Enzymes

The human body is built to break down substances, whether it is food, bacteria, or foreign materials. Within dentin, naturally occurring enzymes, matrix metalloproteinases (MMPs) and cysteine cathepsins, lie dormant until activated by acidic conditions. Once triggered, these enzymes begin degrading the collagen matrix that supports the adhesive bond, gradually weakening the restoration's attachment to the tooth [20].

b) Chemical Warfare: Hydrolysis and Biodegradation

Dental adhesives rely on polymer chemistry to form strong, durable bonds. However, many of these polymers contain ester bonds, which are particularly vulnerable to hydrolytic degradation. In Figure 19.2, the "hydrolysis of dental monomers" refers to this degradation process. Saliva contains enzymes like cholesterol esterase and pseudocholinesterase, which break

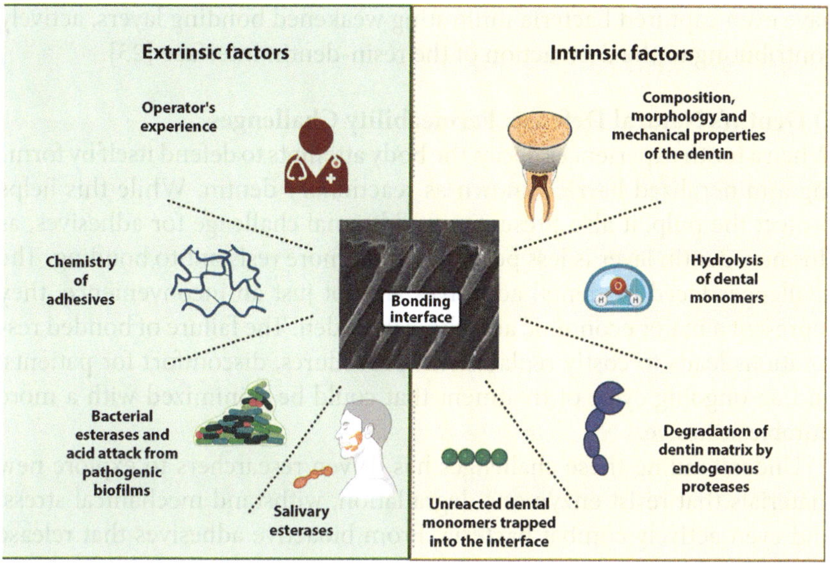

Figure 19.2 An overview of the factors contributing to the deterioration of the resin-dentin interface [13].

down these bonds over time, leading to a gradual disintegration of the adhesive layer. Additionally, the natural moisture in dentin can lead to water uptake by the adhesive material, causing it to expand and gradually weaken the polymer structure [21, 22].

c) The Battle of Forces: Mechanical Stress and Fatigue

Every time you bite, chew, or grind your teeth, you subject your dental fillings to cyclic mechanical forces. Unlike a rigid structure, teeth have a natural elasticity, meaning that restorations and adhesive bonds must endure continuous flexing and compressive forces. Over time, these repetitive stresses lead to microscopic cracks in the adhesive layer, creating pathways for bacteria and fluids to penetrate, a phenomenon known as microleakage [18].

d) Microbial Invasion: Bacteria at the Frontline

The oral cavity is home to over 700 species of bacteria, many of which thrive in the microscopic gaps that form around dental restorations. These bacteria release acidic byproducts that further degrade the adhesive interface, accelerating restoration failure. Studies using high-resolution imaging have even captured bacteria infiltrating weakened bonding layers, actively contributing to the destruction of the resin-dentin interface [23].

e) Dentin's Natural Defense: Permeability Challenges

When a tooth experiences decay, the body attempts to defend itself by forming a mineralized barrier known as reactionary dentin. While this helps protect the pulp, it also presents an additional challenge for adhesives, as this new dentin layer is less permeable and more resistant to bonding. The challenges faced by dental adhesives are not just an inconvenience; they represent a major economic and clinical burden. The failure of bonded restorations leads to costly replacement procedures, discomfort for patients, and an ongoing cycle of treatment that could be minimized with a more durable adhesive.

Understanding these challenges has driven researchers to explore new materials that resist enzymatic degradation, withstand mechanical stress, and even actively combat bacteria. From bioactive adhesives that release antimicrobial agents to smart polymers that self-repair when damaged, the next generation of dental adhesives aims to revolutionize restorative dentistry by making fillings last longer, perform better, and ultimately improve patient outcomes [24, 25].

19.3 A Brief History of Dental Adhesives: From Innovation to Evolution

The quest for strong, durable adhesion between restorative materials and tooth structures has driven the evolution of dental adhesives for decades. The story of dental adhesives is one of continuous refinement—seeking to balance efficiency, simplicity, and long-term performance in the demanding oral environment.

The earliest adhesives relied on a simple yet crucial principle: creating a mechanical bond by infiltrating demineralized dental substrates with monomers that could polymerize in place. These systems typically contained volatile solvents, such as water, alcohol, and acetone, to help carry hydrophilic and hydrophobic monomers into the porous structures of prepared dentin and enamel, forming a micromechanical interlock [26–28].

One of the most significant breakthroughs in adhesive dentistry came in 1978, when Clearfil Bond System-F (Kuraray, Japan) became the first commercially available etch-and-rinse adhesive. This system introduced 40% phosphoric acid to condition the tooth surface, a technique that dramatically improved bonding to enamel. However, as the material gained global adoption, it became apparent that dentists in different regions experienced varying clinical outcomes, particularly concerning pulp sensitivity and bond longevity [29].

Differing etching protocols led to unintended consequences. In Western dentistry, intense drying after acid etching was routinely performed to ensure that the enamel appeared chalky and porous, indicating optimal etching. However, dentin behaves differently than enamel—when overly dried, its collagen fibers collapse, reducing the surface area available for adhesive infiltration. This "collapsed collagen" phenomenon weakened the bond to dentin, causing lower bond strength (approximately 5 MPa compared to 20 MPa for enamel). Additionally, polymerization shrinkage further compromised the adhesive interface, leading to microleakage, bacterial infiltration, and increased postoperative sensitivity [30].

Meanwhile, a technique that kept dentin moist during bonding, preventing collagen collapse and enabling better resin infiltration, was introduced. This method, however, did not become widely recognized until 1992, when John Kanca introduced the concept of "wet bonding", a game-changing approach that dramatically improved bond strength to dentin while reducing patient discomfort [31].

As adhesive technology advanced, researchers sought to refine and simplify bonding procedures, leading to the development of self-etch adhesives. Unlike etch-and-rinse systems, which required a separate acid etching step, self-etch adhesives incorporate acidic functional monomers that could condition the tooth surface without rinsing. This innovation reduced technique sensitivity and improved the integrity of the adhesive interface, especially on dentin [30].

The foundation for self-etch adhesives can be traced back to Oskar Hagger's work in the 1940s, when he developed the first self-curing dental adhesive based on glycerophosphoric acid dimethacrylate [32]. However, it was not until the early 2000s that researchers like Van Meerbeek *et al.* (2003) formalized a widely accepted classification system, grouping adhesives into three major categories [28].

19.4 Classification and Composition of Adhesive Systems

Dental adhesives are considered "amphiphilic" because they contain both hydrophilic (water-attracting) and hydrophobic (water-repelling) monomers [30]. This dual nature allows them to bond effectively to both the moist tooth surface and the typically hydrophobic restorative materials. A typical dental adhesive system consists of three main components: an etchant, a primer, and an adhesive resin [30].

The bonding process begins with applying the etchant, usually a phosphoric acid solution, which chemically removes minerals from the tooth surface. This creates a rough, porous surface and increases the tooth's surface energy, making it more receptive to the adhesive. By doing so, the etchant reduces the contact angle between the adhesive and the tooth surface, thereby improving the adhesive's ability to spread evenly and penetrate the dentin structure.

Next, the primer is applied. This solution typically contains hydrophilic monomers dissolved in solvents such as ethanol or acetone. The primer helps the adhesive resin to penetrate the moist dentin by improving wetting and displacing residual moisture. Finally, the adhesive resin is applied. It contains monomers that can polymerize and create a strong micromechanical and chemical bond between the tooth and the restorative material, such as a resin composite.

The classification can be based on the conditioning method and the number of steps: three-step etch-and-rinse, two-step etch-and-rinse, two-step self-etch, and one-step self-etch. Etch-and-rinse adhesive systems always require phosphoric acid to be applied separately, using a different bottle than the one containing the primer or adhesive. Self-etch adhesive systems use an acid primer instead of phosphoric acid in an individual bottle [33, 34]. In addition to the traditional four types of adhesive systems, modern classifications also include "universal" or "multi-mode" adhesives. These are versatile adhesive systems that can be applied using multiple techniques. Specifically, they can be used either with the etch-and-rinse approach, where the tooth is etched with acid and then rinsed before applying the adhesive—or with the self-etch approach, where no rinsing is needed because the acidic primer is included in the adhesive itself. This flexibility allows clinicians to choose the method that best suits the clinical situation or their personal preference, without needing different products [29, 35].

a. Three-step etch-and-rinse adhesive system

The conventional three-step etch-and-rinse adhesive system consists of 37% phosphoric acid, primer, and adhesive. These three materials are used for dentin. However, for enamel, there is no need to use a primer. Therefore, in a cavity with dentin and enamel, the step-by-step application of this adhesive system consists of:

i. Acid etching: Application of phosphoric acid to the enamel for 15 s. Then, the acid is applied to the dentin for 15 s, for a total conditioning time of 30 s for the enamel and 15 s for the dentin.
ii. Rinsing: Wash the acid etching for at least the same time as it was applied.
iii. Drying: While enamel must be completely dry, dentin cannot be overly dry. This step is critical in the hybridization process using etch-and-rinse adhesive systems. Excessive drying of the dentin leads to the collapse of the exposed collagen network, preventing the following materials (primer and adhesive) from penetrating the surface. Therefore, dentin drying must be moderate and can be performed using sterilized filter paper. The paper is gently positioned over the dentin to protect it from the jet of air directed at the enamel. The enamel must be dried until its opacity is observed, indicating that the surface has been well-conditioned and is porous. After air is applied to the enamel, the paper is removed from the dentin. It must be gently dried with absorbent paper.

Excessive drying of dentin is a problem, as mentioned above. However, leaving too much water can also harm adhesion, especially in the long term. In addition to dissolving the primer applied in the next step, excess water makes the bonding interface more hydrophilic and prone to degradation, especially hydrolytic and polymeric degradation. Therefore, when etch-and-rinse adhesive systems are used, the 'wet adhesion' technique on the dentin is recommended—meaning the surface must remain visibly moist, without excess water—to preserve the collagen network and enable proper infiltration of hydrophilic adhesive monomers.

Etching with 37% phosphoric acid removes the topmost enamel layer (about 10 μm), increasing its surface free energy. Acid also creates pores and increases surface free energy by selectively dissolving enamel prisms to a depth of about 20 μm. In dentin, the etchant penetrates approximately 8 μm, dissolving the hydroxyapatite crystals surrounding the collagen fibers. As a result, the exposed collagen matrix in this region is largely depleted of mineral content [30].

iv. Priming: Using a micro-applicator, the primer is actively scrubbed onto the dentin surface for 20 seconds. The application must be performed gently to prevent collagen fiber breakage.

The primer consists of solvents and hydrophilic monomers. Solvents help transport hydrophilic monomers—such as 2-hydroxyethyl methacrylate (HEMA)—into the exposed collagen network by allowing them to penetrate the moist, demineralized dentin. HEMA is a low molecular weight monomer (130.14 g/mol), with a hydroxyl group and a methacrylate group at the other end. The hydroxyl groups facilitate diffusion through the collagen mesh and can bind to the organic tissue and water in dentin. On the other hand, the methacrylate group covalently bonds with the other monomers of the adhesive, which will be applied in the next step. Therefore, the primer is essential for adequate adhesion between the dentin (a hydrophilic substrate) and the adhesive (a hydrophobic material).

v. Solvent evaporation: Excess solvent from the primer is removed using gentle air. Excess solvent in the dentin creates a more hydrophilic interface, which is more susceptible to degradation processes due to osmotic water penetration into the polymeric matrix. In addition, subsequent adhesive polymerization is impaired when solvents are present in dentin, thereby reducing the adhesive's physical and chemical properties.
vi. Adhesive: A thin film must be applied to both enamel and dentin, covering the entire etched surface. After that, it must be light-cured for the time recommended by the manufacturer (generally 10 to 20 s), depending on the distance from the light and the size of the light-curing unit tip.

The composition of adhesives consists of monomers, initiators, polymerization inhibitors, and, sometimes, inorganic fillers. The most used monomers are Bisphenol A-Glycidyl Methacrylate (Bis-GMA), which is a dimethacrylate aromatic ester with high molecular weight (512 g/mol) and high viscosity. This monomer is responsible for the good mechanical properties of the adhesives, providing a higher modulus of elasticity and greater hydrolytic stability than HEMA, for example. The percentage of Bis-GMA varies from about 50% to 70% by mass, depending on the commercial brand. In addition to Bis-GMA, diluent monomers, such as triethylene glycol dimethacrylate (TEGDMA), with a low molecular weight (286 g/mol), and HEMA are added. These monomers aid material handling, reduce viscosity and increase adhesive flow, and support dentin hybridization, as they are more hydrophilic, especially HEMA [36].

Currently, adhesives are light-cured; therefore, light-sensitive initiators are used to initiate polymerization. Photoinitiators can be either type I or type II. As an example of type I photoinitiators, phosphine oxide stands out. These initiators undergo fragmentation after absorbing electromagnetic energy from the photoactivator, generating free radicals that initiate polymerization. Type I photoinitiators usually exhibit high photobleaching, reducing yellowing after polymerization. However, the spectral range of light absorption is narrower than that of Camphorquinone. Camphorquinone is a type II photoinitiator and is the most common photoinitiator used in Dentistry. Type II photoinitiators require a co-initiator to form free radicals. Co-initiators commonly used with camphorquinone are tertiary amines. The initiating system is based on camphorquinone and amine at a molar mass of 0.5 to 1% [36, 37].

Inorganic fillers may be incorporated into adhesives at low concentrations (1-5%) to maintain low viscosity and ensure proper wetting of the dental substrate. Instead, fillers are added primarily to fine-tune viscosity, improve mechanical properties (such as higher modulus of elasticity and tensile strength), reduce hydrolytic degradation, and increase radiopacity. Examples of fillers include silica, quartz, and glass containing barium, boron, aluminum, and silicon [36].

As an advantage, this adhesive system has a long clinical follow-up period and demonstrates satisfactory *in vivo* results, making it the gold standard among adhesive systems (OptiBond Fl, Kerr, USA). However, these systems have disadvantages, primarily due to their sensitivity to technique. The bonding procedure involves multiple steps, such as etching, rinsing, drying, priming, and bonding, which must be carefully controlled to ensure proper adhesive performance. In addition to the extensive clinical protocol, aggressive demineralization of the dentin occurs as acid dissolves hydroxyapatite crystals, exposing the collagen fibers deep within the tissue. The adhesive resin infiltrates part of the demineralized dentin, but a layer beneath the hybrid layer remains unprotected by polymer and is therefore more susceptible to hydrolytic and enzymatic degradation.

b. Two-step etch-and-rinse adhesive system

The industry has released a simplified version of the three-step etch-and-rinse adhesive system to streamline the clinical procedure. This strategy involves separating the acid component while combining the primer and adhesive into a single step. The step-by-step clinical procedure with this strategy is the same as the previous one, except for applying the primer and the adhesive separately. In the conventional two-step etch-and-rinse adhesive system, wet adhesion will form after carefully drying the dentin.

First, a mixture of primer and adhesive is applied with a microapplicator. After using this mixture, the solvent is volatilized. Next, another layer of the combined primer and adhesive is applied again. The solvent is volatilized again. Finally, the material is photoactivated according to the manufacturer's instructions. The composition of this system is very similar to the unsimplified version with three steps, but many clinical steps are reduced.

Although this strategy seems advantageous for reducing the clinical protocol, it is debated whether the time difference is significant, since this material is applied twice, resulting in a substantial layer of adhesive on the tissues. As a disadvantage, this system still presents difficulties related to operative sensitivity and aggressive dentin conditioning, as with the conventional 3-step adhesive system. Furthermore, this system has fewer hydrophobic monomers and forms a more hydrophilic hybrid layer, which is not sealed by a separate hydrophobic adhesive layer.

Creating a more hydrophilic hybrid layer and using a non-hydrophobic adhesive layer over it simplifies adhesives, making them act as semipermeable membranes [38]. As a result, water is entrapped within the adhesive interface, creating water channels called "water-tree" or "water trees" [39, 40], which make the dental interface prone to hydrolytic degradation.

c. Two-step self-etch adhesive system

In self-etch adhesive systems, phosphoric acid is replaced by acidic functional monomers. As a result, washing and drying steps after acid conditioning are dispensed with. Therefore, self-etch systems have fewer operator errors by reducing the steps required for their application.

The basic composition of the 2-step self-etch adhesive system comprises a primer containing solvents, hydrophilic monomers (HEMA), and acidic monomers, and an adhesive comprising photoinitiators, polymerization inhibitors, and a mixture of hydrophobic monomers and diluent monomers. Water is used as the ionizing medium in self-etch adhesives, as it enables the dissociation of acidic monomers — with rare exceptions among commercial products [33, 36].

For most two-step self-etch adhesive systems, enamel and dentin must be restored by selective enamel etching. First, the dentin must be protected (this can be done with a sterile paper), and 37% phosphoric acid in the gel must be applied to the enamel margins (the same acid used in the etch-and-rinse adhesive system) for 30 s, then washed and dried. Next, the acid primer is applied to the dentin with a micro-applicator in a rubbing motion for 20 s. The solvent must be evaporated using an air jet. Next, a thin adhesive layer should be applied to both enamel and dentin with a micro-applicator. Finally, the adhesive is photoactivated for the time established by the manufacturer. As an advantage, self-etch adhesive systems exhibit lower operating errors and cause less aggressive demineralization of dentin, thereby providing greater benefits to the dentin-adhesive interface compared to etch-and-rinse adhesive systems.

Minimally invasive approaches to dentin bonding involve three key mechanisms: shallow demineralization; simultaneous demineralization and monomer infiltration, which leaves virtually no exposed dentin at the base of the hybrid layer; and enhanced chemical stability between hydroxyapatite, calcium, and functional acidic monomers compared to phosphoric acid.

Before adhesion, hydroxyapatite is decalcified. First, acids bind to calcium in the hydroxyapatite; then, phosphate and hydroxyl ions are released. Calcium and acid binding to hydroxyapatite can occur at different levels, and these ions are quickly released into the medium, altering binding stability and exposing the fibers. On the other hand, phosphoric acid does not affect bonding stability. With acidic monomers, this bond can exhibit varying degrees of stability depending on the monomer's pH and molecular structure. Among the main acidic monomers (HEMA-p, 4-MET, phenyl-P, and 10-MDP), 10-MDP has shown the best clinical results [33, 41, 42].

Some studies suggest that the best performance of 10-MDP adhesive systems occurs due to self-organized nanolayers formed between 10-MDP and calcium in the tooth-hard tissues. The first work to demonstrate the formation of nanolayers at the adhesive interface using commercial adhesive systems with 10-MDP was published in 2012 [43]. The authors suggested that the nanolayers increased the hybrid layer's hydrophobicity, preventing water molecules from diffusing through it and causing collagen and polymer hydrolysis. Furthermore, these systems could create residual hydroxyapatite, which is more resistant to acid dissolution [43]. However, although more recent studies still suggest that nanolayers constitute a critical phase in bond strength, others demonstrate that nanolayers lack high stability on their own and that they may not be the main factor responsible for the good results of adhesive systems with 10-MDP [44].

Self-etch adhesives can also be classified as mild (pH> 1.5), moderately strong (pH ≥ 1.5), or strong (pH < 1.5), according to Peumans *et al.* [45]. The thickness of the hybrid layer dictates the bond strength. For instance, it was found that the hybrid layer thickness in strong bonding is 4 μm, while when it drops to 300 nm, less bond strength is achieved, depending on the pH of the self-etch adhesive [33]. Adhesive systems with a strong pH (<1.5) form a thicker, more densely demineralized hybrid layer, similar to phosphoric acid demineralization. Nevertheless, the disadvantage of this approach is that the demineralization products remain at the interface, with no washing step after using the acidic monomer. When performing a restoration involving enamel and dentin, the ideal procedure is to perform the selective acid etch technique on enamel with phosphoric acid and apply a self-etch adhesive system with pH ≥ 1.5 and 10-MDP on dentin.

d. One-step self-etch adhesive system

To reduce clinical time, the industry created the one-step self-etch adhesive system. Two bottles are used to separate the reagents, maximizing the adhesive's shelf-life. The composition is similar to the two-step self-etch system, but it has a higher quantity of hydrophilic monomers.

The step-by-step restoration involving enamel and dentin with this adhesive system should be performed using selective acid etching of enamel, as described for the two-step self-etch system. A drop from each vial must be mixed with a micro-applicator and applied to the dental tissues to volatilize the solvent and light-activate the material. Therefore, like the two-step self-etch systems, this adhesive system has advantages such as shorter clinical time and lower technical sensitivity. However, like a two-step etch-and-rinse adhesive system, it generates a more hydrophilic hybrid layer. There is no hydrophobic sealing by an adhesive layer over the hybrid layer, forming a more hydrophilic interface and a hybrid layer that acts as a semi-permeable membrane, impairing the longevity of restorative treatments.

e. Universal or multi-mode adhesive system

This most recent class of dental adhesives was created to simplify the technique. This material is a self-etch adhesive system that allows its application to dental tissues with or without phosphoric acid. In addition, some brands indicate their use in adhesive protocols involving metals and ceramics. Universal adhesives have an elementary composition similar to that of one-step self-etch adhesive systems, with the addition of 10-MDP [29, 46]. Universal adhesives present superior chemical stability and less phase separation than one-step self-etch adhesive systems and can be marketed in a single bottle. In addition, polyalkenoic acid, also found in other adhesive systems, is incorporated into some universal adhesive brands, providing more molecules to bind calcium in hydroxyapatite. Silane has also been incorporated into some commercially available universal adhesives.

In 2016, researchers found that the ingredients in single-bottle universal adhesives not only coexist but also interact with each other [47]. The best example in this situation is silane. Manufacturers have added silane to the composition of these adhesive systems to eliminate the need for silane pretreatment on ceramics, as both are contained in a single universal bottle. Surprisingly, research has shown a chemical interaction between the silane and water (required to ionize the acid) in the flask, leading to silane dehydration and reducing its bonding ability to ceramics. Furthermore, although manufacturers sell the material as suitable for application with or without prior phosphoric acid treatment, research shows that selective acid etching of enamel improves clinical performance [48].

19.5 New Approaches Targeting the Longevity of Adhesive-Dentin Interfaces

Significant advancements have been made in the composition of adhesive systems and resin composites to enhance the longevity of adhesive-dentin interfaces and dental restorations. These modifications primarily focus on refining the organic matrix, fillers, and photoinitiation systems to improve durability and performance. In the case of resin composites, key developments aim to enable placement in a single, thick increment rather than in multiple layers, to promote self-adhesion, reduce hydrolytic degradation, and incorporate antimicrobial and bioactive properties. Additionally, emerging research explores self-healing resins and smart materials that can dynamically respond to changes in the intraoral environment [24, 25].

Improvements in adhesive systems center on simplifying clinical application while enhancing bond stability. Introducing more stable acidic monomers, such as 10-MDP, has significantly improved adhesion. This chapter highlights these innovations, particularly those targeting hydrolytic degradation resistance, antimicrobial properties, and bioactive potential.

Bulk-fill composites represent a significant advancement, as they allow the material to be applied and cured in thicker layers—typically 4 to 5 millimeters at a time—instead of the traditional method, which required placing and curing the composite in multiple thinner layers (around 2 mm). This reduces clinical time and helps minimize polymerization shrinkage [49]. Manufacturers have introduced fragmentation monomers, aromatic urethane dimethacrylate, and enhanced photoinitiator systems to optimize these materials. Some formulations also modify initiators, such as combining phosphate oxide with camphorquinone and amine, to increase the degree of polymer conversion.

Additional refinements include adjustments in filler shape, color, and refractive index to improve light penetration and polymerization efficiency [49]. A systematic review of 18 eligible clinical studies concluded that bulk-fill resin composite restorations exhibit clinical performance comparable to incrementally layered composites over follow-up periods ranging from

6 months to 10 years. Despite variations in study design and quality, most studies raised concerns about risk of bias, yet the overall findings support the similar effectiveness of both techniques in clinical outcomes [50].

Researchers have integrated antibacterial and remineralizing agents into resin composites to enhance their bioactivity. To achieve these effects, a variety of compounds have been incorporated into dental materials, including fluoride-releasing agents, chlorhexidine, zinc oxide nanoparticles, quaternary ammonium methacrylate (QAM) monomers, and calcium phosphate-based fillers. These additives help reduce biofilm accumulation, inhibit demineralization, and support the remineralization of adjacent dental tissues [25].

Further advancements in the organic matrix involve the development of self-adhesive resins that utilize acidic monomers, such as glycerol phosphate dimethacrylate (GPDM), to eliminate the need for a separate adhesive step [29]. Additionally, acrylamide-based monomers, which lack ester groups, have been investigated for their potential to enhance hydrolytic stability, thereby extending the longevity of adhesive restorations [25].

Beyond structural improvements, researchers are also exploring self-repairing resins that contain encapsulated initiators and triethylene glycol dimethacrylate (TEGDMA). These microcapsules break open when cracks form in the material. They release monomers, which then polymerize within the crack, helping reinforce and repair the polymer. Meanwhile, smart materials are being engineered to release antimicrobial, remineralizing, or reparative agents in response to environmental triggers such as pH changes or bacterial activity [51, 52]. These advancements represent a promising shift toward longer-lasting, bioactive, and self-sustaining dental materials, ultimately improving the durability and clinical success of adhesive restorations.

19.6 Dental Adhesives Loaded with Antibacterial Properties

Ensuring the longevity of dental restorations requires strong adhesion and resistance to bacterial colonization at the tooth-restoration interface. Secondary caries, caused by bacterial infiltration, remains a leading cause of restoration failure. To address this challenge, researchers have developed antibacterial dental adhesives that actively combat biofilm formation while maintaining mechanical integrity. These materials incorporate a variety of antimicrobial agents, each with distinct mechanisms of action and long-term effectiveness. The antibacterial polymers added to dental materials can be categorized into three primary groups:

(1) Methyl methacrylate-based organic compounds containing cationic quaternary ammonium compounds disrupt bacterial cell membranes through electrostatic interactions, leading to cell lysis and death. Examples include 12-methacryloyloxydodecyl pyridinium bromide (MDPB) and dimethylaminododecyl methacrylate (DMADDM), both of which have demonstrated long-term antibacterial effects against cariogenic bacteria such as Streptococcus mutans by integrating into the polymer matrix without significantly affecting the mechanical properties of the adhesive [53, 54].
(2) Other organic compounds possessing properties that stimulate microorganism death – These include antimicrobial agents such as chlorhexidine, benzalkonium chloride, and epigallocatechin-3-gallate (EGCG). Chlorhexidine is widely studied for its ability to inhibit matrix metalloproteinases (MMPs) while reducing bacterial adhesion. EGCG, a polyphenol from green tea, has shown promise as a natural antibacterial agent by interfering with bacterial biofilm formation. However, these leachable agents may suffer from reduced long-term efficacy due to their potential to be washed out over time [55].
(3) Inorganic compounds such as metallic particles – Silver (Figure 19.3), zinc oxide (Figure 19.4), copper iodide, and titanium dioxide nanoparticles have been incorporated into dental adhesives for their strong, broad-spectrum antibacterial property. Nano-silver and zinc oxide nanoparticles generate reactive oxygen species (ROS) and disrupt bacterial metabolic pathways. Recent advances in nanotechnology have led to the development of antibacterial quantum dots that exhibit enhanced bactericidal activity while maintaining biocompatibility [56].

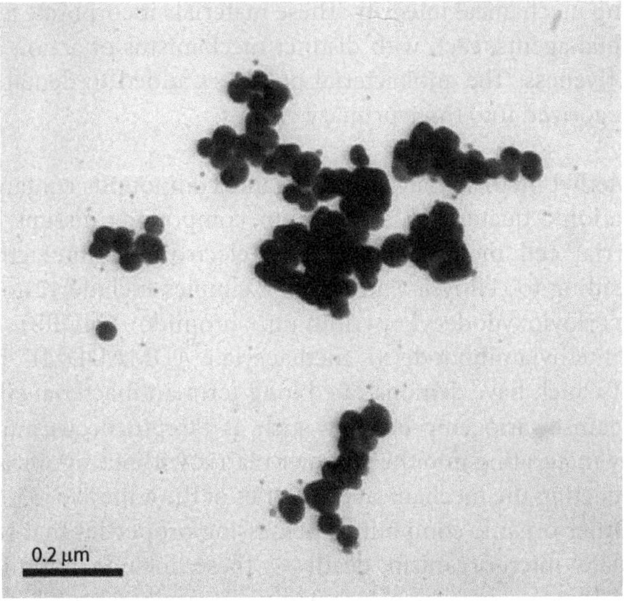

Figure 19.3 Transmission electron microscopy (TEM) image of silver nanoparticles [57].

19.7 Recent Strategies for Antibacterial Dental Adhesives

In addition to these traditional categories, several innovative approaches have emerged to enhance the antibacterial efficacy of dental adhesives further [55]:

- *Antibacterial Peptide-Modified Adhesives* – Short antimicrobial peptides (AMPs) such as GL13K and nisin have been incorporated into dental adhesives. These biomimetic agents selectively target bacterial membranes while being less toxic to human cells. Research suggests that peptides can be covalently linked to polymer matrices, providing sustained antibacterial activity without leaching.
- *pH-Responsive and Smart Release Systems* – Bioactive adhesives containing pH-sensitive microcapsules can release antibacterial agents only when cariogenic bacteria lower the

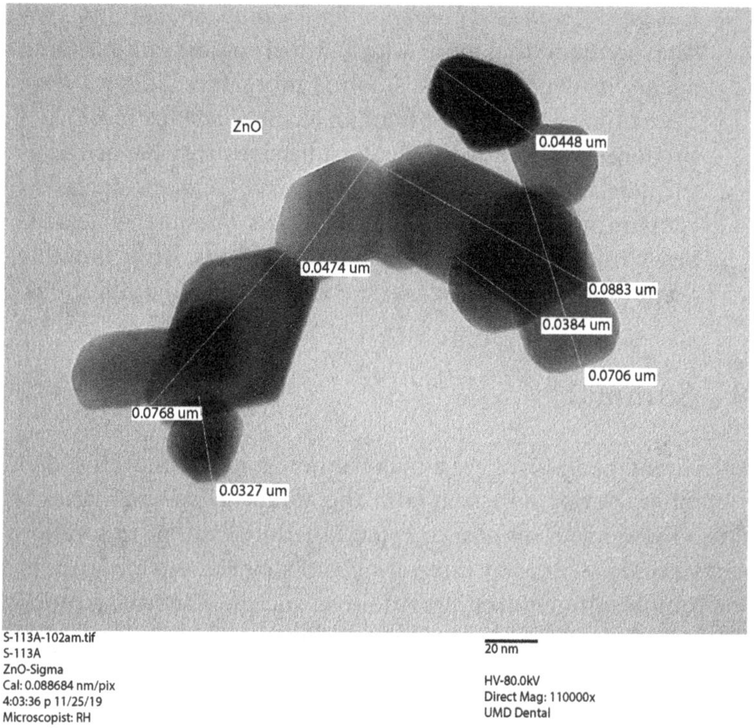

Figure 19.4 Transmission electron microscopy (TEM) image of zinc oxide nanoparticles [58].

local pH. This targeted approach minimizes unnecessary exposure to antimicrobials while maximizing their effectiveness in acidic environments that promote secondary caries.

- *Biofilm-Resistant Surfaces* – Surface modifications using hydrophilic zwitterionic polymers and superhydrophobic coatings have shown potential in preventing bacterial adhesion. These strategies do not rely on bactericidal action but instead make it physically difficult for bacteria to colonize the adhesive interface, thus reducing biofilm formation.
- *Photodynamic and Light-Activated Antibacterial Adhesives* – Some dental adhesives have been modified with light-sensitive agents, such as cationic porphyrins and graphene oxide, which produce ROS upon light activation. These photodynamic systems offer a unique advantage by providing an antibacterial effect on demand, reducing concerns about long-term cytotoxicity.

- *Cationic Quaternary Ammonium Compounds* (QACs) – They remain the most widely incorporated antibacterial agents in dental materials, while inorganic additives have been utilized to a lesser extent. In terms of application, QACs predominantly added to dental adhesives are categorized as primers, adhesives, and resin composites. Composite resins remain the dominant carrier materials due to their extensive clinical use, followed by bonding agents, while flowable composites are the least used for antibacterial modifications.

19.8 Summary

Bioactive dental adhesives are a major innovation in restorative dentistry. They combine strong adhesion with the ability to combat caries-related biofilms. These materials aim to extend the longevity of restorations and prevent secondary caries by integrating antibacterial and bioactive properties to promote remineralization. Additives such as quaternary ammonium compounds, antimicrobial peptides, metallic nanoparticles (e.g., silver and zinc oxide), and bioactive fillers contribute to both antibacterial activity and mechanical reinforcement.

Recent developments include pH-responsive and "smart-release" systems that activate only under acidic conditions, reducing unnecessary antimicrobial activity and improving biocompatibility. Innovations like peptide-modified adhesives and photoactivated antibacterial systems further broaden their potential.

However, most research remains in the early stages, with findings primarily from *in vitro* and preclinical studies. Clinical translation demands thorough validation, especially for long-term safety, durability in the oral environment, and large-scale trials. Looking ahead, multifunctional adhesives that not only bond well but also help prevent caries and support long-term stability could redefine restorative dentistry and improve patient outcomes.

AI disclosure statement: The author acknowledges the use of AI-based tools for minor language editing or grammatical correction. However, all intellectual contributions, analyses, and conclusions presented in this chapter are entirely the author's own.

References

1. GBD 2021 Oral Disorders Collaborators, Trends in the global, regional, and national burden of oral conditions from 1990 to 2021: a systematic analysis for the Global Burden of Disease Study 2021, Lancet 405, 897-910 (2025).
2. R. H. Selwitz, A. I. Ismail, and N. B. Pitts, Dental caries, The Lancet 369, 51-59 (2007).
3. M. T. Martins, F. Sardenberg, C. B. Bendo, M. H. Abreu, M. P. Vale, S. M. Paiva, and I. A. Pordeus, Dental caries remains as the main oral condition with the greatest impact on children's quality of life, PLoS One 12, 1-8 (2017).
4. P. Corrêa-Faria, A. Daher, M. do C. M. Freire, M. H. N. G. de Abreu, M. Bönecker, and L. R. Costa, Impact of untreated dental caries severity on the quality of life of preschool children and their families: a cross-sectional study, Qual Life Res 27, 3191-3198 (2018).
5. J. van Houte, Role of microorganisms in caries etiology, J Dental Res 73, 672-681 (1994).
6. J. D. B. Featherstone, The continuum of dental caries--evidence for a dynamic disease process, J Dental Res 83 Spec No C, C39-C42 (2004).
7. J. D. Featherstone, The science and practice of caries prevention, J Am Dental Assoc 131, 887-899 (2000).
8. K. Papadimitriou and Á. Alegría, P. A. Bron, M. de Angelis, M. Gobbetti, M. Kleerebezem, J. A. Lemos, D. M. Linares, P. Ross, C. Stanton, F. Turroni, D. van Sinderen, P. Varmanen, M. Ventura, M. Zúñiga, E. Tsakalidou, J. Kok, Stress physiology of lactic acid bacteria, Microbiol Mol Biol Rev 80, 837-890 (2016).
9. F. E. Dewhirst, T. Chen, J. Izard, B. J. Paster, A. C. R. Tanner, W.-H. Yu, A. Lakshmanan, and W. G. Wade, The human oral microbiome, J Bacteriol 192, 5002-5017 (2010).
10. Products - Data Briefs - Number 96 - May 2012, https://www.cdc.gov/nchs/products/databriefs/db96.htm.
11. M. Laske, N. J. M. Opdam, E. M. Bronkhorst, J. C. C. Braspenning, and M. C. D. N. J. M. Huysmans, Risk factors for dental restoration survival: a practice-based study, J Dental Res 98, 414-422 (2019).
12. F. F. Demarco, M. B. Corrêa, M. S. Cenci, R. R. Moraes, and N. J. M. Opdam, Longevity of posterior composite restorations: not only a matter of materials, Dental Mater 28, 87-101 (2012).
13. L. S. Mokeem, I. M. Garcia, and M. A. Melo, Degradation and failure phenomena at the dentin bonding interface, Biomedicines 11, 1256-1271 (2023).
14. I. A. Mjör and F. Toffenetti, Secondary caries: a literature review with case reports, Quintessence Int 31, 165-179 (2000).
15. R. L. Sakaguchi, Review of the current status and challenges for dental posterior restorative composites: clinical, chemistry, and physical behavior considerations. Summary of discussion from the Portland Composites Symposium

(POCOS) June 17-19, 2004, Oregon Health and Science University, Portland, Oregon, Dental Mater 21, 3-6 (2005).
16. A. Jokstad, S. Bayne, U. Blunck, M. Tyas, and N. Wilson, Quality of dental restorations FDI Commission Project 2–95*, International Dental Journal 51, 117-158 (2001).
17. T. Beazoglou, S. Eklund, D. Heffley, J. Meiers, L. J. Brown, and H. Bailit, Economic impact of regulating the use of amalgam restorations, Public Health Rep 122, 657-663 (2007).
18. F. F. Demarco, K. Collares, M. B. Correa, M. S. Cenci, R. R. de Moraes, and N. J. Opdam, Should my composite restorations last forever? Why are they failing?, Braz Oral Res 31, 92-99 (2017).
19. N. J. M. Opdam, F. H. van de Sande, E. Bronkhorst, M. S. Cenci, P. Bottenberg, U. Pallesen, P. Gaengler, A. Lindberg, M. C. D. N. J. M. Huysmans, and J. W. van Dijken, Longevity of posterior composite restorations: a systematic review and meta-analysis, J Dental Res 93, 943-949 (2014).
20. A. Mazzoni, L. Tjäderhane, V. Checchi, R. Di Lenarda, T. Salo, F. R. Tay, D. H. Pashley, and L. Breschi, Role of dentin MMPs in caries progression and bond stability, J Dental Res 94, 241-251 (2015).
21. R. M. Carvalho, A. P. Manso, S. Geraldeli, F. R. Tay, and D. H. Pashley, Durability of bonds and clinical success of adhesive restorations, Dental Mater 28, 72-86 (2012).
22. J. L. Ferracane, Hygroscopic and hydrolytic effects in dental polymer networks, Dental Mater 22, 211-222 (2006).
23. J. L. Ferracane, Models of caries formation around dental composite restorations, J Dental Res 96, 364-371 (2017).
24. M. A. S. Melo, I. M. Garcia, L. Mokeem, M. D. Weir, H. H. K. Xu, C. Montoya, and S. Orrego, Developing Bioactive Dental Resins for Restorative Dentistry, J Dental Res 102, 1180-1190 (2023).
25. A. P. P. Fugolin and C. S. Pfeifer, New resins for dental composites, J Dental Res 96, 1085-1091 (2017).
26. M. G. Brackett, N. Li, W. W. Brackett, R. J. Sword, Y. P. Qi, L. N. Niu, C. R. Pucci, A. Dib, D. H. Pashley, and F. R. Tay, The critical barrier to progress in dentine bonding with the etch-and-rinse technique, J Dentistry 39, 238-248 (2011).
27. N. Nakabayashi, M. Nakamura, and N. Yasuda, Hybrid layer as a dentin-bonding mechanism, J Esthet Dentistry 3, 133-138 (1991).
28. B. Van Meerbeek, J. De Munck, Y. Yoshida, S. Inoue, M. Vargas, P. Vijay, K. Van Landuyt, P. Lambrechts, and G. Vanherle, Buonocore memorial lecture. Adhesion to enamel and dentin: current status and future challenges, Oper Dentistry 28, 215-235 (2003).
29. L. Breschi, T. Maravic, C. Mazzitelli, U. Josic, E. Mancuso, M. Cadenaro, C. S. Pfeifer, and A. Mazzoni, The evolution of adhesive dentistry: From etch-and-rinse to universal bonding systems, Dental Mater 41, 141-158 (2025).

30. D. H. Pashley, F. R. Tay, L. Breschi, L. Tjäderhane, R. M. Carvalho, M. Carrilho, and A. Tezvergil-Mutluay, State of the art etch-and-rinse adhesives, Dental Mater 27, 1-16 (2011).
31. J. Kanca, Improving bond strength through acid etching of dentin and bonding to wet dentin surfaces, Journal of the American Dental Association 123, 35-43 (1992).
32. K.-J. M. Söderholm, Dental adhesives .... how it all started and later evolved, J Adhesive Dentistry 9, Suppl 2, 227-230 (2007).
33. B. Van Meerbeek, K. Yoshihara, Y. Yoshida, A. Mine, J. De Munck, and K. L. Van Landuyt, State of the art of self-etch adhesives, Dental Mater 27, 17-28 (2011).
34. M. Giannini, P. Makishi, A. P. A. Ayres, P. M. Vermelho, B. M. Fronza, T. Nikaido, and J. Tagami, Self-etch adhesive systems: a literature review, Braz Dental J 26, 3-10 (2015).
35. L. Hardan, R. Bourgi, N. Kharouf, D. Mancino, M. Zarow, N. Jakubowicz, Y. Haikel, and C. E. Cuevas-Suárez, Bond strength of universal adhesives to dentin: a systematic review and meta-analysis, Polymers (Basel) 13, 1-34 (2021).
36. K. L. Van Landuyt, J. Snauwaert, J. De Munck, M. Peumans, Y. Yoshida, A. Poitevin, E. Coutinho, K. Suzuki, P. Lambrechts, and B. Van Meerbeek, Systematic review of the chemical composition of contemporary dental adhesives, Biomaterials 28, 3757-3785 (2007).
37. A. Kowalska, J. Sokolowski, and K. Bociong, The Photoinitiators used in resin-based dental composite—A review and future perspectives, Polymers (Basel) 13, 1-17 (2021).
38. F. R. Tay, R. Frankenberger, I. Krejci, S. Bouillaguet, D. H. Pashley, R. M. Carvalho, and C. N. S. Lai, Single-bottle adhesives behave as permeable membranes after polymerization. I. *In vivo* evidence, J Dentistry 32, 611-621 (2004).
39. F. R. Tay, C. N. S. Lai, S. Chersoni, D. H. Pashley, Y. F. Mak, P. Suppa, C. Prati, and N. M. King, Osmotic blistering in enamel bonded with one-step self-etch adhesives, J Dental Res 83, 290-295 (2004).
40. F. R. Tay and D. H. Pashley, Water treeing--a potential mechanism for degradation of dentin adhesives, Am J Dentistry 16, 6-12 (2003).
41. J. Fehrenbach, C. P. Isolan, and E. A. Münchow, Is the presence of 10-MDP associated to higher bonding performance for self-etching adhesive systems? A meta-analysis of *in vitro* studies, Dental Mater 37, 1463-1485 (2021).
42. E. Carrilho, M. Cardoso, M. Marques Ferreira, C. M. Marto, A. Paula, and A. S. Coelho, 10-MDP based dental adhesives: adhesive interface characterization and adhesive stability-a systematic review, Materials (Basel) 12, 1-18 (2019).
43. Y. Yoshida, K. Yoshihara, N. Nagaoka, S. Hayakawa, Y. Torii, T. Ogawa, A. Osaka, and B. V. Meerbeek, Self-assembled nano-layering at the adhesive interface, J Dental Res 91, 376-381 (2012).

44. F.-C. Tian, X. Y. Wang, Q. Huang, L. N. Niu, J. Mitchell, Z. Y. Zhang, C. Prananik, L. Zhang, J. H. Chen, L. Breschi, D. H. Pashley, F. Tay, Effect of nanolayering of calcium salts of phosphoric acid ester monomers on the durability of resin-dentin bonds, Acta Biomater 38, 190-200 (2016).
45. M. Peumans, J. De Munck, A. Mine, and B. Van Meerbeek, Clinical effectiveness of contemporary adhesives for the restoration of non-carious cervical lesions. A systematic review, Dental Mater 30, 1089-1103 (2014).
46. K. Yoshihara, N. Nagaoka, A. Sonoda, Y. Maruo, Y. Makita, T. Okihara, M. Irie, Y. Yoshida, and B. Van Meerbeek, Effectiveness and stability of silane coupling agent incorporated in "universal" adhesives, Dental Mater 32, 1218-1225 (2016).
47. F. Murillo-Gómez, F. A. Rueggeberg, M. F. De Goes, Short- and long-term bond strength between resin cement and glass-ceramic using a silane-containing universal adhesive, Operative Dentistry 42, 514-525 (2017).
48. W. L. de O. da Rosa, E. Piva, and A. F. da Silva, Bond strength of universal adhesives: a systematic review and meta-analysis, J Dentistry 43, 765-776 (2015).
49. A. Van Ende, J. De Munck, D. P. Lise, and B. Van Meerbeek, Bulk-fill composites: a review of the current literature, J Adhesive Dentistry 19, 95-109 (2017).
50. A. Sengupta, O. Naka, S. B. Mehta, and S. Banerji, The clinical performance of bulk-fill versus the incremental layered application of direct resin composite restorations: a systematic review, Evidence- Based Dentistry 24, 143-154 (2023).
51. J. Liang, F. Liu, J. Zou, H. H. K. Xu, Q. Han, Z. Wang, B. Li, B. Yang, B. Ren, M. Li, X. Peng, J. Li, S. Zhang, X. Zhou, L. Cheng. pH-Responsive Antibacterial Resin Adhesives for Secondary Caries Inhibition, J Dental Res 99, 1368-1376 (2020).
52. A. V. B. Pintor, C. M. G. Monteiro, L. R. de Menezes, M. A. S. Melo, and L. C. Maia, Trends in pH-triggered strategies for dental resins aiming to assist in preventing demineralization: A scoping review, J Dentistry 153, 1-10 (2025).
53. K. Zhang, S. Wang, X. Zhou, H. H. Xu, M. D. Weir, Y. Ge, M. Li, S. Wang, Y. Li, X. Xu, L. Zheng, L. Cheng, Effect of antibacterial dental adhesive on multispecies biofilms formation, J Dental Res 94, 622-629 (2015).
54. S. Imazato, Y. Kinomoto, H. Tarumi, S. Ebisu, and F. R. Tay, Antibacterial activity and bonding characteristics of an adhesive resin containing antibacterial monomer MDPB, Dental Mater 19, 313-309 (2003).
55. M. S. Ibrahim, I. M. Garcia, A. Kensara, A. A. Balhaddad, F. M. Collares, M. A. Williams, A. S. Ibrahim, N. J. Lin, M. D. Weir, H. H. K. Xu, M. A. S. Melo, How we are assessing the developing antibacterial resin-based dental materials? A scoping review, J Dentistry 99, 1-9 (2020).
56. G. Naguib, A. A. Maghrabi, A. I. Mira, H. A. Mously, M. Hajjaj, and M. T. Hamed, Influence of inorganic nanoparticles on dental materials' mechanical properties. A narrative review, BMC Oral Health 23, 1-18 (2023).

57. F. W. Degrazia, V. C. B. Leitune, I. M. Garcia, R. A. Arthur, S. M. W. Samuel, and F. M. Collares, Effect of silver nanoparticles on the physicochemical and antimicrobial properties of an orthodontic adhesive, J Appl Oral Sci 24, 404-410 (2016).
58. I. M. Garcia, A. A. Balhaddad, M. S. Ibrahim, M. D. Weir, H. H. K. Xu, F. M. Collares, and M. A. S. Melo, Antibacterial response of oral microcosm biofilm to nano-zinc oxide in adhesive resin, Dental Mater 37, 182-193 (2021).

20

Bioadhesives: Recent Advances and Biomedical Applications

Megha Goyal and Nitu Bhatnagar*

Department of Chemistry, Manipal University Jaipur, Jaipur, Rajasthan, India

Abstract
Bioadhesives, materials capable of adhering to biological tissues, have garnered significant attention due to their potential in various biomedical applications, including wound closure, hemostasis, and tissue engineering. This review provides an overview of the types of bioadhesives, their adhesion mechanisms, recent developments, and their applications. Emphasis is laid on natural and synthetic bioadhesives as well as a range of new bio-inspired adhesives, highlighting their advantages, limitations, and future prospects in the biomedical field.

Keywords: Adhesion, bioadhesives, biomedical, natural, synthetic, wound

20.1 Introduction

Bioadhesives have been changing the surgical procedure with their increasing importance and rapid development over the past 30 years. The growing interest in producing adhesives and sealants makes them constitute a market share of $38 billion currently. According to the global boioadhesives market size forecast, the market is projected to grow from USD 6.02 billion in 2024 to USD 15 billion by 2035, representing a CAGR of 8.65%, during the period 2024-2035 [1, 2].

Bioadhesives behave as natural polymeric materials which have an ability to stick to a biological tissue for a long period of time. This is similar to conventional adhesion phenomenon. The ideal polymer characteristics for bioadhesion include sufficient quantity of hydrogen bonding chemical

**Corresponding author*: niturbhatnagar@gmail.com

Anil N. Netravali and Frank M. Etzler (eds.) The Wide World of Adhesion: A Festschrift in Honor of Dr. Kash Mittal on his 80th Birthday, (669–690) © 2026 Scrivener Publishing LLC

groups, anionic surface charges, high molecular weight, high chain flexibility, and cost effectiveness [3, 4].

Bioadhesives are being investigated in a variety of fields such as clinical applications, drug delivery and others [5, 6]. Bioadhesives exhibit properties that closely resemble those of biological soft tissues, making them highly suitable for biomedical applications. These materials demonstrate mechanical properties such as elasticity, flexibility, and strength similar to natural tissues like skin, muscles, or cartilage. Additionally, bioadhesives are biocompatible, minimizing adverse reactions when in contact with living tissues. Their ability to function effectively in wet or aqueous environments, i.e., maintaining adhesive performance in the presence of bodily fluids, further enhances their relevance in medical contexts. Some bioadhesives also possess adaptive and self-healing capabilities, mirroring the dynamic and reparative nature of biological soft tissues. Compared to traditional invasive surgical techniques for wound closure, like sutures, staples, and wires, bioadhesives provide a non-invasive alternative that significantly mitigates pain and minimizes tissue trauma. These advantages have led to their widespread adoption in various medical situations [7, 8]. Most studies have tended to concentrate on either the adhesion mechanism of bioadhesives or their uses for wound closure and leakage prevention [9]. Also focus has been on new biomaterials, treatments, and technology devices like biosensors which leverage bioadhesives for enhanced functionality in monitoring, diagnostics, and therapeutic applications. The majority of bioadhesives find applications in drug delivery systems, leveraging their ability to interact effectively with biological tissues. These adhesives are broadly categorized in two adhesive classes, based on their solubility. Firstly, water-soluble adhesives, such as polylysine, polybrene, alginic acid, and polyvinylpyrrolidone (PVP), are particularly useful in applications where dissolution in aqueous environments is advantageous. Secondly, water-insoluble adhesives, including gelatin, carbopol, and ethyl cellulose, provide stability and durability in wet conditions, making them ideal for applications requiring long-term adhesion in moist environments. This classification, along with examples, is detailed in Table 20.1 [10–12].

This review gives an overview on animal and plant-inspired adhesives. Later, in this review, bio-inspired adhesives with their adhesion mechanism (dry and wet adhesion) are also discussed in detail. Lastly, the discussion of bio-inspired adhesives is complemented by an overview of potential applications, including plywood manufacturing, drug delivery, wound healing, and military uses. Additionally, a forward-looking perspective on future methods is provided, emphasizing innovative strategies such as the development of novel bioadhesive materials with enhanced biocompatibility

Table 20.1 Some water soluble and insoluble adhesives.

Water soluble adhesives			**Water insoluble adhesives**		
Cationic	**Neutral**	**Anionic**	**Cationic**	**Neutral**	**Anionic**
Polylysine	Poly(ethylene glycol) (PEG)	Alginic acid	Gelatin	Ethyl cellulose	Carbopol
Polybrene	Polyvinylpyrrolidone (PVP)	Carrageenan		Hydroxy propylmethyl cellulose	Polycarbophill
Polyvinyl methyl imidazole	Hydroxy propyl cellulose	Carboxy methyl cellulose		Poly(methyl methacrylate)	Cross-linked poly(methacrylic acid)

and self-healing properties, the integration of bioadhesives with advanced technologies like biosensors and smart drug delivery systems, and the exploration of environmentally sustainable and bio-inspired fabrication techniques to advance the field. These methods include the development of stronger, multifunctional adhesives with properties such as self-healing and antimicrobial activity, the use of nanotechnology and AI-assisted design for optimized adhesive systems, and a focus on sustainable and eco-friendly production processes. This outlook underscores the potential for bio-inspired adhesives to address emerging challenges and expand their applications in diverse fields.

20.2 Classification of Bioadhesives [13]

20.2.1 Natural Bioadhesives

Derived from biological sources, natural bioadhesives include:

- **Polysaccharide-based**: Chitosan and alginate, known for their biocompatibility and biodegradability [14].
- **Protein-based**: Gelatin, fibrin, and collagen, which mimic the extracellular matrix and promote cell adhesion.
- **Marine-derived**: Mussel foot proteins (Mfps) and barnacle cement, exhibiting strong adhesion in wet environments.

20.2.2 Synthetic Bioadhesives

Engineered through chemical synthesis, synthetic bioadhesives offer tunable properties:

- **Cyanoacrylates**: Rapid-setting adhesives used in skin closures.
- **Poly(ethylene glycol) (PEG) derivatives**: Hydrophilic polymers forming hydrogels for tissue adhesion.
- **Polyurethanes**: Versatile adhesives with adjustable mechanical properties.
- **Acrylic copolymers**: Used in dental and orthopedic applications.

An ideal bioadhesive requires strong tissue adhesion and mechanical performance, along with excellent biocompatibility and biodegradability, particularly in moist and dynamic physiological environment [15]. However, existing bioadhesives for clinical applications do not fully meet these requirements. Bioadhesives exhibit poor mechanical properties, weak interfacial interactions and incompatible biosafety which refers to their inability to meet essential biomedical safety requirements, including issues such as cytotoxicity, immune reactions (where the body's immune system reacts negatively to the adhesive material, causing inflammation or rejection), poor biocompatibility, and harmful degradation by-products. This has given rise to the creation of new generations of synthetic adhesives based on the principles used by Nature and correlating them with the concepts used in these bio-adhesives. They are categorised under Bio-inspired Adhesives.

20.2.2.1 Bio-Inspired Adhesives

Recently, scientists have been working with artificial adhesives, which are synthetic materials designed to replicate the structure and functionality of natural bio-adhesives, known as bio-inspired adhesives. Bio-inspired adhesives can be considered as a great research innovation. Examples of bio-inspired adhesives include the feet of geckos, beetles, flies, mussels, tree frogs, and arboreal salamanders, which demonstrate exceptional adhesion property; lotus leaves, red rose petals, rice leaves, and water striders, known for their superhydrophobicity and self-cleaning ability; butterfly wings, peacock feathers, blue penguin feathers, and Japanese jewel beetle skin, which exhibit optical engineering (the manipulation of light to produce vibrant colors, iridescence, or structural coloration) and superhydrophobicity; chameleon skin, tropical fish skin (*neon tetra*), brittle stars (*ophiocoma wendtii*), and chameleon hydrangea petals, which showcase color-changing capability; shark skin and blue-crowned parakeet feathers, valued for their antibiofouling, hydrodynamic property, wear resistance, and self-sharpening (maintaining sharpness by natural abrasion or alignment of surfaces); photosynthetic leaf surfaces and polar bear skin, which demonstrate energy scavenging; and mimosa leaves, notable for their sensing and actuation abilities. Bio-inspired adhesives are expected to be nontoxic, bond quickly with high strength, capable of adhering under water, and provide fast reversible, and good adhesion [16–19]. Bio-inspired adhesives originated from biological precursors which is very important for bonding to biological surfaces with low toxicity. Bovine serum albumin (BSA) and glutaraldehyde adhesives are the two examples of bio-inspired medical adhesives approved by the U.S. Food and Drug Administration (FDA) [20, 21]. BSA acts as a biological component that forms an interface between

Table 20.2 Adhesive strengths of bio-inspired adhesives mimicking natural adhesion mechanisms.

Bio-inspired adhesives	Adhesive strength	Key features	Applications
Mussel adhesive protein mimetic hydrogel adhesives	13 kPa	Wet adhesion, biocompatibility	Biomedical adhesives, wound healing
Mimetic sandcastle worm glue	~75–115 kPa	Strong underwater adhesion, injectable	Marine repairs, tissue fixation
Mimetic self-cleaning gecko tape	3 N cm^{-2}	Reversible adhesion, self-cleaning property	Robotics, dry adhesives
Carbon nanotube-based synthetic gecko tape	~6–36 N cm^{-2}	High strength, durability, versatile use	Electronics, industrial adhesives

the adhesive and the tissue, facilitating effective bonding and ensuring compatibility with the biological surface, while glutaraldehyde behaves as the synthetic component for cross-linking [22]. Recently, researchers have discovered that the complex structure formed between natural adhesives and surfaces plays a critical role in enhancing adhesion strength and stability, ensuring effective bonding under various conditions. By this method, new adhesives show high strength, reversibility, and applications in wet and other extreme conditions. Table 20.2 gives an overview of selected bio-inspired adhesives with their adhesive strengths [23].

20.3 Bio-Inspired Adhesives Derived from Plants and Animals

20.3.1 Plant-Inspired Adhesives

The easiest and most economical technique for making plant-derived adhesives is to use the natural materials with little modification. Researchers have used soy-based proteins to create adhesives for prospective uses in the wood sector for strong adhesive property. A corn-germ mixture has also been developed by researchers at the Agricultural Research Service (USA) for use in the production of spray coatings, which provide surface finishing or protection, and glues, which are used to bond materials together [24]. Since the biofuel business has grown recently, non-starch by-products of ethanol production have attracted interest for use in adhesives. *Ruminococcus* bacteria also show good adhesive property. Hiraishi *et al.* [25] have investigated mussel-mimetic synthesized from plant inspired sources. The strong adhesion is caused by interactions between the catechol groups at the ends of the polymer terminal chains and the substrate surface [26]. Gan *et al.* [27] have investigated plant-inspired adhesives and hydrogels based on Ag-Lignin nanoparticles-triggered dynamic redox catechol chemistry. They concluded that the hydrogels show good cell affinity and antibacterial activity. Plant-inspired adhesives, which mimic the natural adhesive property of plants, also contribute to enhancing self-healing adhesion strength, thereby improving the effectiveness and durability of the wound healing process.

20.3.2 Animal-Inspired Adhesives

Bio- inspired adhesives are abundant in animal kingdom. Marine organisms like barnacles, sandcastle worms, oysters, and mussels adapt and bond to different materials in the environment with the protein-based adhesives. With their adhesive nature, bioinspired catechol chemistry is used which helps to develop artificial adhesives for underwater organisms [28]. For the adhesive process, mussels use byssus and plaque, which consist of protein, to secrete a liquid-state protein glue. This secretion is then transformed into a coacervate, achieving rapid setting and robust underwater adhesion [29, 30].

The sandcastle worm adhesive is a notable example of bio-inspired adhesives. It is composed of a combination of synthetic polycations and polyanions with oppositely charged side chains, present in balanced proportions. This adhesive mimic the intricate coacervation model observed in the natural sandcastle worm, allowing it to adhere effectively to surfaces when injected underwater. The coacervate formed is ideal for underwater applications due to its unique physical and chemical properties, enabling a direct delivery of liquid adhesive without dispersion. Beyond the success of chemically bio-inspired adhesives, the principles of dry adhesion have also been replicated through physical imitation. The most common approach for creating dry bio-inspired adhesives involves mimicking natural adhesion mechanisms found in insects, spiders, lizards, and frogs. These efforts have resulted in reversible adhesives with a wide range of applications, leveraging the sophisticated adhesion strategies observed in Nature. Geim *et al.* [31] designed a self-cleaning and re-attachable dry adhesive inspired by gecko hairs, featuring microscopic structures with a scale bar measuring 2 μm, indicating the size of the adhesive's structural features observed under magnification. Similarly, Bhushan and Sayer [32] performed the same experiment by using polyvinylsiloxane to design microscale pillars (about 230/mm^2) to replicate biological attachment system of gecko. Carbon nanotubes (CNTs) are also used to synthesize gecko tapes.

20.4 Mechanisms of Adhesion

There are various mechanisms of adhesion in general which have been proposed in the literature [33–37] which can be categorized in four categories: mechanical interlocking, chemical bonding, diffusion theory, and electrostatic theory [38]. The interactions between an adherend and adhesive are mainly combination of these mechanisms.

In case of plant- and animal- inspired adhesives, there are two types of adhesion involved which are discussed below:

20.4.1 Dry Adhesion

Dry adhesion relies on physical interactions involving van der Waals forces, electrostatic forces, friction, and suction [39, 40]. Examples of organisms that exhibit these adaptable adhesion mechanisms include beetles (*Coleoptera*) [41], flies (*Diptera*), spiders (*Araneae*) [42–44], and geckos (*suborder Gekkota*) [45]. These organisms demonstrate strong, reversible, and repeatable adhesion capability, allowing them to adhere effectively to a variety of rough surfaces. Numerous investigations have shown that their hairy textures at the nano-or micrometre scale are the source of their high stickiness ability. Gecko lizard contains about one million microscale foot hairs having hundreds of nanoscale projections [45, 46]. Hairy structures, extrusion tips, and structural arrangements are the three key characteristics that enhance adhesion in dry adhesive systems by maximizing van der Waals interactions. These structural features collectively improve the adhesion performance observed in natural dry adhesive systems.

20.4.1.1 *Hairy Structure*

One of the key characteristics of dry adhesive systems is hairy texture. A high density of hair-like structures demonstrates excellent adaptation to a rough surface, resulting in large contact area. Additionally, the van der Waals interactions between micro- and nanoscale hairy structures improve as the aspect ratio increases and the diameter decreases. Greiner *et al.* [47] fabricated vertical pillars with heights ranging from 2.5 to 80 μm. They concluded that adhesion strength increases as the pillar radius decreases, correlating with a higher aspect ratio.

20.4.1.2 *Extrusion Tip*

Extrusion tip structure like spatula also has a great influence on dry adhesive performance. It not only increases the contact area of the hairy structures but also equally distributes the stress at contact interface. So, the adhesive get attached to the substrate with good stability [48].

20.4.1.3 *Hierarchical Structure*

Another important factor that can improve adhesion effectiveness is the structure of the gecko foot hair. Numerous studies have shown that the hierarchical structure enhances adaptability to surfaces with different scales of roughness i.e., macro to nano. Furthermore, these structures facilitate an effective distribution of stress across hierarchical levels during repeated loading cycles, significantly enhancing structural stability. Furthermore, several studies have found that hierarchical designs can enhance the adhesion performance. Jeong *et al.* designed the monolithic microscale and nanoscale hierarchical hairs using a 2-step UV-assisted molding technique having robust shear adhesion force also on a rough surface [49]. Murphy *et al.* [50] designed a new adhesive structure that leverages the advantages of hierarchical organization and extruding tip features to enhance adhesion performance.

20.4.2 Wet Adhesion

The adhesion mechanism observed in organisms that live in wet and underwater condition is different from the organisms which live in dry conditions. As a result, octopus, tree frog, and clingfish are capable of reversible adhesion with a powerful attachment to surfaces in wet condition. Micro/nanostructures such as suction cups on the surface of organisms play an important role in exhibiting adhesion in wet condition. The principles of wet adhesion have been the subject of in-depth study in recent years, as well as attempts are being made by the researchers to replicate the adhesion mechanism.

An octopus uses a mechanism that creates a pressure difference through a suction cup in a wet environment to gain adhesion. The clingfish (*gobiesocidae*) gets strongly attached to different wet surfaces through the mechanism of suction and friction. Tree frogs (*hsylidae*) show different type of mechanism. Toe pads of the tree frog have remarkable ability to strongly adhere to the liquid-vapor interface. The wet adhesion originates from capillary forces, where the attraction between the surface and the structure, facilitated by the presence of liquid, is converted into adhesion strength. Based on this principle, an organism can stick to a surface with strong adhesion. Lee *et al.* [51] designed a suction cup made up of thermoresponsive polymers having adhesion strength of 94 kPa. Zhuo *et al.* [52] developed a pneumatic suction device that exhibited strong adhesion with a force of 3.75 N on rough surfaces, supported by a material modulus of 1.29 MPa, thus enabling effective elastic deformation for enhanced grip.

20.5 Applications

Numerous bio-inspired adhesives have been developed to meet consumer and industrial needs. These compounds can be used for tissue engineering, plywood bonding, drug delivery, bio-medical engineering, wound healing, and military purposes. It is believed that in coming days, bio-inspired adhesives will become common in use.

20.5.1 Wound Closure

Bio-inspired adhesives mostly find their application for wound closure. Traditionally, closure of wounds has been achieved using sutures, staples, and metallic wires [38]. However, these conventional methods are often associated with various drawbacks, including scar formation, risk of secondary injuries, foreign body reactions, infection due to fluid wicking, delayed healing, and the need for complex postoperative care [53]. Cyanoacrylate is the synthetic adhesive effectively used in wound healing process, but it can cause erosion, ulceration or necrosis in the surrounding tissues. Also, synthetic adhesives can cause various problems like skin, allergic, and breathing problems. BioGlue®, [22, 54] an albumin–glutaraldehyde adhesive, is a bio-inspired adhesive easy to use. This adhesive can be used in circulatory anastomosis, carotid endarterectomy, aortic dissection, as well as in lung surgery. Bio-inspired adhesives are soft in texture, which decreases the discomfort of patients after application. Traditionally, closure of wounds has been achieved using sutures, staples, and metallic wires [38]. Bioadhesives provide an effective and minimally invasive alternative for wound closure by bonding tissue surfaces together without requiring physical penetration, such as stitches or staples. Wound closure using bioadhesives can generally occur through three primary approaches: adhesion at the surface level to approximate the wound edges, adhesion of sub-surface tissue layers, or a combination of both. All these methods require strong adhesive property to ensure secure tissue bonding. Additionally, for surface applications, the bioadhesives must be biocompatible, biodegradable, and must not interfere with the natural healing process [55].

Surface bioadhesives are typically designed in the form of adhesive tapes, which require high internal cohesion strength to perform effectively [56, 57]. Cohesion, which refers to the internal binding strength of the adhesive material itself, has received relatively little attention as an independent property in previous research. Techniques such as photo-crosslinking have been frequently employed to enhance the cohesion of tape-like bioadhesives. Additionally, double-network strategies have been introduced to further boost the cohesion strength and consequently improve their wound closure performance [58–60]. Among the commercially available tissue adhesives, cyanoacrylate-based bioadhesives are the most commonly used for clinical wound sealing. Initially developed in 1949, cyanoacrylate adhesives are primarily used for bonding sub-surface layers due to their high adhesion strength. However, their application on moist tissues presents challenges, as water-activated curing can be difficult to control. Additionally, their use raises safety concerns, including exothermic reactions during polymerization, the release of cytotoxic by-products, and prolonged degradation time, which may hinder tissue compatibility and healing [53, 61].

To prevent complications like tissue irritation and foreign body responses, care must be taken to avoid allowing cyanoacrylate adhesives to enter the wound itself. It is generally advised to hold the wound edges in contact for a minimum of 30 seconds during application to ensure proper bonding. Due to the brittle nature of the adhesive barrier formed, cyanoacrylates are not ideal for use in high-movement areas like joints (e.g., knees, elbows, or fingers), where skin flexing can easily disrupt the adhesion.

20.5.2 Plywood Manufacturing

As more people become aware of the drawbacks of using petrochemical-derived products, ecologically-friendly materials are being employed more frequently, as is also the case in the plywood business [62, 63]. Since 1940, raw ingredients such as animal blood, casein, and soy proteins have held the largest market for this purpose. But as new technologies have emerged, bio-inspired adhesives have established themselves in the plywood sector. A wood adhesive based on bacterial exopolysaccharide or on waste products from the manufacturing of biofuel is one example that has already been developed for use in industry [64]. Plywood manufacturing has traditionally relied on synthetic adhesives derived from petrochemicals, such as urea-formaldehyde and phenol-formaldehyde resins. While these adhesives provide strong bonding and durability, their use raises environmental and health concerns due to the release of volatile organic compounds (VOCs) and formaldehyde emissions. In response to growing demands for sustainable and eco-friendly materials, bio-inspired adhesives have emerged as a promising alternative in the wood and plywood industry [65, 66].

Bio-inspired adhesives for plywood are typically derived from natural sources, including plant-based proteins, polysaccharides, and microbial byproducts. Soy protein, lignin, starch, and tannin have gained significant attention due to their renewability, biodegradability, and non-toxic nature. For example, soy protein adhesives have been developed by incorporating cross-linking agents or enzymatic treatments to enhance their bonding strength and water resistance, making them suitable for industrial plywood production [67, 68].

Another notable innovation includes the use of bacterial exopolysaccharides and byproducts from biofuel production as adhesive components. These materials not only reduce reliance on fossil-based chemicals but also promote waste valorization, contributing to a circular economy. A key example is the development of adhesives based on corn germ-derived polymers and lignocellulosic residues, which have shown favorable adhesion characteristics for wood composites.

The adhesion mechanism in bio-inspired systems often mimics natural adhesion processes found in marine organisms like mussels, which utilize catechol-functionalized proteins to adhere to wet and rough surfaces. Incorporating similar catechol chemistry into plant-based adhesives has significantly improved their performance under moist conditions, which is a common requirement in plywood and outdoor wood applications [69–71]. Moreover, recent research has demonstrated that combining natural polymers with nanomaterials or employing bio-mimetic cross-linking strategies can further enhance the thermal stability, mechanical strength, and durability of bio-adhesives. These innovations have paved the way for industrial-scale applications of bio-inspired adhesives in plywood manufacturing.

In summary, bio-inspired adhesives offer an environmentally-friendly and health-conscious alternative to conventional wood glues. By leveraging natural bonding mechanisms and renewable raw materials, these adhesives are poised to play a crucial role in the future of sustainable plywood production.

20.5.3 Biomedical Applications

The Australian frog *(notaden bennettii)*, commonly known as the crucifix toad, produces a bio-inspired adhesive with a wide range of potential applications. Millar *et al.* [72] discovered that this adhesive, often referred to as "frog glue," is particularly effective in repairing shoulder injuries, such as rotator cuff tears. Frog glue demonstrates exceptional strength, flexibility, and the ability to set in wet environments, making it a promising adjunct for rotator cuff repair in humans. These unique properties highlight its potential for broader biomedical applications. In recent years, there has been a significant surge in interest toward implantable and wearable medical devices, such as tissue scaffolds, biosensors, and biodetectors, due to their promising clinical potential. One of the critical challenges in this area is achieving a stable and conformal interface between these devices and biological tissues [58, 73]. Traditional fixation methods, like sutures and wires, although widely used, can introduce complications such as tissue damage, infection risk, and even potential structural damage to the implanted scaffolds. As a noninvasive alternative, bioadhesives offer a compelling solution to overcome these challenges. For example, Deng *et al.* [74] introduced an innovative electrically conductive bioadhesive composed of a graphene nanocomposite that enables rapid, robust, and reversible bonding of electronic devices to moist, dynamic tissues. This "e-bioadhesive" was effectively used in animal models to record epicardial electrocardiograms and to electrically stimulate a sciatic nerve, demonstrating its utility in real-time physiological monitoring and therapeutic applications.

One of the significant advantages of using bioadhesives for device integration is their functional adaptability, allowing them to be customized for specific applications, such as incorporating antimicrobial property to reduce infection risk or tailoring mechanical properties for dynamic environments. Specific properties such as antimicrobial activity can be embedded within the adhesive material to minimize device-related infection [75]. Despite these advantages, it is essential to note that bioadhesives must be carefully selected and designed to avoid interfering with the functionality of sensitive medical devices. For instance, Macnab *et al.* [76] observed that Tisseel®, a commonly used surgical adhesive, significantly reduced near-infrared light transmission during spectroscopy, which could impair optical device performance.

Bioadhesives also show potential in improving tissue transplantation techniques. In islet transplantation, a treatment for type I diabetes designed to restore insulin production bioadhesives offer innovative solutions to improve the procedure's efficiency and outcomes. Brubaker *et al.* used a mussel-inspired adhesive to directly anchor pancreatic islets onto internal tissue surfaces [77]. This approach offers several benefits: it simplifies the transplantation process by eliminating the need for vascular infusion, reduces damage from direct blood contact, and supports rapid revascularization of the transplanted islets while still enabling effective insulin delivery into the portal vein.

Overall, bioadhesives represent a promising frontier in medical device stabilization and tissue integration, though continued research is required to optimize their compatibility and effectiveness across various biomedical applications.

20.5.4 Drug Delivery

In the mid 1960s, the idea of drug delivery first emerged. Park and Robinson were the first to combine medication administration with bioadhesives in 1984 [78, 79]. 3,4-Dihydroxy-L-phenylalanine (DOPA) is a key component utilized in bio-inspired adhesives, particularly for repairing tissues. Its effectiveness is especially pronounced in mucus-rich structures with high cellular turnover, where its adhesive property enables robust and efficient bonding. The main benefit of bio-inspired adhesive is to increase the systemic absorption of conventional drugs and polypeptides across the nasal membrane without the use of the absorption enhancing agents. A bio-inspired adhesive depends on plant lectins to bind substances to epithelial cells. Delivering drugs through the skin offers several advantages over traditional methods like injections, inhalation, or oral administration. It helps to bypass issues such as poor drug absorption, first-pass metabolism, and patient discomfort often associated with these routes. Skin-related diseases are highly prevalent, affecting nearly one-third of the global population, yet their overall impact is often overlooked. In fact, skin conditions rank as the fourth most common cause of human illness worldwide [80–84].

Topical drug delivery plays a key role in the management of various skin conditions. It is especially important in treating disorders such as bacterial and fungal infections, skin cancers, dermatitis, burns, chronic wounds, and psoriasis. By delivering therapeutic agents directly to the affected area, this method enhances treatment effectiveness while minimizing systemic side effects.

20.6 Challenges and Limitations [85]

Bioinspired structural adhesives have received a lot of attention and have made considerable progress in the last few decades because of the expanding needs of the real world and the rapid growth of biomimetics. The delicate adhesive structures found in living things are an endless source of ideas for making synthetic adhesives. Some of these have found their application in biomedicine, drug delivery, underwater manipulation, and wearable technologies. But even the best bioinspired structural adhesives still fall short of their synthetic counterparts in many ways, and they face a number of major obstacles in their approach to being used on a large scale. The following are the challenges that limit the advanced usage of these adhesives on a large scale.

(1) Understanding the adhesion mechanisms in these natural structures, such as those found in geckos, mussels, or sandcastle worms, is complex. The dynamic adhesion processes occurring at submicrometer and nanoscale levels are challenging to observe and analyze in real time due to their intricate and extremely small-scale nature [24]. Even the typical adhesion models such as Johnson-Kendall-Roberts, capillary, and Stefan adhesion are not able to explain these complex phenomena in Nature [86].

(2) In bioinspired material processing, it is very difficult to construct artificial structures that perfectly match the modulus, surface properties, and fine micro-/nanostructures of natural adhesive structures due to the limitation of present processing technologies.

(3) As far as assessing the adhesive performance of the bioinspired structural adhesives is concerned, there are still no uniform standards for the test methods and parameters.

(4) Furthermore, artificially developed structural adhesives lack the advanced capabilities of natural systems, such as sensibility (the ability to respond to environmental stimuli), memory, and adaptability, which are inherently exhibited by living organisms.

20.7 Future Directions

The bioinspired adhesive architectures at the present stage still have a long way to go to outperform synthetic adhesive structures and meet the ever-growing demands of the real world. There is a need for development of multifunctional bioadhesives combining antimicrobial, self-healing, and regenerative capabilities which is a promising avenue. Personalized adhesives tailored to patient-specific needs and AI-assisted materials design are expected to revolutionize the field. Thus, there is a need for an interdisciplinary approach to have a thorough understanding of current challenges and future opportunities of bioinspired structural adhesive systems. This may definitely inspire researchers to develop new-generation artificial adhesives with optimized adhesive performance and are stronger and more effective.

AI disclosure statement: The author acknowledges the use of AI-based tools for minor language editing or grammatical correction. However, all intellectual contributions, analyses, and conclusions presented in this chapter are entirely the author's own.

References

1. Y. Liang, X. Zhao, T. Hu, B. Chen, Z. Yin, P. X. Ma and B. Guo, Adhesive hemostatic conducting injectable composite hydrogels with sustained drug release and photothermal antibacterial activity to promote full-thickness skin regeneration during wound healing, *Small* 15, 1900046 (2019).
2. J. Qu, X. Zhao, Y. Liang, T. Zhang, P. X. Ma and B. Guo, Antibacterial adhesive injectable hydrogels with rapid self-healing, extensibility and compressibility as wound dressing for joints skin wound healing, *Biomaterials* 183, 185-199 (2018).
3. H. S. Ch'Ng, H. Park, P. Kelly and J. R. Robinson, Bioadhesive polymers as platforms for oral controlled drug delivery II: synthesis and evaluation of some swelling, water-insoluble bioadhesive polymers, *Journal of Pharmaceutical Sciences* 74, 399-405 (1985).

4. D. Ulkoski and C. Scholz, Bioadhesives: Chemistry and mode of operation In: *Encyclopedia of Polymeric Nanomaterials*, S. Thomas, K. Müllen and J. Hu (Eds.) pp. 98–105, Springer (2015).
5. K. L. Mittal, I. S. Bakshi and J. K. Narang, (Eds.) *Bioadhesives in Drug Delivey*, Wiley-Scrivener, Beverley, MA (2020).
6. M. Dunky and K. L. Mittal, (Eds.) *Biobased Adhesives: Sources, Characteristics, and Applications*, Wiley-Scrivener, Beverley, MA (2023).
7. Y. Ding, L. Yu and Z. Mao, Adhesion mechanisms and design strategies for bioadhesives, *Colloid and Interface Science Communiactions* 63, 100809 (2024).
8. Y. Li, T. Xu, Z. Tu, W. Dai, Y. Xue, C. Tang, W. Gao, C. Mao, B. Lei and C. Lin, Bioactive antibacterial silica-based nanocomposites hydrogel scaffolds with high angiogenesis for promoting diabetic wound healing and skin repair, *Theranostics* 10, 4929 (2020).
9. L. Ge and S. Chen, Recent advances in tissue adhesives for clinical medicine, *Polymers* 12, 939 (2020).
10. V. G. Kadajji and G. V. Betageri, Water soluble polymers for pharmaceutical applications, *Polymers* 3, 1972-2009 (2011).
11. S. Khanlari and M. A. Dubé, Bioadhesives: a review, *Macromolecular Reaction Engineering* 7, 573-587 (2013).
12. K. C. Ugoeze, Bioadhesive polymers for drug delivery applications, In: *Bioadhesives in Drug Delivery*, K.L. Mittal, I.S. Bakshi and J.K. Narang (Eds.) pp 29-56, Wiley-Scrivener, Beverley, MA (2020).
13. S. Nam and D. Mooney, Polymeric tissue adhesives, *Chemical Reviews* 121, 11336-11384 (2021).
14. H. Hamedi, S. Moradi, S. M. Hudson, A. E. Tonelli and M. W. King, Chitosan based bioadhesives for biomedical applications: A review, *Carbohydrate Polymers* 282, 119100 (2022).
15. S. J. Wu and X. Zhao, Bioadhesive technology platforms, *Chemical Reviews* 123, 14084-14118 (2023).
16. Y. Liu, H. Wang, J. Li, P. Li and S. Li, Gecko-inspired controllable adhesive: structure, fabrication, and application, *Biomimetics* 9, 149 (2024).
17. J. Liu, Q. Yang, J. Xu, K. Liu, L. Guo and L. Jiang, Adhesive materials inspired by gecko and mussel, *Progress in Chemistry* 24,1946-1954 (2012).
18. F. Meng, Q. Liu, X. Wang, D. Tan, L. Xue and W. J. P. Barnes, Tree frog adhesion biomimetics: opportunities for the development of new, smart adhesives that adhere under wet conditions, *Philosophical Transactions of the Royal Society A* 377, 20190131 (2019).
19. G. J. Shah and M. Sitti, Modeling and design of biomimetic adhesives inspired by gecko foot-hairs, *IEEE International Conference on Robotics and Biomimetics* (2004).
20. Y. Wang, J. Guo, B. Li, D. Li, Z. Meng and S.-K. Sun, Biocompatible therapeutic albumin/genipin bioglue for postoperative wound adhesion and residual tumor ablation, *Biomaterials* 279, 121179 (2021).

21. S. Tarafder, G. Y. Park, J. Felix and C. H. Lee, Bioadhesives for musculoskeletal tissue regeneration, *Acta Biomaterialia* 117, 77-92 (2020).
22. H. H. Chao and D. F. Torchiana, BioGlue: albumin/glutaraldehyde sealant in cardiac surgery, *Journal of Cardiac Surgery* 18, 500-503 (2003).
23. P. M. Favi, S. Yi, S. C. Lenaghan, L. Xia and M. Zhang, Inspiration from the natural world: from bio-adhesives to bio-inspired adhesives, *Journal of Adhesion Science and Technology* 28, 290-319 (2014).
24. M. Liu, W. Ma, M. Zhang, S. Huang, S. Lin, J. Liao, P. Zhang and L. Zhang, Bio-inspired design of antifouling polymeric coatings with natural extracts: key evidence for resistance to fouling adhesion, *Advanced Composites and Hybrid Materials* 8, 139 (2025).
25. N. Hiraishi, D. Kaneko, S. Taira, S. Wang, M. Otsuki and J. Tagami, Mussel-mimetic, bioadhesive polymers from plant-derived materials, *Journal of Investigate and Clinical Dentistry* 6, 59-62 (2015).
26. H. Zhang, T. Zhao, B. Newland, P. Duffy, A. N. Annaidh, E. D. O'Cearbhaill and W. Wang, On-demand and negative-thermo-swelling tissue adhesive based on highly branched ambivalent PEG–catechol copolymers, *Journal of Materials Chemsitry B* 3, 6420-6428 (2015).
27. D. Gan, W. Xing, L. Jiang, J. Fang, C. Zhao, F. Ren, L. Fang, K. Wang and X. Lu, Plant-inspired adhesive and tough hydrogel based on Ag-Lignin nanoparticles-triggered dynamic redox catechol chemistry, *Nature Communication* 10, 1487 (2019).
28. Y. Ma, B. Zhang, I. Frenkel, Z. Zhang, X. Pei, F. Zhou and X. He, Mussel-inspired underwater adhesives from adhesion mechanisms to engineering applications: a critical review, in: *Progress in Adhesion and Adhesives*, Vol. 6, K.L. Mittal (Ed.) pp 739-759, Wiley-Scrivener, Beverley, MA (2021).
29. T. M. Lutz, C. Kimna, A. Casini and O. Lieleg, Bio-based and bio-inspired adhesives from animals and plants for biomedical applications, Materials Today Bio 13, 100203 (2022).
30. A. Ji, L. Han and Z. Dai, Adhesive contact in animal: Morphology, mechanism and bio-inspired application, *Journal of Bionic engineering* 8, 345-356 (2011).
31. A. K. Geim, S. Dubonos, I. V. Grigorieva, K. S. Novoselov, A. Zhukov and S. Y. Shapoval, Microfabricated adhesive mimicking gecko foot-hair, *Nature Materials* 2, 461-463 (2003).
32. B. Bhushan and R. A. Sayer, Surface characterization and friction of a bio-inspired reversible adhesive tape, *Microsystem Technologies* 13, 71-78 (2007).
33. J. Schulz and M. Nardin, Theories and mechanism of adhesion, In: *Handbook of Adhesive Technology*, A. Pizzi and K.L. Mittal (Eds.), pp. 19-34, Marcel Dekker, New York (1994).
34. D. E. Packham, The mechanical theory of adhesion, In: *Handbook of Adhesive Technology*, Second edition, A. Pizzi and K.L. Mittal (Eds.), pp. 69-93, Marcel Dekker, New York (1994).

35. M. M. Chehimi, A. Azioune and E. Cabet-Deliry, Acid-base interactions: Relevance to adhesion and adhesive bonding, In: *Handbook of Adhesive Technology*, Second edition, A. Pizzi and K.L. Mittal (Eds.), pp. 95-144, CRC Press, Boca Raton FL (2003).
36. K. L. Mittal, (Ed.) *Acid-Base Interactions: Relevance to Adhesion Science and Technology*, Vol. 2, CRC Press, Boca Raton FL (2000).
37. F. M. Fowkes, Acid-base interactions in polymer adhesion, In: *Physiochemical Aspects of Polymer surfaces*, Vol. 2, K.L. Mittal (Ed.) pp. 583-603, Plenum Press, Boca Raton, FL (2003)
38. M. Mehdizadeh and J. Yang, Design strategies and applications of tissue bioadhesives, *Macromolecular Bioscience* 13, 271-288 (2013).
39. Y. Akdogan, W. Wei, K. Y. Huang, Y. Kageyama, E. W. Danner, D. R. Miller, N. R. Martinez Rodriguez, J. H. Waite and S. Han, Intrinsic surface-drying properties of bioadhesive proteins, *Angewandte Chemie* 126, 11435-11438 (2014).
40. P. Kord Forooshani and B. P. Lee, Recent approaches in designing bioadhesive materials inspired by mussel adhesive protein, *Journal of Polymer Science Part A: Polymer Chemistry* 55, 9-33 (2017).
41. G. J. Amador, T. Endlein and M. Sitti, Soiled adhesive pads shear clean by slipping: a robust self-cleaning mechanism in climbing beetles, *Journal of the Royal Society Interface* 14, 20170134 (2017).
42. M. Monier and V. Courtier-Orgogozo, Drosophila glue: A promising model for bioadhesion, *Insects* 13, 734 (2022).
43. J. von Byern, C. Müller, K. Voigtländer, V. Dorrer, M. Marchetti-Deschmann, P. Flammang and G. Mayer, Examples of bioadhesives for defence and predation, *Functional Surfaces in Biology III: Diversity of the Physical Phenomena*, S. N. Gorb (Ed.) pp. 141–191, Springer (2017).
44. B. D. Opell, H. M. Elmore and M. L. Hendricks, Adhesive contact and protein elastic modulus tune orb weaving spider glue droplet biomechanics to habitat humidity, *Acta Biomaterialia* 151, 468-479 (2022).
45. K. Autumn and A. M. Peattie, Mechanisms of adhesion in geckos, *Integrative and Comparative Biology* 42, 1081-1090 (2002).
46. S. Sikdar, M. H. Rahman, A. Siddaiah and P. L. Menezes, Gecko-inspired adhesive mechanisms and adhesives for robots—a review, *Robotics* 11, 143 (2022).
47. C. Greiner, A. Del Campo and E. Arzt, Adhesion of bioinspired micropatterned surfaces: effects of pillar radius, aspect ratio, and preload, *Langmuir* 23, 3495-3502 (2007).
48. M. Kang, K. Sun, M. Seong, I. Hwang, H. Jang, S. Park, G. Choi, S.-H. Lee, J. Kim and H. E. Jeong, Applications of bioinspired reversible dry and wet adhesives: A review, *Frontiers in Mechanical Engineering* 7, 668262 (2021).
49. H. E. Jeong, J.-K. Lee, H. N. Kim, S. H. Moon and K. Y. Suh, A nontransferring dry adhesive with hierarchical polymer nanohairs, *Proceedings of the National Academy of Sciences USA* 106, 5639-5644 (2009).

50. M. P. Murphy, S. Kim and M. Sitti, Enhanced adhesion by gecko-inspired hierarchical fibrillar adhesives, *ACS Applied Materials & Interfaces* 1, 849-855 (2009).
51. H. Lee, D.-S. Um, Y. Lee, S. Lim, H.-j. Kim and H. Ko, Octopus-inspired smart adhesive pads for transfer printing of semiconducting nanomembranes, *Adv. Mater* 28, 7457-7465 (2016).
52. S. Zhuo, Z. Zhao, Z. Xie, Y. Hao, Y. Xu, T. Zhao, H. Li, E. M. Knubben, L. Wen and L. Jiang, Complex multiphase organohydrogels with programmable mechanics toward adaptive soft-matter machines, *Science Advances* 6, 1-10 (2020).
53. K. Harsha and P. Vasudha, Cyanoacrylates: An overview in its application as a tissue adhesive, *Der Pharma Chem* 10, 42-46 (2018).
54. S. A. LeMaire, Z. C. Schmittling, J. S. Coselli, A. Ündar, B. A. Deady, F. J. Clubb Jr and C. D. Fraser Jr, BioGlue surgical adhesive impairs aortic growth and causes anastomotic strictures, *Annals of Thoracic Surgery* 73, 1500-1506 (2002).
55. S. Li, J. Zhou, Y. Huang, J. Roy, N. Zhou, K. Yum, X. Sun and L. Tang, Injectable click chemistry-based bioadhesives for accelerated wound closure, *Acta Biomaterialia* 110, 95-104 (2020).
56. W. G. Bae, D. Kim, M. K. Kwak, L. Ha, S. M. Kang and K. Y. Suh, Enhanced skin adhesive patch with modulus-tunable composite micropillars, *Advanced Healthcare Materials* 2, 109-113 (2013).
57. Y. Yang, X. Zhao, J. Yu, X. Chen, R. Wang, M. Zhang, Q. Zhang, Y. Zhang, S. Wang and Y. Cheng, Bioactive skin-mimicking hydrogel band-aids for diabetic wound healing and infectious skin incision treatment, *Bioactive Materials* 6, 3962 (2021).
58. H. Yuk, C. E. Varela, C. S. Nabzdyk, X. Mao, R. F. Padera, E. T. Roche and X. Zhao, Dry double-sided tape for adhesion of wet tissues and devices, *Nature*, *575*, 169-174 (2019).
59. B. Liu, Y. Wang, Y. Miao, X. Zhang, Z. Fan, G. Singh, X. Zhang, K. Xu, B. Li and Z. Hu, Hydrogen bonds autonomously powered gelatin methacrylate hydrogels with super-elasticity, self-heal and underwater self-adhesion for sutureless skin and stomach surgery and E-skin, *Biomaterials* 171, 83-96 (2018).
60. T. M. Pausch, C. Mitzscherling, S. Abbasi, J. Cui, X. Liu, O. Aubert, M. Weissenberger, H. Johansson, P. Schuisky and C. Buesch, SmartPAN: a novel polysaccharide-microsphere-based surgical indicator of pancreatic leakage, *Journal of Biomaterials Applications* 35, 123-134 (2020).
61. Y.-z. Bu, G.-f. Sun, L.-c. Zhang, J.-h. Liu, F. Yang, P.-f. Tang and D.-c. Wu, POSS-modified PEG adhesives for wound closure, *Chinese Journal of Polymer Science* 35, 1231-1242 (2017).
62. T. Sellers, *Plywood and Adhesive Technology*, Marcel Dekker, New York (1985).

63. R. H. Gillespie and B. H. River, Durability of adhesives in plywood, Forest Products Journal 26, 21-25 (1976).
64. X. Mo and X. S. Sun, Soy proteins as plywood adhesives: formulation and characterization, *Journal of Adhesion Science and Technology* 27, 2014-2026 (2013).
65. R. Dwiyanna, A. T. Zikri and R. Widyorini, Harnessing bio-based adhesives for engineered wood: a density functional theory-guided systematic review of properties and binding mechanisms, *Journal of Adhesion*, 101, 1-27 (2025).
66. W. Zhang, Y. Ma, Y. Xu, C. Wang and F. Chu, Lignocellulosic ethanol residue-based lignin–phenol–formaldehyde resin adhesive, *International Journal of Adhesion and Adhesives* 40, 11-18 (2013).
67. D. Gonçalves, J. M. Bordado, A. C. Marques and R. Galhano dos Santos, Non-formaldehyde, bio-based adhesives for use in wood-based panel manufacturing industry—a review, *Polymers* 13, 4086 (2021).
68. L. A. Heinrich, Future opportunities for bio-based adhesives–advantages beyond renewability, *Green Chemistry* 21, 1866-1888 (2019).
69. J. Chen and H. Zeng, Designing bio-inspired wet adhesives through tunable molecular interactions, *Journal of Colloid and Interface Science 645*, 591-606 (2023).
70. T. S. Pal, S. K. Raut and N. K. Singha, Mussel-inspired catechol-functionalized EVA elastomers for specialty adhesives based on triple dynamic network, *Chemistry of Materials* 37, 2516-2534 (2025).
71. P. M. Costa, D. A. Learmonth, D. B. Gomes, M. P. Cautela, A. C. Oliveira, R. Andrade, J. Espregueira-Mendes, T. R. Veloso, C. B. Cunha and R. A. Sousa, Mussel-inspired catechol functionalisation as a strategy to enhance biomaterial adhesion: a systematic review, *Polymers* 13, 3317 (2021).
72. N. L. Millar, T. A. Bradley, N. A. Walsh, R. C. Appleyard, M. J. Tyler and G. A. Murrell, Frog glue enhances rotator cuff repair in a laboratory cadaveric model, *Journal of Shoulder and Elbow Surgery* 18, 639-645 (2009).
73. G. Schiavone and S. P. Lacour, Conformable bioelectronic interfaces: Mapping the road ahead, *Science Translational Medicine* 11, 1-3 (2019).
74. J. Deng, H. Yuk, J. Wu, C. E. Varela, X. Chen, E. T. Roche, C. F. Guo and X. Zhao, Electrical bioadhesive interface for bioelectronics, *Nature Materials* 20, 229-236 (2021).
75. I. Hwang, H. N. Kim, M. Seong, S. H. Lee, M. Kang, H. Yi, W. G. Bae, M. K. Kwak and H. E. Jeong, Multifunctional smart skin adhesive patches for advanced health care, *Advanced Healthcare Mateials* 7, 1800275 (2018).
76. A. Macnab, R. Pagano, B. Kwon, G. Dumont and B. Shadgan, *In vivo* near infrared (NIRS) sensor attachment using fibrin bioadhesive in: *Optical Diagnostics and Sensing XVIII: Toward Point-of-Care Diagnostics*, SPIE Proc. Vol. 10501, 114-121, SPIE (2018).
77. C. E. Brubaker, H. Kissler, L.-J. Wang, D. B. Kaufman and P. B. Messersmith, Biological performance of mussel-inspired adhesive in extrahepatic islet transplantation *Biomaterials* 31, 420-427 (2010).

78. J. W. Lee, J. H. Park and J. R. Robinson, Bioadhesive-based dosage forms: The next generation, *Journal of Pharmaceutical Sciences* 89, 850-866 (2000).
79. D. A. Pecosky and J. R. Robinson, Bioadhesive polymers and drug delivery, In: *Polymers for Controlled Drug Delivery*, P. J. Tarcha (Ed.), pp. 99–125, CRC Press, Boca Raton FL (2023).
80. P. Kamble, B. Sadarani, A. Majumdar and S. Bhullar, Nanofiber based drug delivery systems for skin: A promising therapeutic approach, *Journal of Drug Delivery Science and Technology* 41, 124-133 (2017).
81. N. Raina, R. Rani, V. K. Thakur and M. Gupta, New insights in topical drug delivery for skin disorders: from a nanotechnological perspective, *ACS Omega* 8, 19145-19167 (2023).
82. C. Flohr and R. Hay, Putting the burden of skin diseases on the global map, *British Journal of Dermatology* 184, 189-190 (2021).
83. U. Badilli, M. Gumustas, B. Uslu, and S. A. Ozkan, Lipid-based nanoparticles for dermal drug delivery In: *Organic Materials as Smart Nanocarriers for Drug Delivery*, A. M. Grumezescu (Ed.), 369–413, Elsevier (2018).
84. A. B. Nair, S. R. Kiran Vaka and S. N. Murthy, Transungual delivery of terbinafine by iontophoresis in onychomycotic nails, *Drug Development and Industrial Pharmacy* 37, 1253-1258 (2011).
85. X. Wan, Z. Wang, M. Liu, F. Zhang and S. Wang, Bioinspired structural adhesives: A decades-old science but emerging materials, *Matter* 7, 1710-1723 (2024).
86. H. Jin, L. Tian, W. Bing, J. Zhao and L. Ren, Bioinspired marine antifouling coatings: Status, prospects, and future, *Progress in Materials Science* 124, 100889 (2022).

21

Surface Matters: The Role of Treatment Techniques in Multi-Material Adhesive Bonding

Mani Mohan Tiwari[1], Saleema Noormohammed[1,2], Dilip Kumar Sarkar[1]* and X.-Grant Chen[1]

[1]Department of Applied Science, Aluminium Research Centre REGAL, University of Quebec at Chicoutimi, Chicoutimi, QC, Canada
[2]Trans-Al, Inc., Chicoutimi, QC, Canada

Abstract

Adhesive bonding of aluminum and polymer materials has emerged as a promising approach for creating lightweight, high-performance multi-material joints in structural applications. This article presents a systematic overview of surface treatments and interfacial engineering approaches aimed at improving the adhesion between similar (metal–metal) and dissimilar (metal–polymer) pairs. Emphasis is placed on both conventional treatments, such as mechanical abrasion and alkaline etching, as well as advanced methodologies including electrochemical anodization, organosilane functionalization, and atmospheric pressure corona discharge plasma. The effects of these treatments on surface morphology, chemical activation, and surface wetting behavior are critically examined, with particular attention to their influence on joint strength and durability. Experimental results revealed that machine-assisted abrasion in similar pairs led to an 80% increase in shear strength over untreated aluminum. In dissimilar pairs, corona discharge treatment of poly(vinyl chloride) (PVC) combined with anodized aluminum resulted in more than a 12-fold improvement, while a novel epoxy–silicone adhesive formulation further enhanced bonding strength by over 400% compared to silicone alone. These findings underscore the importance of surface engineering in achieving reliable, high-strength joints in lightweight, multi-material assemblies.

**Corresponding author*: dsarkar@uqac.ca

Anil N. Netravali and Frank M. Etzler (eds.) The Wide World of Adhesion: A Festschrift in Honor of Dr. Kash Mittal on his 80th Birthday, (691–712) © 2026 Scrivener Publishing LLC

Keywords: Adhesive bonding, surface modification, electrochemical anodization, corona discharge plasma, polymer–metal interfaces, lap shear strength, PVC–aluminum joints

21.1 Introduction

Adhesive bonding has emerged as a transformative technique for joining dissimilar materials in a wide range of industrial applications, including automotive, aerospace, construction, electronics, and biomedical engineering. The demand for sustainable manufacturing, fuel-efficient transportation, and advanced lightweight structures has necessitated the transition from conventional joining methods, such as welding and riveting, to more refined, adaptable, and material-friendly alternatives. Adhesive bonding, particularly structural bonding using engineered polymeric adhesives, is now recognized for its potential to deliver uniform stress distribution, corrosion resistance, reduced component weight, and enhanced fatigue life [1, 2].

In recent decades, the push for energy efficiency and global climate mitigation strategies has catalyzed the adoption of lightweight materials such as aluminum alloys and polymers like PVC, which offer high strength-to-weight ratio, recyclability, and compatibility in multi-material systems [3, 4]. However, the performance of adhesive joints using these materials is highly dependent on surface chemistry, microroughness, and the physicochemical interactions at the adhesive-substrate interface. A key scientific and engineering challenge has been the development of reliable surface treatment strategies to optimize adhesion and ensure the long-term durability of bonded assemblies, particularly under environmental exposure and mechanical stress [5, 6].

Numerous models have been developed to explain the mechanisms of adhesion, leading to a range of widely accepted theories in the field [7–13]. The mechanical interlocking theory suggests that adhesives anchor into surface asperities to resist mechanical separation [8]. The diffusion theory, relevant in polymer-polymer systems, describes the entanglement of macromolecular chains across the interface [11]. Electrostatic and chemical bonding theories, including adsorption theory, describe the formation of interfacial forces, such as hydrogen bonds, van der Waals forces, and covalent linkages that are crucial in polymer-metal systems [9, 12, 13]. The combined insights from these models guide the design of surface preparation techniques to enhance adhesion and joint reliability.

Surface pretreatment remains central to successful adhesive bonding. Conventional mechanical treatments such as grit blasting and sanding are frequently used to increase surface area and promote micromechanical anchorage [14, 15]. On the chemical side, surface oxidation, acid etching, and silanization have proven effective in altering surface chemistry to enhance wettability and promote covalent bonding with reactive groups in adhesives [16, 17]. More advanced treatments, including plasma activation, corona discharge, and laser texturing, offer the advantage of introducing surface polarity and cleaning without involving toxic chemicals [18, 19]. Despite these advances, environmentally benign and industrially scalable surface treatments remain a priority, particularly in the context of regulatory restrictions and ecological concerns [20].

A promising route to achieving this balance involves anodization, an electrochemical process that forms a stable porous oxide layer on metallic substrates. Traditional anodization using sulfuric or chromic acid produces high-performance oxide layers but involves hazardous reagents and toxic waste [5, 21]. Recent innovations in salt anodization, using electrolytes such as sodium phosphate (Na_2HPO_4), offer a more sustainable alternative. These methods generate porous oxide structures that facilitate both mechanical interlocking and chemical bonding with adhesives while reducing environmental impact [6, 22]. Mechanical abrasion using silicon carbide (SiC) grits of varying sizes, chemical etching, and anodization are examined as primary surface modification strategies. Furthermore, hybrid techniques, such as *in-situ* silanization during anodization are explored for their synergistic effects on surface topography and interfacial chemistry [17, 23]. For polymers, like PVC, which pose unique adhesion challenges due to their low surface energy and inertness, physical activation methods such as corona discharge are employed to enhance polar functional group density and adhesion [24, 25]. To evaluate bonding efficacy, single lap shear (SLS) testing is adopted, in accordance with the American Society for Testing and Materials (ASTM) D1002 standard [26]. The performance of various surface-prepared substrates is assessed under both unaged and environmentally aged conditions, including exposure to deionized water and saline solution. Fractographic analysis, optical profilometry, scanning electron microscopy (SEM), and Attenuated Total Reflectance-Fourier Transform Infrared (ATR-FTIR) spectroscopy are employed to characterize surface morphology, chemical composition, and failure modes [6, 27]. These techniques offer insights into the mechanisms of adhesion degradation and the influence of environmental agents such as water and chloride ions on joint integrity.

A significant finding of this study is that mechanically abraded aluminum surfaces, which involve fewer processing steps and reduced treatment time, can attain bonding strengths comparable to those achieved through conventional anodization. In contrast, anodization typically relies on acid-based treatments, raising environmental concerns due to their hazardous nature [15, 28]. Additionally, salt anodization using Na_2HPO_4 is shown to yield porous oxide layers with root mean square (rms) roughness values around 1 μm, correlating with Lap Shear Strength (LSS) exceeding 21 MPa. These values approach those reported for industrial acid-anodized surfaces, confirming the viability of this approach [6, 29]. This study contributes to the advancement of adhesive bonding technologies by systematically investigating the role of surface treatment-particularly mechanical abrasion, electrochemical anodization, and silane modification-in enhancing adhesion at polymer– metal and metal–metal interfaces. Hybrid techniques such as *in-situ* silanization during anodization and corona discharge activation of PVC are also explored for their synergistic effects on surface chemistry and adhesion performance. These strategies are designed to optimize interfacial compatibility, improve bonding strength, and provide durable adhesion in lightweight, multi-material assemblies. The implications of this research extend beyond the laboratory. As industries strive to meet emissions targets, improve energy efficiency, and comply with environmental legislation, structural adhesive bonding enabled by clean and scalable surface treatments will be a cornerstone of next generation manufacturing. The integration of porous oxide layers, hybrid silane treatments, and sustainable surface modification strategies presents a promising and scientifically grounded approach to developing robust, durable, and environmentally-compliant adhesive bonding technologies for both metals and polymers.

21.2 Materials and Methods

21.2.1 Materials Studied

The study employed AA6061 and AA3031 aluminum alloys, as well as Type 1 PVC, as base materials for the preparation of adhesively bonded joints. The aluminum substrates were cut into rectangular specimens with dimensions of 101.6 mm × 25.4 mm × 3.0 mm, in accordance with ASTM D1002 standard specifications for SLS joints. PVC sheets were similarly cut and prepared for dissimilar material bonding tests. All substrates were initially cleaned to remove surface contaminants.

Aluminum surfaces underwent a three-step ultrasonic cleaning protocol comprising degreasing in acetone (15 min), rinsing in isopropanol (15 min), and final deionized water rinsing (15 min). The cleaned samples were dried on a hotplate at 70°C to eliminate residual moisture. PVC substrates were ultrasonicated in 2-propanol for 10 minutes and allowed to air dry under ambient conditions.

To ensure consistent bonding conditions and enable comparison across various surface treatments, four structural adhesives were used in this study: Adhesive #1 (2C-epoxy, LOCTITE EA L9460), Adhesive #2 (2C-methyl methacrylate, LOCTITE AA H8003), Adhesive #3 (2C-epoxy, LOCTITE EA E-20HP), Adhesive #4 (organic 2C-epoxy), and Adhesive #5 (silicone, LOCTITE SI 595). A mixed formulation of epoxy and silicone was also prepared for bonding dissimilar PVC/Al substrates. The adhesives were selected to represent a range of curing times and mechanical properties.

21.2.2 Morphological, Chemical, and Mechanical Analyses

Surface roughness measurements were conducted using a MicroXAM 3D optical profilometer equipped with MapVue and SPIP software. RMS roughness values were calculated from the acquired topographic data to assess changes in microroughness resulting from surface treatments.

Surface morphology and microstructure were examined using a JEOL JSM-6480LV scanning electron microscope (SEM). Energy Dispersive X-ray Spectroscopy (EDS) was performed in tandem to assess the elemental composition of the treated surfaces, particularly for the detection of aluminum oxide, silane residues, and elemental chlorine from corrosion.

ATR-FTIR analysis (Cary 630, Agilent Technologies) was performed to identify functional groups introduced by surface treatments. Spectra were recorded in the 4000–450 cm^{-1} range to confirm the presence of silanol, hydroxyl, epoxy, and organosilane bonding groups.

SLS testing was performed using an INSTRON 8801 universal testing machine at crosshead speeds of 0.3–1.0 mm/min. The maximum load at failure was used to calculate shear strength. Fracture surfaces were analyzed using optical imaging to classify failure modes (adhesion, cohesive, or substrate failure) and correlate them with surface treatments and environmental exposure.

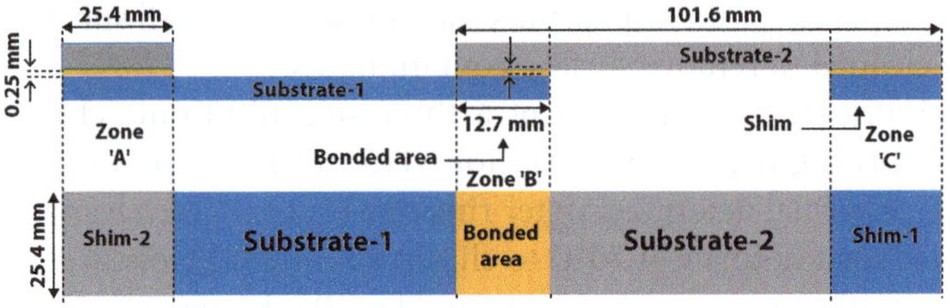

Figure 21.1 Schematic illustration of the adhesive joint configuration used for single lap shear (SLS) testing, following ASTM D1002 standard. All dimensions are in accordance with standard specimen design used for adhesive strength evaluation [30].

21.2.3 Preparation of Adhesive Joints and Adhesives

Figure 21.1 schematically illustrates the test configuration used, described in more detail below. It shows the schematic of a general SLS specimen of two substrates according to the ASTM D1002 standard [26]. The blue part depicts substrate-1 and gray portion represents substrate-2 bonded with adhesive (shown as yellow portion) with defined bondline thickness of 0.25 mm.

To facilitate a clear understanding, the schematic can be divided into three distinct zones, where zone 'A' represents the first end of SLS specimen prepared by mounting a square shaped shim-2 (made of substrate-2) over substrate-1 with a bondline thickness of 0.25 mm. Zone 'B' depicts the adhesively bonded area of two substrates 25.4 mm × 12.7 mm, with the maintained bondline thickness of 0.25 mm. Zone 'C' represents the last end of SLS specimen prepared by mounting a square-shaped shim-1 (made of substrate-1) over substrate-2 with the same dimension. The adhesively joined substrates are allowed to cure according to the manufacturer's recommendation prior to mechanical testing. The LSS was determined using an INSTRON 8801 at a recommended traction speed according to the used adhesive.

21.3 Structural Adhesive Bonding

As outlined in the Introduction section, adhesive bonding is a widely utilized technique for joining both similar and dissimilar materials, offering significant benefits such as uniform stress distribution and lightweight construction. Nevertheless, achieving strong and durable adhesive joints requires careful surface preparation to ensure optimal adhesion. The present investigation has been developed into two sections. The first section addresses surface preparation and adhesive bonding of metal/metal substrates, and the second section deals with the same on polymer/metal substrates. Aluminum (Al) was selected for the metal/metal system, while Al and PVC were chosen as substrates for the polymer/metal system.

21.3.1 Adhesive Bonding of Similar Materials (Metal-Metal)

21.3.1.1 Treatment of Surfaces Using Mechanical Abrasion

One of the simplest and most effective methods to improve adhesive bonding is mechanical abrasion. The roughening of aluminum surfaces using silicon carbide (SiC) abrasives introduces micro-roughness, which enhances mechanical interlocking between the adhesive and the substrate.

Mechanical surface abrasion was investigated as a practical and scalable method to improve the bond strength of aluminum-to-aluminum adhesive joints. Different grit sizes (G-80 to G-600) of SiC abrasive paper were applied in both manual and machine-assisted modes. The structural adhesives used for bonding aluminum joints are summarized as follows:

Figure 21.2(a) shows the Lap Shear Strength (LSS) of the as-received and mechanically abraded specimens, allowing a comparison of the joint strengths achieved with the two different adhesives. Adhesive #1 is a two-component epoxy (LOCTITE EA L9460) requiring 72 hours of curing at room temperature, while Adhesive #2 is a two-component methyl methacrylate (LOCTITE AA H8003) that cures within 1 hour under ambient conditions. These adhesives were used to evaluate bond performance under different surface preparation and aging scenarios. The highest surface roughness (rms ~4.01 ± 0.64 μm) was achieved using grit-80 under machine mode, correlating with the highest LSS (22.4 ± 1.23 MPa using Adhesive #2). Figure 21.2(b) shows the fracture surfaces of aluminum substrates bonded with Adhesive #1 and Adhesive #2 following mechanical testing, accompanied by an analysis of the observed failure modes. All specimens exhibited adhesion failure, with the exception of the machine-assisted and manually prepared samples bonded with Adhesive #2, which demonstrated cohesive failure.

Surface profilometry and SEM analysis confirmed the development of a textured topography capable of supporting mechanical interlocking. Lower grit abrasion and manual mode treatments resulted in reduced bonding performance due to insufficient surface profiling [14].

SLS tests were conducted on Adhesive #2 based aluminum joints after exposure to air (0 h), deionized (DI) water, and 3.5 and 7.0 wt.% NaCl solutions for 24, 360, and 720 hours, as shown in Figure 21.3. The durability

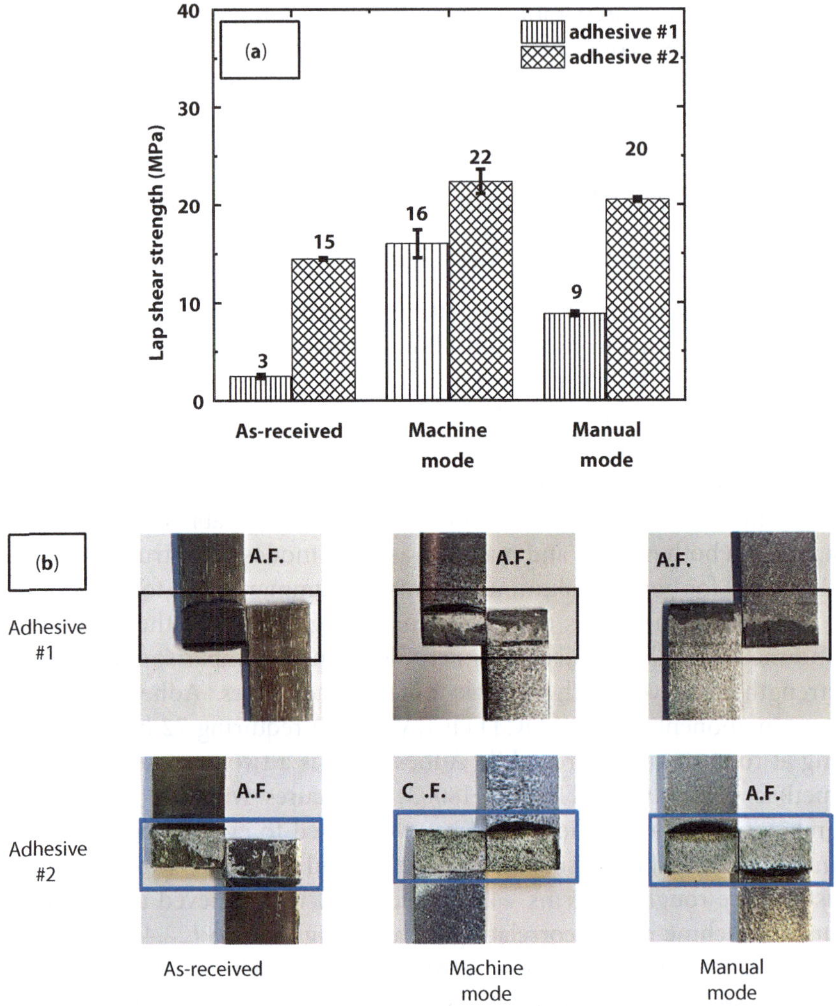

Figure 21.2 Influence of surface preparation method and adhesive type on the shear strength of bonded aluminum joints. (a) Lap shear strength (LSS) of as-received and abraded AA6061 aluminum surfaces, bonded using Adhesive #1 and Adhesive #2. Abrasion was carried out using 80-grit SiC paper in machine and manual modes. Higher LSS was consistently observed with Adhesive #2, particularly under machine abrasion conditions. (b) Fracture surface images of ruptured SLS specimens for each configuration. The observed visual differences indicate a shift in failure behavior from adhesion failure (A.F.) to cohesive failure (C.F.) [30].

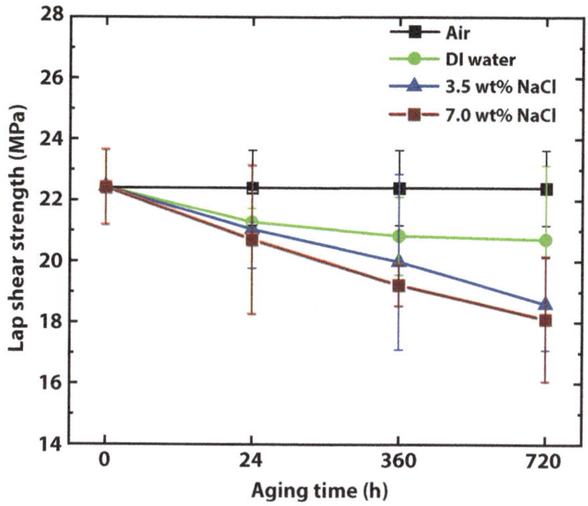

Figure 21.3 Evolution of lap shear strength (LSS) of adhesively bonded AA6061 joints over time under different environmental aging conditions. SLS tests were conducted after exposure to air (0 h), deionized (DI) water, and 3.5 and 7.0 wt.% NaCl solutions for 24, 360, and 720 hours. A progressive decrease in joint strength is observed with increasing exposure time, particularly in NaCl solutions, indicating the detrimental effect of salt-induced degradation on adhesive bonding durability [30].

assessments revealed that DI water exposure had minimal impact on shear strength (8% drop), whereas 3.5wt.% NaCl and 7.0wt.% NaCl immersions led to reductions of 17% and 19.6%, respectively. A progressive decrease in joint strength is observed with increasing exposure time, particularly in NaCl solutions, indicating the detrimental effect of salt-induced degradation on adhesive bonding durability. EDS and FTIR analyses showed that hydrolysis and chloride-induced corrosion were responsible for adhesion loss. Fracture mode shifted from cohesive to adhesion failure in saline-aged specimens, validating the importance of surface protection under environmental stress.

21.3.1.2 Electrochemical Anodization as an Adhesion-Enhancing Strategy

Beyond mechanical methods, electrochemical strategies such as anodization were also explored to improve adhesion. This technique enables the formation of a porous oxide layer on the aluminum surface, which enhances surface area and mechanical interlocking with adhesives. The anodization process in an aqueous sodium phosphate solution facilitates the formation of a porous oxide layer, enhancing surface roughness and improving chemical compatibility with the adhesive. Salt-based anodization using Na_2HPO_4 electrolyte at 30 mA·cm^{-2} and 5°C for 60 minutes created a porous oxide layer with average rms roughness of ~1.04 μm. Figures 21.4(a, b) show SEM images of the surfaces of the as-received aluminum and the 60-minute anodized counterpart, respectively. The SEM image of the as-received aluminum shows a smooth surface, whereas the image obtained after 60 minutes of anodization reveals visible open pores highlighted by blue circles. EDS spectrum (Figure 21.4c) confirms the presence of aluminum and oxygen on the anodized surface, indicating oxide formation. Figure 21.4 (d) shows ATR-FTIR spectrum, showing characteristic Al_2O_3 absorption bands that confirm the formation of aluminum oxide after anodization. SEM and EDS analyses confirmed the formation

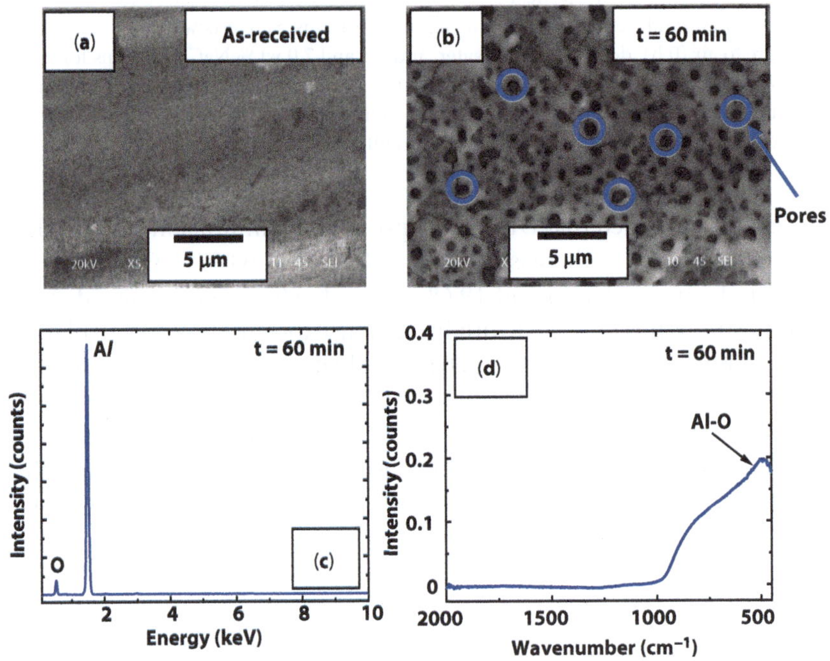

Figure 21.4 Morphological and chemical characterization of aluminum surface before and after 60 minutes of salt anodization. (a, b) The SEM image of the as-received aluminum shows a smooth surface, whereas the image obtained after 60 minutes of anodization reveals visible open pores highlighted by blue circles. (c) EDS spectrum confirming the presence of aluminum and oxygen on the anodized surface, indicating oxide formation. (d) ATR-FTIR spectrum showing characteristic Al_2O_3 absorption band, confirming the formation of aluminum oxide after anodization [30].

of a homogeneous oxide layer, and FTIR spectrum exhibited strong Al–O bond signature.

SLS specimens were prepared using Adhesive #3 according to the ASTM D1002 standard and allowed to cure according to the manufacturer's recommendation prior to mechanical testing. This surface treatment significantly improved shear strength, reaching up to 21 ± 1 MPa with Adhesive #3—nearly twice the value observed for the untreated substrate, as shown in Figure 21.5(a). The improvement is attributed to increased surface area and the presence of hydroxyl groups that facilitate hydrogen bonding with the adhesive matrix [16, 22].

Furthermore, the rupture modes of the failed specimens, shown in Figure 21.5(b), presented 100% cohesive failure on specimens prepared with surfaces treated for 60 minutes, indicating that the adhesion at the adhesive/surface interface was strong enough to make the failure within the adhesive. All other specimens showed 100% adhesion failure at the interface between

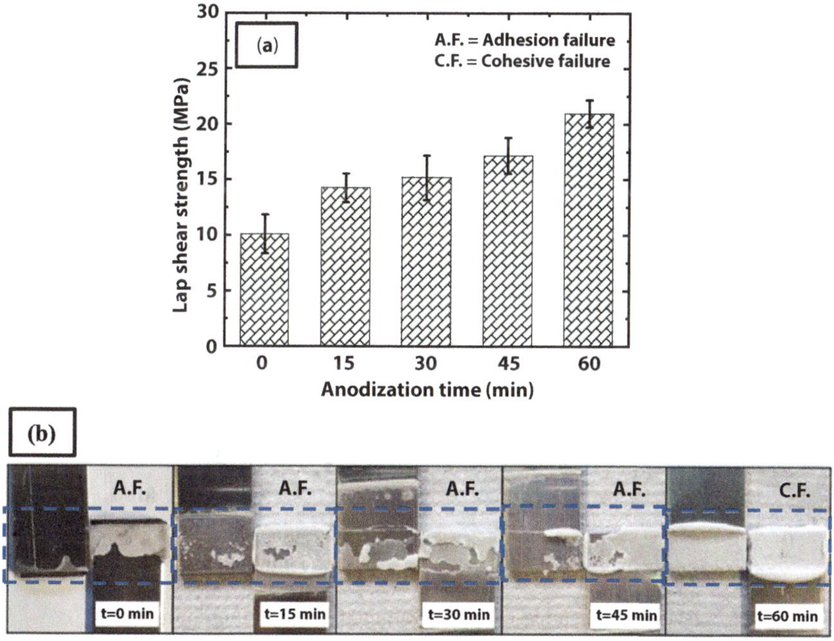

Figure 21.5 Evolution of lap shear strength (LSS) and failure mode as a function of anodization time for adhesively bonded joints. (a) Single lap shear (SLS) test results for anodization times ranging from 0 to 60 minutes, showing a steady increase in strength. A transition from adhesion failure (A.F.) to cohesive failure (C.F.) is observed at 60 minutes, indicating improved interfacial bonding. (b) Corresponding images of fracture surfaces illustrating the failure mechanisms [30].

the adhesive and the surface, indicating that the porous texture and dimensions with lower surface roughness values on the surfaces anodized for less than 60 minutes were not sufficient to obtain a strong interfacial bond between the adhesive and the surface and thereby failing at the interface.

21.3.1.3 Incorporation of Organosilane for Enhanced Adhesion

To further enhance adhesion properties, hybrid approaches combining chemical functionalization with anodization were investigated. In particular, the integration of [3-glycidoxypropyl] trimethoxysilane (GPS) into the anodization process was employed to chemically modify the oxide layer and enhance adhesive compatibility. By anodizing aluminum in a mixed electrolyte of zinc nitrate and GPS, the oxide layer is chemically modified to improve wettability and adhesive compatibility. Figures 21.6 (a, b) shows

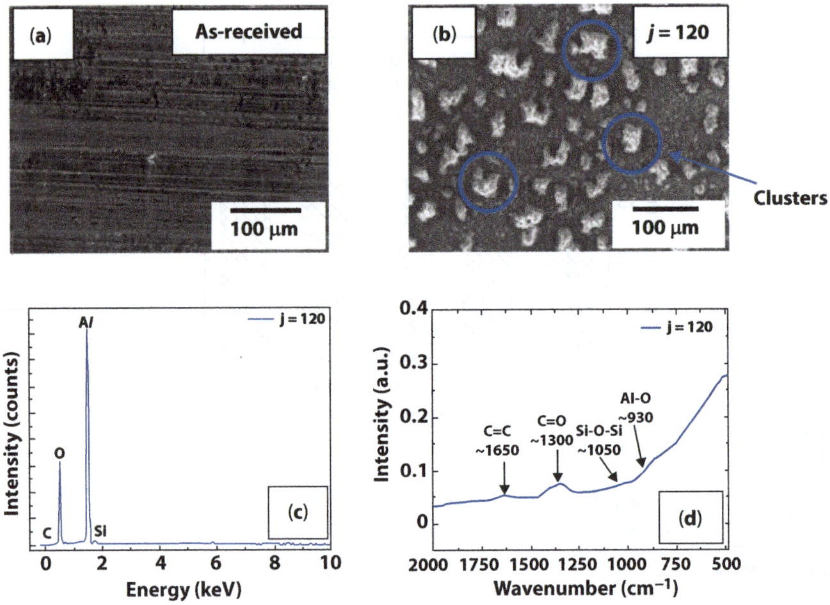

Figure 21.6 Surface and chemical characterization of AA3031 aluminum sheets before and after anodization at current density (j) 120 mA·cm^{-2}. (a) SEM image of the as-received surface showing a smooth, striated morphology. (b) SEM image after anodization, revealing the formation of micro-clusters (highlighted by blue circles), indicating surface modification. (c) EDS spectrum of the anodized surface showing peaks for Al, O, C, and minor peak of Si. (d) ATR-FTIR spectrum indicating the presence of Al_2O_3 and other functional groups (C=O, C–O–C), confirming oxide layer formation and potential for enhanced adhesion [30].

SEM images of as-received clean aluminum and aluminum anodized at a current density of 120 mA·cm^{-2}.

This dual treatment enhances both surface hydrophilicity and chemical interactions between the adhesive and the aluminum surface, resulting in superior bonding performance. A hybrid anodization-silanization process was developed using $Zn(NO_3)_2$ electrolyte containing 1% GPS, executed at current densities between 10 and 150 mA·cm^{-2}. At an optimal current density of 120 mA·cm^{-2}, dual-scale microstructures were formed with rms roughness of ~6.95 μm. As shown in Figures 21.6(c, d), the ATR-FTIR spectrum revealed characteristic signatures of Si–O–Si and epoxy groups, while EDS analysis confirmed the presence of silicon- and oxygen-rich layers.

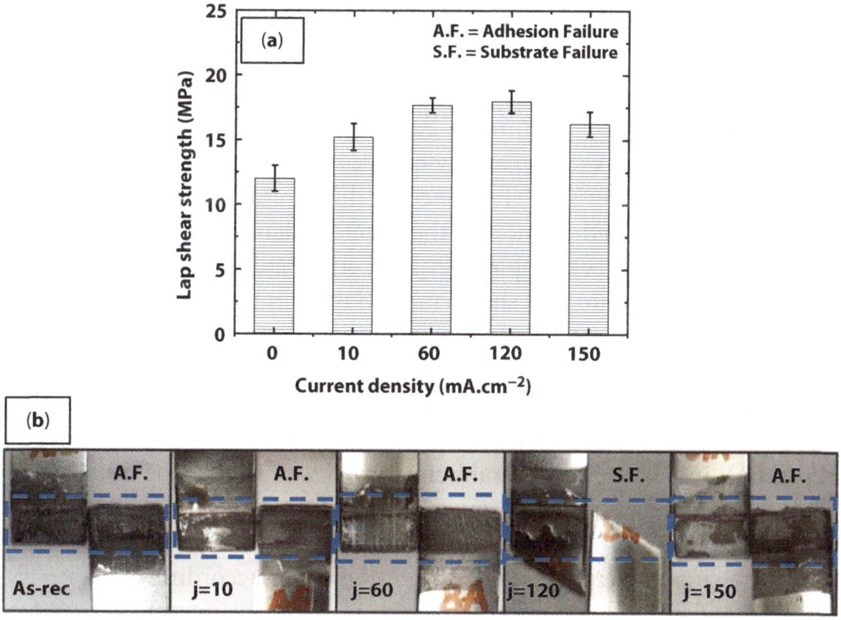

Figure 21.7 Effect of anodizing current density on the shear performance and failure mode of adhesively bonded AA3031 aluminum joints. (a) Lap shear strength (LSS) as a function of anodizing current density (j = 10 to 150 mA·cm^{-2}), showing a peak in strength at j = 120 mA·cm^{-2}, where failure shifts from adhesion (A.F.) to substrate failure (S.F.), indicating improved interfacial bonding. (b) Fracture surface images corresponding to each condition. The shift in failure mode with increasing j demonstrates the critical influence of anodizing on adhesion [30].

SLS specimens were prepared using Adhesive #4 according to the ASTM D1002 standard and allowed to cure according to the manufacturer's recommendation prior to mechanical testing.

Figure 21.7(a) shows the results of the Single Lap Shear (SLS) test conducted on as-received and electrochemically treated specimens. The test revealed a maximum bonding strength of 18 ± 1 MPa, representing an approximate 50% improvement compared to untreated aluminum substrates. Notably, the failure mode at 120 $mA \cdot cm^{-2}$ was complete substrate failure, indicating superior bonding strength, as shown in Figure 21.7(b). This behavior can be attributed to microstructural evolution and chemical changes occurring on the treated surfaces at different current densities. A critical current density (j) of 120 $mA \cdot cm^{-2}$ was found to be optimal for achieving the highest Lap Shear Strength (LSS), validating the contribution of both mechanical interlocking and chemical bonding mechanisms [16, 31].

21.3.2 Adhesive Bonding of Multi-Material (Polymer-Metal)

21.3.2.1 Treatment of Surfaces of Polymer Using Corona Discharge Atmospheric Air Plasma

Unlike aluminum, PVC is naturally hydrophobic and exhibits poor adhesive bonding characteristics. To overcome this limitation, corona discharge treatment is employed to modify the surface chemistry of PVC. Corona discharge treatment was applied to PVC substrates to improve bonding with aluminum. Figure 21.8 shows the variation in the static water contact angle of PVC surfaces treated with corona discharge atmospheric plasma for 0, 5, 10, 20, and 30 seconds. Surface analysis *via* ATR-FTIR revealed a progressive increase in –OH group intensity with treatment time up to 20 seconds. Water contact angles dropped from 98° (untreated) to 36° (20 s exposure), indicating considerably improved wettability. The relationship among wettability, surface energy, and bond strength is discussed by Mittal [32].

SLS specimens were prepared using Adhesive #3 according to the ASTM D1002 standard and allowed to cure according to the manufacturer's recommendation prior to mechanical testing. Figure 21.9 shows that the LSS of PVC/Al joints increased from less than 1 MPa for untreated samples to 12.4 MPa when anodized aluminum was joined with PVC treated by corona discharge for 20 seconds. These findings demonstrate the role of surface polarity in polymer-metal bonding [33].

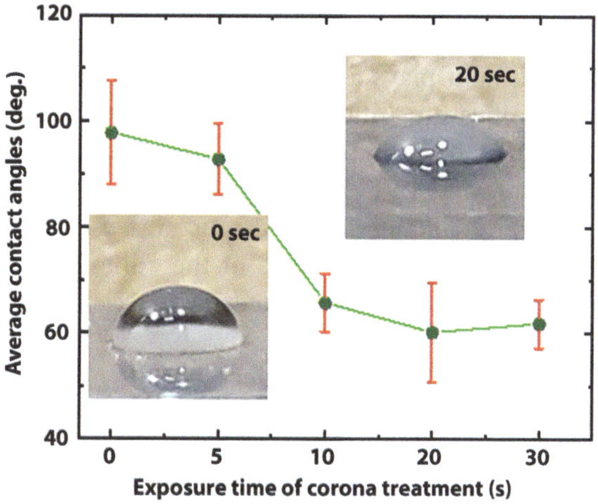

Figure 21.8 Variation in water contact angle on PVC surfaces treated by corona discharge atmospheric plasma, as a function of exposure time. Increasing treatment time leads to a progressive reduction in contact angle, indicating improved surface wettability and activation. The effect stabilizes beyond 10 seconds. Inset photographs show representative water droplet profiles at 0 and 20 seconds. Enhanced hydrophilicity supports the suitability of corona treatment for improving adhesive bonding performance [30].

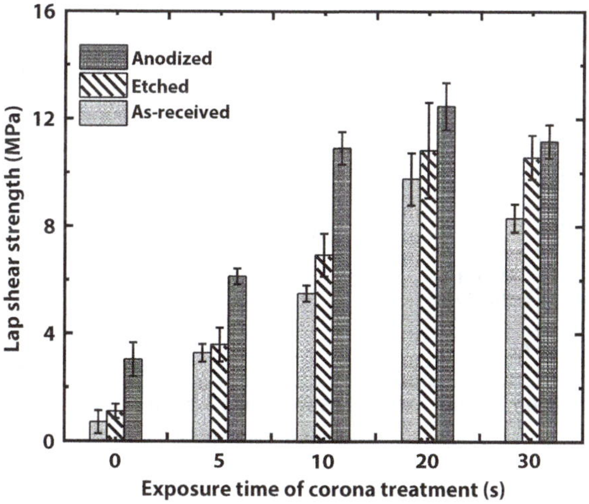

Figure 21.9 Effect of corona discharge treatment time on lap shear strength (LSS) of PVC/aluminum joints using different aluminum surface preparations. Single lap shear (SLS) specimens were prepared with PVC bonded to as-received, etched, and anodized AA6061 aluminum. A consistent increase in LSS is observed with increased corona treatment time, with anodized Al showing the highest performance at all intervals. These results highlight the synergistic effect of surface texturing and corona activation on joint integrity [30].

21.3.2.2 Formulation of Mixed Adhesive

While surface treatment plays a key role, the chemistry of adhesive formulation also significantly influences bond quality. A novel approach explored in the current research involves the use of mixed Adhesive #3 and Adhesive #5 to bond PVC/Al joints. A mixed adhesive formulation comprising epoxy and silicone in equal ratio was used to improve compatibility across PVC/Al joints. Details of the adhesives used, including the epoxy–silicone mixture, are provided in materials and methods (see Section 21.2.1).

Figure 21.10 shows the ATR-FTIR spectra of the mixed adhesive of the epoxy and silicone as well as of the PVC and Al substrate surfaces. The Al surface showed no IR absorptions, except for a small band at around ~950 cm^{-1} due to the possibility of the presence of an ultrathin oxide layer. On the other hand, the PVC substrate displayed multiple strong characteristic IR bands, such as 600 cm^{-1} (C-Cl stretching) and a broad absorption peak at 1425 cm^{-1} corresponding to $-CH_2$ bending, typical of PVC surfaces. Small bands of $-CH_2$ and $-CH_3$ between 2800–3000 cm^{-1} were also observed.

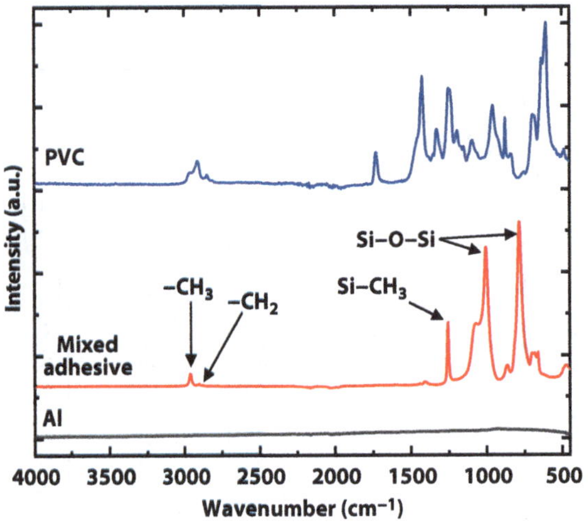

Figure 21.10 ATR-FTIR spectra of aluminum, PVC, and a mixed epoxy–silicone adhesive used in single lap shear (SLS) joint formation. Characteristic peaks in the adhesive spectrum correspond to methyl ($-CH_3$), methylene ($-CH_2$), and siloxane (Si–O–Si) groups, confirming the presence of silicone functionalities. PVC exhibits multiple sharp bands associated with its polymer backbone, while aluminum shows a featureless spectrum in the organic region, as expected. These spectral features support the chemical compatibility of the adhesive with both substrates and help explain the observed interfacial bonding behavior [30].

Upon analyzing the epoxy-silicone adhesive mixture, overlapping peaks with PVC, namely (i) $-CH_2$, $-CH_3$ around 2800–3000 cm^{-1}, (ii) $Si-CH_3$ at 1260 cm^{-1} were observed. The presence of a Si–O–Si stretching mode at ~1050 cm^{-1} and a bending mode at 800 cm^{-1} was also observed, confirming functional groups associated with the silicone adhesive in the mixed formulation. The presence of these components effectively enhanced the bonding with the Al substrate due to their chemical affinity to PVC that has an inherently low surface energy chemical structure.

The presence of siloxane (Si–O–Si) in the mixed adhesive shows promise for enhancing bonding with the mixed adhesive compared to those bonded with silicone alone. Further mechanical tests were carried out on the PVC/Al adhesive joints prepared using the pure silicone adhesive and mixed adhesive. The LSS of the PVC/Al bonded joints with pure silicone was found to be 0.43 MPa, while the joint strength using the mixed adhesive increased to 2.21 MPa, showing 413% enhancement. This behavior can be attributable to the compatibility of low surface energy PVC with silicone.

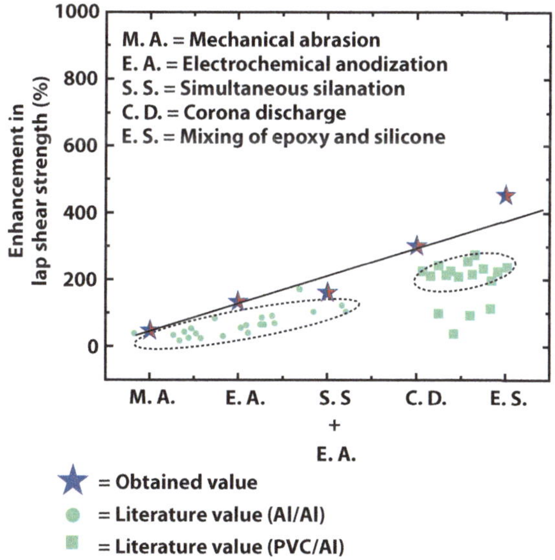

Figure 21.11 Comparative analysis of lap shear strength (LSS) enhancement for various surface treatment methods applied to similar (Al/Al) and dissimilar (PVC/Al) adhesive joints. Experimental results (★) are benchmarked against literature values (▪: Al/Al, •: PVC/Al). Surface treatments include mechanical abrasion (M.A.), electrochemical anodization (E.A.), simultaneous silanation (S.S.), and corona discharge (C.D.). The combination of treatments and adhesive systems led to significant improvements in joint strength, particularly with adhesive mixing strategy, exceeding literature benchmarks [30].

The results show that the mixing of epoxy with silicone in an equal proportion provides an increment in the joint strengths, with the mixed adhesive having excellent compatibility with PVC/Al. Figure 21.11 provides a comparative overview of adhesive joint performance, consolidating the key findings from the various surface treatments and adhesive formulations explored in this study.

Figure 21.11 demonstrates the significant impact of various surface treatment methods on the enhancement in LSS. A clear trend is observed, where mechanical abrasion results in the lowest improvement, while advanced treatments such as electrochemical anodization, simultaneous silanation with anodization, and corona discharge progressively enhance bond strength. Notably, surface treatment using corona discharge and modifying adhesive chemistry through mixing achieves the highest enhancement in the adhesive bonding of dissimilar materials. The distribution of data points indicates that surface modifications not only improve adhesion but also enhance the overall structural integrity of the bonded materials, highlighting the importance of selecting an appropriate treatment technique for optimizing adhesive performance in structural applications.

21.4 Conclusions

In summary, the present investigation provides a comprehensive analysis into the effects of surface treatment—particularly electrochemical anodization—and adhesive formulation on the mechanical performance of adhesively bonded similar (Al/Al) and dissimilar (PVC/Al) joints. Through a combination of mechanical testing and surface characterization techniques (SEM, FTIR, EDS), the findings demonstrate that surface activation, especially when paired with appropriate adhesive selection, significantly enhances bond strength and alters failure mode from adhesion to cohesive or even substrate failure.

Notably, anodized aluminum bonded with an epoxy–silicone mixture achieved the highest LSS. Corona-treated PVC exhibited improved wettability and chemical functionality, as evidenced by water contact angle reduction and FTIR spectrum. Mixed adhesive formulations proved particularly effective in bridging the mechanical and chemical disparities between aluminum and PVC, yielding strong and durable dissimilar material joints. These findings highlight the influence of surface treatment on interfacial adhesion mechanisms, offering practical guidance for optimizing bonding strategies in lightweight, multi-material assemblies.

Future research should focus on evaluating the long-term durability of these joints under cyclic loading and environmental aging, as well as exploring the industrial scalability of the proposed surface treatments.

21.5 Acknowledgments

We acknowledge the financial support provided by the Natural Sciences and Engineering Research Council of Canada (NSERC). The authors are grateful to Mr. Marc-André Hudon and Mr. Marc Veillette of Henkel, Inc. for their valuable support in providing the adhesive samples used in this study.

AI disclosure statement: The author acknowledges the use of AI-based tools for minor language editing or grammatical correction. However, all intellectual contributions, analyses, and conclusions presented in this chapter are entirely the author's own.

References

1. R.D. Adams, J. Comyn, and W.C. Wake, *Structural Adhesive Joints in Engineering*, Springer Science & Business Media, London, UK (1997).
2. W.C. Wake, *Adhesion and the Formulation of Adhesives*, Applied Science Publishers, London, UK (1978).
3. O. Faruk, J. Tjong, and M. Sain (Eds.), *Lightweight and Sustainable Materials for Automotive Applications*, CRC Press, Boca Raton, FL (2017).
4. J. Fan and J. Njuguna, An introduction to lightweight composite materials and their use in transport structures, in: *Lightweight Composite Structures in Transport*, J. Njuguna (Ed.), pp. 3–34, Woodhead Publishing, Cambridge, UK (2016).
5. G.W. Critchlow, K.A. Yendall, D. Bahrani, A. Quinn, and F. Andrews, Strategies for the replacement of chromic acid anodising for the structural bonding of aluminium alloys, *International Journal of Adhesion and Adhesives* 26, 419–453 (2006).
6. E. Dayss, G. Leps, and J. Meinhardt, Surface modification for improved adhesion of a polymer–metal compound, *Surface and Coatings Technology* 116, 986–990 (1999).
7. J. Schulz and M. Nardin, Theories and mechanisms of adhesion, in: *Handbook of Adhesive Technology*. A. Pizzi and K.L. Mittal (Eds.) pp. 19–34, Marcel Dekker, New York (1994).

8. D.E. Packham, The mechanical theory of adhesion, in: *Handbook of Adhesive Technology, Second edition,* A. Pizzi and K.L. Mittal (Eds.) pp. 69–93, CRC Press, Boca Raton, FL (2003).
9. M.M. Chehimi, A. Azioune, and E. Cabet-Deliry, Acid-base interactions: relevance to adhesion and adhesive bonding, in: *Handbook of Adhesive Technology, second edition.* A. Pizzi and K.L. Mittal (Eds.) pp. 95–144, CRC Press, Boca Raton, FL (2003).
10. F.M. Fowkes, Acid-base interactions in polymer adhesion, in: *Physicochemical Aspects of Polymer Surfaces,* Vol. 2, K.L. Mittal (Ed.) pp. 583-603, Plenum Press, New York (1983)
11. K.L. Mittal (Ed.), *Acid-Base Interactions: Relevance to Adhesion Science and Technology. Vol. 2,* CRC Press, Boca Raton, FL (2000).
12. S.S. Voyutskiĭ and V.M. Zamaziĭ, Self-adhesion (autohesion) of polyisobutylene. II, *Rubber Chemistry and Technology* 30, 544–547 (1957).
13. B.V. Derjaguin and V.P. Smilga, Electronic theory of adhesion, *Journal of Applied Physics* 38, 4609–4616 (1967).
14. A. Ghumatkar, R. Sekhar, and S. Budhe, Experimental study on different adherend surface roughness on the adhesive bond strength, *Materials Today: Proceedings* 4, 7801–7809 (2017).
15. S.G. Prolongo and A. Ureña, Surface treatments for aluminum alloy bonding, *International Journal of Adhesion and Adhesives* 29, 23–31 (2009).
16. N. Saleema, D.K. Sarkar, R.W. Paynter, D. Gallant, and M. Eskandarian, A simple surface treatment and characterization of AA 6061 aluminum alloy surface for adhesive bonding applications, *Applied Surface Science* 261, 742–748 (2012).
17. M.-L. Abel, A.N.N. Adams, A.J. Kinloch, S.J. Shaw, and J.F. Watts, The effects of surface pretreatment on the cyclic-fatigue characteristics of bonded aluminium-alloy joints, *International Journal of Adhesion and Adhesives* 26, 50–61 (2006).
18. M. Akram, K.M.B. Jansen, L.J. Ernst, S. Bhowmik, G. Ajeesh, S. Ahmed, and D. Chakraborty, Effect of atmospheric pressure plasma modification on polyimide and adhesive joining with titanium, *Metallurgical and Materials Transactions A* 46, 4680–4687 (2015).
19. M. Strobel and C.S. Lyons, Plasma surface modification of polymers—relevance to adhesion-Guest Editorial, *Journal of Adhesion Science and Technology* 8, 303–304 (1994).
20. A.S.H. Makhlouf, *Handbook of Smart Coatings for Materials Protection,* Woodhead Publishing, Cambridge, UK (2014).
21. B. Wielage, G. Alisch, T. Lampke, and D. Nickel, Anodizing–a key for surface treatment of aluminium, *Key Engineering Materials* 384, 263–281 (2008).
22. M.M. Lohrengel, Anodic oxide layers on aluminum and their applications, *Materials Science and Engineering: R: Reports* 11, 243–294 (1993).
23. L. Pan, A. Zhang, Z. Zheng, L. Duan, L. Zhang, Y. Shi, and J. Tao, Enhancing interfacial strength between AA5083 and cryogenic adhesive *via* anodic

oxidation and silanization, *International Journal of Adhesion and Adhesives* 84, 317–324 (2018).

24. A. Kruse, G. Krüger, A. Baalmann, and O.-D. Hennemann, Surface pretreatment of plastics for adhesive bonding, *Journal of Adhesion Science and Technology* 9, 1611–1621 (1995).
25. A. Suganya, G. Shanmugavelayutham, and C. Serra Rodríguez, Study on structural, morphological and thermal properties of surface modified polyvinylchloride (PVC) film under air, argon and oxygen discharge plasma, *Materials Research Express* 3(9), 095302 (2016).
26. Standard test method for apparent shear strength of single-lap-joint adhesively bonded metal specimens by tension loading (metal-to-metal), *ASTM D1002* (2005).
27. J.S. Ferguson, Measurement of delamination crack tip displacement in G/E laminates using scanning electron microscopy, Ph.D. thesis, University of British Columbia, Canada (1992).
28. X. Zhan, J. Chen, C. Gu, Q. Peng, J. Chen, and Y. Wei, Study on effects of pre-treatment and surface roughness on tensile-shear strength of 2060 Al–Li alloy adhesive joints, *Journal of Adhesion* 93, 613–625 (2017).
29. W. Lee and S.-J. Park, Porous anodic aluminum oxide: anodization and templated synthesis of functional nanostructures, *Chemical Reviews* 114, 7487–7556 (2014).
30. M.M. Tiwari, Studies of surface preparation and structural adhesive bonding of aluminum and other materials, Ph.D. thesis, Université du Québec à Chicoutimi, Canada (2024).
31. A.J. Sturgeon, Glass-ceramic coatings for metals, Ph.D. thesis, University of Warwick, United Kingdom (1987).
32. K.L. Mittal, The role of the interface in adhesion phenomena, *Polymer Engineering and Science* 17, 467–473 (1977).
33. G. Yang, T. Yang, W. Yuan, and Y. Du, The influence of surface treatment on the tensile properties of carbon fiber-reinforced epoxy composites-bonded joints, *Composites Part B: Engineering* 160, 446–456 (2019).

22

Properties of Adhesives for High-Temperature Applications: A Critical Review

Momin Urooj Ahmed and S.K. Panigrahi*

Department of Mechanical Engineering, Defence Institute of Advanced Technology, Pune, Maharashtra, India

Abstract

A critical and comprehensive review on properties of adhesives for high-temperature applications is presented in the present study. The mechanical performance of epoxy adhesives exhibits significant sensitivity to temperature variations, with a general decline in tensile, compressive, flexural, and impact properties as temperature increases, particularly near or above the glass transition temperature (Tg). Elevated temperatures induce softening and increased ductility, leading to reduced strength and stiffness. While nanofillers enhance mechanical properties at room temperature and offer moderate improvements under thermal stress, their effectiveness diminishes beyond the Tg. Functionally Graded Adhesives (FGAs) demonstrate superior thermo-mechanical stability and resistance to degradation by effectively distributing stress and delaying failure. Composite adhesives reinforced with fibers or nanoparticles retain better mechanical integrity under thermal loading by impeding crack propagation and enhancing energy dissipation. Lap shear strength benefits from increased ductility up to the Tg but deteriorates beyond it; nanoparticle incorporation mitigates this effect by enhancing modulus and interfacial bonding. Peel stress analysis indicates improved stress transfer at elevated temperatures with nanofillers. Furthermore, rising temperatures improve adhesive wettability due to reduced surface tension, with FGAs showing the most pronounced benefits due to their tailored gradient structures. This comprehensive evaluation highlights the complex interplay between thermal softening and mechanical performance, emphasizing the need for advanced adhesive formulations tailored for high-temperature structural applications.

**Corresponding author*: panigrahi.sk@gmail.com

Anil N. Netravali and Frank M. Etzler (eds.) The Wide World of Adhesion: A Festschrift in Honor of Dr. Kash Mittal on his 80th Birthday, (713–748) © 2026 Scrivener Publishing LLC

Keywords: Adhesive, adhesion, nanofillers, wettability, surface tension

22.1 Introduction

The study of adhesives under elevated temperature conditions constitutes a fundamental aspect of modern materials science and structural engineering. This area of research holds substantial relevance across a wide array of industrial applications, including but not limited to aerospace, automotive, electronics, marine, and civil construction. In these sectors, adhesives are increasingly favored for their ability to bond dissimilar materials with minimal added weight, uniform stress distribution, and improved fatigue resistance compared to traditional mechanical fastening methods such as bolts, rivets, or welds [1]. The lightweight and design flexibility offered by adhesive bonding also enable the manufacture of complex assemblies and contribute to overall system efficiency. However, the performance of adhesive bonds under varying environmental conditions, particularly high temperatures, is critical, as many structural components are expected to operate under thermal loads for extended periods. A thorough understanding of how adhesives behave at elevated temperatures is therefore vital for ensuring safety, reliability, and longevity in high-performance systems [1].

Despite their numerous advantages, adhesives are inherently sensitive to environmental factors, with temperature being one of the most influential. Elevated temperatures can significantly alter the mechanical behavior of bonded joints, leading to reductions in strength, stiffness, and durability. This vulnerability becomes especially critical in the context of single-lap joints (SLJs), a commonly employed configuration in structural bonding, particularly in aerospace and automotive designs. Experimental and numerical investigations have demonstrated that when SLJs are subjected to high temperatures, thermal expansion mismatches between the adhesive and the adherends can generate internal stresses. When these thermal stresses are combined with external mechanical loads such as tension, shear, or bending they can lead to stress concentrations at the bondline edges, resulting in premature failure or delamination [2]. Furthermore, adhesive degradation at elevated temperatures, often due to polymer chain scission or softening near or above the glass transition temperature (Tg), can further reduce joint integrity. Understanding these mechanisms is critical for designing joints capable of sustaining loads in demanding thermal environments.

The performance of adhesives is strongly influenced by temperature, which can significantly alter their mechanical properties. As temperature changes, adhesives may transition between brittle and ductile behavior, affecting their strength and deformation characteristics. When toughening agents are used, adhesives often exhibit nonlinear stress-strain responses [3], which are effectively modeled using plasticity theory. In addition to temperature sensitivity, adhesives also respond to changes in strain rate [4], making their dynamic behavior a critical consideration under varying loading conditions, such as impacts. Moreover, the fracture behavior of adhesive joints under mixed-mode loading and different loading rates has become a key focus of recent research, especially for improving the safety and durability of bonded structures in automotive and other high-performance applications [1, 3, 4].

Adhesively bonded joints are increasingly being used as an alternative to traditional mechanical fasteners in various engineering applications due to their numerous benefits. These include reduced overall weight of the structure, lower manufacturing costs, and enhanced resistance to damage. When designing adhesive joints, it is essential to consider how temperature variations affect joint strength. Key factors influencing the performance of adhesive joints across a broad temperature range include the shrinkage that occurs during curing, the adhesive's coefficient of thermal expansion (CTE) particularly in relation to that of the bonded substrates and the temperature-dependent changes in the adhesive's mechanical properties [5].

Adhesive materials typically exhibit temperature-dependent mechanical behavior. At elevated temperatures, properties such as yield strength and Young's modulus tend to decrease, making the adhesive softer and less stiff. In contrast, at lower temperatures, adhesives become more rigid and stronger. Ductility follows an inverse trend, increasing with temperature and decreasing when exposed to colder conditions [6]. These temperature-induced changes can significantly influence the durability and effectiveness of adhesive joints in real-world applications. The ability to withstand high temperatures is especially critical in sectors like aerospace, electronics, and the oil and gas industry, where thermal stability is often a key indicator of an adhesive's performance under demanding environmental conditions.

22.2 Properties of Adhesives at Elevated Temperature

Adhesive properties are profoundly influenced by temperature variations, particularly at elevated levels. As temperature rises, the tensile strength of adhesives generally shows a marked decrease, a behavior that becomes especially pronounced once the adhesive reaches or surpasses its glass transition temperature (Tg). At this critical point, the adhesive undergoes a significant change in molecular mobility, transitioning from a rigid or glassy form to a more flexible and rubbery-like form. This transformation results in a considerable reduction in both stiffness and load-bearing capability. Studies across various polymer-based adhesives have consistently demonstrated that Young's modulus and tensile strength decline steadily with temperature [7], reflecting a loss of structural integrity. Conversely, tensile strain tends to increase, indicating that although the adhesive becomes weaker, it can stretch more before breaking. This balance between strength loss and strain gain must be carefully evaluated when selecting adhesives for thermally demanding applications such as in electronics, aerospace, and automotive industries [8].

An increase in temperature typically leads to enhanced ductility in adhesive materials. When the matrix of an adhesive softens, its tensile strength typically diminishes, but at the same time, the material often becomes more capable of elongating before breaking. This phenomenon is particularly beneficial in applications where flexibility is required to accommodate thermal expansion, contraction, or movement between bonded components. The increased elongation-at-break signifies a transition in material behavior from brittle to ductile, allowing it to absorb more deformation energy without fracturing. Such behavior is crucial in dynamic environments or under thermal cycling, where rigid adhesives may fail prematurely due to their inability to accommodate movement. This ductile transformation, however, is a double-edged sword while it provides improved strain tolerance, it often comes with a trade-off in mechanical strength, particularly under sustained or cyclic loading [9].

Prolonged exposure to elevated temperatures can induce significant aging effects in adhesive systems, primarily due to irreversible chemical changes within the polymer structure. These changes may include oxidation, chain scission, loss of volatile components, or thermal degradation, all of which progressively weaken the adhesive over time. As materials age, both their cohesive strength (in the adhesive itself) and adhesion strength (at the interface with the substrate) can weaken, leading to diminished performance and eventual failure of the bond. This degradation process can be accelerated in the presence of oxygen, UV radiation, or humidity, making environmental stability a critical factor in adhesive formulation. Conducting long-term thermal aging tests is crucial for accurately estimating the service life and maintaining structural integrity in high-temperature environments, such as in power electronics, engine compartments, and heat-exposed structural components [10].

The thermal response of adhesives can vary significantly depending on their chemical composition and classification. Thermosetting adhesives, such as epoxies and phenolics, generally maintain their mechanical integrity better at elevated temperatures due to their highly crosslinked structures, which resist flow and deformation. In contrast, thermoplastic adhesives tend to soften more readily when heated, leading to a faster decline in mechanical properties such as tensile and shear strengths. Specific formulations, such as polyurethane-based systems, may show intermediate behavior depending on their hard and soft segment composition. Moreover, the presence of fillers, curing agents, and additives can greatly influence an adhesive's thermal resistance. Therefore, selection of adhesive systems for high-temperature use must be guided by a detailed knowledge of their thermal transitions (e.g., Tg), stability under thermal cycling, and their long-term degradation mechanisms [8, 11].

Elevated temperatures typically result in a noticeable reduction in the compressive strength of adhesives, especially as the adhesive approaches or surpasses its glass transition temperature. This temperature-induced transition causes the material to shift from a glassy, hard form to a soft, rubber-like consistency, significantly reducing its ability to resist compressive loads. As a consequence, the compressive modulus also declines, meaning the adhesive deforms more easily under pressure. This softening effect undermines the adhesive's load-bearing capability, which is critical in structural bonding applications subjected to compressive forces. In multilayer assemblies or joint configurations where compression is a primary mode of loading, such thermal softening could result in permanent deformation, buckling, or delamination over time [9, 10].

Similarly to compressive properties, the flexural strength of adhesives also diminishes with increasing temperature. This is primarily due to the same softening mechanisms that reduce tensile and compressive strengths, namely increased molecular mobility and reduced intermolecular forces within the adhesive matrix. However, a noteworthy outcome of this softening is that the material becomes more flexible, allowing for greater deformation before breaking. While this can be advantageous in scenarios where bending or flexing is expected, it also reduces the adhesive's resistance to mechanical failure under bending loads. This trade-off between increased flexibility and decreased strength must be carefully balanced in design applications, particularly in joints subjected to bending stresses or mechanical fatigue over time [10, 12].

At elevated temperatures, adhesives often exhibit increased ductility, which can improve their ability to absorb energy during impact events, which is beneficial in applications where impact resistance is crucial, such as crash-resistant structures or shock-absorbing mounts. However, this gain in ductility is frequently accompanied by a loss in overall impact strength. While the material may deform more before failing, it absorbs less energy before rupture, making it more vulnerable to failure under high-energy impacts. This is due to the adhesive's softened state and reduced internal resistance to crack initiation and propagation at high temperatures. Therefore, when designing for thermal and mechanical impact loading, the reduced impact resistance of adhesives at high temperatures must be carefully considered in material selection and joint design [8, 10].

The resistance of adhesives to crack initiation and propagation typically decreases as temperature increases. At higher temperatures, the adhesive material becomes softer and more compliant, which facilitates the growth of existing cracks and flaws within the adhesive layer. This is associated with a reduction in fracture toughness, a critical property that determines a material's ability to resist crack growth under stress. Increased molecular motion at elevated temperatures also contributes to lower energy dissipation around the crack tip, allowing fractures to propagate more easily. This makes high-temperature performance a vital concern in critical structural applications, where failure due to crack growth could compromise the safety or integrity of the bonded system [12, 13].

Lap shear stress is a commonly encountered stress mode in adhesive joints, especially in structural bonding applications. The valuable insights into its load-bearing capacity, durability, and overall performance in service environments can be obtained from lap shear strength [14]. Since these joints are frequently subjected to both mechanical and thermal stresses, maintaining adequate shear strength under temperature fluctuations is essential. A reduction in lap shear strength at high temperatures can signal potential failure points and impact the service life of the joint. Engineers often rely on lap shear testing as a standardized metric to compare adhesives, evaluate aging behavior, and ensure the joint can withstand operational loads over extended periods [14].

22.3 Methodology

22.3.1 Experimental Methods

22.3.1.1 Tensile Test

The tensile properties of epoxy adhesives are typically determined in accordance with the international standard ISO 527-1 [15], which outlines the methodology for testing plastics and polymer materials, including thermosetting adhesives. The standard prescribes the use of a dogbone-shaped

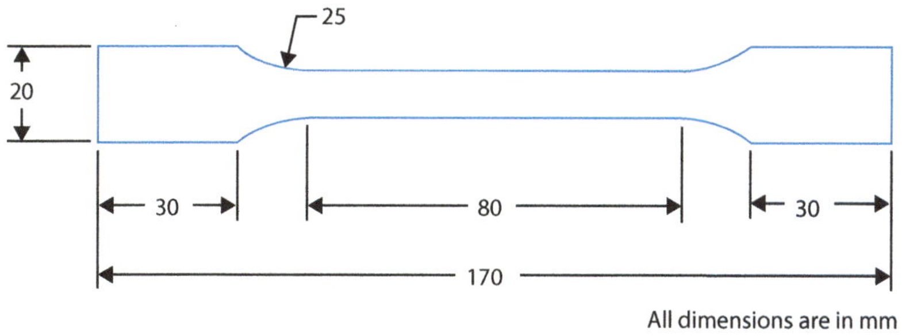

Figure 22.1 Tensile test specimen according to ISO 527-1.

specimen as shown in Figure 22.1 to achieve a uniform stress distribution across the gauge section during testing. The dimensions of this specimen, including the gauge length, width, and thickness, are detailed in the ISO 527-1 document. The tensile test is usually performed under displacement control by displacement control method in which the crosshead speed should be kept as 1 mm/min, which provides a controlled strain rate suitable for brittle and semi-brittle materials like epoxy. This test allows for the determination of several important mechanical properties, such as Poisson's ratio, Young's modulus (elastic modulus), ultimate tensile strength, and yield strength, which are critical for assessing the performance of the adhesive under uniaxial loading. These properties are used in numerical simulations and design assessments of bonded joints subjected to tensile loads.

Additionally, ASTM D638 [16] is another widely recognized standard that outlines similar procedures for evaluating the tensile behavior of plastics and adhesive materials. Both standards provide comprehensive guidelines on specimen preparation, test setup, and data interpretation, ensuring reliable and reproducible results across various laboratories and research facilities.

22.3.1.2 Compression Test

The EN 196-1 [17] standard is a well-established European specification used primarily for testing the compressive strength of hydraulic cement, but it can also be effectively applied to determine the compressive properties of epoxy adhesives, particularly in standardized research and quality control environments. According to the guidelines provided in this standard, the specimen used for compression testing should be a prism with a square base of 40×40 mm^2 and a height of 80 mm as shown in Figure 22.2.

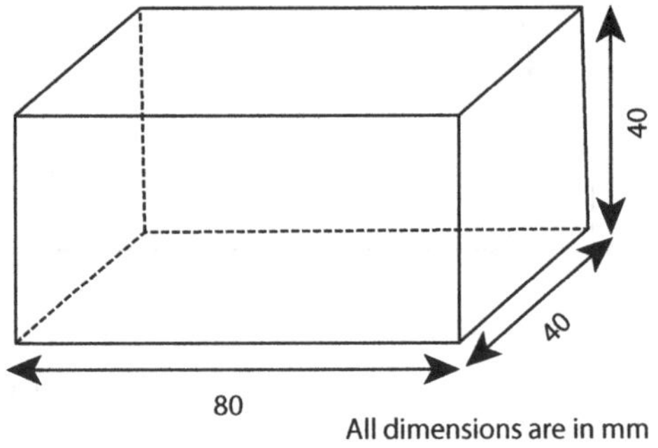

Figure 22.2 Compression test specimen according to EN 196-1.

This geometry ensures consistent stress distribution and minimizes the risk of buckling or misalignment during testing. The compression test is carried out under force-controlled conditions [18], where the application of the load is at a steady and smooth rate of 2400 ± 200 N/s until the specimen fails or fractures. This controlled loading rate is essential to accurately assess the material's resistance to compressive forces without introducing dynamic effects or abrupt failures that might skew the results. The test setup is designed to simulate real-world compressive stresses that adhesives may encounter in structural applications such as composite assemblies and column bonding. The compression test yields important mechanical properties such as compressive strength, modulus, and deformation characteristics, which are vital for assessing the adhesive's structural performance under different loading scenarios. Figure 22.2 depicts the specimen geometry in accordance with the standard guidelines.

Besides EN 196-1, the ASTM D695 [19] standard is also a widely accepted international method for evaluating the compressive properties of rigid plastics and adhesives, including those based on epoxy formulations. While ASTM D695 may use cylindrical or rectangular specimens, its test principles are similar, focusing on precise control of loading conditions and specimen alignment to obtain reliable, reproducible results. Both EN 196-1 and ASTM D695 are essential tools for materials scientists and engineers in assessing how adhesives will perform under compressive stress, particularly in demanding environments such as aerospace, automotive, civil infrastructure, and electronics packaging. Using both standards together can offer a broader comparative insight into material performance across different international test quality and test protocols.

22.3.1.3 Flexural Test

The flexural behavior of epoxy adhesives and polymeric materials can be accurately assessed using the standard ASTM D790 [20], which is designed to evaluate the flexural strength and flexural modulus of plastics as well as reinforcement like nanofillers added in plastics [21], including insulating materials. This is a three-point bend test, where a rectangular bar-shaped specimen is supported at its ends and at the midpoint the load is applied. The standard recommends a support span-to-width ratio of 16:1, unless specific requirements dictate otherwise. The specimen is loaded progressively until failure occurs at the outer surface or until it reaches a maximum strain of 5.0%, depending on which happens first. The test provides valuable data on the specimen's stiffness and resistance to bending loads, these properties are very important for applications involving flexural stress in service. ASTM D790 outlines two test procedures: Procedure A employs a slower strain rate of 0.01 mm/mm/min [22], ideal for evaluating brittle materials, while Procedure B utilizes a faster rate of 0.1 mm/mm/min, making it more appropriate for ductile materials. The specific procedure chosen depends on the expected behavior of the test material and the required resolution of the stress-strain response. The specimen dimensions are depicted in Figure 22.3 in accordance with the test standard. Flexural testing is crucial for understanding how the adhesive responds to bending forces in practical structural assemblies, such as curved joints or surfaces exposed to torsion or flexure.

22.3.1.4 Impact Test

ASTM D256 [23] is an internationally accepted method for assessing impact resistance through the Izod pendulum test [24]. It evaluates the amount of energy a notched sample absorbs when impacted by a swinging

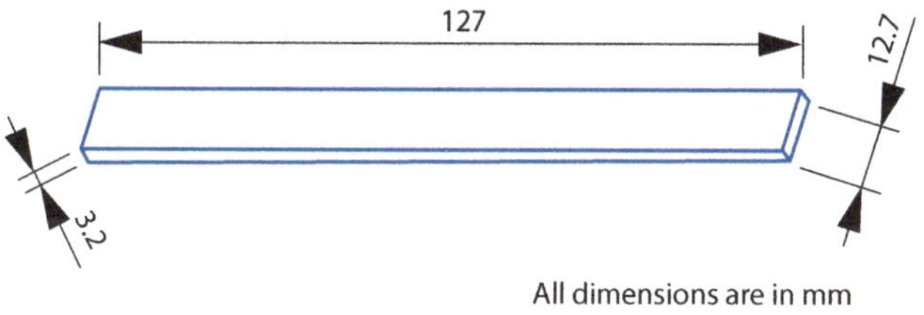

Figure 22.3 Flexural test specimen according to ASTM D790.

pendulum, offering essential insights into the material's toughness and sensitivity to notches under abrupt loading. The test entails hitting a notched specimen with a pendulum hammer and recording the energy needed to cause fracture. The standard specifies dimensions for the test specimen, typically 63.5 mm (2.5 inches) in length and 12.5 mm (0.5 inches) in height as shown in Figure 22.4 with a notched section to concentrate stress and promote brittle fracture.

The test is designed to simulate real-world impact conditions, helping manufacturers to assess the durability and resilience of plastics used in various industries, such as consumer electronics and automobile industry [25]. The notched specimen is clamped vertically with the notch facing the pendulum. The pendulum is dropped from a predetermined height and impacts the notched side of the specimen. The energy required to fracture the specimen is determined by measuring the difference in the pendulum's height before and after the impact. The standard pendulum hammer has an initial potential energy of 2.7 J and an impact velocity of approximately 3.46 m/s.

ISO 11343:2019 [26] is also a standard to evaluate the impact resistance of adhesives. It outlines a dynamic impact wedge test designed to evaluate the cleavage resistance of high-strength adhesive joints when subjected to impact forces.

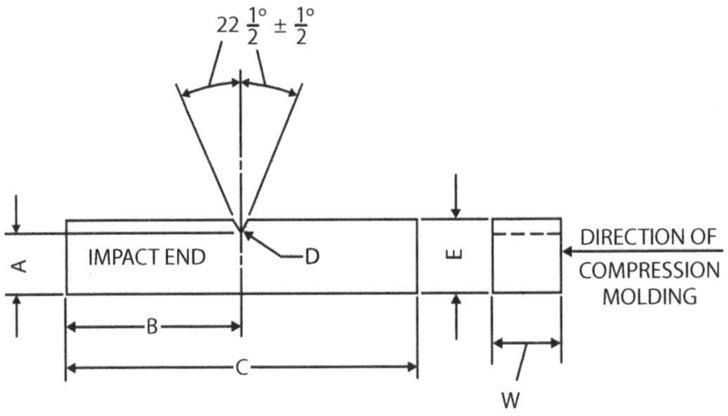

Figure 22.4 Standard Izod impact test specimen according to ASTM D256 [23]. The dimensions are as follows. A = 10.16 ± 0.05 mm, B = 31.8 ± 1.0 mm, C = 63.5 ± 2.0 mm, D = 0.25 ± 0.05 mm, E = 12.70 ± 0.20 mm. The specimen width should be according to material specification or defined by the supplier.

22.3.1.5 *Peel Tests*

Peel tests are used to determine the force needed to separate an adhesive from a substrate at defined angles, typically 90° or 180°. These tests are commonly employed to evaluate the adhesion strength and toughness of materials [27[. The angle at which peeling occurs significantly affects the peel force. According to existing studies, as the peel angle increases, the required force generally decreases up to a certain point. Eitner and Rendler [28] conducted peel tests on soldered or bonded ribbons used in solar cell metallization and observed that lower angles below 90° required greater peel forces, while the weakest forces were recorded at an angle of 135° [29]. DIN 53282 is a German standard that specifies the method for testing adhesives used for metals and adhesively bonded metal joints, specifically through the T-peel test. The T-peel test measures the force required to peel apart two metal adherends that have been bonded together with an adhesive, providing a quantitative assessment of the adhesive's performance in peel loading situations. ASTM D1876 – 08 [30] is also a standard test method for T-peel test. The standard specimen and T-peel panel for T-peel test are given in Figure 22.5. The T-peel test panel is 152 mm by 305 mm with unbonded length of 76 mm whereas test specimen is 25 mm wide. The standard test rate of loading (peel rate) in the T-peel test under ASTM D1876-08 is 254 mm/min (10 inches/min). Other rates, including

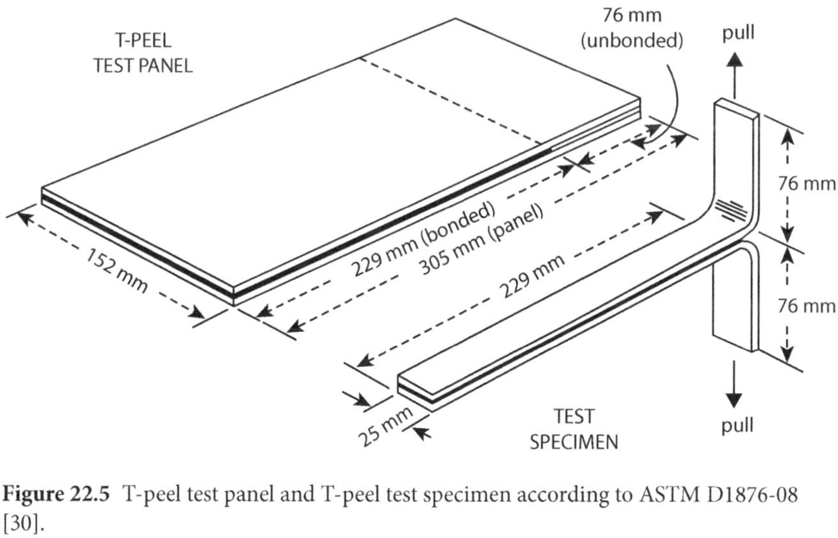

Figure 22.5 T-peel test panel and T-peel test specimen according to ASTM D1876-08 [30].

50.8 mm/min (2 in/min) and 25.4 mm/min (1 in/min), are occasionally used in special applications, but 254 mm/min is the default test rate in the standard.

22.3.1.6 Shear Test

The shear testing of adhesives can be done using single-lap joints. The single-lap joints are made according to the ASTM D1002 standard [31]. It outlines a standardized procedure for evaluating the apparent shear strength [32] of adhesives using a single-lap joint specimen setup. In this method, two metal plates are bonded using the adhesive and cured before being machined into uniform specimens, typically 25.4 mm in width with an overlap of 12.7 mm. The adhesive layer is maintained at a thickness of 0.2 mm. The metal plates should ideally be 1.62 mm thick, and the overall specimen length is 177.8 mm, as illustrated in Figure 22.6. Testing is performed using a universal testing machine, where the specimen is pulled at a constant rate usually 1.3 mm/min (0.05 in/min) until failure. The load is applied along the centerline to promote even stress distribution across the bonded area.

22.3.2 Numerical Methods

Non-linear geometric analysis can be carried out using widely available finite element (FE) software tools such as ABAQUS, ANSYS, and others. To perform the finite element analysis, the geometry of the sample which needs to be tested is replicated in the software. After importing

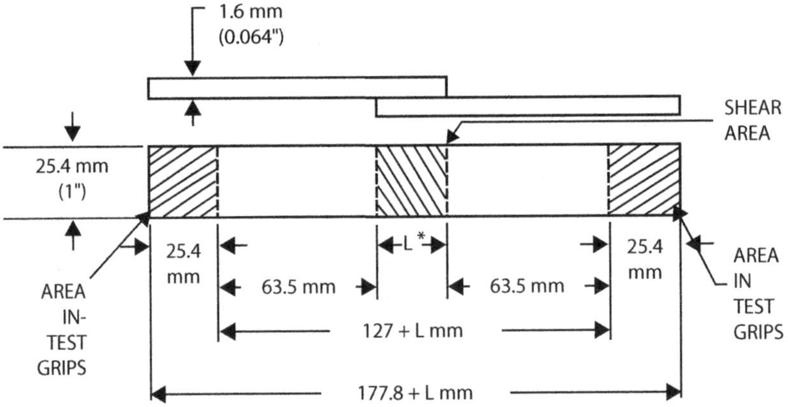

Figure 22.6 Standard single lap joint configuration according to ASTM D1002 [31].

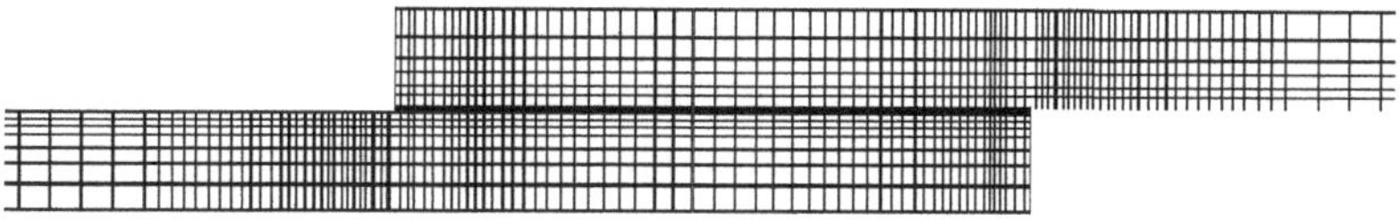

Figure 22.7 Meshing a single-lap joint.

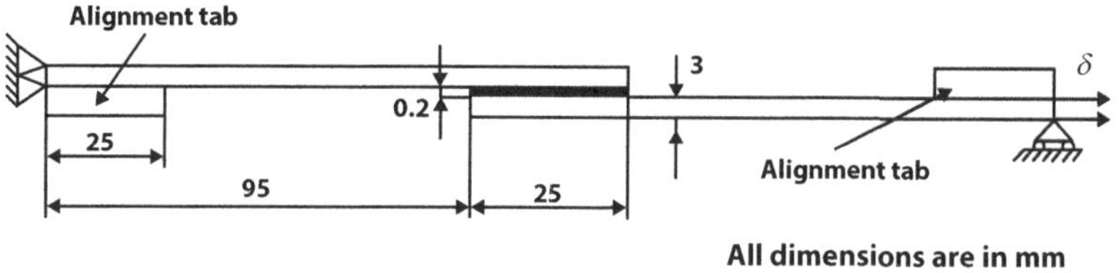

Figure 22.8 Boundary conditions in a single-lap joint [34].

the geometry of the specimen in the software the geometry needs to be meshed. A meshing is a process in which the geometry of the specimen is divided into smaller elements called finite elements. There are multiple types of elements like tetrahedral elements, hexahedral elements etc. These different types of elements can be used in different types of geometries depending on the complexity of the geometry. The mesh for the SLJs is shown in Figure 22.7.

After performing meshing, boundary conditions need to be applied on the meshed specimen. The boundary conditions are fixed supports, roller supports, applied load, displacement etc. The boundary conditions for a single-lap joint, illustrated in Figure 22.8, involve fixing one end of the joint while applying a vertical constraint along with a tensile displacement (δ) at the opposite end. This setup is intended to accurately replicate the conditions observed during experimental testing [33, 34].

22.3.2.1 *Cohesive Zone Modelling*

Cohesive zone models (CZMs) are used to simulate the elastic behavior, onset of damage, and progressive failure in materials due to local degradation. These models are built upon a relationship between relative displacements and corresponding stresses at initially coincident nodes within cohesive elements. This framework captures both the initial elastic response up to the maximum load and the subsequent softening that leads to full material failure. The shape of the softening curve can be tailored to reflect the actual behavior of the material or interface being modeled.

Traction–separation laws, which govern the separation in different loading modes (such as tension and shear), are associated with specific fracture energies. In purely tensile or shear modes, damage initiates at integration points when stresses begin to decline according to the relevant traction–separation law. Under mixed-mode conditions, energy-based approaches are commonly used to combine tension and shear effects, enabling accurate modeling of the complex failure behavior in bonded structures [35]. The traction–separation model typically begins with a linear elastic phase followed by a linear damage evolution phase. The elastic behavior is described using a constitutive matrix that relates interface stresses to strains, as detailed in the following equation.

$$t = \begin{Bmatrix} t_{\mathrm{n}} \\ t_{\mathrm{s}} \end{Bmatrix} = \begin{bmatrix} K_{\mathrm{nn}} & K_{\mathrm{ns}} \\ K_{\mathrm{ns}} & K_{\mathrm{ss}} \end{bmatrix} \begin{Bmatrix} \varepsilon_{\mathrm{n}} \\ \varepsilon_{\mathrm{s}} \end{Bmatrix} = K\varepsilon.$$

Here t is traction vector, representing force per unit area acting on a surface. It has components t_n (normal traction) and t_s (shear traction). Also K is a constitutive matrix that relates the traction vector (stresses) to the displacement jump vector (relative displacements) across the cohesive interface, before damage initiates, which characterizes the material's resistance to deformation. Its components are K_{nn}, K_{ns}, and K_{ss}, representing normal-normal, normal-shear, and shear-shear stiffness terms, respectively and ε is strain vector, representing the deformation of the material. It has components ε_n (normal strain) and ε_s (shear strain). For a thin adhesive layer, the relationship between the adhesive's elastic properties and its thickness can be simplified using an appropriate approximation. In this approach, the stiffness parameters are defined as K_{nn}=E, K_{ss}=G, and K_{ns}=0 where E and G represent the elastic modulus in the normal and shear directions, respectively. Several methods can be employed to characterize damage initiation. In the referenced study [34], the quadratic nominal stress criterion was adopted for this purpose, as it has been shown to provide accurate predictions. The expression for this criterion is given as follows,

$$\left\{\frac{\langle t_{\mathrm{n}}\rangle}{t_{\mathrm{n}}^{0}}\right\}^{2} + \left\{\frac{t_{\mathrm{s}}}{t_{\mathrm{s}}^{0}}\right\}^{2} = 1.$$

In this context, t^0_n and t^0_s represent the maximum nominal stresses in the pure normal and shear modes, respectively. The use of Macaulay brackets around t_n indicates that only tensile stresses contribute to damage initiation, as compressive stresses are excluded from triggering failure. Complete separation of the adhesive layer is often modeled using a linear power law that combines the fracture energies associated with pure tensile and shear modes.

$$\frac{G_n}{G_n^c}+\frac{G_s}{G_s^c}=1.$$

The terms G_n and G_s correspond to the energy dissipated due to the tractions and relative displacements in the normal and shear directions, respectively. The critical fracture energies required to cause failure under pure normal and pure shear loading conditions are denoted as Gn^c and Gs^c, respectively [35].

22.4 Results and Discussion

22.4.1 Tension Test

Jahani [36] has studied the influence of test temperature on the tensile behavior of the adhesive. The adhesive examined is a solvent-free, thixotropic, two-part epoxy, commercially known as S&P 220 HP, designed for bonding carbon-fiber reinforced polymer (CFRP) to concrete surfaces. Analysis of the stress–strain curves reveals that an increase in test temperature leads to a noticeable reduction in both tensile strength and modulus [37]. This reduction is attributed to the adhesive becoming softer and more ductile at elevated temperatures [38]. A pronounced drop in mechanical performance is observed as the temperature approaches the glass transition point of the adhesive. Detailed values for tensile strength and modulus are provided in Table 22.1 and illustrated in Figure 22.9, while the corresponding stress–strain curves are shown in Figure 22.10.

Nanofillers can be added to the adhesive to enhance its properties. Banea *et al.* [39] used a two-component epoxy adhesive Betamate 2098 and nanofiller was Expancel 031 DU 40 particles. The nanofillers were added in the epoxy adhesive and tested at different temperatures. The neat epoxy had a tensile strength of 25 MPa at room temperature, and at 80 °C it went down to 2.5 MPa. When nanofiller was added 10 and 20 percent in the epoxy its strength was 17 MPa and 14 MPa at room temperature while its strength was reduced to 2.3 MPa, and 2.4 MPa, respectively, as shown in Figure 22.11.

Table 22.1 Tensile strength and elastic modulus of adhesives samples tested at different temperatures. The sample tested at 20°C is named as TT-TEN-20, similarly for samples tested at 40°C, 50°C, 60°C, 70°C, 85°C are named as TT-TEN-40, TT-TEN-50, TT-TEN-60, TT-TEN-70, TT-TEN-85.

Specimen	Tensile strength (MPa)	Elastic modulus (MPa)
TT-TEN-20	28±0.1	8102.4±67.0
TT-TEN-40	23±0.2	5520.8±179.0
TT-TEN-50	19.6±0.7	4289.6±100.7
TT-TEN-60	6.9±0.2	673.2±24.5
TT-TEN-70	2.7±0.1	271.5±12.9
TT-TEN-85	1.9±0.0	247.5±1.8

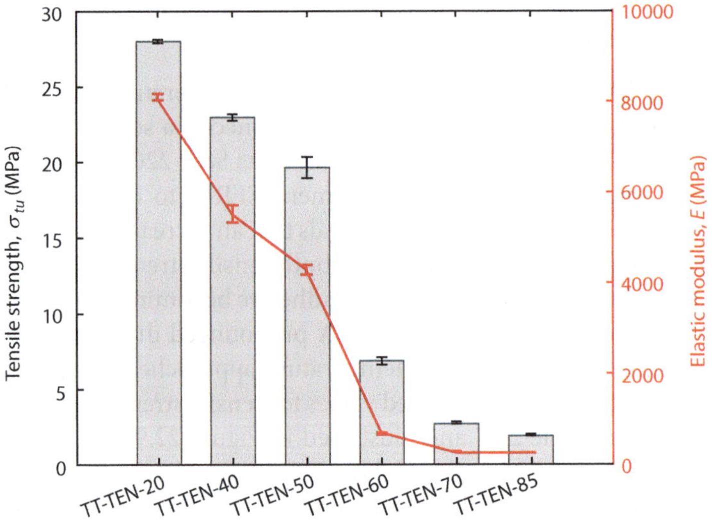

Figure 22.9 Tensile strength and elastic modulus of the adhesive at different testing temperatures. The sample tested at 20°C is named as TT-TEN-20, similarly for samples tested at 40°C, 50°C, 60°C, 70°C, 85°C are named as TT-TEN-40, TT-TEN-50, TT-TEN-60, TT-TEN-70, TT-TEN-85, respectively.

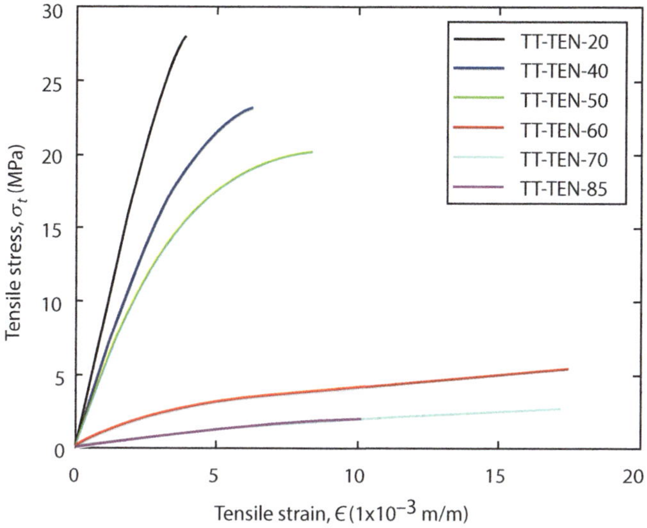

Figure 22.10 Stress-strain graph of the adhesive at different test temperatures. The sample tested at 20°C is named as TT-TEN-20, similarly the samples tested at 40°C, 50°C, 60°C, 70°C, 85°C are named as TT-TEN-40, TT-TEN-50, TT-TEN-60, TT-TEN-70, TT-TEN-85, respectively.

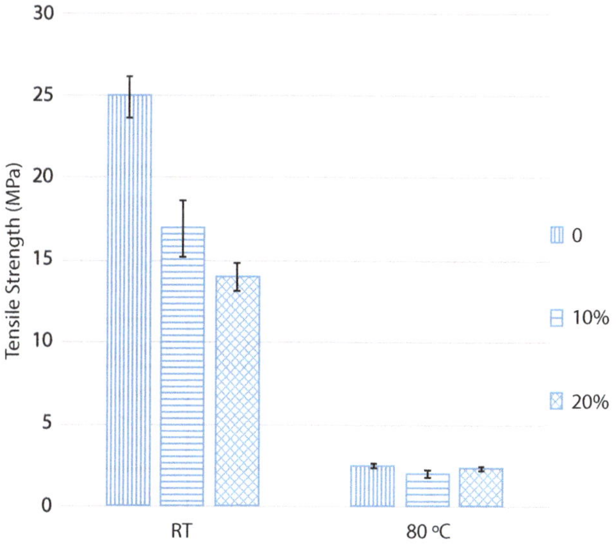

Figure 22.11 Tensile strength of epoxy adhesive at room temperature (RT) and 80°C with addition of nanofiller.

22.4.2 Compression Test

Along with tensile properties, the paper [38] also provides the compression properties of the S&P HP 220 adhesive under various temperatures. Similarly to the tensile behavior, the compressive properties of the adhesive are also influenced by increasing test temperatures [29]. As the test temperature rises, the adhesive becomes increasingly ductile and flexible, leading to a reduction in the ultimate compressive strength with higher temperatures [37]. Table 22.2 shows the values of compressive strength and load. The compressive strength appeared to be half at the test temperature around 60°C. Figure 22.12 shows the compressive load versus the deflection of the adhesive at different temperatures. Figure 22.13 shows the compressive strength and ultimate compressive force at different temperatures.

22.4.3 Flexural Test

With rising temperature, the flexural strength of adhesives typically declines. This is a result of thermal softening, which diminishes the material's resistance to deformation and fracture under stress [40]. At lower temperatures, adhesives tend to be stronger but more brittle, showing higher flexural strength but with a greater likelihood of brittle failure [41]. The relationship between temperature and strength is often nonlinear, with a marked drop in flexural strength as the adhesive approaches or exceeds its glass transition temperature (Tg). Rothenhäusler and Ruckdaeschel [42]

Table 22.2 Compressive strength of adhesives samples tested at different temperatures alongside their compressive load. The sample tested at 20°C is named as TT-COM-20, similarly for samples tested at 40°C, 50°C, 60°C, 70°C, 85°C are named as TT-COM-40, TT-COM-50, TT-COM-60, TT-COM-70, TT-COM-85, respectively.

Specimen	Compressive strength (MPa)	Compressive load (kN)
TT-COM-20	79.1±0.1	128.4±0.4
TT-COM-40	60.5±2.6	96.9±4.7
TT-COM-50	49.5±0.6	79.3±1.7
TT-COM-60	44.2±1.0	70.6±1.8
TT-COM-70	24.4±0.1	38.9±0.7
TT-COM-85	9.9±0.7	15.9±1.2

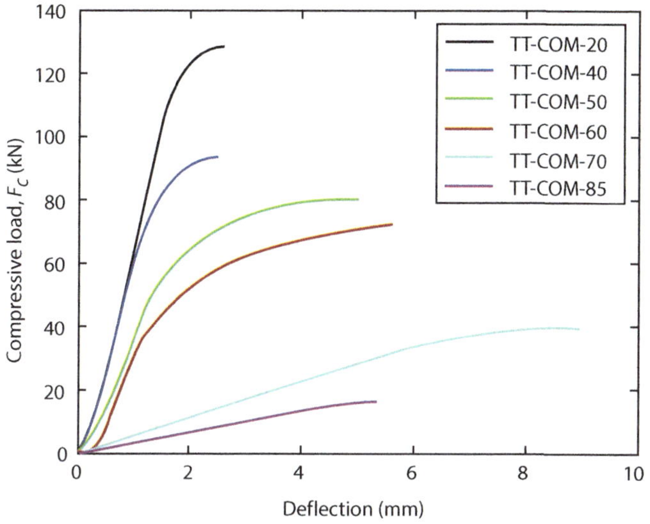

Figure 22.12 Compressive load-deflection graph of the adhesive at different test temperatures. The sample tested at 20°C is named as TT-COM-20, similarly for samples tested at 40°C, 50°C, 60°C, 70°C, 85°C are named as TT-COM-40, TT-COM-50, TT-COM-60, TT-COM-70, TT-COM-85, respectively.

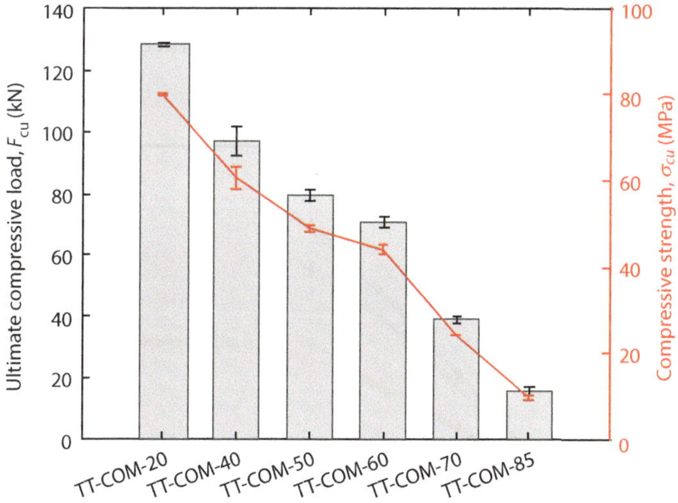

Figure 22.13 Ultimate compressive load and compressive strength of the adhesive at different test temperatures. The sample tested at 20°C is named as TT-COM-20, similarly for samples tested at 40°C, 50°C, 60°C, 70°C, 85°C are named as TT-COM-40, TT-COM-50, TT-COM-60, TT-COM-70, TT-COM-85, respectively.

performed a three-point bending test to determine the flexural strength of diglycidyl ether of bisphenol A (DGEBA) resin [43], which was cured using L-arginine, referred to as Argopox in the study. The test was conducted at various temperatures, with stress–strain curves for Argopox shown in Figure 22.14 at temperatures of −40°C, 22°C, and 80°C. Similarly to the results observed in tensile and compression tests, the flexural modulus of the thermoset decreases with rising temperature, from 3.3 GPa at −40°C to 2.2 GPa at 80°C. As a result, the flexural strength diminishes and the flexural strain at failure increases with higher temperatures (refer to Table 22.3). Failure typically initiates at the side of the specimen where tensile stresses are present, as the tensile strength of the thermoset is much lower than its compressive yield strength [42]. Furthermore, at room temperature, the tensile and compression yield strengths of Argopox (84.7 MPa) are lower than those of petroleum-based amine-cured DGEBA (95 MPa to 123 MPa) [44].

Doping adhesives with nanoparticles (e.g., zinc oxide) greatly increases their flexural strength, stiffness, and thermal stability, particularly at room and moderately high temperatures. Optimum nanoparticle loading enhances interfacial bonding and stress transfer, resulting in higher flexural strength and modulus. For instance, incorporation of 2 phr zinc oxide nanoparticles enhanced the modulus by 37% and tensile strength by 60% compared to neat epoxy. At elevated nanoparticle concentrations, agglomeration is possible, and this can minimize the mechanical advantage.

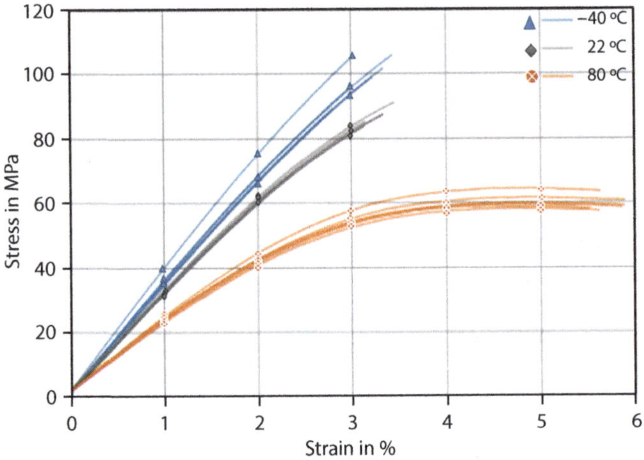

Figure 22.14 Stress versus strain graph of 3-point bend test on Agropox at different temperatures.

Table 22.3 Flexural properties of Agropox at different temperatures.

Temperature (°C)	Flexural modulus (GPa)	Flexural strength (MPa)	Flexural strain (%)
80	2.2 ± 0.1	60.6 ± 2.1	5.81 ± 0.14
22	2.9 ± 0.1	84.7 ± 6.9	3.17 ± 0.36
−40	3.3 ± 0.2	100.5 ± 5.9	3.18 ± 0.22

Nanoparticles also enhance thermal stability, enabling the adhesive to retain a greater percentage of its flexural strength at high temperatures than unmodified adhesives [45].

Functionally graded adhesives (FGAs) are particularly engineered to have more stable thermo-mechanical properties at high temperatures than homogeneous adhesives. The graded structure enables more efficient thermal stress distribution and postpones the beginning of property degradation that usually happens at high temperatures [46]. Though all polymer adhesives suffer from some extent of stiffness and flexural strength reduction with temperature increase (as a result of matrix softening), FGAs do so at a reduced rate. The gradual variation of material properties in functionally graded adhesives (FGAs) helps reduce stress concentrations caused by higher temperatures, which are a common cause of premature failure in conventional adhesives, especially under flexural (bending) loads at elevated temperatures [47, 48].

22.4.4 Impact Test

At higher temperatures (up to 80°C), monolithic structural adhesives typically show a reduction in dynamic resistance to cleavage. This is due to softening of the adhesive matrix, which lowers the force required to propagate a crack. For epoxy-based adhesives, increasing test or cure temperature (e.g., curing at 180°C) can enhance initial crosslinking and toughness, but testing at elevated temperatures still leads to decreased absorbed energy and peel strength compared to room temperature. In carbon-fiber reinforced thermoplastic composites (e.g., Carbon fiber reinforced polyetheretherketone called as CF/PEEK), wedge peel strength increases with processing temperature up to a certain point (e.g., 220–240°C), likely due to improved matrix flow and fiber wetting. However, at temperatures above this range, peel strength can decrease, attributed to changes in fiber distribution and matrix crystallinity. For example, wedge peel strength was observed to increase to approximately 3.2 N/mm at 220–240°C, but further temperature increases led to a decline in peel strength [49].

At higher temperatures, the impact strength of monolithic adhesives (such as pure epoxy) decreases significantly, primarily due to softening as the material reaches or passes its glass transition temperature (Tg). This reduction in strength and toughness limits the ability of monolithic adhesives to absorb and dissipate impact energy under thermal stress [6].

In contrast, composite adhesives formed by adding reinforcements like fibers or fillers to the resin matrix demonstrate improved impact resistance and mechanical integrity at elevated temperatures. The reinforcing components in composite adhesives help maintain structural performance by restricting crack propagation and providing additional energy dissipation mechanisms.

22.4.5 Lap-Shear Test

The lap-shear strength in single-lap joints (SLJs) has a very strong temperature dependence that is centered mainly on adhesive's glass transition temperature (Tg). Tests on aluminum alloy 6082-T651 (AlSi1MgMn - Aluminium as base material which contains Si1 –1% Silicon, Mg – Magnesium, Mn –Manganese) single-lap adhesive joints reveal that while tensile stress diminishes with increasing temperature, lap-shear strength shows an increase with increasing temperature up to Tg. The apparently contradictory effect can be accounted for by the increased ductility of the adhesive, which enables better stress distribution across the bonded area. But beyond Tg, the lap-shear strength falls significantly. The test was conducted according to ISO standard 11003-2:1999 [50]. Adhesive selected was XN1244 adhesive (XN1244 is a one-component high temperature paste epoxy adhesive). Prior to adhesive application, the joint surfaces were grit-blasted and degreased with acetone [51]. After applying the adhesive, the joints were cured, forming a bondline with a thickness of 0.2 mm and an overlap length of 25 mm. The adherends had a width of 25 mm and a thickness of 3 mm. Figure 22.8 illustrates the boundary conditions used for a single-lap joint (SLJ) test [52]. Mechanical testing was conducted using an Instron 8801 universal testing machine at both room temperature (RT) and elevated temperatures (100, 125, 150, 175, and 200°C), with a constant crosshead speed of 1 mm/min. An environmental chamber integrated with the testing machine was used to achieve and maintain the desired temperatures.

Below the adhesive's glass transition temperature (Tg), the SLJ lap-shear strength improved showing approximately a 9% increase at 100°C and 30%

at 125°C. However, above Tg, there was a marked decline in shear strength dropping by about 54% at 175°C and approximately by 75% at 200°C [34], as presented in Figure 22.15. The corresponding load-displacement curves for various temperatures are shown in Figure 22.16. A numerical simulation was also carried out for the same material system. The adhesive layer was modeled using a cohesive zone model (CZM) with a traction–separation law, while the adherends were considered elastic-perfectly plastic. Fracture was driven by cohesive crack propagation, initiating at the edges of the overlap and progressing rapidly toward the bond interior. Table 22.4 lists the CZM material parameters for the XN1244 adhesive.

At room temperature, the predicted failure load using the CZM approach was around 25% higher than the experimental value. This discrepancy was partly attributed to residual stresses, which arise when high-temperature-cured adhesives (such as XN1244, cured at 140°C) cool down to room temperature. These stresses may result from differential thermal contraction between the adhesive and adherends or from chemical shrinkage during curing. Such residual stresses can influence the energy available for crack propagation.

Furthermore, the finite element (FE) model tended to exhibit stiffer (less strain or displacement) responses compared to experimental data. At 100°C, the simulation closely matched the experimental behavior, while at 150°C and 200°C i.e. near or exceeding Tg the simulation showed reasonable agreement in terms of load but under-predicted fracture displacement. This indicates that some failure mechanisms may not have been fully captured by the CZM at elevated temperatures. Figure 22.17 compares the load–displacement curves for XN1244 adhesive SLJs obtained experimentally and numerically [34].

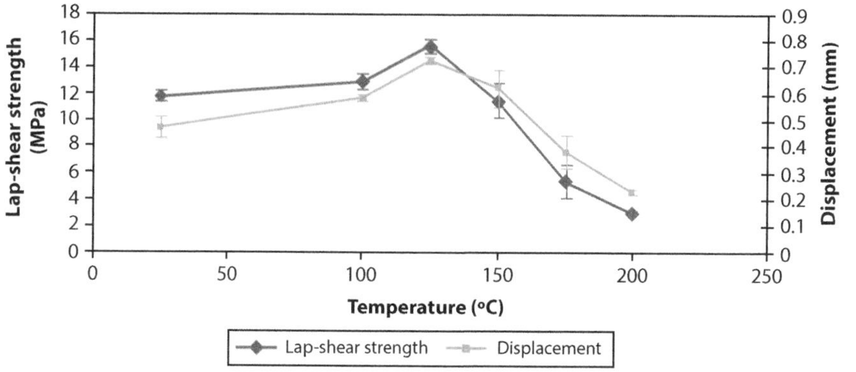

Figure 22.15 Lap-shear strength and displacement of SLJ vs temperature [34].

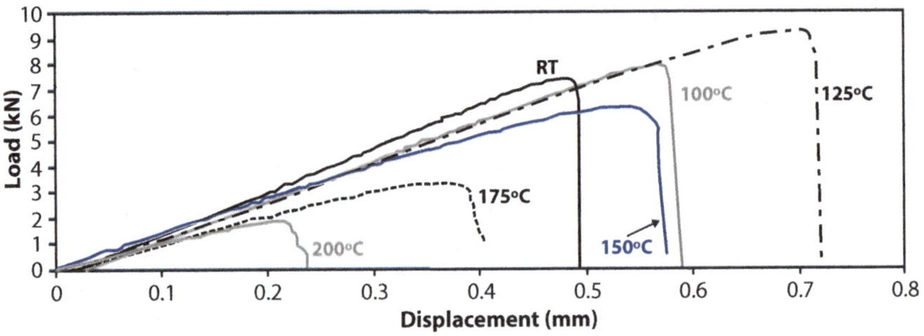

Figure 22.16 Load versus displacement of an SLJ at different temperatures [34].

Table 22.4 The properties of XN1244 adhesive for CZM modelling [34].[1]

Temperature (°C)	G^c_n (N/mm)	G^c_s (N/mm)	t^0_n (MPa)	t^0_s (MPa)	E (MPa)	G (MPa)
RT	0.47	2.2	68	32	5872	2150
100	0.5	2.6	6.45	25	4173	1527
150	0.42	1.7	6.5	12	72	25
200	0.07	0.3	1.4	3	40	14.6

[1]The critical fracture energies under pure normal and pure shear loading conditions are denoted as G^c_n and G^c_s. t^0_n and t^0_s represent the maximum nominal stresses in the pure normal and shear modes, respectively and E and G represent the elastic modulus in the normal and shear directions, respectively.

The influence of incorporating Al_2O_3 nanoparticles into the Epocast 50-A1/946 epoxy adhesive (Epocast 50-A1/946 is a high-performance, two-part epoxy system developed by Huntsman Advanced Materials) has been investigated [53]. The numerical simulation is done to find the shear behavior of the adhesive with Al_2O_3 inclusion in the adhesive at different temperatures. The adherend used in this study was AL 5083 aluminium and carbon fiber reinforced polymer (CFRP). The temperatures were 25°C, 50°C, and 75°C. The results were found at four different locations in the overlap region. Location A which is interface of aluminium and adhesive. Location B refers to the central plane within the adhesive layer, while Location C corresponds to the midpoint along the length at the edge of the overlap region. Location D was along the length of overlap. The addition of Al_2O_3 increases the strength of adhesive at room temperature and higher

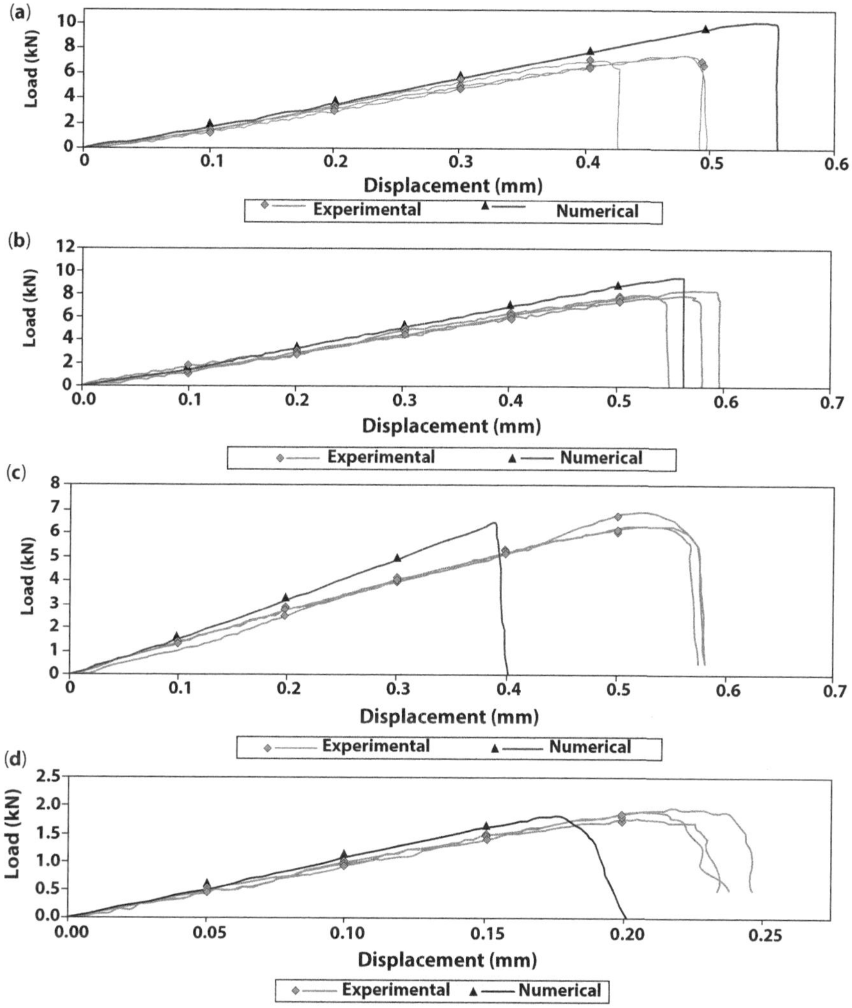

Figure 22.17 Representative experimental and numerical load-displacement curves for XN1244 adhesive SLJs at (a) RT, (b) 100°C, (c) 150°C and (d) 200°C [34].

temperature also. The geometry is shown in Figure 22.18, Figure 22.19 illustrates the shear stress distribution for the unmodified (neat) epoxy adhesive, while Figure 22.20 presents the corresponding distribution for the adhesive enhanced with Al_2O_3 nanoparticles.

Adding Al_2O_3 nanoparticles to epoxy adhesive enhances its stiffness by increasing the elastic modulus and overall strength [54]. The impact of incorporating 1.5% Al_2O_3 into the epoxy adhesive is detailed in Table 22.5.

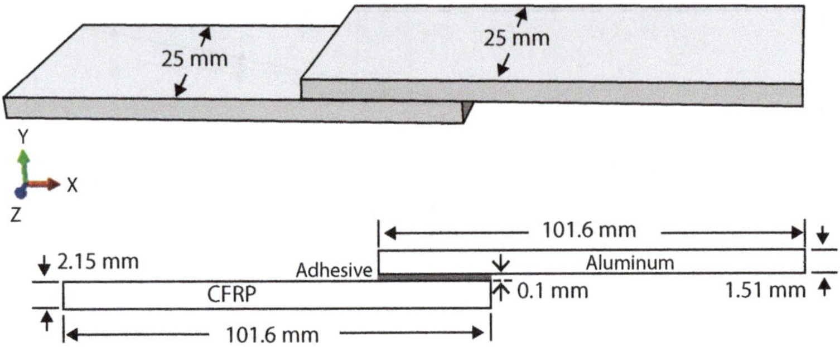

Figure 22.18 Geometry of SLJ of aluminium and CFRP [53].

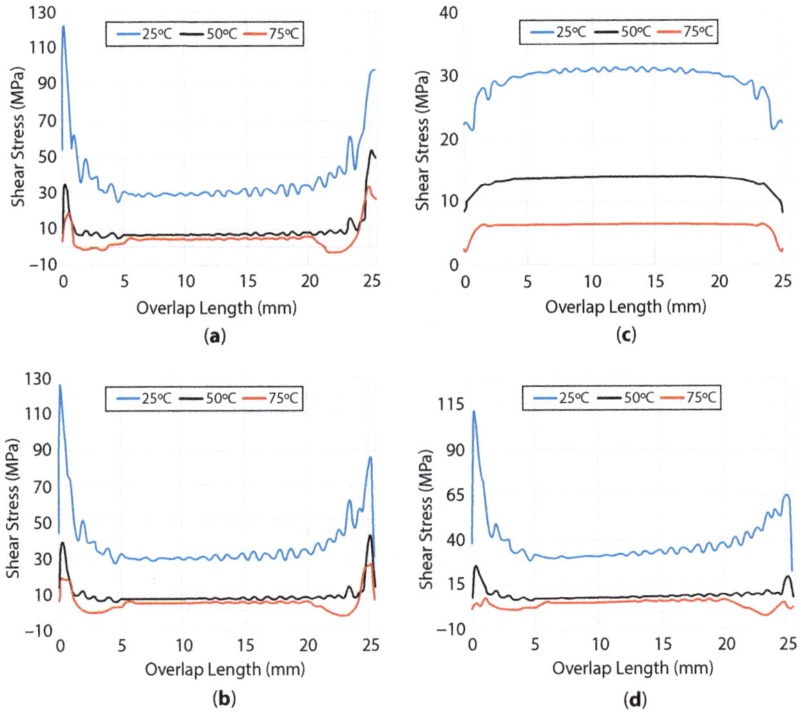

Figure 22.19 Shear stress distribution in SLJ for neat epoxy at different temperatures [53].

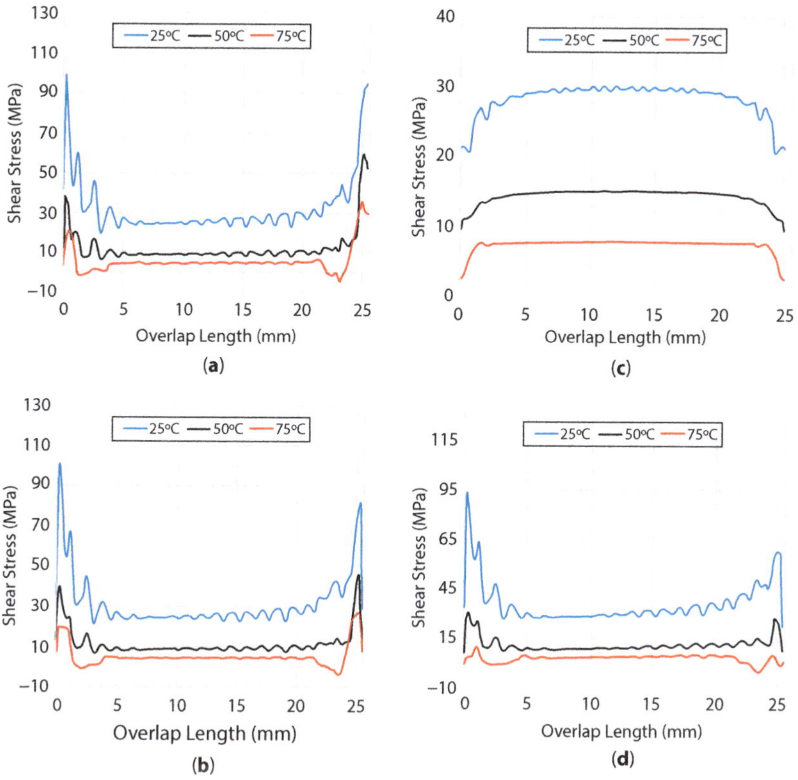

Figure 22.20 Shear strength of epoxy adhesive with 1.5% Al_2O_3 added at different temperatures [53].

Table 22.5 The percentage change in shear stress of epoxy with 1.5% of Al_2O_3 at different locations, and different temperatures [53].

Location	25°C	50°C	75°C
A	−19.66%	7.71%	8.72%
B	−19.45%	4.96%	9.21%
C	−4.16%	7.15%	14.42%
D	−20.72%	9.06%	23.83%

22.4.6 Peel Test

The peel stress distribution is studied at different temperatures for Epocast 50-AL/946 epoxy adhesive with incorporating Al_2O_3 nanoparticles in the epoxy. The study examines the influence of Al_2O_3 nanoparticles on peel stress distribution in epoxy adhesive at various temperatures. Peel stress was evaluated at four distinct locations: Location A (the interface between the aluminum substrate and adhesive), Location B (the central plane within the adhesive layer), Location C (the midpoint along the length at the edge of the overlap), and Location D (across the width of the overlap). Incorporating nanofillers into the adhesive led to a reduction in peak peel stress by 1.3% at 25°C. However, at elevated temperatures of 50°C and 75°C, the peak peel stress increased by 2.7% and 10.7%, respectively. Figure 22.21 displays the peel stress distribution of the unmodified adhesive at different temperatures, and Table 22.6 summarizes the variations in peel strength at

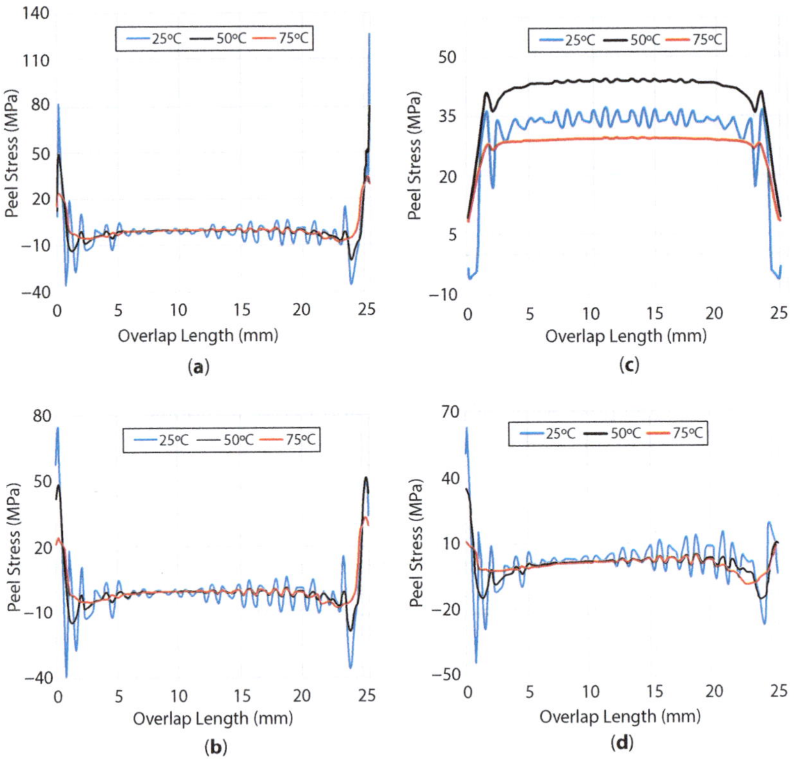

Figure 22.21 Peel stress distributions in SLJ of neat epoxy at different temperatures [53].

Table 22.6 The percentage change in peel stress of epoxy with 1.5% of Al_2O_3 at different locations and different temperatures [53].

Location	25°C	50°C	75°C
A	−1.34%	2.72%	10.74%
B	−2.70%	5.77%	11.44%
C	1.70%	−4.75%	−0.29%
D	−0.80%	6.77%	21.24%

the specified locations after the addition of 1.5% Al_2O_3 nanoparticles. Also Figure 22.22 shows the effect of adding 1.5% Al_2O_3 in epoxy on peel stress distribution of SLJ at various different temperatures [54].

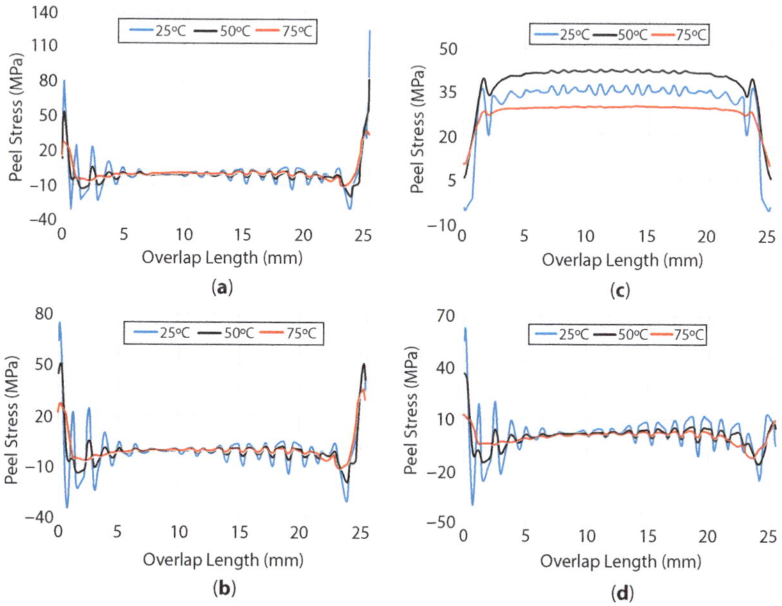

Figure 22.22 Peel stress of epoxy adhesive with 1.5% Al_2O_3 added at different temperatures [53].

22.4.7 Surface Tension

At higher temperatures, the surface tension of adhesives generally decreases. This occurs because increasing temperature leads to greater molecular motion. This leads to a decrease in cohesive forces at the adhesive surface [55, 56]. Consequently, the adhesive's capacity to effectively spread across the substrate known as wettability increases initially which allow the adhesive to spread more easily however higher temperature can start premature curing of adhesives which may potentially weakening the resulting bond [57].

22.4.7.1 Impact on Adhesion and Wetting

As surface tension decreases, the adhesive generally spreads more easily; however, if the adherend has a low surface energy, effective spreading may still be limited [57, 58]. Lower surface tension of the adhesive means it is less likely to form strong molecular attractions with the substrate, which can lead to weaker adhesion. A reduction in surface tension influences the contact angle formed between the adhesive and the substrate. As temperature increases [59], this contact angle generally becomes larger, signifying a decline in wetting capability.

22.4.8 Wettability and Contact Angle

With rising temperature, the surface tension of most adhesives decreases, causing a decrease in cohesive forces in the adhesive. This tends to enhance wettability, i.e., the adhesive is more easily spread on the surface of the substrate.

The measure of wettability is contact angle (lower contact angle = improved wetting) which generally decreases with increasing temperature for most adhesive systems. This trend is found in a wide range of materials, including hot-melts and polymeric adhesives applied to monolithic, composite, and functionally graded joints [60].

Monolithic adhesives are uniform in composition. At higher temperatures, their viscosity drops and surface tension decreases, resulting in a lower contact angle and improved wetting of the substrate. This enhanced wettability at elevated temperatures facilitates better adhesive spreading and potentially stronger interfacial bonding.

Composite adhesives contain fillers or reinforcements. The temperature effect on wettability and contact angle is similar to monolithic adhesives, but the extent may vary depending on the filler type and distribution.

At higher temperatures, the matrix phase (often a polymer) softens, reducing the contact angle and improving wetting. However, the presence of fillers could slightly modify the temperature dependence due to differences in thermal expansion and surface energy.

Functionally-graded adhesives have a gradual change in composition or properties across their thickness. This grading can be designed to optimize wetting and adhesion at the interface. At elevated temperatures, the graded structure may further enhance wettability by matching the adhesive's surface energy to that of the substrate, leading to a more pronounced decrease in contact angle compared to uniform adhesives [61].

22.5 Summary and Conclusions

The mechanical performance of epoxy adhesives declines significantly with increasing temperature, particularly near or beyond the glass transition temperature (Tg). Elevated temperatures cause thermal softening, resulting in reduced tensile strength, compressive strength, elastic modulus, and flexural strength, while increasing ductility and flexibility. Nanofillers improve strength and stiffness at room temperature but provide limited benefits at higher temperatures. However, they do help in partially retaining mechanical integrity under thermal stress.

Flexural strength and impact resistance also diminish at elevated temperatures, although the use of nanoparticle doping (e.g., zinc oxide) and fiber reinforcements enhances thermal stability and crack resistance. Functionally Graded Adhesives (FGAs) further improve thermal performance due to their ability to distribute stresses more evenly, delaying structural failure.

Lap-shear strength initially increases up to the Tg due to enhanced ductility but deteriorates sharply beyond this point. Nanoparticles such as Al_2O_3 enhance interfacial strength and modulus, thereby improving shear performance at elevated temperatures. Peel stress distribution studies reveal improved stress transfer with nanofillers, particularly at high temperatures.

Thermal conditions also influence adhesive wettability. At elevated temperatures, the surface tension of adhesives generally decreases due to increased molecular motion. This reduction in surface tension typically enhances the adhesive's ability to wet and spread over substrates. Consequently, weaker molecular interactions form at the adhesive-substrate interface, leading to diminished adhesion strength. The associated decrease in surface tension further contributes to a decrease in contact angle, a clear indicator of reduced wettability.

A complex interplay between increased ductility and decreased strength across various mechanical properties characterizes the performance of adhesives at higher temperatures. Understanding these changes is crucial for selecting appropriate adhesives for applications subjected to fluctuating thermal environments.

AI disclosure statement: The author confirms that no artificial intelligence (AI)-generated content or AI-assisted writing tools were used in the preparation of this chapter (including summaries and the abstract). All text, analysis, and conclusions are the result of the author's original work.

References

1. H. Oguma and K. Naito, Effect of stress ratio on the fatigue fracture mechanism of adhesive single-lap joints. Procedia Structural Integrity 19, 224-230 (2019).
2. P. Hu, X. Han, W.D. Li, L. Li and Q. Shao, Research on the static strength performance of adhesive single lap joints subjected to extreme temperature environment for automotive industry. International Journal of Adhesion and Adhesives 41, 119-126 (2013).
3. P. C. Mohapatra and L. V. Smith, Adhesive hardening and plasticity in bonded joints. International Journal of Adhesion and Adhesives 106, 102821 (2021).
4. C.S.P. Borges, P.D.P. Nunes, A. Akhavan, E.A.S. Marques, R.J.C. Carbas, L. Alfonso and L.F.M. da Silva, Influence of mode mixity and loading rate on the fracture behavior of crash resistant adhesives. Theoretical and Applied Fracture Mechanics 107, 102508 (2020).
5. M. D. Banea , F. S. M. de Sousa, L. F. M. da Silva, R. D. S. G. Campilho and A. M. de Bastos Pereira, Effects of temperature and loading rate on the mechanical properties of a high temperature epoxy adhesive. Journal of Adhesion Science and Technology 25, 2461–2474 (2011).
6. G. Viana, M. Costa, M.D Banea and L.F.M da Silva. A review on the temperature and moisture degradation of adhesive joints, Proceedings of the Institution of Mechanical Engineers, Part L: Journal of Materials: Design and Applications, 231, 488-501 (2016).
7. M.N.M Zafri, N. Mohamad, M.E.A. Manaf, T.A. Rahim, H.E.A. Mailod, Q. Ahsan, M.A. Ani and T.M. Ming, Vulcanization bonded natural rubber-aluminum by chemically modified graphene nanoplatelets-epoxy adhesive without primer. Journal of Advanced Research in Fluid Mechanics and Thermal Sciences 110, 182-199 (2023).

8. M. Banea, L.F.M. da Silva and R. Campilho, Effect of temperature on tensile strength and mode I fracture toughness of a high temperature epoxy adhesive. Journal of Adhesion Science and Technology 26, 939-953 (2012).
9. W. Tan, J. Na and Z. Zhou, Effect of service temperature on mechanical properties of adhesive joints after hygrothermal aging. Polymers 13, 3741 (2021).
10. W. Luo, W. Chen, D. Liu, X. Huang and B. Ma. Effect of temperature and humidity on mechanical properties and constitutive modeling of pressure-sensitive adhesives. Scientific Reports 14, 14634 (2024).
11. C.T. Mitchell, C.B. Dayan, D.M. Drotlef, M. Sitti and A.Y. Stark. The effect of substrate wettability and modulus on gecko and gecko-inspired synthetic adhesion in variable temperature and humidity. Scientific Reports 10, 19748 (2020).
12. R. D. Adams and V. Mallick, The effect of temperature on the strength of adhesively-bonded composite-aluminium joints. Journal of Adhesion 43, 17–33. (1993).
13. S. Akarsu, S.K. Buyuk and A.S. Kucukekenci. Effects of adhesive systems at different temperatures on the shear bond strength of orthodontic brackets. Journal of Dental Research, Dental Clinics, Dental Prospects 13, 103-108 (2019).
14. Masterbond.com. How Heat Affects the Lap Shear Strength of Structural Adhesives. https://www.masterbond.com/articles/how-heat-affects-lap-shear-strength-structural-adhesives.
15. Plastics-Determination of tensile properties Part 1: General principles, ISO 527-1 (2012).
16. Standard test method for tensile properties of plastics, ASTM D638 (2014).
17. Methods of testing cement – Part 1: Determination of strength, EN 196-1 (2016).
18. S. Ebnesajjad and A.H. Landrock, Durability of adhesive bonds, in: *Adhesives Technology Handbook (Third Edition)*, pp 297-338, William Andrew Publishing (2015).
19. Standard test method for compressive properties of rigid plastics, ASTM D695 (2015).
20. Standard test methods for flexural properties of unreinforced and reinforced plastics and electrical insulating materials, ASTM D790 (2003).
21. ADMET Materials testing system manufacturer, Bend testing equipment guide. https://www.admet.com/blog/bend-testing-equipment-guide (2018).
22. O. Devaux, Analysis and optimization of mixed-mode conical adhesively bonded joints under thermo-mechanical loadings, Ph.D. thesis, Université de Bretagne Occidentale, Brest, France (2015).
23. Standard test methods for determining the Izod pendulum impact resistance of plastics, ASTM D256 (2018).
24. C.J. Tzeng, Y.K, Yang, Y.H. Lin and C.H. Tsai, A study of optimization of injection molding process parameters for SGF and PTFE reinforced PC

composites using neural network and response surface methodology. International Journal of Advanced Manufacturing Technology 63, 691–704 (2012).

25. T.L.S. Ming, E. Jayamani and S.K. Heng. A review of microwave curing technique to fabricate natural fiber reinforced polymer composites, Jurnal Teknologi (Sciences & Engineering) 85, 113-123 (2023).
26. Adhesives determination of dynamic resistance to cleavage of high-strength adhesive bonds under impact wedge conditions - wedge impact method, ISO 11343 (2019).
27. A. Prathuru, Structural and residual strength analysis of metal-to-metal adhesively bonded joints. PhD. Thesis, Robert Gordon University, Aberdeen, Scotland (2019).
28. U. Eitner and L. C. Rendler, The impact of ribbon properties on measured peel forces, in: Proceedings of 6th International Conference on Silicon Photovoltaics, pp. 500 – 504 (2016).
29. Y.J. Awalekar, A.S. Takalkar and S.S. Shinde, Investigation of peel resistance of adhesives materials: A review, Proceedings of Engineering and Technology Innovation 10, 19-28 (2018).
30. Standard test method for peel resistance of adhesives (T-Peel Test), ASTM D1876-08 (2023).
31. Standard test method for apparent shear strength of single-lap-joint adhesively bonded metal specimens by tension loading, ASTM D1002 (2019).
32. Y. Wei, X. Jin, Q. Luo, Q. Li and G. Sun. Adhesively bonded joints – A review on design, manufacturing, experiments, modeling and challenges, Composites Part B: Engineering 276, 111225 (2024).
33. R. Campilho, M.F.S.F. de Moura, J.J.M.S. Domingues and J.J.L. Morais. Computational modelling of the residual strength of repaired composite laminates using a cohesive damage model. Journal of Adhesion Science and Technology 22, 1565–1591 (2008).
34. M.D. Banea, L.F.M. da Silva and R.D.S.G. Campilho. Effect of temperature on the shear strength of aluminium single lap bonded joints for high temperature applications, Journal of Adhesion Science and Technology 28, 1367-1381 (2014).
35. R. Campilho, M.F.S.F. de Moura and J.J.M.S. Domingues .Using a cohesive damage model to predict the tensile behaviour of CFRP single-strap repairs. International Journal of Solids and Structures 45, 1497–1512 (2008).
36. Y. Jahani, Effect of temperature on the flexural behavior of NSM CFRP strengthened RC beams under time-dependent loading and fatigue, Ph.D. Thesis, University of Girona, Girona, Spain (2022).
37. M.D. Banea, High temperature adhesives for aerospace applications, Ph.D. Thesis, University de Porto, Porto, Portugal (2011).
38. Y. Jahani, M. Banea, C. Barris, R. Perera, and L. Torres .Influence of curing temperature, post-curing and testing temperatures on mechanical properties

of a structural adhesive. Construction and Building Materilas 324, 126698 (2022).
39. M. Banea, L.F.M. da Silva, R. Carbas and S. de Barros. Effect of temperature and moisture on the tensile properties of a TEPs-modified adhesive. Materiale Plastice, 55. 478-481 (2018).
40. A. Nema, C. N. Mallineni, P. K. Penumakala, R. Adusumalli, K. Tejasvi, and M. K. Buragohain, Effect of temperature on flexural and interlaminar shear strength properties of carbon-epoxy composites: Experiment and modeling, Polymer Composites 45, 9139–9155 (2024).
41. T. Walander, Cohesive modelling of the temperature dependence of epoxy based adhesives in Mode I and Mode II loading, Licentiate of engineering thesis, Chalmers University of Technology, Gothenburg, Sweden (2013).
42. F. Rothenhäusler and H. Ruckdaeschel, l-arginine as bio-based curing agent for epoxy resins: Temperature-dependence of mechanical properties. Polymers 14, 4696. (2022)
43. P. K. Balasubramani, and J. O. Iroh, Mechanism and kinetics of curing of diglycidyl ether of bisphenol A (DGEBA) resin by chitosan, Polymer Engineering & Science 57, 865-874 (2017).
44. F. G. Garcia, B. G. Soares, V.J.R.R Pita, R. Sánchez and J. Rieumont, Mechanical properties of epoxy networks based on DGEBA and aliphatic amines. Journal of Applied Polymer Science 106, 2047–2055 (2007).
45. S.M. Mirmohammadi, O. Moini Jazani and A. Gharieh, Thermomechanical analyses and ANN modeling of novel epoxy adhesives with CSR particles and zinc oxide nanoparticles in structural bonding. Scientific Reports 15, 11201 (2025).
46. D. Muniraj, S. Vignesh and V.M. Sreehari, Mechanical characterization and impact damage assessment of Al/SiC functionally graded coating under elevated temperatures. Scientific Reports 14, 29134 (2024).
47. M.Q. dos Reis, E.A.S. Marques, R.J.C. Carbas and L.F.M. da Silva, Functionally graded adherends in adhesive joints: An overview, Journal of Advanced Joining Processes 2, 100033 (2020).
48. J.F. Durodola, Functionally graded adhesive joints – A review and prospects, International Journal of Adhesion and Adhesives 76, 83-89 (2017).
49. C.M. Stokes-Griffin, A. Kollmannsberger, P. Compston and K. Drechsler, The effect of processing temperature on wedge peel strength of CF/PA6 laminates manufactured in a laser tape placement process, Composites Part A: Applied Science and Manufacturing 121, 84-91 (2019).
50. I. Kocabas, Strength of glue joints of metallic materials in connection with design of joints, Ph.D. Thesis, University of Pardubice, Pardubice, Czech Republic (2016).
51. G.M.S.O. Viana, Development of a cohesive zone model for adhesive joints that includes environment degradation, Ph.D. Thesis, Universidade do Porto, Portugal (2018).

52. X.C. He, Effect of supports on stress distributions in the adhesive joints. Applied Mechanics and Materials 130–134, 1495–1498 (2011).
53. M. Hassan, A. Mubashar, M. Masud, A Zafar, M. Umair Ali and Y.S.Rim, Effect of temperature and Al_2O_3 nanofiller on the stress field of CFRP/Al adhesively bonded single-lap joints. Coatings 12, 1865 (2022).
54. H. Nassiraei and P. Rezadoost, Stress concentration factors in tubular T/Y-joints strengthened with FRP subjected to compressive load in offshore structures, International Journal of Fatigue 140, 105719 (2020).
55. Productos Colcar, How do surface tension and surface energy influence bonding with hotmelt adhesives and water based adhesives. https://productoscolcar.com/en/how-do-surface-tension-and-surface-energy-influence-bonding-with-hotmelt-adhesives-and-water-based-adhesives/ (2020).
56. BYJU'S the learning app, Angle of contact. https://byjus.com/physics/relation-between-angle-of-contact-and-temperature/ (2021).
57. K.L. Mittal, The role of the interface in adhesion phenomena, Polymer Engineering & Science, 17, 467–473 (1977).
58. JTAPE, Surface Energy & Adhesion: Explained https://www.jtape.com/news/surface-energy-and-adhesion-explained/ https://www.jtape.com/news/surface-energy-and-adhesion-explained/ (2024).
59. J.D. Glover, X. Yang, R. Long and J.T. Pham, Creasing in microscale, soft static friction. Nature Communication 14, 2362 (2023).
60. B. Duchemin, G. Cazaux, M. Gomina and J. Bréard, Temperature-dependence of the static contact angle: A transition state theory approach. Journal of Colloid and Interface Science 592, 215-226 (2021).
61. KRUSS, Wetting and Adhesion of Hotmelts. https://www.sanyo-si.com/wp-content/uploads/kruss-ar223-en.pdf#:~:text=The%20measurements%20of%20the%20surface,combination%20with%20the%20High%20Temperature (2001).

23

Methods to Enhance the Mechanical Performance of FRP Composite Adhesive Joints

Henrique F.M. de Queiroz[1] **and Mariana D. Banea**[1,2*]

[1]*Federal Center of Technological Education of Rio de Janeiro (CEFET/RJ), Av. Maracanã, Rio de Janeiro, Brazil*
[2]*CICECO - Aveiro Institute of Materials, Department of Materials and Ceramic Engineering, University of Aveiro, Aveiro, Portugal*

Abstract
Fibre-reinforced polymer (FRP) composites are widely used in various industries due to their high strength-to-weight ratio, corrosion resistance, and durability. However, the mechanical performance of FRP composite adhesive joints is often limited by factors such as debonding, crack propagation, and stress concentration. This chapter explores various methods proposed in the literature to enhance the mechanical performance of FRP composite adhesive joints, providing a comprehensive overview of advancements and techniques aimed at addressing these limitations. These methods include geometrical optimizations and adherend and adhesive modifications. The advantages and limitations of each method are discussed and a critical evaluation of their effectiveness based on available experimental and numerical studies is provided. In addition, current research trends and areas that require further investigation are identified. Ultimately, this chapter serves as a comprehensive guide for researchers and engineers involved in the design and optimization of FRP composite adhesive joints, providing insights into the most effective methods for enhancing their mechanical performance.

Keywords: Composite materials, adhesive joints, mechanical properties

**Corresponding author*: mbanea@ua.pt

Anil N. Netravali and Frank M. Etzler (eds.) The Wide World of Adhesion: A Festschrift in Honor of Dr. Kash Mittal on his 80th Birthday, (749–774) © 2026 Scrivener Publishing LLC

23.1 Introduction

Fibre-reinforced polymers (FRPs) are highly versatile structures that present a wide array of advantages when compared to more classical structural materials (e.g., steel and aluminium). Since their introduction to the industry, FRPs have boomed into an ever-increasing envelope of applications, from automotive to aerospace, marine, sports, medical among many others [1]. Their main advantages are their lightweight, high strength-to-weight ratio, and an incredibly high design flexibility as a function of many factors such as type of reinforcement fibres, resin matrices, filler materials, chemical treatments of the fibre surface, etc [2, 3]. In industries such as automotive and aerospace, the ability of materials like carbon fibre-reinforced plastics (CFRPs) to meet or exceed structural requirements while significantly reducing weight is unparalleled and highly valuable. As a result, CFRPs have become a transformative innovation among FRPs, marking a paradigm shift in numerous industrial applications. On the other hand, glass fibre-reinforced plastics (GFRPs) also play a significant role in industrial applications within the same sectors as CFRPs, offering similar benefits, such as high strength-to-weight ratio, but at a much lower cost. Furthermore, natural fibre-reinforced composites (NFRCs) have been utilized for decades across various industries, primarily in non-structural applications. These materials stand out for their minimal environmental impact and low production costs, making them an increasingly attractive option for sustainable design [4–6]. However, there are several drawbacks to the use of FRPs, such as longevity concern, environmental resistance, non-destructive structural integrity analysis among others [7, 8]. Additionally, the fabrication of complex shapes often requires the development of specialized moulds, which can significantly increase production costs. Nonetheless, despite these drawbacks, FRPs are a staple of modern industry and will continue to be used in an ever-increasing variety of industrial applications.

FRP materials, like any material used in the industry as a part of a larger structure, need to be joined with the rest of the structure. Traditionally, spot welding, drilling/bolting, and riveting have been the most commonly used joining methods for steel and aluminum. However, these techniques are unsuitable for joining FRP materials to a structure due to their inherently destructive nature. For instance, drilling a hole can damage the surrounding matrix and fibers, induce microcracks, and cause delamination of the outer layers [6]. Therefore, structural adhesive bonding has emerged as the leading technology for joining FRPs to a variety of materials, and it is now recognized as the optimal joining method [9–11]. However, the mechanical performance of FRP composite adhesive joints is often limited by factors such as debonding, crack propagation, and stress concentration. This chapter presents an overview of the main methods used to increase the mechanical performance of FRP composite adhesive joints. First, the main factors that affect the performance of composite adhesive joints are briefly discussed. Further, the main methods used to increase the mechanical performance of FRP adhesive joints are presented. Finally, some conclusions and an outlook on critical challenges and future perspectives are summarized.

23.2 Factors Affecting the Performance of Composite Adhesive Joints

There are several key factors that can significantly impact the performance of FRP composite adhesive joints, including: adhesive properties (the adhesive material's mechanical properties, such as strength, stiffness, toughness, and durability, are critical factors that can affect the joint's overall performance), adherend (type of resin and fibre, fibre architecture and orientation and stacking sequence) properties [9, 11]. The compatibility of the adhesive with the composite material, including its resin matrix and fiber type, can also significantly affect the bond strength and durability of the joint.

The joint geometry, including the overlap length, thickness, and shape, can affect the load-carrying capacity and the distribution of stress within the joint. Therefore, the main methods of joint load-bearing capacity improvements are focused on tailoring these parameters to the design necessities.

Proper surface preparation is essential for achieving a strong bond between the FRP composite material and the adhesive. Surface contaminants, roughness, and surface free energy can significantly impact the joint's strength and durability. An improper surface preparation can easily lead to interfacial failures [12]. The failure mode must also be analysed and reported carefully, since it can give crucial insight into the crack propagation behaviour and stress concentrations.

Another design choice that significantly affects the structure is the type of bonded joint. The geometry of the joint plays a crucial role in the mechanical performance of FRP composite adhesive joints as different joint geometries can have different load-bearing capabilities and failure modes.

There are many different joint geometries, such as double lap joints, scarf joints, among many others [13]. The choice of joint geometry depends on a variety of factors, such as the materials being joined, the loading conditions, and the desired performance characteristics.

The single lap joint (SLJ) is the most used and studied joint geometry due to its fabrication simplicity. However, this very common design choice introduces a stress state that is unique to the SLJ and a significant issue for the structure (i.e., the peel stress). Due to the misalignment of the load axis during service, a dual stress state emerges, where over the course of the overlap peel and shear stress states exist simultaneously (see Figure 23.1) [14]. Because of this, a serious concern arises in SLJs, i.e., the peel stress concentrations at the overlap edges. This is significantly detrimental for the load bearing capability of the joint, since due to this peeling tendency, the adhesive/adherend interface will tend to de-bond rather than initiating crack formation within the adhesive that propagates up to a fully cohesive failure. Specifically, for FRP materials, this is worsened by the fact that in a classically laminated composite, the layers are stacked one by one in the design specification stacking sequence and there will be nothing between the fibre layers other than a resin-rich zone (see Figure 23.2). This means that the transverse strength of the material is significantly lower than its in-plane strength. This is especially true for long fibre unidirectional reinforced composites. In fact, for CFRP composites, more than an order of

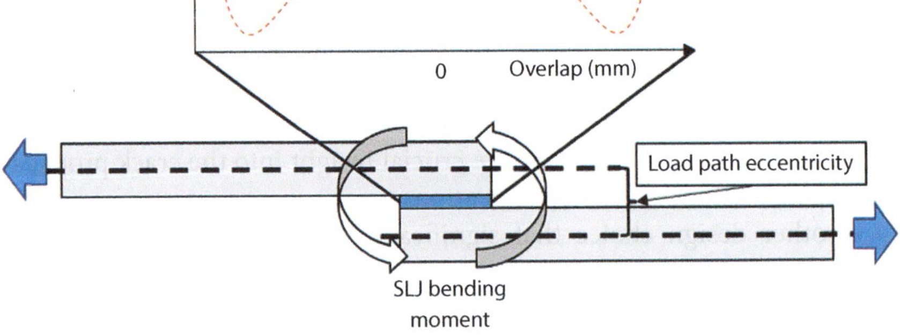

Figure 23.1 A simplified view of the prevailing stress states in an SLJ.

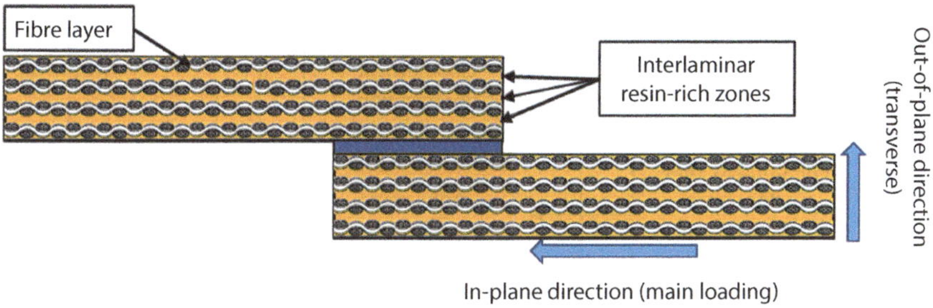

Figure 23.2 Schematic of an FRP SLJ.

magnitude difference between these direction-dependant material properties has been reported, where the transverse tensile strength of the composite is even lower than the tensile strength of the neat resin matrix [15].

The primary challenge facing the structural bonding of FRP materials utilizing the SLJ geometry is the phenomenon of delamination. Delamination is characterized by the formation of a crack at the location of the maximum peel stress, which penetrates the adherend and propagates through the interlaminar resin-rich region, resulting in the detachment of a fibre layer. This mode of failure typically results in a reduction in the joint's overall load-bearing capacity [16, 17]. Different methods have been proposed in the literature to improve the mechanical performance of FRP composite adhesive joints and are briefly discussed in the next section.

23.3 Methods to Increase the Mechanical Performance of FRP Composite Adhesive Joints

In the context of the inherent issues that arise owing to the stress state, and the issue of delamination of FRPs that is exacerbated by the peak peel stresses, various solutions have been proposed in the literature to tackle these issues. Generally, they fall in three main categories: geometrical optimizations, adhesive modifications and adherend modifications (see Figure 23.3).

These material modifications and geometry optimization techniques all attempt to solve the same issue: the maximization of the joint load-bearing capacity by avoiding delamination failures facilitated by the peak peel stresses at the overlap along with the significant orthotropic FRP properties. The local geometrical optimization aims at lowering the sharp peel stresses at the overlap edges by generating a smoother stress transfer

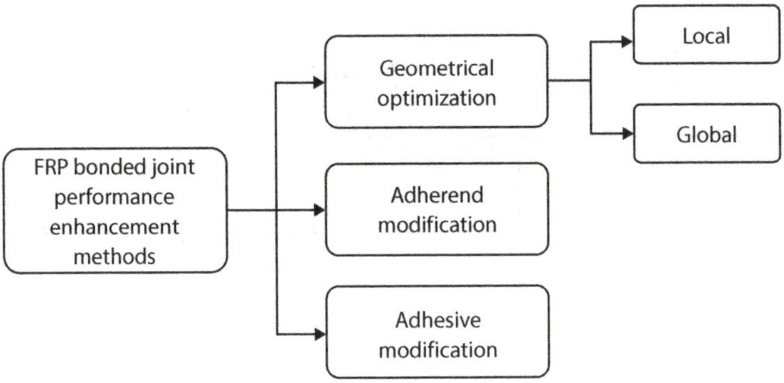

Figure 23.3 Schematic of possible methods for performance enhancement of FRP bonded joints.

as a function of material geometry optimization at the edges (e.g., adhesive spew fillets and adherend tapering/rounding). The global geometrical optimization attempts to alter the stress state from a dominantly peel state to a mostly compressive state at the edges by altering the overall overlap geometry (e.g., wavy joints). The adherend modifications alter the

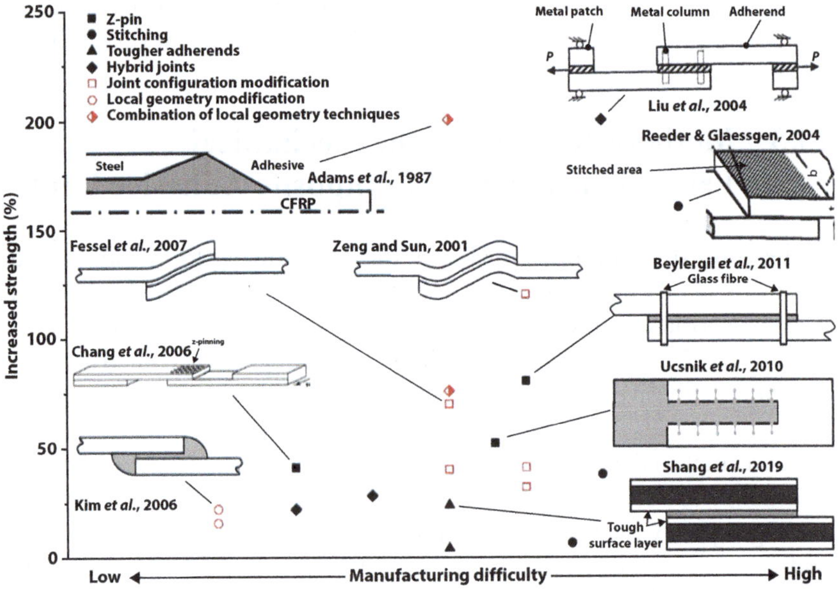

Figure 23.4 A comparison of different techniques in terms of strength improvement and manufacturing difficulty increase [27].

properties of the adherend material in ways that benefit the final load bearing capacity of the joint. For example, a more rigid adherend will rotate less during SLJ loading, leading to a decrease in peak peel stresses at the overlap edge. Another possibility is to toughen the surface by increasing the fracture toughness, therefore deflecting the crack from propagating within the adherend, altering the crack propagation, which occurs within the adhesive instead. Finally, the adhesive modifications mainly focus on tailoring the adhesive properties to maximize energy spent by the crack propagation and increase the failure load of the joint [1]. The following topics will address these methodologies and their corresponding state of the art. Shang *et al.* [17] described the various methodologies used in the literature to enhance the strength of composite bonded joints (see Figure 23.4). It was noted that while very significant improvements were possible, the fabrication difficulty sharply increased, and therefore the cost [17–26]. The design of the joint must take into consideration these variables so that it meets the requirements without unnecessary expenses.

23.3.1 Geometrical Optimizations

It is well known that a bonded joint must be designed to distribute the stress evenly and avoid stress concentrations. This can be achieved by using tapered or stepped-lap joints, which help to reduce stress concentrations [8]. Global and local geometrical optimization techniques focus on altering the governing stress states by controlling the load paths and where and how the stress concentrations will occur. Global geometries include the many different types of joints (i.e., double-lap joint (DLJ), scarf joints, non-linear overlaps, etc., see Figure 23.5) as well as other geometrical features such as overlap length/width and adhesive layer and adherend thickness [28–32]. Local geometry optimization focuses more specifically on the overlap edges by tailoring the local geometry of the adhesive and/or the adherend by spew fillets or rounding/tapering of the adherend, see Figure 23.6) [23, 33–40]. Table 23.1 presents a summary of some recent research efforts in the field.

Qin *et al.* [32] studied the effects of geometrical dimensions on the failure of composite-to-composite adhesively bonded joints. Tri-axial glass fabrics as well as unidirectional fabrics were used along with an epoxy resin system to fabricate the composite plates. A two-component epoxy adhesive paste was employed to fabricate the bonded joints. The overlap length and bonded area width were varied (i.e., overlap lengths of 12.7, 25.4, 38.1, 50.8, and 63.5 mm and overlap widths of 25.4, 38.1, 50.8, 63.5, and 101.6 mm, respectively). The adhesive thickness was also varied from 1 to 5 mm

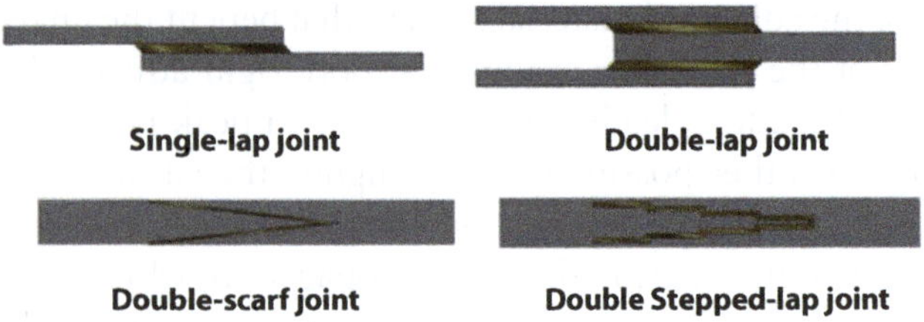

Figure 23.5 Adhesively bonded joints configurations.

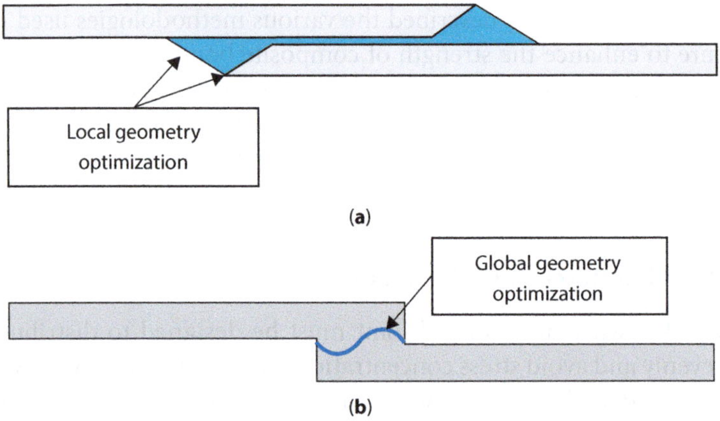

Figure 23.6 Examples of (a) local and global (b) geometry optimizations.

(1, 2, 4 and 5 mm) as well as the adherend thickness (4.5, 3 and 2.25 mm). The joints were characterized mechanically under uniaxial tension at room temperature. It was observed that the failure load of the bonded joints decreases and then slightly increases when the adhesive thickness increases from 2.0 to 5.0 mm. The main failure mode reported for the bonded joints is an alternating combination of interfacial failure and thin-layer cohesive failure (i.e., TLC as defined by a removal of adhesive from one of the surfaces with a few adhesive particles left) around the bonded surface. The failure load of bonded joints increases overall with overlap lengths from 12.7 to 63.5 mm, and the slope of the failure load decreases with the overlap length. The failure load of bonded joints increases overall with adherend thickness for different overlap lengths. As the adherend thickness decreases, fibre-tear failure in the adherends occurs in the thinnest bonded joints.

Table 23.1 Recent research efforts in geometrical modifications of bonded joints.

Adhesive	Adherend	Modification technique	Joint failure load/ Shear strength	Ref
Epoxy	Tri-axial and unidirectional glass fibre fabric/epoxy	A thick layer, large overlap length and width, and laminate layups were evaluated.	It was reported that failure load decreases with adhesive thickness, while it increases rapidly with overlap length and increases at a slightly faster rate than is proportional with overlap width.	[32]
Epoxy	GFRP	Nonflat interfaces were produced by embedding different material reinforcements during vacuum infusion (i.e., nylon, bamboo, glass fibre, aluminium, and iron).	Results showed that the ultimate load and average shear stress increased with the increasing distance from the singularity point to the center of reinforcements. The embedding of reinforcements with high Young's moduli and low diameters can significantly improve the ultimate load of the nonflat surface specimens.	[40]

(*Continued*)

Table 23.1 Recent research efforts in geometrical modifications of bonded joints. (*Continued*)

Adhesive	Adherend	Modification technique	Joint failure load/ Shear strength	Ref
Epoxy/ Araldite 2015	GFRP	Two types of global nonflat overlaps were investigated: straight wavy and reverse wavy joints.	The reverse wavy joint presented a significant enhancement in joint failure load when compared to the flat counterpart.	[35]
Epoxy	Poly(lactic acid) (PLA)	A wavy nonflat overlap joint with 3D printed PLA adherends was studied.	It was reported that significant joint failure load enhancement is possible *via* global geometry optimization.	[41]
Epoxy	GFRP	A flat-joggle-flat joint geometry was fabricated and analysed.	An increase in strength of 90% is observed over flat joints.	[42]
Epoxy	GFRP	Bonded joints with nonflat sinusoidal interfaces were analyzed.	A significant improvement in bonded joint failure load was reported for the wavy joint whose edge is predominantly under compression.	[37]

Chen *et al.* [40] investigated the strength and failure modes of adhesively bonded joints with easily fabricated nonflat surfaces. Glass fabric $[0/90]_4$ and epoxy resin were used to fabricate the composite panels *via* the Seemann Composite Resin Infusion Molding Process (SCRIMP). To ensure the nonflat surface geometry of the adherends, reinforcements were embedded within a glass fibre composite laminae, and a nonflat surface was formed under vacuum pressure. Various edge distances, reinforcement diameters and materials were investigated (i.e., 2, 4, 6, 8 and 12 mm reinforcement diameters and nylon, bamboo, glass fibre, aluminium and iron as adherends). A two-component epoxy adhesive was used to fabricate the bonded joints and the adhesive thickness was kept at 1.6 mm. The joints were assessed mechanically *via* uniaxial tension tests. Results showed that the ultimate load and average shear stress increased with the increasing distance from the singularity point to the center of reinforcements. The embedding of reinforcements with high Young's modulus and low diameters can significantly improve the ultimate load of the nonflat surface specimens.

In summary, global or local or even a combination of both techniques have been shown to significantly enhance the failure load of composite bonded joints. However, these techniques invariably come with increased fabrication complexity and costs, therefore care must be taken to avoid overdesign of the bonded joints.

23.3.2 Adherend Modifications

The adherend properties are crucial for the overall efficiency of an FRP composite bonded joint. Both the in-plane and transverse mechanical properties come into play in affecting the prevailing stress states as well as crack propagation, failure modes and ultimately failure load [43]. In an SLJ, a stiffer adherend for example, will require a higher peel stress state for a given rotation of the overlap edge compared to a less rigid one. This effect has a limit, i.e., continuous increases in rigidity do not yield linear enhancements in joint failure load. In a classical composite layup of bidirectionally or unidirectionally oriented fibre fabrics, during bending, a crack initiation site will form somewhere on the overlap edge due to the stress concentrations [44, 45]. After that, the crack will progress following the path of lowest fracture toughness. This is a function of the material toughness. Since most ductile and high-strength modern adhesives are also relatively high in toughness, (i.e., when compared to most used epoxy matrix materials) the crack will enter the adherend. Its propagation will then either be very close to the interface, generating a light-fibre tear (LFT) or thin-layer cohesive (TLC) failure mode or within the adherend and inside the resin-rich zone between laminas, yielding a delamination (fibre tear) of the first layer of the composite adherend [16]. The failure load will be lower if the failure is either LFT/TLC or delamination when compared to a pure cohesive failure, as the crack will require less energy to propagate through a medium of lower toughness. In this context, the surface toughness, and transverse properties of composite adherends are a vital area of study and there have been a number of techniques employed by researchers in the current literature to solve this issue. The surface toughness can be enhanced either by the use of a different material at the critical overlap edge zones [46, 47], by the hybridization technique [9–11, 16, 17, 48, 49], by the use of natural fibres at the adherend/adhesive interface [50] or by using a layer of metal around a composite core, creating a fibre-metal laminate (FML) [51]. Another possibility for enhancement of the bonded joint *via* adherend property tailoring is the improvement of the transverse adherend material properties. This can be done *via* z-pins [52], stitching [53], and micro/nanofiller materials [54, 55]. A schematic of these possible techniques can be seen in Figure 23.7. Table 23.2 presents some recent research efforts in the adherend modification of bonded joints.

Recently, de Queiroz *et al.* [48] studied experimentally the behaviour of SLJs made of natural, synthetic and interlaminar hybrid fibre-reinforced polymer composites. The effect of the number of the interlaminar synthetic layers needed to match the bonded joint efficiency of a 100% synthetic GFRP bonded joint was studied by testing SLJs bonded with a modern tough structural adhesive used in the automotive industry. A significant enhancement to the joint load bearing capacity as a function of adherend material was found. The failure load of the hybrid adherend joints increased by increasing the number of external synthetic layers (i.e., the failure load of hybrid 3-layer joint increased by 28.6% compared to hybrid 2-layer joint) and reached the pure synthetic adherends joint failure load (i.e., CFRP and GFRP bonded joint failure loads). The authors state that this is due to the optimum compromise between the adherend material properties (i.e., stiffness and strength) and a diminished bondline peel stress state as a result of lower rotation of the adherend during loading, which leads to lower stresses and delayed delamination failure onset. Schollerer *et al.* [46] investigated the effect of a novel local adherend surface toughening concept *via* a thermoplastic poly(vinylidene fluoride) (PVDF) layer and compared it to classical local geometrical modifications. A reduction of peak stresses at the overlap edges while maintaining a simple fabrication methodology so that additional expenses are avoided was the goal. A prepreg adherend made from Hexcel Hexply® 8552 was used along with Hysol® EA 9695 adhesive system. PVDF thermoplastic was added to the prepreg before curing. The PVDF layer length effect was studied, and an ideal dimension was chosen. The adherend received atmospheric pressure plasma surface treatment. The results from four PVDF length joint specimen groups (i.e., 5, 10, 17.5 and 30 mm) were compared to the unmodified (neat) case along with local adhesive geometry modifications (i.e., half triangle, arc and chamfering). Significant improvements were observed for the 5, 10 and 17 mm PVDF layer cases, where an approx. 28, 84% and

Table 23.2 Research efforts in adherend modification of bonded joints.

Adhesive	Adherend	Modification technique	Joint failure load/ Shear strength	Ref
Epoxy	Natural/synthetic fibres	Intralaminar hybridization of a jute bidirectional fabric with natural and synthetic fibres.	Significant joint failure load enhancement was reported as a function of hybridization.	[16]
Epoxy	Natural/synthetic fibres	Interlaminar hybridization of a jute bidirectional fabric core with fabric outer layers.	Significant joint failure load was observed as a function of hybridization. Optimum load was found for the symmetric 3 glass layers around a 5-layer jute core (i.e., G3J5G3) specimen, matching the failure load of GFRP.	[48]
Epoxy	Hexcel HexPly® 8552 IM7 UD	Surface toughening by a layer of thermoplastic poly(vinylidene fluoride) (PVDF).	Significant improvements were reported for the shear strength as a function of the toughening of the edges. Improvements were also reported when compared to classical local geometrical modifications.	[46]

(*Continued*)

Table 23.2 Research efforts in adherend modification of bonded joints. (*Continued*)

Adhesive	Adherend	Modification technique	Joint failure load/ Shear strength	Ref
Epoxy	Aluminum 6082T6/ Polyphthalamide	Discrete aluminum patches were used to decrease the peak peel stresses at the edges of the overlap of dissimilar bonded joints.	It was reported that the technique significantly enhances the lap shear strength (LSS) of the dissimilar joints by increasing the interfacial stiffness of the joints.	[47]
Epoxy	3D GFRP	Adherends reinforced with non-crimp 3D orthogonal woven glass fibre fabrics.	Single-lap and double butt-strap joints were fabricated *via* co-curing and adhesive bonding. Stitching the co-cured single-lap joints resulted in significant increase in joint strength. Moreover, the delamination failure was not observed within any of the 3D woven composite adherends.	[53]
Epoxy	Mild carbon steel/ GFRP	Stainless steel pins were used as mechanical fasteners near the overlap edges of the adhesively bonded joints.	It was reported that the pins increase the strength of the joints *via* a bridging mechanism as a function of number and location of pins by up to 58%.	[52]

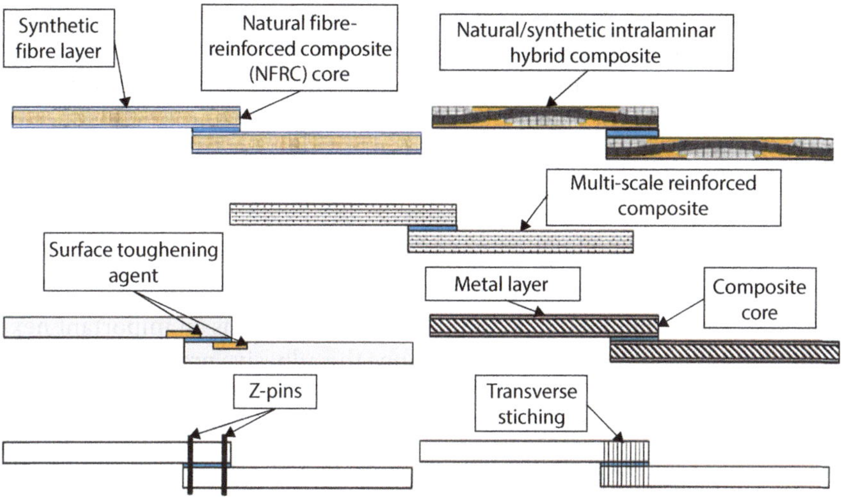

Figure 23.7 A schematic of possible techniques for adherend modification.

30% joint failure load increases were reported compared to the unmodified joint, respectively. Moreover, this PVDF case also presented significant improvements when compared to the classical geometrical modifications. It was noted that only the overlap edges are affected from a more ductile material in the stress transfer zone, which acts to transfer peel stresses from one adherend to the other without a crack up to joint failure.

Islam and Tong [52] studied experimentally the effect of through-the-thickness pin reinforcement on the static joint strength of hybrid steel/glass fibre joints. Stainless steel pins were applied as mechanical fasteners close to the overlap edge of the adhesively bonded joints. It was reported that the pins enhance the strength of the joints through a bridging mechanism as a function of number and location of pins by up to 58%.

In summary, the adherend modification techniques have great potential for significantly improving the load-bearing capabilities of composite bonded joints. However, similarly to the geometrical modifications, the cost of fabrication, quality assurance and control, non-destructive damage assessment and life prediction of the structures are still significant hindrances to wider industrial applications of these methods.

23.3.3 Adhesive Modifications

Similarly to the adherend material properties, the properties of the adhesive material are also a highly significant parameter in the joint design.

Material properties such as the classical mechanical characteristics as well as fracture toughness and the type of adhesive (i.e., ductile or brittle) significantly alter the performance and behaviour of a bonded joint in service. For example, for short overlaps, a high-strength and brittle adhesive will yield a higher failure load than a ductile one. On the other hand, a more ductile adhesive will present an almost linear relationship between higher failure loads and overlap lengths, while for a brittle one a diminishing return will be observed [32]. Assuming that for a given composite bonded joint, delamination is not occurring, and the failure is exclusively cohesive, the enhancement of the adhesive properties is an important next step. The most used technique to enhance the adhesive properties is by the use of filler materials. The more difficult it is for the crack to propagate, the higher the bonded joint load will be at total failure. This is a function of the fracture toughness of the adhesive and the objective of the multi-scale reinforcement of the adhesive (i.e., micro or nano) is to enhance this property. The idea is that the crack tip will be blunted by the filler particles and will either have to shear through them or have its path diverted, ultimately spending more energy than in the neat adhesive [56]. The interface between the filler and the adhesive is very important here, since it directly affects the efficiency of the technique. Many particles have been used for adhesive reinforcement such as: carbon nanotubes (CNTs), metallic particles (e.g., zinc oxide, titanium dioxide, iron oxide), silicon dioxide, nano-clays, cork, rubber, glass among others [56–58]. However, the enhancements are a function of the adhesive/filler mixture, where thorough homogenization is key. Clusters of agglomerated filler materials, especially for nanoparticles, are very likely to create zones of stress concentrations and crack nucleation sites. Some advances in the adhesive modification technique are described in Table 23.3. Razavi *et al.* [59] investigated single lap joints with structural adhesives reinforced with a mixture of silica nanoparticles and multi-walled carbon nanotubes. A mixture of silica spherical nanoparticles (SNPs) along with multi-walled carbon nanotubes (MWCNTs) in a ratio of 1:1 by weight was fabricated *via* mechanical stirring followed by ultrasonication to ensure homogeneity. Three weight filler fractions were studied: 0.2, 0.5 and 0.8 wt% of combined SNPs and MWCNTs. The adherends were made of 7075-T6 aluminium and received mechanical abrasion surface preparation as well as acid etching and acetone degreasing. The bonded joints were characterized in uniaxial quasi-static tensile testing. It was reported that the mechanical properties of the joint were enhanced including the shear strength and elongation at failure. No degradation was reported for all the weight fractions of the mixed filler reinforced joints. This was in contrast to the incorporation of a single type of nanoparticle,

Table 23.3 Recent research efforts in adhesive modification of bonded joints.

Adhesive	Adherend	Filler material	Joint failure load/ Shear strength	Ref
Epoxy	7075 T6 Aluminum	Silica nanoparticles/ MWCNTs Neat and mixed in 1:1 ratio (0.2, 0.5, 0.8 wt.%)	Significant enhancements were reported especially for the mixed filler specimens.	[59]
Epoxy	GFRP	Nano-clay 1, 3 and 5 wt.%	Lap shear strength (LSS) and joint stiffness increased for the 1 wt% case.	[60]
Acrylic	Steel/ glass-epoxy	TiO_2 nanoparticles up to 3.5 wt.%	Significant enhancements were reported for the shear and tensile strengths of composite joints. A reduction was reported in the peel strength.	[61]
Acrylic	Steel/ glass-epoxy	Silica nanoparticles up to 2.5 wt.%	Significant improvements in shear and tensile strengths of the joints were reported up to 1.5 wt.%. After that, the properties tended to decrease.	[62]

(*Continued*)

Table 23.3 Recent research efforts in adhesive modification of bonded joints. (*Continued*)

Adhesive	Adherend	Filler material	Joint failure load/ Shear strength	Ref
Polyether ketoneketone (PEKK)	Stainless steel	Hydroxyl functionalized MWCNTs, up to 1 wt.%	The highest improvement for the LSS was reported for the 0.3 wt.% (i.e., 152% when compared to the neat unreinforced adhesive).	[63]
Epoxy	High strength steel, Aluminum and CFRP	Thermally Expandable Particles (TEPs), up to 25 wt.%	It was reported that debonding of structurally similar and dissimilar joints is possible in 60 seconds to 2 minutes without any damage to the adherends.	[64]

where at certain weight percentages, a mechanical properties decrease was observed. It was also reported that the failure mode tended to be more cohesive for the filler reinforced specimens.

Another possibility for the use of particles in the adhesive layer is to tackle the issue of the dismantling difficulty of the bonded structures. Unlike fastened structures, bonded joints cannot be easily de-bonded. The adhesive layer would have to be thermally degraded and significant thermal damage would occur on the FRP adherends as well. Thermally expandable particles (TEPs) are microparticles made of a thermoplastic shell filled with a liquid hydrocarbon. Upon heating, the shell softens, and the liquid becomes a gas, causing the TEPs to expand up to 100 times their original size and after cooling the shells harden and the final volume is maintained. Recently, it has been reported that TEPs can be used to successfully de-bond structural SLJs bonded with high performance adhesives [64, 65]. It is important to note that the effect of the particles in the adhesive material properties is deleterious as a function of increasing the TEP content (i.e., higher weight fractions of TEP will result in lower LSS) [66, 67].

The modification of the adhesive material is a well-defined area for enhancing the efficiency of bonded joints usually through the use of particles by improving the fracture toughness and crack blunting characteristics of the adhesive and therefore, generating a higher load at failure. Fibres can also be used as reinforcement materials for the adhesive and can improve the mechanical performance of FRP composite adhesive joints by enhancing the stiffness and strength of the adhesive, improving the overall strength of the joint.

23.4 Summary and Future Prospects

This chapter has presented the available methodologies for enhancing the bonded joint efficiency of FRP joints. Several techniques have been applied in the literature to varying degrees of success. Generally, there are three potential areas for possible improvement: geometrical optimizations, adherend modifications and adhesive modifications. Each has a range of possible techniques, and all have their advantages and disadvantages. A common theme of the most efficient methodologies such as reinforcing the transverse direction of the composite adherend, or the wavy global joint geometry is that significant fabrication costs are to be expected as well as difficult quality control and assurance. Therefore, more work needs to be done in order to control these variables in order for these complex and efficient techniques to be industrially applicable. While existing methods have demonstrated significant improvements in joint performance, further research is necessary to maximize their potential for industrial applications. Some critical challenges and future research directions are summarized below:

- Optimizing the geometry of FRP bonded joints is critical for improving performance in specific applications. Techniques like topography optimization, combined with finite element analysis (FEA), can be employed to design geometries that distribute stresses more effectively, enhancing joint durability and strength.
- Develop optimization methods tailored to specific materials and applications. These methods should focus on improving adhesive properties, surface treatments, and bonding

strategies to maximize joint performance under varying operational conditions.

- Additive manufacturing (AM), including 3D printing, offers great potential for fabricating composite adhesive joints with complex geometries, which can improve the joint strength and efficiency and decouple the part complexity from its price. Further research into hybrid AM techniques, combining traditional and additive methods, could lead to significant advancements.
- Research into advanced adhesives, including those with nano-enhancements (e.g., carbon nanotubes or graphene-based fillers), could improve bonding strength, thermal stability, and environmental resistance. Adaptive or self-healing adhesives may also offer long-term durability and reliability.
- Embedding sensors or smart materials within joints to monitor performance in real-time and predict failures.

Acknowledgements

This work was developed within the scope of the project CICECO Aveiro Institute of Materials, UID/50011/2025 (DOI 10.54499/UID/50011/2025) & LA/P/0006/2020 (DOI 10.54499/LA/P/0006/2020), financed by national funds through the FCT/MCTES (PIDDAC).

AI disclosure statement: The author confirms that no artificial intelligence (AI)-generated content or AI-assisted writing tools were used in the preparation of this chapter (including summaries and the abstract). All text, analysis, and conclusions are the result of the author's original work.

References

1. H.F.M. de Queiroz and M.D. Banea, Methods to increase the mechanical performance of composite adhesive joints: An overview with focus on joints with natural fibre composite adherends. *J. Composite Mater.* 56, 3993-4010 (2022).
2. M.D. Banea, M. Rosioara, R.J.C. Carbas, and L.F.M. da Silva, Multi-material adhesive joints for automotive industry. *Composites Part B-Eng.* 151, 71-77 (2018).

3. F. Ramezani, B.D. Simões, R.J.C. Carbas, E.A.S. Marques, and L.F.M. da Silva, Developments in laminate modification of adhesively bonded composite joints. *Materials* 16, 568 (2023).
4. K.L. Pickering, M.G.A. Efendy, and T.M. Le, A review of recent developments in natural fibre composites and their mechanical performance. *Composites Part A-Appl. Sci. Manuf.* 83, 98-112 (2016).
5. O. Akampumuza, P.M. Wambua, A. Ahmed, W. Li, and X.-H. Qin, Review of the applications of biocomposites in the automotive industry. *Polym. Composites* 38, 2553-2569 (2017).
6. Rampal, G. Kumar, S.M. Rangappa, S. Siengchin, and S. Zafar, A review of recent advancements in drilling of fiber-reinforced polymer composites. *Composites. C: Open Access* 9, 100312 (2022).
7. J. Qureshi, A review of fibre reinforced polymer structures. *Fibers* 10, 27 (2022).
8. G. de Seixas, H. de Queiroz, J. Neto, and M. Banea, Effect of water on the mechanical and thermal properties of natural fibre reinforced hybrid composites. *J. Composite Mater.* 57, 1941-1958 (2023).
9. H.F.M. de Queiroz, J.S.S. Neto, D.K.K. Cavalcanti, and M.D. Banea, A novel 3D fibre-reinforcement architecture for high performance natural fibre reinforced composite adhesively bonded joints. *J. Adhesion* 100, 118-138 (2024).
10. H.F.M. de Queiroz, R.H.A. Rosa, J.S.S. Neto, D.K.K. Cavalcanti, and M.D. Banea, Mechanical characterization of short Curauá/glass fiber-reinforced interlaminar hybrid composite adhesive joints. *J. Adhesion Sci. Technol.* 38, 1473-1490 (2024).
11. H.F.M. de Queiroz, V. Pastor, A.C. de Mendonça, J.S.S. Neto, D.K.K. Cavalcanti, and M.D. Banea, Effect of hybridization and composite architecture symmetry in the bonded joint performance of jute/carbon fibre-reinforced composites. *Int. J. Adhesion Adhesives* 133, 103745 (2024).
12. M.D. Banea, Dissimilar joining of PMCs to metals – Adhesive bonding, in: *Encyclopedia of Materials: Composites*, Vol. 3, D. Brabazon (Ed.), pp. 324-333, Elsevier, Oxford (2021).
13. M.D. Banea and L.F.M. da Silva, Adhesively bonded joints in composite materials: An overview. *Proc. Inst. Mech. Eng. L: J. Mater.: Design Appl.* 223, 1-18 (2009).
14. J.P.A. Valente, R.D.S.G. Campilho, E.A.S. Marques, J.J.M. Machado, and L.F.M. da Silva, Adhesive joint analysis under tensile impact loads by cohesive zone modelling. *Composite Struct.* 222, 110894 (2019).
15. L. Liu, X. Zhang, Z. Wang, Y. Wang, and J. Guo, Micromechanics modeling of transverse tensile strength for unidirectional CFRP composite. *Materials* 15, 8577 (2022).
16. H. de Queiroz, M. Banea, and D. Cavalcanti, Experimental analysis of adhesively bonded joints in synthetic- and natural fibre-reinforced polymer composites. *J. Composite Mater.* 54, 1245-1255 (2020).

17. X. Shang, E. Marques, J. Machado, R. Carbas, D. Jiang, and L.F.M. da Silva, A strategy to reduce delamination of adhesive joints with composite substrates. *Proc. Inst. Mech. Eng. L: J. Mater.: Design Appl.* 233, 521-530 (2019).
18. S. Ucsnik, M. Scheerer, S. Zaremba, and D.H. Pahr, Experimental investigation of a novel hybrid metal–composite joining technology. *Composites Part A-Appl. Sci. Manuf.* 41, 369-374 (2010).
19. G. Fessel, J. Broughton, N. Fellows, J. Durodola, and A. Hutchinson. A numerical and experimental study on reverse-bent joints for composite substrates. in 48th AIAA/ASME/ASCE/AHS/ASC Structures, Structural Dynamics, and Materials Conference, pp. 2192 (2007).
20. R.D. Adams and J.A. Harris, The influence of local geometry on the strength of adhesive joints. *Int. J. Adhesion Adhesives* 7, 69-80 (1987).
21. K.-S. Kim, J.-S. Yoo, Y.-M. Yi, and C.-G. Kim, Failure mode and strength of uni-directional composite single lap bonded joints with different bonding methods. *Composite Struct.* 72, 477-485 (2006).
22. P. Chang, A.P. Mouritz, and B.N. Cox, Properties and failure mechanisms of pinned composite lap joints in monotonic and cyclic tension. *Composites Sci. Technol.* 66, 2163-2176 (2006).
23. Q.-G. Zeng and C.T. Sun, Novel design of a bonded lap joint. *AIAA J.* 39, 1991-1996 (2001).
24. J. Liu, J. Liu, and T. Sawa, Strength and failure of bulky adhesive joints with adhesively-bonded columns. *J. Adhesion Sci. Technol.* 18, 1613-1623 (2004).
25. J.R. Reeder and E.H. Glaessgen, Debonding of stitched composite joints under static and fatigue loading. *J. Reinf. Plast. Composites* 23, 249-263 (2004).
26. B. Beylergil, Y. Cunedioglu, and A. Aktas, Experimental and numerical analysis of single lap composite joints with inter-adherend fibers. *Composites Part B-Eng.* 42, 1885-1896 (2011).
27. X. Shang, E.A.S. Marques, J.J.M. Machado, R.J.C. Carbas, D. Jiang, and L.F.M. da Silva, Review on techniques to improve the strength of adhesive joints with composite adherends. *Composites Part B-Eng.* 177, 107363 (2019).
28. J.A.B.P. Neto, R.D.S.G. Campilho, and L.F.M. da Silva, Parametric study of adhesive joints with composites. *Int. J. Adhesion Adhesives* 37, 96-101 (2012).
29. P.N.B. Reis, F.J.V. Antunes, and J.A.M. Ferreira, Influence of superposition length on mechanical resistance of single-lap adhesive joints. *Composite Struct.* 67, 125-133 (2005).
30. J.-I. Choi, S.M. Hasheminia, H.-J. Chun, and J.-C. Park, Experimental study on failure mechanism of hybrid composite joints with different adhesives. *Fibers Polymers* 18, 569-574 (2017).
31. M.-S. Seong, T.-H. Kim, K.-H. Nguyen, J.-H. Kweon, and J.-H. Choi, A parametric study on the failure of bonded single-lap joints of carbon composite and aluminum. *Composite Struct.* 86, 135-145 (2008).

32. Z. Qin, K. Yang, J. Wang, L. Zhang, J. Huang, H. Peng, and J. Xu, The effects of geometrical dimensions on the failure of composite-to-composite adhesively bonded joints. *J. Adhesion* 97, 1024-1051 (2021).
33. L. F M da Silva and R. D Adams, Techniques to reduce the peel stresses in adhesive joints with composites. *Int. J. Adhesion Adhesives* 27, 227-235 (2007).
34. T.P. Lang and P.K. Mallick, The effect of recessing on the stresses in adhesively bonded single-lap joints. *Int. J. Adhesion Adhesives* 19, 257-271 (1999).
35. F. Nosouhi, M. Farahani, and M. Ansari, Experimental and numerical study on the composite adhesive joint reinforcement using wavy edge. *J. Adhesion Sci. Technol.* 32, 1007-1017 (2018).
36. A.F. Ávila and P.O. Bueno, Stress analysis on a wavy-lap bonded joint for composites. *Int. J. Adhesion Adhesives* 24, 407-414 (2004).
37. M. Ashrafi, A. Ajdari, N. Rahbar, J. Papadopoulos, H. Nayeb-Hashemi, and A. Vaziri, Adhesively bonded single lap joints with non-flat interfaces. *Int. J. Adhesion Adhesives* 32, 46-52 (2012).
38. M.Y. Tsai and J. Morton, The effect of a spew fillet on adhesive stress distributions in laminated composite single-lap joints. *Composite Struct.* 32, 123-131 (1995).
39. A.R. Rispler, L. Tong, G. P. Steven, and M.R. Wisnom, Shape optimisation of adhesive fillets. *Int. J. Adhesion Adhesives* 20, 221-231 (2000).
40. P.-I. Chen, K.-Y. Wang, and H.-H. Huang, Strength and failure modes of adhesively bonded composite joints with easily fabricated nonflat interfaces. *Composite Struct.* 225, 111162 (2019).
41. D. Cavalcanti, M. Banea, and H. Queiroz, Mechanical characterization of bonded joints made of additive manufactured adherends. *Ann. Dunarea Jos Univ. Galati Fascicle XII Weld. Equip. Technol.* 30, 27-33 (2019).
42. A. Nanda Kishore and N. Siva Prasad, An experimental study of Flat-Joggle-Flat bonded joints in composite laminates. *Int. J. Adhesion Adhesives* 35, 55-58 (2012).
43. B.R.K. Blackman, A.J. Kinloch, F.S. Rodriguez-Sanchez, and W.S. Teo, The fracture behaviour of adhesively-bonded composite joints: Effects of rate of test and mode of loading. *Int. J. Solids Struct.* 49, 1434-1452 (2012).
44. M.D. Banea, Influence of adherend properties on the strength of adhesively bonded joints. *MRS Bull.* 44, 625-629 (2019).
45. M.D. Banea, L.F.M. da Silva, R. Carbas, and R.D.S.G. Campilho, Effect of material on the mechanical behaviour of adhesive joints for the automotive industry. *J. Adhesion Sci. Technol.* 31, 663-676 (2017).
46. M.J. Schollerer, J. Kosmann, O. Völkerink, D. Holzhüter, and C. Hühne, Surface toughening – a concept to decrease stress peaks in bonded joints. *J. Adhesion* 95, 495-514 (2019).
47. A.Y. Kanani, X. Hou, and J. Ye, A novel dissimilar single-lap joint with interfacial stiffness improvement. *Composite Struct.* 252, 112741 (2020).

48. H.F.M. de Queiroz, M.D. Banea, and D.K.K. Cavalcanti, Adhesively bonded joints of jute, glass and hybrid jute/glass fibre-reinforced polymer composites for automotive industry. *Appl. Adhesion Sci.* 9, 2 (2021).
49. D.K.K. Cavalcanti, H.F.M. de Queiroz, and M.D. Banea, Performance enhancement of adhesive joints of additive manufactured parts by using different types of fibre reinforcements. *Int. J. Adhesion Adhesives* 124, 103371 (2023).
50. F. Delzendehrooy, M.R. Ayatollahi, A. Akhavan-Safar, and L.F.M. da Silva, Strength improvement of adhesively bonded single lap joints with date palm fibers: Effect of type, size, treatment method and density of fibers. *Composites Part B-Eng.* 188, 107874 (2020).
51. M.A. Morgado, R.J.C. Carbas, E.A.S. Marques, and L.F.M. da Silva, Reinforcement of CFRP single lap joints using metal laminates. *Composite Struct.* 230, 111492 (2019).
52. M.S. Islam and L. Tong, Influence of pinning on static strength of co-cured metal-GFRP hybrid single lap joints. *Composites Part A-Appl. Sci. Manuf.* 84, 196-208 (2016).
53. A.E. Bogdanovich, M. Dannemann, J. Döll, T. Leschik, J.N. Singletary, and W.A. Hufenbach, Experimental study of joining thick composites reinforced with non-crimp 3D orthogonal woven E-glass fabrics. *Composites Part A-Appl. Sci. Manuf.* 42, 896-905 (2011).
54. J. Kumar, V. Kumar, I. Singh, and P.K. Rakesh, Joining behavior of polymeric composites fabricated using agricultural waste as fillers. *J. Adhesion Sci. Technol.* 35, 1652-1663 (2021).
55. M.R. Bagheri, S.A. Sadough Vanini, N. Kordani, and M. Sadighi, The effect of nanoparticles in single-lap composite joints studied by experimental and numerical analyses. *Iran. Polym. J* 24, 629-640 (2015).
56. A. Nemati Giv, M.R. Ayatollahi, S.H. Ghaffari, and L.F.M. da Silva, Effect of reinforcements at different scales on mechanical properties of epoxy adhesives and adhesive joints: a review. *J. Adhesion* 94, 1082-1121 (2018).
57. H. Ejaz, A. Mubashar, E. Uddin, Z. Ali, and N. Arif, Influence of MWCNTs on strength properties of high viscous epoxy adhesive and fracture behavior of adhesively bonded joints. *Theoretical and Applied Fracture Mechanics* 120, 103412 (2022).
58. M. Jouyandeh, O. Moini Jazani, A.H. Navarchian, and M.R. Saeb, High-performance epoxy-based adhesives reinforced with alumina and silica for carbon fiber composite/steel bonded joints. *J. Reinf. Plast. Composites* 35, 1685-1695 (2016).
59. S.M.J. Razavi, M.R. Ayatollahi, A. Nemati Giv, and H. Khoramishad, Single lap joints bonded with structural adhesives reinforced with a mixture of silica nanoparticles and multi walled carbon nanotubes. *Int. J. Adhesion Adhesives* 80, 76-86 (2018).

60. S.M.R. Khalili, M. Tavakolian, and A. Sarabi, Mechanical properties of nano-clay reinforced epoxy adhesive bonded joints made with composite materials. *J. Adhesion Sci. Technol.* 24, 1917-1928 (2010).
61. A. Tutunchi, R. Kamali, and A. Kianvash, Steel-epoxy composite joints bonded with nano-TiO2 reinforced structural acrylic adhesive. *J. Adhesion* 91, 663-676 (2015).
62. A. Tutunchi, R. Kamali, and A. Kianvash, Adhesive strength of steel–epoxy composite joints bonded with structural acrylic adhesives filled with silica nanoparticles. *J. Adhesion Sci. Technol.* 29, 195-206 (2015).
63. U. Marathe, M. Padhan, and J. Bijwe, Carbon nanotubes- A powerful nano-filler for enhancing the performance properties of polyetherketoneketone composites and adhesives. *Composites Sci. Technol.* 210, 108813 (2021).
64. M.D. Banea, L.F.M. da Silva, R.J.C. Carbas, and S. de Barros, Debonding on command of multi-material adhesive joints. *J. Adhesion* 93, 756-770 (2017).
65. M.D. Banea, L.F.M. da Silva, and R.J.C. Carbas, Debonding on command of adhesive joints for the automotive industry. *Int. J. Adhesion Adhesives* 59, 14-20 (2015).
66. M.D. Banea, L.F.M. da Silva, R.J.C. Carbas, and R.D.S.G. Campilho, Mechanical and thermal characterization of a structural polyurethane adhesive modified with thermally expandable particles. *Int. J. Adhesion Adhesives* 54, 191-199 (2014).
67. M.D. Banea, L.F.M. da Silva, R.J.C. Carbas, and R.D.S.G. Campilho, Structural adhesives modified with thermally expandable particles. *J. Adhesion* 91, 823-840 (2015).

Index